An Introduction To Statistics

Carmine De Santo
Michael Totoro

Nassau Community College

Fourth Edition

whittier publications inc.

Published by Whittier Publications Inc.
Lido Beach N.Y. 11561

Cover by Les Schacter

Copyright © 1993 by Carmine De Santo and Michael Totoro
All Rights Reserved.

ISBN   1-878045-15-6

No part of this publication may be reproduced,
stored in a retrieval system, or transmitted, in any
form or by any means, electronic, mechanical,
photocopying, recording, or otherwise, without the
prior permission of the publisher.

Printed in the United States of America

10 9 8 7 6 5 4 3 2 1

TABLE OF CONTENTS

CHAPTER 1: INTRODUCTION TO STATISTICS
§ 1.1 Introduction   1
§ 1.2 Uses of Statistics   10
§ 1.3 Misuses of Statistics   12
§ 1.4 Overview and Summary   17
Glossary   19
Exercises   20

CHAPTER 2: ORGANIZING AND PRESENTING DATA
§ 2.1 Introduction   28
§ 2.2 Classifications of Data   31
§ 2.3 Exploring Data Using The Stem-and-Leaf Display   34
§ 2.4 Frequency Distribution Tables   43
§ 2.5 Graphs: Bar Graph, Histogram, Frequency Polygon And Ogive   65
§ 2.6 Specialty Graphs: Pie Chart and Pictograph   81
§ 2.7 Identifying Shapes and Interpreting Graphs   86
§ 2.8 Using Minitab   95
Glossary   97
Exercises   98

CHAPTER 3: MEASURES OF CENTRAL TENDENCY
§ 3.1 Introduction   116
§ 3.2 The Population Mean   117
§ 3.3 The Median   124
§ 3.4 The Mode   126
§ 3.5 The Sample Mean   129
§ 3.6 The Relationship of the Mean, Median and Mode   131
§ 3.7 Using Minitab   135
Glossary   139
Exercises   140

CHAPTER 4: MEASURES OF VARIABILITY
§ 4.1 Introduction   151
§ 4.2 The Range   152
§ 4.3 The Population Variance and The Population Standard Deviation   154
§ 4.4 The Sample Standard Deviation, s   166
§ 4.5 Applications of the Standard Deviation   171
§ 4.6 Using Minitab   180
Glossary   184
Exercises   185

CHAPTER 5: z-SCORES AND PERCENTILES

§ 5.1 Introduction	198
§ 5.2 z Scores	199
§ 5.3 Converting z Scores To Raw Scores	204
§ 5.4 Percentile Rank and Percentiles	206
§ 5.5 Box-and-Whisker Plot: An Exploratory Data Analysis Technique	212
§ 5.6 Using Minitab	220
Glossary	223
Exercises	224

CHAPTER 6: PROBABILITY

§ 6.1 Introduction	235
§ 6.2 Some Terms Used in Probability	236
§ 6.3 Permutations and Combinations	240
§ 6.4 Probability	251
§ 6.5 Fundamental Rules and Relationships of Probability	257
§ 6.6 Conditional Probability	278
§ 6.7 Binomial Probability	292
§ 6.8 Using Minitab	309
Glossary	314
Exercises	315

CHAPTER 7: THE NORMAL DISTRIBUTION

§ 7.1 Introduction	331
§ 7.2 Properties of the Normal Distribution	332
§ 7.3 Using the Normal Curve Area Table: TABLE II	335
§ 7.4 Applications of the Normal Distribution	340
§ 7.5 Percentiles	347
§ 7.6 Probability	360
§ 7.7 The Normal Approximation to the Binomial Distribution	365
§ 7.8 Using Minitab	379
Glossary	383
Exercises	384

CHAPTER 8: SAMPLING DISTRIBUTIONS

§ 8.1 Introduction	395
§ 8.2 The Sampling Distribution of The Mean	398
§ 8.3 The Effect of Sample Size on the Sampling Distribution of The Mean	407
§ 8.4 Using Minitab	408
Glossary	413
Exercises	414

CHAPTER 9: INTRODUCTION TO HYPOTHESIS TESTING

§ 9.1 Introduction	423
§ 9.2 Hypothesis Testing	424
§ 9.3 The Development of a Decision Rule	440
§ 9.4 p-Values for Hypothesis Testing	466
§ 9.5 Using Minitab	473
Glossary	478
Exercises	479

CHAPTER 10: HYPOTHESIS TESTING INVOLVING A POPULATION MEAN

§ 10.1 Introduction	490
§ 10.2 Hypothesis Testing Involving a Population Mean	490
§ 10.3 Introduction to the t Distribution	498
§ 10.4 Using TABLE III: Critical Values For The t Distribution	501
§ 10.5 Hypothesis Testing: Population Standard Deviation Unknown	504
§ 10.6 Using Minitab	517
Glossary	520
Exercises	521

CHAPTER 11: HYPOTHESIS TESTING INVOLVING TWO POPULATION MEANS

§ 11.1 Introduction	531
§ 11.2 The Sampling Distribution of the Difference Between Two Means	532
§ 11.3 Hypothesis Testing Involving Two Population Means And Unknown Population Standard Deviations	535
§ 11.4 Hypothesis Tests Comparing Treatment And Control Groups	545
§ 11.5 Using Minitab	552
Glossary	557
Exercises	558

CHAPTER 12: HYPOTHESIS TESTING INVOLVING A POPULATION PROPORTION

§ 12.1 Introduction	568
§ 12.2 The Sampling Distribution of The Proportion	572
§ 12.3 Hypothesis Testing Involving A Population Proportion	575
§ 12.4 Using Minitab	588
Glossary	591
Exercises	592

CHAPTER 13: ESTIMATION

§ 13.1	Introduction	600
§ 13.2	Point Estimate of The Population Mean	600
§ 13.3	Interval Estimation	604
§ 13.4	Interval Estimation: Confidence Intervals For The Population Mean	606
§ 13.5	Interval Estimation: Confidence Intervals For The Population Proportion	619
§ 13.6	Determining Sample Size And The Margin of Error	624
§ 13.7	Using Minitab	634
	Glossary	637
	Exercises	638

CHAPTER 14: CHI-SQUARE

§ 14.1	Introduction	647
§ 14.2	Properties of the Chi-Square Distribution	649
§ 14.3	Chi-Square Hypothesis Test of Independence	650
§ 14.4	Assumptions Underlying the Chi-Square Test	662
§ 14.5	Test of Goodness-of-Fit	664
§ 14.6	Using Minitab	671
	Glossary	675
	Exercises	676

CHAPTER 15: LINEAR CORRELATION AND REGRESSION ANALYSIS

§ 15.1	Introduction	689
§ 15.2	The Scatter Diagram	690
§ 15.3	The Coefficient of Linear Correlation	698
§ 15.4	More on the Relationship Between Correlation Coefficient And The Scatter Diagram	702
§ 15.5	Testing the Significance of the Correlation Coefficient	703
§ 15.6	The Coefficient of Determination	710
§ 15.7	Linear Regression Analysis	712
§ 15.8	Assumptions for Linear Regression Analysis	720
§ 15.9	Using Minitab	722
	Glossary	727
	Exercises	728

Answers	739
Appendices	784
Index	801

Preface To The Fourth Edition

Today, courses in elementary statistics are being presented at all levels of education, reaching an audience with increasingly diverse mathematical backgrounds and goals. Many texts require better than average mathematical ability and sophistication. Unfortunately, a large segment of today's audience does not possess the necessary skills needed to succeed with the advanced level of presentation in currently available texts. This text is written in a style and at a level that addresses the needs of this diverse audience. It is intended for college freshmen with no background in statistics and minimal computational skill. Our pedagogy emphasizes a step-by-step problem solving method while encouraging the student to tie together concepts for the purpose of drawing conclusions. This structure will enable the student to successfully use statistics and appreciate its value. The writing style "talks" to the student making it easy for the student to understand and comprehend the basic concepts.

Our text is a comprehensive presentation of topics covered in a standard first course in statistics. The text begins with an overview of statistics, its uses and misuses. A full discussion of descriptive measures and graphing techniques provide the student with basic skills to summarize and interpret data.

Besides the presentation of the descriptive topics of elementary statistics, this text offers an extensive chapter on elementary probability. Although students often find probability difficult we have presented this topic in a manner that is easily understood. You will find, however, that the rigor of the material in our chapter is comparable to probability chapters in texts that require a significantly higher level of mathematical ability.

An extensive chapter is devoted to the purpose and use of the normal distribution and its applications, while a chapter on estimation introduces the concepts of inferential statistics, confidence intervals, and the estimation of population parameters for the mean and proportion.

We have been involved in the teaching of elementary statistics for over twenty years and have found the hypothesis testing concept difficult for students to comprehend. This text provides the student and instructor with a unique presentation of the hypothesis testing concept. It offers a presentation that is independent of a particular sampling distribution. This enables the student to concentrate on the concepts related to hypothesis testing and not be distracted by the mathematical characteristics of the specific sampling distribution. This general approach focuses on the logic and structure of the hypothesis testing procedures. Furthermore, it allows the instructor the flexibility to choose hypothesis testing applications related to means, or proportions.

Special attention has been given to include an introduction to Exploratory Data Analysis(EDA) approaches, which in recent years, has seen a strong movement within the area of statistical education. In our discussion of EDA techniques, we have devoted attention to the techniques called the "stem and leaf" display and the "box-and-whisker" plot. These two visual techniques can yield valuable information about the shape of a distribution, and an indication about the presence of outliers.

FEATURES OF THE FOURTH EDITION

In this fourth edition we have made changes and additions that emphasize the use of calculators with statistical functions, the inclusion of real life applications of statistics, a section within each chapter covering the use of the computer statistics software package MINITAB, and two exercise sections for thoughtful and practical applications of statistics. To accomplish these objectives the following new material has been included in this edition:

- **Case Studies:** Case studies which appear in all the chapters provide actual applications of statistical concepts based on articles, statistical snapshots, or research published in newspapers, magazines, or journals. All case studies are based on real data and allow the student to see the relevance of statistics and why a student should study statistics. We believe the students will enjoy discussing the appropriateness of these case studies as well as enhance their appreciation of statistics.

- **Using MINITAB:** At the end of every chapter is a section entitled Using MINITAB, which provides a detailed instruction on the use of MINITAB. Each chapter contains a description of MINITAB commands that describe how to perform the calculations and analysis for the concepts contained within the chapter. In addition, each chapter contains an example of an actual MINITAB printout with a discussion of the MINITAB results.

- **What Do You Think? Problems:** Each chapter contains a What Do You Think? section in the chapter exercises. This section provides the student with a set of realistic exercises which enable the student to explore diverse applications. In most cases, the problems will present applications which will require the student to analyze and interpret the information relating to some real practical situation.

- **Exploring DATA With MINITAB:** Every chapter includes a section within the chapter exercises entitled Exploring DATA With MINITAB. These exercises encourage the student to investigate the concepts within each chapter using the software package MINITAB.

- **Applications of the Standard Deviation:** A new section entitled Applications of the Standard Deviation was added to Chapter 4. In this section, Chebyshev's Theorem and the Empirical Rule demonstrate how the standard deviation provides a measure of variability for different shaped distribution. The coefficient of variation is described as a relative measure of variability and demonstrated with a few meaningful comparisons.

- **Box-and-Whisker Plot:** A new section entitled Box-and-Whisker Plot: An Exploratory Data Analysis Technique was included within Chapter 5. This section introduces the important aspects of the Box-and-Whisker Plot as an effective exploratory data analysis technique in displaying, through a visual presentation, the major characteristics of a distribution.

- **p-Values For Hypothesis Testing:** A new section entitled p-Values For Hypothesis Testing has been added to Chapter 9 which presents an explanation on the use of p-values when performing a hypothesis test using a computer software package, or when interpreting the results of statistical tests of hypotheses in research journals and reports.

ADDITIONAL FEATURES OF THE TEXT

The main features of the text have been maintained that we believe distinguish it from the other introductory statistics texts in enhancing a student's appreciation of statistics and an understanding of its application in the real-life situations. These features are:

- **Presentation of the Material:** Our informal writing style in the explanation of the statistical concepts is simple and very clear for students to read and learn statistics. The pedagogical approach of learning statistics through discovery and applications makes the subject relevant to the beginning students, and shows them why the study of statistics is valuable as a tool to analyze information and draw conclusions.

- **Design and Format:** All definitions, procedures, theorems, formulas, and other important concepts are highlighted within boxes with appropriate headings. This will help students to focus in on the important aspects of each chapter. In particular, all procedures are outlined with a step-by-step approach. The general hypothesis testing procedure has been outlined in Chapter 9 using a five step procedure which provides a model that is applied to hypothesis tests for all chapters.

- **Examples:** Every chapter contains an abundant number of examples which cover the concept of each section topic. The format for each example is the statement of the problem followed by its solution. Our belief is that a student will better understand the concepts, techniques, theorems and definitions of the chapter after practicing these ideas through many relevant examples.

- **Solutions:** The solution to every example is presented with a very clear and readable style. Our approach is built on the premise that **each step is not trivial**. Consequently, we have provided a step-by-step approach in solving each problem.

- **Numerous Diagrams:** The text uses a large number of diagrams to picture many of the concepts visually. In particular, the applications of the normal distribution, and the hypothesis testing procedure involve numerous use of diagrams to illuminate the statistical concepts.

- **Exercises:** The exercises are a special feature of this text. Each chapter contains numerous problems that cover a wide range of applications in many diverse fields. The content of the problems will interest both the student and instructor. Most of the problems are real data which are taken from sources such as periodicals, reports, newspapers, and journals. These problems provide an insight into the many diverse applications of the use of statistics. Each exercise set consists of fill-in the blank, multiple choice, true-false questions, word problems, what do you think? problems, exploring data using MINITAB, projects and computer applications. The answers to most chapter problems appear in the answer section at the back of the text.

- **Glossary:** A glossary of key terms and concepts listing the appropriate section in which the term appears is printed at the end of each chapter.

- **Using the Calculator:** General instructions for using a statistical calculator are given within appropriate chapters. A special calculator instruction section appears in Appendix B which contains specific instructions for some brands of calculators.

- **Using the Computer:** The use of computer software packages to perform statistical computations and analysis has become a common tool in learning and using statistics. These computer packages allow you to concentrate on understanding and applying statistical techniques and concepts while enabling you to avoid complex data manipulation and tedious calculations. At the end of each chapter a section entitled: **Using MINITAB** will appear. Within these sections we will be using the Student Edition of MINITAB as our software package. A description of the MINITAB commands are explained and illustrated that are required to perform the appropriate statistical analysis for that chapter. In addition, MINITAB is also used to investigate some of the important concepts introduced within the chapter. Actual MINITAB output are presented to illustrate the usefulness of the computer for numerical calculations and to explain how to interpret the computer output. The main objectives of these sections will be:

(1) to illustrate the usefulness of computers.
(2) to provide you with a computer output of selected text exercises.
(3) to familiarize you with reading and interpreting computer outputs.

A MINITAB computer supplement entitled:
<u>Learning to Use Minitab</u> (F. Ripps and B. Smith) is available for students using this text.

This Minitab supplement provides an introduction to the use of the student edition of MINITAB and is linked to this text through text exercises, examples, discussions and data sets.

- **Database:** A database representing 75 student records appears in Appendix A. This is available for computer assignments using the Student Edition of MINITAB. At the end of each chapter an exercise set entitled Database, computer assignments are presented so students can practice using MINITAB on real data.

Acknowledgements

We express our thanks to our colleagues at Nassau Community College who have made helpful criticisms and suggestions. We name some of them below as a sign of our gratitude but emphasize that they are not responsible for any weaknesses that may remain in the text. Special gratitude is reserved for Professors Joe Altamura, Alice Berridge, Mauro Cassano, Anthony Catania, Dennis Christy, John Earnest, Larry Kaufer, Ken Lemp, George Miller, Faith Ripps, John Schreiber, Richard Silvestri, Barbran Smith and Thomas Timchek. Furthermore, we appreciate the special attention that Grace Totoro has given to the project.

In addition to the above individuals we would like to express our heartfelt thanks to Professor William Porreca whose years of contributions to this project are deeply appreciated.

Finally, and most importantly, we would like to thank our wives Debbie and Kathleen, and our children Craig, Matthew, and Allison for their continuous support, understanding, patience, and encouragement during the preparation of this project, and for bringing love and happiness to our lives.

Carmine De Santo
Michael Totoro

CHAPTER 1
INTRODUCTION TO STATISTICS

❄ 1.1 INTRODUCTION

What is Statistics?

Your life is a *statistic*!

Can you describe a typical day in your life?

For example:

>What do you have for breakfast?
>- eggs or egg beaters?
>- pastry or oat bran muffins?
>- coffee, tea or decaffeinated drink?
>- orange or grapefruit juice?
>
>Are you a full-time or part-time student?
>
>How many hours a week do you work?
>
>Consider the following fascinating facts that have been used to describe an "average" day in America.

Did you know that on an average day in America. . .
- 9,077 babies are born. (1,282 are illegitimate.)
- 5,962 couples will wed; 1,986 will divorce.
- 56,000 animals are turned over to animal shelters.
- 3 million people will go to the movies.
- a car is stolen every 33 seconds.
- 2,466 children are bitten by dogs.
- 500 million cups of coffee are drunk.
- $54,794 is spent to fight dandruff.
- Amateurs take 19,178,000 snapshots.
- 679 million telephone conversations occur, of which 90 million are long distance.
- People drink 90 million cans of beer.
- 2,740 teenagers get pregnant.
- 438 immigrants become citizens.
- 10,930,000 cows are milked.
- Tobacco chewers chew up to 1.3 million packages of the stuff.
- The U.S. Postal Service sells 90 million stamps, handles 230 million pieces of mail and delivers 834,000 packages.
- The snack bar at Chicago's O'Hare Airport sells 5,479 hot dogs, covered with 12 gallons of relish and nine gallons of mustard, washed down with 890 gallons of coffee![1]

---

[1] These facts appeared in *American Averages: Amazing Facts of Everyday Life*, by M. Feinsilber and W. B. Mead (Doubleday, New York)

The previous statements represent statistical descriptions of a large collection of information or observations for some phenomenon or characteristic. The information or observations before they are arranged or analyzed are called **raw data**. The term **raw** is used because statistical techniques have not been applied to analyze the data.

After the data is collected, we are usually confronted with a massive list of numbers, names, opinions, dates, measurements, percentages, etc. To make this data collection more meaningful and useful we need to examine the data set in a formal way.

This formal **exploration** of the data can include many different forms: For example,

- graphs can be used to visually display the data.

- numerical techniques can be used to describe the main characteristics of the data.

These data explorations can help you discover patterns or relationships which help you generate hypotheses or draw inferences from the data set.

Before we formally define statistics let us consider some case studies that illustrate applications of statistics.

CASE STUDY 1.1

An opinion poll conducted by **The New York Times/CBS News** describes the results of a survey of 1,283 adults who responded to the question:

Do you think the typical major league baseball player generally makes too much money, too little money, or do you think their incomes are about right?

Referring to the owners of the teams a similar question was asked:

Do you think owners generally make too much money, too little money, or do you think their incomes are about right?

The **pie graphs** in the following figure summarize the results of the opinion poll.

Bucks Too Big, a Majority Tells Poll

By ROBERT McG. THOMAS Jr.

If you're convinced that, say, Dwight Gooden is worth every penny of the $1,850 or so he will make for every pitch he throws next season, or that Jose Canseco deserves the $5,000 he earns every time he bats, boy, do you have an argument on your hands.

By a substantial margin, Americans apparently believe that baseball players are paid too much, although most people don't seem to hold it against the players, in part perhaps because a majority also believes baseball owners make too much, too.

Those, anyway, are among the findings of The New York Times/CBS News Poll conducted on Monday, Tuesday and Wednesday of last week. To be fair, the 1,283 adults reached in the survey were not asked about high-paid superstars like Met pitcher Gooden or Oakland slugger Canseco. Instead, they were asked whether the "typical" major leaguer makes too much or too little money, or whether his income is about right.

More Than a 3-1 Margin

The people surveyed agreed by more than a three-to-one margin (71 percent to 19 percent) that the typical player is paid too much rather than just the right amount. Only 1 percent thought the typical player is paid too little. The poll's margin-of-sampling error was plus or minus three percentage points.

The responses were pretty much the same for baseball fans and non-fans alike. But one variable produced a striking difference: the respondent's income. The richer the person, it seems, the greater the resentment that baseball players are even richer.

Although a majority of people in all income categories said the players make too much, just 59 percent of those with incomes under $15,000 a year expressed that view. The figure jumped to 70 percent for the $15,000-to-$30,000 group, to 74 percent for those making between $30,000 and $50,000 and to 79 percent for those earning more than $50,000.

If the average American believes the typical player makes too much, how accurate is the average American's estimate of how much the average player makes? Not very.

The typical player makes nowhere near as much as the $5.15 million Gooden will average when his three-year contract extension kicks in next year. Nor is the typical player's salary very close to the average salary of the 708 major leaguers, $890,844, a figure inflated by the impact of the few dozen multimillion-dollar incomes. Actually, a typical player — one at the midpoint of the salary ranking — makes $500,000.

Yet of the respondents who provided an answer (and 36 percent did not) only 14 percent came within $100,000 of the actual $500,000 figure, with 3 percent picking within the $400,000-to-$500,000 range and 11 percent choosing a number in the $500,000-to-$600,000 range. The others were about evenly divided between those who named a higher figure (44 percent) and those who picked a lower one (42 percent).

The most popular estimates — between $900,000 and $1 million, picked by 17 percent of the respondents who answered the question — were more than $400,000 too high, and the next most common, between $100,000 and $200,000, named by 14 percent, were more than $300,000 too low.

Figure 1.1

The pie graphs in Figure 1.1 are being used to summarize and describe the data obtained from the opinion poll. ■

Case Study 1.2

The results of a survey appearing in USA Today indicated what secretaries felt was most stressful in their jobs.

Figure 1.2

The graph in Figure 1.2 describes the opinions of 1,000 administrative assistants within Fortune 500 companies. This case study, like Case Study 1.1, is another illustration of how statistics is used to summarize and describe data.

Case Study 1.3

Due to an increase in the number of children contracting measles and rubella, as shown in Figure 1.3, medical researchers from Madigan Army Medical Center in Tacoma, Washington, conducted a study to determine why children who have been vaccinated for measles contracted the disease. The research team vaccinated two groups of infants, 51 who were perfectly healthy when they got their shots and 47 who had bad colds when they were immunized. Follow-ups showed that 98 percent of the kids who are healthy when vaccinated were fully protected against measles. But the vaccine protected only 79 percent of those who had colds when they got their shots.

In this case study, statistics is used to explore the data for any patterns or relationships that might exist that will enable the researchers to generate hypotheses or draw inferences.

Colds May Hinder Measles Shots

Researchers study vaccine's effectiveness in kids with virus

By Laurie Garrett
STAFF WRITER

Measles vaccines may not be effective in children who have bad colds when they get their shots, according to the first study of its kind appearing in yesterday's Journal of the American Medical Association.

The researchers say their finding could explain why 2 to 5 percent of all people vaccinated during early childhood against measles develop the disease later in life.

The United States is in its third year of a serious measles epidemic, and New York City leads the nation in cases of the disease.

Last year more than 27,000 measles cases and 100 deaths were reported to the Centers for Disease Control in Atlanta, the largest number since the government's measles eradication program began in 1978. About 800 of those officially reported cases were in New York City, according to the CDC's Dr. William Atkinson, but city health officials say the real number, not yet reported to CDC, was 2,479 cases and nine deaths for 1990. So far this year New York City has nearly equalled its total number of 1990 cases, with more than 2,000 illnesses and eight deaths in children, according to the city health department.

Between 1980 and 1988, the United States averaged 3,000 measles cases a year, most of which occurred in children and young adults who had been vaccinated, Dr. Georges Peter, chair of the infectious disease committee of the American Academy of Pediatrics, said in an interview.

In an effort to understand why the vaccine sometimes fails, Dr. Marvin Krober and a medical team from Madigan Army Medical Center in Tacoma, Wash., vaccinated two groups of infants; 51 who were perfectly healthy when they got their shots, and 47 who had bad colds when they were immunized. Follow-ups showed 98 percent of the kids who are healthy when vaccinated were fully protected against measles. But the vaccine took in only 79 percent of those who had colds when they got their shots.

Such an association between viral illnesses at the time of vaccination and subsequent immunization failure has not been found with any other disease, Atkinson said.

"This is not sufficient data to recommend broad sweeping changes in immunization practices. I would wait for further studies before making policy changes," Krober said. "But individually, if a physician were seeing a child with a cold and was comfortable that the parents would, indeed, bring the child back two weeks later, I would delay vaccination in the presence of a cold. But 80 percent protection is better than zero percent, which is what you'll get if the parents don't bring that kid back in."

"In the mid-80s, vaccine failure was the problem, but the epidemic we're having now has clearly changed. It is undoubtedly due to a failure to vaccinate at all," Peter said. In some inner-city areas — notably in New York City — only half of all school-age children have received their proper vaccines, he said.

The Krober study is, "a preliminary finding, an important finding," Peter said. "And the concern is that it could be misinterpreted as a reason not to vaccinate in the presence of other viral infections. The danger is this will shift attention to vaccine failure instead of health access and the need to be immunized."

The CDC's Atkinson noted that diphtheria and pertussis, which are covered by DPT shots in the first year of life, are not increasing in incidence. Parents, he said, are fairly scrupulous about taking their newborns to doctors. But measles vaccines are, ideally, given at 15 months of age and boosted before starting kindergarten. "Many parents have simply dropped their kids out of the health care system by then," he said.

The majority of New York City measles cases, "are kids who are three or four years old who have had zero measles vaccinations. That's the problem," Atkinson said.

An as-yet-unreleased U.S. Health and Human Services vaccination report calls for a massive nationwide effort to immunize all pre-school children at an increased federal cost of $40 million to $50 million a year, Peter said.

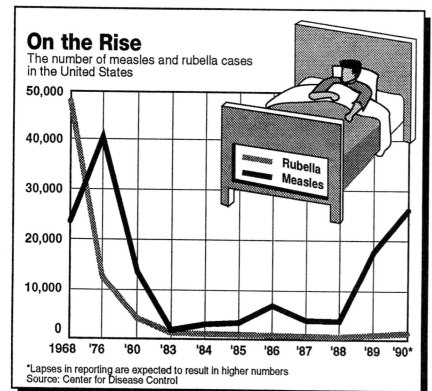

Copyright 1991, Newsday. Reprinted with permission. Newsday/Steve Madden

Figure 1.3

CASE STUDY 1.4

In an article appearing in the April 24, 1991 issue of USA Today, researchers from the American Health Foundation in Valhalla, N.Y., claimed that wheat is the best bran to eat to protect against colon cancer. The researchers studied 75 women who ate up to 30 grams of fiber a day from either oat bran, corn bran, or wheat bran muffins for a period of eight weeks. One of their findings was: **Wheat bran eaters had lower amounts of colon cancer inducing compounds.**

In this case study the researchers are making an **inference** about the effects of eating wheat bran muffins and its link to the risk of colon cancer. ■

In learning about statistics we will be mostly concerned with the application of statistical techniques. The applications of the statistical techniques will help us to summarize and describe the data, to search for patterns or relationships in the data, and to draw inferences from the data.

Statistics can be defined as:

> **Definition: Statistics.** Statistics is the collection, presentation, analysis and interpretation of data.

The study of statistics can be separated into two areas:
Descriptive Statistics and *Inferential Statistics*.

> **Definition: Descriptive Statistics.** Descriptive Statistics uses numerical and/or visual techniques to summarize data.

Case studies 1.1, 1.2 and 1.3 are examples of descriptive statistics because the data was summarized using numerical and visual techniques.

In Case Study 1.4, the researchers are making an inference from the data collected on 75 women that ate up to 30 grams of fiber per day for a period of eight weeks. The researchers **inferred** from the data pertaining to the 75 women in their study to **all people** that eat the same amount of wheat bran fiber per day. This is the **essence of Inferential Statistics** because the researchers are drawing **a conclusion or inference about *all* people** who eat wheat bran **using only the results of a portion**, (i.e. the 75 women used in their study), of these people.

As indicated in Case Study 1.4, the researchers' conclusions are being generalized to **all** people who eat wheat bran. In statistics this entire group is referred to as the **population**.

> **Definition: Population.** The total collection of individuals or objects under consideration.
>
> **Definition: Sample.** The portion of the population that is selected for study.

In Case Study 1.4, the researchers selected a sample of 75 women to conduct their study. This sample represented a portion of the population of all people who eat wheat bran.

> **Definition: Inferential Statistics.** Inferential Statistics is the process of using **sample information** to draw inferences or conclusions about the **population**.

Inferential Statistics is used to help researchers and decision makers make inferences or draw conclusions about a population. The following statements represent situations where inferential statistics can be applied:

- A medical researcher wants to determine if large doses of vitamin C are effective in combating colds.

- Pollsters want to estimate what percentage of the American public approve of the President's economic program.

- A quality control engineer for a video tape manufacturer must determine whether their high grade video tape meets the manufacturers' specifications before shipment to their regional distributors.

- The Food and Drug Administration would like to determine if women who use oral contraceptives have undesirable side effects.

- A market researcher wants to know in what quantities the American consumer is willing to purchase a new product.

These situations require the use of inferential statistics because the researchers and decision makers will use sample data selected from the population to make inferences or draw conclusions about the population. In most applications of statistics, researchers use only a sample rather than the entire population, because it is usually impractical or impossible to obtain all the population observations or measurements.

In such situations, we will examine samples to describe populations. For example, suppose an electrical company wanted to determine the average life of their new 40 watt light bulb. All the 40 watt bulbs the company manufactures would constitute the population. To obtain the average life of the population of light bulbs it would be necessary to compute the life of each bulb. Obviously, this would be impractical since they would not have any bulbs left to sell! In such instances, it becomes necessary to select a sample, with similar pertinent characteristics as all the new 40 watt light bulbs, that we can use to make inferences about the entire population.

Thus, the electrical company would try to design a procedure to select a representative sample of these new light bulbs and use this sample data to estimate the population average life of their new 40 watt light bulbs.

> A **representative sample** is a sample that has the pertinent characteristics of the population in the same proportion as they are included in that population.
>
> For example, if a population has 60% females, then a representative sample of the population with regard to sex should also have 60% females.

Since the **primary objective of inferential statistics** is to use sample information to estimate a population characteristic, it is imperative that the researcher try to design a procedure to select a **sample which is representative of the population**. Thus, the electrical company that wants to **estimate** the population average life of their new light bulb would take all possible measures to select a **representative** sample from the population. If the sample is **not representative** of the population, then inferences about the population characteristics **may not** be reasonable. The idea of ensuring sampling representativeness is similar to the "toothpick" technique used to determine if a cake is completely baked. A toothpick is inserted in **several areas** of the cake, and if the toothpick comes out free of cake batter each time, we conclude that the **entire** cake is done.

In the previous example about the new 40 watt light bulb, a sample was selected and the **sample average** was used **to estimate** the **average life** for *all* the new **40 watt light bulbs**. That is, a sample average was used to estimate the population average. Whenever a number is used to describe a characteristic of a sample, such as sample average, this number is called a **statistic**.

Definition: Statistic. A number that describes a characteristic of a sample.

On the other hand, a number like the population average, that describes a characteristic of a population is called a **parameter**.

Definition: Parameter. A number that describes a characteristic of a population.

Thus, a population average is an example of a parameter while a sample average is an example of a statistic. The sample statistic (i.e., sample average for the 40 watt light bulb) is being used to estimate the population parameter (i.e., the average life of the population of new light bulbs). We can say that the concept of inferential statistics is to use a sample statistic to make inferences about a population parameter.

Examples 1.1 and 1.2 illustrate the difference between a parameter and a statistic.

Example 1.1

A statistician computes the batting average of the American League batters to be .285 and the batting average of the New York Yankees, an American League team, to be .267. If the population is defined to be the baseball players in the American League, then determine which batting average represents a parameter and which one represents a statistic?

Solution:

Since the population is defined to be the baseball players in the American League, then the batting average of .285 is an example of a parameter because it describes a characteristic of the population. The New York Yankees represent a sample of the players in the American League. Therefore, the Yankees batting average of .267 is an example of a statistic because it describes a characteristic of the sample. ■

Example 1.2

A politician is interested in determining the proportion of voters among the 20,000 voters in her district who will vote for her. An opinion poll of 1500 potential voters will estimate how **all** the voters in her district will vote in the upcoming election.

The results of the opinion poll indicate for the upcoming election that 52% will vote for the politician.

Determine:
a) the population.
b) the sample.
c) whether the result, 52%, is a parameter or statistic.

Solution:

a) The 20,000 voters in her district comprise the population.
b) The opinion poll of 1500 potential voters represents the sample.
c) The value of 52% is a statistic since it describes a characteristic of the sample of 1500 potential voters.

CASE STUDY 1.5

The following appeared in an edition of *The New York Times*.

How the Poll Was Conducted

The latest New York Times/CBS News Poll is based on telephones interviews conducted June 3 to 6 with 1,424 adults around the United Stated, excluding Alaska and Hawaii.

The sample of telephone exchanges called was selected by a computer from a complete list of exchanges in the country. The exchanges were chosen so as to assure that each region of the country was represented in proportion to its population. For each exchange, the telephone numbers were formed by random digits, thus permitting access to both listed and unlisted numbers. The numbers were then screened so that calls would be placed to residences only. Within each household one adult designated by a random procedure to be the respondent for the poll.

The results have been weighted to take account of household size and number of telephone lines into the residence, and to adjust for variations in the sample relating to region, race, sex, age and education.

In theory, in 19 cases out of 20 the results based on such samples will differ by no more than three percentage points in either direction from what would have been obtained by seeking out all American adults.

The potential sampling error for smaller subgroups is larger. For example, for those adults who are between 18 and 44 years of age and single it is plus or minus about five percentage points.

In addition to sampling error, the practical difficulties of conducting any survey of public opinion may introduce other sources of error into the poll.

New York Times © 1991 Reprinted by Permission

Notice the great care taken by the pollsters in selecting the sample. **Identify** the population and the number of people selected for the sample. **Describe** the sampling techniques used by the pollsters in their attempt to obtain a representative sample.

In the article the following statement appears:

> **In theory, in 19 cases out of 20 the results based on such samples will differ by no more than three percentage points in either direction from what would have been obtained by seeking out all American adults.**

If a second sample were selected using the same procedure, do you believe the results of the second sample would be identical to the first? Explain your answer in relation to the previous statement.

This case study uses terms that identify some important concepts that will be studied in later chapters of this text. What do you think the terms **random digits** and **sampling error** mean?

1.2 Uses of Statistics

The role of statistics in our lives is increasing and some may even find it disturbing. Statements such as, "Your family's socio-economic index is too high to qualify for financial aid," or "Your cholesterol level is too high, therefore you are advised to change your diet immediately," are all by-products of the advances in data collection and processing which has occurred during the twentieth century.

Today, even at an elementary level, it is impossible to understand economics, finance, psychology, sociology, or the physical sciences without some understanding of the meaning and interpretation of average, variability, correlation, inference, charts or tables.

Extensive data collection and distribution activities are performed by the federal and other governmental and private agencies in areas such as education, employment, health, crime, prices, housing, medical care, manufacturing, agriculture, construction, transportation, etc.

Furthermore, the development of the electronic computer has led to a revolution in the area of data collection and analysis. With the aid of the computer, statistical analysis can be applied to vast amounts of data accurately and quickly. Such diverse problems as weather forecasting, economic stabilization, and disease control are today being solved using statistical analysis.

The following examples represent some applications of statistics.

Example 1.3 *"Auto Thefts On The Rise"*

FBI crime figures stated in *Uniform Crime Reports* indicate that the number of vehicle thefts in 1989 went up by 9%. This descriptive statistic represents an increase of the largest annual percentage gain for any of the eight major crimes regularly tracked by the FBI. ∎

Example 1.4 *"1990 Census Rate Of Return Low"*

The Census Bureau reports that the national rate of return is 57%. This rate is unprecedentedly low as compared to an 83.9% rate of return ten years ago.

A consequence of this vital descriptive statistic may effect the disbursement of federal and state aid as well as the boundaries of the voting districts. ∎

Example 1.5 *"Dewey Defeats Truman"*

In the 1948 presidential election the Gallup Poll incorrectly predicted the defeat of Truman by Dewey. The Gallup Poll uses statistical methods in selecting samples for their public opinion surveys and applies inferential techniques to make predictions. Their method of sampling has produced presidential election predictions which have had an error averaging around 2% since 1948. (They failed to pick the winner in 1948 but have been correct ever since.) ∎

Examples 1.6 and 1.7 discuss descriptive economic measures to reflect changes in the economy.

Example 1.6 *"This past month The Consumer Price Index rose 1.2%"*

What does this mean? The Bureau of Labor Statistics uses a collection of over 400 goods and services called a fixed market basket to reflect changes in the purchasing power of the dollar. An increase of 1.2% in the Consumer Price Index indicates that prices for the 400 goods and services have increased 1.2% during the past month as compared to the prices for these goods and services in the base period (currently 1982-84). ∎

Example 1.7 *"Stocks Soar, as Dow Gains 63.86, to 2,945.05"*

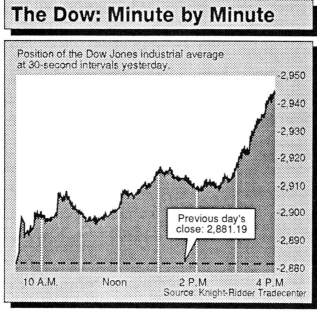

Figure 1.4

The Dow Jones Industrial Average climbed a record 63.86 points in a day representing a 2.2 percent increase in the Dow. The graph in Figure 1.4 records the position of the Dow at 30 second intervals.

The Dow Jones Industrial Average represents an average of a composite of 30 blue chip stocks. This average is used as a yardstick of the stock market performance. That is, it reflects the direction of the stock prices in general. ∎

Example 1.8 *"New Fat-Free Veggie Burger is Marketed"*

The vice president of marketing for a fast food chain wants to study the sales performance of a new fat-free veggie burger. Before mass marketing this new product nationwide, the vice president will examine the sales performance from a representative sample of their fast food chain stores. The results of this sample will help the vice president to decide whether to mass market this new product.

In essence, the vice president will use the sample results to infer how the new product will be accepted nationwide. This is an example of inferential statistics. ∎

Example 1.9 *"College Projects Increase in Fall Enrollment"*

For next year's planning and decision making at a community college, a college administrator needs to predict the number of students that will enroll for the upcoming fall semester. The administrator's predicting procedure relies on the number of students graduating from high school for the past five years and the corresponding number of students who enrolled in the college for the past five years. Based upon this information, the administrator will predict how many of this year's graduating high school seniors will enroll in the college. ∎

Example 1.10 "There is a Tornado Watch in Effect Tonight from 5 to 9 PM"

Such forecasts are issued by the U.S. Weather Bureau which has the ability to gather weather data from all over the world, quickly analyze this vast amount of information, and produce relatively accurate predictions. Early warnings issued by the bureau are now responsible for the savings of thousands of lives and millions of dollars in property. ■

In Examples 1.11 and 1.12 the researchers analyzed the data for possible patterns or relationships that might exist. This type of analysis is called **Exploratory Data Analysis**.

Example 1.11 "Caution: Cigarette Smoking May be Hazardous to Your Health"

In the early 1950s, Dr. Richard Doll and Professor Austin Bradford studied the relationship between cigarette smoking and lung cancer. Their conclusion that "smoking is a factor, and an important factor, in the production of carcinoma of the lung" shook the foundation of the tobacco industry. ■

Example 1.12 "Rise in Rate of Breast Cancer on Long Island"

Medical researchers are trying to determine what factors have led to an abnormally high incidence of breast cancer in women on Long Island. To aid in this study, the researchers will look for any patterns or relationships which may exist among the vast amounts of data. For example, the researchers examine such factors as race, diet, age, occupation, body fat, family history, age at first child's birth, etc, and the relationship of these factors to the incidence of breast cancer. ■

Example 1.13 "Morning Workout Burns More Fat"

A study conducted at Kansas State University concluded that exercise done before breakfast burns more fat than exercise done at any other time of the day. The researcher hypothesized that the burning of fat is related to the level of insulin. Since the level of insulin is known to be low before breakfast, the researcher studied a group of runners before breakfast and after lunch.

This study illustrates an example of Inferential Statistics since the researcher used a sample to draw a conclusion about the general population. ■

Example 1.14 "Study States 5% of Young Adults Are Illiterate"

The Department of Education found that 5% of young adults between the ages of 21 to 25 years old read below the fourth grade level. The report is based on a survey of 3600 people.

This example illustrates another use of inferential statistics because the Department of Education is using a sample to infer about the general population of young adults aged 21 to 25 years old. ■

The previous examples have illustrated the many uses of statistics. Some of the examples emphasized the use of descriptive statistics and exploratory data analysis, while others illustrated the use of inferential statistics in forecasting, predicting and the testing of hypotheses.

❄ 1.3 Misuses of Statistics

Disraeli, a British Statesman, was quoted as saying, *"There are three kinds of lies: lies, damned lies, and statistics."* Disraeli's statement attests to the fact that there are many instances where statistics are abused.

In this section, our objective is to briefly discuss some of the misuses of statistics. Hopefully, the illustrations presented will make you aware of the need to become statistically literate. After reading this section, we believe you will be more inclined to agree with H. G. Wells' statement that:

"Statistical thinking will one day be as necessary for efficient citizenship as the ability to read and write."

We will examine four techniques that illustrate how information can be presented in a misleading fashion. These techniques are: (I) Graphs, (II) Non-Representative Samples, (III) Inappropriate Comparisons and (IV) The Omission of Variation About an Average. Let's illustrate how graphs can be misleading.

I GRAPHS

A graph is a descriptive tool used to visually describe the characteristics and relationships of collections of data quickly and attractively. However, graphs also provide an excellent opportunity to distort the truth.

Example 1.15

An investment analyst could use a graph to show that the rate of return on retirement annuities are falling. This is shown in Figure 1.5.

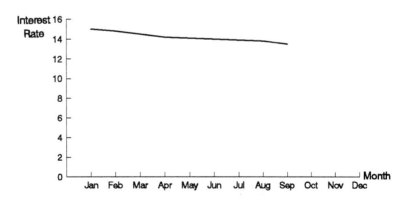

Figure 1.5

However, if the analyst wanted to make the decrease in the rate of return appear more dramatic, she could simply change the units of the vertical scale and delete its lower portion. This is illustrated in Figure 1.6.

Although Figures 1.5 and 1.6 present the same information, Figure 1.6 presents the distorted impression that the rate of return is dropping at an incredible rate. ■

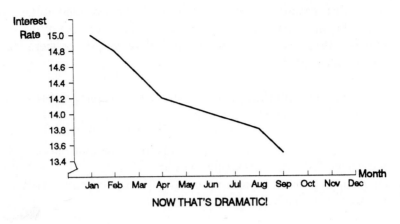

Figure 1.6

Another graph commonly used is the *bar graph* as illustrated in Example 1.16.

Example 1.16

In order to attract new advertisers, a magazine that wishes to depict dramatic growth in its circulation produced the following bar graph.

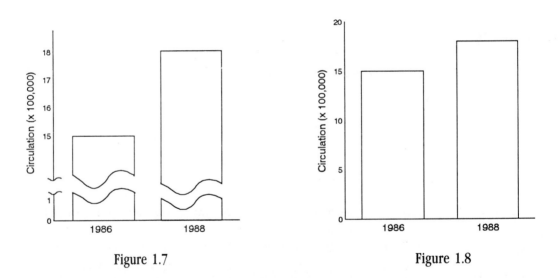

Figure 1.7   Figure 1.8

Notice the bars in Figure 1.7 are presented with a portion missing. If not examined carefully this gives the visual impression that the magazine's circulation for 1988 is twice the circulation for 1986. However, when the bars are completely drawn as illustrated in Figure 1.8, one sees only a slight increase in circulation. ■

A *pictograph* is another popular graphing technique used to distort the truth.

Definition: Pictograph. A pictograph is a graph involving pictures of objects in which the size of the object in the picture represents the relative size of the quantity being represented by the object.

Example 1.17

During a recent contract negotiation, the owner of a baseball team decided to dramatize the fact that his players' salaries have doubled from 1988 to 1991 by placing an advertisement in a local newspaper as illustrated by Figure 1.9.

1988   1991

Figure 1.9

This pictograph is misleading because it creates a visual impression that the baseball players' salaries have more than doubled since 1988. This impression was created by doubling both dimensions of the 1988 picture. This doubling of both dimensions quadrupled the area of the 1991 picture. ∎

II NON-REPRESENTATIVE SAMPLES

Since a sample is only a portion of a population, great care must be taken to insure that the sample is representative of the population. Drawing a conclusion from a non-representative sample (or a biased sample) is another device often used to distort the truth.

Let's take a look at an instance where a non-representative sample led to an inappropriate inference about the population.

The 1936 Literary Digest Poll predicted that the Republican Presidential Candidate Alfred M. Landon would defeat the Democratic incumbent Franklin D. Roosevelt by an overwhelming margin. However, to the surprise of the Literary Digest pollsters, Roosevelt won the election by a landslide. The Literary Digest made a staggering error when it used an unrepresentative sample. The Digest's sampling procedure involved mailing sample ballots to ten million prospective voters of which only 2.3 million ballots were returned.

However, the 2.3 million ballots that were returned represented only that portion of the population with a "relatively intense interest" in expressing their opinion on the election. Thus, this sample was non-representative (biased). In the 1936 election, it seems clear that the minority of anti-Roosevelt voters felt more intense about expressing their opinions about the election than did the pro-Roosevelt majority. In essence, the Digest's error was to predict the outcome of the election based on a *biased* voluntary response to its mailed survey. (Incidentally, the Literary Digest went bankrupt soon after).

Examples 1.18, 1.19 and 1.20 contain conclusions that are based upon sample information where the information regarding the sampling selection process was not mentioned. In particular, information such as sample size and representativeness of the population have been conveniently omitted. In such instances it is important to question the procedures used to gather the sample information. For each of the examples we have suggested a possible deception that could have been used in the selection of the sample that led to the conclusion.

Example 1.18

One recent study stated that "90% of the people surveyed were in favor of nuclear power."

What if this survey was based on only 10 workers whose livelihood is dependent upon the construction of nuclear power plants, would you consider the 90% statistic startling?
Probably not, because the study was based on a small non-representative sample! ∎

Example 1.19

An advertisement in a video magazine stated that "8 out of 10 VCR owners recommend the use of a wet head cleaner."

What if the ad agency trying to market this cleaner polled many groups of 10 VCR owners until they obtained a result favorable to their product, never mentioning the unfavorable results, would you then consider this advertisement to be misleading?
Yes, because the ad agency's advertisement is only using the sample result that supports the use of the product. ∎

Example 1.20

An independent laboratory test showed that "People who use DPT toothpaste have 30% fewer cavities."

This laboratory made an impressive statement about DPT toothpaste. **But, there are some questions which need to be answered.**
First, how many people were sampled?
Second, was this the only sample selected?
Third, what does the statement "...30% fewer cavities" mean? Fewer than what?! Could this mean fewer than those people who did not use any toothpaste at all? ∎

III *Inappropriate Comparisons*

The statement "Do Not Compare Apples to Oranges," conveys the essence of this next deceptive technique which is illustrated in Example 1.21.

Example 1.21

During the gasoline shortage of the 70's, people started buying foreign cars because they were more fuel efficient. An advertisement agency for an American automobile manufacturer trying to persuade potential buyers to purchase an American-made auto used a television commercial showing two automobiles, the American-made auto and the foreign import, side by side. The commercial emphasized the Environmental Protection Agency's (EPA) fuel efficiency rating of the two autos. In fact, on the screen, the EPA rating figures were shown in bold print as illustrated in Figure 1.10.

Figure 1.10

The EPA ratings represented two different driving conditions which were inconspicuously written in small print. The fact that the EPA rating of the American-made auto was based on highway driving while the import's EPA rating was *not* based on highway driving made this an inappropriate comparison. ∎

IV THE OMISSION OF VARIATION ABOUT AN AVERAGE

The omission of pertinent statistical figures concerning variation about an average can result in the public drawing inaccurate conclusions.

Example 1.22

New mothers usually read many books about child development. However, these books can cause many unpleasant moments, especially if the reader misinterprets the author's statements concerning a child's development rate.

For instance, a new mother may interpret the statement, "A child begins to walk on the average around 12 months," to mean that if her child does not walk by 12 months, then her baby is abnormal or that the baby is a late developer.

This type of misinterpretation can be avoided if the author provided a range of months for the developmental walking stage. For example, the statement, "A child begins to walk between 11 to 15 months," is more helpful than a statement giving just the average walking age. ∎

❄ 1.4 OVERVIEW AND SUMMARY

In this chapter we introduced the two branches of statistics, descriptive and inferential, and showed ways statistics are used and misused. Within the next few chapters, we will explore ways to describe and summarize collections of data. We will then consider the concepts and techniques associated with inferential statistics. These include the study of probability, the notions of sampling, and the statistical models of inference.

Throughout the text, we will present real life examples, case studies, and articles to help you develop an appreciation of the use of statistics. Also, we will include computer applications using the statistical package MINITAB and instructions on the use of a calculator to perform statistical computations. Throughout the text, we will display examples of computer output. An example of computer output that has been generated by MINITAB is shown below. MINITAB was used to analyze and graph the following distribution of 50 IQ scores.

IQ Scores

95	115	120	105	116	127	101	125	99	104	114	130
130	107	122	110	145	103	103	135	118	105	115	128
131	118	117	113	104	126	104	107	129	134	108	112
104	98	123	105	134	138	121	99	112	127	136	121
102	111										

In MINITAB, the command DESCRIBE is used to generate the following analysis of the distribution of IQ scores. We will explain this output and analysis as we study the appropriate concepts within the following chapters.

MTB > DESCRIBE C8

	N	MEAN	MEDIAN	TRMEAN	STDEV	SEMEAN
IQ	50	116.12	115.00	115.80	12.47	1.76

	MIN	MAX	Q1	Q3
IQ	95.00	145.00	104.75	127.00

The command HISTOGRAM is used to generate the following graph of the distribution of IQ scores. Histograms will be discussed in Chapter 2.

MTB > HISTOGRAM C8

Histogram of IQ   N = 50

Midpoint	Count	
95	1	*
100	5	*****
105	11	***********
110	5	*****
115	6	******
120	6	******
125	5	*****
130	5	*****
135	4	****
140	1	*
145	1	*

GLOSSARY

TERM	SECTION
Raw Data	1.1
Data	1.1
Inference	1.1
Statistics	1.1
Descriptive Statistics	1.1
Inferential Statistics	1.1
Population	1.1
Sample	1.1
Representative Sample	1.1
Parameter	1.1
Statistic	1.1
Graph	1.3
Bar Graph	1.3
Pictograph	1.3
Non-representative Sample	1.3

20 • CHAPTER 1 INTRODUCTION TO STATISTICS

Exercises

Part I Fill in the blanks.

1. Information or observations before they are arranged or analyzed are called _____ data.
2. The collection, presentation, analysis and interpretation of data is called _____.
3. The study of statistics can be separated into two areas:
   _____ Statistics and _____ Statistics.
4. The branch of Statistics that uses numerical and/or visual techniques to summarize data is called _____.
5. The total collection of individuals or objects under consideration is called the _____.
6. The portion of the population that is selected for study is called a _____.
7. Inferential Statistics is the process of using a **sample** to draw inferences or conclusions about a _____.
8. The use of sample information to estimate a population characteristic is the primary objective of _____ statistics. Thus, the researcher must take care to insure that the sample is _____ of the population.
9. A number that is used to describe a characteristic of a sample, such as a sample average, is called a _____.
10. A number such as the population average that describes a characteristic of a population is called a _____.
11. A graph involving pictures of objects in which the size of the object in the picture represents the relative size of the quantity being represented by the object is called a _____.
12. The 1936 Literary Digest Poll is an instance where a _____ sample led to an inappropriate inference about the population.

For questions numbered 13 to 17 use one of the following words:
SAMPLE, POPULATION, STATISTIC, PARAMETER, DESCRIPTIVE, INFERENTIAL

13. The average IQ score of a sample of 100 college students represents a _____.
14. The population of basketball players at UCLA has an average height of 6'3". This is an example of _____ statistics.
15. On the basis of an independent poll, it is estimated that 57% of the voters will vote for politician DePortoro. This is an example of _____ statistics.
16. A survey of 1,000 undergraduate students at a midwestern university of 40,000 undergraduate students was taken to determine the percentage of undergraduates living at home during the current semester. The survey indicated that 20% of the students live at home.
    a) The 1,000 undergraduate students surveyed represents the _____.
    b) The 40,000 undergraduate students attending the midwestern university represents the _____.
    c) The percentage value, 20%, represents a _____.
17. During the month of December, a county hospital recorded 450 live births. Of these births, 52% were boys.
    a) The percentage value, 52%, represents a _____ of the population of December births.
    b) If the data on December births is used to estimate the percentage of boy births for the past year, then the percentage value of 52% would represent a statistic, since the December births would represent a _____ of the population.

For problems 18 to 20 use the information in each problem statement to fill in the blanks.

18. A study of all working women in San Francisco, California, was made to determine their average annual salary. A sample of 250 working women was selected, and their salaries were recorded.
    a) The population is _____.
    b) The sample is _____.
    c) The population parameter being studied is _____.
19. A survey was conducted to determine the opinion of people in a Miami suburb on a proposed law to legalize casino gambling. A sample of 400 people were questioned, and it was determined that 58% of the people surveyed were in favor of legalizing casino gambling.
    a) The population is _____.
    b) The sample is _____.
    c) The percentage value, 58%, is a statistic because it describes a characteristic of a _____.
20. A sociologist was interested in determining the

socio-economic level of families living in a suburban community.

The sociologist used the average annual income as a measure of the socio-economic level of the 10,000 families living in this community. To estimate the average annual income, the sociologist randomly selects 100 families.

Determine:
a) The population._____.
b) The sample._____.
c) The parameter being studied._____.
d) The statistic used to estimate the parameter. _____.

PART II Multiple choice questions.

1. A research medical team, using data collected from a sample of 4000 sport-related injuries, reported that the average recovery time for a broken ankle is eight weeks. This average is an example of a:
   a) parameter   b) statistic
2. Most psychology texts report that the average IQ score for an adult male is between 90 and 110. This average is an example of a:
   a) parameter   b) statistic
3. The results of a survey indicate that 23% of the district will support the proposition. This is an example of:
   a) descriptive statistics b) inferential statistics
4. The average age of those surveyed is 26.73 years. This is an example of
   a) descriptive statistics b) inferential statistics
5. Statistics describe the characteristics of a _____.
   a) population b) sample
6. Parameters describe the characteristics of a _____.
   a) population b) sample
7. The primary reason for the error in prediction made by *Literary Digest* in predicting the 1936 Presidential winner is _____.
   a) insufficient sample size
   b) the sample of 10 million was unrepresentative
   c) the 2.3 million voluntary responses was biased
8. Which of the following is an example of a parameter?
   a) In a sample of 10,000 children under age fifteen, 2,466 play soccer.
   b) The average male SAT verbal score is 460.
   c) Sixty-seven out of 100 pollsters surveyed predict Mayor DePortoro will win the election.
9. Which of the following is the most important criteria for a sample to be considered unbiased?
   a) representativeness   b) size   c) inferential

PART III Answer each statement True or False.

1. Data exploration techniques are primarily used to help discover errors in the collected information.
2. Inferential statistics helps researchers make generalizations about large populations by using data from a sample of the population.
3. Population characteristics are called parameters.
4. Sample characteristics are called statistics.
5. In most applications of inferential statistics all the observations of the population should be obtained.
6. A sample represents a portion of the population.
7. Predictions made from a straw poll are an example of inferential statistics.
8. If a sample is not representative of the population, inferences made from the sample information may not be correct.
9. The incorrect prediction made by the *Literary Digest* for the 1936 presidential election was primarily due to the mailing of the ballots to a non-representative sample of 10 million prospective voters.
10. A pictograph is a popular technique used to distort the truth because it can create a visual impression that does not accurately represent the relative size of the quantities being represented.

PART IV Problems.

1. For each of the following examples of statistics that have appeared in the media, determine which illustrate the use of inferential statistics.

   For those examples of inferential statistics, identify what you believe is the population being sampled and the sample size. State one conclusion or inference being drawn and whether or not you are in agreement. Explain.

1(a)

POLL: AMERICANS WANT TOUGHER LAWS

A nationwide survey of 1,047 people shows that two out of three favor the death penalty for murder—and long prison sentences for those convicted of violent crimes.

Half of those queried support forced sterilization of habitual criminals and the insane, according to A-T-O Inc., an Ohio business conglomerate which sponsored the study.

Sixty-three percent of the people questioned in the poll said they favor giving police more power to question suspects.

1(b)

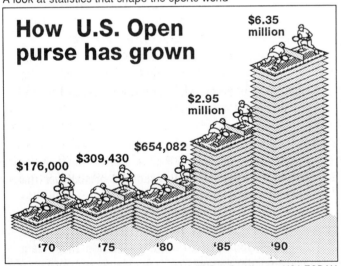

1(c)

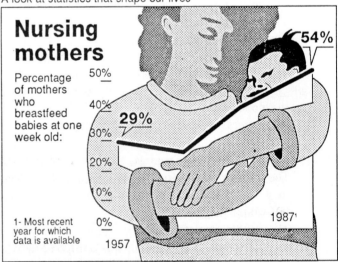

1(d)

2. The following data represents the unemployment rates from a large northeast city. Using this data sketch:
   a) a graph making the unemployment rate appear to be stable.
   b) a graph that exaggerates the rise in the unemployment rates.

Month	Unemployment
Jan	6.5
Feb	6.6
Mar	6.6
Apr	6.7
May	6.8
Jun	6.9
Jul	6.8
Aug	6.9
Sep	7.0
Oct	7.1
Nov	7.2
Dec	7.3

   rate in %

3. The following data represents the trend of U.S. imports of foreign automobiles from 1980 to 1990. Using this data construct:
   a) a graph exaggerating the rise in U.S. imports of foreign autos due to the gasoline shortage.
   b) an accurate graph depicting the trend of U.S. imports of foreign autos.

Year	80	81	82	83	84	85	86	87	88	89	90
Auto Imports per million Population	45	46	50	60	65	66	70	75	78	80	83

4. Discuss how the presentation of the statistics within each of the following statements could be misleading.
   a) Ivory Soap is 99.44 % pure.
   b) Four out of five doctors prescribe the pain reliever found in aspirin for headache pain.
   c) Nine out of ten dentists recommend chewing sugarless gum for their patients who chew gum.
   d) A brand of chewing tobacco claims that it is 40% juicier.
   e) One recent study claimed that 55% of all insured homes in America are underinsured, that is, insured to less than 80% of actual replacement cost.
   f) The profits of a foreign owned oil company have doubled in the last two years.

   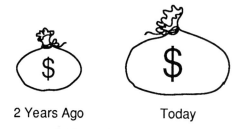
   2 Years Ago   Today

   g) The bar graph indicates how the top five insurance companies pay off on auto accidents. Company X is using the graph to express just the *plain truth*.

   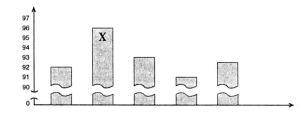

   h) A prominent mathematics association journal used the following graph to illustrate its increase in circulation. The ad read:
   **"In Five Years, We'll Have
   10 GREAT Years BEHIND us!"**

i) A New Mexico land developer, Rio Oro, used the following ad to influence people to purchase land in Arizona.

A GOOD Investment!

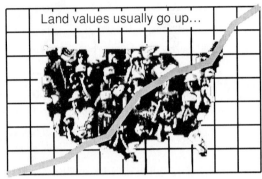

AS POPULATION INCREASES

j) Rio Oro used the following pictograph to explain why investors would prefer Arizona land. The ad was stated as:

"Here's why so many investors PREFER Arizona land."

  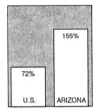

POPULATION GROWTH   EMPLOYMENT GROWTH   BANK DEPOSIT GROWTH

k) **"Join The Army – It's Safer."**

An Army recruiting center in Texas coined the above slogan based on the following statistical data.

The death rate in the Army is 5 per thousand while the death rate for civilians is 20 per thousand during the same period.

The Army recruiting center tried to persuade individuals to join the Army by explaining it was safer in the Army than out of it.

l) **"The book value of this automobile has increased through the years."**

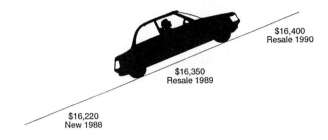

5. Read the following satiric essay entitled: **I'm Tired**, which comments on how many people work in the United States. Discuss the author's misuse of data and facts within the essay. Within your discussion, comment on:

1) the author's biases on who works and who doesn't work.
2) the author's misuse of facts. For example, are there some groups listed in the nonworking categories who do work? Is it possible to belong to two or more of these categories?
3) what the author used as the population of the labor force and why it is not as useful as the Census Bureau's definition. (Note: the Census Bureau calculates labor force statistics for persons who are 16 years and over.)
4) the percentage of people that work as stated by the author.
5) what suggestions would you give the anonymous author that would help him/her to use the data more accurately.

I'm Tired

I'm tired. For several years I've been blaming it on middle age, iron poor blood, lack of vitamins, air pollution, obesity, dieting, and a dozen other maladies that make you wonder if life is really worth living.

But now I find out, tain't that. I'm tired because I'm overworked.

The population of this country is 200 million. Eighty-four million are retired. That leaves 116 million to do the work. There are 74 million in school, which leaves 42 million to do the work. Of this total, there are 23 million employed by the Federal Government. That leaves 19 million to do the work.

Four million are in the Armed Forces, which leaves 15 million to do the work. Take from that total the 14,800,000 people who work for States and Cities. That leaves 200,000 to do the work. There are 188,000 in hospitals, so that leaves 12,000 to do the work.

Now, there are 11,998 people in prisons. That leaves just two people to do the work. You and me. And you're sitting there reading this.

No wonder I'm tired!!

6. The article entitled "How Ads Can Bend Statistics" appeared in the *NY Times* on January 17, 1981.

CONSUMER SATURDAY

HOW ADS CAN BEND STATISTICS

"Doctors recommend" "Four out of five people preferred...." "Twice as many repairmen choose...." The slogans and claims are survey advertisements, a technique of selling products that has grown in use over the last five years and one in which the Federal Trade Commission is increasingly interested.

"It is an important technique because consumers give it credibility and they use the information contained in them to make important purchasing decisions," Wallace S. Synder, assistant director for the F.T.C.'s advertising practices division, said in Washington this week. Regulation of advertising is a traditional function of the F.T.C.

Survey advertisements are an effective tool increasingly used in the 60 billion-a-year advertising trade, but such advertisements are not always as scientific or responsible as they may sound.

During the last two weeks, three manufacturers have made consent agreements with F.T.C. to discontinue the use of less-than-substantive claims for their products.

In one, Litton Industries agreed not to advertise its microwave oven as superior to others unless there is a "reasonable basis" for such claims. The company had been claiming that independent technicians preferred their product over competing products, but their survey used only authorized Litton service agencies' technicians who could possibly have a bias toward their product. In addition, there was no effort to make sure that the technicians who preferred Litton microwave ovens were experienced at repairing ovens other than Litton's.

Under F.T.C. regulation, "reasonably basis" means that competent and reliable surveys or tests are conducted by qualified persons. The results, say the regulation, must be objectively evaluated "using procedures that ensure accurate and reliable results." The Litton advertisements, the agency said, did meet such standards.

In a similar case, Standard Brands and Ted Bats & Company, formerly the advertising agency for Standard Brand's Fleischmann's Margarine, agreed not to make claims based on surveys or tests unless they were scientifically conducted and the claims accurately reflected the results.

The agreements between the companies and the F.T.C. cover all the products of Standard Brands, one of the nation's largest food and beverage companies, and every advertisements created by the advertising agency.

At issue were advertisements used between 1974 and 1977. They claimed the doctors preferred and recommended Fleischmann's Margarine. Two of the four advertisements, allegedly based on a national survey of doctors stated:

"When a doctor chooses margarine, chances are it's Fleischmann's."

The F.T.C. alleged that the claims were unproven, charging that only 15.5 percent of the doctors surveyed recommended any specific name brand.

The companies have now agreed that any surveys or tests featured in advertisements will be designed and analyzed in a reliable scientific manner and must clearly establish any superiority claims.

In another case, Teledyne Inc. and Teledyne Industries Inc. makers of Water Pik, and J. Walter Thompson, its advertising agency, agreed that their advertising claims about the effectiveness of the dental-care product in preventing gum diseases would have a "reasonable basis." The Water Pik is an "oral irrigator" that sprays water into the mouth and that, advertisements say, flushes away debris from beneath the gum line and "reduces the causes of gum disease."

The F.T.C. further alleged that Teledyne falsely claimed that the Water Pik is approved by the American Dental Association. To make such a claim, the endorsement must be made in writing, according to the F.T.C.

C. Collot Guerard, deputy assistant director of the F.T.C.'s advertising division, offered pointers for consumers when they read or see an advertisement:

"Ask yourself," she said, "how many doctors were actually surveyed when you see something like "five out of six doctors." Consumers tend to think of big, nationally projected surveys, but it might not be so."

Learn to recognize the "dangling comparatives" in advertisements such as "Products X will get your teeth whiter than ever" and ask yourself just what does that mean. Whiter than the same product did a year ago? Whiter than other products.

Watch for parity claims. "If an advertisements states that "no hair dryer will dry your hair quicker,'" she said "it implies that it dries faster than other brands. Usually it means that it dries just as quick as other products."

Maintain a healthy skepticism.

New York Times © 1981
Reprinted by permission

This article discusses various problems that arise when claims are made within advertisements. Enumerate several of these problems and discuss some of the possible solutions.

7. A cereal manufacturer used the following graph within its advertisement to convince prospective buyers to purchase their new cereal *CRUNCH-O*.

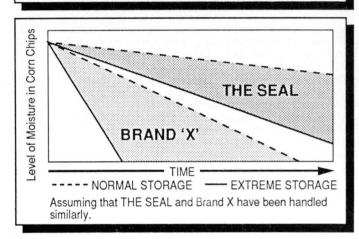

THE SEAL™ retains the perfect moisture level in an unopened pack. In fact, it retains moisture ten times better than conventional packaging. Ten times better! And it's 1000 times better at keeping out air.

The advertisement claimed that their new packaging technique called "THE SEAL" keeps the cereal fresher.
a) Examine the graph and explain why it is misleading.
b) The advertisement also states the following:

"THE SEAL" retains the perfect moisture level in an unopened box. In fact, it retains moisture 10 times better than conventional packaging. *Ten times better! And it's 1000 times better at keeping out air.*

Comment on why this statement can be misleading. Select the words or phrases that need to be further explained.

8. The following advertisement appeared in a wildlife magazine:

Environmental T-Shirts
Support wildlife by wearing environmental t-shirts.
(10% of profits go to environmental groups)
a) If each t-shirt costs $15 and the manufacturer's profit per t-shirt is $5, then how much of your $15 is being contributed to the environmental groups?
b) Calculate the percent of the $15 that would be contributed to the environmental groups.
c) Do you feel this advertisement can be misleading?

PART V What Do You Think?

1. **What Do You Think?**

The following statistic appeared in the periodical *USA Today*:

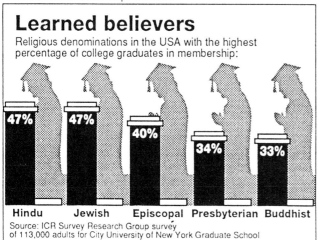

a) Identify the population being surveyed.
b) Since there is a greater number of Roman Catholics in the US than Hindus, why don't the Roman Catholics appear as one of the top percentages?
c) Why are Hindus one of the top percentages?
d) What information would you like to see that was not included?

2. **What Do You Think?**

The following statistic appeared in the periodical USA Today:

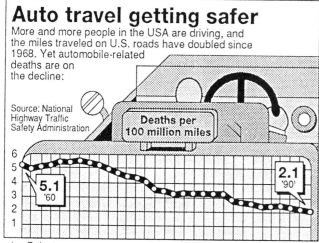

a) Identify the population being surveyed.
b) What factors might contribute to the comment "Auto travel getting safer"?
c) Approximately between what two years has the death rate remained relatively stable?
d) Can you conclude from the graph that the number of automobile-related accidents have also decreased?

PART VI **Projects.**

1. While watching TV pay close attention to the commercials.
   a) Determine the type of statistics (descriptive or inferential) being presented.
   b) Comment as to what information (if any) has been left out that might lead you to a different conclusion.
2. Look in magazines or newspapers for ads using statistics. Comment on the presentation and state if you think any deceptive tricks might have been utilized.

PART VII **Database.**

The following exercises refer to the file DATABASE listed in Appendix A. We have indicated the appropriate MINITAB commands that are necessary to answer each exercise.

1. Using the MINITAB commands:
   RETRIEVE and PRINT
   Retrieve the file DATABASE.MTW and PRINT the data for the variables:
   a) AGE   e) STY
   b) AVE   f) GPA
   c) SEX   g) GFT
   d) MAJ

   For the variables SEX and MAJ identify what each data value represents.

   For the other variables except SEX and MAJ determine the lowest and hightest data values.

2. Using the MINITAB commands:
   RETRIEVE and PRINT
   Retrieve the file DATABASE.MTW and PRINT the data for the variables:
   a) ID   d) SEX   g) GPA
   b) AGE   e) MAJ   h) GFT
   c) AVE   f) STY

   Discuss the characteristics of the student whose ID number is:
   1) 8573   3) 8809
   2) 7679   4) 3529

CHAPTER 2
ORGANIZING AND PRESENTING DATA

❋ 2.1 INTRODUCTION

The student government association is interested in exploring some characteristics of the student body. A survey was developed and administered to a sample of 150 students. After the data was collected, a graph comparing a student's major area of concentration to the student's sex appeared in the school's newspaper. This graph is shown in Figure 2.1.

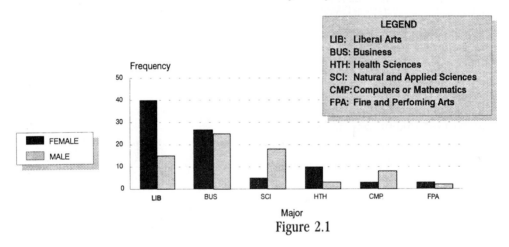

Figure 2.1

Notice how this graph is a very convenient way to convey the number of males and females that belong to the five listed majors. From the graph, you can see instantly that there is a greater number of males than females majoring in Computers or Mathematics, and the Natural and Applied Sciences. On the other hand, the graph also shows there is a greater number of females than males majoring in Liberal Arts, Business, Health Sciences and Fine and Performing Arts.

Graphs are commonly used to organize, summarize, and analyze collections of data. Although you frequently see graphs in the media, have you ever wondered about the procedures that are used to generate them?

One important step is data collection. This is often accomplished by using a survey. The graph in Figure 2.1 was constructed from the data collected in the student government's survey. The survey contained the following questions:

SURVEY QUESTIONS

1. How many credits are you currently enrolled in?

2. How many hours per week do you spend studying for your courses?

3. What is your current cumulative Grade Point Average?

4. **Select the area of concentration (Major) that best indicates your field of study:**

   a) **Liberal Arts**
   b) **Business**
   c) **Health Sciences**
   d) **Computers or Mathematics**
   e) **Natural and Applied Science**
   f) **Fine and Performing Arts**

5. **Sex: (M) male (F) female**

After the data was collected, a listing representing the **raw data** was compiled as shown in Table 2.1.

The data set in its present form, as displayed in Table 2.1, makes it very difficult to observe the characteristics of the data. In this chapter we will present techniques that will help us organize, summarize, and explore the data to determine if there exists patterns or relationships among the data. For the given data set, these techniques will be able to help us answer questions such as:

1) How many credits do most students take for a semester?
2) What is the most popular major?
3) How many hours do students usually study per week?

or to explore meaningful comparisons such as:

4) the number of hours that females study to the number of hours that males study.
5) the relationship of the type of major to a student's sex.
6) the number of study hours per student to the student's GPA.

Within this chapter, we will present techniques that involve the use of tables and graphs to display the important characteristics of the data. These methods are dependent upon the types of data collected. In Section 2.2, we will discuss the different classifications and characteristics of data.

Table 2.1
Student Government Survey Data of 150 Students

The column abbreviations are defined as follows:

SEX: student sex
M: male
F: female

HR: study hours per week
GPA: student grade point average
CR: number of credits this semester

MAJ: college major
LIB: liberal arts
BUS: business

CMP: computers or mathematics
SCI: natural and applied sciences
HTH: health sciences
FPA: fine and performing arts

SEX	MAJ	HR	GPA	CR	SEX	MAJ	HR	GPA	CR	SEX	MAJ	HR	GPA	CR	SEX	MAJ	HR	GPA	CR
M	LIB	15	2.79	14.0	M	BUS	24	3.61	18.0	F	LIB	19	3.05	17.0	F	LIB	8	2.47	15.5
M	BUS	18	2.87	12.5	M	BUS	20	3.45	19.0	M	BUS	31	3.62	15.0	M	LIB	10	2.66	13.0
F	LIB	17	2.71	15.0	F	BUS	15	2.42	15.0	F	CMP	33	3.67	18.0	M	BUS	12	3.28	15.0
M	CMP	20	3.42	17.0	F	LIB	12	2.49	14.0	F	BUS	23	2.25	10.5	F	LIB	15	3.41	17.0
M	BUS	19	2.88	6.0	M	CMP	24	3.41	13.0	M	LIB	20	2.70	15.0	M	LIB	31	3.88	18.0
F	LIB	23	2.91	12.0	F	BUS	29	3.18	12.0	F	LIB	17	2.30	7.5	F	LIB	11	2.32	10.5
M	LIB	8	1.76	18.0	M	CMP	30	3.65	7.5	M	LIB	16	2.51	8.0	F	HTH	20	2.97	16.0
M	SCI	10	2.95	15.0	F	LIB	32	3.55	10.5	M	BUS	26	3.54	18.0	F	BUS	29	3.75	15.5
F	LIB	27	2.82	4.5	F	LIB	30	3.45	12.0	M	BUS	12	1.92	4.5	F	LIB	20	2.78	12.0
M	SCI	28	2.96	12.0	M	LIB	19	2.35	16.0	F	LIB	20	3.15	13.0	F	HTH	15	2.84	16.0
M	BUS	20	2.48	17.0	M	BUS	20	2.37	15.0	F	BUS	11	2.00	3.0	F	HTH	10	3.10	17.0
F	BUS	27	2.24	15.0	F	LIB	21	2.54	17.0	F	LIB	9	1.95	15.0	F	LIB	16	2.49	10.0
F	LIB	36	3.02	14.0	F	BUS	21	2.00	10.0	M	BUS	9	2.88	13.0	F	HTH	10	3.24	15.5
F	LIB	17	2.45	15.5	M	LIB	22	2.59	17.0	F	LIB	9	2.65	15.0	M	BUS	21	2.95	7.5
F	HTH	24	3.39	15.0	F	LIB	17	1.94	14.0	F	LIB	8	1.98	12.0	M	BUS	23	2.81	8.0
F	SCI	26	2.75	17.0	F	BUS	16	2.59	15.5	F	BUS	19	3.21	16.0	F	HTH	10	2.36	17.0
M	BUS	28	2.82	16.5	M	LIB	19	3.16	16.5	M	BUS	7	1.95	12.0	M	CMP	17	3.10	15.0
F	BUS	22	2.29	15.0	F	LIB	18	2.66	7.5	F	LIB	10	2.96	15.0	F	LIB	28	3.78	16.0
F	BUS	16	3.56	12.0	M	BUS	22	3.05	13.0	M	LIB	8	2.95	16.5	M	LIB	15	3.08	17.0
M	LIB	27	3.28	8.0	M	LIB	21	3.16	15.0	M	BUS	18	2.00	10.5	M	BUS	11	3.07	15.0
F	BUS	20	3.06	14.0	M	LIB	24	3.29	12.0	F	LIB	16	2.82	16.0	F	LIB	12	3.15	13.0
M	SCI	24	3.56	7.5	F	LIB	20	2.01	12.0	F	BUS	10	1.95	11.0	F	LIB	15	3.65	12.0
M	BUS	27	3.54	4.5	M	BUS	27	3.51	7.5	F	LIB	12	3.01	13.0	M	BUS	14	3.70	17.0
F	BUS	18	3.07	12.0	F	BUS	18	2.84	15.0	F	HTH	15	2.70	14.0	F	LIB	21	2.73	15.0
M	SCI	28	3.05	12.0	F	LIB	29	3.59	16.0	F	LIB	11	2.67	17.0	M	LIB	17	3.75	17.0
M	BUS	24	2.92	14.0	M	CMP	23	3.25	17.0	M	CMP	9	2.81	16.0	F	LIB	20	3.17	14.0
F	HTH	12	2.48	15.0	M	LIB	25	3.65	7.5	M	LIB	27	3.76	15.5	M	BUS	25	3.56	12.0
M	SCI	24	3.23	13.0	M	CMP	31	3.89	12.0	F	CMP	10	2.67	14.0	F	FPA	30	3.75	17.0
F	LIB	25	3.17	7.5	F	BUS	23	2.96	15.0	F	HTH	7	2.32	14.0	F	BUS	33	3.66	18.0
M	BUS	26	3.57	10.0	F	BUS	18	2.77	15.5	F	BUS	8	2.09	16.0	M	LIB	28	3.85	7.5
F	BUS	28	3.19	11.0	F	LIB	26	3.28	12.0	F	LIB	9	2.42	17.0	M	LIB	9	2.74	17.0
F	LIB	30	3.08	12.0	M	LIB	27	3.13	10.0	F	LIB	10	1.90	6.0	F	BUS	9	2.64	15.0
M	CMP	15	2.47	6.0	F	BUS	33	2.75	7.5	M	LIB	7	1.98	12.0	F	FPA	12	2.93	12.0
F	LIB	12	1.88	7.5	M	CMP	20	2.76	17.0	F	LIB	12	3.15	16.0	M	LIB	24	2.47	16.0
M	BUS	11	2.40	15.0	M	LIB	17	3.17	6.0	F	HTH	14	3.40	17.0	M	FPA	8	2.18	14.0
F	BUS	24	2.89	17.0	M	HTH	15	1.86	16.0	F	BUS	10	2.87	15.0	M	LIB	6	2.43	13.0
F	LIB	25	3.25	14.0	F	BUS	17	2.81	13.0	F	LIB	28	3.51	17.0					
F	LIB	16	2.65	14.0	M	LIB	11	1.87	6.0	F	BUS	8	1.95	12.0					

2.2 CLASSIFICATIONS OF DATA

An important consideration in the collection of data is the type of information about a person or thing you are interested in obtaining. For example, a person's height, an individual's sex, a student's GPA, the circumference of a redwood tree, the closing price of a stock, the pollen count for a given day, or one's preference in the choice of an automobile illustrate types of information pertaining to a person or thing. These types of information are called **variables**.

> **Definition: Variable.** A variable is a type of information, usually a property or characteristic of a person or thing, that is measured or observed.

A **specific measurement or observation** for a variable is called the **value** of the variable. Suppose you are interested in the variables height, sex and GPA for the students in your class. If a female student in the class stands 63 inches tall and has a GPA of 3.23, then female would be the value for the variable sex, 63 inches represents the value for the variable height and 3.23 is the value for the variable GPA. A collection of pieces of data pertaining to one or more variables is called a **data set**. Table 2.1 represents a data set. In Example 2.1, we will examine the variables of this data set.

Example 2.1

Using Table 2.1,
a) identify the variables.
b) determine the value of each of these variables for the first student listed in the table.

Solution:

a) The variables are Sex, Major, Study Hours, GPA, and Credits.
b) The first student is a Male with a Major in Liberal Arts, who studies 15 hours per week, has a GPA of 2.79 and is taking 14 credits.

Therefore, the variable:

sex has a value of male;
major has a value of Liberal Arts;
study hours has a value of 15 hours per week;
GPA has a value of 2.79; and
Credits has a value of 14. ■

CATEGORICAL AND NUMERICAL VARIABLES

There are basically two types of variables: **Categorical** and **Numerical**. **A categorical variable yields values that denote categories.** For example, the variable sex is a categorical variable since sex is classified by the categories male or female. The variable major is a categorical variable where the values have been classified by the categories liberal arts, business, health sciences, computers or mathematics, natural and applied sciences, or fine and performing arts. **The values of these categorical variables are referred to as categorical data.**

> **Definition: Categorical Variable.** A categorical variable represents categories, and is non-numerical in nature. Its values are referred to as categorical data.

A numerical variable yields numerical values that are the result of a measurement. For example, the variable GPA is a numerical variable because it represents a numerical measurement of a student's academic performance. The variable study hours is a numerical variable since it numerically measures the number of hours per week a student studies. The specific values of numerical variables are called **numerical data**.

> **Definition: Numerical Variable.** A numerical variable is a variable where the values are the result of a measurement process. Its values are referred to as numerical data.

Numerical data can be further classified as either continuous or discrete depending upon the numerical values it can assume.

Continuous Data

Numerical data such as height, weight, temperature, distance, and time are examples of continuous data because they represent measurements that can take on any value between two numbers. The values of continuous data are usually expressed as a rounded-off number.

A person's weight is an example of continuous data since it is a measurement that can assume any value between two numbers. For example, a person's true weight could be any value between two numbers such as 150.5 and 151.5 pounds. Usually, we represent this person's weight as the *rounded-off weight* of 151 pounds.

A meteorologist may report that the temperature in Central Park at 2 p.m. is 83 degrees. This temperature of 83 degrees is a rounded off measurement usually representing a value between 82.5 degrees and 83.5 degrees. Thus, temperature is an example of continuous data.

Discrete Data

Numerical data such as the number of full-time students attending Yale Law School, the number of automobile accidents in Arizona per year, or the number of unemployed people in New York City are examples of discrete data because these data can only take on a limited number of values. Usually discrete data values represent **count** data and are expressed as whole numbers. However, the main characteristic of discrete data is the **break** between successive discrete values.

The closing price of a stock selling between $24 and $25 is considered discrete data since it can only take on a limited number of price values such as: $24, $24 1/8, $24 1/4, $24 3/8, $24 1/2, $24 5/8, $24 3/4, $24 7/8, $25.

Example 2.2

For each of the following:
a) Identify the variable and classify the variable as either categorical or numerical.
b) If the variable is numerical, then further classify it as either continuous or discrete.

1) A person's height.
2) An individual's political affiliation.
3) The number of rainy days in July.
4) The number of gallons of fuel used to travel from New York to Chicago.
5) A license plate number.
6) A person's hair color.
7) An opinion to the question: "Is the President of the United States doing a good job?"
8) A shoe size.
9) The number of inches of snowfall per year.
10) The number of vowels in a paragraph.
11) The time of day.
12) The amount of soda in a 2 liter bottle.
13) A student's credit load for the semester.

Solution:

1) A person's height represents a numerical variable. Since height is a measurement that has been rounded off, it is continuous.
2) Political affiliation represents a category. Therefore, this is a categorical variable.
3) The number of rainy days in July is a numerical variable. This number represents count data and is therefore discrete.
4) The number of gallons of fuel used to travel from New York to Chicago is a numerical variable. This represents a measurement that has been rounded off and is continuous.
5) License plate numbers represent a categorical variable since they identify the owner of the automobile.
6) A person's hair color is a categorical variable.
7) One's opinion to a question represents a categorical variable.
8) Shoe size is a numerical variable. Since shoe sizes can only take on a limited number of values, it is discrete.
9) The number of inches of snow is a numerical variable. Since this number is a measurement that has been rounded off, it is continuous.
10) The number of vowels in a paragraph represents a numerical variable. Since this number is count data, it is discrete.
11) The time of day is a numerical variable. It represents a measurement that has been rounded off. Therefore, it is continuous.
12) The amount of soda in a 2 liter bottle is a numerical variable. The amount represents a measurement that has been rounded off. Therefore, it is continuous.
13) The number of credits is a numerical variable. Since the number of credits can only take on a limited number of values, it is discrete.

2.3 EXPLORING DATA USING THE STEM-AND-LEAF DISPLAY

Statistical studies, marketing surveys, scientific and medical experiments all involve collections of data. Before any conclusions can be drawn from these collections of data, the data must be carefully explored to discover any useful aspects, information, or unanticipated patterns that may exist in the data. This approach to the analysis of data is called **Exploratory Data Analysis**. The idea of exploratory data analysis is to learn as much as possible about the data before conducting any statistical testing of hypotheses or relationships, or drawing any conclusions about the data. The assumption of exploratory data analysis is that the more you know about the data, the more effectively the data can be used to perform any subsequent statistical analysis. The emphasis of exploratory data analysis is to use visual displays of the data to reveal vital information about the data. One visual technique used to explore data is the **stem-and-leaf display** developed by John W. Tukey of Princeton University. The stem-and-leaf display helps to provide information about the shape or pattern of the data values of the variable. To help examine the shape or pattern of the data values, it is necessary to organize the data values to form a **distribution**.

> **Definition: Distribution.** A distribution represents the numerical data values of a variable from the lowest to the highest value along with the number of times each data value occurs. The number of times each data value occurs is called its *frequency*.

A visual display of a distribution of data values helps to yield valuable information about the shape of the distribution.

> **Definition: Stem-and-Leaf Display.** The stem-and-leaf display is an exploratory data analysis technique that visually shows the shape of a distribution. The display uses the actual values of the variable to present the shape of the distribution of data values.

To see how the stem-and-leaf display is constructed, let's analyze the following data values representing the cholesterol levels of 50 middle-aged men on a regular diet.

Cholesterol Levels
(expressed in milligrams of cholesterol per 100 milliliters)

263	258	240	233	225	222	199	282	239	236	232
283	200	212	225	235	240	258	263	274	250	259
241	237	226	213	269	199	253	201	265	226	238
242	259	233	238	229	215	202	319	277	229	239
243	248	245	219	276	246					

To construct a stem-and-leaf display for this distribution of 50 cholesterol levels, we will use the actual data values to display the shape of the distribution. This is accomplished by rewriting each data value into two components called a stem and a leaf. Let's discuss how to do this. The first data value from the list of cholesterol levels is 263, and we will write this data value within the stem-and-leaf display as: **26 | 3**. The first two leading digits (26), which appear to the left of the vertical line, are referred to as the **stem**. The last digit of the data value (3), which appears to the right of the vertical line, is called the **leaf**. This is illustrated as follows:

STEM LEAF
the data value of 263 is displayed as: 26 | 3

For each cholesterol level, the stem will pertain to the first two digits of the data value, while the leaf of the data value will indicate the last digit of the data value. Thus, the second data value of 258 would be represented as: **25 | 8** within the stem-and-leaf display, where the 25 represents the stem and 8 is the leaf of this data value. The stem portion of the third data value of 240 is 24, while the leaf portion is 0. Thus, the cholesterol level of 240 is expressed as: **24 | 0** within the stem-and-leaf display. To construct a stem-and-leaf display for the 50 cholesterol levels, we will use the following procedure:

Step 1. List all the possible stems, regardless of whether there exists a data value having such a stem, in a vertical column starting with the smallest stem on top, and ending with the largest stem at the bottom of the column.

Since the smallest data value is 199 and the largest data value is 319, then the top stem value will be 19 (which corresponds to the smallest cholesterol value of 199) and the bottom stem value will be 31 (which corresponds to the largest cholesterol value of 319). Thus, the vertical column representing the stem values will include all stem values starting with 19 and ending with 31. This is illustrated in Figure 2.2.

Step 2. Draw a vertical line to the right of the column of stems. This is illustrated in Figure 2.2. The vertical line simply serves to separate the two parts (stem and leaf) of each data value.

```
Stem |
-----+------------------
  19 |
  20 |
  21 |
  22 |
  23 |
  24 |
  25 |
  26 |
  27 |
  28 |
  29 |
  30 |
  31 |
```

Figure 2.2

Step 3. Place the leaf of each data value in the row of the display corresponding to the stem of the data value and to the right of the vertical line that separates the stem portion (or left side) of the display from the leaf portion (or right side) of the display.

Thus, for the first data value of 263, the leaf portion 3 will be written within the stem row 26, but the 3 will appear to the right of the vertical line which separates the stem from the leaves of the display. This procedure of placing the leaf of each data value within the appropriate stem row is continued until all the data values are included within the stem-and-leaf display. The completed stem-and-leaf display is illustrated in Figure 2.3.

```
Stem | Leaf
19   | 9 9
20   | 0 1 2
21   | 2 3 5 9
22   | 5 2 5 6 6 9 9
23   | 3 9 6 2 5 7 8 3 8 9
24   | 0 0 1 2 3 8 5 6
25   | 8 8 0 9 3 9
26   | 3 3 9 5
27   | 4 7 6
28   | 2 3
29   |
30   |
31   | 9
```

Figure 2.3

Step 4. Arrange the leaves within each stem row in increasing order from left to right. This provides a more informative stem-and-leaf display. In practice, it is not necessary to name the stem and leaf portions within the display, so we've omitted them in Figure 2.4.

Stem-and-leaf display of cholesterol levels
of 50 middle-aged men on a regular diet

```
19 | 9 9
20 | 0 1 2
21 | 2 3 5 9
22 | 2 5 5 6 6 9 9
23 | 2 3 3 5 6 7 8 8 9 9
24 | 0 0 1 2 3 5 6 8
25 | 0 3 8 8 9 9
26 | 3 3 5 9
27 | 4 6 7
28 | 2 3
29 |
30 |
31 | 9
```

Figure 2.4

The stem-and-leaf display arranges the data in a very convenient form since the number of leaves for each stem will indicate the number of data values pertaining to that stem. For example, in the second row of Figure 2.4, we have: **20 | 0 1 2** This row represents the three data values: 200, 201, and 202. Also, rotating the page sideways, we can also see the shape of the distribution of cholesterol levels for the 50 men since the stem-and-leaf display viewed in this manner has a likeness to a bar graph. From Figure 2.4, we notice that the greatest number of cholesterol levels are concentrated between the values 222 to 248. **We can see that a major advantage of a stem-and-leaf display is that it shows the shape of the distribution by displaying the number of data values pertaining to each stem using the actual values from the distribution. We can also examine the leaves within each stem to determine the distance between the data values, thus providing more information on the shape of the distribution.** For example, upon examining the stem-and-leaf display in Figure 2.4, and in particular, the two rows pertaining to the stems 28 and 31, **28 | 2 3** and **31 | 9**, notice that the largest distance between any two data values within the distribution occurs between data values 283 and 319. In fact, you should notice that the cholesterol level of 319 stands apart from the pattern of all the other cholesterol levels. This cholesterol level is called an **outlier** since it lies well above all the other cholesterol levels within the distribution.

> **Definition: Outlier.** An outlier is an individual data value which lies far (above or below) from most or all of the other data values within a distribution.

In Figure 2.4, we have seen that the stem-and-leaf display presents the data values of a distribution in a more convenient form that will make it easier to study the shape and important characteristics of a distribution while preserving the actual data values. As we explained before, the stem-and-leaf display is an exploratory data analysis technique that assists one in looking at the data. When examining the data, it is important to know what to look for in the data. The important aspects of the data are called the characteristics of a distribution. Let's discuss some of these characteristics.

Characteristics of a Distribution

- **Center of the Distribution**
  Determine where the middle value of all the data values is located.

- **Determine the Overall Shape of the Distribution**
  Look for the number of peaks within the distribution:
  Is there one peak or are there several peaks?
  How are the data values clustered? Are they **symmetric about the center or skewed in one direction**?
  1) **A distribution is symmetric** when the data values greater than the center of the display and the data values less than the center of the display are mirror images of each other.
  2) **A distribution is skewed** when the data values tend to be concentrated toward one end of the display or tail of the distribution, while the data values in the other tail are spread out through extreme values resulting in a longer tail.

- A **positively skewed** or **skewed to the right distribution** is when the data values are concentrated in the left tail (or end of the display containing the smaller data values) with the longer tail of the display on the right.

- A **negatively skewed** or **skewed to the left distribution** is when the data values are concentrated in the right tail (or end of the display containing the larger data values) with the longer tail of the display on the left.

- **Look at the Spread of the Data Within the Distribution**
  Examine the distribution for any **gap(s)** within the distribution or for any individual data value that falls well outside the overall range of the data values, that is, look for any **outliers**.

Let's now summarize the procedure to be used to construct a stem-and-leaf display.

Procedure To Construct a Stem-And-Leaf Display

1. Define the stem and leaf you will use for your data values. In general, the stem can have as many digits as needed to represent the data values, but the leaf should only contain one digit. The stem represents the beginning digit(s) of the data values, while the leaf represents the last or terminating digit of the data values.

2. List all the possible stems, regardless of whether there are any data values having such a stem, in a vertical column starting with the smallest stem on top and ending with the largest stem at the bottom. Include all possible stems in this column, even if there are stems with no corresponding leaves.

3. Draw a vertical line to the right of the column of stems. This line will be used to separate the two parts of each data value, the stem and the leaf.

4. For each data value within the distribution, record the leaf within the corresponding stem row and to the right of the vertical line.

5. Arrange the leaves within each stem row in increasing order from left to right. This provides a more informative stem-and-leaf display.

6. If the display contains too many data values and begins to look too cramped and narrow, you can stretch the display by using two lines per stem. To stretch the display, subdivide each stem into two rows: a first row, symbolized by an "*" attached to the stem portion, containing all the leaf digits from zero through four, and a second row, symbolized by a "·" attached to the stem portion, containing all the leaf digits from five through nine. This type of stem-and-leaf display is called a stretched stem-and-leaf display.

Let's illustrate the use of this procedure in Example 2.3.

Example 2.3

For the following data values representing the cholesterol levels of 50 middle-aged men on a low-fat diet and exercise program:
  a) Construct a stem-and-leaf display.
  b) Describe the characteristics of the distribution. Is it symmetric, skewed to the right, or skewed to the left?
  c) Determine the center of the distribution and then decide between what two values on a stem that the greatest concentration of data values are located?

Cholesterol Levels
(expressed in milligrams cholesterol per 100 milliliters)

161, 172, 193, 190, 205, 214, 217, 205, 168, 176, 188, 195, 209, 219, 189, 209, 220, 211, 200, 195, 184, 175, 198, 209, 211, 228, 199, 188, 172, 165, 177, 197, 202, 211, 198, 184, 171, 200, 218, 220, 210, 189, 231, 176, 180, 193, 230, 200, 181, 190

Solution:

a) To construct a stem-and-leaf display for the 50 cholesterol levels, we will use the previously outlined procedure:

Step 1. List all the possible stems in a vertical column starting with the smallest stem on top and ending with the largest stem at the bottom of the column.

Since the smallest data value is 161 and the largest data value is 231, then the stem values will range from 16 to 23.

Step 2. Draw a vertical line to the right of the column of stems. This is illustrated in Figure 2.5.

```
Stem |
-----+
  16 |
  17 |
  18 |
  19 |
  20 |
  21 |
  22 |
  23 |
```

Figure 2.5

Step 3. For each data value within the distribution, record the leaf within the corresponding stem row and to the right of the vertical line.

```
Stem | Leaf
-----+----------------------
  16 | 1 8 5
  17 | 2 6 5 2 7 1 6
  18 | 8 9 4 8 4 9 0 1
  19 | 3 0 5 5 8 9 7 8 3 0
  20 | 5 5 9 9 0 9 2 0 0
  21 | 4 7 9 1 1 1 8 0
  22 | 0 8 0
  23 | 1 0
```

Figure 2.6

Step 4. Arrange the leaves within each stem row in increasing order from left to right. This provides a more informative stem-and-leaf display.

Stem-and-leaf display of cholesterol levels
of 50 middle-aged men on a low-fat diet and exercise program

```
16 | 1 5 8
17 | 1 2 2 5 6 6 7
18 | 0 1 4 4 8 8 9 9
19 | 0 0 3 3 5 5 7 8 8 9
20 | 0 0 0 2 5 5 9 9 9
21 | 0 1 1 1 4 7 8 9
22 | 0 0 8
23 | 0 1
```

Figure 2.7

b) Rotating the stem-and-leaf display in Figure 2.7 sideways, we see that the shape of the distribution of cholesterol levels of men on a low-fat diet and exercise program is approximately symmetrical.

c) Starting from the lowest data value and counting to the twenty-fifth data value (or to the middle of the data values), we can determine that the center of the distribution is between 197 and 198. The greatest concentration of cholesterol levels on a stem falls between 190 and 199, with the majority (more than 50%) of cholesterol levels falling between 171 and 219. ■

❊ BACK-TO-BACK STEM-AND-LEAF DISPLAY

At times, it might be of interest to compare two distributions using a stem-and-leaf display with a common stem. This type of display is called a **back-to-back stem-and-leaf display**. In the next example, we will examine how to construct a back-to-back stem-and-leaf display to compare the distribution of cholesterol levels for middle-aged men who are on a regular diet to those on a low-fat diet and exercise program.

Example 2.4

Using the stem-and-leaf display for the cholesterol levels for middle-aged men on a regular diet (shown in Figure 2.4) and the stem-and-leaf display for middle-aged men on a low-fat diet and exercise program (shown in Figure 2.7): a) construct a back-to-back stem-and-leaf display.
b) use the back-to-back stem-and-leaf display to compare the two distributions.

Solution:

a) A back-to-back stem-and-leaf display is constructed by writing **one common** stem for both displays. This **common stem** is placed between the leaves of both displays. **This stem must contain all the possible stem values that can exist for both displays.** If you examine Figures 2.4 and 2.7, you will see that the possible stem values range from 16 to 31. Thus, the back-to-back display is constructed using the form shown in Figure 2.8.

Back-to-Back Stem-and-Leaf Display

Leaf	Stem	Leaf
	16	
	17	
	18	
	19	
	20	
	21	
	22	
	23	
	24	
	25	
	26	
	27	
	28	
	29	
	30	
	31	

Figure 2.8

Figure 2.9 shows the back-to-back display comparing the cholesterol levels for middle-aged men on a regular diet (shown to the left side of the stems) to the cholesterol levels of middle-aged men on a low-fat diet and exercise program (shown to the right of the stems).

Leaves for regular diet	Stem	Leaves for low-fat diet
	16	1 5 8
	17	1 2 2 5 6 6 7
	18	0 1 4 4 8 8 9 9
9 9	19	0 0 3 3 5 5 7 8 8 9
2 1 0	20	0 0 0 2 5 5 9 9 9
9 5 3 2	21	0 1 1 1 4 7 8 9
9 9 6 6 5 5 2	22	0 0 8
9 9 8 8 7 6 5 3 3 2	23	0 1
8 6 5 3 2 1 0 0	24	
9 9 8 8 3 0	25	
9 5 3 3	26	
7 6 4	27	
3 2	28	
	29	
	30	
9	31	

Figure 2.9

b) Comparing the two distributions shown in the back-to-back stem-and-leaf display of Figure 2.9, we can see that there is a greater number of lower cholesterol levels for the men on the low-fat diet and exercise program. Both distributions are essentially symmetric, but they have different data values representing their centers (239 for the regular diet distribution and 197 or 198 for the low-fat distribution). For the regular diet distribution, the greatest concentration of data values falls between 232 and 239, while the low fat distribution has the greatest concentration from 190 to 199. The cholesterol level of 319 in the regular diet distribution represents an outlier. ■

Example 2.5

Fifty college students, consisting of twenty-five males and twenty-five females, were surveyed to determine the number of hours that they spend talking on the telephone per week.

The survey results are recorded in the back-to-back stem-and-leaf display shown in Figure 2.10. In this display the stem represents the tens digit and the leaf represents the units digit.

Construct a **stretched** stem-and-leaf display from this information, and compare the two distributions.

Back-to-Back Stem-and-Leaf Display
Hours spent on telephone per week

Male Leaves	stem	Female Leaves
7 7 6 6 5 4 4 3 3 3 3 3 2 1 1 1 1 1 0	0	0 2 2 2 2 3 3 5 5 5 6 7 8
4 0 0 0 0 0	1	0 0 2 2 4 5 5 6 7 7 8 9

Figure 2.10

Solution:

A stretched stem-and-leaf display would be appropriate for the back-to-back stem-and-leaf display in Figure 2.10, since there are **too many** data values being **cramped** into the two stems, 0 and 1. For example, there are 19 males who spend from 0 to 7 hours on the phone per week, and they are being represented by the values: 0 | 0 to 0 | 7 in the stem-and-leaf display of Figure 2.10. However, there are 13 females who spend from 0 to 8 hours on the phone per week, and they are represented by the values: 0 | 0 to 0 | 8 in the display of Figure 2.10.

Since the display of Figure 2.10 shows too many data values cramped into one row, we can stretch the display by subdividing each stem into two rows. One row will be symbolized using an "*" attached to the stem portion. This row will contain all data values having the leaf digits from zero to four. The second row for the same stem portion will be symbolized by an "•" attached to the stem portion. This row will contain all data values having the leaf digits from five to nine. Thus, the stem 0^* will pertain to all data values from 0 to 4 hours while the stem $0^•$ will pertain to all the data values from 5 to 9 hours. The stem 1^* will pertain to the data values from 10 to 14 hours, and the stem $1^•$ will pertain to the data values from 15 to 19 hours. Using this notation for the stems, we can construct a stretched stem-and-leaf display for the back-to-back male and female distributions of hours spent on the telephone per week. The stretched stem-and-leaf display is shown in Figure 2.11.

Back-to-Back Stretched Stem-and-Leaf Display
Hours spent on telephone per week

Male Leaves	Stem	Female Leaves
4 4 3 3 3 3 3 2 1 1 1 1 1 1 0	0*	0 2 2 2 2 3 3
7 7 6 6 5	0•	5 5 5 6 7 8
4 0 0 0 0 0	1*	0 0 2 2 4
	1•	5 5 6 7 7 8 9

Figure 2.11

The stretched stem-and-leaf display of Figure 2.11 provides a better grouping of the data values. Examining the stretched display of Figure 2.11, one can conclude that the male students have a distribution that has a greater concentration of smaller data values than the female students. The female students have a more uniform distribution of data values. Thus, one can conclude that the female students have a tendency to spend more time on the phone per week than do the male students. ■

2.4 FREQUENCY DISTRIBUTION TABLES

In many statistical studies, a large quantity of data is usually collected and the statistician is interested in summarizing the large data set by placing it in a more manageable form. The stem-and-leaf display that was developed in Section 2.3 is an example of one way that data can be put into a more manageable form. In this section we will present another form of representing data called a **frequency distribution**. A frequency distribution indicates the values of a variable(s) and how often each value occurs. The frequency distribution of a variable can be displayed by a table or graph. In either form we will be able to determine such things as: the highest and lowest values, apparent patterns of a variable, apparent patterns between variables, what values the data may tend to cluster or group around, or which values are the most common. We will first examine the **frequency distribution table** to help determine these characteristics of the data.

> **Definition: Frequency Distribution Table.** A frequency distribution table is a table in which a data set has been divided into distinct groups, called classes, along with the number of data values that fall into each class, called the frequency.

FREQUENCY DISTRIBUTION TABLE FOR CATEGORICAL DATA

The first step in constructing a frequency distribution table is to organize the data into distinctive groups. These groupings are called **classes** (or **class intervals**). If the data collected are categorical, then **each category** will be used to represent a **class** (or **class interval**).

For example, if a biology student is investigating the prevalence of eye color within his college and records the data as either hazel, brown, blue, or black for each student, then each of these four categories of eye color will represent a class.

The second step is to determine the frequency of each class by counting the number of data values that fall within each class. For categorical data, the frequency is the number of data values that fall within each category or class.

In Example 2.6, we will use this procedure to construct a **frequency distribution table for the categorical data** contained within Table 2.1.

Example 2.6

Using the raw data pertaining to the Student Government Survey Data of 150 Students listed in Table 2.1,
a) construct a frequency distribution table for the categorical variable **major**, and
b) determine the most popular major for the 150 students.

Solution:

a) To construct a frequency distribution table for the categorical variable **major**, we will need to organize the data into classes and then determine the frequency for each class.

Step 1. Organize the data into distinct groups called classes. For a categorical variable, we will use the categories of the variable to represent the classes.

For the variable **major**, these categories or classes are:
**liberal arts
business
computers or mathematics
health sciences
natural and applied sciences
fine and performing arts.**

Step 2. Determine the frequency of each class by counting the number of data values that fall within each class. For the categorical variable major, the frequency is the number of data values that fall within each category of the variable major. From Table 2.1, we count 54 students that are liberal arts majors; therefore, the class labeled liberal arts has a frequency of 54. Continuing this procedure for the remaining five classes, we can construct the frequency distribution table for the categorical variable **major** which is shown in Table 2.2.

Table 2.2
Frequency Distribution Table For The Variable Major

CLASSES (Major)	FREQUENCY
Liberal Arts	54
Business	51
Computers or Mathematics	11
Health Sciences	11
Natural and Applied Sciences	20
Fine and Performing Arts	3

b) The frequency distribution table in Table 2.2 displays the data pertaining to major in a more concise and convenient form and makes it easier to interpret the data than the original data set contained in Table 2.1. From Table 2.2, one is able to see at a glance that Liberal Arts and Business are the most popular majors for the students within the data set. ■

FREQUENCY DISTRIBUTION TABLE FOR NUMERICAL DATA

Organizing categorical data into a frequency distribution table is a relatively simple task since the classes represent the categories of the variable while the frequencies are determined by counting the number of times the data values fall within each category.

On the other hand, constructing a frequency distribution table for numerical data is not as obvious. There are two problems that need to be addressed:
(1) How many classes should be used for the data set?
(2) How should these classes be defined?

The choice of the number of classes representing the data set is arbitrary. However, if there are too many classes used, then it would be difficult to see any apparent pattern within the distribution of the data set. For example, examining the data in Table 2.1 for the **numerical** variable **study hours**, you should notice that the values for study hours range from 6 hours to 36 hours. In constructing a frequency distribution for study hours, if we were to consider using **each study hour as a separate group or class,** then this procedure would give us too many classes (31 classes, in fact) as illustrated in Table 2.3. This would not accomplish our objective of summarizing and examining any patterns that may exist for the study hours.

Table 2.3
Frequency Distribution of Study Hours

study hours	frequency	study hours	frequency	study hours	frequency	study hours	frequency
6	1	14	2	22	3	30	4
7	3	15	9	23	5	31	3
8	7	16	6	24	9	32	1
9	7	17	8	25	4	33	3
10	10	18	6	26	4	34	0
11	6	19	5	27	7	35	0
12	9	20	12	28	7	36	1
13	0	21	5	29	3		

On the other hand, if we were to use too few classes, say for example, if we were to use only *two* classes to group the data set, then although we could conclude that more students study less than twenty-one hours, the data would be so clustered together within these two classes that we would lose much of the information and not be able to discover any existing patterns.

Table 2.4
Frequency Distribution of Study Hours

CLASSES (Study Hours)	FREQUENCY
6 - 21	96
22 - 37	54

Thus, although there are no hard and fast rules for making a choice on the number of classes to use, we will use the following guidelines to help us in our decision.

Guidelines For Constructing a Frequency Distribution Table for Numerical Data

1. The number of classes should be neither too small nor too large. Generally, the number of classes should be between 5 and 15. The number of classes we will use will depend on the number of data values within the data set as well as how much the data values vary.

2. The classes should be constructed so that every data value will fall into one class and only one class.
   That is, we need to decide how to define the classes so that they cover the entire range of the data values and that each data value will fall into one and only one class.

3. Whenever possible it is desirable to construct all classes with the same width.
   A frequency distribution with unequal class widths is more difficult to interpret than one with equal class widths.

We will now illustrate how to use these guidelines to construct a frequency distribution table for the variable study hours using the data set in Table 2.1.

Step 1. Decide on the number of classes into which the data are to be grouped.

Using the guideline of a number between 5 and 15, we will use 7 classes for the variable study hours in Table 2.1. This selection of 7 classes is arbitrary.

Step 2. Define the individual classes.

If we want the 7 classes to accommodate the entire data set and each class to have equal width, then we must decide upon a class width large enough to include all the data within these 7 classes. You should realize that the number of classes and the class width are dependent upon one another. Since the fewer the number of classes, the larger the class width, and vice versa. To determine an approximate class width, we will use the rule:

$$\text{approximate class width} = \frac{\text{largest data value} - \text{smallest data value}}{\text{number of classes}}$$

Therefore, for the variable study hours in Table 2.1, the approximate width for 7 classes is:

$$\text{approximate class width} = \frac{36 - 6}{7}$$

$$= \frac{30}{7}$$

$$\approx 4.29$$

Thus, the class width must be **at least 4.29 to accommodate all the data into 7 classes**. However, **the actual class width we will use will be dependent upon the number of significant digits used to measure the data values.** Since the data values are all expressed as *whole numbers*, we will round this result of 4.29 **up to 5**. Thus, **our choice for the width of each class will be 5**.

Step 3. Construct each class using the appropriate class width.

Now, we must define the limits of each of the classes. Each class must have an upper and lower class limit. The **class limits** represent the **smallest and largest data values** that can fall into each given class. The **lower class limit is the smallest data value of the class** while the **upper class limit is the largest data value of the class**. These class limits must be stated so that each data value will fall into one and only one class. The way the class limits are written will depend upon the type of data that is being grouped and the way the data was recorded or rounded off.

For example, the study hour data in Table 2.1 represent continuous data that have been rounded off to the nearest hour. Thus, the data values are recorded as whole numbers so the class limits will be written using whole numbers. Beginning with the smallest data value of 6 hours as the lower class limit of the first class, we will add one less than the class width to obtain the upper class limit. Thus, the upper class limit will be:

$$6 + (5 - 1) = 10.$$

Thus, **the first class represents the data values 6 to 10 and is written as: 6 - 10**.

We can now form the remaining classes by adding the class width to the previous lower and upper class limits. That is:

lower class limit = lower class limit of previous class plus class width

upper class limit = upper class limit of previous class plus class width

Following this procedure for each successive class until the 7 classes have been constructed for our study hour data set, we have:

CLASSES (Study Hours)
6 – 10
11 – 15
16 – 20
21 – 25
26 – 30
31 – 35
36 – 40

Notice all 7 classes have the same class width of five; they do not overlap so every data value will fall into one, and only one class and the classes cover the entire range of data values for the variable study hour. The first class contains the study hours 6 to 10, where the lower class limit is 6 hours and the upper class limit is 10 hours. The second class contains the study hours 11 to 15 where 11 is the lower class limit and 15 is the upper class limit. The last class has a lower class limit of 36 hours and an upper class limit of 40 hours.

Step 4. Determine the frequency of each class.

After the classes have been constructed, the number of data values falling within each class is tallied. This number is called the **frequency of the class** and is recorded under the frequency column of the frequency distribution table.

Thus, Table 2.5 represents the **frequency distribution table** for the study hour data of Table 2.1.

Table 2.5
**Frequency Distribution Table
for the Variable Study Hour**

CLASSES (Study Hours)	FREQUENCY
6 – 10	28
11 – 15	26
16 – 20	37
21 – 25	26
26 – 30	25
31 – 35	7
36 – 40	1

Examining Table 2.5, we see that the data values for the distribution of study hours have been organized into a more compact and useful form that tends to highlight the important characteristics of the data on study hours. From the table, we notice that the majority of students study from 6 to 30 hours with the greatest concentration of students within the class of 16 to 20 study hours. Also, rarely does any student study more than 36 hours. Although, the frequency distribution shown in Table 2.5 summarizes the data in a convenient form, we have sacrificed information about the original study hour data set. **Unlike a stem-and-leaf display, we can see from Table 2.5 that when the data values of a distribution are organized into a frequency distribution table, the actual individual values of the data are no longer apparent.** Notice that there was one student who studied from 36 to 40 hours, however, from Table 2.5 we cannot determine the exact number of hours that this particular student studied. Therefore, for each class of a frequency distribution table, we will **identify a particular value to represent all the data values within that class.** The **midpoint of the class** is used for this purpose, and it is called the **class mark**. The class mark of a class is obtained by calculating the average of the class limits for the class.

Definition: Class Mark. The class mark is the midpoint of a class and is used as the representative value for the class. The formula to calculate the class mark is:

$$\text{Class Mark} = \frac{\text{upper class limit} + \text{lower class limit}}{2}$$

For the frequency distribution in Table 2.5, the class mark of the first class, 6 - 10, is 8. This class mark is determined by adding the class limits of 6 and 10 and dividing by 2. That is,

class mark of first class is: $\frac{10 + 6}{2} = 8$

The class mark of each class is obtained using this class mark formula. In Table 2.6, the class marks of the frequency distribution of study hours have been added to the frequency distribution table under the column class mark.

Table 2.6
Frequency Distribution Table for the Variable Study Hour

CLASSES (Study Hours)	FREQUENCY	CLASS MARK
6 - 10	28	8
11 - 15	26	13
16 - 20	37	18
21 - 25	26	23
26 - 30	25	28
31 - 35	7	33
36 - 40	1	38

In examining the class mark column of Table 2.6, you should notice that the difference between each successive class mark is 5. This difference is the class width. **Within a frequency distribution table, the difference between successive class marks will always be equal to the class width.**

Although Table 2.6 represents the frequency distribution for the continuous variable study hours, it is not apparent from Table 2.6 that the data represents a continuous variable. We see from Table 2.6 that 28 students studied 6 to 10 hours, while there were 26 students who studied 11 to 15 hours. Since the number of hours a student studies represents continuous data, it is possible for a student to have studied 10.7 hours. In which class would such a student be included? Although the number of hours a student studies represents continuous data, the study hours were recorded and rounded off to the nearest hour. Thus, the 10.7 study hours would be rounded-off to 11 hours, and the student should be included in the second class of 11-15 study hours. You should understand that the class limits shown in Table 2.6 do not represent the "**real class limits**" or the so-called "class boundaries" of the data. For example, consider the first class: 6 - 10. Since the data values were originally rounded to the nearest hour, any actual study time that was greater than or equal to 5.5 hours but less than 10.5 hours is included in this first class. Thus, the numbers 5.5 and 10.5 are referred to as the **class boundaries** or **real class limits** of the first class. The class boundaries of the second class are: 10.5 and 15.5. The class boundaries of the last class are 35.5 and 40.5.

Definition: Class Boundary. The class boundary of a class is the midpoint of the upper class limit of one class and the lower class limit of the next class. The formula to determine the class boundary is:

$$\text{Class Boundary} = \frac{\text{Upper Class Limit of one class} + \text{Lower Class Limit of the next class}}{2}$$

If we add an additional column representing the class boundaries to the frequency distribution table of Table 2.6, we obtain Table 2.7. The purpose of including the class boundaries in the frequency table is to present study hours as continuous data and to identify the real class limits of each class.

Table 2.7
Frequency Distribution Table for the Variable Study Hour

CLASSES (Study Hours)	FREQUENCY	CLASS MARK	CLASS BOUNDARIES
6 - 10	28	8	5.5 - 10.5
11 - 15	26	13	10.5 - 15.5
16 - 20	37	18	15.5 - 20.5
21 - 25	26	23	20.5 - 25.5
26 - 30	25	28	25.5 - 30.5
31 - 35	7	33	30.5 - 35.5
36 - 40	1	38	35.5 - 40.5

Notice the classes show a gap of one unit between successive class limits. That is, the first class ends at 10 hours while the second class begins at 11 hours. However, **the class boundaries have eliminated this gap of one unit** by taking one-half of this unit gap and adding it to the upper class limit and subtracting one-half of this unit gap from the lower class limit of the following class. Thus, **class boundaries are numbers midway between consecutive upper and lower class limits that are used to eliminate the gap between the class limits to signify that the data is continuous**. In actual practice, class boundaries are numbers created when a measurement of a continuous variable has been rounded off. Therefore, **when constructing a frequency distribution table for a continuous variable we will include class boundaries as a column within the frequency distribution table**.

On the other hand, **when constructing a frequency distribution table for discrete data, a column for the class boundaries is not required**. Let's now summarize the steps required for constructing a frequency distribution table for numerical data.

Procedure for Constructing a Frequency Distribution Table for Numerical Data

1. Decide on the number of classes for the data set.
2. Determine an approximate class width, using the formula:

$$\text{approximate class width} = \frac{\text{largest data value} - \text{smallest data value}}{\text{number of classes}}$$

3. Round up the approximate class width **result to the next significant digit used to measure the data. This rounded up number will represent the width of each class.**

4. Construct the class limits for the first class by:
   a) selecting the smallest data value within the data set as the lower class limit of the first class, and
   b) using the formula to obtain the upper class limit of the first class:

upper class limit = lower class limit plus the class width minus one unit

The remaining class limits of the frequency distribution table can be constructed using the formulas:

lower class limit = lower class limit of previous class plus class width
upper class limit = upper class limit of previous class plus class width

5. To determine the class marks for each class, use the formula:

$$\text{Class Mark} = \frac{\text{Upper Class Limit} + \text{Lower Class Limit}}{2}$$

List the class marks for each class in the frequency distribution table under the class mark column.

6. For continuous data, determine the class boundaries by finding the midpoint between two successive class limits. This is determined using the formula:

$$\text{Class Boundary} = \frac{\text{Upper Class Limit of one class} + \text{Lower Class Limit of the next class}}{2}$$

* Class boundaries are not required for discrete data. *

7. Determine the frequency for each class by finding the number of data values that fall into each class. List the frequency for each class in the frequency distribution table under the column labelled frequency.

Example 2.7

The following data represent the weekly earnings (in dollars) of 50 university students.

180	221	163	52	69	189	172	163	164	78	148
122	139	144	168	114	88	122	153	93	106	168
193	68	119	129	133	149	168	83	198	165	123
148	144	118	115	109	153	138	152	129	138	143
111	119	140	122	145	156

a) For this data set, construct a frequency distribution table containing five classes using the previous procedure outlined for numerical data.
b) What is the class mark (midpoint) of the weekly earnings class that has the greatest number of university students? What purpose does the class mark serve?

Solution:

a) To construct a frequency distribution table, the steps outlined in the procedure for numerical data are used.

Step 1. Decide on the number of classes.

For this data set, we will use *five classes*.

Step 2. Determine an approximate class width by using the formula:

$$\text{approximate class width} = \frac{\text{largest data value} - \text{smallest data value}}{\text{number of classes}}$$

Since the largest data value is 221 and the smallest data value is 52, the approximate class width for the five classes is:

$$\text{approximate class width} = \frac{221 - 52}{5}$$

$$= \frac{169}{5}$$

$$= 33.8$$

Step 3. Round up the approximate class width result to the next significant digit that the data was measured to. This rounded up number will be the width of each class.

Since the university salaries are represented as **whole** numbers, the approximate class width, 33.8, will be rounded up to the next whole number, 34. Thus, **the width of each class will be 34**.

Step 4. Construct the class limits for the first class by:
   a) **selecting the smallest data value within the data set as the lower class limit of the first class.**

The smallest data value in the data set is the salary 52 dollars. This will be the lower class limit of the first class.

b) using the formula for the upper class limit, obtain the upper class limit for the first class:

upper class limit = lower class limit + the class width − 1

Since, the lower class limit for the first class is 52, and the class width is 34, then the upper class limit is:
upper class limit = 52 + (34 −1)
= 52 + 33
= 85.

Thus the class limits for the first class are: 52 and 85. So the first class is written as: **52 − 85.**

The remaining class limits of the frequency distribution table can be constructed using the formulas:

lower class limit = lower class limit of previous class plus class width

upper class limit = upper class limit of previous class plus class width

The lower class limit for the first class is 52. To obtain the lower class limit of the next class we add the class width of 34 to 52. That is,

lower class limit of the next class = 52 + 34
= 86

To obtain the upper class limit of the next class, we add the class width of 34 to the first class's upper class limit of 85. That is,

upper class limit of the next class = 85 + 34
= 119

Thus, the limits of the second class are 86 and 119.
The second class is written as: **86 − 119.**

Continuing this procedure for the remaining three classes, we get:

the limits of the third class: **120 − 153,**
the limits of the fourth class: **154 − 187,** and
the limits of the fifth class: **188 − 221.**

Table 2.8 represents the five classes for the 50 university salaries.

Table 2.8
Weekly Earnings of the University Students

CLASSES (Earnings)
52 - 85
86 - 119
120 - 153
154 - 187
188 - 221

Step 5. **To determine the class marks for each class, use the formula:**

Class Mark = $\dfrac{\textit{upper class limit + lower class limit}}{2}$

List the class marks for each class in the frequency distribution table under the class mark column.

To determine the class marks for each of the classes of Table 2.8, we must find the midpoint of these classes. Using the class mark formula for the first class, 52 - 85, we have:

$$\text{class mark} = \frac{85 + 52}{2}$$
$$= 68.5$$

The class mark for the second class, 86 - 119, is:

$$\text{class mark} = \frac{119 + 86}{2}$$
$$= 102.5$$

Continuing this procedure for the remaining classes, we can now construct a table that includes the class marks. Table 2.9 contains the five classes representing the weekly earnings of the university students that includes the class marks.

Table 2.9
Weekly Earnings of the University Students

CLASSES (Earnings)	CLASS MARK
52 - 85	68.5
86 - 119	102.5
120 - 153	136.5
154 - 187	170.5
188 - 221	204.5

Examining Table 2.9, you should notice that if you were to subtract any two successive class marks, the difference will equal 34. For example, the difference between the class marks representing the classes 120 - 153 and 154 - 187 is the difference between the class marks 170.5 and 136.5, or 170.5 - 136.5 = 34.

This difference between any two successive class marks of 34 represents the class width for the weekly earnings of the university students.

Step 6. Since weekly earnings represent discrete data, class boundaries are not required and are not included in the frequency distribution table.

Step 7. Determine the frequency for each class by finding the number of data values that fall within each class. List the frequency for each class in the frequency distribution table under a column labelled frequency.

The frequency for each class is represented in Table 2.10. This table is the frequency distribution table for the weekly earnings of the fifty university students.

Table 2.10
Frequency Distribution Table
Weekly Earnings of the University Students

CLASSES (Earnings)	CLASS MARK	FREQUENCY
52 - 85	68.5	5
86 - 119	102.5	10
120 - 153	136.5	21
154 - 187	170.5	10
188 - 221	204.5	4

b) The greatest number of students are in the class representing the weekly earnings $120 to $153. The class mark which represents the class midpoint of this class is $136.50.

Since the frequency distribution table doesn't indicate the actual weekly earnings of the 21 students within the class 120 - 153, the class mark of $136.50 serves as a representative weekly earning for these 21 students. ■

RELATIVE FREQUENCY DISTRIBUTIONS

We have discussed how to construct a frequency distribution table which displays the frequency of each class as the number of data values that fall within that class. At times, however, it is useful to report the **proportion** or the **percentage** of data values that belong to each class. This is referred to as the **relative frequency** of the class.

> **Definition: Relative Frequency.** The proportion of data values within a class is called the **relative frequency** of the class. The formula used to calculate the relative frequency of each class is:
>
> Relative Frequency of a Class = $\dfrac{\textit{class frequency}}{\textit{total number of data values}}$

The sum of the relative frequencies of all the classes will always be one except for possible round-off error. If the relative frequencies are multiplied by 100, then they are expressed as percents and are referred to as the **relative percentage frequencies, or relative percentages**.

Definition: Relative Percentage Frequency. The relative percentage frequency of a class is the percent of data values within a class. The formula used to calculate the relative percentage frequency of each class is:

Relative Percentage Frequency = (relative frequency)(100%)

The sum of the relative percentage frequencies of all the classes will always equal 100 except for possible round-off error.

A frequency distribution table that displays the relative frequencies or the relative percentage frequencies of each class is called a **relative frequency distribution table**.

The purpose of using the relative frequencies within a distribution table is to help compare the frequency distribution tables of two data sets with a different number of data values within each data set.

Let's construct a relative frequency distribution table for the frequency distribution data set of Example 2.7.

Example 2.8

Using the frequency distribution table representing the weekly earnings of the university students shown in Table 2.11,
a) construct a relative frequency distribution table.
b) determine the sum of the relative frequencies and the sum of the relative percentages.
c) identify the class that has the greatest relative percentage.

Table 2.11
Frequency Distribution Table
Weekly Earnings of the University Students

CLASSES (Earnings)	FREQUENCY
52 - 85	5
86 - 119	10
120 - 153	21
154 - 187	10
188 - 221	4

Solution:

a) The **total number of students** within the data set is determined by summing the frequencies within each class: 5 + 10 + 21 + 10 + 4 = 50. Since there are 50 students that comprise the data set, and since there are 5 students that earn from 52 to 85 dollars weekly, the relative frequency of this class is: $\frac{5}{50} = 0.10$. If we multiply this result by 100%, we get: (0.10)(100%) = 10%. Thus, we can state that 10% of the university students earn from 52 to 85 dollars.

The relative frequency of the second class, 86 - 119, is: $\frac{10}{50} = 0.20$. Multiplying this result by 100%, we obtain: (0.20)(100%) = 20%. Therefore, 20% of the university students earn from 86 to 119 dollars per week.

The relative frequency of the third class, 120 - 153, is: $\frac{21}{50} = 0.42$. In terms of a percent, we have: (0.42)(100%) = 42%. Thus, 42% of the university students earn from 120 to 153 dollars per week.

The relative frequency of the class: 154 - 187 is:

$$\frac{10}{50} = 0.20 \text{ or } 20\%$$

While, the relative frequency of the last class, 188 - 221, is:

$$\frac{4}{50} = 0.08 \text{ or } 8\%$$

The relative frequencies along with the relative percentages corresponding to each class are shown in Table 2.12.

This table is referred to as a **Relative Frequency Distribution Table**.

Table 2.12
Relative Frequency Distribution Table
Weekly Earnings of the University Students

CLASSES (Earnings)	FREQUENCY	RELATIVE FREQUENCY	RELATIVE PERCENTAGES
52 - 85	5	0.10	10
86 - 119	10	0.20	20
120 - 153	21	0.42	42
154 - 157	10	0.20	20
188 - 221	4	0.08	8

sum = 1.00   sum = 100%

b) **The sum of the relative frequencies is:**
relative frequencies sum = 0.10 + 0.20 + 0.42 + 0.20 + 0.08
or,
relative frequencies sum = 1.00

The sum of the relative percentages is:
relative percentages sum = 10% + 20% + 42% + 20% + 8%
or,
relative percentages sum = 100%

c) The greatest percentage of students, which is 42%, earn from 120 to 153 dollars per week. Thus, the class 120 - 153 represents the largest relative percentage. This type of information is very useful in making a comparison between classes or making a comparison of the 50 university students to another sample of students. ∎

CUMULATIVE FREQUENCY DISTRIBUTIONS

There are times when we are interested in presenting a frequency distribution table that shows the number of data values that are **less than** or **greater than** a particular data value. Suppose, in considering the information within Table 2.12, we were interested in knowing how many students are earning **less than** a particular weekly salary or how many students are earning **more than** a particular weekly salary. This type of information is displayed in a frequency distribution table called a **cumulative frequency distribution table**. There are two types of cumulative frequency distributions, one is referred to as a **"less than" cumulative frequency distribution** while the other is known as a **"more than" cumulative frequency distribution**.

In a **"less than" cumulative frequency distribution table**, the cumulative frequency of a particular class indicates the number of data values within the class plus the sum of all the data values **less than** this particular class. Such a distribution is constructed by starting with the lowest class and successively summing up all the frequencies within each of the other classes.

In constructing a cumulative frequency distribution table, the manner in which each class is defined is changed. Each class is defined by a single class limit and is written as **less than the lower class limit**.

Let's examine the technique to construct a "less than" cumulative frequency distribution table for the data in Table 2.13.

Example 2.9

Construct a "Less Than" Cumulative Frequency Distribution for the data in Table 2.13.

Table 2.13
Frequency Distribution Table Weekly Earnings of the University Students

CLASSES (Earnings)	FREQUENCY
52 - 85	5
86 - 119	10
120 - 153	21
154 - 187	10
188 - 221	4

Solution:

Let's first begin by constructing the classes for the "less than" cumulative frequency distribution. Except for the last class of a "less than" cumulative frequency distribution, **each class is constructed by taking the lower class limit within Table 2.13 and expressing the class of the cumulative frequency table in the form**:

less than the lower class limit

To construct the last class of a "less than" cumulative frequency distribution table, **the lower class limit of the previous class**, which is 188, **is added to the class width**, which is 34, to obtain 222 (188 + 34). Using this value of 222, write the last class using the same form as in the previous classes. That is, the last class would be written as **less than 222**. This procedure is illustrated in Table 2.14.

Table 2.14
"Less Than" Cumulative Frequency Distribution Table
for the Weekly Earnings of the University Students

CLASSES (Earnings)	CUMULATIVE FREQUENCY
less than $52	
less than $86	
less than $120	
less than $154	
less than $188	
less than $222	

In comparing Table 2.13 to Table 2.14, Table 2.14 has two changes:
(1) there is one additional class.
(2) the class limits have been changed to allow for the frequencies to be summed or accumulated. That is, the classes are now written as: "less than 52" dollars, "less than 86" dollars and so on.

To determine the frequencies for each of these classes, we need to sum up the frequencies of each class that falls below the class limit of these new classes. These summed frequencies can now be listed under the column labelled cumulative frequency. To determine the cumulative frequencies for each of these classes, we simply sum the class frequencies of Table 2.13 as we move down the table. That is, the first class "less than 52" has a frequency of 0 since there are no students who earned less than 52 dollars. The second class "less than 86" has a frequency of 5 since there are 5 students who earned less than 86 dollars. These 5 students represent all the students in the class 52-85 of Table 2.13. The third class "less than 120" has a frequency of 15, which represents the sum of the 5 students of the class 52-85 and the 10 students of the class 86-119 of Table 2.13. Continuing in this manner, the "less than" cumulative frequency distribution is shown in Table 2.15.

Table 2.15
"Less Than" Cumulative Frequency Distribution Table
for the Weekly Earnings of the University Students

CLASSES (Earnings)	CUMULATIVE FREQUENCY
less than $52	**0**
less than $86	**5** (0 + 5)
less than $120	**15** (0 + 5 + 10)
less than $154	**36** (0 + 5 + 10 + 21)
less than $188	**46** (0 + 5 + 10 + 21 + 10)
less than $222	**50** (0 + 5 + 10 + 21 + 10 + 4)

In a "Less Than" Cumulative Frequency Distribution, the first class will always have a cumulative frequency of zero since there are no data values smaller than the lower class limit of the first class. On the other hand, the last class will always equal the total number of data values within the data set, since all the data values will be smaller than the lower class limit of this last class.

Some observations that can be made from Table 2.15 are:

- all the 50 students made at least $52 per week.
- no student made more than $222 per week.
- the 15 students of the class that earned less than $120 per week includes the 5 students who earned less than $86 per week.
- thirty-six students earned less than $154 per week.

In some instances, we are interested in determining the number of data values that are greater than a specific data value. For example, we might be interested in knowing the number of university students that earned $154 or more per week. This type of information is contained within a distribution referred to as a **"more than" cumulative distribution**. A **"More Than" Cumulative Distribution** is similar to a "less than" cumulative distribution except that the frequencies are summed in the reverse manner, that is, from highest class to the lowest class, and the classes are written in the form: **lower class limit or more**. Thus, in a **"more than" cumulative frequency distribution table**, the cumulative frequency of a particular class indicates the number of data values within the class plus the sum of all the data values **greater than** this particular class. Let's construct a "more than" cumulative frequency distribution table for the data listed in Table 2.16.

Example 2.10

Construct a "More Than" Cumulative Frequency Distribution Table for the weekly earnings data of the university students in Table 2.16.

Table 2.16
Frequency Distribution Table
Weekly Earnings of the University Students

CLASSES (Earnings)	FREQUENCY
52 - 85	5
86 - 119	10
120 - 153	21
154 - 187	10
188 - 221	4

Solution:

Let's first begin by constructing the classes for the "more than" cumulative frequency distribution. Except for the last class, **each class of the cumulative distribution is constructed by taking the lower class limit of the class within Table 2.16 and writing the class for the cumulative frequency distribution in the form:**

lower class limit or more

To construct the last class of a "more than" cumulative frequency distribution table, **the lower class limit of the previous class**, which is 188, **is added to the class width**, which is 34, to obtain 222 (188 + 34). Using this value of 222, write the last class in the same form as the previous classes. Thus, the last class is written as: **222 or more**. This is illustrated in Table 2.17.

Table 2.17
"More Than" Cumulative Frequency Distribution Table
Weekly Earnings of the University Students

CLASSES (Earnings)	CUMULATIVE FREQUENCY
52 or more	
86 or more	
120 or more	
154 or more	
188 or more	
222 or more	

In Table 2.17, you will notice that the class limits have been changed to allow for the frequencies to be summed. These summed frequencies will now be listed under the column labelled "cumulative frequency".

To determine the cumulative frequencies for each of these classes, we simply sum the class frequencies of Table 2.16 as we move up the table. Since all fifty students earned more than 52 dollars, for the first class "52 or more" dollars the frequency is 50. The second class "86 or more" has a frequency of 45 since there are 45 students who earned 86 or more dollars. These 45 students represent all the students in the classes 86-119, 120-153, 154-187, and

188-221 of Table 2.16. The third class "120 or more" has a frequency of 35, which represents the sum of the 21 students of the class 120-153, the 10 students of the class 154-187, and the 4 students of the class 188-221 of Table 2.16. Continuing in this manner, the "more than" cumulative frequency distribution is shown in Table 2.18.

Table 2.18
"More Than" Cumulative Frequency Distribution Table
Weekly Earnings of the University Students

CLASSES (Earnings)	CUMULATIVE FREQUENCY
52 or more	50
86 or more	45
120 or more	35
154 or more	14
188 or more	4
222 or more	0

■

In a "More Than" Cumulative Frequency Distribution, the first class will always have a cumulative frequency that is equal to the total number of data values within the data set, since all the data values are greater than the lower class limit of the first class. On the other hand, the last class will always have a cumulative frequency of zero, since all the data values will be smaller than the lower class limit of this last class.

Some observations that can be made examining Table 2.18 are:
- all the 50 students made at least $52 per week.
- no student made $222 or more per week.
- the 14 students of the class that earned $154 or more per week includes the 4 students who earned $188 or more per week.
- forty-five students earned $86 or more per week.

Sometimes cumulative frequency distributions are represented using **relative percentages** of the cumulative frequencies. To represent a cumulative frequency value as **a relative percentage**, **divide** each cumulative frequency value by the **total** number of data and **multiply by 100 percent**. We can express this definition using the formula:

$$\text{Relative Percentage Cumulative Frequency} = \frac{(\textit{Cumulative Frequency})}{(\textit{Total Number of Data})} (100\%)$$

Table 2.19 represents the relative percentages for the "less than" cumulative frequency distribution in Table 2.15.

Table 2.19
"Less Than" Cumulative Percentage Distribution Table
Weekly Earnings of the University Students

CLASSES (Earnings)	CUMULATIVE FREQUENCY	RELATIVE PERCENTAGE CUMULATIVE FREQUENCY $\frac{\text{Cumulative Frequency}}{\text{Total number of data}}(100\%)$
less than 52	0	$\frac{0}{50}(100\%) = 0\%$
less than 86	5	$\frac{5}{50}(100\%) = 10\%$
less than 120	15	$\frac{15}{50}(100\%) = 30\%$
less than 154	36	$\frac{36}{50}(100\%) = 72\%$
less than 188	46	$\frac{46}{50}(100\%) = 92\%$
less than 222	50	$\frac{50}{50}(100\%) = 100\%$

Using the "less than" cumulative percentage distribution in Table 2.19, we can conclude that 72% of the students earned less than $154 per week.

Example 2.11

Construct a "more than" cumulative percentage distribution for the "more than" cumulative frequency distribution in Table 2.20. Draw a conclusion regarding the percent of students earning $120 or more using the "more than" cumulative percentage distribution.

Table 2.20
"More Than" Cumulative Frequency Distribution Table
Weekly Earnings of the University Students

CLASSES (Earnings)	CUMULATIVE FREQUENCY
52 or more	50
86 or more	45
120 or more	35
154 or more	14
188 or more	4
222 or more	0

Solution:

To represent a cumulative frequency value as a relative percentage, each cumulative frequency value must be divided by the total number of data and multiplied by 100 percent. Since Table 2.20 represents a "more than" cumulative frequency distribution, the first class in the table corresponds to the **total frequency** of the distribution. Thus, the total frequency of the distribution is fifty. To represent a cumulative frequency value as a relative percentage, each frequency value must be divided by 50 and then multiplied by 100 percent.

The formula for the relative percentage cumulative frequency is:

$$\text{Relative Percentage Cumulative Frequency} = \frac{(\text{Cumulative Frequency})}{(\text{Total Number of Data})}(100\%)$$

Using this formula, the relative percentage for each cumulative frequency value is determined and listed in Table 2.21.

Table 2.21
"More Than" Cumulative Percentage Distribution Table
Weekly Earnings of the University Students

CLASSES (Earnings)	PERCENT CUMULATIVE FREQUENCY
52 or more	100%
86 or more	90%
120 or more	70%
154 or more	28%
188 or more	8%
222 or more	0%

Using the "More Than" Cumulative Percentage Distribution in Table 2.21, we can conclude that 70% of the students earned $120 or more per week.

2.5 Graphs: Bar Graph, Histogram, Frequency Polygon and Ogive

A **graph** is a descriptive tool used to visualize the characteristics and the relationships of the data quickly and attractively. A well-constructed graph will reveal information about a distribution that may not be apparent from a quick examination of a frequency distribution table. There are many different types of graphs that are used to display information. One such graph is a **Bar Graph**.

Bar Graph

Figure 2.12 displays examples of bar graphs that are commonly seen in a newspaper or magazine. What kind of summary statements can you draw from the graph in Figure 2.12a? What kind of information can you draw from the graph in Figure 2.12b?

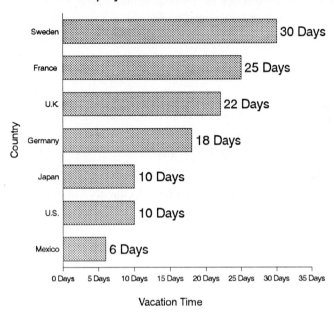
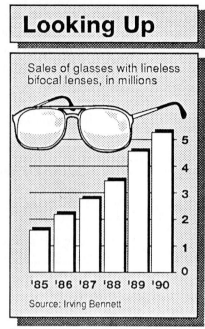

Figure 2.12a   Figure 2.12b

Notice that these graphs in Figures 2.12a and 2.12b are using bars to represent data. This type of graph is appropriately referred to as a **Bar Graph**. Let's examine how to construct and use a bar graph to represent the data.

Bar Graph

A **bar graph** is generally used to depict discrete or categorical data. A bar graph is displayed using two axes. One axis represents the discrete or categorical variable while the other axis represents the frequency of the variable(s).

We will consider two types of bar graphs: **horizontal** and **vertical**. A **horizontal bar graph** entitled "The Leisure Lag" is illustrated in Figure 2.12a. The bar graph shows the vacation time, in days, for employees with one year of service for the seven listed countries. The vertical axis represents the different countries, while the horizontal axis depicts the number of vacation days. Furthermore, the longer the horizontal bar, the greater the vacation time. Since this type of graph is used to represent a categorical variable, such as countries, preferences, sex, etc., the categorical values are placed along the vertical axis. The frequencies or percentages for each of the categories are placed along the horizontal axis, where the length of each bar represents the frequency or percentage of categorical value.

A **vertical bar graph** entitled "Looking Up" is displayed in Figure 2.12b. In this vertical bar graph, one variable is **time**, represented in years, and is scaled along the horizontal axis. The vertical axis represents the variable **sales of glasses with lineless bifocal lenses**. The sales, which are measured in millions, corresponding to each year are placed along the vertical axis. In the vertical bar graph, the height of each bar represents the frequency or percentage for the discrete data.

The bar graph in Figure 2.12b is a **Time-Series Graph**, since it represents the value of a variable, sales, over different periods of time, years.

Time-Series Graph

A time-series graph contains information about two variables, where one of the variables is time.

Example 2.12

The frequency distribution in Table 2.22 represents the categorical data resulting from a wine merchant's preference survey of 500 customers in the San Francisco Bay Area. The customers were asked to respond to the question: "Which type of wine do you prefer?" The survey responses are shown in Table 2.22. Construct a horizontal bar graph using the data in the frequency distribution table.

Table 2.22
Wine Drinker's Preference Survey

CLASSES (Wine Types)	FREQUENCY
Fume Blanc	35
Chablis	65
Chardonnay	80
Chenin Blanc	45
Cabernet Sauvignon	110
Zinfandel	50
Other	115

Solution:

To construct a horizontal bar graph the vertical axis is labelled "wine type" and the horizontal axis is labelled frequency. The categories of the variable "wine type" are listed in Table 2.22. These wine categories are represented on the vertical axis as illustrated in Figure 2.13. The frequencies in Table 2.22 range from 35 to 115, and the horizontal axis is scaled in units of 20 to represent these frequency values.

Figure 2.13 represents a horizontal bar graph for the categorical data of Table 2.22.

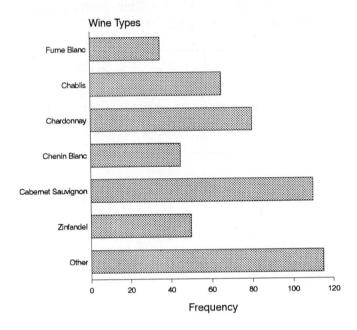

Figure 2.13

Notice that the wine types have been placed along the vertical axis while the number of customers preferring each wine type have been placed along the horizontal axis. The length of each horizontal bar represents the number of customers preferring that wine type. The space between the bars emphasizes that the data is discrete.

The second bar graph we will construct is the **vertical** bar graph. This type of bar graph is generally used to represent numerical data that is discrete or data that is classified by time.

Example 2.13

The frequency distribution in Table 2.23 represents the number of marriages per 10,000 population in the United States for selected years from 1960 to 1990. Construct a vertical bar graph.

Table 2.23
U.S. Marriages, 1960 – 1990

Year	Number of Marriages (per 10,000 population)
1960	85
1965	93
1970	106
1975	100
1980	106
1985	102
1990	98

Solution:

To construct a vertical bar graph, the vertical axis is labelled "frequency". This vertical axis represents the number of marriages per 10,000. Since the largest number of marriages is 106, the vertical axis is scaled in units of twenty. The horizontal axis depicts the selected years from 1960 to 1990.

Figure 2.14 is a vertical bar graph representing the numerical data in Table 2.23 where the length of each bar corresponds to the number of marriages per 10,000 for each selected year. This vertical bar graph is a time-series graph since one of the variables is time.

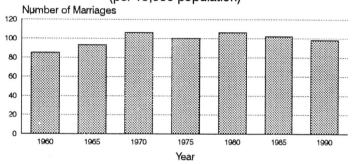

Figure 2.14

The next graph we will consider is the histogram. It is the most common graphical representation of a frequency distribution.

HISTOGRAM

A **histogram** is a graph of a **continuous** variable. It consists of rectangles connected to each other **without any gaps or breaks** between two adjacent rectangles.

Essentially, a histogram is a vertical bar graph where the width of each class of the frequency distribution corresponds to the width of the rectangle and the height of the rectangle corresponds to the frequency or relative frequency of the class.

In constructing a histogram, the vertical sides of each rectangle corresponds to the class boundaries of each class, and consequently, there are no breaks or gaps between the rectangles. This illustrates the continuous nature of the data.

Let us examine how to construct a histogram using the data in Table 2.24 which represents the frequency distribution of average daily temperatures of a south pacific island during the month of December.

Table 2.24
Average Daily Temperatures of a South Pacific Island

CLASSES (Temperatures)	FREQUENCY	CLASS BOUNDARIES
74 - 77	1	73.5 - 77.5
78 - 81	7	77.5 - 81.5
82 - 85	15	81.5 - 85.5
86 - 89	6	85.5 - 89.5
90 - 93	1	89.5 - 93.5
94 - 97	1	93.5 - 97.5

In constructing a histogram, the horizontal axis will be scaled using the class boundaries of the data to show that the variable is continuous. In this case, since the variable temperature is continuous, the class boundaries of each of the classes of daily temperatures have been determined and included within the frequency distribution table.

Column three of Table 2.24 displays the class boundaries of the daily temperatures. The horizontal axis will be scaled from the lowest class boundary of 73.5 to the highest class boundary of 97.5, with all the other class boundaries marked off in equal distances between 73.5 to 97.5 along the horizontal axis. The class frequencies are scaled along the vertical axis. Since the greatest frequency is 15, the vertical axis will be scaled from 0 to 16 in increments of 2.

Rectangles are constructed representing each class of the frequency distribution. The width of each rectangle equals the class width, and the height of each rectangle corresponds to the frequency of each class. In constructing each rectangle, the lower class boundary of each class is used as the beginning point of each rectangle, and the lower class boundary of the next class is the ending point. Thus, the first class, which has the class boundaries 73.5 - 77.5 and a class width of 4, would be represented by a rectangle with a width of 4 that begins at the lower class boundary of 73.5 and ends at the class boundary of 77.5. The second class, which has the class boundaries 77.5 - 81.5 and a class width of 4, would be represented by a rectangle with a width of 4 that begins at the lower class boundary of 77.5 and ends at the class boundary of 81.5. This process continues until the last class 93.5 - 97.5 is represented by a rectangle with a width of 4 that begins at the lower class boundary of 93.5 and ends at the class boundary of 97.5.

Figure 2.15 illustrates the histogram for the frequency distribution in Table 2.24. Notice that this histogram is basically a vertical bar graph where each rectangle or bar has equal widths and there are no breaks or gaps between the adjacent rectangles or bars. Thus, the histogram portraying the daily temperatures of the South Pacific Island is conveying the fact that the temperature data is **continuous**.

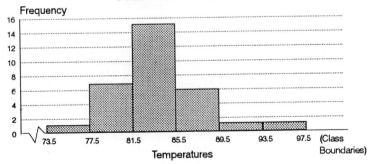

Figure 2.15

The histogram in Figure 2.15 indicates that the greatest frequency of days during the month of December had an average daily temperature from 81.5 to 85.5 degrees, and rarely does the average daily temperature drop below 77.5 degrees or greater than 89.5 degrees on this South Pacific Island.

Notice that a **squiggle (-/\\-)** has been placed on the horizontal axis in Figure 2.15. This indicates that some values have been omitted. Specifically, the temperatures 0 to 73 have been omitted. This is often done for artistic balance in the presentation of the graph. **In general, a squiggle (-/\\-) is used to indicate the omission of values on either the horizontal or vertical axes.**

The following procedure summarizes the steps to construct a histogram.

Procedure to Construct a Histogram

To construct a histogram, perform the following steps.

1. Organize the data into a frequency distribution table using the procedure outlined in Section 2.4 for continuous variables.
2. Label the horizontal axis with the name of the variable. Scale the horizontal axis starting from the lowest class boundary to the highest class boundary, while marking off all the other class boundaries in increments equal to the class width of the frequency distribution table.
3. Label the vertical axis as the frequency, relative frequency, or relative percentage frequency. Examine the frequency of the frequency distribution table to determine the largest value and scale the vertical axis using an appropriate increment to represent these frequencies.
4. Construct a rectangle for each class where the class boundaries correspond to the vertical sides of the rectangle, and the height of the rectangle corresponds to the frequency, relative frequency, or relative percentage frequency. Each rectangle should have the same width, and this width is the class width.

Example 2.14

Table 2.25 represents the frequency distribution of systolic blood pressure values for 100 career women aged 20 to 49 years,
a) construct a histogram.
   From the histogram,
b) determine the class that represents the greatest number of women's blood pressures, and
c) determine within which three blood pressure classes the majority of the blood pressures fall?

Table 2.25
**Systolic Blood Pressure Values
of 100 Career Women Aged 20 to 49 Years**

CLASSES (Blood Pressures)	FREQUENCY
100 – 109	4
110 – 119	10
120 – 129	26
130 – 139	19
140 – 149	15
150 – 159	7
160 – 169	10
170 – 179	9

Solution:

a)

Step 1. To construct a histogram for the continuous variable blood pressure, we need to determine the class boundaries and include the boundaries in the frequency distribution table.

The class boundaries are determined using the formula:

$$\text{Class Boundary} = \frac{\text{Upper Class Limit of one class} + \text{Lower Class Limit of the next class}}{2}$$

Table 2.26 lists the class boundaries for the systolic blood pressure values.

Table 2.26
**Systolic Blood Pressure Values
of 100 Career Women Aged 20 to 49 Years**

CLASSES (Blood Pressures)	FREQUENCY	CLASS BOUNDARIES
100 – 109	4	99.5 – 109.5
110 – 119	10	109.5 – 119.5
120 – 129	26	119.5 – 129.5
130 – 139	19	129.5 – 139.5
140 – 149	15	139.5 – 149.5
150 – 159	7	149.5 – 159.5
160 – 169	10	159.5 – 169.5
170 – 179	9	169.5 – 179.5

Step 2. Label the horizontal axis with the variable name Systolic Blood Pressure. Scale the horizontal axis starting from the lowest class boundary of 99.5 to the largest class boundary of 179.5 while marking off all the other class boundaries in increments equal to the class width of 10. This is illustrated in Figure 2.16.

Step 3. The vertical axis will represent the frequencies of Table 2.26 and is labelled "Frequency". To decide what would be an appropriate scale for the axis, examine the frequencies of Table 2.26 to determine the greatest frequency. Since 26 is the greatest frequency, the vertical axis will be scaled in increments of five. Examine the vertical axis in Figure 2.16

Step 4. Construct a rectangle for each class where the class boundaries correspond to the vertical sides of the rectangle, and the height of the rectangle represents the frequency of the class. The use of the class boundaries on the horizontal axis will depict the variable blood pressure as being continuous since the histogram in Figure 2.16 has no breaks or gaps between consecutive rectangles.

Notice the histogram in Figure 2.16 does illustrate blood pressure as a continuous variable.

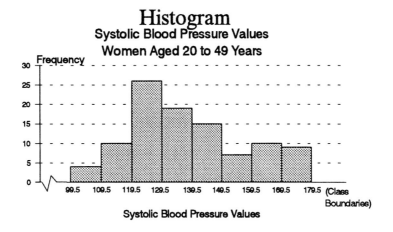

Figure 2.16

b) In examining the histogram in Figure 2.16, the rectangle with the class boundaries 119.5 - 129.5 has the greatest frequency, which is 26. Since the class boundaries of 119.5 - 129.5 represents the class 120 - 129, this class has the greatest number of blood pressures.

c) From the histogram in Figure 2.16 we can see that the majority of the blood pressures fall between the class boundaries starting at 119.5 and ending at 149.5, since these three consecutive rectangles represent more than half the blood pressures. The class boundary 119.5 is the lower class boundary for the class limit 120 and 149.5 is the upper class boundary for the class limit 149. Thus the majority of the blood pressures fall between the blood pressures 120 and 149. ∎

Example 2.15

In Table 2.1 at the beginning of this chapter, survey data was given for 150 students responding to a survey developed by a student government association. The frequency distribution presented in Table 2.27 represents the number of hours studied for the sample of 150 students.

a) Change the frequencies in Table 2.27 to relative percentage frequencies and construct a histogram using the relative percentage frequencies.

b) From the histogram determine the percentage of students who study from 11 to 25 hours.

Table 2.27
Study Hours for Students

CLASSES (Study Hours)	FREQUENCY	CLASS BOUNDARIES
6 - 10	28	5.5 - 10.5
11 - 15	26	10.5 - 15.5
16 - 20	37	15.5 - 20.5
21 - 25	26	20.5 - 25.5
26 - 30	25	25.5 - 30.5
31 - 35	7	30.5 - 35.5
36 - 40	1	35.5 - 40.5

Solution:

a)

Step 1. Since the data is already summarized in a frequency distribution table and the class boundaries for each class are included, we need only to determine the relative percentage frequencies for each class. To determine the relative percentage frequencies for each class, the frequency within each class is divided by the total number of data values and multiplied by 100. That is:

$$\text{relative percentage frequency} = \frac{\text{class frequency}}{\text{total number of data}} (100\%)$$

Since this distribution represents 150 students, the total number of data is 150. Thus, each class frequency is divided by 150 to determine the relative percentage frequency for each class.

For example, the first class representing 6 - 10 study hours has a frequency of 28. Thus, the relative percentage frequency for the first class is:

$$\text{relative percentage frequency for the first class} = \frac{28}{150}(100\%)$$

$$= 18.67\%.$$

The relative percentage frequencies for the remaining classes are determined using this formula and are included in Table 2.28.

Step 2. The horizontal axis is labelled with the data name Study Hours. Scale the axis starting with the lowest class boundary of 5.5 to the largest class boundary of 40.5 while marking off all the other class boundaries in increments equal to the class width of 5. This is illustrated in Figure 2.17.

Table 2.28
Study Hours for Students

CLASSES (Study Hours)	RELATIVE PERCENTAGE	CLASS BOUNDARIES
6 - 10	18.67%	5.5 - 10.5
11 - 15	17.33%	10.5 - 15.5
16 - 20	24.67%	15.5 - 20.5
21 - 25	17.33%	20.5 - 25.5
26 - 30	16.67%	25.5 - 30.5
31 - 35	4.67%	30.5 - 35.5
36 - 40	0.67%	35.5 - 40.5

Step 3. The vertical axis will represent the relative percentages of Table 2.28 and is labelled as relative percentage frequency. To decide an appropriate scale for these percentages, examine the percentages of Table 2.28 to determine the largest percentage. Since the largest percent is 24.67%, the vertical axis will be scaled in increments of 5%. Examine the scaling of the vertical axis in Figure 2.17.

Step 4. Construct a rectangle for each class where the class boundaries correspond to the vertical sides of the rectangle, and the height of the rectangle represents the relative percentage frequency of the class. The class boundaries of the histogram depicts the variable study hours as continuous. This is illustrated in Figure 2.17.

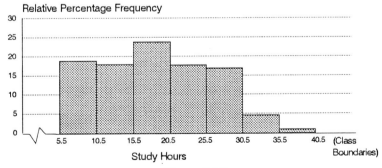

Figure 2.17

b) The study hours 11 to 25 are being represented by the three rectangles starting with the class boundary of 10.5 and ending with the boundary of 25.5. To determine the percentage of students corresponding to these study hours we need to sum up the relative percentages which pertain to these rectangles. This sum is:

sum = 17.33% + 24.67% + 17.33%
sum = 59.33%

Thus, 59.33% of the students study from 11 to 25 hours. ■

We now will examine a graph similar to a histogram that can be used to graphically depict a frequency distribution. This new type of graph uses line segments to connect the midpoints of the bars in a histogram. This line graph is called a **frequency polygon**.

FREQUENCY POLYGON

A **Frequency Polygon** is another type of graph that is used to represent continuous data. A frequency polygon is a line graph representing a frequency distribution using the class marks of the distribution.

The frequency polygon is constructed by using straight line segments to connect the frequency that corresponds to each class by positioning a dot above each class mark or midpoint. Frequency polygons are very useful when comparing two or more distributions on the same set of axes.

Let's examine how to construct a frequency polygon using the data in Table 2.29.

Table 2.29 is the frequency distribution table for the 150 student responses listed in Table 2.1 representing the variable Study Hours. A column for **class marks** has been included in the table since class marks are needed to construct a frequency polygon. Remember, the class mark represents the **midpoint** of each class.

Table 2.29
Study Hours for 150 student Responses

CLASSES (Study Hours)	CLASS MARK	FREQUENCY
6 - 10	8	28
11 - 15	13	26
16 - 20	18	37
21 - 25	23	26
26 - 30	28	25
31 - 35	33	7
36 - 40	38	1

To construct a frequency polygon for the data in the frequency distribution table, a horizontal axis is constructed listing the class marks from the table. A dot is placed directly above each class mark corresponding to the class mark's frequency (see Figure 2.18a). After the dots are plotted, line segments are used to connect the adjacent dots. Notice that the frequency polygon is not connected to the horizontal axis. In order to connect the frequency polygon to the horizontal axis, an **additional class mark must be added to each end of the axis**. These additional class marks have **zero frequency** since there are no students that study less than 6 or more than 40 hours.

After the additional class marks are inserted the graph can be connected to the horizontal axis. Figure 2.18b represents the graph of the frequency polygon.

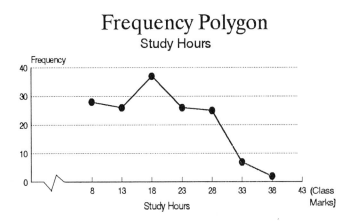

Figure 2.18a   Figure 2.18b

The following steps summarize the procedure to construct a frequency polygon.

Procedure to Construct a Frequency Polygon

To construct a frequency polygon, perform the following steps.

1. Organize the data into a frequency distribution table, including the class marks, using the procedure outlined in Section 2.4 for continuous variables.

2. Label the horizontal axis with the name of the variable. The horizontal axis will represent the class marks. To scale this axis, mark off the class marks from the lowest class mark to the highest class mark. Include two additional class marks, one to the left of the lowest class mark and one to the right of the highest class mark.

3. Place a dot representing the frequency of each class above the class mark or midpoint of each class. Remember, the two additional class marks have zero frequency, and their dots are placed on the horizontal axis.

4. Connect the adjacent dots using straight line segments. The resulting line graph is called a frequency polygon.

Example 2.16

Construct a Frequency Polygon using the data in Table 2.30.

Table 2.30
Cars on the Road by Age

CLASS MARKS (Age in Years)	Number
2	9,400,000
4	10,070,000
6	10,131,000
8	9,081,000
10	5,631,000
12	5,317,000
14	4,989,000
16	2,503,000

Solution:

To construct a frequency polygon for the data in Table 2.30, we will follow the previously outlined procedure.

1. Construct a horizontal axis listing the class marks of 2 to 16 years of Table 2.30. Notice that the distance between each class mark is 2 years. Since two additional class marks are needed at each end of the axis, these class marks will also be 2 years apart ranging from the lowest class mark of 2 years to the highest class mark of 16. Thus, the additional lower class mark is obtained by subtracting 2 years from 2 years to get 2 - 2 = 0 years, while the additional upper class mark is determined by adding 2 years to the upper class mark of 16 to get 16 + 2 = 18. This is illustrated in Figure 2.19a.

2. Place a dot directly above each class mark corresponding to the class mark's frequency. Remember, the two additional class marks have zero frequency, and their dots are placed on the horizontal axis.

3. Connect the dots using line segments.

Figure 2.19b represents the frequency polygon for the data in Table 2.30.

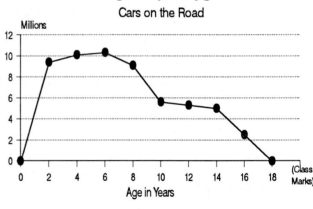

Figure 2.19a Figure 2.19b ∎

Line graphs similar to frequency polygons are especially useful when you wish to compare two or more distributions on the same set of axes. This idea is illustrated in Case Study 2.1.

Case Study 2.1

The two line graphs in the USA Snapshot of Figure 2.20 compares meat versus poultry consumption in pounds per person for the selected years 1970 to 1991.

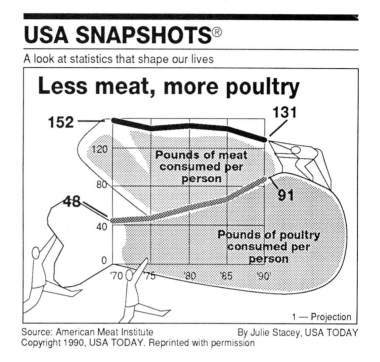

Figure 2.20

Notice these graphs are time-series line graphs. For these types of graphs it would be inappropriate to extend each line graph to the horizontal axis, since this would indicate that the meat and poultry consumption per person is zero for the years prior to 1970 and after 1991.

In examining the line graph pertaining to poultry consumption, we notice that the consumption per person has been **increasing** for the years 1970 to 1991. Furthermore, for this 21 year period, the number of pounds of poultry consumed per person has increased from 48 lbs to 91 lbs. This represents an increase of 43 lbs of poultry consumed per person over the 21 year period, or **an approximate increase in consumption of 2.05 lbs of poultry per year**.

On the other hand, the line graph pertaining to meat consumption per person has been **decreasing** for the years 1970 to 1991. For this same 21 year period, the number of pounds of meat consumed per person has decreased from 152 lbs to 131 lbs. This represents a decrease of 21 lbs of meat consumed per person over the 21 year period, or **a decrease in consumption of 1 lb of meat per year**.

If the rate of poultry consumption continues to increase at the rate of 2.05 lbs per year, while the rate of meat consumption continues to decrease at 1 lb per year, then estimate the year that the poultry and meat consumption per person will be the same. ∎

OGIVE

A graph of a cumulative frequency distribution is called an **ogive** (pronounced "oh-jive"). It is a graph of a cumulative frequency distribution that represents the left end of each class on the horizontal axis and the corresponding cumulative frequency on the vertical axis. The ogive of a "less than" cumulative frequency distribution is always increasing.

Table 2.32 is a cumulative frequency distribution representing the weekly earnings of 50 university students.

Table 2.32
"Less Than" Cumulative Frequency Distribution
Representing Weekly Earnings of 50 University Students

CLASSES (Earnings)	CUMULATIVE FREQUENCY
less than 52	0
less than 86	5
less than 120	15
less than 154	36
less than 188	46
less than 222	50

Let's develop a procedure to construct an ogive using the data in the cumulative frequency distribution of Table 2.32.

Step 1. List the class values on the horizontal axis.

We will scale the horizontal axis from 52 to 222.

Step 2. Scale the vertical axis using an appropriate increment and list the cumulative frequencies.

Examining Table 2.32, we see that the greatest frequency is 50. An appropriate way to scale the vertical axiswould be in units of 10.

Step 3. Place a "dot" above the value which represents the left end of each class and corresponds to the cumulative frequency of the class.

Since the first class, less than 52, has a frequency of zero, a dot is placed on the horizontal axis at 52. For the second class, less than 86, a dot is placed above 86 at a frequency height of 5. The remaining dots are placed at a frequency height that corresponds to the frequencies listed in Table 2.32.

Step 4. To construct the ogive, the adjacent dots are connected using line segments.

Figure 2.21 represents the ogive for the "less than" cumulative frequency Table 2.32.

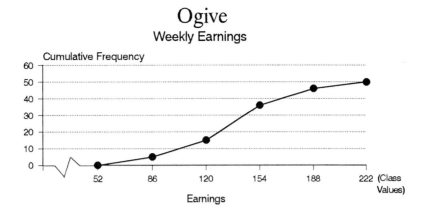

Figure 2.21

By examining the ogive in Figure 2.21, determine between what two successive classes the change in cumulative frequencies is the greatest?

The procedure to construct an ogive is summarized below.

Procedure to Construct an Ogive

To construct an ogive, perform the following steps.

1. Organize the data into a Cumulative Frequency Distribution Table, using the procedure outlined in Section 2.4 for continuous variables.

2. Label the horizontal axis with the name of the variable. Scale the horizontal axis starting from the lowest class value which represents the left end of each class to the highest class value, while marking off all the other class values in equal increments.

3. Label the vertical axis as the cumulative frequency. Examine the frequencies of the cumulative frequency distribution table to determine the largest value and scale the vertical axis using an appropriate increment to represent these frequencies.

4. Place a dot above the value which represents the left end of each class and that corresponds to the cumulative frequency for each class.

5. Use line segments to connect the adjacent dots. The resulting graph is called an ogive.

2.6 Specialty Graphs: Pie Chart and Pictograph

Let's examine two specialty graphs called the **pie chart** and the **pictograph**.

PIE CHART

A **pie chart** is useful in representing categorical data in a circular format. In the pie chart, each category is represented by a sector of the circle or pie that corresponds to the percent or proportion of the data within each category.

Case Study 2.2

The pie chart in Figure 2.22 represents **the reasons why the 309 Los Angeles Marathon registrants say they would participate in the Los Angeles Marathon**.

Figure 2.22

Notice that the pie chart represents the different reasons why the runners are participating in the marathon. These reasons have been categorized as: **sense of accomplishment, fun, challenge, fitness,** and **other**. The **area** of the different sectors of the circle reflects the **percent** of the runners that have indicated their reason for running in the marathon. For example, **40% of the runners** indicated that their main reason for participating in the marathon was a sense of accomplishment. Thus, **40% of the area of the circle** has been sectioned to represent the sense of accomplishment response. ∎

To construct a pie chart, we draw a circle and then subdivide the circle into sectors, or parts, that correspond to the percentages or proportions within each category. Examine Figure 2.23. You will see a circle that has been divided into 10 equal parts. Each part represents thirty-six degrees. A circular ruler called a protractor was used to divide the circle into the ten equal parts.

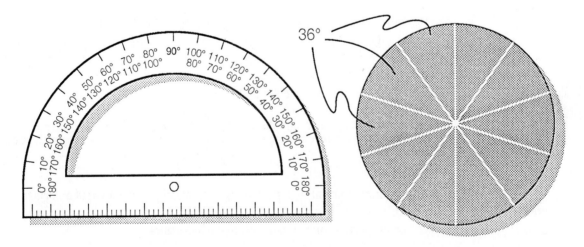

Figure 2.23

Table 2.33 represents a percentage breakdown of the sales budget of DPT Manufacturing Company.

Let's discuss how to construct a pie chart to represent this information.

Table 2.33
Sales Budget for DPT Manufacturing

Category	Percent
Clerical	10%
Consultants	30%
Entertainment	40%
Travel	20%

A pie chart to depict the sales budget of DPT Manufacturing Company is shown in Figure 2.24b. To construct this pie chart, we need to divide a circle into equivalent sectors that represent the percents which are listed in Table 2.33. Since a circle represents 360 degrees, it is necessary to determine the equivalent number of degrees that correspond to each percent within Table 2.33.

To convert a percentage into degrees, the following procedure is used:
    number of degrees = percentage of 360 degrees

For example, to convert 10% into degrees we need to determine 10% of 360 degrees using the procedure:

number of degrees = 10% of 360 degrees
             = (0.10)(360)
             = 36 degrees

Thus, 10% within a pie chart is represented by a sector of the circle which is 36 degrees. To convert 30% into degrees, we must find 30% of 360 degrees. This is calculated by the procedure:

number of degrees = 30% of 360 degrees
             = (0.30)(360)
             = 108 degrees

Therefore, within a pie chart, 30% is represented by a sector of the circle which is 108 degrees. The other percentages are determined using this same procedure. Table 2.34 lists the degrees that correspond to the percentages for DPT manufacturing data.

Table 2.34
Sales Budget for DPT Manufacturing

Category	Percent	Degrees
Clerical	10%	36
Consultants	30%	108
Entertainment	40%	144
Travel	20%	72

In Figure 2.24a we have listed **both** the percent and equivalent degree measurement for the data in Table 2.34. From this information, the pie chart in Figure 2.24b can be drawn.

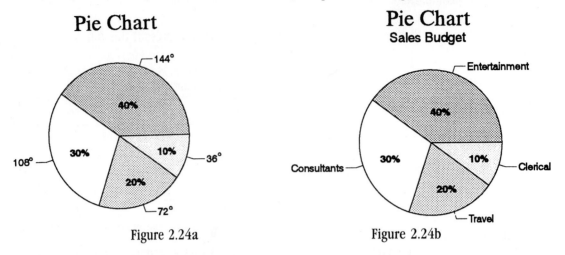

Figure 2.24a

Figure 2.24b

Procedure to Construct a Pie Chart

1. Determine the percent corresponding to each category of the variable.
2. Convert each percentage into its equivalent degree measurement by finding the percentage equivalent of 360 degrees.
3. Use a protractor to construct each sector of the pie chart that corresponds to the degree measurement for each category of the variable.
4. Label each sector with the category name and its corresponding percentage.

Case Study 2.3

The pie chart in Figure 2.25 was constructed from a survey that sampled 8,450 women aged 15-44 regarding the outcomes of unmarried couples who lived together. To add impact, figures of couples were drawn in each of the sectors of the chart.

Figure 2.25

What percent of the unmarried couples are now not living together? Is there enough information in this pie chart to draw any conclusions regarding whether living together results in a successful marriage? ■

PICTOGRAPH

A **pictograph** uses pictures to represent data in a more attractive and eye catching manner.[1]

Case Studies 2.4 and 2.5 are examples of pictographs.

CASE STUDY 2.4

In the graph of Figure 2.26, the **length of an arm holding money** is used to represent the average life of the currency. It is interesting to note which currency amount first catches your eye. Most of us will notice the $100 bill. Upon further observation your eye glances at all the currency amounts. Which currency lasts longest? Which lasts shortest? Do these facts surprise you?

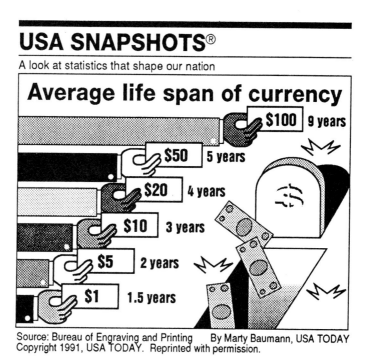

Figure 2.26

■

---

[1] A word of caution: Pictographs can be used to misrepresent data. See Section 1.3 for more information on the misuses of statistics.

86 • CHAPTER 2   ORGANIZING AND PRESENTING DATA

CASE STUDY 2.5

The pictograph in Figure 2.27 is an eye-catching graph that uses a student holding a pencil which reflects the pattern of verbal SAT scores for a twenty-two year period! The use of a pencil emphasizes the *downward trend* for verbal SAT scores.

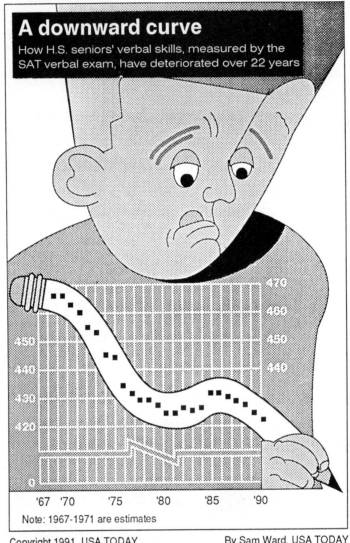

Figure 2.27

■

❄ 2.7 IDENTIFYING SHAPES AND INTERPRETING GRAPHS

SHAPES OF DISTRIBUTIONS

A frequency distribution can have a variety or an unlimited number of different shapes. We will discuss and examine the standard shapes and characteristics of a few different types of distributions. The shapes of these distributions will be represented by a smooth curve which is intended to represent a frequency distribution using very small classes.

The **symmetric bell-shaped distribution** as shown in Figure 2.28 is a familiar distribution that we often encounter in statistics. This type of distribution has its highest frequency or peak at the center with the frequencies steadily decreasing but identically distributed on both sides of the center. If the symmetric bell-shaped distribution were folded along the center, the right and left side would be mirror images of itself as seen in Figure 2.28. The bell-shaped distribution indicates that there are as many data values in both the low and high classes of the distribution. Data representing adult characteristics such as IQ scores, heights, or weights usually have a symmetric bell-shaped distribution since most adults have IQs, heights or weights that are bunched near the center value with few adults falling at the extreme values.

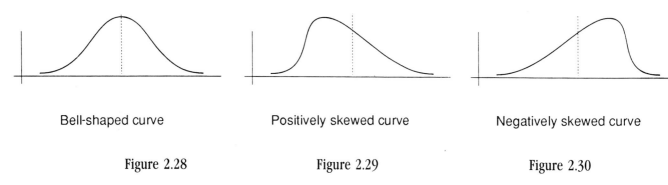

Bell-shaped curve   Positively skewed curve   Negatively skewed curve

Figure 2.28   Figure 2.29   Figure 2.30

A distribution which is not symmetric is said to be a **skewed distribution**. A skewed distribution is a type of distribution that has most of the data values falling into one side of the distribution with very few but extreme data values falling in the other side. There are two types of skewed distributions. A distribution of incomes represents one type of a skewed distribution since usually most of the people have incomes on the lower end of the income scale with a relatively few people who are millionaires falling into the higher end of the income scale. This type of skewed distribution is called **a positively skewed or skewed to the right distribution**.

On the other hand, a person's age at death represents a second type of skewed distribution since there are a relatively few people who die at a very young age with the frequency of death increasing as a person ages. This type of skewed distribution is called **a negatively skewed or skewed to the left distribution**. Let's discuss the characteristics of these two skewed distributions.

A positively skewed or a skewed to the right distribution is illustrated in Figure 2.29. This type of distribution has a greater number of relatively low scores and a few extremely high scores. **The shape of a skewed to the right distribution has a longer tail on the right side which represents the few extremely high scores** as shown in Figure 2.29.

A negatively skewed or a skewed to the left distribution is a distribution that has a larger number of relatively high scores and a few extremely low scores. The shape of a skewed to the left distribution is illustrated in Figure 2.30. From Figure 2.30, we notice that **the distribution has a longer tail in the left side which represents the few extremely low scores.**

Notice that the skewness of a distribution results from the extreme data values falling into one of the tails of the distribution.

If the extreme values occur in the upper or right tail, the distribution has a longer right tail and is referred to as a positively skewed or skewed to the right distribution.

When the extreme values occur in the lower left tail, then the distribution has a longer left tail and is referred to as negatively skewed or skewed to the left distribution.

Let's discuss some other distribution shapes which are illustrated in Figure 2.31.

A **Uniform** or **Rectangular Distribution** is a distribution where all the classes contain the same number of data values or frequencies. This type of distribution is illustrated in Figure 2.31a.

A uniform distribution usually occurs when tossing a fair die say 600 times. We would expect the frequency distribution representing the outcome of the die to be the same for each number 1 to 6.

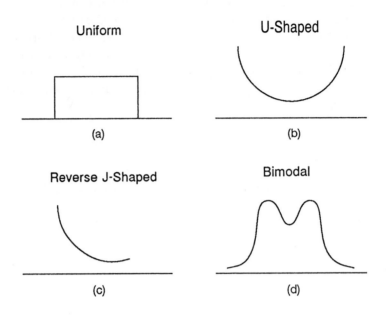

Figure 2.31

A **U-shaped Distribution** is a distribution, as the name suggests, that has a U-shape. That is, it has its two greatest frequencies occurring at each extreme end of the distribution with the lower frequencies of the distribution occurring in the center. Figure 2.31b illustrates a U-shaped distribution. Death rates by age within the United States form a distribution that is roughly U-shaped.

A **Reverse J-Shaped Distribution** has a shape similar to its name. A reverse J-shaped distribution has its greatest frequency of data values occurring at one end of the distribution and then tails off gradually in the opposite direction. This is illustrated in Figure 2.31c. If a pair of fair dice were tossed 720 times, and the outcomes were classified as no sixes, one six, or two sixes, the resulting distribution would resemble a reversed J-shape.

A **Bimodal Distribution** is a distribution that has two greatest frequencies or peaks as illustrated in Figure 2.31d. A bimodal distribution usually indicates that the distribution represents the data values of two different groups or populations. For example, a distribution containing both the weights of males and females would be a bimodal distribution where each peak within this distribution would represent the most frequent weight for each sex.

Example 2.17

For the four distributions pictured in Figure 2.32 identify the shape of each distribution and explain what the shape indicates for the variable represented.

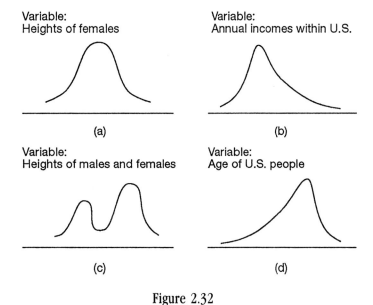

Figure 2.32

Solution:

a) The shape of the distribution in Figure 2.32a is bell-shaped. This indicates most of the females have heights that are clustered in the center of the distribution with very few females having extreme heights, i.e. too short or too tall.
b) The shape of the distribution in Figure 2.32b is skewed to the right. This indicates that few incomes are very high with a greater number of incomes being relatively low.
c) The shape of the distribution in Figure 2.32c is bimodal. Each peak within the distribution represents the most frequent height for each sex.
d) The shape of the distribution in Figure 2.32d is skewed to the left. This indicates that there are a greater number of older people and relatively fewer younger people.

INTERPRETING GRAPHS

Besides identifying the shape of a graph, it is essential that one is able to read and interpret the information in a graph. The following examples illustrate how to read and interpret information within graphs.

Example 2.18

Examine the histogram in Figure 2.33 and answer the following questions.

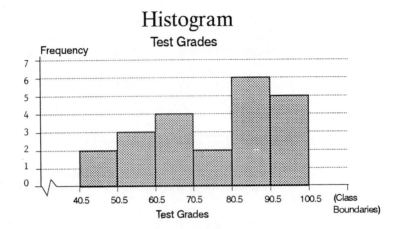

Figure 2.33

a) How many test grades are represented by the histogram?
b) How many test grades are greater than 70?
c) How many test grades are between 61 and 80 (inclusive)?
d) How many test grades are less than 81?
e) Which class has a frequency of 4?
f) Which classes have the lowest frequency?

Solution:

To answer these questions, it's helpful to construct a frequency distribution table as shown in Table 2.35, using the information from the histogram given in Figure 2.33.

Table 2.35
Frequency Distribution of Test Grades

Test Grades	FREQUENCY
41 - 50	2
51 - 60	3
61 - 70	4
71 - 80	2
81 - 90	6
91 - 100	5

a) To calculate the number of test grades represented by the histogram add the frequencies of all the classes. We obtain: **2 + 3 + 4 + 2 + 6 + 5 = 22**
Figure 2.33 represents a distribution of 22 test grades.
b) Any grade greater than 70 must be in the classes 71-80, 81-90, or 91-100. Therefore to calculate how many grades are greater than 70, add the frequencies of these 3 classes. This total indicates 13 test grades are greater than 70.
c) A test grade between 61 and 80 inclusive must be either in the class 61-70 or 71-80. Therefore, the number of grades between 61 and 80 is 6.
d) A test grade less than 81 must be in the classes 71-80, 61-70, 51-60, or 41-50. Therefore the number of grades less than 81 is 11.
e) The class that has a frequency of 4 is 61-70.
f) The lowest frequency is 2. There are two classes with this frequency. They are 41-50 and 71-80. ■

Example 2.19

Use the information in the histogram in Figure 2.34 to answer the following questions.

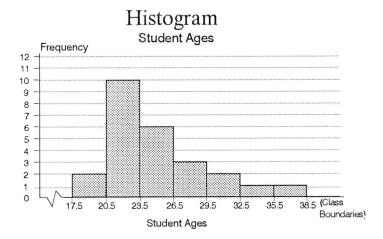

Figure 2.34

a) How many students are represented by this histogram?
b) How many students are older than 26?
c) What percent of the students are older than 26?
d) What percent of the students are younger than 21?
e) What percent of the students are older than 20 but younger than 30?
f) What percent of the students are younger than 40?

Solution:

Table 2.36 has been constructed using the histogram of student ages as indicated in Figure 2.34.

Table 2.36
Frequency Distribution of Student Ages

Student Ages	FREQUENCY
18 - 20	2
21 - 23	10
24 - 26	6
27 - 29	3
30 - 32	2
33 - 35	1
36 - 38	1

a) To calculate the number of students represented by the graph, add the frequencies of all the classes. The total number of students is 25.

b) Any student older than 26 must be in one of the classes: 27-29, 30-32, 33-35, and 36-38. To calculate the number of students older than 26, add the frequencies of these four classes. Therefore, 7 students are older than 26.

c) To calculate the percent of students older than 26:

Step 1: Divide the number of students older than 26 by the total number of students in the distribution. That is,

$$\frac{\text{The number of students older than 26}}{\text{Total number of students in the dist.}} = \frac{7}{25} = 0.28$$

Step 2: Multiply this decimal by 100% to convert the decimal to a percent.
0.28 x 100% = 28%

Therefore, 28% of the students in the distribution are older than 26.

d) To calculate the percent of students younger than 21:

Step 1: Calculate the number of students younger than 21.
The number of students younger than 21 is 2.

Step 2: Divide the number of students younger than 21 by the total number of students in the distribution:

$$\frac{\text{Number of students younger than 21}}{\text{Total number of students in the dist.}} = \frac{2}{25} = 0.08$$

Step 3: Multiply this decimal by 100% to convert the decimal to a percent.
0.08 x 100% = 8%

Therefore, 8% of the students in the distribution are younger than 21.

e) Calculate the percent of the students older than 20 but younger than 30.

Using the same procedure illustrated in part d (above), 76% of the students in the distribution are older than 20 but younger than 30.

f) Since **everyone** is younger than 40, the percent of students younger than 40 is 100%. ■

CASE STUDY 2.6

Examine the pie chart in Figure 2.35 which represents the source of state governments' revenue.

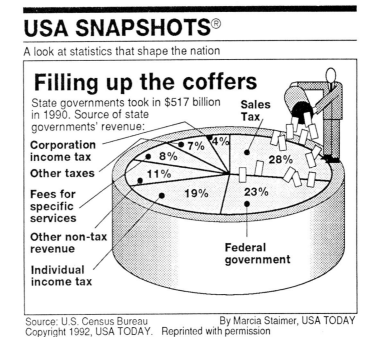

Figure 2.35

Using this information, answer the following questions:

a) How much state revenue was generated from the sales tax?
b) Suppose we assume that the state sales tax is 8%. What was the base sales amount use to generate the sales tax portion of the state revenue?
c) If the state was planning to increase the sales tax to 9% next year, estimate the revenue due to sales tax using the results obtained in question #2.
d) Determine the percentage due to sales tax revenue if the state increased the sales tax to 9%. (Assume that all other conditions remained the same.)

Solution:

a) According to the graph, the state government received $517 billion. Twenty-eight percent was from sales tax. Thus, 28% of $517 billion = $144.76 billion.

That is, the state tax revenue generated from sales tax is $144.76 billion.

b) If we assume that the state sales tax is 8% and this tax generated $144.76 billion then:
8% of (**a base amount due to sales**) = $144.76 billion.

This can be expressed as:

(0.08)(base amount due to sales) = $144.76 billion,
or,

$$\text{base amount due to sales} = \frac{\$144.76 \text{ billion}}{(0.08)}$$

Therefore, base amount due to sales = $1,809.5 billion.
Thus, an amount of sales revenue of $1,809.5 billion yielded $144.76 billion of sales tax revenue.

c) If the amount due to sales revenue remained the same and the sales tax increased to 9%, then the state could expect to generate:
9% of $1,809.5 billion = $162.855 billion

That is, the state tax revenue that could have been generated from a 9% sales tax is $162.855 billion.

d) Since a 9% sales tax would have produced a revenue of $162.86 billion, this revenue represents an overall increase in the state revenues of $18.095 billion. Adding this to the initial $517 billion yields $535.095 billion.

The projected $162.86 billion from a 9% sales tax represents an increase to 30.44% revenue generated from sales tax.

CASE STUDY 2.7

Examine the histogram in Figure 2.36 which represents the percentage of men and women by age who live alone in the USA.

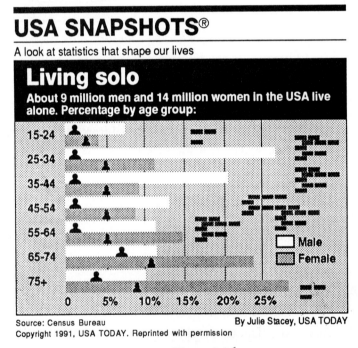

Figure 2.36

Using the information in Figure 2.36, answer the following questions:
a) For which age class do you find the greatest number of men living alone?
b) For which age class do you find the greatest number of women living alone?
c) Are there more men or women living alone in the age class 45-54?
d) Describe the shape of the graph representing the women living alone.
e) Describe the shape of the graph representing the men living alone.
f) Why do you think the shapes of the graphs for parts d and e are different?

Solution:

a) In Figure 2.36, the longest length for the males corresponds to the class pertaining to the 25-34 year old males. Thus, the age class 25-34 would contain the greatest number of men living alone.

b) In Figure 2.36, the longest length for the females corresponds to the class pertaining to the 75+ year old women. So, the age class 75+ would contain the greatest number of women living alone.

c) In the 45-54 age class, it appears that 13% of the 9 million males are represented. Thirteen percent of the 9 million males is 1.17 million males.
   In the 45-54 age class, it appears that 9% of the 14 million females are represented. Nine percent of the 14 million females is 1.26 million females.
   Thus, although, relative to the male population, more of the percent of the males live alone in the 45-54 age class than the female population, the number of females living alone is greater than the number of males living alone in that age class.

d) The shape of the graph for the female population living alone tends to be skewed to the left.

e) The shape of the graph for the male population living alone tends to be skewed to the right.

f) Women tend to live longer than males.

2.8 Using MINITAB

At the beginning of this chapter, a large data set was presented representing 150 responses to a student survey. In MINITAB, the Stem-and-Leaf command for the variable hours studied per week produced the output presented in Figure 2.37.

```
MTB > stem-and-leaf 'hr'

Stem-and-leaf of hr      N = 150
Leaf Unit = 1.0

    5    0  66777
   19    0  88888889999999
   35    1  0000000000111111
   44    1  222222222
   55    1  44555555555
   68    1  6666677777777
  (11)   1  88888899999
   71    2  00000000000011111
   54    2  22233333
   46    2  4444444445555
   33    2  66667777777
   22    2  8888888999
   12    3  0000111
    5    3  2333
    1    3
    1    3  6
```

Figure 2.37

The numbers in the first column represent the cummulative frequency up until the row containing the middle value of the distribution is reached. After the row containing the middle value, the row frequencies represent the number of remaining data values within the distribution. The middle value of the distribution is contained within the row identified by a set of parentheses around the row frequency.

MINITAB can quickly develop a Stem-and-Leaf Display that was discussed in this chapter. From examining Figure 2.37 you should notice that since the data values ranged from 6 to 36 hours studied per week and that the frequency of the data values contained in each stem was very large, MINITAB separates the leaves many times. For example, the stem 10 is separated into leaves 0 and 1, then 2 and 3, 4 and 5, 6 and 7, and, finally 8 and 9.

The output also indicates the line that marks the half-way position for the data, if the data were arranged from lowest to highest. Since the line marked by (11) contained 1 88888899999, the data values 18 and 19 represent the middle of the distribution. Can you determine which of these data values is eactly in the middle of the distribution?

When using MINITAB, there are options to control the output. The output in Figure 2.38 represents a stem-and-leaf display that *forces* the display to group the data values in increments of 10.

```
MTB > Stem-and-Leaf 'hr';
SUBC>    Increment 10.

Stem-and-leaf of hr        N  = 150
Leaf Unit = 1.0

   19    0 6677788888889999999
  (60)   1 000000000011111122222222244555555555666677777777888889999
   71    2 000000000000111112223333344444444455566667777778888888999
   12    3 000011123336
```

Figure 2.38

MINITAB can be used to develop a histogram. Figure 2.39, represents MINITAB's version of a histogram for the same data represented in Figure 2.38. Notice that the midpoint and frequency count is indicated on the output.

```
MTB > Histogram 'hr'.

Histogram of hr    N = 150

Midpoint    Count
      8       19    *******************
     12       25    *************************
     16       24    ************************
     20       28    ****************************
     24       21    *********************
     28       21    *********************
     32       11    ***********
     36        1    *
```

Figure 2.39

Glossary

TERM	SECTION	TERM	SECTION
Variable	2.2	Less Than Frequency Distribution	2.4
Data Set	2.2	More Than Frequency Distribution	2.4
Categorical Variable	2.2	Graph	2.5
Numerical Variable	2.2	Bar Graph	2.5
Discrete Data	2.2	Horizontal Bar Graph	2.5
Exploratory Data Analysis	2.3	Vertical Bar Graph	2.5
Stem-and-Leaf Display	2.3	Time-Series Graph	2.5
Back-to-Back Stem-and-Leaf Display	2.3	Histogram	2.5
Stretched Stem-and-Leaf Display	2.3	Frequency Polygon	2.5
Distribution	2.3	Ogive	2.5
Outlier	2.3	Pie Chart	2.6
Skewed Distribution	2.3	Pictograph	2.6
Frequency Distribution	2.4	Symmetric Bell-Shaped Distribution	2.7
Frequency Distribution Table	2.4	Positively Skewed or Skewed to the Right	2.7
Class or Class Interval	2.4	Negatively Skewed or Skewed to the Left	2.7
Class Limit	2.4	Uniform or Rectangle Distribution	2.7
Class Mark	2.4	U-Shaped Distribution	2.7
Class Boundary or Real Class Limit	2.4	Reverse J-Shaped Distribution	2.7
Relative Frequency Distribution	2.4	Bimodal Distribution	2.7
Cumulative Frequency Distribution	2.4		

EXERCISES

PART I **Fill in the blanks.**

1. A _____ is a type of information, usually a property or characteristic of a person or thing, that is measured or observed.
2. A specific measurement or observation for a variable is called the _____ of the variable.
3. There are two types of variables: _____ and _____.
4. A categorical variable is a variable where the values represent _____.
5. A variable where the values are the result of a measurement process is called a _____ variable. These values are referred to as numerical ___.
6. Numerical data can be classified as either ___ or ___.
7. Numerical data such as weight, height, and temperature are examples of _____ data.
8. Numerical data such as the number of sunny days in August, the number of heads in five tosses of a coin, and the number of people on social security are examples of _____ data.
9. The approach to carefully examining a data set to discover any useful aspects, information, or unanticipated patterns that may exist in the data is called _____.
10. A _____ represents the numerical data values of a variable from the lowest to the highest value along with the number of times each data value occurs.
11. A relatively simple exploratory data analysis technique that helps to visually show the shape of a distribution by using the actual values of the variable is called a _____ display.
12. If a data value within a stem-and-leaf display is written as: 34 | 6, then the stem value is ___ and the leaf value is ___.
13. An individual data value which lies far from most or all of the other data values within a distribution is called an _____.
14. When two distributions are compared using a back-to-back stem-and-leaf display, a common ___ is used to display each distribution.
15. In a stretched stem-and-leaf display, the stem value 3* is used to represent the leaf values from _____ to _____, while the stem value 3· is used to represent the leaf values from _____ to _____.
16. A frequency distribution table is a table that is used to help describe the general characteristics of a distribution by dividing the data into groups, called ___ along with the number of data that fall into each class, called the _____.
17. In constructing a frequency distribution table for a categorical variable, the classes represent the _____ of the variable.
18. In constructing a frequency distribution table, the actual class width is dependent upon the number of __ digits used to measure the data values.
19. The midpoint of a class is called a class ___ and is calculated by taking one-half of the sum of the ___ and ___ class limits.
20. The representative value of a class is called a class _____.
21. The difference between two successive class marks is equal to the class ___.
22. The class boundary of a class is the ___ of the upper class limit of one class and the lower class limit of the next class.
23. The proportion of data values within a class is called the ___ frequency of the class and is calculated by dividing the class frequency by the ___ ___ of data values.
24. The sum of all the relative frequencies of all the classes will always be ___, except for possible round off error.
25. The relative percentage frequency of a class is the ___ of data values within a class. To determine the relative percentage frequency of each class, multiply the relative frequency of each class by ___%. The sum of all the relative percentage frequencies will always be _____ except for possible round off error.
26. In a "less than" cumulative frequency distribution table, the cumulative frequency of a particular class indicates the number of data values within the class plus the ___ of all the data values ___ ___ this particular class.
27. In a "more than" cumulative frequency distribution table, the cumulative frequency of a particular class indicates the number of data values within the class plus the ___ of all the data values ___ ___ this particular class.
28. A bar graph is generally used to depict ___ or ___ data.
29. A time-series graph contains information about two variables, where one of the variables is ___.
30. Generally, in a horizontal bar graph, the categories of the categorical data are placed along the ___ axis, while the frequencies are placed along the ___ axis.
31. In a vertical bar graph, the height of each bar represents the ___.
32. A squiggle is used to represent the ___ of values in either a horizontal or vertical axis.

33. A histogram is a graph of a continuous ____. It consists of rectangles connected to each other without any ____ or ____ between two adjacent rectangles thus showing continuity.
34. A frequency polygon is a line graph that uses the class____ of a frequency distribution to represent a ____variable.
35. A graph of a "less than" cumulative frequency distribution is called an ____.
36. A graph used to represent categorical data in a circular format is called a ____ chart.
37. A graph that uses pictures to represent data in an interesting and eye catching way is called a ____.
38. A symmetrically bell-shaped distribution has its highest frequency or peak at the ____ with the frequencies steadily ____ and are identically distributed on both sides of the center.
39. A distribution which is not symmetric is a ____ distribution.
40. A positively skewed or skewed to the ____ distribution has a greater number of relatively ____ scores and a few extremely ____ scores.
41. A negatively skewed or skewed to the ____ distribution has a larger number of relatively ____ scores and a few extremely ____ scores.
42. A uniform or rectangular distribution is a distribution where all the classes contain the ____ number of data values or frequencies.
43. A U-shaped distribution has its greatest frequencies occurring at each ____ of the distribution with the lower frequencies occurring in the ____.
44. A reverse J-shaped distribution has its greatest frequency of data values occurring at ____ end of the distribution and tails off gradually in the ____ direction.
45. A bimodal distribution is a distribution that has ____ greatest frequencies or peaks.

PART II Multiple choice questions.

To answer questions 1-7 use the following histogram.

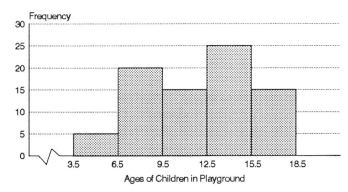

1. Find the total number of children represented in the graph.
   a) 25 b) 80 c) 100 d) 50
2. How many children in the playground are older than 12?
   a) 40 b) 25 c) 15 d) 30
3. What percent of the children are in the last two classes?
   a) 40% b) 50% c) 25% d) 15%
4. Which class has the largest frequency?
   a) 4-6 b) 10-12 c) 13-15 d) 16-18
5. What percent of the children are older than 9 but younger than 16 years old?
   a) 40% b) 50% c) 25% d) 15%
6. What percent of the children are younger than 4 years old?
   a) 40% b) 50% c) 25% d) 0%
7. How many children in the playground are older than 3?
   a) 25 b) 80 c) 100 d) 50
8. The class boundaries for the class 20-26 are:
   a) 20.5-25.5   b) 19.5-25.5
   c) 20.5-26.5   d) 19.5-26.5
9. The class width for the class 25-30 is:
   a) 7 b) 6 c) 5 d) 4
10. If the lower class limit is 17 and the class width is 7, then the upper class limit is:
    a) 23 b) 24 c) 22 d) 25
11. What is the class mark for the class 10 - 16?
    a) 10 b) 13 c) 12.5 d) 16

Use the following information for questions 12 and 13:
In constructing a Stem-and-Leaf display, a collection of anxiety scores are represented as whole numbers ranging from 15 to 40.

12. What would be the range of values used to construct a Stem?
    a) 1 to 4 b) 15 to 40 c) 10 to 100 d) 20 to 35
13. How many digits are used to represent a leaf?
    a) 1 b) 2 c) 5 d) 9
14. If a class has a lower class limit of 34 and an upper class limit of 40, what would be the class mark?
    a) 33.5 b) 34 c) 36 d) 37
15. In a relative frequency distribution, the sum of the percentages should equal:
    a) 1% b) 25% c) 50% d) 100%

PART III Answer each statement True or False.

1. Graphs are used to accurately depict distributions.
2. Numerical data are usually represented as a rounded-off numbers.
3. The values of a variable characterized by descriptive words such as good, poor, or excellent are referred to as categorical data.
4. An example of a continuous variable is the number of

classrooms in a school building.
5. A discrete variable can be represented using a frequency distribution table.
6. A graph of a "less than" cumulative frequency distribution table is an ogive.
7. When constructing a frequency distribution table, the number of classes is always 10.
8. Bar graphs are usually used to represent categorical variables.
9. Histograms are usually used to represent continuous variables.
10. Pie charts are usually constructed using percentages.
11. The frequency polygon is a better graphical representation of data than a histogram.
12. A histogram is essentially a bar graph drawn without space between the bars.
13. Class boundaries are used in the construction of a histogram but not in the construction of a bar graph.
14. If two distributions have the same histogram, then the percent of terms greater than the center of the distribution is the same.
15. Stem-and-Leaf Displays are visual displays that are primarily used to represent categorical data.
16. Exploratory Data Analysis is used to learn as much as possible about data before conducting statistical tests.

PART IV Problems.

1. Classify the following data as either categorical or numerical. If the data is numerical, then classify it as either continuous or discrete data.
   a) The life of an automobile tire.
   b) The number of children in a family.
   c) A person's educational degree.
   d) A person's weight.
   e) The number of registered conservatives in a county.
   f) A person's eye color.
   g) The annual income of a family.
   h) The breaking strength of a fishing line.
   i) The amount of time needed to go to work.
   j) A person's occupation.
   k) The number of bushels of apples produced by a farmer.
   l) The volume of water in a reservoir.
   m) An individual's IQ.
   n) A person's opinion on the type of job Congress is doing.

2. The following table represents a sample of data collected on college students. Each column abbreviation represents the following information:
   sex: student's sex
   status: fr= freshman, sp= sophomore, jr= junior, sr=senior
   age: student's age
   ht: student's height in inches
   wt: student's weight in lbs.
   frat: fraternity or sorority membership: N= No or Y=Yes
   havg: high school average
   gpa: grade point average

student#	sex	status	age	ht	wt	frat	havg	gpa
3748	m	jr	20	68	170	N	77	2.89
9857	f	fr	18	62	105	N	93	3.45
7194	f	sr	21	65	135	Y	82	2.76
6203	m	sp	19	75	195	Y	91	2.94
5924	f	jr	21	60	124	Y	87	3.12

   a) Identify each variable and type of variable within the data set.
   b) For the student identified by the number 9857, specify the value of each of the variables.

3. A health club is interested in the distribution of ages for its members that use the club on Friday evenings. The club's personnel director surveys the membership one Friday evening and collects the following data:

<u>Club Member's Ages in Years</u>

19	31	52	34	84	63	52	37	24	29
33	46	19	32	41	49	26	32	46	44
28	76	49	34	73	63	56	35	74	66
59	39	61	50	37	29	30	51	54	41

   a) Construct a stem-and-leaf display for the distribution of ages using the tens digit for the stem portion.
   b) Describe the shape of the distribution.
   c) In which age decade are most people?
   d) In which fifteen year age difference are most people?

4. A road and highway agency is studying the commuting habits of drivers on a local parkway. Drivers were asked the number of miles they drive to work. The following distribution represents the survey findings.

<u>Number of Miles Travelled to Work</u>

12	34	35	20	4	38	23	10	3	5
40	5	13	19	30	40	45	8	21	5
25	18	15	15	7	64	25	20	50	8

   a) Construct a stem-and-leaf display for the number of miles travelled to work using the tens digit for the stem portion.
   b) Using the stem-and-leaf display, describe the shape of the distribution.
   c) Within what 10 mile range do most people drive?

5. A statistics instructor has recorded the amount of time students need to complete the final examination. The times, stated to the nearest minute, are:

50, 70, 62, 55, 38, 42, 49, 75, 80,
79, 48, 45, 53, 58, 64, 77, 48, 49,
50, 61, 72, 74, 10, 95, 120, 79, 75,
48, 72, 37, 35, 79, 72, 75, 77, 32,
30, 77, 75, 79, 45, 70, 39, 75, 72,
45, 47, 73, 72, 71

a) Construct a stem-and-leaf display using the hundreds and tens digits for the stem portion.
b) Describe the shape of the distribution.
c) Are there any outlier(s) in the data set? If so, give an explanation for the outlier(s)?
d) What percentage of the class required from 40 to 75 minutes to complete the test?
e) If the statistics instructor only allowed one hour and 20 minutes for the exam, what percentage of the class would not have had enough time to complete the exam?

6. The following stem-and-leaf display represents the fuel consumption in miles per gallon for 45 vehicles. The stem represents the tens and unit digit of the mpg data value, while the leaf represents the tenths digit of the mpg value.

```
13 | 8 9
14 | 0 3 5 7
15 | 1 1 4 4 8
16 | 1 2 9
17 | 0 2
18 | 0 1 3
19 | 0 1 4 4 5
20 | 0 5 6 6 8
21 | 1 1 5 5
22 | 0 1 2 2 3
23 | 1 4 7 8
24 | 1 3
25 |
26 |
27 |
28 | 4
```

a) Write the data values that are being represented by this stem-and-leaf display.
b) Is there a data value that falls outside the range of all the other data values? If so, what is that value? What is such a value called?
c) Construct a frequency distribution table using four classes for the data values in the stem-and-leaf display.
d) Compare the frequency distribution table to the stem-and-leaf display. In particular, what information is contained in the stem-and-leaf display that is not contained in the frequency distribution table?

7. Given the following distribution of statistics test grades:

73 92 57 89 70 95 75 80 47
88 47 48 64 86 79 72 71 77
93 55 75 50 53 75 85 50 82
45 40 82 60 55 60 89 79 65
54 93 60 83 59

a) Construct a stem-and-leaf display using the tens digit for the stem portion.
b) Describe the shape of the distribution.
c) Construct a frequency distribution table and a histogram using six classes.
  Using the histogram, answer the following questions:
d) How many test grades are greater than 89?
e) What percent of the test grades are greater than 79?
f) What percent of the test grades are lower than 70?
g) What percent of the test grades are between 70 and 79 (inclusive)?

8. The following data set represents the ages of 110 runners who competed in a city marathon.

44 23 19 51 22 69 31 30 41 21
52 27 20 32 23 71 37 60 29 54
27 31 23 27 44 29 58 43 49 38
60 32 34 50 28 18 29 54 41 22
37 36 37 35 33 37 20 26 22 27
31 29 41 43 67 55 35 29 48 19
25 69 60 43 26 33 24 22 35 34
18 28 65 58 21 60 59 45 27 26
42 34 52 61 56 72 19 29 19 28
44 56 29 43 50 49 19 22 23 27
30 58 64 59 24 55 48 18 40 21

a) Construct a stem-and-leaf display using the tens digit for the stem portion.
b) Construct a frequency distribution table using 8 classes. Within the table include class marks, class boundaries, and relative percentages.
c) Construct a histogram and frequency polygon for the distribution.
d) Describe the shape of the polygon and classify the shape as either symmetric bell-shaped, skewed to the right, or skewed to the left.

9. A medical study was performed to determine if a special new drug could lower a person's blood pressure. The study was conducted with two groups an experimental group and a control group. The experimental group received the new drug, while the control group was given a dummy pill called a placebo. The systolic blood pressure for each participant was measured while the participant was at rest before the

study began. Both groups had similar blood pressure distributions before the experiment began. The following data represents the blood pressures of the participants within each group after the study was conducted.

Experimental Group

109	119	128	131	105	113	108	113	125
130	120	123	120	119	126	125	118	107
114	117	121	112	105	127	115		

Control Group

117	127	135	142	124	137	113	109	126
135	143	138	145	136	126	136	147	138
128	138	137	139	148	149	148		

a) Construct a back-to-back stem-and-leaf display to compare these two groups using the hundreds and tens digits for the stem portion.
b) Does it appear that the new drug had an effect on lowering the blood pressures of the participants, if both groups began with similar blood pressure distributions?
c) Describe the shape of each distribution for the experimental and control groups?

10. A medical researcher surveyed one hundred mothers who recently gave birth. The researcher determined whether the mothers smoked during pregnancy and the babies' birth weights. The birth weights were separated into two groups, smoking mothers and non-smoking mothers. The following data represents the researchers findings.

Smoking Mothers

5 lbs 7 ozs	6 lbs 6 ozs	7 lbs 2 ozs	9 lbs 2 ozs	7 lbs 3 ozs
8 lbs 2 ozs	7 lbs 5 ozs	6 lbs 2 ozs	5 lbs 3 ozs	3 lbs 4 ozs
5 lbs 3 ozs	7 lbs 6 ozs	6 lbs 1 oz	5 lbs 4 ozs	5 lbs 2 ozs
6 lbs 8 ozs	7 lbs 4 ozs	7 lbs 2 ozs	9 lbs 3 ozs	6 lbs 5 ozs
7 lbs 7 ozs	8 lbs 2 ozs	7 lbs 6 ozs	3 lbs 1 oz	5 lbs 12 ozs
6 lbs 2 ozs	7 lbs 5 ozs	6 lbs 5 ozs	8 lbs 8 ozs	9 lbs 1 oz
8 lbs 3 ozs	7 lbs 8 ozs	6 lbs 1 oz	5 lbs 9 ozs	6 lbs 5 ozs
6 lbs 6 ozs	8 lbs 3 ozs	7 lbs 8 ozs	6 lbs 6 ozs	5 lbs 12 ozs
5 lbs 9 ozs	6 lbs 9 ozs	6 lbs 8 ozs	5 lbs 2 ozs	7 lbs 4 ozs
6 lbs 9 ozs	7 lbs 5 ozs	7 lbs 0 ozs	8 lbs 0 ozs	8 lbs 0 ozs

Non-smoking Mothers

8 lbs 2 ozs	7 lbs 1 oz	6 lbs 6 ozs	5 lbs 9 ozs	9 lbs 4 ozs
7 lbs 0 ozs	8 lbs 5 ozs	6 lbs 9 ozs	5 lbs 13 ozs	8 lbs 4 ozs
7 lbs 6 ozs	6 lbs 6 ozs	5 lbs 11 ozs	9 lbs 0 ozs	8 lbs 9 ozs
7 lbs 1 oz	6 lbs 8 ozs	7 lbs 9 ozs	9 lbs 3 ozs	8 lbs 8 ozs
6 lbs 6 ozs	7 lbs 3 ozs	7 lbs 14 ozs	8 lbs 1 oz	7 lbs 5 ozs
6 lbs 5 ozs	7 lbs 6 ozs	6 lbs 12 ozs	7 lbs 8 ozs	7 lbs 9 ozs,
9 lbs 3 ozs	8 lbs 4 ozs	7 lbs 11 ozs	6 lbs 14 ozs	5 lbs 11 ozs
8 lbs 6 ozs	5 lbs 15 ozs	6 lbs 10 ozs	7 lbs 3 ozs	9 lbs 8 ozs
7 lbs 15 ozs	7 lbs 2 ozs	6 lbs 14 ozs	7 lbs 12 ozs	7 lbs 10 ozs
6 lbs 8 ozs	6 lbs 14 ozs	11 lbs 2 ozs	10 lbs 1 oz	10 lbs 5 ozs

a) Construct a back-to-back stem-and-leaf display for the birth weights of the babies for the smoking and non-smoking mothers using lbs. for the stem portion.
b) Describe the shape of each distribution of birth weights for smoking and non-smoking mothers.
c) Does it appear from the stem-and-leaf display that smoking may have an effect on the birth weight of a newborn? Support your conclusion.

11. A bank executive compared the length of time required to service its customers using the auto teller to the customers using the regular banking service. To obtain information on these two distributions of customer service times, two samples of fifty service times were recorded to the nearest second.

Auto Teller Service
(time measured to the nearest second)

195	123	126	178	168	132	180	204	228
216	276	228	138	286	150	192	156	186
268	222	204	132	198	208	204	258	228
428	274	162	156	288	270	186	144	192
144	350	378	207	195	193	170	181	276
265	306	282	194	197				

Regular Banking Service
(time measured to the nearest second)

173	309	264	310	174	428	317	246	204
262	426	306	234	298	161	156	175	168
198	179	210	228	240	174	312	288	360
216	210	282	252	322	246	366	252	192
186	246	304	234	216	228	252	378	198
216	348	123	310	318				

a) Construct a "less than" cumulative frequency distribution table for both samples using 8 classes.
b) Determine what percentages of the service times for the customers using the auto teller are less than or equal to 4 minutes.
c) Determine what percentages of the service times for the customers using the regular banking service have times less than or equal to 4 minutes.

d) Compare the waiting times of both services using the distribution tables.

12. The following blood pressures are for men and women within the age group 19 - 25 years.
    a) Construct a back-to-back stem-and-leaf display for these blood pressures where the stem portion represents the first two digits of the blood pressures.
    b) Construct a "less than" cumulative frequency distribution table for both the men and women blood pressure readings using 5 classes.
    c) Determine the percentage of men who have blood pressures between 104 and 136 inclusive.
    d) Determine the percentage of women who have blood pressures between 104 and 136 inclusive.
    e) Construct an ogive for the blood pressure readings for both men and women.

Men Blood Pressures

146	130	115	120	104	143	120	146	116
118	130	130	145	132	118	136	104	126
156	145	150	110	130	126	106	112	120
130	136	134	156	120	110	150	105	130
124	128	130	122	142	130	106	110	112
120	126	126	112	110	120	120	120	120
130	110	106	112	134	110	120	120	128
120	140	118	120	130	114	118	120	116
120	106	118						

Women Blood Pressures

134	124	108	112	126	114	118	114	110
112	122	128	135	118	110	110	105	106
122	129	105	110	110	134	130	140	116
110	112	110	110	116	114	110	107	130
139	126	138	120	118	116	118	112	104
116	104	114	124	156	120	118	120	128
116	126	128	112	128	108	108	130	116
118	122	126	128	120	119	130	125	120
126	112	138						

13. The test scores on a statistics exam ranged from 44 to 94. Determine the class limits, class marks, and class boundaries of a frequency distribution if these test scores were grouped using:
    a) 8 classes
    b) 7 classes
    c) How many classes should you use if you want the class marks to be a whole number?

14. A survey on the number of hours a college student works per week showed that the hours varied from 1 to 40. Determine the class limits, class marks, and class boundaries of a frequency distribution table if the work hours were grouped using:
    a) 5 classes.
    b) 6 classes.

15. A frequency distribution table representing the number of live births to U.S. mothers for the past year was grouped according to the following classes: 15 - 19 years, 20 - 24 years, 25 - 29 years, 30 - 34 years, 35 - 39 years, 40 - 44 years, and 45 - 49 years. For this frequency distribution table, is it possible to determine the exact number of live births to U.S. mothers having an age:
    a) 25 years or older?
    b) more than 25 years?
    c) more than 24 years?
    d) exactly 25 years?
    e) at least 24 years?

16. The following distribution represents the IQ scores of 120 high school students.

Classes (IQ Score)	Frequency
81 - 90	2
91 - 100	13
101 - 110	44
111 - 120	40
121 - 130	17
131 - 140	4

Determine, if possible, the number and percent of the students who have IQ scores that are:
a) less than 101.
b) at most 110.
c) 111 or less.
d) exactly 121.
e) between 100 to 120 inclusive.
f) at least 121.

17. The following data represents the number of members within each household of forty high school students.

7	4	5	5	6	5	10	8	5	5
6	9	4	5	4	4	7	6	8	5
6	5	3	4	2	5	4	5	4	3
3	4	5	3	4	5	4	2	3	6

Construct a frequency distribution table using:
a) five classes.
b) three classes.

18. The following data set represents the IQ Scores of 120 high school students.

    ```
    145 126 118  95 110 150 108 128 107 123 100 113
    109 130  98 118 110 121 105 109 136 123 112 117
    113  92 118 104 117  97 127 108 122 105 117 121
    109 138 125 118  99  97 149  92 113 100 109 117
    113 116 116  98 125 123 124 129  94 148 144 127
    119 101 127  95 107 112 121 125 113  99 120 122
    123 127 128 117 114 129  95 113 145 112 122 106
    119  93 132 122 110 117  98 121 110  97 121 131
    109 145 103 135  98 142 127 116 111 104 112 116
    125 109 134 110 107  97 101 131  96 113 107 115
    ```

    a) Construct a frequency distribution table using 10 classes. Within the table include class marks, class boundaries, relative frequencies and percentages.
    b) Find the sum of the relative frequencies and percent ages.
    c) Using this table, determine the percentage of students that have an IQ Score which is:
    (1) less than 116
    (2) between 128 to 133 inclusive
    (3) greater than 121
    (4) at most 109
    (5) at least 134

19. The following frequency distribution table represents information collected by a local service station owner on the number of gallons of gasoline, recorded to the nearest gallon, purchased per customer.

Gallons of Gasoline	Frequency
5 - 7	17
8 - 10	33
11 - 13	42
14 - 16	88
17 - 19	61
20 - 22	38
23 - 25	21

    a) What is the class width of each class?
    b) Determine the class mark, class boundaries, and relative percents for each class.
    c) Construct a "less than" cumulative frequency distribution table and an ogive for this "less than" cumulative frequency distribution table.
    d) How many customers purchased less than 11 gallons of gasoline? What percent of the customers does this represent?
    e) What percent of the customers purchased less than 17 gallons of gasoline?
    f) How many customers purchased from 11 to 19 gallons? What percent of the customers does this represent?

20. Fifty people were given a blind taste test to determine their preference for the soft drink colas Sipeps and Coaks. Their responses to the taste test were classified as: S for Sipeps, C for Coaks, and N for Neither. The responses to the taste test were:

    S, S, C, S, C, N, N, C, S, C, C, N, C, S, C, C, C, C, S, S, S, S, C, C, S,
    N, S, C, C, S, S, C, C, S, N, C, S, N, C, S, C, S, C, S, C, S, C, C, S, C

    a) Construct a frequency distribution table representing the responses to the taste test.
    b) Calculate the relative frequencies and relative percents for each of the responses.
    c) Use the percentages to construct a pie chart representing the results of the taste test.

21. After the holiday season, 748 people were polled about when they finished their holiday shopping. The results of the poll were as follows:

Finished Shopping By:	Number of Shoppers
November 30	45
December 15	389
December 23	127
December 25	187

    a) Determine the percentage of shoppers corresponding to each date.
    b) Construct a pie chart for this information.
    c) By what date are the majority of the shoppers finished shopping?
    d) What percent of the shoppers finish by December 23?
    e) What percent of the shoppers are NOT finished by December 23?

22. The following table gives the number of deaths per 100,000 contributed to coronaries and cardiovascular diseases by age group for men and women.

    # RISKS RISE WITH AGE
    Deaths per 100,000 by age group

Age	Men	Women
25-34	2.2	0.8
35-44	18.7	4.4
45-54	80.4	22.0
55-64	231.2	87.4
65-74	523.0	256.1
75-84	1132.0	686.6

a) Determine the class width for each class.
b) Find the class mark and class boundaries for each class.
   Determine the relative percentages for the males and females for each class.
c) Find the sum of the relative percents for the males and females.
d) Construct two frequency polygons one representing the males and one representing the females. Compare the two polygons and comment about the differences between the two groups.

23. A national park foundation allocates funds to individual parks according to park size. In preparing a report, an official must summarize the information regarding park size. He prepares the frequency table that follows:

Park Size in Acres	Number of Parks
1 - 9	17
10 - 49	34
50 - 99	52
100 - 199	47
200 - 499	32
500 - 749	25
750 - 999	68

Using the frequency table:
a) Construct a "less than" cumulative frequency distribution table.
b) Construct an ogive.
c) Represent the cumulative frequency values as relative frequency percentages.
d) What percentage of parks are:
   (1) less than 10 acres?
   (2) less than 100 acres?
   (3) less than 750 acres?
e) Within what range of acres does 50% of the distribution fall?

24. The results of Glenville High School's chemistry regents exam are given below.
44, 76, 82, 80, 83, 84, 48, 80, 73, 75, 80, 77, 100, 62, 65, 82, 66, 64, 60, 86, 90, 88, 92, 95, 78, 53, 59, 52, 89, 75
a) Construct a frequency distribution table using five classes include the class mark and class boundaries.
b) Construct a histogram.
c) Construct a frequency polygon.

25. A business magazine randomly surveyed 210 business owners to determine the age at which they first began their own business. The following distribution represents the survey results:

Age	Frequency
18 - 22	27
23 - 27	54
28 - 32	55
33 - 37	31
38 - 42	21
43 - 47	10
48 - 52	7
53 - 57	3
58 - 62	2

a) Find the class mark, class boundaries, and relative frequencies for each class.
b) Construct a frequency polygon and describe the shape of the distribution.
c) Construct a "less than" cumulative frequency distribution.
d) Construct a "more than" cumulative frequency distribution.
e) Construct an ogive for the "less than" cumulative frequency distribution.
f) What percentage of the business owners started their first business when they were younger than 33 years old?
g) What percentage of the owners started their first business when they were 43 years or older?
h) What percentage of the business owners started their first business when they were 23 to 32 years old?

26. Pollution indices for a city in the northeast for 100 consecutive days are given in the following table.
```
59 54 63 60 69 66 42 47 48 41
43 33 48 52 41 48 44 42 40 47
62 70 54 83 69 57 46 58 52 54
58 54 67 49 61 62 54 57 30 44
33 41 52 43 47 54 59 62 68 66
60 57 57 61 58 59 54 48 39 61
60 53 50 44 58 46 57 63 69 72
42 70 54 77 81 84 33 37 44 52
41 43 37 70 72 59 54 61 50 73
61 73 74 72 67 58 49 82 79 68
```
Construct:
a) a frequency distribution table using seven classes; include class mark and class boundaries within the table.
b) a histogram.
c) a frequency polygon.

27. The student government recently surveyed 300 students over a period of five consecutive registration days as to how long it takes students to register for next semester's classes.

    The following table represents the registration time for the surveyed students:

Number of Minutes to Register for Class	Number of Students
0 - 15	20
16 - 30	45
31 - 45	90
46 - 60	110
61 - 75	30
76 - 90	5

    a) For the given frequency table, add the class boundaries and construct a histogram.
    b) Construct a "less than" cumulative frequency table and an ogive.
    c) From the ogive, determine how long 75% of the students would take to register?

28. The following table contains 150 test scores representing the results of an IQ test that was administered to the third graders at an elementary school in Merrick, New York.

Classes (IQ Scores)	BOYS Frequency	GIRLS Frequency
60 - 79	8	2
80 - 99	27	35
100 - 119	23	15
120 - 139	10	25
140 - 159	2	3

    a) Construct a relative frequency percentage distribution table for:
       (1) the IQ scores of the third grade boys.
       (2) the IQ scores of the third grade girls.
       (3) all IQ scores of the third graders at the school.

    Using the relative frequency percentage distribution tables of part a:

    b) Construct a histogram for:
       (1) the IQ scores of the third grade boys.
       (2) the IQ scores of the third grade girls.
       (3) all IQ scores of the third graders at the school.
    c) Calculate the following:
       (1) the percent of girls that have an IQ of at least 100.
       (2) the percent of boys that have an IQ of at least 100.
       (3) the percent of third graders that have an IQ of at least 100.
       (4) the percent of third graders that have an IQ between 80 and 159 (inclusive).
       (5) the percent of third graders that have an IQ of less than 80.

29. At the High Roller High School in Kings Park, California, all seniors are required to take the Scholastic Aptitude Test (SAT). The following table summarizes last year's test scores.

Classes (SAT Scores)	VERBAL		MATHEMATICS	
	# of Males	# of Females	# of Males	# of Females
200-299	15	10	10	25
300-399	45	30	30	55
400-499	35	30	30	35
500-599	45	60	55	45
600-699	10	15	15	10
700-799	25	50	35	25

    a) Construct a relative frequency percentage distribution table representing the:
       (1) males
       (2) females

    Using the classes within the table:

    b) Construct a histogram for the verbal SAT scores representing the:
       (1) males
       (2) females
    c) Construct a histogram for the mathematics SAT scores representing the:
       (1) males
       (2) females
    d) Construct frequency polygons comparing the:
       (l) verbal SAT scores representing the males to the verbal SAT scores representing the females.

(2) mathematics SAT scores representing the males to the mathematics SAT scores representing the females.

e) Calculate the percent of:
(1) females that have a verbal SAT score of at least 400.
(2) males that have a verbal SAT score of at least 400.
(3) females that have a mathematics SAT score of a least 600.
(4) males that have a mathematics SAT score of at least 600.

30. The manager of Fast Food Grocery Store is preparing a marketing report to present to his district manager. Construct a horizontal bar graph from the information in the following table.

Cereal	Number of Cases
Bran Chex	20
Raisin Bran	35
Wheat Bran	14
Corn Bran	16

31. Using the data given in the following table, construct a horizontal bar graph.

SURGICAL OPERATIONS WITHIN THE UNITED STATES DURING 1991	
Surgery	Number of Operations (in thousands)
Appendectomy	784
Hemorrhoidectomy	200
Hysterectomy	787
Tonsillectomy	724
Prostatectomy	223
Herniorrhaphy	508

32. Using the data given in the following table, construct a horizontal bar graph.

WHERE THE MILLIONAIRES LIVE	
State	Number of Millionaires
New York	56,096
California	38,691
Illinois	35,545
Ohio	31,202
Florida	29,523
New Jersey	28,613
Indiana	24,880
Idaho	24,738
Minnesota	23,381
Texas	23,002

33. The following data was compiled by an Education Commission regarding the question:
In different countries, how long do kids go to school each year?

The average number of days per year for the listed countries were as follows:

Country	Average Number of School Days
England	192
France	185
Netherlands	200
Israel	216
Japan	243
Germany	210
Thailand	200
USA	180
Sweden	180

Arrange the data from highest to lowest and construct a horizontal bar graph.

34. Using the data given in the following table, construct a vertical bar graph.

The Cable Picture	
Year	Number of Subscribers (in millions)
1986	13.0
1987	14.1
1988	16.0
1989	18.3
1990	21.0

35. The following table represents the frequency distribution of the number of children per family for 400 families surveyed in a small community.

Number of Children	Frequency
0 - 1	102
2 - 3	228
4 - 5	57
6 - 7	13

a) Construct a bar graph for this frequency distribution table.
b) Construct a "less than" cumulative frequency distribution table.
c) What percent of the families have two or more children?
d) What percent of the families have less than four children?
e) Using this frequency distribution table, if you were a builder planning to build 80 new homes near this community, how many 4 bedroom homes might you consider building?

36. Central General Hospital in Phoenix has the following data representing weight in pounds at birth for 500 premature babies:

Class	Frequency
0.5 - 0.9	26
1.0 - 1.4	48
1.5 - 1.9	60
2.0 - 2.4	66
2.5 - 2.9	73
3.0 - 3.4	85
3.5 - 3.9	100
4.0 - 4.4	42

a) Construct a "less than" cumulative frequency table.
b) Use the "less than" cumulative frequency table constructed in part (a) to construct an ogive.
c) If babies that weigh less than 3.0 pounds need incubators, then what percentage of the premature babies will need incubators?

37. Construct two pie charts to represent the results of a poll conducted by the American Communications Group of 1,000 households to determine **how often people watch movies**.

The following frequency distribution table represents the results of the poll.

How Often People Watch Movies	Percentage of People Who Watch Movies	
	AT HOME	AT THEATER
Less Than Once a Month	13%	46%
Once a Month	12%	24%
Twice or More a Month	75%	30%

38. A financial advisor has prepared model portfolios on how her clients might divvy up their investments based upon the type of risk that an investor is willing to take. Construct two pie charts to compare these two types of portfolios.

Investment Category	TYPE OF PORTFOLIO	
	LOW RISK	HIGH RISK
Cash	30%	10%
High-Grade Bonds	20%	10%
Blue-Chip Stocks	20%	25%
Income Stocks	10%	5%
Aggressive Growth Stocks	10%	25%
Foreign Stocks	10%	25%

39. The Department of Education surveyed seventeen year-old students pertaining to the amount of time the students spend doing homework. The results of their survey are given in the following table.

Number of Hours Spent Doing Homework	Number of students
Less than 1 hour	196
1 to 2 hours	182
More than 2 hours	84
None assigned	147
Didn't do it	91

a) Using the table information, convert the number of students to a percentage and construct a pie chart.
b) What percentage of the students spend at least one hour doing homework?
c) What percentage of the students spend at most two hours doing homework?

40. The following pictograph shows the percentage of injuries for different levels of football. Using this information, construct a frequency distribution table to display the number of injuries per level of football for the sample of 400.

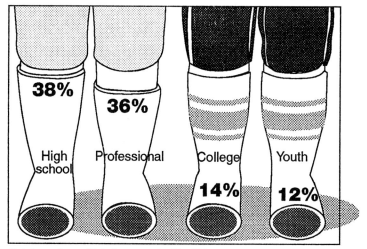

Sports Injuries
Level of football considered to have the highest risk of injury.

Total number of injuries reported is 400.

41. A writing sample (composition) was submitted by each of 50 students applying for admission to Wrighter's College. Each composition was rated by 3 readers and a composite grade was assigned. The following histogram represents the 50 grades.

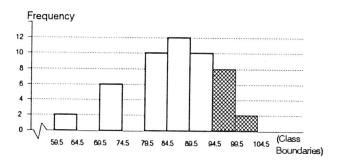

a) How many compositions were rated between 90 and 99 inclusive?
b) What percent of students scored between 80 and 84 inclusive?
c) What percent of the total area is in the criss-crossed section?
d) What percent of the students scored between 90 and 99 inclusive?
e) What percent of the total area is contained under the graph in the 7th and 8th classes combined?
f) Did anyone receive the highest grade of 100?
g) What percent of the total area is contained under the graph in the last class?

110 • CHAPTER 2 ORGANIZING AND PRESENTING DATA

42. Each of the following tables or graphs has a newspaper heading. Using this heading as your main theme, write a paragraph that supports this theme using the information contained in the table or graph.

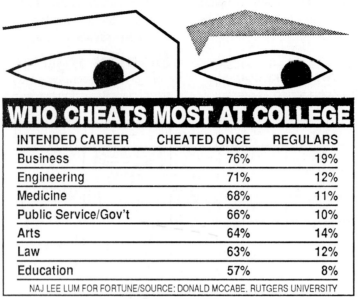

WHO CHEATS MOST AT COLLEGE

INTENDED CAREER	CHEATED ONCE	REGULARS
Business	76%	19%
Engineering	71%	12%
Medicine	68%	11%
Public Service/Gov't	66%	10%
Arts	64%	14%
Law	63%	12%
Education	57%	8%

NAJ LEE LUM FOR FORTUNE/SOURCE: DONALD MCCABE, RUTGERS UNIVERSITY

Copyright 1991, FORTUNE. Reprinted with permission.

(a)

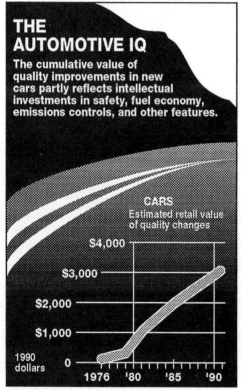

Copyright 1991, Fortune Magazine Reprinted with permission.

(b)

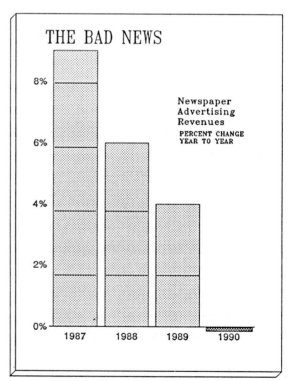

Source: Smith Barney Harris Upham & Co.

(c)

HOW JAPAN IS CHANGING

AVERAGE LIFE EXPECTANCY
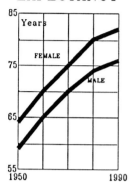

AVERAGE FAMILY SIZE
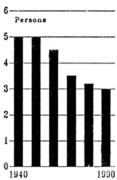

AVERAGE FAMILY INCOME
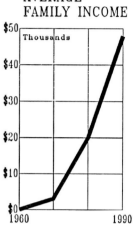

PERCENT OF FAMILY INCOME SAVED
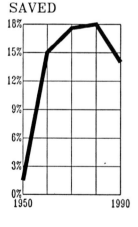

HOURS WORKED PER YEAR
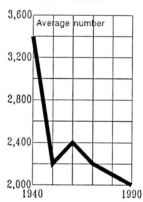

NUMBER OF PERSONS PER CAR
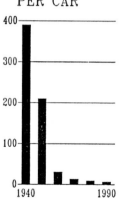

COCA-COLA CONSUMPTION
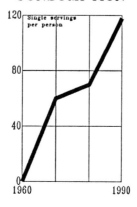

(d)

On The Rise: Prison Population and Violent Crimes

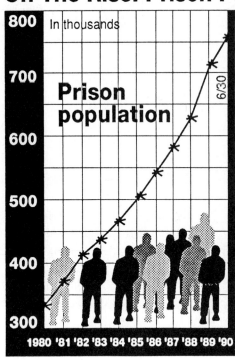
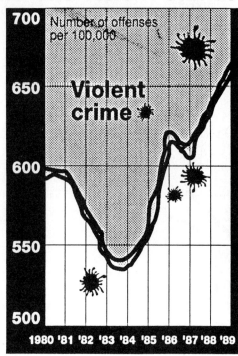

(e)

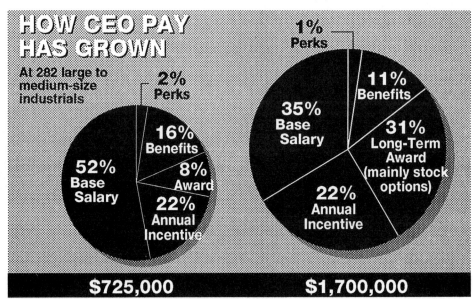

Copyright 1991, Fortune Magazine. Reprinted with permission.

(f)

APARTHEID AMERICAN STYLE

Family Income Distribution

BLACK		WHITE
14.5%	Over $50,000	32.5%
15%	$35,000–$50,000	20.8%
14%	$25,000–$35,000	16.5%
19.5%	$15,000–$25,000	16%
37%	Under $15,000	14.2%
$21,423	1990 MEDIAN INCOME	$36,915

Source: Andrew Hacker

(g)

PART V What Do You Think?

1. What Do You Think?

Examine the following four Time-Series Graphics that represent the distributions of campaign contributions for the 1992 elections.

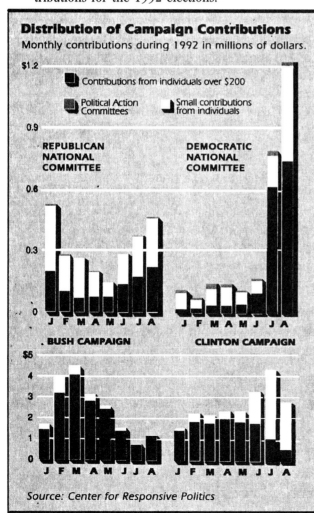

Source: Center for Responsive Politics

Copyright 1992, The New York Times. Reprinted with permission.

a) Describe the shape of each of the graphs.
b) For each graph, *estimate* the amount of money contributed per month. Compare the total amount of money contributed to the Republican National Committee, the Democratic National Committee, the Bush Campaign, and the Clinton Campaign.
c) Which Committee received more money:
   from individuals making small contributions?
   from individuals making over $200?
d) Which Campaign received more money:
   from individuals making small contributions?
   from individuals making over $200?
e) For the months of July and August, compare for each graph, the proportion of money received from small contributions to individuals contributing over $200.
f) In each graph, discuss the change in the amount of dollars contributed per month from January to August.
g) For each graph, determine which two month period had the greatest increase in the amount of dollars contributed?
h) For each graph, determine which two month period had the greatest decrease in the amount of dollars contributed?
i) In each graph, discuss the change in the percentage of small contributions per month. Overall, which graph(s) indicates a general decrease in the amount of dollars contributed from small contributors. Which graph(s) indicates a general increase in the amount of dollars contributed from small contributors.

2. What Do You Think?

Examine the following seven Time-Series Graphics which present an outlook on the business travel picture that appeared in the September 8, 1992, issue of USA TODAY.

For each graphic, describe the situation being presented regarding that one aspect of business travel. That is, are the hotel room rates increasing or decreasing or staying relatively stable? What about the car-rental rates, food expenses and airfares? What is happening to the number of business trips, business travelers, and the number of miles traveled? Is the overall cost of business travel increasing or decreasing over this period? What has been happening to the number of business trips over this same period? During the later part of this period, the economy has been in a recession. Do You Think that this has attributed to any noticeable patterns shown within the graphics? Explain your answer.

Copyright 1992, USA TODAY. Reprinted with permission. Source: Airline Economics, Runzheimer International, Smith Travel Research, U.S. Travel Center. By Marcy E. Mullins, USA TODAY

PART VI Exploring DATA With MINITAB.

1. Select twenty data values for the variable **STUDY HOURS** from the 150 values listed in Table 2.1. Using MINITAB, enter the 20 data values using the Set command and construct a STEM-AND-LEAF display using increments of 1 and 10.
   a) Compare your results to the Stem-and-Leaf displays in Figure 2.37 and Figure 2.38. Are the shapes the same?
   b) If they are different, what do you think contributes to this difference?

2. The following data set represents the test grades of 60 statistics students.

```
22 25 28 29 31 33 35 36 37 39 40 41 43 44
45 45 47 48 49 50 52 53 54 54 55 55 56 57
58 60 61 63 64 65 65 65 66 68 68 69 70 71
73 75 76 76 78 79 80 82 83 84 85 87 87 89
90 90 92 95
```

Using MINITAB, construct three histograms where each histogram has a different number of classes. To enter the data, use the SET or READ command, and to construct the histograms the HISTOGRAM command with the subcommands START and INCREMENT. Use the following number of classes for each of the histograms:

a) two classes
b) seven classes
c) twenty-five classes

Examine each of the histograms and describe:

d) what happens to the class width as the number of classes is increased from two to twenty-five?
e) the shape, and what happens to the shape of each histogram as the number of classes changes.
f) which histogram seems to present the data in a manner that will be more helpful in revealing any useful characteristics or patterns that may exist within the data set?

PART VII Projects.

1. Inspect recent issues of your local tabloids and/or business magazines.
   Find *one* of each of the following:
   a) a frequency distribution table
   b) a bar graph
   c) a histogram
   d) a frequency polygon
   e) a pie chart
   f) a pictograph
   For each example determine:
   a) the variables and the variable values
   b) the type of data (categorical, discrete or continuous)

c) comment as to whether or not you believe the example has been depicted appropriately.

2. Survey one hundred students and record for each student the responses to the following questions:
   Ques. #1: How many hours of television do you watch each week?
   Ques. #2: Is a Cable TV service installed in your home?
   a) Construct a stem-and-leaf display for the student responses to Ques. #1.
      (1) Describe the shape of the display.
      (2) Are there outliers present? If so, try to explain why the outlier(s) occurred.
   b) Construct a back-to-back stem-and-leaf display for the responses to Ques. #1 comparing those students that have Cable TV service to those students that do not have Cable TV service.
   c) Describe the differences (if any) of each category represented in the back-to-back stem-and-leaf display.
   d) Construct a "less than" cumulative frequency table for the responses to Ques. #1.
   e) Construct an ogive for the information in part d.
   f) Formulate three observations using the ogive.

PART VIII **Database.**

The following exercises refer to the file DATABASE listed in Appendix A. We have indicated the appropriate MINITAB commands that are necessary to answer each exercise.

1. Using the MINITAB commands:
   RETRIEVE and STEM-and-LEAF
   Retrieve the file DATABASE.MTW and construct a Stem-and-Leaf display for the variables:
   a) AGE
   b) AVE
   c) WRK
   d) STY
   Describe the general shape of each of the distributions using the terms developed in Section 2.7.

2. Using the MINITAB commands:
   RETRIEVE and HISTOGRAM
   Retrieve the file DATABASE.MTW and construct a Histogram for the variables:
   a) AGE
   b) AGE using increments of 5
   c) AGE by SEX using increments of 5
   How do the shapes of the graphs differ?
   Compare the shapes of the graphs for males verses females.

3. Using the MINITAB commands:
   RETRIEVE and HISTOGRAM
   Retrieve the file DATABASE.MTW and construct a Histogram for the variables:
   a) GPA using increments of 0.5 by MAJ.
   b) Which majors have the highest GPAs?
   c) Which majors have the lowest GPAs?

CHAPTER 3
MEASURES OF CENTRAL TENDENCY

❄ 3.1 INTRODUCTION

*"Are you an **Average** American?"*

If you are, you...

...would be a 29-year old hermaphrodite (slightly more female than male) who stands about 5-feet-4, weighs about 150 pounds, earns close to $17,000 a year, and eats a hamburger three times a week. He (she? it?) has 1.4 children and watches television about 2.5 hours a day.[1]

The previous statement is trying to describe an "**average**" or a "**typical**" **person** for the population of American people. In this chapter, we will be discussing statistical methods used to try to depict a **typical score** for all the scores within a distribution. These methods used in determining the average or typical score are trying to give some idea about where the **center** of all of the scores lie. Let's examine this concept using the following quote from a recent baseball negotiation issue.

> **After five years as a professional, a player would become a free agent if earning less than the "average" salary. (Owners estimate this at $1,095,000; the players at $1,092,000).**

This proposal was one of the principal issues in the negotiations between the baseball club owners and the major league players association. Both groups are trying to describe the salary that is typical. However, one might ask why there is a discrepancy of $3,000 between the owners' and players' estimate of the **average** or **typical** salary if both groups are using the same data? To most people the word "average" has just one meaning. However, in statistics, there are different ways of computing the average or typical score of a distribution. We will examine three such ways. They are the **mean**, the **median**, and the **mode**. These measures are referred to as **Measures of Central Tendency**.

Measures of central tendency are important tools of descriptive statistics because each measure, in its own way, enables a statistician to describe with a single number, the center or typical score of all the scores. However, when we use a single number to represent an **average** or a **typical score**, we are not telling everything about **all the scores** within the distribution. We are trying to give some idea about where the center or middle value lies. We will now investigate the three measures of central tendency and examine what each of these measures really describes.

---

[1] This appeared in <u>The Average American</u>, by Barry Tarshis.

3.2 THE POPULATION MEAN

The first measure of central tendency that we will discuss is the population mean. The mean is the most common and useful measure of central tendency and is usually referred to as the "average".

Definition: Population Mean. The population mean is the number obtained by summing up all the data values of the population and dividing this sum by the total number of data values.

$$\text{Population Mean} = \frac{\text{sum of all the data values}}{\text{total number of data values}}$$

Example 3.1

Suppose in an art appreciation class you received the following quiz grades:

$$8, 9, 7, 6, 8, 10$$

What is your mean quiz grade?

Solution:

The sum of all the data values is found by adding all the quiz grades. The total number of data values is found by counting the number of quiz grades. Thus, the mean is:

$$\text{mean} = \frac{8 + 9 + 7 + 6 + 8 + 10}{6} = \frac{48}{6} = 8$$

Therefore, your mean quiz grade is 8. ■

Let's now develop a formula for the population mean.

A Formula for the Population Mean

The **population mean** is symbolized by the Greek letter μ (pronounced "mu"). In general, the data is represented by $x_1, x_2, x_3, x_4, \ldots, x_N$ where x_1 represents the first data value, x_2 represents the second data value, and x_N represents the last [or Nth] data value. The total number of data values is represented by the letter N. Using this notation, the definition of the population mean, μ, can be written as:

$$\mu = \frac{\text{sum of all the data values}}{\text{total number of data values}}$$

or,

$$\mu = \frac{x_1 + x_2 + x_3 + x_4 + \ldots + x_N}{N}$$

In statistics a more concise representation of $x_1 + x_2 + x_3 + x_4 + \ldots + x_N$ is written using the capital Greek letter Σ (pronounced "sigma") which means SUM.

Thus, $x_1 + x_2 + x_3 + x_4 + \ldots + x_N$ can be represented by Σx.

The formula for the population mean can now be written as:

$$\mu = \frac{\Sigma x}{N}$$

Example 3.2

Let $x_1 = 4$, $x_2 = 7$ and $x_3 = 7$. Calculate the population mean, μ.

Solution:

1. Calculate the sum of the data values:

$$\Sigma x = x_1 + x_2 + x_3$$
$$= 4 + 7 + 7$$
$$\Sigma x = 18$$

2. Determine N.

   N = 3, since there are three data values.

   Thus, the population mean is:

$$\mu = \frac{\Sigma x}{N} = \frac{18}{3} = 6 \quad \blacksquare$$

Example 3.3

Let $x_1 = 12$, $x_2 = 25$, $x_3 = 21$ and $x_4 = 8$. Calculate the population mean, μ.

Solution:

1. Calculate Σx.

$$\Sigma x = x_1 + x_2 + x_3 + x_4$$
$$= 12 + 25 + 21 + 8$$
$$\Sigma x = 66$$

2. Determine N.

   N = 4

   Thus, the population mean is:

$$\mu = \frac{\Sigma x}{N} = \frac{66}{4} = 16.5 \quad \blacksquare$$

OTHER CALCULATIONS USING Σ

There are many formulas in statistics involving the use of Σ, and it is important to know how to interpret expressions containing Σ. Let's consider these expressions and discuss how to evaluate them.

We will first consider the expression: $(\Sigma x)^2$, the square of the sum of the data values.

Procedure to calculate $(\Sigma x)^2$

$(\Sigma x)^2$ means $(x_1 + x_2 + x_3 + x_4 + ... + x_N)^2$

1. Sum the data values, (Σx).
2. Square the sum, $(\Sigma x)^2$.

The next example will illustrate the procedure used to calculate $(\Sigma x)^2$.

Example 3.4

Let $x_1 = 3$, $x_2 = 4$, $x_3 = 1$ and $x_4 = 6$.
Determine $(\Sigma x)^2$.

Solution:

Notice the sum of the data values, Σx, is being squared. This means we must first find the sum of all the data values, and then **square** this sum.
That is,

1) sum of all the data values: (Σx)

$\Sigma x = x_1 + x_2 + x_3 + x_4$
$\Sigma x = 3 + 4 + 1 + 6$
$\Sigma x = 14$

2) square this sum: $[(\Sigma x)^2]$

$(\Sigma x)^2 = (14)^2$

$(\Sigma x)^2 = 196$ ∎

Now, we will consider the meaning of the expression Σx^2, the sum of the squared data values.

Procedure to Calculate Σx^2

Σx^2 means $(x_1)^2 + (x_2)^2 + (x_3)^2 + ... + (x_N)^2$

1. Square each data value.
2. Sum these squared data values.

The next example illustrates the calculation of Σx^2.

Example 3.5

Let $x_1 = 2$, $x_2 = 4$, $x_3 = 1$ and $x_4 = 3$. Find Σx^2.

Solution:

First, square each data value.

$$(x_1)^2 = (2)^2 = 4 \qquad (x_2)^2 = (4)^2 = 16$$
$$(x_3)^2 = (1)^2 = 1 \qquad (x_4)^2 = (3)^2 = 9$$

Second, sum these squared data values.

$$\Sigma x^2 = (x_1)^2 + (x_2)^2 + (x_3)^2 + (x_4)^2$$
$$= 4 + 16 + 1 + 9 = 30 \blacksquare$$

Summary For Calculating $(\Sigma x)^2$ and Σx^2

When calculating $(\Sigma x)^2$ you first **add** all the data values and then **square** this sum. When calculating Σx^2 you first **square** each individual data value and then **add** these squared data values.

Statistical calculators have built-in programs to enable you to calculate the expressions Σx and Σx^2. You should become familiar with the operation of your calculator and review the examples in this section that refer to these expressions.

At the end of this chapter we have outlined a general procedure to help you calculate these expressions.

Example 3.6

Suppose $x_1 = 2$, $x_2 = 0$, $x_3 = 5$, and $x_4 = -3$.
Compute and compare the value of $(\Sigma x)^2$ to Σx^2.

Solution:

Let's first find $(\Sigma x)^2$.

$$\Sigma x = 4 \text{ and } (\Sigma x)^2 = 16.$$

Now find Σx^2.

$$\Sigma x^2 = 2^2 + 0^2 + 5^2 + (-3)^2 = 38.$$

From this example, we can conclude that $(\Sigma x)^2$ and Σx^2 are in general, **not** equal. $\blacksquare$

Now let's examine the meaning of the expression $\Sigma(x - \mu)$, the sum of the deviations from the mean, and how to evaluate it. The procedure to calculate this expression is given by:

Procedure to calculate $\Sigma(x - \mu)$

$\Sigma(x - \mu)$ means $(x_1-\mu)+(x_2-\mu)+(x_3-\mu)+...+(x_N-\mu)$

1. Compute the mean.
2. Subtract the mean from each data value.
3. Sum these differences.

Example 3.7

Let $x_1 = 5$, $x_2 = 3$, $x_3 = 7$. Find $\Sigma(x - \mu)$.

Solution:

1) Determine μ.
   $\mu = 5$
2) Subtract the mean, μ, from each data value. These differences are referred to as **deviations from the mean**.
   $(x_1-\mu) = (5-5) = 0$
   $(x_2-\mu) = (3-5) = -2$
   $(x_3-\mu) = (7-5) = +2$
3) Find the sum of the deviations from the mean.
   $\Sigma(x - \mu) = 0 + (-2) + 2$
   $= 0$ ■

Example 3.8

Given the following ages of seven children at a playground:
   5, 4, 4, 8, 4, 9, 8
a) Compute the mean age, μ.
b) Evaluate $\Sigma(x - \mu)$.

Solution:

a) Calculate the mean, μ.
   $\mu = 6$
b) To evaluate $\Sigma(x - \mu)$:
   1) Subtract the mean, μ, from each data value.

   $(5-\mu)=-1$ $(4-\mu)=-2$ $(4-\mu)=-2$ $(8-\mu)=2$ $(4-\mu)=-2$ $(9-\mu)=3$ $(8-\mu)=2$

   2) Find the sum of the deviations from the mean.
   [Remember, the sum of the deviations from the mean is denoted as: $\Sigma(x - \mu)$].

   $\Sigma(x - \mu) = -1 + -2 + -2 + 2 + -2 + 3 + 2$
   $= 0$ ■

From the previous two examples we might guess that the **sum of the deviations from the mean is always zero**. That is, $\Sigma(x - \mu) = 0$. In fact, this is true, and it is **a property of the mean. This property indicates that the mean is the balance point within a distribution**. The mean is like the fulcrum on a seesaw. Let's explain this idea using the ages of the seven children from Example 3.8. If the ages of the seven children were marked on a board, and a disk for each child is placed on the board corresponding to the child's age, then this board would have its balance point at the mean value of 6. That is, if the fulcrum of the board is placed at the value of 6, then the board is balanced! This is illustrated in Figure 3.1.

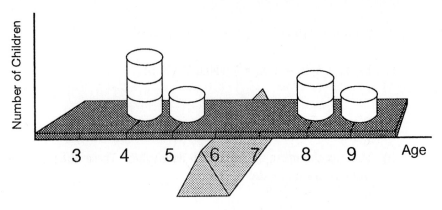

Figure 3.1

From Figure 3.1, since the fulcrum of the board is at the mean value of 6, this serves to illustrate that the mean is the point that balances all the values on either side of it. In other words, the sums of the distances from the mean must be the same on each side of the mean. In symbols, this is written:
$\Sigma(x - \mu) = 0$.

Property of the Mean

The sum of the deviations from the mean is always zero.

It is expressed symbolically as: $\Sigma(x - \mu) = 0$.

This property of the mean states that the mean is the point within a distribution that balances all the data values on either side of it. That is, the sums of the distances from the mean must be the same on each side of the mean.

Example 3.9

Given the distribution of data values:

2, 3, 4, 6, 10

a) Find the mean of this distribution.
b) Add the constant 4 to each data value to form a new distribution.
c) Find the mean of the new distribution.

Solution:

a) $\mu = \dfrac{\Sigma x}{N} = \dfrac{25}{5} = 5$ ∎

b) Add the constant 4 to each data value to form a distribution of new data values.

original data values x	new data values $x + 4$
2	(2+4)= 6
3	(3+4)= 7
4	(4+4)= 8
6	10
10	14

Thus, the distribution of new data values is:
6, 7, 8, 10, 14

c) Calculate the mean of the distribution of new data values.

$$\text{mean of the new data values} = \frac{45}{5}$$

$$\text{new mean} = 9$$

Notice that the new mean of 9 is four more than the old mean of 5. This increase is exactly the amount we had added to each data value of the distribution. This example illustrates a useful property of the mean. ■

**Property of the Mean:
Adding a Constant to the Data Values**

If a constant C is added to each data value of a distribution whose mean is μ, then the new mean would become $\mu + C$. That is,

new mean = original mean + C

In Example 3.9, we added the constant 4 to each data value in the original distribution with a mean of 5. The new mean can be determined by:

new mean = original mean + C
new mean =   5   + 4
new mean =   9

In a similar manner if a constant were subtracted from each data value in a distribution, the mean of the new distribution will be decreased by the constant.

**Property of the Mean:
Subtracting a Constant From the Data Values**

If a constant C is subtracted from each data value within a distribution whose mean is μ, then the new mean would become $\mu - C$. That is,

new mean = original mean - C

Let's consider what effect multiplying (or dividing) each data value by a constant has on the mean.

**Property of the Mean:
Multiplying or Dividing by a Constant**

If each data value within a distribution, whose mean is μ, is multiplied (or divided) by a constant C, then the mean of the new distribution is equal to the original mean multiplied (or divided) by the constant C.

❄ 3.3 THE MEDIAN

Another measure of central tendency which we will consider is the median. The median is defined to be the middle value of a distribution. The median is dependent upon the number of data values within the distribution, and not on the actual value of the data.

> **Definition: Median.** The median is the middle value of a distribution after the data values have been arranged in numerical order. Thus, half (or 50%) of the remaining data values are less than the value at the position of the median and half of the remaining data values are greater than the value at the position of the median.

Let's consider a procedure that we can use to determine the median or middle value of a distribution.

> **Procedure to Determine the Median**
>
> The procedure to determine the median depends upon the number of data values within the distribution. **After the data values have been arranged in numerical order, there are two possible cases:**
>
> a) For an **odd number** of data values, the median is the middle score of the distribution.
>
> b) For an **even number** of data values, the median is the mean value of the two middle data values of the distribution.

Example 3.10

Using the quiz grades:

8, 7, 9, 9, 9, 6, 10, 7, 5

find the median quiz grade.

Solution:

1) Arrange the data values in numerical order.

   5, 6, 7, 7, 8, 9, 9, 9, 10

2) Since there is an odd number of data values, the median is the middle score, which is 8. ∎

In Example 3.10, **8 was the median because it was positioned in the middle of all the data values**. That is, 8 was the fifth or middle data value of the nine data values. In this example, the median was positioned at the fifth data value within the nine data values of the distribution. **The position of the median within a distribution can be determined by adding one to the number of data values and dividing by 2.** Thus, in Example 3.10, if one is added to the number of data values, 9, to get 10 and divide this result by 2, then the median would be the fifth data value, which is 8.

Median's Position Within a Distribution

In general, the median's position within a distribution of N data values that have been arranged in numerical order can be determined by using the following formula:

$$\text{position of median} = \frac{N + 1}{2} \text{ th data value}$$

Example 3.11

Find the median of the following distribution:

3, -2, -4, 5, 8, -6, 0

Solution:

1) Arrange the data values in numerical order.

   -6, -4, -2, 0, 3, 5, 8

2) Since there are 7 data values, the median is the value of the middle data value. To determine the position of the middle data value, use the formula:

$$\text{position of median} = \frac{N + 1}{2} \text{ th data value}$$

Thus, **the position of the median is the (7 + 1)/2 or 4th data value.** Since the fourth data value is 0, then the median is 0. ∎

Example 3.12

Find the median of the data values:

3, 7, 5, 5, 0, 2

Solution:

1) Arrange the data values in numerical order.

   0, 2, 3, 5, 5, 7

2) Since there is an even number of data values, the median is the mean value of the two middle values.

$$\text{median} = \frac{3 + 5}{2} = \frac{8}{2} = 4 \quad \blacksquare$$

3.4 THE MODE

The last measure of central tendency we will examine is the mode. The mode is the easiest measure to compute and the simplest to interpret.

> **Definition: Mode.** The mode of a distribution is the data value within the distribution which occurs most frequently. It is possible that a distribution may **not** have a mode or may have **more** than one mode.

Example 3.13

Find the mode of the following distribution representing eight test scores.

65, 75, 45, 90, 75, 68, 85, 60

Solution:

Since 75 appears most frequently, it is the mode. ∎

Example 3.14

Find the mode for the following distribution of screw lengths (measured in inches).

0.25, 0.33, 0.5, 0.33, 0.5, 0.33, 0.25, 0.16, 0.5, 0.5, 0.33

Solution:

Notice there are two data values which appear most frequently. Therefore, this distribution has *two* modes. The modes for the distribution are 0.33 and 0.5. ∎

In the previous example, there were two modes. This type of distribution is called bimodal and leads us to the following definition.

> **Definition: Bimodal Distribution.** If a distribution has two data values which appear with the greatest frequency, then the distribution is a bimodal distribution.

Example 3.15

Find the mode for the following distribution:

5, 4, 6, 10, 6, 4, 10, 3, 8, 7

Solution:

Since there are more than two data values which appear most frequently, (i.e. 4, 6 and 10 appear twice), we will refer to such a distribution as having no mode. ∎

In the previous example, the distribution had no mode and is called a nonmodal distribution.

> **Definition: Nonmodal Distribution.** If a distribution has no mode or more than two modes, then the distribution is nonmodal.

Example 3.16

A lifeguard at the beach observed the following bathing suit colors that were worn by male bathers:

black, red, white, red blue, black, blue, blue, white, green, yellow, green, black, red, blue, white, orange, blue, yellow, black, red, gray, blue, white, blue, blue, gray, green, white, red, black, gray, blue, red, black, blue, red.

The lifeguard wants to determine the typical bathing suit color. Which measure of central tendency should the lifeguard use and what is the typical suit color?

Solution:

Since the mean and median would not be appropriate [WHY?], the lifeguard would need to use the mode. The most frequent bathing suit color is blue. Therefore, the typical bathing suit color would be blue. ■

Example 3.17

The small community of Poorich is applying for federal aid. One question on the application requests the *average* family income for the community.

The following data are the family incomes for the five community residents:

$25,000; $20,000; $200,000; $30,000; $15,000.

If the town uses the mean income of $58,000 to represent the average family income, do you believe this number accurately represents the income level of the community?
If not, which measure of central tendency would you use to represent the income level of this community?

Solution:

Since the mean income of $58,000 is greater than four of the five family incomes, it does not represent the typical income level of the community. The median income of $25,000 more accurately depicts the typical income level of the residents since it is not sensitive to the extreme values in the distribution. ■

CASE STUDY 3.1

Examine the three USA Snapshots shown in Figure 3.2. In each snapshot a measure of central tendency is being used to describe an aspect of a person's life. For each snapshot, state which measure of central tendency is being used and explain why this is or is not an appropriate measure to use in each instance.

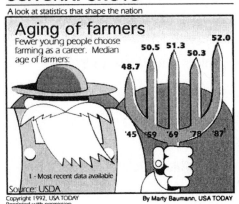
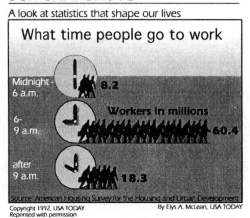
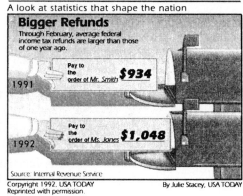

Figure 3.2

CASE STUDY 3.2

Read the two articles *The Average Man* and *Welcome to the Most Normal City in America* displayed in Figure 3.3. Discuss how the word *average* and *typical* are being used in each of the articles. For each statement within each article that contains a statistic, state which measure of central tendency you believe was used or could have been used to determine that statistic. Also, explain why you chose that particular measure of central tendency.

The Average Man

From *Men's Health*:

- Is 5 feet 9 inches tall and 173 pounds.
- Is married, 1.8 years older than his wife and would marry her again.
- Has not completed college.
- Earns $28,605 per year.
- Prefers showering to taking a bath.
- Spends about 7.2 hours a week eating.
- Does not know his cholesterol count, but it's 211.
- Watches 26 hours and 44 minutes of TV a week.
- Takes out the garbage in his household.
- Prefers white underwear to colored.
- Cries about once a month — one-fourth as much as Jane Doe.
- Falls in love an average of six times during his life.
- Eats his corn on the cob in circles, not straight across, and prefers his steak medium.
- Can't whistle by inserting his fingers in his mouth.
- Prefers that his toilet tissue unwind over, rather than under, the spool.
- Has sex 2.55 times a week.
- Daydreams mostly about sex.
- Thinks he looks okay in the nude.
- Will not stop to ask for directions when he's in the car.

Welcome to the most normal city in America

EXPERTS have pinpointed the most ordinary city in the U.S. - and because of this it will become one of the biggest test markets in the country.

According to 1990 census data, Tulsa is now the most typical city in the nation.

Marketers who used to ask, "Will it play in Peoria?" can now turn to Tulsa, which came closest of all cities with populations over 50,000 to matching the national average for ages, racial mix and housing prices, according to Donnelley Marketing Information Services.

What makes living in Tulsa such a typical experience?

•Tulsa is 79 percent white and 14 percent black - very close to the national average of 80 percent white and 12 percent black.

•Tulsa's median home value is $60,500 - compared to the national median of $79,100.

•Tulsa's age distribution - of children, baby boomers and seniors - is almost identical to the nation's.

•Tulsans, like the rest of Americans, are spending more conservatively now than they did in the designer-label 80's.

Businesses look to such typical towns to test out new products or measure public opinion.

"Tulsa is middle America to the max," says Rick Bahlinger, president of Golden Eagle Distributing.

Other typical cities, according to Donnelley, are Charleston, W. Va.; Midland, Texas; Springfield, Ill., Wichita, Kan. and Oklahoma City.

Figure 3.3

3.5 THE SAMPLE MEAN

In Section 3.2, we defined the population mean to be $\frac{\Sigma x}{N}$ and symbolized it by μ. However, as mentioned in Section 1.1 it is often impractical to obtain all the data necessary to compute the population mean. In such instances, a sample is selected from the population and the mean of the sample is calculated. The **sample mean** can be used to estimate the population mean, μ.

Definition: Sample Mean. The mean of a sample is determined by summing up all the data values of the sample and dividing this sum by the total number of data values.

The symbol for the sample mean is $\bar{x}$ and the formula to calculate the sample mean is:

$$\bar{x} = \frac{\Sigma x}{n}$$

The **sample mean** is used to estimate the mean of a population, μ.

Example 3.18

A sample of twenty 9 volt batteries were tested to determine their average life. The following data represents the battery life in months:

20, 11, 15, 18, 24, 17, 19, 12, 19, 22
18, 15, 19, 21, 20, 15, 17, 24, 16, 18

a) Calculate the sample mean, $\bar{x}$.
b) Give an estimate of the population mean, μ.

Solution:

a) The sample mean is calculated by using the formula:

$$\bar{x} = \frac{\Sigma x}{n}$$

$$\bar{x} = \frac{360}{20}$$

$$\bar{x} = 18$$

b) Therefore, an estimate of the population mean life for all such batteries is 18 months. ∎

Case Study 3.3

For each graphic in Figure 3.4, identify which measure of central tendency is used and determine how each descriptive figure can be interpreted as a statistic or a parameter.

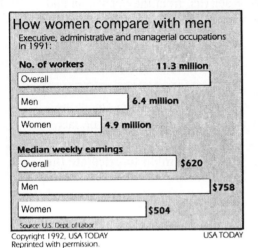

Figure 3.4

Using a Calculator to Evaluate Σ Expressions and to Find the Mean

Statistical calculators have built-in programs that enable users to evaluate the expressions: $(\Sigma x)^2$ and Σx^2 as well as to the find the mean. The procedure to obtain these results require the user to simply enter the data values using special keys and then retrieve the results from the statistical memories through a combination of keystrokes.

The procedure to determine these quantities is dependent upon the calculator manufacturer and the model you own.

The general procedure is:

Step 1. Set the calculator in statistical mode.
Step 2. Enter the data values one at a time using the appropriate key.
For Texas Instruments Calculators, this key is usually the Σ+ key, while for Casio and Sharp Calculators the key is usually the M+ or Data Key. Depending on the calculator, as the data value is entered, the display will either show the number of data values entered or the last data value entered.
Step 3. Retrieve the statistical results from memory using the appropriate keystrokes outlined in the calculator manual.

Although the previous procedure is generic and not dependent upon any specific calculator, Appendix B contains specific calculator instructions of some models for the calculators manufactured by Texas Instruments, Casio, Sharp, and Radio Shack.

3.6 THE RELATIONSHIP OF THE MEAN, MEDIAN AND MODE

In this chapter, we've discussed three ways to describe the center or identify the "typical" value of a distribution. To help decide which measure of central tendency would be appropriate in describing the center of a specific distribution type, we will examine the relationship of the mean, median, and mode within three common distribution shapes: *symmetric bell-shaped*, *skewed to the left*, and *skewed to the right*. We will first discuss how to determine where each measure of central tendency will be positioned within each of these distribution shapes.

The Mode and its Location within the Distribution Shapes

Within each of the distribution shapes illustrated in Figure 3.5, you will notice that the mode, which represents the most frequent value, is located at the point on the curve corresponding to the highest value.

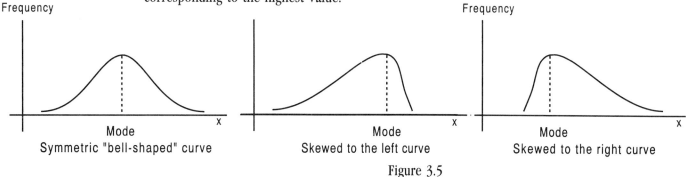

Figure 3.5

The Median and its Location within the Distribution Shapes

The median is located at the point on the curve which divides the area under the curve exactly in half. That is, half or 50% of the area lies to the left of the median value while half or 50% of the area lies to the right of the median value. This is illustrated in Figure 3.6.

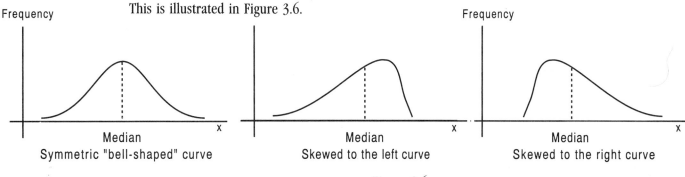

Figure 3.6

The Mean and its Location within the Distribution Shapes

The mean of a distribution is located at the value which represents the "balance point" of the curve. To understand the idea of a balance point, try to think of the curve as representing a solid object and determining the point beneath the curve which will balance the solid curve. Thus, the balance point is the position under the solid curve where half the weight of the solid curve is to the left of this point and half the weight is to the right of this point. The idea of the balance point for each distribution shape is illustrated in Figure 3.7

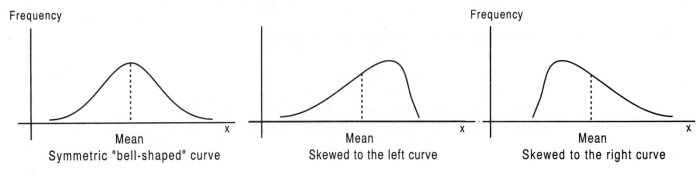

Figure 3.7

Now let's examine the position of each measure of central tendency with respect to each other within the three distribution shapes: symmetric bell-shaped, skewed to the right, and skewed to the left.

Symmetric Bell-Shaped Distribution

Within a symmetric bell-shaped distribution, the mean, median, and mode are all located at the center of the distribution as illustrated in the symmetric bell-shaped curves in Figures 3.5 to 3.7. The mean is located at the center of a symmetric bell-shaped distribution since every data value to the left of the center is balanced by a corresponding data value to the right of the mean. Thus, the mean or the balance point is at the center of a symmetric bell-shaped distribution. The median is located at the center of a symmetric bell-shaped distribution since this is the point that divides the area under the curve into two equal halves (50% of the area to the left and 50% of the area to the right). The mode is located at the center of a symmetric bell-shaped distribution since this is the point at which the curve attains its highest value.

Skewed to the Left Distribution

Within a distribution which is skewed to the left, the mode occurs at the highest point and it has a value greater than both the median or the mean as illustrated in Figure 3.8.

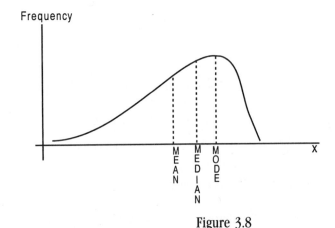

Figure 3.8

Within a skewed to the left distribution, the mean has the smallest value of all the measures of central tendency because the mean is greatly influenced by the extreme data values that occur in the left tail of the distribution and is *pulled in the direction of the extreme values in the left tail*. The median has a smaller value than the mode since the median is *pulled to the left by the extreme values in the left tail* of the distribution. However, the median will be greater than the mean since the median is not as sensitive or greatly influenced as the mean to the extreme data values in the left tail.

Skewed to the Right Distribution

Within a distribution which is skewed to the right, the mode occurs at the highest point and the mode has a value smaller than the median or the mean as illustrated in Figure 3.9.

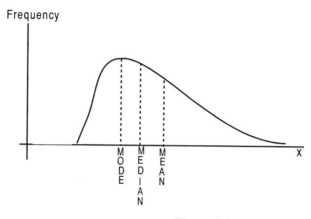

Figure 3.9

The mean has the largest value of all the measures of central tendency because the mean is greatly influenced by the extreme data values that occur in the right tail of the distribution. The median has a value greater than the mode since the median is effected by the extreme values within the distribution and *pulled in the direction of the right tail*. However, the median has a value smaller than the mean since the median is not as sensitive or greatly influenced as the mean to the extreme data values in the right tail.

Example 3.19

Using Figure 3.10, determine the value for:
a) the mode
b) the mean
c) the median

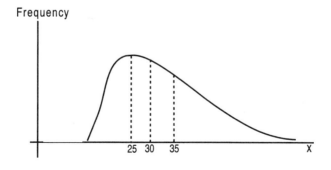

Figure 3.10

Solution:

a) Within a distribution, the mode occurs at the highest point. Therefore, the mode is 25.
b) Since the distribution in Figure 3.10 is skewed to the right, then the mean is greater than both the median and the mode. Thus, the mean is 35.
c) In a skewed to the right distribution, the median is greater than the mode but less than the mean. Therefore, the median is 30.

COMPARING THE MEAN, THE MEDIAN AND THE MODE

From the examples and discussions within this chapter, you should have noticed that there are advantages and disadvantages for each measure of central tendency. Let's discuss some of these advantages and disadvantages for each measure.

As illustrated in Example 3.17, the mean is a measure of central tendency that is affected by the extreme data values or outliers within a distribution. Thus we say that **the mean is not a resistant measure of central tendency**, because it is not resistant to the influence of the extreme values or outliers.

On the other hand, **the median is a resistant measure of central tendency.** Since the median is resistant to the influence of extreme data values or outliers, its value does not respond strongly to the changes of a few extreme data values regardless of how large the change may be. Thus, if a distribution has an outlier or is skewed, then the mean and median may differ greatly in value, and the mean, which is sensitive to the extreme data values, will move toward the direction of the long tail in a skewed distribution as illustrated in Figures 3.8 - 3.9.

However, as illustrated in the symmetric bell-shaped curves in Figures 3.6 - 3.7, the mean and median will have the same value. Therefore, in the presence of outliers, the mean will be sensitive to outliers while the median is resistant, or unaffected, to outliers or extreme data values.

Since the mean is sensitive to outliers or extreme data values, **the mean can be used to provide us with information that an outlier does exist within a distribution**. Once an outlier is detected, the statistician can investigate the reason for such an extreme data value. In such instances when the statistician is made aware of the existence of outliers, he/she can decide whether to use a measure of central tendency that is resistant to these outliers or to give individual attention to the outliers. Notice in Example 3.17, the existence of the outlier $200,000 was detected by the fact that the mean ($58,000) was much larger than the median ($25,000). In this particular instance, the Poorich Community decided to use the median to represent the "average" or center value to ignore the presence of the outlier, since the median value of $25,000 is resistant to the outlier $200,000.

The mode has an advantage over both the mean and median when the data is categorical, since it is not possible to calculate a mean or median for this type of data. This is illustrated in Example 3.16. In addition, the mode usually indicates the location within a large distribution where the data values are concentrated. A shoe retailer would be interested in modal shoe size when placing an order for new shoes since the shoe sales will be concentrated around this modal size. While, a discount consumers club is interested in the brands of items that most people are purchasing, since these items represent the modal brands.

However, the mode has some disadvantages too. One disadvantage is that the mode cannot always be calculated. For example, if a distribution has all different data values or has more than two modes, then the distribution is nonmodal.

❄ 3.7 USING MINITAB

Fifty-two students enrolled in an elementary statistics course responded to the question: "How many hours per week do you spend studying for your courses?". The data values for their responses along with a recording of their gender were entered into a data file using the READ command of MINITAB. A printout of the file using the PRINT command of MINITAB produced the following output:

For the output, 'hrsty' stands for the number of hours a student studies per week. Within the output labelled 'sex', "0" represents a "male" response and "1" represents a "female" response.

MTB > PRINT 'hrsty' 'sex'

ROW	hrsty	sex	ROW	hrsty	sex
1	5	1	27	9	1
2	3	0	28	13	1
3	90	0	29	8	0
4	40	0	30	6	1
5	23	1	31	35	0
6	21	1	32	7	0
7	10	0	33	10	0
8	3	0	34	7	1
9	9	1	35	14	0
10	10	0	36	14	1
11	10	1	37	7	0
12	12	1	38	20	0
13	12	0	39	4	1
14	5	0	40	13	1
15	3	0	41	10	0
16	7	0	42	35	1
17	3	0	43	14	0
18	12	0	44	8	1
19	3	0	45	12	1
20	15	0	46	3	0
21	10	0	47	14	1
22	10	1	48	5	1
23	5	0	49	5	1
24	10	1	50	20	1
25	10	1	51	10	1
26	5	0	52	15	1

Notice that the printout *does not* summarize the distribution. For example, by examining the printout it would be very difficult to determine the distribution's shape

or

to estimate the typical number of hours studied per week by a student enrolled in elementary statistics?

In Chapter 2, many techniques were introduced that help summarize data. One such technique was the Stem-and-Leaf Display.

Using the STEM-AND-LEAF command of MINITAB for the distribution of study hours, we obtain:

MTB > STEM-AND-LEAF 'hrsty'

```
Stem-and-leaf of hrsty    N = 52         Leaf Unit = 1.0

      22      0    3333334555555677778899
     (22)     1    0000000000222233444455
       8      2    0013
       4      3    55
       2      4    0
       1      5
       1      6
       1      7
       1      8
       1      9    0
```

Using this stem-and-leaf display, what can you conclude about the *distribution's shape*?

In MINITAB, the stem-and-leaf display also indicates the row that contains the median by enclosing the row count within parentheses. For example, in this distribution of 52 data values, the median is contained within the second row of the stem-and-leaf display. To determine the value of the median, it is necessary to find the middle data value of the 52 values. Thus, the median is 10, which is located in the second row.

In addition, since the actual data are presented, the display is helpful in identifying the modal number of study hours per week.

MINITAB will calculate the MEAN and MEDIAN along with other useful statistics using the DESCRIBE command.

MTB > DESCRIBE 'hrsty'

	N	MEAN	MEDIAN	TRMEAN	STDEV	SEMEAN
hrsty	52	12.77	10.00	10.65	13.53	1.88

	MIN	MAX	Q1	Q3
hrsty	3.00	90.00	5.25	14.00

Why do you think the MEAN is different than the MEDIAN? In answering this question, consider the shape of the distribution, the relationship of the mean and median for this type of distribution, and the existence of outliers along with the concept of a resistant measure.

When you use the DESCRIBE command, the *trimmed mean*, indicated on the output as the TRMEAN, is also given. The trimmed mean calculates the mean of the distribution after the upper and lower 5% of the scores have been taken out of the distribution. Essentially this eliminates the influence of *outliers* on the value of the trimmed mean. You should notice that the STEM-AND-LEAF display indicated that one student studied 90 hours per week. Certainly, the data value of 90 hours is considered an outlier. Although it is possible for a student to study 90 hours per week, a responsible data exploration should question the validity of this data response. Was it recorded correctly? Or was it truly an unusual occurrence?

If the data value, 90, was incorrectly recorded then the **trimmed mean** would be a better measure to use than the mean since the trimmed mean **does not** include the data value 90 within its calculation. Explain why the TRMEAN and the Median are close in value?

MINITAB can sort the data and produce STEM-AND-LEAF displays for further comparisons. For example, if a comparison of the number of study hours per week for males verses females was desired, then this can be accomplished by using the command STEM-AND-LEAF with a subcommand to identify sex. This results in the following output:

MTB >   STEM-AND-LEAF 'hrsty';
SUBC>   BY 'sex'.

Stem-and-leaf of hrsty   C2 = male   N = 27   Leaf Unit = 1.0

```
   13    0   3333335557778
  (10)   1   0000022445
    4    2   0
    3    3   5
    2    4   0
    1    5
    1    6
    1    7
    1    8
    1    9   0
```

Stem-and-leaf of hrsty   C2 = female   N = 25   Leaf Unit = 1.0

```
    9    0   455567899
  (12)   1   000002233445
    4    2   013
    1    3   5
```

What can you conclude about the shape of each STEM-AND-LEAF?

In addition, the DESCRIBE command with a subcommand to identify sex produces the following output where:

"0" represents male and "1" represents female.

MTB > DESCRIBE C1;
SUBC> BY C2.

	sex	N	MEAN	MEDIAN	TRMEAN	STDEV	SEMEAN
hrsty	0	27	13.48	10.00	10.84	17.72	3.41
	1	25	12.00	10.00	11.35	6.89	1.38

	sex	MIN	MAX	Q1	Q3
hrsty	0	3.00	90.00	5.00	14.00
	1	4.00	35.00	7.50	14.00

Notice that although the medians are the same for both males and females, the means are different. What do you attribute this to? Examine the TRMEANs for each. Why do you think the females' TRMEAN is greater than the males'?

Glossary

TERM	SECTION
Measures of Central Tendency	3.1
Population mean, μ	3.2
Σ notation	3.2
Median	3.3
Mode	3.4
Bimodal Distribution	3.4
Nonmodal Distribution	3.4
Sample mean, $\bar{x}$	3.5
Resistant Measure	3.6
Trimmed Mean	3.7

Exercises

PART I Fill in the blanks.

1. Three measures of central tendency are _____ , _____ and _____ .
2. The measures of central tendency usually describe the _____ of a distribution.
3. The population mean is found by _____ up all the data values of a distribution and dividing by the _____ of data values in the distribution.
4. The Greek symbol used for the population mean is _____.
5. The Greek symbol used for the command "sum" is _____.
6. The symbol used for the total number of data values in a population is _____.
7. The formula for the population mean is _____.
8. For a distribution of data values, $\mu = 7$. Find the mean of a new distribution formed by:
   a) adding 6 to each data value data._____
   b) subtracting 8 from each data value._____
   c) dividing 14 into each data value._____
   d) multiplying each data value by 3._____
9. If $\mu = 5$, and $N=2$, then $\Sigma x =$ _____.
10. A distribution $x_1, x_2, ..., x_N$ has a mean of 15. Then for the distribution:
    a) $3x_1, 3x_2, ... , 3x_N$, the mean is _____.
    b) each data value is divided by 5, the mean is _____.
    c) $x_1 - 10, x_2 - 10, ... , x_N - 10$, the mean is _____.
    d) formed by adding 5 to each data value, the mean is _____.
11. If the data values $x_1, x_2, ... ,x_N$ have a mean of 13, then $\Sigma(x -13)=$_____.
12. If a distribution of 5 data values has $\mu = 11$, and if four of the data values are 5, 10, 20, 25 then the fifth data value is _____.
13. If $\Sigma(x - 3)=0$, then 3 is _____.
14. If $x_1, x_2, ... ,x_N$ are the data values of a distribution, then:
    a) $\Sigma x = x_1 + x_2 + ... + x_N$
    b) $\Sigma x^2 = (x_1)^2 + (x_2)^2 + ... + (x_N)^2$
    c) $(\Sigma x)^2 =$ _____.
    d) $\Sigma(x - 2)=$ _____.
    e) $\Sigma(x - 2)^2 =$ _____.
    f) $[\Sigma(x - 2)]^2 =$ _____.
15. When a distribution with an odd number of data values is arranged in numerical order, the median is the _____ data value of the distribution.
16. When a distribution with an even number of data values is arranged in numerical order, the median is the _____ of the two middle data values of the distribution.
17. In a distribution having an odd number of data values, the number of data values less than the median _____ the number of data values greater than the median.
18. In a distribution which has an odd number of data values, the mean is 100, and the median is 75. The middle data value is _____.
19. In a distribution, the data value that appears most frequently is called the _____.
20. A distribution containing two modes is called a _____ distribution.
21. A distribution containing more than two modes is called a _____ distribution.
22. The mean of a sample is denoted by _____.
23. The sample mean formula is: _____.
24. The sample mean is used as an estimate of the _____ mean.
25. If a mean is computed using only a portion of the population data, then this statistic is called the _____ mean.
26. If the entire population data is used to compute the mean, then this parameter is called the _____ mean.
27. A sample mean describes a characteristic of a sample, therefore it is not a parameter but a _____.
28. The mean is a measure of central tendency that is affected by the _____ data values or _____ within a distribution.
29. Since the mean is sensitive to outliers, then it _____ a resistant measure of central tendency.
30. In a skewed distribution, since the mean is affected by the extreme values of the distribution, the mean will move toward the direction of the long _____ of the distribution.
31. Since the median is unaffected by the extreme data values or outliers of a distribution, it is said to be _____ to the extreme data values.
32. If the data values are categorical, the _____ is the best measure of central tendency to use.
33. One advantage of the mean is that it helps to detect the presence of an _____ within the distribution.
34. In a skewed to the right distribution, mean is located to the _____ of the median, because it is affected by the _____ data values.
35. In a symmetric bell-shaped distribution, the measures of central tendency are located at the _____ of the distribution.

PART II **Multiple choice questions.**

1. Which of the following is commonly referred to as the "average"?
   a) mean   b) median   c) mode
2. The mode of a distribution is the data value which appears:
   a) least   b) most   c) in the middle

   For questions 3 – 7 consider the following distribution:
   25, 20, 8, 7, 9, 6, 8, 4, 5, 8
3. The mean of this distribution is:
   a) 12   b) 16   c) 10   d) 7
4. The median is:
   a) 7.5   b) 8   c) 9   d) 10
5. The mode is:
   a) 10   b) 8   c) 20   d) 4
6. Which of the three measures of central tendency will change if we remove one of the data values from the above distribution?
   a) mean   b) median   c) mode
7. If we removed the extreme data values (25 and 20) from the distribution, which measure of central tendency would not change?
   a) mean   b) median   c) mode
8. The symbol used to represent the mean of a sample is:
   a) n   b) x   c) μ   d) $\bar{x}$
9. $\bar{x}$ is often used to estimate the _____ of a population.
   a) mean   b) median   c) mode
10. Which measure of central tendency is less resistant to outliers?
    a) mean   b) median
11. In a skewed to the left distribution, the mean is located to the _____ of the median.
    a) left   b) right
12. Which measure of central tendency is more resistant to extreme data values or outliers?
    a) mean   b) median
13. If a distribution has outliers, then the value of the mean and median will be:
    a) the same   b) different   c) unknown

PART III **Answer each statement True or False.**

1. The mean, median, and mode are known as measures of central tendency.
2. When adding the same constant to each data value in a distribution, the mean is increased by the value of that constant.
3. When adding the same constant to each data value in a distribution, the mode in not affected.
4. If each data value in a distribution is multiplied by a different constant, the mean of the new distribution is equal to the mean of the original distribution multiplied by one of the constants.
5. If 15 is subtracted from every data value in a distribution where the mean is 45, then the original mean was 30.
6. The mean, median, and mode are never the same value.
7. If there are an even number of data values in a distribution, the median is never the value of a data value in the distribution.
8. If there are 15 data values in a distribution and they are arranged in numerical order, then the median is the value of the data value in the 8th position.
9. If the mean of a distribution is equal to 5 and the mode equals 3, then the median must equal 4.
10. The mean is always one of the data values of a distribution.
11. Both the population mean and the sample mean represent parameters of a distribution.
12. A sample mean, $\bar{x}$, is usually equal to the population mean, μ.
13. The mean is a more resistant measure of central tendency.
14. In a skewed to the right distribution, the median is located to the left of the mean.
15. The mean is less resistant to outliers than the median.

PART IV **Problems.**

1. a) For the distribution: 6, 2, 4, 4, 5, 3, 5, 3, 1, 7, find:
      1) the mean
      2) the median
      3) the mode
      4) Σ(x - 4)
      5) Σx²
      6) (Σx)²
      7) Σ(x - 4)²
      8) [Σ(x - μ)]²
      9) Σx² - μ²
   b) For the distribution:
      12, 9, 7, 13, 9, 8, 10, 9, 11, 12, find:
      1) the mean
      2) the median
      3) the mode
      4) Σ(x - 10)
      5) Σx²
      6) (Σx)²

7) $\Sigma(x - \mu)^2$
8) $[\Sigma(x - \mu)]^2$
9) $\Sigma x^2 - \mu^2$

2. Given:
x y
1 7
2 8
3 9
4 10

Compute the following:

(a) Σx (b) Σy (c) $\Sigma x \Sigma y$ (d) $(\Sigma x)^2$
(e) $(\Sigma y)^2$ (f) Σx^2 (g) Σy^2 (h) Σxy

(i) $\dfrac{4(\Sigma xy) - \Sigma x \Sigma y}{\sqrt{4\Sigma x^2 - (\Sigma x)^2}\sqrt{4\Sigma y^2 - (\Sigma y)^2}}$

3. Given:
x y
1 3
2 5
2 4
3 4
4 5

Compute the following:

(a) Σx (b) Σy (c) $\Sigma x \Sigma y$ (d) $(\Sigma x)^2$
(e) $(\Sigma y)^2$ (f) Σx^2 (g) Σy^2 (h) Σxy

(i) $\dfrac{5(\Sigma xy) - \Sigma x \Sigma y}{\sqrt{5\Sigma x^2 - (\Sigma x)^2}\sqrt{5\Sigma y^2 - (\Sigma y)^2}}$

4. In order to determine a fair market price for a particular CD player, a consumer obtained the following selling prices:
$160, $165, $190, $145, $170, $165, $195
   a) Calculate the mean, median and mode.
   b) Which measure(s) of central tendency, if any, do you think best illustrates a fair price?

5. A travel agent, trying to promote travel to a particular Caribbean island, began an intensive ad campaign. The campaign stressed the good weather of the island. For this island, the number of sunny days per month for the last year were:
27, 24, 28, 22, 26, 28, 25, 28, 23, 29, 24, 28.
   a) Calculate the mean, median, and mode.
   b) Which one of the three measures would you use for the ad campaign? Why?

6. The following distribution represents real estate listings for homes in southern California:
$375,000, $525,000, $425,000, $1,486,000, $510,000, $655,000, $425,000, $490,000, $575,000 and $505,000.
   a) Determine the mean, median, and mode for these prices.
   b) Why are the mode and mean not good measures of central tendency for this distribution?

7. The table below represents a summary of the final test grades of a calculus class.

Final Test Grade	Number of Students
90	2
87	1
86	2
85	5
84	2
83	6
82	8
79	1
76	3
66	7
60	5
20	4
10	5
0	1

   a) Determine the mean, median, and modal final test grades.
   b) Which measure of central tendency is less resistant to the extreme data values?

8. At a recent golf tournament the following data were reported for the 10th hole.

10th hole data

strokes to hole	number of golfers
3	30
4	220
5	120
6	5
7	5

   a) Find the mean, median, and mode number of strokes for the 10th hole.
   b) Which measure of central tendency best is not appropriate in describing the typical number of strokes for the 10th hole? Explain.

9. Twenty-four baseball players earn the following annual salaries: 2 players at $4.6 million; 4 at $2.45 million; 4 at $1.93 million; 5 at $960,000; 2 at $790,000; and 7 at $640,000.
   a) Determine the mean, median, and modal salary.
   b) Which measure is a poor measure of central tendency? Explain.

10. A statistics instructor gave the following final semester grades to his students:
86, 82, 100, 96, 91, 94, 85, 95, 78, 80, 87, 92,
85, 81, 97, 98, 89, 90, 84, 92, 91, 87, 98, 90
   a) Compute the mean grade.
   b) Comment on the remark of one of the instructor's statistics students: " On the average, this instructor does not give grades above 90."

11. Given the distribution: 10, 20, 30, 40, 50, 60, 70, 80, 900.

Question # 11 continued:
  a) Calculate the mean and the median.
  b) Which measure seems to be a better indicator of the central tendency of this distribution?
  c) For this distribution, change the last data value (900) to 90 and recalculate the mean and the median.
  d) Now which measure seems to be a better indicator of central tendency?
  From your observation, which measure of central tendency was:
11e) more resistant to the extreme data values?
11f) less resistant to the extreme data values?
12. The student government association is in the process of planning a special awards dinner. To try to determine the menu for the awards dinner the association surveyed the number and types of meals purchased at the campus cafeteria. The following meal table was recorded during the past week.

Meals	Monday	Tuesday	Wednesday	Thursday	Friday
Fried Chicken	125	45	80	100	80
Meat Loaf	40	60	95	70	75
Veal Cutlets	50	35	55	75	45
Spag. & Meatball	100	135	90	125	135
Beef Stew	120	90	80	135	90
Pizza	30	80	70	45	160

12. a) What are the most popular daily meals?
    b) What is the most popular weekly meal?
    c) Based on the survey, which meal do you believe the association should choose for the awards dinner?
13. The following data represents a psychiatrist's rating of fifty severely depressed mental patients' responses to drug therapy.

response to drug therapy	number of patients
excellent	12
very good	17
good	16
no response	5

   The psychiatrist claimed that "very good" was the typical response to this drug. Explain which measure of central tendency she used, and why it is appropriate in this situation.
14. The following data represent estimated miles per gallon figures for a new front wheel drive car under city driving conditions:
    30, 28, 29, 28, 28, 42, 40, 28, 29, 30, 42, 40, 40, 42
    a) Calculate the three measures of central tendency.
    b) Which measure of central tendency will yield the maximum mpg rating for this type car?
15. The following data represents the shoe sizes of a popular brand of women's shoes sold at Sole City during the past week:
    5, 5½, 6½, 7, 5½, 6½, 7, 8, 9, 5½, 6, 7, 7½, 8, 5, 6, 6½, 9, 5½, 5, 5½, 6, 6½, 8, 5½, 6½, 5½, 5, 5½, 7½

    Use the data to compute the:
    a) mean shoe size sold.
    b) median shoe size sold.
    c) modal shoe size sold.
    d) Which measure(s) of central tendency would be most beneficial to the store manager when she places her order for purchasing more of these shoes?
16. DPT Motors, Inc. would like to estimate the miles per gallon rating for its new economy car. To estimate the mpg rating, fifteen of the cars were test driven and the results of the sample are indicated.
    35, 29, 27, 28, 29, 21, 34, 19,
    27, 26, 24, 34, 35, 33, 29
    a) Calculate the mean mpg rating for this sample.
    b) Which symbol would be used to represent this mean?
17. A cigarette manufacturer wants to determine the "average" tar content per cigarette for their new brand of licorice flavored cigarettes.

"average" tar content per cigarette for their new brand of licorice flavored cigarettes.

a) Calculate the mean and median tar content using the following sample data.
b) If the manufacturer wishes to emphasis a low tar content in an advertisement, what would you guess the manufacturer will state as the "average" tar content per cigarette for their new licorice flavored cigarettes?

sample data (TAR content in milligrams)
5.0, 5.7, 4.9, 5.3, 4.8, 6.1, 5.0, 4.8, 4.8, 5.9, 5.1, 6.0, 4.9, 4.9, 4.9, 6.1, 4.9, 4.8, 6.0, 4.9, 5.9, 6.0, 4.9, 5.0, 4.9, 4.7

18. A mutual fund company bases its quarterly management and investment service charge on the fund's average market value per share.
   If the following stocks represent the fund's portfolio for the first quarter, determine the mean market value per share.

number of shares	company	market value
50,000	Continental Oil	1,900,000
35,000	Exxon	2,736,875
40,000	Houston Natural Gas	900,000
60,000	Tenneco	2,455,000
75,000	Foremost-McKesson	1,575,000
60,000	Gillette	1,527,500
70,000	Walgreen	1,929,750
35,000	Control Data	1,964,425
20,000	Data General	1,540,000
50,000	Modular Computer Sys.	600,000
20,000	Eastman Kodak	950,000
40,000	Boeing	1,740,000
45,000	General Dynamics	1,300,000
50,000	Lockheed	945,000
13,000	Intel	800,000
115,000	City Investing	1,725,000

19. A leading manufacturer of radial tires claims that their X-60 radial tire will last an average of 60,000 miles. To test this claim, a consumer protection group purchases and tests 32 of these tires.
   The number of miles each tire lasted before needing replacement is recorded below:
   60,900 59,500 58,900 59,650 60,350 60,100
   60,500 58,500 58,800 58,200 60,000 60,400
   60,500 58,000 59,000 58,900 59,500 60,000
   61,000 61,000 59,700 59,800 59,600 60,100
   60,500 61,500 59,950 59,750 59,750 60,150
   60,200 62,000

a) Using the sample information, determine the mean tire life of the X-60 radial tire. Comment on the manufacturer's claim.
b) Assume you are a statistician working for the tire manufacturer. Using the same information, determine which measure of central tendency you would use to support the manufacturer's claim.

20. A manufacturer needs to maintain a close check over the performance of their production process. For example, a sugar refinery needs to control the weight of their 5 lb bag of sugar because the weight of the bags may fluctuate due to gradual changes in the production process. This controlling process is referred to as *quality control*. Suppose the sugar refinery decides to use the mean weight of 20 bags of the sugar that have been randomly selected from the production line to study the performance of the production process.
a) Using the following sample data, calculate the mean weight.
b) Examine the 20 weights used for part a and comment as to whether you believe the estimate is sufficient to check the quality control of the production process.

actual weight in lbs for the five pound bag of sugar
4.8750 5.2500 4.8750 4.8750 4.8750 4.8750
5.2500 4.8750 4.8750 4.9375 5.1875 4.9375
5.1875 4.9375 5.2500 4.9375 5.2500 4.9375
4.8750 4.9375

21. A sociology professor wants to get an estimate of the "average" age of the students in his large lecture. The professor randomly selects 10 students from his class and records their ages. The ages are as follows:
   19, 18, 18, 22, 19, 58, 20, 23, 21, 22
a) Compute the mean age for this sample.
b) At the next class meeting the professor is startled at the response to his question: "Who is 24 or older?" Almost no one responded. Can you explain this?

22. A health insurance company has conducted a survey that determines the length of hospital stay (in days) for three hospitals in Cincinnati. The results of the survey are given below.

Length of hospital stay for patients (in days)		
hospital I	hospital II	hospital III
2	2	11
1	14	3
13	5	14
4	40	5
4	3	43
15	13	31
3	5	1
6	7	9
10	2	4
4	16	12

a) Determine the mean length of hospital stay at each hospital.
b) Compute the mean length of hospital stay combining all the survey data.
c) What do you think is the insurance company's best estimate for the average length of hospital stay for a patient?

23. Two baseball veterans are going to compute their respective lifetime batting average by adding all their season averages and dividing by the number of seasons. The following table contains the data for the two players.

	Player A			Player B		
season	AB	Hits	ave	AB	Hits	ave
1	203	54	.266	310	90	.290
2	392	120	.306	204	64	.314
3	151	37	.245	426	106	.246
4	435	148	.340	126	46	.365
5	215	59	.274	515	145	.281
6	574	181	.315	224	74	.330
7	263	75	.285	374	110	.294
8	271	71	.261	335	93	.278
9	380	116	.305	180	56	.311

Is the procedure the veterans are using to compute their lifetime batting averages correct? If not, how should they compute their lifetime batting averages? Justify your answer.

24. A student states that his cumulative grade point average for the past three semesters is 3.0. The student's reasoning is as follows: "I earned a 3.5 average for 4 credits during the summer semester, a 3.0 average for 12 credits in the fall and a 2.5 average for 16 credits in the spring."
Is the student correct in his reasoning? Justify your answer.

25. Fourteen psychology exam scores and their respective deviations from the mean of a class of 15 students are listed in the following table. The exam score for student number 7 is missing. Using the information in the table, find the missing psychology score. (Hint: Use the property $\Sigma(x - \mu) = 0$).

student #	psychology score	deviation from mean
1	125	2
2	130	7
3	115	-8
4	126	3
5	132	9
6	127	4
7		
8	118	-5
9	128	5
10	134	11
11	126	3
12	119	-4
13	118	-5
14	115	-8
15	125	2

26. The hair color, eye color and sex of twenty new born babies which were born at a county hospital are listed in the following table.

sex	hair color	eye color
M	BROWN	BROWN
M	BROWN	BROWN
F	BLONDE	BLUE
F	RED	BROWN
M	BROWN	BROWN
M	BLONDE	BLUE
F	BROWN	BROWN
F	BROWN	BLUE
F	BROWN	BROWN
M	BLACK	BROWN
F	BLONDE	BLUE
M	BLACK	BROWN
M	BROWN	BLUE
M	BROWN	BROWN
F	BROWN	BLUE
F	BROWN	BLUE
F	BLACK	BROWN
F	RED	BLUE
M	BLACK	BROWN
M	BROWN	BLUE

a) Estimate the typical:
 1) hair color.
 2) eye color.
 3) characteristics for a male baby.
 4) characteristics for a female baby.
b) Which measure of central tendency did you use to calculate the answers to the part (a), and expalin why you chose this measure?

27. An investor decides to diversify his $450,000 investment by opening a $150,000 one year CD with an effective yield of 9.25% and purchasing 2000 shares

at $50 per share of a blue chip stock paying a 7.5% yearly dividend and two hundred triple A bonds at $1000 per bond yielding 9.75% per year.

Determine:
a) the total interest earned on the CD for the year.
b) the total stock dividend earned for the year.
c) the total bond interest earned for the year.
d) the mean monthly return on the total investment.

28. An exercise equipment manufacturer states that their new electronic treadmill will burn an average of 900 calories for a thirty minute workout. The following sample data represents the amount of calories burned per individual for twenty individuals.
930, 890, 920, 910, 850, 870, 895, 880, 940, 830, 825, 890, 925, 855, 960, 860, 915, 830, 920, 875
a) Calculate the sample mean number of calories burned.
b) If an individual loses one pound for every 3500 calories burned, based on the sample mean calculated in part (a), determine the mean number of pounds lost per individual for a thirty minute workout.
c) If an individual works out 30 minutes a day for 30 days on this treadmill, determine the expected number of pounds lost.

29. Eastinghouse, an electric company, advertises that their new improved energy saving light bulb has a mean life of 2500 hours. A consumer group decides to sample 30 of these bulbs to determine if the company's advertisement is correct. The following data represents the consumer group's sample.
2540 2340 2325 2300 2510 2380 2520 2370
2390 2540 2400 2430 2500 2520 2320 2550
2410 2535 2510 2405 2315 2515 2415 2540
2305 2570 2580 2415 2445 2305
a) Calculate the sample mean life of these bulbs.
b) Calculate the difference between the sample mean life and the company's advertised population mean life.
c) The consumer group decides to dispute the company's claim if the difference between the sample and the population mean life is greater than 50 hours. Does the consumer group have reason to dispute the company's claim?

30. Comment on how the word "**average**" is being used in the following statements:

- The **average** physician earned $164,300 in 1990 while working 59.9 hours.
- The starting salary for the **average** graduate of N.Y.U. Law School is $76,424.
- Pediatricians **averaged** $106,500 - and that's 170 % of what an **average** newspaper reporter earns for a 35-hour week.
- An **average** American eats 50 bananas a year, drinks 374 beers, goofs off 45 minutes a day at work, swallows 215 aspirins a year, and owes more than $5,000.

31. Using the following figure which represents a skewed to the right distribution, identify the letter which represents the position of:
a) the mode
b) the mean
c) the median

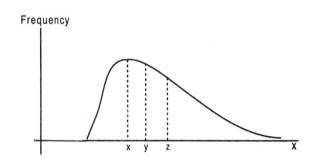

32. Using the following figure which represents a skewed to the left distribution, determine the value for:
a) the mode
b) the mean
c) the median

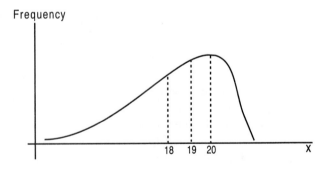

33. For each of the following distribution of data values,

a) identify the general shape that you believe best describes each distribution, and
b) identify whether the mean or median would have the largest value within this type of distribution.

(1) The annual income of people living in the United States.

(2) The number of calories consumed weekly by twenty-five year old Las Vegas women weighing 125 pounds.

(3) The test results of an easy exam.

(4) The test results of a difficult exam.

(5) The heights of all the female students at a University.

(6) The heights of all the male college students playing basketball within the state of North Carolina.

(7) The number of words typed per minute by executive secretaries working in the World Trade Center.

PART V **What Do You Think?**

1. **What Do You Think?**

   Examine the USA Snapshot entitled *How to live longer*.
   How many more years can the average person expect to live by including exercise in his/her lifestyle? What would you interpret to be the number of years the average person would expect to gain by watching his/her diet and including exercise within his/her lifestyle?

2. **What Do You Think?**

   For each of the following graphics, discuss which measure of central tendency might have been used or was used within the graphic. Explain why you believe this measure is or is not the appropriate measure to use to describe the data. If possible, draw a conclusion using the information within the graphic.

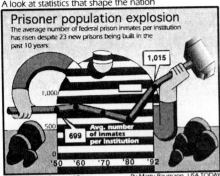

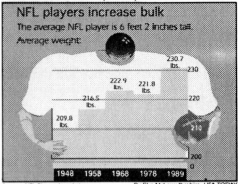

PART VI **Exploring DATA With MINITAB.**

1. A statistics instructor at a community college administers the same exam to two groups of students. The first group represents full-time day students, while the second group is composed of part-time evening students. The results of the exam for each group are:

   Full-Time Day Students
   22, 32, 41, 50, 62, 74, 80, 90, 95, 86, 86, 74, 75, 62, 62, 53, 53, 43, 43, 35, 35, 27, 29, 29, 38, 39, 39, 87, 87, 75, 79, 69, 69, 59, 58, 49, 48, 53, 59, 59, 64, 62, 62, 68, 76, 79, 87, 96, 76, 97, 79, 76, 98, 62, 65, 54, 59, 44, 49

   Part-Time Evening Students
   32, 47, 58, 17, 64, 77, 89, 86, 99, 92, 80, 71, 63, 55, 46, 49, 57, 65, 74, 86, 94, 92, 86, 89, 96, 74, 61, 70, 84, 85, 92, 97, 78, 65, 56, 64, 84, 86, 95, 98, 86, 72, 52, 94, 96, 84, 94

   Use the SET or READ command in MINITAB to enter the data values for each distribution. For each distribution:
   a) construct a stem-and-leaf display using the tens digit as the stem.
   b) discuss the shape of the distribution.
   c) using the DESCRIBE command in MINITAB, determine the mean, median, mode, and the trimmean for the distribution, and interpret these measures.
   d) locate the value of the mean, median, and mode on each stem-and-leaf display.
   e) what does the relationship between the location of the measures of central tendency indicate about the skewness or symmetry of the two distributions.
   f) determine if there is a measure of central tendency that is very sensitive to the extreme data values?
   If there is such a measure, then identify the measure, the shape of the distribution, and the direction in which this measure is shifted. Explain why the measure has shifted in that direction.
   g) compare the value of the mean to the trimmean. In which distribution is the difference larger? Can you explain what might attribute to this larger difference?
   h) determine, overall, which group you believe did better on the exam? Explain your answer.

2. The following MINITAB output represents a stem-and-leaf display for 42 freshman enrolled at a small community college. The display represents the number of hours worked per week for both male and female students.

   MTB > STEM-AND-LEAF 'hrswrk';
   SUBC> INCREMENT 10.

   Stem-and-leaf of hrswrk   N = 42   Leaf Unit = 1.0

   ```
    4   0  0000
   12   1  00005555
   21   2  000555555
   21   3  0005555
   14   4  0000005555
    4   5  005
    1   6  5
   ```

   a) Draw the shape of the display using a smooth curve.
   b) Indicate, by drawing a line, where the mean,

median and mode fall.
c) Discuss the shape of the distribution in terms of the relationship between the location of the mean, median and mode within the distribution.
d) What conclusions can you draw regarding the differences in the work habits of the male and female students?

3. The following MINITAB outputs indicate two stem-and-leaf displays representing 42 freshman enrolled at a small community college.

The first display indicates the number of hours worked per week for the female students, while the second display indicates the number of hours worked per week for the male students.

OUTPUT for Female Students.

MTB > STEM-AND-LEAF 'hrswrk';

SUBC> BY sex;

SUBC> INCREMENT 10.

Stem-and-leaf of hrswrk   sex = 1   N = 20
                          LeafUnit = 1.0

```
  20    0   0
   8    1   000555
  (7)   2   0005555
   5    3   005
   2    4   05
```

OUTPUT for Male Students.

Stem-and-leaf of hrswrk   sex = 2   N = 22
                          Leaf Unit = 1.0

```
   2    0   00
   4    1   05
   6    2   55
  10    3   0555
  (8)   4   00000555
   4    5   005
   1    6   5
```

For each display:
a) Draw the shape using a smooth curve.
b) Indicate, by drawing a line, where the mean, median, and mode fall.
c) Discuss the shape of the distribution in terms of the relationship between the location of the mean, median, and mode within the distribution.
d) Compare the shapes of each of the distributions.

What are the differences between the work habits of the male and female students?

PART VII **Projects.**

1. Choose 10 stocks arbitrarily form the New York Stock Exchange and compute the mean stock price for these stocks per day for one week. Compare your results per day with the Dow Jones Industrial Stock Average per day by using the chart below:

day	10 stock average	dollar change per day	Dow Jones	dollar change per day
Monday				
Tuesday				
Wednesday				
Thursday				
Friday				

2. Consider the table below:

Ages

	16-25	26-35	36-45	46-55
1.				
2.				
3.				
4.				
5.				
6.				
7.				
8.				
9.				
10.				

For each age group, survey ten people and record their answer to the question, "How many years should a married couple wait before having children?"
a) According to your survey, what is the mean number of years a couple should wait before having children as answered by each group?
b) What is the modal number of years a couple should wait before having children as answered by each group?
c) Consider all the responses as one group. What are the mean and mode of this distribution?
d) Do you think people in different age groups responded with a significantly differently waiting period?

3. The following data represents personal information for fifty students.

student ID	hair color	eye color	ht inchs	wt lbs	# hours worked weekly	sex	IQ score	# of sib	age
8476	BR	HAZEL	66	118	22	M	95	2	19
3834	BK	BROWN	67	107	32	F	115	2	18
3165	BL	BROWN	69	154	35	M	120	5	18
1207	BL	BROWN	72	176	23	M	105	3	23
6784	BR	BROWN	60	103	20	F	116	1	19
6780	BR	HAZEL	67	132	18	M	127	5	20
5380	BK	BLUE	66	123	13	M	101	1	18
4227	BR	BROWN	66	121	40	M	125	1	18
8625	BR	HAZEL	64	105	0	F	99	0	19
0817	BR	BLUE	71	170	35	M	104	3	32
0882	BR	HAZEL	71	175	24	M	114	4	18
0638	BR	HAZEL	67	135	30	M	121	2	19
4390	BK	BROWN	74	206	35	M	130	2	20
5345	BK	BROWN	63	98	0	F	107	2	19
6819	BK	BLUE	74	198	0	M	122	0	20
4002	RD	HAZEL	59	99	10	F	110	2	18
1252	BR	BROWN	68	143	28	F	145	5	18
1031	BL	BROWN	61	115	35	F	109	0	43
7873	BR	HAZEL	60	93	19	F	103	3	19
0194	BR	BROWN	70	160	15	M	135	5	20
9189	BR	BLUE	56	144	23	F	118	5	18
3695	BR	BROWN	64	115	22	F	105	3	18
2444	BK	BROWN	71	172	16	M	115	2	17
2467	BK	BROWN	61	113	19	F	128	2	19
7286	BR	HAZEL	67	98	25	F	131	1	20
7896	BR	BROWN	61	105	24	F	118	2	18
8905	BR	HAZEL	65	135	25	F	117	0	19
7299	BR	BROWN	65	122	0	F	113	3	18
9638	BK	BROWN	67	129	26	M	104	1	29
6798	BR	BROWN	66	118	20	F	126	4	21
4442	BL	BROWN	69	143	44	F	104	1	26
0132	BL	BROWN	73	194	16	M	107	3	18
1884	RD	BROWN	70	157	31	M	129	2	19
7345	BK	BROWN	63	126	9	F	134	1	20
0655	BK	HAZEL	69	149	7	M	108	2	18
4982	BR	HAZEL	73	193	36	M	112	1	24
8825	BL	HAZEL	73	188	17	M	104	3	37
6696	BR	HAZEL	61	122	44	F	98	1	18
9498	BR	BROWN	70	164	10	M	123	3	19
6217	RD	HAZEL	66	133	37	F	105	0	20
6014	BK	BLUE	66	123	24	M	134	1	18
1394	RD	BROWN	69	184	9	M	138	2	17
3199	BR	BROWN	69	190	21	M	121	0	19
2426	BR	HAZEL	66	122	15	M	99	2	19
2033	BR	BROWN	74	207	0	M	112	2	22
2435	BK	HAZEL	68	157	33	M	127	5	20
3173	RD	HAZEL	69	176	28	M	136	2	19
3361	BR	BROWN	70	192	14	M	121	2	36
8886	BR	BROWN	70	125	31	F	102	3	21
3593	BR	BROWN	74	197	41	M	111	2	24

Compute the suitable measure of central tendency for each characteristic and construct a composite description of the typical student at this school. Use the composite of the average American given in the introduction to this chapter as a guide.

PART VIII **Database.**

The following exercises refer to the file DATABASE listed in Appendix A. We have indicated the appropriate MINITAB commands that are necessary to answer each exercise.

1. Using the MINITAB commands:
   RETRIEVE, MEAN, MEDIAN and TALLY
   Retrieve the file DATABASE.MTW and determine the measures of central tendency for the variables:
   a) AGE
   b) AVE
   c) STY
   d) GPA
   e) GFT

2. Using the MINITAB commands:
   RETRIEVE, MEAN, LET, NAME and PRINT
   Retrieve the file DATABASE.MTW and display a table representing the effect of subtracting the mean of each variable from the data for the variables:
   a) AGE
   b) GPA
   c) GFT
   The table should include columns representing:
   1) the variable name along with the data values
   2) the data value minus the mean, as well as the *mean* of the variable and the *mean* of the column for the data value minus the mean.
   Examine the results within each column. Did you expect these results? Comment.

CHAPTER 4
MEASURES OF VARIABILITY

❄ 4.1 Introduction

A favorite story told in the math department is about the late department chairperson, a nonswimmer, who recently drowned in a lake whose **average depth is only two feet!**

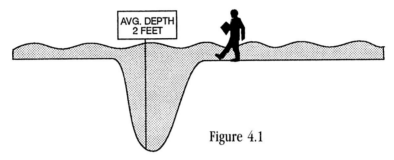

Figure 4.1

If the chairperson had been aware of the tremendous variations in depth as illustrated in Figure 4.1, he would have realized that knowing *only* **the lake's average depth does not adequately describe its depth everywhere**.

As illustrated in this story the Measures of Central Tendency alone are sometimes inadequate to fully describe a distribution. Additional descriptive measures, such as the measures of variability will further enhance our ability to describe the characteristics of a distribution.

Before we discuss the measures of variability, let's discuss the notion of variability by examining the two distributions given in Table 4.1.

Table 4.1

Distribution 1	Distribution 2
4	1
4	2
4	89
4	100
4	

In Distribution 1, notice that the data values are all the same, that is, there is **no variability**. In Distribution 2, there is **variability** since the data values vary from 1 to 100.

4.2 THE RANGE

The first measure of variability we will discuss is the range. The range provides a quick method of describing the variability of a distribution.

> **Definition: Range.** The range of a distribution is the number representing the difference between the largest value and smallest value in the distribution.
>
> **Range = largest value − smallest value**

Example 4.1

The following data represents the closing prices of a high tech stock during an active week of trading (stock prices expressed in dollars): 15, 21, 19, 25, 32

Compute the range of the closing prices of the stock.

Solution:

Notice that the closing price of the high tech stock varied from a low of 15 dollars to a high of 32 dollars. Therefore, the range of the stock for the week was 17 dollars. The range of 17 dollars indicates that the closing price of the high tech stock varied by 17 dollars during the week. ■

Example 4.2

Listed below are the math test grades for two students:
  student A: 50, 70, 70, 70, 70, 70, 90
  student B: 50, 55, 62, 70, 78, 85, 90

Compute the range of test grades for each student.

Solution:

For student A, the range is 40. For student B, the range is 40. ■

Case Study 4.1

The graphic in Figure 4.2 represents the performance of the Dow Jones Industrial Average for the six day period from Friday, June 4, 1993 to Friday, June 11, 1993.

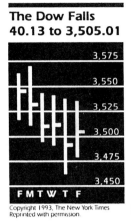

Figure 4.2

The Dow Jones Average, a widely quoted average of 30 big company stocks, is a barometer of the U.S. stock market performance. The bar for each day within Figure 4.2 provides information about the high, low and closing value of the Dow Jones Average as illustrated in Figure 4.3.

Figure 4.3

a) In Figure 4.2, estimate the closing, high and low value of the Dow Jones for Wednesday?
b) From Figure 4.2, which day of the week had the shortest bar? the longest bar? What does the length of the bar in Figure 4.2 indicate about the variability of the stock market's performance?
c) Observe the **closing** value of the Dow Jones for Tuesday and Wednesday. How much did the closing value of the Dow Jones vary from Tuesday to Wednesday?
d) Examine the closing value of the Dow Jones for Wednesday and Thursday. How much did the closing value of the Dow Jones vary from Wednesday to Thursday?
e) In examining the distance between the daily closing values of the Dow Jones Average, what does this indicate about the stock market for the week?
f) What variability is being presented within the headline: *The Dow Falls 40.13 to 3,505.01* for Figure 4.2? Explain.

In Example 4.2, the range for both students is the same, since the calculation of the range involves only the two extreme data values (the largest and the smallest). Overall, the test grades of student B appear to **vary** more than the test grades of student A. Table 4.2 highlights the two distributions with a box separating the extreme test grade values.

Table 4.2

student A:	50	70	70	70	70	70	90
student B:	50	55	62	70	78	85	90

Examining the test grades within the box, you should notice that the variability of the two distributions is different. In this instance, the range may not be a completely adequate way of distinguishing between the variability of these two test grade distributions. Thus, **one disadvantage of the range is that it does not take into consideration the value of all the data values when measuring the variability of a distribution.**

Let's examine two additional measures of variability, where *every* **data value within the distribution is taken into consideration when computing variability.**

4.3 THE POPULATION VARIANCE AND THE POPULATION STANDARD DEVIATION

The population variance and population standard deviation are two measures of variability that are defined using all the data values of a distribution. Before we define these measures, let's examine the development of the variance and standard deviation so we can get a better understanding of their definitions and how to compute their formulas.

To develop these new measures of variability, we will choose a value that depicts the **typical value** of the distribution and determine how far each data value varies from this value. **The mean will be used to represent this typical value.**

For the following data, find the mean:

$$1, 2, 3, 4, 5, 6, 7$$

$$\mu = \frac{\Sigma x}{N} = 4$$

Now let's determine how far each data value varies from the mean. To do this we begin by subtracting the mean from each data value in the distribution. These differences are called **deviations from the mean** and are illustrated in Table 4.3.

Table 4.3

Data Value x	Deviation from the Mean $x - \mu$	Explanation
1	1 - 4 = -3	-3 indicates that the value 1 is 3 units *below* the mean
2	2 - 4 = -2	-2 indicates that the value 2 is 2 units *below* the mean
3	3 - 4 = -1	-1 indicates that the value 3 is 1 unit *below* the mean
4	4 - 4 = 0	0 indicates that the value 4 *is* the mean
5	5 - 4 = 1	1 indicates that the value 5 is one unit *above* the mean
6	6 - 4 = 2	2 indicates that the value 6 is two units *above* the mean
7	7 - 4 = 3	3 indicates that the value 7 is three units *above* the mean

Since we want these new measures to reflect the variability of all the data values, let's calculate the mean of the deviations from the mean that were obtained in Table 4.3.

The mean of these deviations is:

$$\frac{(-3)+(-2)+(-1)+(0)+(1)+(2)+(3)}{7} = 0$$

Let's take time to interpret this result. Did we expect the mean of the deviations to be zero? Remember in Chapter three we showed that **the sum of the deviations from the mean**, expressed as $\Sigma(x-\mu)$, is always zero. You can see that finding the mean of these deviations is not a good way to define a measure of variability since the result will always equal zero.

To remedy this problem we must consider a technique which will change each deviation to a nonnegative number before calculating the mean. One method of making the deviations nonnegative is to square each deviation. This is represented in Table 4.4.

Table 4.4

Data Value x	Deviation from the Mean $x - \mu$	SQUARED Deviation from the Mean $(x - \mu)^2$
1	-3	$(-3)^2 = 9$
2	-2	$(-2)^2 = 4$
3	-1	$(-1)^2 = 1$
4	0	$(0)^2 = 0$
5	1	$(1)^2 = 1$
6	2	$(2)^2 = 4$
7	3	$(3)^2 = 9$

We will now use the squared deviations, symbolized as $\Sigma(x-\mu)^2$, to define these new measures of variability. Let's now calculate **the mean of these squared deviations**, by adding up the squared deviations and dividing this sum by the number of squared deviations. The **mean of the squared deviations** is written as:

$$\frac{\Sigma(x-\mu)^2}{N}$$

This result is:

$$\frac{\Sigma(x-\mu)^2}{N} = \frac{9+4+1+0+1+4+9}{7} = 4$$

Therefore, the mean of the squared deviations is 4.
This result is called the ***population variance.***

> **Definition: Population Variance.** The population variance is the mean of the squared deviations from the mean. The variance of the population is symbolized by the square of the lower case Greek letter sigma, written as σ^2.
> Thus, the formula for the population variance is:
>
> $$\text{population variance} = \frac{\Sigma(x-\mu)^2}{N}$$
>
> OR
>
> $$\sigma^2 = \frac{\Sigma(x-\mu)^2}{N}$$

The population variance is a measure of variability that considers all the data values of a distribution. Although the variance is an important measure of variability, there are instances when the variance is extremely difficult to interpret as a measure of variability. We will explore this disadvantage in Example 4.3.

Example 4.3

The following data represents the hourly wages five employees are paid in a small cannoli bakery in Boston.

$$\$6, \$10, \$9, \$12, \$8$$

Compute the variance of the hourly wages.

Solution:

Step 1. Compute the mean.

$$\mu = \$9$$

Step 2. Subtract the mean of $9 from each wage and then square the deviations from the mean. These results are shown in Table 4.5.

Table 4.5

Data Value x	Deviation from the Mean $x - \mu$	Squared Deviation from the Mean $(x - \mu)^2$
$6	$6 - $9 = -$3	$(-\$3)^2$ = 9 squared dollars
$10	$10 - $9 = $1	$(\$1)^2$ = 1 squared dollar
$9	$9 - $9 = $0	$(\$0)^2$ = 0 squared dollars
$12	$12 - $9 = $3	$(\$3)^2$ = 9 squared dollars
$8	$8 - $9 = -$1	$(-\$1)^2$ = 1 squared dollar

Step 3. Add the squared deviations.

$$\Sigma(x - \mu)^2 = (9 + 1 + 0 + 9 + 1) \text{ squared dollars}$$

$$= 20 \text{ squared dollars}$$

Step 4. To calculate the variance, divide the sum of the squared deviations from the mean by the number of data values.

$$\text{variance} = \frac{\Sigma(x - \mu)^2}{N}$$

$$= \frac{20 \text{ squared dollars}}{5}$$

$$\text{variance} = 4 \text{ squared dollars} \quad \blacksquare$$

One disadvantage of the variance is that the variance is measured in squared units, while the data values within the distribution are not measured in squared units.

Since the unit measurements for the variance and the data will always be different, it may be difficult to relate the variance as a measure of variability to the data values within the distribution.

Examining Example 4.3, we see that the variance is expressed in squared dollar units. We would like to define a new measure of variability, which is expressed in the same unit as the original data values. This can be accomplished by taking the positive square root of the variance. This new measure of variability is called the population standard deviation, and is symbolized by the lower case Greek letter sigma, written as σ.

Let's now define this new measure of variability called the population standard deviation.

Definition: Population Standard Deviation. The population standard deviation, σ, is the positive square root of the variance.

population standard deviation = $\sqrt{\text{variance}}$

Therefore, the formula for the population standard deviation is symbolized as:

$$\sigma = \sqrt{\frac{\Sigma(x-\mu)^2}{N}}$$

We will refer to this formula for the standard deviation as the ***definition formula for the population standard deviation***.

To calculate the population standard deviation for Example 4.3, take the positive square root of the variance. Since the variance is 4 squared dollars, then the population standard deviation is:

population standard deviation = $\sqrt{4 \text{ squared dollars}}$

$$= 2 \text{ dollars}$$

This is symbolized as:

$$\sigma = \$2$$

Notice that the standard deviation expresses the variability in the *same* unit of measure as the original data values. That is, the hourly wages and the standard deviation are both expressed in dollars.

Intuitively, we can interpret the standard deviation to be the "average" or "typical" deviation from the mean. Thus, in Example 4.3 we can interpret a standard deviation of $2.00 as indicating that the hourly wages of 6, 10, 9, 12, and 8 dollars vary by an "average" of $2.00 from the mean hourly wage of 9 dollars.

Example 4.4

Calculate the standard deviation for the following distribution of ages (represented in years) using the definition formula for the population standard deviation.

$$1, \; 3, \; 5, \; 18, \; 8$$

Solution:

Step 1. Calculate the mean.

$$\mu = \frac{\Sigma x}{N} = 7 \text{ years}$$

Step 2. Subtract the mean of 7 from each data value and then square these deviations. These results are shown in Table 4.6.

Table 4.6

Data Value x	Deviation from the Mean x - μ	Squared Deviation from the Mean (x - μ)²
1	-6	(-6)² = 36
3	-4	(-4)² = 16
5	-2	(-2)² = 4
18	11	(-11)² = 121
8	1	(-1)² = 1

Step 3. Add the squared deviations.

$$\Sigma(x - \mu)^2 = 178$$

Step 4. Divide the squared deviations by the number of data values. This result is the variance.

$$\text{variance} = \frac{\Sigma(x - \mu)^2}{N} = \frac{178}{5} = 35.6 \text{ squared years}$$

In symbols, we write:
$$\sigma^2 = 35.6 \text{ squared years}$$

Notice **the variance is expressed in *squared years*.**

Step 5. Take the square root of the variance. This is the standard deviation.

$$\sigma = \sqrt{\text{variance}}$$

$$\sigma = \sqrt{35.6}$$

$$\sigma \approx 5.97$$

A standard deviation of 5.97 years indicates that the ages within this distribution vary by an "average" of 5.97 years from the mean age of 7 years. ■

There is an equivalent formula for the population standard deviation which is better suited for calculator usage. This formula is called the ***computational formula for the population standard deviation***.

The Computational Formula for the Population Standard Deviation is:

$$\sigma = \sqrt{\dfrac{\Sigma x^2 - \dfrac{(\Sigma x)^2}{N}}{N}}$$

Example 4.5

Using the computational formula, find the population standard deviation for the following distribution of data values.

$$2, 2, 3, 4$$

Solution:

Step 1. List all the data values and the squares of each value. These results are shown in Table 4.7.

Table 4.7

x	x²
2	4
2	4
3	9
4	16

Step 2. Find the sum of the squared data values, Σx^2.

$$\Sigma x^2 = 4 + 4 + 9 + 16 = 33$$

Step 3. Find the sum of the data values, Σx.

$$\Sigma x = 2 + 2 + 3 + 4 = 11$$

Step 4. Square the sum of the data values, $(\Sigma x)^2$.

$$(\Sigma x)^2 = (11)^2 = 121$$

Step 5. Divide $(\Sigma x)^2$ by N.

$$\frac{(\Sigma x)^2}{N} = 30.25$$

Step 6. Subtract $\dfrac{(\Sigma x)^2}{N}$ from Σx^2.

We have:

$$\Sigma x^2 - \frac{(\Sigma x)^2}{N} = 33 - 30.25 = 2.75$$

Step 7. Divide $\Sigma x^2 - \dfrac{(\Sigma x)^2}{N}$ by N.

This result is the variance.

$$\text{variance} = \frac{\Sigma x^2 - \dfrac{(\Sigma x)^2}{N}}{N} = 0.6875$$

$$\sigma^2 \approx 0.69$$

Step 8. Take the square root. This result is the standard deviation.

$$\sigma = \sqrt{\frac{\Sigma x^2 - \dfrac{(\Sigma x)^2}{N}}{N}}$$

$$\sigma = \sqrt{0.69}$$

$$\sigma \approx 0.83 \quad \blacksquare$$

Using a Calculator to Evaluate the Population Standard Deviation

Statistical calculators have built-in programs that enable users to evaluate the population standard deviation. The procedure to obtain this result requires the user to simply enter each data value using special keys, and then retrieve the population standard deviation from the statistical memory through a combination of keystrokes. The procedure to determine the standard deviation is dependent upon the calculator manufacturer and the model you own.

The general procedure for the population standard deviation is:

Step 1. Place the calculator in statistical mode.

Step 2. Enter each data values using the appropriate key. For Texas Instruments Calculators, this key is usually the $\Sigma+$ key, while for Casio and Sharp Calculators the key is usually the M+ or Data Key. Depending on the calculator, as each data value is entered, the display will show either: the number of data values presently entered or the last data value entered.

Step 3. Retrieve the statistical results from memory using the appropriate keystrokes outlined in the manual for the calculator.

Although, the previous procedure is generic and not dependent upon any specific calculator, Appendix B contains specific calculator instructions for some models of the calculator manufacturers: Texas Instruments, Casio, Sharp and Radio Shack.

Example 4.6

Two high school women basketball players are being considered for an award representing the most consistent outstanding woman's scorer for the season. The number of points scored/game has been collected and is given in Table 4.8.

Table 4.8

Points per game for player A	Points per game for player B
5	18
32	20
45	22
25	15
8	35
10	24
35	29
12	19
37	25
21	23

a) Calculate the mean and standard deviation for each player.
b) Using the results of part (a) determine which player should receive the award.

Solution:

a) Using your statistical calculator and following the general procedure outlined below compute the mean and standard deviation for each player.

Step 1. Place the calculator in statistical mode.
Step 2. Enter each data values using the appropriate key.
Step 3. Retrieve the statistical results from memory using the appropriate keystrokes outlined in the manual for the calculator.

The mean and standard deviation for each player is shown in Table 4.9.

Table 4.9

	Mean	Standard deviation
Player A	23.00	13.23
Player B	23.00	5.47

b) Since both players had the same average number of points per game, the award for the most consistent player should be given to player B because player B has the smaller standard deviation. ■

Let us examine what will happen to the value of the standard deviation if we change each data value of a distribution by a constant. First we will consider adding a constant to each data value of a distribution.

Example 4.7

The standard deviation for the distribution: 1, 2, 3, 4, 5, 6, 7 is 2. Let us add 5 to each data value of this distribution and see what effect this will have on the standard deviation.
When the constant 5 is added to each data value of the distribution: 1, 2, 3, 4, 5, 6, 7, the new distribution becomes 6, 7, 8, 9, 10, 11, 12.
Use a statistical calculator to compute the standard deviation, σ, for this new distribution.

Solution:

To evaluate the standard deviation, σ, using a statistical calculator, use the following steps.

Step 1. Place the calculator in statistical mode.

Step 2. Enter each data value using the appropriate key.
For Texas Instruments Calculators, this key is usually the $\Sigma+$ key, while for Casio and Sharp Calculators the key is usually the M+ or Data Key.
Depending on the calculator, as each data value is entered, the display will show either: the number of data values presently entered or the last data value entered.

Step 3. Retrieve the standard deviation from memory using the appropriate keystrokes outlined in the manual.
The standard deviation is:
$$\sigma = 2$$

Notice that the standard deviation of the new distribution is 2, which is the same value as the standard deviation of the original distribution. ∎

This leads us to the following property of the standard deviation.

Property of the Standard Deviation For Adding a Constant

If a constant is added to each data value in a distribution whose standard deviation is σ, then the standard deviation of the new distribution will also equal the same value σ.

The reason the value of the standard deviation remains the same when adding a constant to each data value is because the differences or variations between the data values remain the same. This is illustrated in Figure 4.4.

Variations between data values of original distribution

original distribution:	1	2	3	4	5	6	7
differences between data values:		1	1	1	1	1	1

A new distribution is formed by adding the constant 5 to each data value in the original distribution.

Variations between data values of new distribution

new distribution:	6	7	8	9	10	11	12
differences between data values:		1	1	1	1	1	1

Figure 4.4

What do you think will happen to the standard deviation of a distribution if we subtract a constant from each data value in a distribution? Answer: The standard deviation will remain the same, since the variation between the data values is not changed.

The next example will show the effect on the value of the standard deviation when multiplying each data value in a distribution by the same constant.

Example 4.8

Let's take the distribution 1, 2, 3, 4, 5, 6, 7, where $\sigma = 2$, and multiply each data value of the distribution by the constant 3.
Compute the standard deviation of the new distribution and compare it to the standard deviation of the original distribution.

Solution:

Multiplying each data value of the distribution: 1, 2, 3, 4, 5, 6, 7 by the constant 3, the new distribution becomes:

3, 6, 9, 12, 15, 18, 21

Using your statistical calculator the standard deviation for the new distribution is:

$$\sigma = 6$$

Comparing this new standard deviation of 6 to the standard deviation of the original distribution, which is 2, we notice that this new standard deviation is three times the value of the standard deviation of the original distribution. That is,

standard deviation of = 3(standard deviation of original distribution)
new distribution
= 3(2)
= 6 ∎

Using the result of this example, we can state the following property of the standard deviation.

Property of Standard Deviation For Multiplying by a Constant

If each data value of a distribution, whose standard deviation is σ, is multiplied by a positive constant, C, then the standard deviation of the new distribution will be C times σ, expressed $C(\sigma)$.

This property is true because when multiplying each data value by the constant C, the differences or variations between the data values are also multiplied by this same constant C. This is illustrated in Figure 4.5.

Variations between original data values

| original distribution: | 1 | 2 | 3 | 4 | 5 | 6 | 7 |

| differences between data values: | 1 | 1 | 1 | 1 | 1 | 1 |

A new distribution is formed by multiplying each data value in the original distribution by the constant 3. This results in the following variation.

Variation between new data values

| new distribution: | 3 | 6 | 9 | 12 | 15 | 18 | 21 |

| differences between data values: | 3 | 3 | 3 | 3 | 3 | 3 |

Figure 4.5

A similar result holds true if a negative number is used as the constant factor provided the negative sign of the constant is ignored. That is, if a number like (-3) was used as the constant factor, the resulting standard deviation of the new distribution would still be 6. Finally, let's consider the effect on the standard deviation if each data value of a distribution is divided by a positive constant.

Property of Standard Deviation For Dividing by a Constant

If each data value of a distribution, whose standard deviation is σ, is divided by a positive constant C, then the standard deviation of the new distribution will be σ divided by C, expressed σ/C.

Case Study 4.2

Examine the two USA Snapshots in Figure 4.6. The USA Snapshot entitled "Teens at Loose Ends" in Figure 4.6a presents the highest and lowest percentages of teenagers who are neither attending school nor working. In Figure 4.6b, the USA Snapshot entitled "What's Spent on Car Insurance" shows the highest and lowest average car insurance premiums for some states.

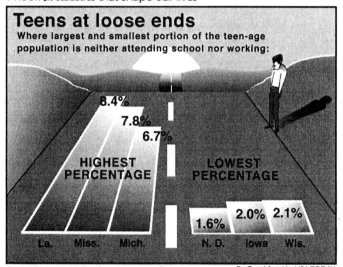

(a)

(b)

Figure 4.6

(1) Explain how the information contained in each USA Snapshot is useful in describing the variability of all states.
(2) What additional descriptive measure is contained in the USA Snapshot pertaining to the car insurance premiums in Figure 4.6b? What does this additional measure describe and how is it helpful in describing the data on car insurance premiums within the USA?
(3) Which measure of variability can be calculated using the information contained within each snapshot of Figure 4.6? What disadvantage(s) is/are there in using this measure to indicate the variability of all the data values?

CASE STUDY 4.3

Examine the verbal and mathematical SAT score averages for college-bound high school seniors shown in Figure 4.7. In your observation, which data set of SAT scores seems to have the **more consistent** SAT scores? Would you expect this data set to have a smaller or a larger standard deviation?

SAT SCORE AVERAGES FOR COLLEGE-BOUND HIGH-SCHOOL SENIORS

School Year	Verbal	Mathematical
1983-84	426	471
1984-85	431	475
1985-86	431	475
1986-87	430	476
1987-88	428	476
1988-89	427	476
1989-90	424	476
1990-91	422	474
1991-92	423	476

SOURCE: COLLEGE ENTRANCE EXAMINATION BOARD
Copyright 1993, NEWSWEEK. Reprinted with permission.

Figure 4.7

Calculating the mean and standard deviation for the verbal and mathematical SAT averages, we obtain:

SAT	VERBAL	MATHEMATICAL
MEAN	426.89	475
STANDARD DEVIATION	3.21	1.66

Using this information, we can see that the more consistent mathematical SAT scores had the smaller standard deviation than the verbal scores.

What do you think would happen to the standard deviation of the verbal and mathematical SAT scores, if the score for the 1983-84 school year is omitted? Compare the SAT score for the 1983-84 school year to the other scores within each data set and determine whether you believe that the standard deviation of each would decrease, increase or stay the same? ∎

4.4 THE SAMPLE STANDARD DEVIATION, s

There are many situations when the entire population data is either unattainable or impractical to collect. In such instances, a *representative sample* is selected from the population and is used to *estimate* the main characteristics of the population, such as the population standard deviation, σ. An estimate of the population standard deviation can be obtained by using the sample standard deviation, denoted by s. The following formula for the sample standard deviation will be referred to as the definition formula for s.

The Definition Formula for the Sample Standard Deviation, s

$$s = \sqrt{\frac{\Sigma(x-\bar{x})^2}{n-1}}$$

Notice the definition formula for s is similar to the definition formula for the population standard deviation except the sample mean, $\bar{x}$, replaces the population mean, μ, and the denominator is n-1 rather than N. Using a denominator of n-1 gives a **better estimate** for the population standard deviation, σ. Consequently, ***s is a good estimate of the population standard deviation, σ, when the sample size, n, is large***.

An equivalent formula that can be used to calculate the sample standard deviation, s, which serves as an estimate of the population standard deviation, is called the computational formula for s. This formula is better suited for use with a calculator.

The Computational Formula for the Sample Standard Deviation, s, is:

$$s = \sqrt{\frac{\Sigma x^2 - \frac{(\Sigma x)^2}{n}}{n-1}}$$

Example 4.9

Professor Candy Hart, a Health instructor, decides to use her class as a sample to estimate the mean and standard deviation of the pulse rate of all the students at the university. She collected the following 15 pulse rates (measured in heart beats per minute while resting) from her students.

$$48, 74, 76, 80, 85, 45, 70, 50, 78, 65, 73, 78, 74, 73, 66$$

Compute the sample mean, $\bar{x}$, as an estimate of the population mean and the sample standard deviation, s, as an estimate of the population standard deviation.
Use the computational formula to compute s.

Solution:

We use the sample mean formula: $\bar{x} = \frac{\Sigma x}{n}$.

The sample mean is:

$$\bar{x} = \frac{1035}{15} = 69$$

Therefore, 69 heart beats per minute is an estimate for the population mean pulse rate.

To compute an estimate for the population standard deviation, use the computational formula for the sample standard deviation, s, which is outlined in the following 7 steps.

Step 1. List all the data values of the sample and their squares. These results are listed in Table 4.10.

CHAPTER 4 MEASURES OF VARIABILITY

Table 4.10

x	x²
48	2304
74	5476
76	5776
80	6400
85	7225
45	2025
70	4900
50	2500
78	6084
65	4225
73	5329
78	6084
74	5476
73	5329
66	4356

Step 2. Find the sum of the squared data values, Σx^2.

$$\Sigma x^2 = 73{,}489$$

Step 3. Find the sum of the data values, Σx.

$$\Sigma x = 1035$$

Step 4. Square the sum of the data values, $(\Sigma x)^2$.

$$(\Sigma x)^2 = (1035)^2 = 1{,}071{,}225$$

Step 5. Divide $(\Sigma x)^2$ by n.

$$\frac{(\Sigma x)^2}{n} = \frac{1{,}071{,}225}{15} = 71{,}415$$

Step 6. Subtract $\frac{(\Sigma x)^2}{n}$ from Σx^2.

$$\Sigma x^2 - \frac{(\Sigma x)^2}{n} = 73{,}489 - 71{,}415$$
$$= 2{,}074$$

Step 7. Divide $\Sigma x^2 - \frac{(\Sigma x)^2}{n}$ by n-1 and take the square root of this quotient.

$$\frac{\Sigma x^2 - \frac{(\Sigma x)^2}{n}}{n-1} = \frac{2074}{14} \approx 148.14$$

Therefore, $s = \sqrt{\dfrac{\Sigma x^2 - \dfrac{(\Sigma x)^2}{n}}{n-1}}$

$$s = \sqrt{148.14}$$

$$s \approx 12.17$$

Therefore, the sample standard deviation serves as an estimate for the population standard deviation pulse rate is 12.17 heart beats per minute.

USING A CALCULATOR TO EVALUATE THE POPULATION STANDARD DEVIATION AND THE SAMPLE STANDARD DEVIATION

Statistical calculators have built-in programs that enable users to evaluate the population standard deviation and the sample standard deviation. The procedure to obtain these results require the user to simply enter the data values using special keys, and then retrieve the results from the statistical memories through a combination of keystrokes.

The procedure to determine these quantities is dependent upon the calculator manufacturer and the model you own.

The general procedure is:

Step 1. Set the calculator in statistical mode.

Step 2. Enter the data values one at a time using the appropriate key. For Texas Instruments Calculators, this key is usually the Σ+ key, while for Casio and Sharp Calculators the key is usually the M+ or Data Key.
Depending on the calculator, as the data value is entered, the display will either show the number of data values entered or the last data value entered.

Step 3. Retrieve the statistical results from memory using the appropriate keystrokes outlined in the manual for the calculator.

Although, the previous procedure is generic and not dependent upon any specific calculator, Appendix B contains specific Calculator Instructions for some models of the Calculator manufacturers: Texas Instruments, Casio, Sharp and Radio Shack.

Example 4.10

The annual report of a natural gas company compares the average sale prices for its natural gas sold within the United States and International for the five year period from 1988 to 1992. Table 4.11 contains the price of the natural gas in dollars per thousand cubic feet.

Table 4.11

Average Sale Price of Natural Gas (dollars per thousand cubic feet)					
Year	1992	1991	1990	1989	1988
United States	$1.60	$1.57	$1.61	$1.62	$1.57
International	$1.77	$2.18	$1.82	$1.43	$1.48

a) Using the information within Table 4.11, compute the sample mean, $\bar{x}$, and sample standard deviation, s, for the United States and International natural gas prices.

b) Using the results of part (a), are the natural gas sale prices in the United States more or less consistent than the International sale prices over this five year period? Explain why this is true?

Solution:

a) To obtain the sample mean, and sample standard deviation, let's use the statistical calculator. Let's use the previously outlined procedure for the statistical calculator. The procedure to determine these quantities is dependent upon the calculator manufacturer and the model you own. The general procedure is:

Step 1. Set the calculator in statistical mode.

Step 2. Enter the data values one at a time using the appropriate key. For Texas Instruments Calculators, this key is usually the Σ+ key, while for Casio and Sharp Calculators the key is usually the M+ or Data Key.

On a Texas Instrument Calculator, you would do the following to enter the data for the United States gas prices:

key in 1.60 then press Σ+
key in 1.57 then press Σ+
key in 1.61 then press Σ+
key in 1.62 then press Σ+
key in 1.57 then press Σ+

On either a Casio or Sharp, you would do the following to enter the data for the United States gas prices:

key in 1.60 then press M+ or Data Key
key in 1.57 then press M+ or Data Key
key in 1.61 then press M+ or Data Key
key in 1.62 then press M+ or Data Key
key in 1.57 then press M+ or Data Key

Depending on the calculator, as the data value is entered, the display will either show the number of data values entered or the last data value entered.

Step 3. Retrieve the statistical results from memory using the appropriate keystrokes outlined in the manual for the calculator.

The results are shown in Table 4.12.

Table 4.12

	United States	International
Sample mean, $\bar{x}$	$1.59	$1.74
Sample standard deviation, s	$0.02	$0.30

a) Not only is the mean gas price in the United States smaller than the International gas price, but the variability of gas price, as measured by the standard deviation, for the United States is only 2 cents. This is much smaller than the standard deviation of the International gas price of 30 cents. Consequently, the United States gas price has been more consistent from 1988 to 1992. The dramatic increase in International gas prices in the years 1990 and 1991 resulted in a greater mean and standard deviation for the International gas price, while the United States prices for gas remained relatively stable. ∎

4.5 APPLICATIONS OF THE STANDARD DEVIATION

In this section, we will look at some rules that illustrate the relationship between the value of the standard deviation and the percentage of data values within a distribution that fall within a certain distance from the mean.

Two measures that help to describe a distribution of data values are the mean and the standard deviation. Remember that the mean is a measure that is used to describe the center of the distribution, while the standard deviation measures the spread of the data values within the distribution. But how can one use the mean and standard deviation to interpret the spread of data values within a distribution? There are two general rules that establish a relationship between the standard deviation and the distribution of data values. These rules are *Chebyshev's Theorem* and the *Empirical Rule*.

CHEBYSHEV'S THEOREM

For a distribution of data values, regardless of shape, Chebyshev's Theorem provides the percentage of data values that **may** lie at various standard deviations from the mean.

> **Chebyshev's Theorem states for any distribution, regardless of its shape,:**
>
> **at least** $1 - \dfrac{1}{k^2}$ **of the data values will lie within k standard deviations of the mean, where k is greater than one.**

For example, the percent of data values in a distribution which lie within **two standard deviations of the mean** (k=2) must be at least:

$$1 - 1/k^2 = 1 - 1/2^2 = 1 - 1/4 = 3/4 \text{ or } 75\%$$

This is interpreted to mean that at least 75% of the data values within a distribution will lie within plus and minus 2 standard deviations from the mean.

The percent of data values in a distribution which lie within **three standard deviations of the mean** (k=3) must be at least:

$$1 - 1/k^2 = 1 - 1/3^2 = 1 - 1/9 = 8/9 \text{ or } 89\%.$$

This is interpreted to mean that at least 89% of the data values within a distribution will lie within plus and minus 3 standard deviations from the mean.

Keep in mind that Chebyshev's Theorem only indicates the **minimum** percentage of data values that will lie within a specific number of standard deviations of the mean. In many distributions, this percentage will be greater than the minimum guaranteed by Chebyshev's Theorem.

In the Figure 4.8, the results of **Chebyshev's Theorem** are shown for the shapes of the following distributions:
a) symmetric bell-shaped shown in Figure 4.8(a).
b) skewed to the left shown in Figure 4.8(b).
c) skewed to the right shown in Figure 4.8(c).

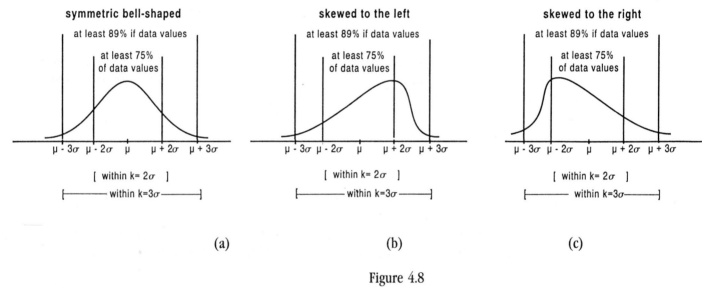

Figure 4.8

Example 4.11

A statistics instructor has recorded the amount of time students need to complete the final examination. The times, stated to the nearest minute, are:

50, 70, 62, 55, 38, 42, 49, 75, 80, 79, 48, 45, 53, 58, 64, 77,
48, 49, 50, 61, 72, 74, 10, 95, 120, 79, 75, 48, 72, 37, 71, 35,
79, 72, 75, 77, 32, 30, 77, 75, 79, 45, 70, 39, 75, 72, 45, 47, 73, 72

Using this data, determine:
a) μ and σ to the nearest minute.
b) the percent of data values that are within 2 standard deviations of the mean.
c) the percent of data values that are within 3 standard deviations of the mean.
d) compare the results of parts (b) and (c) to the results of Chebyshev's Theorem.
e) Is there any data value within the distribution that falls outside 3 standard deviations from

the mean? If there is such a data value, then can this data value be considered an outlier? Explain.

Solution:

a) The mean and standard deviation, to the nearest minute, are
$\mu = 61$ minutes and $\sigma = 19$ minutes.

b) The time which represents the value that is 2 standard deviations below the mean is:
$\mu - 2\sigma = 61 - 2(19) = 23$ minutes,
while the time which represents the value that is 2 standard deviations above the mean is:
$\mu + 2\sigma = 61 + 2(19) = 99$ minutes.

The number of times that are from 23 to 99 minutes, that is within 2 standard deviations of the mean, is 48. Therefore, the percent of times that are within 2 standard deviations of the mean is:
$$\frac{48}{50}(100\%) = 96\%$$

c) The time which represents the value that is 3 standard deviations below the mean is:
$\mu - 3\sigma = 61 - 3(19) = 4$ minutes,
while the time which represents the value that is 3 standard deviations above the mean is:
$\mu + 3\sigma = 61 + 3(19) = 118$ minutes.

The number of times that are from 4 to 118 minutes, that is within 3 standard deviations of the mean, is 49. Therefore, the percent of times that are within 3 standard deviations of the mean is:
$$\frac{49}{50}(100\%) = 98\%$$

d) Chebyshev's Theorem asserts that the percent of data values that are within 2 standard deviations from the mean is **at least 75% or a minimum of 75%**. From part (b), we determined that the percent of actual times that are within 2 standard deviations of the mean is 96%. This is consistent with Chebyshev's Theorem since 96% is **more than the minimum of 75%.**

According to Chebyshev's Theorem, the percent of data values that are within 3 standard deviations from the mean is **at least 89% or a minimum of 89%**. From part (c), we determined that the percent of actual times that are within 3 standard deviations of the mean is 98%. This is consistent with Chebyshev's Theorem since 98% is **more than the minimum of 89%.**

e) The data value 120 minutes is outside the range of values representing 3 standard deviations from the mean, 4 to 118 minutes. The data value 120 minutes can be considered to be an outlier since it is considered to lie far from most of the other data values within the distribution.

Example 4.12

A machine is set to produce an auto part with a mean size of 2.50 inches with a standard deviation of 0.07 inches. If any auto parts smaller than 2.29 inches or greater than 2.71 inches are defective, then according to Chebyshev's Theorem at most what percentage of the auto parts would one expect to be defective?

Solution:

To determine the number of standard deviations that 2.29 and 2.71 are from the mean, subtract the mean from each of the data values, 2.29 and 2.71 and divide each result by the standard deviation. That is,

$$\frac{2.29 - 2.50}{0.07} = -3 \quad \text{and} \quad \frac{2.71 - 2.50}{0.07} = 3$$

Thus the values 2.29 inches and 2.71 inches represent data values that are 3 standard deviations from the mean.

According to Chebyshev's Theorem, one would expect the percentage of auto parts falling within three standard deviations to be at least 89%. Therefore, the percentage of parts that one would expect to be defective, or fall outside the interval from 2.29 to 2.71 inches, is:

100% − at least 89% = at most 11%. ∎

Another rule that measures with more precision the percentage of data values that fall within a certain distance from the mean for a symmetric bell-shaped distribution is described by the Empirical Rule.

EMPIRICAL RULE

Although Chebyshev's Theorem applies to any shaped type of distribution, the Empirical Rule provides an explanation of how the standard deviation measures the spread of the data values from the mean only for a distribution of data values having an approximately bell-shaped curve.

Empirical Rule states for a bell-shaped distribution that:

1. **approximately 68% of the data values will lie within one standard deviation of the mean.**

2. **approximately 95% of the data values will lie within two standard deviations of the mean.**

3. **approximately all (99% to 100%) of the data values will lie within three standard deviations of the mean.**

Figure 4.9 illustrates the Empirical Rule for a symmetric bell-shaped distribution.

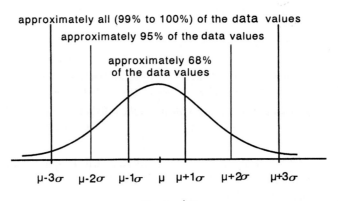

Figure 4.9

The Empirical Rule is a widely used rule in real life applications that provides a reasonable estimate for distributions that are approximately bell-shaped. However, you may find that the percentage of data values may not be exactly equal to the stated percentages for the intervals within the Empirical Rule. The Empirical Rule is especially useful when the mean and standard deviation of a bell-shaped distribution is known while all the data values of the distribution are not known. This type of application of the Empirical Rule is illustrated in Example 4.13.

Example 4.13

The results of a standardized achievement test are approximately bell-shaped with $\mu = 350$ and $\sigma = 85$.

According to the Empirical Rule, approximately what percentage of test scores would you expect to fall between:

a) 265 to 435?

b) 180 to 520?

c) 95 to 605?

d) If there was a test score equal to 625, could you consider such a score to be an outlier? Explain.

Solution:

a) Since the values 265 and 435 are one standard deviation from the mean, that is, $\mu - 1\sigma = 350 - 85 = 265$ and $\mu + 1\sigma = 350 + 85 = 435$, then according to the Empirical Rule the percentage of test scores within 265 to 435 is approximately 68%.

b) Within the interval 180 to 520 which represents test scores that are two standard deviations from the mean, that is, $\mu - 2\sigma = 180$ and $\mu + 2\sigma = 520$, then approximately 95% of the test scoresshould lie within this interval.

c) Almost all the test scores, or 99% to 100%, are expected to lie in the interval 95 to 605 because these test scores are three standard deviations from the mean. That is, $\mu - 3\sigma = 95$ and $\mu + 3\sigma = 605$.

d) The test score 625 is considered an outlier, since an outlier is a test score that lies far from the other test scores. According to the Empirical Rule, the test scores that may lie more that three standard deviations from the mean occurs less than 1%. This makes the test score 625 a very rare score, therefore, it should be considered an outlier. ■

CASE STUDY 4.4

The newspaper clipping in Figure 4.10 presents the *Home Prices By Region* for typical homes within the United States.

Figure 4.10

a) Which measure of central tendency is being used to represent the typical home price? What advantage is there to use this measure of central tendency over the mean?
b) Compare and interpret the median price for a home in Nashville, Tenn. to a home in Orange Co., Calif..
c) Determine the μ and σ of the median home prices.
d) Find the percentage of rates that fall within two standard deviations of the mean, three standard deviations of the mean, and four standard deviations of the mean.
e) Calculate the minimum percent of data values that are expected to fall within two, three and four standard deviations from the mean according to Chebyshev's Theorem.

f) Compare the results of parts (d) and (e). Do the results differ or are they equal? If the results differ, explain why they are different.

g) Is there any home price that falls outside the range of four standard deviations from the mean? If such a value exists, identify this value, and might you consider this value to be an outlier? Explain. ■

CASE STUDY 4.5

The time-series vertical bar graphs in Figure 4.11 display the Securities and Exchange Commissions (SEC) staff (Figure 4.11a) and budget (Figure 4.11b) for the five selected years from 1976 to 1992. Based on these bar graphs, which one has more variability over the selected years?

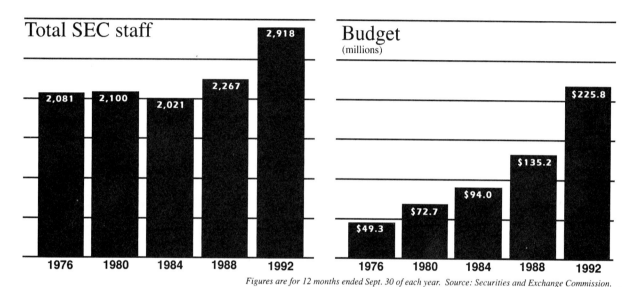

(a)   (b)

Figure 4.11

From the figure, you should observe that the budget bar graph indicates more variability than the SEC staff graph. Using the information in the bar graphs, the mean and standard deviation for the SEC staff and budget are given in Table 4.13.

Table 4.13

SEC	STAFF	BUDGET
Sample Mean	2277.4	115.4
Sample Standard Deviation	369.55	69.34

Notice that in Table 4.13 the standard deviation for the SEC staff data is greater than the standard deviation for the SEC budget. Why is there a discrepancy between the variability indicated by the graphs verses the variability contained within the table?

Let's compare this information about the mean and standard deviation to the heights of the bars within the graphs.

For each graph, draw a horizontal line on the bar graph which represents the value of the mean. Within which bar graph does there appear to be a greater fluctuation among the heights of the bars with respect to the mean? Explain why the standard deviation reflects the variability among the heights of the bars. ■

In Case Study 4.5 we observed that when comparing the variability of the distributions where the size or magnitude of the data values are very different, the standard deviation **doesn't** take into consideration the magnitude of the data values. That is, the standard deviation is an absolute measure of variability. In such situations we need a **relative measure** of dispersion, that is, a measure that takes into consideration the magnitude and the variability of the data values. This objective can be accomplished by comparing the standard deviation relative to the mean of the distribution. Such a relative measure of variability is called the **coefficient of variation.**

Definition: Coefficient of Variation. The Coefficient of Variation, is the ratio of the standard deviation to the mean, expressed as a percent.

$$Coefficient\ of\ Variation = \frac{standard\ deviation}{mean}(100\%)$$

WHEN TO USE THE COEFFICIENT OF VARIATION

The coefficient of variation is a very effective measure to use when one needs to make a meaningful comparison of the dispersion of two distributions under the following two situations.

1. The data values of the two distributions are in different units.
2. The data values of the two distributions are in the same unit, but the distance between their means is large.

Let's reinterpret the information in Table 4.13 by calculating the coefficient of variation for the staff and budget data.

For SEC staff data, the coefficient of variation is

$$Coefficient\ of\ Variation = \frac{standard\ deviation}{mean}(100\%)$$

$$= \frac{369.55}{2277.4}(100\%)$$

$$\approx 16\%$$

For SEC budget data, the coefficient of variation is

$$= \frac{69.34}{115.4}(100\%)$$
$$\approx 60\%$$

We find that the SEC staff, which has the greater **absolute** variation, (i.e. standard deviation = 369.55), has the smaller **relative** variation as measured by the coefficient of variation. In this particular situation, the coefficient of variation was an appropriate measure to use because the data values were expressed in different units. The SEC staff data values represent the number of staff members, while the data values of the SEC budget is dollars.

Example 4.14

An investor has studied the performance of two stocks for one year. The following data summarizes their performance:
Stock A: mean = $98.38   standard deviation = $9.78
Stock B: mean = $23.50   standard deviation = $6.95

a) Which stock has the greater absolute variation?
b) Which stock has the greater relative variation?
c) If the investor decides to use the standard deviation as a measure of the **risk** of the investment, explain why the coefficient of variation is a more useful measure of risk in comparing these two stocks.

Solution:

a) Since the standard deviation is an absolute measure of dispersion, then the larger standard deviation of $9.78 has the greater absolute standard deviation.
b) Using the coefficient of variation as a relative measure of dispersion, we need to calculate the coefficient of variation for each stock. For stock A we have:

$$\text{Coefficient of Variation} = \frac{\text{standard deviation}}{\text{mean}}(100\%)$$
$$= \frac{9.78}{98.30}(100\%)$$
$$\approx 10\%$$

For stock B we have:

$$\text{Coefficient of Variation} = \frac{\text{standard deviation}}{\text{mean}}(100\%)$$
$$= \frac{6.95}{23.50}(100\%)$$
$$\approx 30\%$$

Stock B has the greater relative measure of variability, since stock B has the larger coefficient of variation.
c) Notice the standard deviation of the two stocks are relatively close in value. Yet the mean for stock A is about four times larger than the mean of stock B. Since the distance between the means of each stock is large, then the coefficient of variation is a very useful

measure to make a meaningful comparison of the dispersion between the two stocks. In particular, a change of one standard deviation from the mean price of either stock will have a more profound effect on the total investment in stock B than in stock A because the standard deviation of stock B represents 30% of mean as compared to 10% of the mean for stock A. Therefore stock B represents a riskier investment. ■

❄ 4.6 Using MINITAB

Exploring the Effects of a Constant on the Variability of a Distribution.

Since MINITAB can quickly compute the standard deviation of a distribution, let's use MINITAB to explore the effects of ADDING and SUBTRACTING a constant to each DATA value, as well as MULTIPLYING and DIVIDING each DATA value by a constant on the **variability** of the resulting distributions.

Consider the distribution of Reading Test Scores (RTS).

Reading Test Scores
45 35 48 52 35 65 68 90 75 53 47 44 68 35 24 78
55 68 64 47

The SET command of MINITAB is used to place the reading scores in column C1. Once the data values are entered the NAME command is then used to name column C1 as RTS (i.e. Reading Test Scores).

Using the LET command, a constant can be added, subtracted, multiplied or divided with respect to each reading test score.

Suppose we try this using the constant 16. To add 16 to each of the reading test scores in C1 and form a new distribution C2, the command
MTB ⟩   LET C2 = C1 + 16 is used. If we name C2 as 'RTS+16', and print C2 we have
MTB ⟩   LET C2 = C1 + 16
MTB ⟩   NAME C2 'RTS+16'
MTB ⟩   PRINT C2

RTS+16
61 51 64 68 51 81 84 106 91 69 63 60 84 51 40 94
71 84 80 63

A similar process can be performed on C1 to form columns C3, C4 and C5 to represent subtraction, multiplication and division respectively. The columns are named as RTS-16, RTS*16 and RTS/16.

MTB ⟩   LET C3 = C1 - 16
MTB ⟩   LET C4 = C1*16
MTB ⟩   LET C5 = C1/16
MTB ⟩   NAME C3 'RTS-16'
MTB ⟩   NAME C4 'RTS*16'
MTB ⟩   NAME C5 'RTS/16'
MTB ⟩   PRINT C1-C5

ROW	RTS	RTS+16	RTS-16	RTS*16	RTS/16
1	45	61	29	720	2.8125
2	35	51	19	560	2.1875
3	48	64	32	768	3.0000
4	52	68	36	832	3.2500
5	35	51	19	560	2.1875
6	65	81	49	1040	4.0625
7	68	84	52	1088	4.2500
8	90	106	74	1440	5.6250
9	75	91	59	1200	4.6875
10	53	69	37	848	3.3125
11	47	63	31	752	2.9375
12	44	60	28	704	2.7500
13	68	84	52	1088	4.2500
14	35	51	19	560	2.1875
15	24	40	8	384	1.5000
16	78	94	62	1248	4.8750
17	55	71	39	880	3.4375
18	68	84	52	1088	4.2500
19	64	80	48	1024	4.0000
20	47	63	31	752	2.9375

Notice that for row 10, the Reading Test Score (RTS) is 53, while:
the RTS+16 is 69 (i.e. 53 + 16 = 69),
the RTS-16 is 37 (i.e. 53 - 16 = 37),
the RTS*16 is 848 (i.e. 53*16 = 848)
and,
the RTS/16 is 3.3125 (i.e. 53/16 = 3.3125).

A Histogram for each of the distributions is given in Figures 4.12a to 4.12e.

```
MTB >   HIST C1
Histogram of RTS    N = 20
Midpoint   Count
   20       1    *
   30       0
   40       4    ****
   50       6    ******
   60       2    **
   70       4    ****
   80       2    **
   90       1    *
```

Figure 4.12a

```
MTB >  HIST C2
Histogram of RTS+16   N = 20
Midpoint  Count
   40      1    *
   50      3    ***
   60      5    *****
   70      3    ***
   80      5    *****
   90      2    **
  100      0
  110      1    *
```

Figure 4.12b

```
MTB >  HIST C3
Histogram of RTS-16   N = 20
Midpoint  Count
   10      1    *
   20      3    ***
   30      5    *****
   40      3    ***
   50      5    *****
   60      2    **
   70      1    *
```

Figure 4.12c

```
MTB >  HIST C4
Histogram of RTS*16   N = 20
Midpoint  Count
  400      1    *
  500      0
  600      3    ***
  700      2    **
  800      5    *****
  900      1    *
 1000      2    **
 1100      3    ***
 1200      2    **
 1300      0
 1400      1    *
```

Figure 4.12d

```
MTB >  HIST C5
Histogram of RTS/16   N = 20
Midpoint   Count
   1.5      1    *
   2.0      3    ***
   2.5      0
   3.0      5    *****
   3.5      3    ***
   4.0      2    **
   4.5      4    ****
   5.0      1    *
   5.5      1    *
```

Figure 4.12e

By examining the graphs ESTIMATE:

a) which distribution has the greatest variability.
b) which distribution has the least variability.
c) if any of the distributions have the same variability.

Using the STDEV command we compute the standard deviation of each of the distributions.

```
MTB >  STDEV C1
ST.DEV. = 16.879
MTB >  STDEV C2
ST.DEV. = 16.879
MTB >  STDEV C3
ST.DEV. = 16.879
MTB >  STDEV C4
ST.DEV. = 270.07
MTB >  STDEV C5
ST.DEV. = 1.0549
```

By examining the standard deviation values you should notice that the distribution formed by **multiplying** the data values of C1 by the constant 16 (i.e. RTS*16) produced the **greatest variability**, whereas, the distribution formed by **dividing** the data values of C1 by the constant 16 (i.e. RTS/16) produced the **least variability**. Furthermore, notice that the standard deviation of column RTS*16 is equal to 16 times the standard deviation of the original distribution RTS. Similarly, the standard deviation of column RTS/16 is equal to the standard deviation of the original distribution divided by the constant 16.

The effect of **adding** or **subtracting** the constant 16 from each of the reading test scores has **no effect on the variability**. By comparing these results with the graphs in Figures 4.12a to 4.12e you should notice the effects of the constants on the histograms.

Glossary

TERM	SECTION
Measures of variability	4.1
Range	4.2
Variance of the population, σ^2	4.3
Population standard deviation, σ	4.3
Sample standard deviation, s	4.4
Chebyshev's Theorem	4.5
Empirical Rule	4.5
Coefficient of Variation	4.5

Exercises

PART I Fill in the blanks.

1. Three measures of variability are: _____, _____, and _____.
2. Measures of variability describe the _____ of a distribution.
3. The range is calculated by subtracting the _____ value of the distribution from the _____ value of the distribution.
4. If a distribution has a range of 10, and the largest value in the distribution has a value of 12, then the smallest value in the distribution has a value of _____.
5. For a distribution, the range is 5. If we add a constant 6 to each data value, then the range of the new distribution is _____.
6. For a distribution, $\mu = 18$ and the range is 0, then the median is _____.
7. The variance of the population, σ^2, is the mean of the squared deviations from the _____.
8. The formula for variance of the population, σ^2, is _____.
9. If a distribution of five data values has a variance of 100, then the sum of the squared deviations from the mean, $\Sigma(x - \mu)^2$, is _____.
10. If the variance of a distribution is 25, and N = 8 then $\Sigma(x - \mu)^2 =$ _____.
11. The population standard deviation of a distribution is the square root of the _____.
12. The Greek symbol used to represent the population standard deviation is _____.
13. The definition formula for the population standard deviation is _____. The computational formula for the population standard deviation is _____.
14. In a distribution, if the variance is 144, then $\sigma =$ _____.
15. In a distribution, if all the data values have the same numerical value of 14, then:
    $\mu =$ _____
    Mode = _____
    Median = _____
    Range = _____
    Variance = _____
    Standard deviation = _____
16. The expression "the sum of the squared deviations from the population mean" is symbolized as _____.
17. For a distribution, if $x - \mu$ has a negative value, then x is positioned _____ the mean. If $x - \mu$ has a positive value, then x is positioned _____ the mean. If $x - \mu = 0$ then $x =$ _____.
18. For a distribution of 7 data values, if $\sigma = 0$ and one of the data values is 13, then all the other six data values are _____.
19. For a distribution, $\sigma = 3$. Calculate the standard deviation of a new distribution formed by:
    a) adding 10 to each data value. _____
    b) subtracting 3 from each data value. _____
    c) dividing 6 into each data value. _____
    d) multiplying each data value by 5. _____
    e) multiplying each data value by (-4). _____
20. For a distribution, if the variance is 1/4 then $\sigma =$ _____.
21. Given the following two distributions:
    Distribution x: 1, 11, 11, 11, 20
    Distribution y: 1, 6, 10, 15, 20
    Using your intuition, the standard deviation of distribution ___ will be larger than the standard deviation of distribution ___.
22. For a distribution where $\mu = 50$ and $\sigma = 5$, calculate:
    $\mu + \sigma$ _____
    $\mu + 2\sigma$ _____
    $\mu + 3\sigma$ _____
    $\mu - \sigma$ _____
    $\mu - 2\sigma$ _____
    $\mu - 3\sigma$ _____
23. A good estimate of a population standard deviation is denoted by the symbol _____.
24. The definition formula for the sample standard deviation is s = _____. The computational formula for the sample standard deviation is: s = _____.
25. Two general rules that establish a relationship between the standard deviation and the distribution of data values are _____ Theorem and the _____ Rule.
26. For a distribution of data values, regardless of shape, the percentage of data values that **may** lie at various standard deviations from the mean is given by _____.
27. The Empirical Rule provides an explanation of how the standard deviation measures the spread of the data values from the mean only for a distribution of data values having an approximately _____ curve.
28. Chebyshev's Theorem states for any distribution, regardless of its shape, at least $1 - \dfrac{1}{k^2}$ of the data values will lie within k _____ _____ of the mean, where k is greater than one.
29. Empirical Rule states for a bell-shaped distribution that:

a. approximately _____ % of the data values will lie within one standard deviation of the mean.
b. approximately _____ % of the data values will lie within two standard deviations of the mean.
c. approximately _____ or _____ % to _____ % of the data values will lie within three standard deviations of the mean.

30. According to Chebyshev's Theorem, at least 75% of the data values will lie within _____ standard deviations of the mean.

31. The coefficient of variation is a _____ measure of dispersion whereas the standard deviation is an _____ measure of dispersion.

32. The formula to compute the coefficient of variation is _____.

33. The coefficient of variation indicates the what percent of the mean is represented by the _____.

34. When comparing the variability of two distributions the coefficient is a meaningful measure of variability when the data values are measured in _____ units or when the _____ of the distributions are far apart.

PART II **Multiple choice questions.**

1. Which measure of variability is determined only by the extreme values of a distribution?
   a) range   b) variance   c) standard deviation

2. Which measure of variability is always expressed in squared units?
   a) range   b) variance   c) standard deviation

3. If the standard deviation of distribution I is σ_1, and every score in this distribution is multiplied by five to form a new distribution II with a standard deviation represented symbolically by σ_2, then _____.
   a) σ_1 is larger than σ_2.
   b) σ_1 equals σ_2.
   c) σ_1 is smaller than σ_2.

4. If the range of distribution x is denoted symbolically by R_x and every data value in this distribution is increased by adding seven to form a new distribution y with a range denoted symbolically as R_y, then _____.
   a) R_x is larger than R_y.
   b) R_x equals R_y.
   c) R_y is larger than R_x.

5. Which measure of variability is the mean of the squared deviations from the mean?
   a) range   b) variance   c) standard deviation

6. Which measure of variability is the square root of another measure of variability?
   a) range   b) variance   c) standard deviation

7. The standard deviation of a distribution is _____.
   a) never negative
   b) always a whole number
   c) always positive

8. The range of a distribution is _____.
   a) never negative
   b) always a whole number
   c) always positive

9. The symbol used to represent an estimate of the population standard deviation is _____
   a) R^2 b) s^2 c) s   d) σ

10. For the same set of data, which formula will yield a larger value:
    a) σ b) s

11. Which of the following can be applied to a skewed shaped distribution?
    a) Chebyshev's Theorem   b) Empirical Rule
    c) Schwartz' Rule

12. According to the Empirical Rule, the approximate percentage of data values that are within 2 standard deviations is:
    a) 68%   b) 95%   c) 99%   d) 100%

13. For a bell-shaped distribution, which of the following gives a conservative estimate for the percentage of data values that are within two standard deviations of the mean?
    a) Chebyshev's Theorem   b) Empirical Rule
    c) Schwartz' Rule

14. The coefficient of variation represents the percent of the standard deviation to the _____.
    a) median   b) mean   c) range   d) variance

15. Which of the following is a measure of absolute variability?
    a) standard deviation   b) mean
    c) coefficient of variation

PART III **Answer each statement True or False.**

1. The range, variance and standard deviation are all known as measures of variability.

2. When you add the same constant to each data value in a distribution the range does not change.

3. When you add the same constant to each data value in a distribution, the standard deviation is changed by the value of that constant.

4. The standard deviation is sensitive to the extreme data values.

5. If a distribution consists of all negative data values, the standard deviation is also negative.

6. If a distribution consists of all negative data values then the range is negative.

7. When computing the standard deviation it is important to first arrange the data values in numerical order.

8. In a distribution whose data values all have the same numerical value, the standard deviation is always zero.

9. If you multiply every data value in a distribution by the same constant, the standard deviation is sometimes affected.
10. Two distributions that have the same range are not necessarily the same.
11. Two distributions that have the same standard deviation are always the same.
12. A distribution that has a range equal to zero must also have a standard deviation equal to zero.
13. A distribution that has a range of one must also have a standard deviation of one.
14. The formula used to the sample standard deviation, s, is the same as the formula that is used to compute the population standard deviation, σ.
15. The estimate of the population standard deviation, s, is always equal to the population standard deviation, σ.
16. For the same set of data, if both σ and s are calculated, then the value of s would be slightly larger than the value of σ.
17. Chebyshev's Theorem can only be applied to a bell-shaped distribution.
18. Chebyshev's Theorem states that at least 89% of the data values within a distribution will lie within 3 standard deviations from the mean.
19. According to the Empirical Rule, a bell-shaped distribution will have 68% of the data values within two standard deviations of the mean.
20. The Empirical Rule can be applied to either a bell-shaped or a skewed distribution.
21. When comparing two distributions, the distribution with the larger standard deviation will have the greater value for the coefficient of variability.
22. If the coefficient of variability is 20% then this indicates that the standard deviation is 20% of the mean.

PART IV Problems.

1. For the distribution: 2, 3, 7, 8, 15 compute the:
   a) range
   b) variance, σ^2
   c) population standard deviation, σ
   d) estimate of the population standard deviation, s, assuming the data represents a sample.
2. For the distribution: 3, 5, 8, 13, 21 compute the:
   a) range
   b) variance, σ^2
   c) population standard deviation, σ
   d) estimate of the population standard deviation, s, assuming the data represents a sample.
3. A travel agency, trying to promote travel to two ski resorts, base its ad campaign on the following temperature data: (data measured in degrees Fahrenheit)

   Resort 1   34   37   15   38   42   -16
   Resort 2   25   32   16   20   27   30

   The ad emphasizes the mean temperature. Observe that the two temperature distributions are not identical. Use the population standard deviation to determine which resort has the more consistent weather.

4. Three candidates applied for a legal secretarial opening in the office of I. M. Guilty. A four hour standardized test was administered to each candidate to determine their typing skills, writing skills, word processing skills, stenographic skills and knowledge of the law.
   The scores on each part are as follows:

Candidate	typing	writing	steno	word proc.	law
Ms. Libel	7.9	7.9	7.8	7.7	88
Ms. Sue	8.6	8.1	7.9	7.5	80
Mr. Will	7.8	8.4	8.0	7.4	85

   The candidates are told that the sole selection criterion will be based on the highest total of the five skill scores. If this criterion is unable to separate the candidates, then the candidate who has the highest total with the most consistent scores will be hired. Based on this criteria, which candidate should be hired for the job? (Hint: The measures of variability should be used to measure consistency among the scores.)

5. Consider the following distribution:
   2, 9, 16, 23, 30, 37, 44
   Calculate the range and population standard deviation for each of the following:
   a) the original distribution.
   b) the distribution formed by subtracting 10 from each data value in the original distribution.
   c) the distribution formed by multiplying each data value in the original distribution by (-4).

6. Craig has recorded prices for a statistical calculator at two discount electronic stores during the last three months.
   Based on this data, Craig calculated the mean and standard deviation of the calculator prices for each store.

Discount store	mean price	standard deviation
store 1	$25.00	$6.25
store 2	$25.00	$2.50

   During the past 3 months, in which store did Craig have a better chance to purchase the calculator at a cheaper price?

7. The following data represents annual salaries of real estate agents from two different offices located in Palmetto Beach, Florida (salaries are expressed in thousands of dollars).

Real Estate Office	Agent					
	One	Two	Three	Four	Five	Six
Sarabell	400	42	100	300	75	37
20th Century	200	50	410	47	150	250

In which real estate office are the salaries more varied?

8. For the distribution: 14, 10, 8, 16, 20, 12, 18
   a) Find the μ and σ.
   b) Find the data value that is one standard deviation above the mean. (i.e. μ + σ).
   c) Find the data value that is 1σ below the mean, (i.e. μ − σ).
   d) Find the data values that are within one standard deviation of the mean, (i.e., from μ − σ to μ + σ inclusive).
   e) How many data values are at least two standard deviations above the mean, (i.e. at least μ + 2σ)?
   f) How many data values are between μ − σ and μ + 2σ ?

9. An award is going to be given to the player on a women's college basketball team who has shown the most consistent performance over the entire season. The coach feels that consistent play is a very important factor in team performance. To measure consistency, she decides to use the standard deviation for those players that have scored an mean average of 10 points or more per game. What follows are the scoring distributions for all the players on the team:

player	games									
	1	2	3	4	5	6	7	8	9	10
Joann	7	5	0	16	X	7	4	0	7	8
Amy	12	17	X	10	5	12	0	20	X	7
Maryann	16	18	19	X	15	20	20	14	17	14
Kathleen	9	12	7	14	8	5	4	16	18	7
Debbie	28	18	24	19	22	17	24	12	20	16
Ellen	15	3	9	8	9	13	7	18	9	19
Grace	11	8	20	25	17	16	8	5	10	10

X denotes the player did not play in that game.

Calculate the population mean and standard deviation of each player and then decide which player gets the award for the most consistent performance.

10. A union leader for a small group of workers in Benito's Restaurant of Little Italy wants to make them an offer they can't refuse. His goal is to reduce the variation of wages paid to members of his union. Their current wages and three proposals offered by management are listed as follows:
   Current Wages (in dollars per hour)
   28.50, 27.25, 29.00, 27.75, 28.00
   Proposals
   PLAN A: Each union member receives a raise of $2.80/hr.
   PLAN B: Each union member receives a 10% raise.
   PLAN C: Each union member receives a raise equal to $80.00 divided by their current hourly wage.
   a) Determine the salary for each union member for plans A, B, and C.
   b) Determine which proposal would best achieve the union leader's goal by calculating the population standard deviation and range for each wage plan.

11. Matthew, an employee of the utility company ILLCO, enrolls in the company's reinvestment plan to purchase 10 shares of ILLCO stock per month. The following dollar figures represent Matthew's purchase price per share for the ILLCO stock during the past year:
   (price per share in dollars)
   15.25, 16.00, 17.50, 16.50, 15.00, 15.75,
   15.50, 16.25, 16.75, 16.00, 15.25, 15.50
   a) Compute the mean price per share that Matthew paid for the stock.
   b) Compute the population standard deviation of the stock prices per share.

12. The points allowed by two Los Angeles pro football teams in games against their common opponents are given in the following table. Using the population standard deviation as a measure of consistency, determine which team has the *more* consistent defense? (Hint: Defense can be measured by the points allowed in a game)
   team x: 35 29 17 30 20
   team y: 27 31 29 25 19

13. Fifteen economists were asked to estimate the discount rate that will be in effect at the start of the new year. The following figures represent their estimates:
   0.065   0.080   0.060   0.090   0.105   0.080
   0.075   0.095   0.080   0.095   0.100   0.090
   0.085   0.075   0.070
   Compute (to three decimal places):
   a) the sample mean, $\bar{x}$
   b) the sample standard deviation, s.
   c) determine how many of the estimated discount rates fall within the range: $\bar{x} - s$ and $\bar{x} + s$.
   d) determine the percent of estimated rates that are within one standard deviation of the mean.

e) determine the percent of estimated rates that are within two standard deviations of the mean.

14. The closing prices of three common stocks traded on the New York Stock Exchange for two weeks during the past month are given in the following table:

	Stock X	Stock Y	Stock Z
First Week	66	70	74
	68	64	78
	67	70	69
	66	68	64
	66	69	60
Second Week	68	67	65
	68	66	70
	70	70	66
	71	72	70
	70	64	64

a) Compare the three stocks using the sample mean.
b) Compute the sample standard deviation for each of the three stocks.
c) Based upon this information which stock is most speculative (i.e. most varied) and which one is most conservative (least varied)?

15. A beverage company samples 10 soda bottles from two dispensing machines to check the output quality of the production process. The dispensing machines are set to fill the bottles with two liters of soda (67.6 oz). The actual amounts of soda which was dispensed into the bottles per machine are recorded in the following table.

output (in ounces) for machine one	output (in ounces) for machine two
67.8	67.9
67.5	67.4
67.7	67.8
67.6	67.3
67.9	67.2
67.5	67.9
67.4	67.3
67.7	67.8
67.4	67.6
67.5	67.8

a) Using these results, compare the output quality of the two dispensing machines by computing the sample mean amount of soda dispensed per bottle by each machine as your standard.
b) Compute the sample standard deviation, s, for each machine's output. Using this additional information, compare the output quality of the two dispensing machines.

16. A study was conducted by a consumer protection group to determine the effectiveness of two leading pain relievers, Afferin and Banacin. The pain relievers were administered to a sample of 30 subjects suffering from headache pain. Fifteen subjects were administered Afferin while 15 were administered Banacin. The amount of time it took the drug to become effective in relieving the pain was recorded and the results are listed in the following table:

Number of minutes for relief of headache pain

Afferin Group	Banacin Group
45	48
63	54
54	42
37	38
42	42
58	53
47	64
36	44
41	44
39	50
48	52
54	42
55	51
49	52
52	44

Compute the sample mean and sample standard deviation, s, for each group and determine which drug has the more consistent time for relieving headache pain.

17. During twenty randomly selected days, a measurement of the pollutant level of a toxic chemical is made in a stream running adjacent to a steel mill that has complied with the established Environmental Protection Agency's (EPA) water pollution standards. The results are listed in the following table:

pollutant level of the toxic chemical (in parts per million)

17.1	18.1
16.8	15.8
17.3	17.5
16.4	16.6
16.8	15.2
18.1	18.3
15.7	17.5
16.3	16.9
17.4	17.1
15.2	16.2

a) Using this information, compute $\bar{x}$, and s.

b) If the EPA's safety level standard is $\bar{x} + 2s$, determine the percentage of days sampled that exceeded this standard.

18. The following table shows the amount spent for research, development and marketing for the past year by eight drug and eight cosmetic firms. (note: expenditures are measured in millions of dollars.)

Expenditures for

Research and Development	Expenditures for Marketing	Type of Company
2.64	1.74	Drug
3.79	2.59	Drug
1.84	1.97	Drug
2.36	3.64	Drug
3.59	2.18	Drug
4.61	2.64	Drug
3.17	2.55	Drug
2.89	1.90	Drug
1.17	3.75	Cosmetic
2.38	4.69	Cosmetic
0.87	2.75	Cosmetic
1.24	3.15	Cosmetic
1.75	3.56	Cosmetic
0.75	2.94	Cosmetic
1.29	4.75	Cosmetic
1.67	3.87	Cosmetic

a) Compute the mean and s for each type of company and for each type of expenditure.
b) Based on the given data, what conclusions can you draw concerning these expenditures?

19. The Student Government Association (SGA) at a community college conducted a survey of faculty and students to determine their opinions about instituting a preferred parking area for faculty members only. The responses to the poll were ranked from one to ten, where 1 meant strongly against the preferred parking for faculty and 10 meant strongly in favor of the preferred parking for faculty.

A random sample of 15 student and faculty responses were:

1, 10, 1, 10, 1, 2, 9, 8, 1, 10, 2, 1, 9, 10, 10

a) Compute the sample mean and standard deviation of the responses to the opinion poll.

A second opinion poll conducted by the SGA asked for student and faculty to express their views about whether the college should conduct a seminar on Literacy.

The following responses reflect the opinions of 15 student and faculty members, where 1 meant strongly against the seminar and 10 meant strongly in favor of the seminar.

5, 6, 5, 5, 4, 5, 5, 6, 5, 5, 5, 6, 5, 5, 6

b) Compute the sample mean and standard deviation of the responses to the opinion poll.
c) Compare the results of parts a and b. Do the means of the opinion polls distinguish a difference between the two issues? Do the standard deviations reflect a difference between the two issues? Explain your answer.

20. Use Chebyshev's theorem to answer the following two questions:
a) A bus company's records indicate that its buses complete their routes 9.3 minutes late on the average with a standard deviation of 2.1 minutes. At least what percentage of its buses complete their routes anywhere between:
 (1) 5.1 minutes and 13.5 minutes late?
 (2) 3.0 minutes and 15.6 minutes late?
 (3) 0.9 minutes and 17.7 minutes late?
b) A machine is set to produce screws with a mean size of 0.42 mm and a standard deviation of 0.003 mm. If screws smaller than 0.411mm or greater than 0.429mm are unusable, at most what percentage of the screws would one expect to be unusable?

21. The nation's 15 largest bank holding companies have the following assets in billions of dollars:
$194.4 $98.9 $96.9 $78.4 $74.7
$73.8 $64.7 $54.7, $51.8 $44.7
$41.9 $41.7 $34.4 $33.4 $33.2

Using this data, calculate:
a) μ and σ.
b) the percent of assets within 2 standard deviations of the mean.
c) the percent of assets within 3 standard deviations of the mean.
d) compare the results of parts (b) and (c) to the results of Chebyshev's Theorem (see problem #20).

22. Assuming the distribution of IQ scores of 8,000 college-aged students is bell-shaped with a mean of 120 with a standard deviation of 8. Using the Empirical rule, determine the approximate percentage and number of students who are expected to have IQ scores from:
a) 104 to 136.
b) 112 to 128.
c) 112 to 136.
d) 104 to 144.

23. A radial tire manufacturer claims that their new radial Z-80 tire has an average life of 80,000 miles with a standard deviation of 2,500 miles. Assuming the tire life is approximately bell-shaped, use the Empirical rule to determine the percentage of tires that will last:
a) from 75,000 to 85,000 miles.
b) more than 77,500 miles.
c) less than 72,500 miles.
d) If you purchased one of these radial tires and it lasts 71,250 miles, what would you infer about the manufacturer's claim?

24. During last years Marathon on Long Island, the

average time to run the marathon was 225 minutes with a standard deviation of 25 minutes.
a) What can be said about the percentage of runners that took more than 275 minutes to finish the Marathon, if you have no information about the shape of this distribution?
b) What is the approximate percentage of runners that took more than 275 minutes to finish the Marathon, if the distribution has an approximate bell-shaped distribution?

25. A fair die is tossed 40 times and the outcome is recorded. The resulting outcomes are:
3, 3, 5, 1, 2, 2, 4, 5, 2, 3, 6, 6, 4, 1, 2, 3, 4, 3, 2, 1, 4, 4, 5, 5, 2, 6, 6, 4, 6, 2, 3, 5, 3, 1, 1, 6, 1, 2, 1, 5
Determine:
a) μ and σ.
b) the percent of data values that are within 2 standard deviations of the mean.
c) the percent of data values that are within 3 standard deviations of the mean.
d) construct a Stem-and-Leaf display for the outcomes and describe its shape.
e) compare the results of parts (b) and (c) to the results of Chebyshev's Theorem.

26. A **pair** of dice is tossed 40 times and the **sum** of the face values are displayed in the following stem-and-leaf.

```
Stem-and-leaf of sum of dice   N = 40
         Leaf Unit = 0.10
    1      2  | 0
    2      3  | 0
    6      4  | 0000
   12      5  | 000000
   17      6  | 00000
  (10)     7  | 0000000000
   13      8  | 000
   10      9  | 000
    7     10  | 000
    4     11  | 000
    1     12  | 0
```

Determine:
a) μ and σ.
b) the percent of data values that are within 2 standard deviations of the mean.
c) the percent of data values that are within 3 standard deviations of the mean.
d) compare the results of parts (b) and (c) to the results of Chebyshev's Theorem.

27. Ralphy, a trolley car operator in San Francisco, records the number of hours of sleep per night he has received for the past 20 evenings. A stem-and-leaf display representing the data values is depicted as follows:

```
Stem-and-leaf of hours of sleep   N = 20
         Leaf Unit = 0.10
    2      4  | 00
    2      5  |
    4      6  | 00
    9      7  | 00000
   (5)     8  | 00000
    6      9  | 0000
    2     10  | 0
    1     11  | 0
```

Determine:
a) μ and σ.
b) the percent of data values that are within 2 standard deviations of the mean.
c) the percent of data values that are within 3 standard deviations of the mean.
d) compare the results of parts (b) and (c) to the results of Chebyshev's Theorem.

28. The exam scores for a biology midterm have been represented in the histogram below:

Histogram of C5   N = 100

```
Midpoint  Count
   66       2   **
   68       2   **
   70       7   *******
   72       9   *********
   74      13   *************
   76      11   ***********
   78      14   **************
   80      11   ***********
   82      13   *************
   84       8   ********
   86       6   ******
   88       0
   90       2   **
   92       2   **
```

For this distribution the mean is 78 and the standard deviation is 5.5.
a) Describe the shape of the histogram.
b) Does Chebyshev's Theorem apply? If so, determine the minimum percent of the exam scores, according to Chebyshev's Theorem, that are within 2 standard deviations of the mean?
c) Calculate the actual percentage of exam scores within 2 standard deviations of the mean. Compare this actual result to the theoretical result stated by Chebyshev's Theorem.
d) Does the Empirical Rule apply? If so, how does the results for the Empirical Rule compare to part (c)?

29. Each day, employment officer A reviews 40 resumes with a standard deviation of 5. Employment officer B completes 150 resumes with a standard deviation of 15. Compare the absolute variability and relative variability of each employment officer to determine which officer shows less variability in output?

30. The president of Debbie's Daytime Secretarial school is trying to determine which computer processing training program to use. Two groups were trained for the same task each using a unique training program. Group 1 used training program A; group 2 used training program B. It took the first group an average of 25 hours with a standard deviation of 6 hours to train each student. Whereas, the second group took an average of 16 hours with a standard deviation of 5 hours to train each student. Which training program has less relative variability in its performance?

31. An investor is examining the past performance of three growth mutual funds which invest primarily in small-sized companies. During the past 5 years, the average return on investment and standard deviation of the investment return for each mutual fund are given in the following table.

Mutual Fund	Average Five Year Return	Standard Deviation
Fund A	27.6%	6.47%
Fund B	25.7%	5.91%
Fund C	31.8%	6.92%

a) Which fund had a greater absolute variability in return over this 5 year period?
b) If this investor considers risk to be associated with a greater relative dispersion, then which of these three mutual funds has pursued a riskier strategy?
c) If this investor decides to narrow the selection down to mutual fund B and C because they both belong to excellent mutual fund families, then which fund should the investor choose if he is interested in a less riskier investment strategy?

32. Ellen, a stock broker, is studying the performance of two stocks over a three month period. Stock D had a mean selling price of $13.38 with a variance of 5.62, while Stock E had a mean selling price of $98.13 with a variance of 61.62.
a) Which stock has a greater relative variation in the selling price per share?
b) Why should you consider using a measure of relative dispersion to compare these two stocks?

33. Linda, a sports guru, is interested in comparing the careers of the New York Yankee players Mickey Mantle and Whitey Ford to determine which player had a more consistent career. Mickey Mantle's career batting average was .298 with a standard deviation of 0.033, while Whitey Ford had a career ERA of 3.14 with a standard deviation of 0.46.
a) If you use the coefficient of variation to compare the consistency of both players' career, which player had the more consistent career?
b) Why is the coefficient of variation an appropriate measure to use in making this comparison?

34. Scientists have been studying data dealing with the earth's weather to determine whether there is a **Global Warming Effect** going on. However, there are conflicting conclusions drawn from the data. Some scientists are proclaiming that there is a dangerous warming in progress due to a **greenhouse effect**. But, there are other scientists who believe that there is little evidence within the data to conclude that there is a greenhouse effect in progress.

Examine the following statements made by scientists regarding the earth's climate and *write your interpretation* of this global warming phenomena in response to the question:

"Do you believe that the 1980's is the start of a global warming trend or is this just the start of a natural variability of the weather pattern?"

- Every annual high (temperature) since 1950 falls below the trend of the natural increase.
- About 1950, emissions of greenhouse gases rose very rapidly. Yet, the warming rate, which should have escalated, promptly slowed.
- A net advance of about one third of 1 degree is what made the 1980's, the hottest decade "on record." Some researchers believe that this amount is within the **standard deviation**, or natural variability, of the weather.
- Through the 1980's one warm year followed another. Some scientists attributed this increase in temperatures to a greenhouse effect. They speculate that an increase of carbon dioxide in the atmosphere is the culprit and the future can only get hotter.
- The greenhouse effect has been concluded based upon the fact that there has been a slight increase in temperature from past data records. However, scientists state that the data records are fuzzy by a few shades of a degree, plus or minus. And, this increase in temperature is at best only slightly larger than the margin of error for the data.

PART V What Do You Think?

1. **What Do You Think?**

   Examine the vertical bar graph in Figure A. The graph represents the number of passengers expected to travel by air each day of the Christmas holidays, in millions.

   a) For which of the following three day periods: Dec. 19 - 21st, Dec. 23 - 25th, or Dec. 28 - 30th, do you believe the variability of the number of passengers is the largest? the smallest variability? Explain.
   b) From the bar graph, estimate the variability of all the data values using the range. What is a disadvantage in using the range as a measure of variability?
   c) For which consecutive three day period, is the variability the greatest? the smallest? Explain.

2. **What Do You Think?**

   Examine the weather map of the United States in Figure B on page 194. The bar graphs for selected cities indicates the normal monthly total precipitation, in inches.

   Locate the following cities: Buffalo, Duluth, Miami, and Seattle.

   a) For each of these cities, describe the shape of the distribution of monthly total precipitation.
   b) For which of these four cities, is the variability of the precipitation the smallest? Explain.
   c) For which of these cities, do you believe the variability of the precipitation is the greatest? Explain.

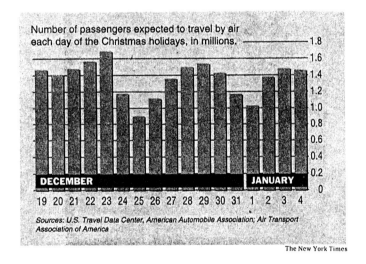

FIGURE A

194 • CHAPTER 4 MEASURES OF VARIABILITY

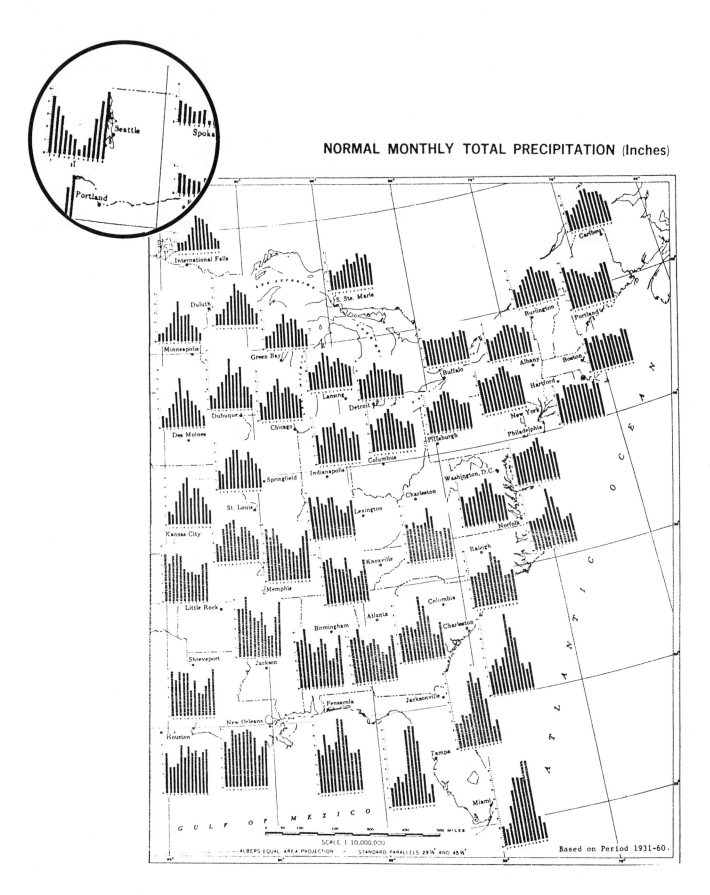

FIGURE B

3. **What Do You Think?**

The following information represents the listing of stock quotations for six stocks. Examine the information contained within this listing. Notice that the listing includes the high and low price for each stock during the past 52 weeks, as well as the high, low and closing price for each stock for the previous trade day. The last column represents the net change in the stock price from the previous day's quotation. The stock's name and symbol are also included within the listing.

52 weeks							net
Hi	Lo	Stock	Sym	Hi	Lo	Close	chg
13 1/2	4	Aileen	AEE	5 1/2	5 1/8	5 1/8	-1/2
43 3/4	25 1/8	AutoZone	AZO	40 7/8	39 5/8	40 5/8	-1/2
15 1/2	12 5/8	WestcstEngy	WE	15 1/4	15	15 1/8	...
28 1/4	11 1/8	Marvel Entn	MRV	28 1/8	24 7/8	27 5/8	+2 5/8
15	9 5/8	Homestake	HM	16 1/4	14 7/8	16	+1 5/8
9 5/8	3 3/4	Ft Nt Film	FNAT	8 9/16	8 7/16	8 7/16	-1/16

a) Which measure of variability is being used to reflect the variability in the stock prices for the year? for the day?

b) Over the year, which stock has been less volatile? more volatile?

c) From the previous trading day, which stock has been less volatile? more volatile?

d) Which stock had the greatest percentage gain in its price from the previous day's trading price? the smallest percentage gain?

e) Which stock had the greatest percentage loss in its price from the previous day's trading price? the smallest percentage loss?

f) The following graph represents the daily prices for the stock First National Film Corp (Symbol: FNAT) for the period from January, 1993 to April, 1993.

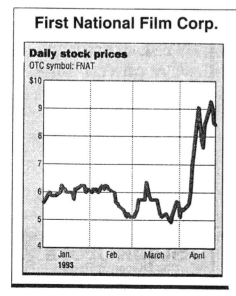

f) cont'd (i) Within which month, was the variability in First National Film Corp. stock's price the smallest? the greatest?

(ii) What information is contained within the graph for First National Film Corp. that is not represented by the stock quotation previously listed above? Explain how this shows a disadvantage of the range as a measure of variability as compared to the graphic information for First National Film Corp.

4. **What Do You Think?**

The following distribution list represents the ages of the 400 USA's wealthiest people, according to Forbes Magazine, for the year 1992.

<u>The Ages of the 400 Wealthiest</u>

71, 46, 81, 43, 73, 44, 48, 70, 43, 62, 43, 69, 65, 47, 80, 67, 67, 58, 60, 79, 75, 64, 73, 69, 72, 49, 61, 40, 63, 69, 64, 61, 48, 64, 59, 75, 62, 75, 68, 49, 70, 68, 52, 65, 47, 64, 71, 52, 63, 69, 40, 35, 38, 77, 60, 79, 79, 70, 54, 46, 52, 93, 67, 91, 72, 45, 64, 66, 55, 50, 85, 50, 76, 75, 57, 83, 82, 66, 90, 69, 68, 78, 75, 43, 69, 50, 52, 68, 71, 76, 86, 56, 75, 58, 46, 79, 72, 77, 91, 81, 86, 80, 62, 74, 49, 56, 68, 50, 54, 64, 85, 73, 52, 67, 86, 65, 82, 73, 53, 92, 74, 56, 51, 87, 75, 72, 54, 63, 50, 85, 69, 60, 74, 69, 86, 83, 70, 66, 67, 67, 70, 81, 51, 55, 64, 55, 86, 54, 61, 51, 66, 64, 64, 61, 66, 75, 56, 64, 76, 92, 75, 77, 72, 60, 69, 62, 80, 48, 67, 54, 55, 66, 68, 69, 67, 55, 63, 69, 86, 67, 56, 84, 68, 78, 77, 64, 63, 76, 59, 58, 69, 70, 72, 70, 63, 58, 62, 83, 86, 67, 69, 34, 38, 55, 50, 52, 55, 79, 74, 74, 54, 47, 64, 63, 64, 49, 61, 63, 69, 66, 74, 52, 53, 77, 79, 83, 84, 59, 48, 90, 82, 76, 84, 54, 62, 84, 53, 83, 56, 78, 67, 82, 52, 58, 67, 48, 79, 84, 77, 66, 76, 83, 82, 81, 49, 71, 89, 84, 51, 50, 44, 88, 76, 64, 78, 57, 31, 60, 47, 49, 53, 36, 82, 60, 77, 51, 60, 55, 55, 86, 79, 70, 59, 72, 80, 78, 71, 78, 57, 51, 67, 73, 50, 54, 50, 80, 73, 84, 42, 60, 64, 64, 59, 77, 74, 54, 64, 73, 61, 69, 54, 53, 50, 52, 76, 58, 76, 74, 63, 60, 59, 66, 54, 52, 49, 58, 62, 36, 49, 49, 44, 47, 76, 61, 66, 63, 87, 73, 57, 65, 64, 64, 75, 69, 48, 54, 87, 74, 66, 53, 48, 56, 67, 67, 58, 27, 72, 67, 48, 50, 44, 68, 71, 55, 69, 97, 78, 61, 57, 85, 79, 65, 88, 79, 70, 59, 35, 36, 36, 39, 36, 60, 43, 46, 41, 39, 56, 64, 59, 72, 48, 59, 76, 61, 60, 45, 68, 57, 57

a) Construct a histogram for these ages using the classes: 20-29, 30-39, 40-49, 50-59, 60-69, 70-79, 80-89, and 90-99.

b) Describe the shape of the histogram.

c) Using Chebyshev's Theorem, determine the minimum percentage of ages that fall within two standard deviations of the mean? within three standard deviations of the mean?

d) Using the Empirical rule, determine the approximate percentage of ages that fall within one standard deviation of the mean? within two standard deviations of the mean? within three standard deviations of the mean?

e) Calculate the actual percentage of ages that fall within one, two or three standard deviations of the mean.

f) Compare the actual results of part (e) to the results found by using Chebyshev's Theorem and the Empirical rule. Which rule provides a better approximation? Which rule gives a more conservative estimate?

g) For this type of shaped distribution, which rule would be more appropriate to use to approximate the percentage of ages that fall within k standard deviations of the mean? Explain.

PART VI Exploring DATA With MINITAB.
1. The test results for the MIDTERM EXAM SCORES for two statistics classes are given as follows:

CLASS 1	CLASS 2
33	9
34	33
24	28
32	15
19	28
21	13
20	16
17	25
21	25
9	23
23	19
27	17
25	16
28	28
25	31
17	17
24	15
26	34
24	22
34	19
22	24
25	22
32	23
34	20
26	14
20	18
22	22
22	20
19	19
33	20

a. Use MINITAB to construct a histogram for each class' MIDTERM EXAM SCORES.
b. Using the graphs, estimate which class' exam scores are more variable.
c. Use MINITAB to compute the standard deviation for each class' exam scores.
d. Which class' scores indicate the greater variability?
e. Is your estimate for part b and the calculation for part d consistent?

2. The Systolic blood pressures for a sample of twenty 45 year old females and twenty 45 year old males are represented as follows:

FEMALES	MALES
105	105
120	95
110	140
95	130
100	125
125	120
95	150
100	120
105	125
110	140
105	130
95	125
120	110
125	120
110	120
115	115
115	125
120	140
105	120
95	110

a. Use MINITAB to construct a histogram for each group of data values.
b. Using the graphs, estimate which group's blood pressures are more variable.
c. Use MINITAB to compute the standard deviation for each group of data values.
d. Which group's blood pressures indicate the greater variability?
e. Is your estimate for part b and the calculation for part d consistent?

PART VII Projects.
1. Select 25 male students and 25 female students and administer the following arithmetic test. Allow 10 minutes for the examination.
The correct answers are (1) b (2) c (3) d (4) d (5) a. Grade each of the tests and determine which sex had more consistent grades (Calculate the range and standard deviation for the grades for each sex).

The directions: Answer all questions by choosing the letter of the answer of your choice.

1) $\dfrac{4}{3} \div \dfrac{1}{2} = ?$ a) $\dfrac{2}{3}$ b) $\dfrac{8}{3}$ c) $\dfrac{4}{5}$ d) $\dfrac{3}{8}$

2) $18 + 24 + ? = 60$ a) 28 b) 8 c) 18 d) 38

3) $(0.03\%) = \dfrac{3}{?}$ a) 10 b) 100 c) 1000 d) 10,000

4) $\dfrac{2}{5} + \dfrac{3}{4} = ?$ a) $\dfrac{9}{5}$ b) $\dfrac{5}{9}$ c) $\dfrac{6}{20}$ d) $\dfrac{23}{20}$

5) $7 - 2/5 + 0.12 = ?$ a) 6.72 b) 5.22 c) 6.52 d) 6.12

2. Select a well known brand name food product for which there is a comparable store brand. For example, a brand name peanut butter and a store brand peanut butter (same size jar).

   Get the prices for the brand name product and the store brand product from as many different food stores as you can (at least five). Form two distributions, one from the name brand prices and the other from the store brand prices. Use the standard deviation to determine which distribution is more variable.

PART VIII **Database.**

The following exercises refer to the file DATABASE listed in Appendix A. We have indicated the appropriate MINITAB commands that are necessary to answer each exercise.

1. Using the MINITAB commands:
   RETRIEVE, MAXIMUM, MINIMUM, LET, STDEV and PRINT
   Retrieve the file DATABASE.MTW and determine the range and sample standard deviation for the variables:
   a) AGE
   b) AVE
   c) STY
   d) GPA
   e) GFT

2. Using the MINITAB commands:
   RETRIEVE, and DESCRIBE
   Retrieve the file DATABASE.MTW and use the DESCRIBE command to determine the mean and standard deviation of the variables:
   a) AVE
   b) STY
   c) GPA
   for (1) the male students (Males are coded 0)
       (2) the female students (Females are coded 1)

CHAPTER 5
z SCORES AND PERCENTILES

❄ 5.1 Introduction

John Gulliver, the famous fictitious British traveler, in his fantastic journeys to faraway lands encountered many strange situations. Two of these situations are pictured in Figures 5.1 and 5.2.

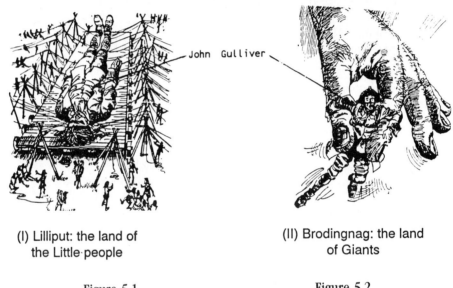

(I) Lilliput: the land of the Little people

(II) Brodingnag: the land of Giants

Figure 5.1   Figure 5.2

As can be seen from the pictures Gulliver's size is extreme relative to the inhabitants of the kingdoms he visits. In the land of Lilliput, where the inhabitants are only as large as Gulliver's thumb, he is considered a giant (see Figure 5.1). On the other hand, in the land of Brobdingnag, where Gulliver is only as large as an acorn, he is a dwarf among giants (see Figure 5.2).

Throughout his travels, Gulliver's height remains constant. However depending upon the height of the people in the strange lands he visits, his height is interpreted or ranked differently.

For example, we were able to rank Gulliver as a giant in Lilliput by observing that the Lilliputians were, in comparison to his size, extremely small. Similarly, by observing that the Brobdingnagians were all giants, we were able to rank Gulliver as a dwarf in Brobdingnag.

In statistics, one encounters numerous situations where it will be necessary to describe an individual data value within a distribution. In the previous two chapters we discussed the statistical measures of central tendency and variability. So far we have used these measures to describe characteristics of entire distributions. Now we will examine a technique which uses these measures to describe and rank an individual data value within a distribution and more importantly explore methods which will enable us to compare data values taken from different distributions.

5.2 z SCORES

In Example 5.1 we will consider a statistical measure that will rank an individual data value within a distribution by measuring how far the data value lies from the mean.

Example 5.1

Debbie earned a grade of 87 on her History exam, and a grade of 19 on her English exam. She would like to determine on which exam she did better, relative to the students in each class. How can this be done?

Solution:

Using all the class test scores, Debbie calculated the mean and standard deviation for each class and summarized the information in Table 5.1.

Table 5.1

Exam	Mean	Standard Deviation	Debbie's Grade
History	83	4	87
English	15	2	19

Using the mean, μ, as a typical grade, she calculated the deviation from the mean for each of her exams. A deviation from the mean is called a deviation score, and is calculated by the formula:

deviation score = data value - mean.

On the History exam, Debbie's deviation score is:

Test Grade - Mean = Deviation Score

87 - 83 = 4

On the English exam, Debbie's deviation score is:

Test Grade - Mean = Deviation Score

19 - 15 = 4

Debbie noticed that in both of her classes, her test grades were four units greater than the mean, thereby producing no meaningful comparison. To obtain a meaningful comparison Debbie decided to take into consideration the different variations in the test grades for each of her classes. Since the standard deviation reflects the typical deviation from the mean, Debbie decided to reevaluate her test grades using the standard deviation. Table 5.2 shows this information for Debbie's history test grade.

Table 5.2

History Exam		
test grade − mean = deviation score	standard deviation	number of standard deviations from the mean
87 − 83 = 4	4	1

We can interpret Debbie's history grade of 87 to be **one standard deviation** (1σ) greater than the mean of 83. That is,

mean + 1(standard deviation) = Debbie's History grade

$$83 + 1(4) = 87$$

Table 5.3 contains the information for Debbie's English grade.

Table 5.3

English Exam		
test grade − mean = deviation score	standard deviation	number of standard deviations from the mean
19 − 15 = 4	2	2

Debbie's English grade of 19 is interpreted to be **two standard deviations** (2σ) greater than the mean of 15. That is,

mean + 2(standard deviations) = Debbie's English grade

$$15 + 2(2) = 19$$

Debbie concluded that the English test grade of 19 showed a better achievement relative to her English class since it is two standard deviations greater than the mean English grade, whereas her History exam grade of 87 is only one standard deviation greater than the mean History grade. ∎

The statistical measure used in Example 5.1 to describe **how far each of Debbie's test grades were from the mean grade in *standard deviation units*** was the **z score**.

> **Definition: z score.** The z score of a data value indicates the number of standard deviations that the data value is away from the mean. The formula for the z score of a data value is:
>
> $$\text{z score of a data value} = \frac{\text{data value} - \text{mean}}{\text{standard deviation}}$$
>
> $$\text{z score of a data value} = \frac{\text{deviation score}}{\text{standard deviation}}$$
>
> In symbols, the z score formula is written:
>
> $$z = \frac{X - \mu}{\sigma}$$

Let's use the z score formula to determine the z score for each of Debbie's exam grades from Example 5.1, and interpret these results. The pertinent data of Example 5.1 is summarized in Table 5.4.

Table 5.4

Exam	Mean (μ)	Standard Deviation (σ)	Debbie's Grade (X)
History	83	4	87
English	15	2	19

The z score for the History test grade of 87 is computed using the formula:

$$z = \frac{X - \mu}{\sigma}$$

and the information contained in Table 5.4.
Thus, the z score for the data value 87 is:

$$z = \frac{87 - 83}{4} = +1 \ .$$

Thus, the test grade of 87 has a z score of one. This indicates that **87 is one standard deviation greater than the mean**.
The z score for the English test grade of 19 is:

$$z = \frac{19 - 15}{2} = +2$$

Thus, the test grade of 19 has a z score of +2. This indicates that **19 is two standard deviations greater than the mean**. In comparing the two z scores as illustrated in Figure 5.3, we see that Debbie's English grade has a greater z score.

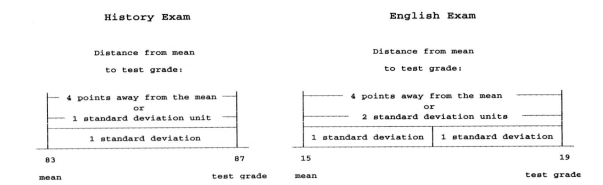

Figure 5.3

Therefore, we can interpret this result to mean that Debbie's English test grade has a **better ranking** in the English class than her test grade in the History class. Thus, Debbie can conclude that her English test grade is considered to be a better performance relative to each class than her History test grade.

Example 5.2

Calculate the z scores for each of the following four test grades that were selected from a Psychology class where the mean is 75 and the standard deviation is 10.

Psychology test grades: 65, 80, 90, 75

Solution:

To calculate the z score of 65, use $z = \dfrac{X - \mu}{\sigma}$.

Since $X = 65$, $\mu = 75$ and $\sigma = 10$, then z of 65 is:

$$z = \frac{65 - 75}{10} = \frac{-10}{10} = -1$$

To compute the z score of 80, we have:

$$z = \frac{80 - 75}{10} = +0.5$$

Using the z score formula, the test score $X = 90$ has a z score of 1.5. While the mean test score $X = 75$ has a z score of zero. ■

Notice, in Example 5.2 the test grade of 65 is less than the mean and has a negative z score. The test grade of 75 is equal to the mean and has a z score of 0. While, the test grade of 80 is greater than the mean and has a positive z score. Notice from the previous results, the sign of the z score indicates the position of the test score in relation to the mean. These results are summarized in the Table 5.5.

INTERPRETING THE SIGN OF A z SCORE

The sign of the z score of a data value will indicate the position of the data value in relation to the mean. This is indicated in Table 5.5.

Table 5.5

z Score	Position of Data Value
z score is positive	Data Value is greater than the mean
z score is zero	Data Value is equal to the mean
z score is negative	Data Value is less than the mean

Example 5.3

Matthew received grades of 85 in sociology, 82 in communication and 77 in philosophy. Using the data in Table 5.6 determine on which exam Matthew did best and on which exam Matthew did worst, relative to each class.

Table 5.6

Exam	Mean (μ)	Standard Deviation (σ)	Matthew's Grade (X)
Sociology	90	5	85
Communication	72	5	82
Philosophy	77	10	77

Solution:

Step 1. Calculate the z scores for each test grade using the formula:

$$z = \frac{X - \mu}{\sigma}$$

for sociology: $z = \frac{85 - 90}{5} = -1$

for communication: $z = +2$

for philosophy: $z = 0$

Step 2. Using the z score of each test grade, compare the test grades.

On the sociology exam, Matthew's grade of 85 has a z score of -1. Thus, his sociology grade is one standard deviation less than the mean grade.

On the communication exam, Matthew's grade of 82 has a z score of +2. Therefore, his communication grade is two standard deviations greater than the mean grade.

Matthew's philosophy grade is equal to the mean grade, since his z score is zero.

Step 3. Determine on which exam Matt did best relative to each class.

Since Matthew's communication grade has the highest z score, Matthew's performance on the communication exam was the best grade relative to his class.

Step 4. Determine on which exam Matt did worst relative to each class.

Since Matthew's sociology grade has the lowest z score, Matthew's performance on the sociology exam was the worst grade relative to his class.

Example 5.4

Last season, the American League's leading hitter batted .337 whereas the National League's leading hitter batted .362. Which hitter was more outstanding with respect to his league if the American League's mean batting average was .247 with a standard deviation of .030 and the National League's mean batting average was .262 with a standard deviation of .040?

Solution:

Step 1. Calculate the z score for each league's leading hitter.

For the American League's leading hitter batting .337 his z score is:

$$z = \frac{.337 - .247}{.030} = +3$$

For the National League's leading hitter:

$$z \text{ of } 0.362 = +2.5$$

Step 2. Compare the z scores and determine which hitter was more outstanding with respect to his league.

Since the American league's hitter had the higher z score, he was the more outstanding hitter relative to his league. ∎

5.3 CONVERTING z SCORES TO RAW SCORES

In this section, we will develop a formula to convert a z score back to a raw score or data value. In Examples 5.5 and 5.6, we will begin the development of this formula.

Example 5.5

A recent medical report claimed that if an individual's weight is more than one standard deviation from the mean weight for the individual's height, then the individual is either overweight or underweight. Find the two weights that determine overweight and underweight for a height of 5ft 8in. if the mean weight for this height is 155 lb and the standard deviation is 10 lb.

Solution:

Step 1. Overweight: To determine the overweight we need to find the weight which is one standard deviation greater than the mean. That is, overweight is any weight greater than $\mu + 1\sigma$.

Since $\mu = 155$ lbs and $\sigma = 10$ lbs then $\mu + 1\sigma$ would be $155 + 1(10)$ or 165 lbs.

Therefore, the overweight for a height of 5ft 8in. is any weight greater than 165 lbs.

Step 2. Underweight: To determine the underweight we need to find the weight which is one standard deviation less than the mean. That is, underweight is any weight less than $\mu - 1\sigma$.

Thus, μ − 1σ = 155 − 1(10)
= 145 lbs.

Therefore, the underweight for a height of 5ft 8in. is any weight less than 145 lbs.

Consequently, if a 5ft 8in person has a weight outside the range of 145 − 165 (lbs), the person is either underweight or overweight. ∎

Example 5.6

An Army regulation states:

"...no recruit's height can have a z score greater than 2 or less than −1.5."

If the mean height of recruits is 68 inches with a standard deviation of 6 inches, determine which of the recruits in Table 5.7 are ineligible for induction according to the above regulation.

Table 5.7

Recruit	Height (in inches)
Abe	72
Larry	81
Lou	79
Tony	57

Solution:

Step 1. Calculate the maximum and minimum heights allowable, using the mean height, μ = 68 inches, and the standard deviation, σ = 6 inches.

Since maximum height is two standard deviations above the mean (i.e. z = 2), then the maximum height is:

$$\mu + 2(\sigma) = 68 + 2(6) = 80.$$

Thus, the maximum height = 80 inches.

Since the minimum height is 1.5 standard deviations less than the mean (i.e. z = −1.5), then the minimum height is:

$$\mu - 1.5(\sigma) = 68 - 1.5(6) = 59.$$

Thus, the minimum height = 59 inches

Step 2. Determine which of the recruits are ineligible.

Since Tony's and Larry's heights fall beyond the allowable limits of 59 inches and 80 inches, they are ineligible for recruitment. ∎

In the previous two examples, we converted a z score to a raw score, X. This procedure can be expressed in the following formula.

> **Converting a z Score to a Raw Score**
>
> To convert a z score to a raw score or data value, we use the **raw score formula**:
> Raw Score = Mean + (z Score)(Standard Deviation)
>
> In symbols, the **raw score formula** is written as:
> $$X = \mu + z\sigma$$

Example 5.7

If $\mu = 50$ and $\sigma = 12$, find the raw score, X, for:
a) a z score of 2.
b) $z = 0.5$.
c) $z = 0$.
d) $z = -1$.

Solution:

a) Using $X = \mu + z\sigma$, where $\mu = 50$, $\sigma = 12$ and $z = 2$.

$X = 50 + (2)(12)$
$X = 74$

Therefore, if $z = 2$, then the raw score is 74.

b) For $z = 0.5$, $\mu = 50$ and $\sigma = 12$, we have:

$X = 50 + (0.5)(12)$
$X = 56$

Therefore, if $z = 0.5$, then the raw score is 56.

c) For $z = 0$, we have: $X = 50$. Therefore, the raw score that corresponds to a $z = 0$ is the mean score, $X = 50$.

d) For $z = -1$, we have $X = 38$. ■

5.4 Percentile Rank and Percentiles

A common statistical technique used to rank a data value relative to the other data values in a distribution is called the **Percentile Rank**. The percentile rank of a data value can be thought of as a measure that indicates the percent of values within the distribution that are less than the given data value.

For example, in a senior class, Julie's class standing is 125. How is this interpreted? Is 125 a high or a low class standing? In order to interpret this class standing more information is needed. Further investigation of Julie's records shows that her class standing has a percentile rank of 60. We interpret this to mean that 60 percent of the seniors are below Julie's class standing.

Let's define this percentile rank concept and examine the formula to calculate the percentile rank.

> **Definition:** **Percentile Rank.** The percentile rank of a data value, X, within a distribution, is equal to the percentage of data values less than the given data value, X, plus one-half the percentage of data values equal to the data value, X.
>
> **Formula to Compute The Percentile Rank of a Data Value**
>
> After arranging the data values in numerical order, the formula used to compute the percentile rank is given by:
>
> $$\text{PR of the data value } X = \frac{[B + (\frac{1}{2})E]}{N}(100)$$
>
> where: PR = percentile rank
> B = the number of data values less than X
> E = the number of data values equal to X
> N = the number of data values in the distribution
>
> **Percentile ranks are always expressed as whole numbers**

Percentile ranks are meaningful when applied to large distributions. Scores from college boards, I.Q. tests and family incomes in the U.S. represent examples of large populations where the data values are reported both as a score and as a percentile.

Example 5.8

On an exam, 25 college students receive scores of:
102, 85, 109, 83, 112, 105, 98, 115, 91, 117, 88, 95, 116,
105, 130, 80, 115, 84, 95, 121, 85, 89, 94, 111, 106.

Compute the percentile ranks of the following scores:
a) 102   b) 84   c) 115

Solution:

Step 1. First *arrange* the data in numerical order. We get:

80, 83, 84, 85, 85, 88, 89, 91, 94, 95, 95, 98, 102, 105,
105, 106, 109, 111, 112, 115, 115, 116, 117, 121, 130

Step 2. a) To find the percentile rank of the score 102, use the formula:

$$\text{PR of the data value } X = \frac{[B + (\frac{1}{2})E]}{N}(100)$$

To use the formula, we need to arrange the scores in numerical order, and determine the values of B, E, and N. After placing the scores in numerical order, we observe that there are twelve scores less than 102, therefore B = 12. Since there is only one 102, then E = 1. Finally, since there are 25 scores, N = 25.

Substituting the values of B, E and N into the formula for the percentile rank we get:

$$\text{PR of the score } 102 = \frac{[12 + (\frac{1}{2})1]}{25}(100)$$

$$\text{PR of the score } 102 = \frac{12.5}{25}(100)$$

PR of the score 102 = 50. This indicates that there are approximately 50% of the scores below the score of 120.

b) To use the percentile rank formula in computing the percentile rank of 84, determine the values of B, and E for 84. For the score 84, B = 2 and E = 1. Therefore,

$$\text{PR of the score } 84 = \frac{[2 + (½)1]}{25}(100)$$

$$= \frac{(2.5)}{25}(100)$$

PR of the score 84 = 10. This indicates that there are approximately 10% of the scores below the score of 84.

c) To use the formula to compute the percentile rank of 115, the values for B and E are: B = 19 and E = 2.
Thus,

$$\text{PR of the score } 115 = \frac{[19 + (½)2]}{25}(100)$$

$$= \frac{20}{25}(100)$$

PR of the score 115 = 80. This indicates that there are approximately 80% of the scores below the score of 115.

Another concept closely related to percentile rank is that of **percentile**. Let's examine this concept.

Definition: Percentile. The pth percentile is the data value within the distribution which has p percent of the data values less than this value and the rest of the data values greater than this data value.

In Example 5.8 part a, the score 102 has a percentile rank of 50. This indicates that 102 is the median because 50% of the scores are less than 102 and 50% of the scores are greater than 102. Using percentiles, we state that 102 is the 50th percentile. This is written as:

$$P_{50} = 102$$

In Example 5.8 part b, the score 84 has a percentile rank of 10. Therefore, we can state that 84 is the tenth percentile, written as:

$$P_{10} = 84.$$

In Example 5.8 part c, the score 115 has a percentile rank of 80. 115 is the 80th percentile, expressed:

$$P_{80} = 115.$$

Meaning of Percentile and Percentile Rank

There is a difference in the use of the words **percentile** and **percentile rank**. In general, the **percentile** is a **data value** of the distribution while a **percentile rank** is a **percentage** value ranging from 0 to 100.

Example 5.9

Identify the percentile and the percentile rank in each statement.
a) Exactly 62 percent of the students in a given school system received IQ scores less than 112.
b) In Nina's third grade class, 70% of the students have heights less than 54 inches.

Solution:

a) 112 is the 62nd percentile and 62 is the percentile rank of 112.
b) 54 is the 70th percentile of the student's heights in the third grade class and 70 is the percentile rank of the height of 54 inches.

DECILES AND QUARTILES

Some important percentiles have individual names. Two such names are *deciles* and *quartiles*. There are nine deciles, and three quartiles. We will first examine the relationship between percentiles and deciles.

Deciles

The nine deciles are:

- The **10th percentile** is called the **1st decile**
- The **20th percentile** is called the **2nd decile**
- The **30th percentile** is called the **3rd decile**
- The **40th percentile** is called the **4th decile**
- The **50th percentile** is called the **5th decile**
- The **60th percentile** is called the **6th decile**
- The **70th percentile** is called the **7th decile**
- The **80th percentile** is called the **8th decile**
- The **90th percentile** is called the **9th decile**

The relationship between the nine deciles and the corresponding percentiles are shown in Figure 5.4

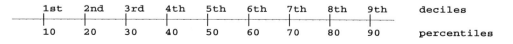

Figure 5.4

Let's now examine the relationship between the three quartiles and the corresponding percentiles.

> **Quartiles**
>
> There are **three quartiles.**
>
> - The **25th percentile** is called the **1st quartile, Q_1**.
> - The **50th percentile** is called the **2nd quartile, Q_2**.
> - The **75th percentile** is called the **3rd quartile, Q_3**.

The relationship between the three quartiles and the corresponding percentiles are shown in Figure 5.5.

Figure 5.5

CASE STUDY 5.1

Examine the statistical tables shown in Figure 5.6. The table in Figure 5.6a presents the percentile ranks of SAT-verbal and SAT-mathematical scores while the table in Figure 5.6b shows the standard error of measurement for the different educational tests given by the Admissions Testing Program(ATP) for the years 1992-93.

Statistical Tables

Percentile ranks of SAT scores
(for national college-bound seniors)

Score	Verbal	Mathematical
800	99+	99+
750	99+	99
700	99	95
650	97	90
600	93	82
550	86	71
500	75	57
450	60	43
400	42	29
350	26	17
300	13	6
250	5	1
200		
Average score	422	474

(a)

Standard errors of measurement
of ATP scores*

Test	SEM
SAT	
Verbal	31
Reading Comprehension	4.0
Vocabulary	4.3
Mathematical	36
Test of Standard Written English	3.7
Achievement Tests	
American History &	
Social Studies	28
Biology	27
Chemistry	29
English Composition	29
English Comp. with Essay	35**
European History &	
World Cultures	27
French	28
French with Listening†	
German	27
Modern Hebrew	30
Italian	28
Japanese with Listening†	
Latin	34
Literature	37
Mathematics Level I	33
Mathematics Level II	31
Mathematics Level IIC	31
Physics	29
Spanish	27

(b)

Figure 5.6

a) Interpret the percent of students that scored below a score of 600 on the SAT-verbal.
b) Interpret the percent of students that scored below a score of 650 on the SAT-mathematical.
c) What percent of students scored above a score of 400 on the SAT-verbal? above the score of 500 on the SAT-mathematical?
d) What SAT-verbal score represents the 6th decile? the 3rd quartile?
e) What SAT-mathematical score represents the 9th decile?
f) What is the average score for the SAT-verbal? Examining the table, what do you think is the percentile rank of this score?
g) What is the average score for the SAT-mathematical? Examining the table, what do you think is the percentile rank of this score?
h) Using the standard errors of measurement of the ATP scores as an estimate of the standard deviations of the different tests, what is the standard deviation of the SAT-verbal and the SAT-mathematical?
i) Using the results of parts f, g and h, determine the z score of the SAT-verbal scores: 500, 400, and 450? What is the percentile rank for each of these scores? Interpret these results.
j) Using the results of parts f, g and h, determine the z score of the SAT-mathematical scores: 500, 400, and 450? What is the percentile rank for each of these scores? Interpret these results.
k) On which SAT test would a score of 500 be considered to be a better test score? Explain your answer using the information in part (j).

Example 5.10

Some results from the mathematics reasoning part of the Scholastic Aptitude Test (SAT) are given in Table 5.8.

Table 5.8

Score	Percentile Rank
320	10
365	20
380	25
400	30
410	35
474	50
510	60
570	75
595	80
650	90
750	99

Using Table 5.8 determine:
a) the score which is the 1st decile.
b) the median.
c) the score which has a percentile rank of 90.
d) the percentile rank of the score 380, and state what quartile it represents.
e) the 3rd quartile.
f) the score which is the 6th decile.

Solution:

a) The first decile is another name for the 10th percentile. Thus, since 320 is the 10th percentile, it is also the 1st decile.
b) The median is the 50th percentile. Thus, the median score is 474.
c) A score that has a percentile rank of 90 is the 90th percentile. Thus, the 90th percentile is the score 650.
d) The score 380 is the 25th percentile and has a percentile rank of 25. Thus, 380 represents the 1st quartile, since it has a percentile rank of 25.
e) The 3rd quartile is another name for the 75th percentile. The 75th percentile is 570 and thus, 570 is the 3rd quartile.
f) The sixth decile is the 60th percentile. Since 510 is the 60th percentile and thus the sixth decile is 510. ■

❄ 5.5 BOX-AND-WHISKER PLOT: AN EXPLORATORY DATA ANALYSIS TECHNIQUE

The Box-and-Whisker Plot is an Exploratory Data Analysis technique developed by John Tukey to visually display several summary statistics in order to describe the location of the center of the data, the spread and shape of a distribution, and to identify any outliers that may exist within the data. It is also helpful in comparing **two** or **more** distributions.

> **Definition: Box-and-Whisker Plot.** A Box-and-Whisker Plot or Boxplot is a visual device that uses a 5-number summary consisting of: smallest data value, the first quartile (Q_1), the median, the third quartile (Q_3), and the largest data value to reveal the characteristics of a distribution.

It is a graphing technique that can be used when there is a small number of data values in a distribution and therefore the construction of either a stem-and-leaf display, a bar chart, or a histogram may not be appropriate.

An example of a Box-and-Whisker Plot is shown in Figure 5.7.

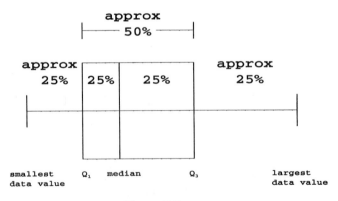

Figure 5.7

In Figure 5.7, the five values used to construct a Box-and-Whisker plot are called the *5-Number Summary* for a distribution and they are often reported as part of the descriptive statistics for a distribution. As illustrated in Figure 5.7, these five numbers are used in the construction of the rectangular Box and the Whiskers for the Boxplot. Specifically, the rectangular box is constructed by using Q_1, the 1st quartile, and Q_3, the 3rd quartile, with a vertical line drawn within the rectangle which represents the position of the median. The horizontal lines extending from the quartiles to the smallest and largest data values are called the whiskers of the Boxplot.

The Boxplot is particularly useful in conveying the location of the center of the distribution as well as the distribution's variability. Let's discuss some observations that can be made by examining a Box-and-Whisker Plot.

Box-and-Whisker Plot Interpretations

(1) The median represents a measure of the center and is denoted by a vertical line within a rectangular box.
(2) Approximately 50% of the data values are contained within the box.
(3) Within each component of the box, separated by the median are approximately 25% of the data values.
(4) Approximately the lower 25% of the data values is represented by the whisker connecting the first quartile to the smallest data value.
(5) Approximately the upper 25% of the data values is represented by the whisker connecting the third quartile to the largest data value.

To construct a Boxplot for a distribution the **5-number summary** consisting of the smallest data value, Q_1, the median, Q_3 and the largest data value must first be determined.

In Section 3.3 of Chapter 3, we developed a procedure to determine the median for a distribution. In essence the median is the middle value of a distribution **after** the data values have been arranged in numerical order.

In a similar way the first and third quartiles can be determined. That is, the first quartile, Q_1, is essentially the *median* of the data values below the **position** of the median, while the third quartile, Q_3, is the *median* of the data values above the **position** of the median.

Example 5.11

Determine the 5-number summary for the following distribution of annual salaries.

Annual Salaries (in Thousands of Dollars)
55 57 60 62 63 64 65 66 67 67 67 70 72 75 78 83 86 87 90

Solution:

The 5-number summary of a distribution consists of the smallest data value, Q_1: the first quartile, the median, Q_3: the third quartile, and the largest data value.
First, determine the median.

Since the data are arranged in numerical order the median is the $\frac{n+1}{2}$ th data value.

For this distribution of 19 data values, the median is the $\frac{19+1}{2}$ or 10th data value which is 67.
Second, determine the 1st quartile, Q_1.

The 1st quartile, Q_1, is the median of the data values which are below the position of the median. In Figure 5.8, the median's position is enclosed within a box, and the values below are shown in bold print.

                                        67 at the 10th data value
                                    represents the median's position

55 57 60 62 63 64 65 66 67 | 67 | 67 70 72 75 78 83 86 87 90

Figure 5.8

To determine Q_1, we need to find the median of the data values which are below the position of the median. The values below the median are:

55 57 60 62 63 64 65 66 67

The median of these data values is 63, since it is the middle value of these nine numbers. Thus, 63 represents the first quartile, that is $Q_1 = 63$.

Third, determine the 3rd quartile, Q_3.

The third quartile, Q_3, is the median of the data values which are above the position of the median. In Figure 5.9, the median's position is enclosed within a box. The values above the median's position are shown in bold print.

                                        67 at the 10th data value
                                    represents the median's position

55 57 60 62 63 64 65 66 67 | 67 | **67 70 72 75 78 83 86 87 90**

Figure 5.9

To determine Q_3, we need to find the median of the data values which are above the position of the median. The values above the median are:

67 70 72 75 78 83 86 87 90

The median of these data values is 78, since it is the middle value these numbers. Thus represents the third quartile, that is $Q_3 = 78$.

Fourth, find the smallest and largest data values.

The smallest data value is 55 while the largest data value is 90. Therefore, the 5-number summary for the distribution of annual salaries is:

    smallest data value = 55
    1st quartile: $Q_1 = 63$
    median = 67
    3rd quartile: $Q_3 = 78$
    largest data value = 90 ∎

Let's summarize the procedure used to determine the 5-number summary.

Procedure to Determine the 5-number Summary

Find:

Step 1. the median.
Step 2. the 1st quartile, Q_1.
Step 3. the 3rd quartile, Q_3.
Step 4. the smallest and largest data values.

Example 5.12

For the following 5-number summary
  smallest data value = 55
  the 1st quartile, Q_1 = 63
  median = 67
  the 3rd quartile, Q_3 = 78
  largest data value = 90
construct a Box-and-Whisker plot.

Solution:

To construct a Box-and-Whisker plot, the five numerical values from the 5-number summary are graphed and labelled on a horizontal axis.

The **first step** is to draw a horizontal axis and scale it appropriately from the smallest to the largest values of the 5-number summary. This is illustrated in Figure 5.10(a).

The **second step** is to draw a rectangle above the horizontal scale by drawing vertical lines at the 1st quartile, Q_1, and the 3rd quartile, Q_3, to represent the sides of the rectangle. Thus, the left side of the rectangle is placed at the value of Q_1 which is 63, while the right side of the rectangle is placed at the value of Q_3 which is 78. The height of the rectangle is unimportant. This is illustrated in Figure 5.10(b).

The **third step** is to draw a vertical line within the rectangle to represent the position of the median. Thus, the vertical line representing the median is drawn at 67. This is illustrated in Figure 5.10(c).

The **fourth step** is to draw two horizontal lines. One line connects the smallest data value to the left side of the box. The second horizontal line is drawn connecting the largest data value to the right side of the box. These horizontal lines connecting the 1st quartile, Q_1, to the smallest data value and the 3rd quartile, Q_3, to the to largest data value are called the Whiskers. The completed Box-and Whisker Plot is illustrated in Figure 5.10(d).

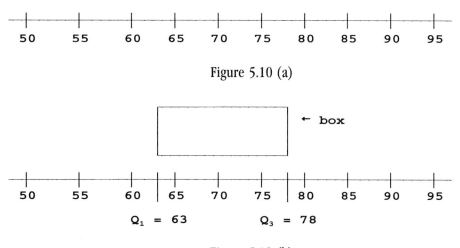

Figure 5.10 (a)

Figure 5.10 (b)

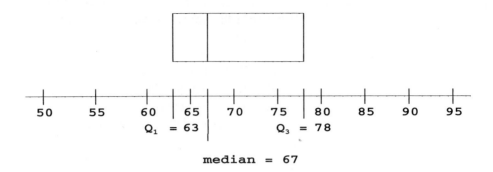

Figure 5.10 (c)

Box-and-Whisker Plot

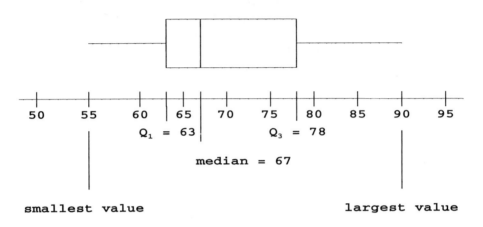

Figure 5.10 (d) ■

Let's summarize the procedure to construct a Box-and-Whisker Plot.

Procedure to Construct a Box-and-Whisker plot

Step 1. Draw a horizontal line and scale it from the smallest to the largest data values.
Step 2. Draw a rectangular box above the horizontal line with the sides at the 1st quartile, Q_1, and the 3rd quartile, Q_3.
Step 3. Draw a vertical line within the box to represent the position of the median.
Step 4. Draw horizontal lines from the box to the smallest and largest data values. These are called the Whiskers of the Boxplot.

A Box-and-Whisker Plot is useful in identifying whether a distribution's shape is symmetric or skewed. Notice that in Figure 5.10(d), the median is not centered within the box. It is positioned closer to Q_1 than to Q_3. This is an indication that the distribution is not symmetric. In fact it is **skewed** to the right. Furthermore, notice that the whisker to the right of the third quartile is longer than the whisker to the left of the first quartile. This suggests that there may be outliers present that are greater than Q_3.

In Figure 5.11 three box plots have been constructed. In Figure 5.11(a) notice that the whisker to the left of Q_1 is longer than the whisker to the right of Q_3. Perhaps there are outliers less than Q_1. Also notice that the median line is positioned closer to Q_3 than to Q_1. This suggests that the distribution is skewed to the left.

In Figure 5.11(b) notice that the whisker to the left of Q_1 is the same length as the whisker to the right of Q_3 and that the median line is positioned in the middle of the box. Although this distribution may contain outliers we can certainly conclude that the distribution is symmetric.

In Figure 5.11(c), as was the case in Figure 5.10(d) the whisker to the left of Q_1 is shorter than the whisker to the right of Q_3 and that the median line is closer to Q_1 than to Q_3. This distribution is skewed right.

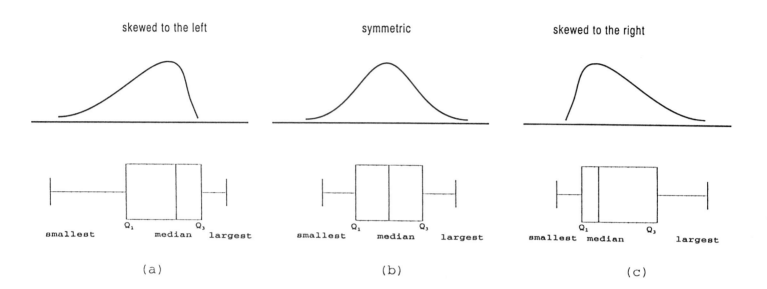

Figure 5.11

Example 5.13

The 5-number summaries of the distributions of the calories per serving of many brands of ice cream verses frozen yogurt are shown in Table 5.9.

Table 5.9

Dessert	Calories per serving				
	smallest value	Q_1	median	Q_3	largest value
Ice Cream	155	165	175	195	210
Yogurt	125	145	165	185	205

a) Construct a Boxplot for each of the 5-number summaries.
b) Examine the shape of each distribution, and determine what can you conclude about the calories per serving of ice cream compared to yogurt?

Solution:

a) Using the 5-number summary per each frozen dessert the Boxplots are shown in Figure 5.12:

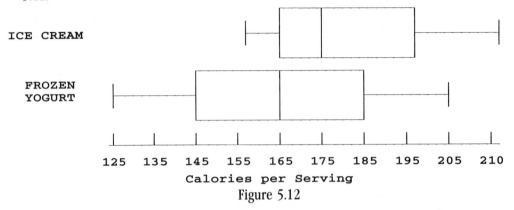

Figure 5.12

b) In examining the Boxplots of Figure 5.12, it appears that ice cream tends to have more calories per serving than frozen yogurt. Notice that the median calories per serving for ice cream is 175 while the median calories per serving for frozen yogurt is 165. The Boxplot for ice cream indicates the shape of the distribution is skewed to the right, since Q_1 is closer to the median, whereas the Boxplot for yogurt has a symmetric shaped distribution. ■

Boxplots are also useful in detecting the presence of an outlier. Data values that are outside the first and third quartiles, Q_1 and Q_3, may be considered somewhat extreme. In fact, the difference between Q_1 and Q_3, is often used as a guide for determining **outliers**. The difference between Q_1 and Q_3 is called *The Interquartile Range (IQR)*.

Definition: Interquartile Range (IQR). The Interquartile Range is the difference between the first quartile, Q_1, and the third quartile, Q_3. It is symbolized as IQR,
$$IQR = Q_3 - Q_1$$

The Interquartile Range, IQR, is a measure of spread within a distribution. **Data values that are at least 1.5 x IQR above the third quartile or below the first quartile are considered to be possible outliers of a distribution.** Tukey recommends using the IQR to detect outliers. In fact, his rule uses the distance (1.5)IQR. **Any data falling below the value Q_1 - (1.5)IQR or above the value Q_3 + (1.5)IQR is considered a potential outlier. Furthermore, any data falling below the value Q_1 - (3)IQR or above the value Q_3 + (3)IQR is an outlier.**

An outlier is an extreme data value that may be a valid data value or it can be a data value due to a measurement error, a recording error, etc. Therefore, when a potential outlier has been detected it is necessary to determine whether the outlier is due to error or is an unusual data value. If the outlier is due to error, then it should be removed. If the outlier represents an unusual but valid data value, then it should be investigated further since it may reveal useful information about the distribution of data values and may play an important role in the further statistical analysis of the data.

CASE STUDY 5.2

Examine the graphic in Figure 5.13 which represents a Salary Survey of U.S. Colleges and Universities offering degrees in statistics. Table 1 shows the salary information from 77 departments on statisticians in departments with programs in statistics for the academic year 1992-93.

TABLE 1—Salary Survey, Statisticians in Departments with Programs in Statistics

Salaries for the Academic Year 1992-93

Assistant Professors

		Number	First Quartile	Median	Third Quartile
Years in Rank	1 or Less	29	38,000	40,000	41,700
	2	41	37,500	40,000	43,500
	3	32	37,400	40,200	44,600
	4	20	38,800	40,500	44,600
	5 or More	45	37,150	39,800	44,700

Associate Professors

		Number	First Quartile	Median	Third Quartile
Years in Rank	2 or Less	35	39,900	45,300	49,700
	3-4	44	40,900	44,850	52,450
	5-6	21	41,450	45,200	49,450
	7 or More	75	43,000	46,100	51,600

Professors

		Number	First Quartile	Median	Third Quartile
Years In Rank	3 or Less	53	49,550	54,500	63,400
	4-6	51	53,400	62,100	69,000
	7-9	40	52,600	63,400	73,500
	10-12	43	55,700	62,600	73,100
	13-15	43	55,900	65,500	76,100
	16-18	35	61,000	70,100	81,500
	19-21	27	62,800	71,200	80,500
	22-24	19	63,900	72,000	84,000
	25 or More	31	58,000	66,000	80,500

Figure 5.13

a) What salary represents the 25th percentile for assistant professors with 3 years in rank?
b) What percent of the associate professors with 7 or more years in rank have a salary which is:
   1) at most $51,600?
   2) at least $51,600?
   3) between $43,000 and $51,600 inclusive?
c) For professors with 25 or more years in rank:
   1) what salary represents the 75th percentile?
   2) the 25th percentile?
   3) the 50th percentile?
   4) What is the interquartile range? What percent of the salaries fall within the interquartile range? How many professors have salaries within the interquartile range?
   5) Approximately how many of these professors have salaries between the first quartile and the median? between the median and the 3rd quartile?
   6) Calculate the range in salaries between the first quartile and the median? between the median and the 3rd quartile? Which has the smaller range?
   7) Associate with each range of part 5, the number and percent of professors that fall within that range. Comment on these results. What does the smaller range indicate regarding the variability of the salaries?
d) For professors with 7-9 years in rank, and associate professors with 7 or more years in rank, make a box representing the 1st, 3rd, and median salary on the same scale. Compare these boxes and draw a conclusion regarding this comparison. ∎

5.6 Using MINITAB

Among the many MINITAB commands, the two that apply most readily to this chapter are the BOXPLOT and DESCRIBE command.

A Boxplot can be constructed using the MINITAB command

BOXPLOT C

The "C" in the command refers to the column where the data values are contained.

Example 5.14

Consider the following data values:
22 44 15 56 34 34 25 16 45 56 75 80 100 24 19 33 28 85 34 30

a) Enter the data values into a MINITAB work sheet in column C_1 and print the column of data values.
b) Construct a Boxplot of the data values in C_1.

Solution:

a) Using the MINITAB command PRINT C1 the printed data values are displayed as:

MTB > PRINT C1

```
C1
22   44   15   56   34   34   25   16   45   56
75   80  100   24   19   33   28   85   34   30
```

b) To the data in C_1 the MINITAB command BOXPLOT C1 is applied.

MTB > BOXPLOT C1

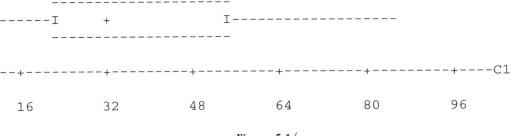

Figure 5.14

The Boxplot produced by MINITAB in Figure 5.14 is essentially the same Boxplot we would have produced using the 5-number summary discussed in Section 5.5. From Figure 5.14 you should notice that the smallest data value appears to be about 15, Q1 about 24, the median 34, Q3 about 56, and the largest data value is 100.

You may find it difficult to read the data values from MINITAB's Boxplot. When using the BOXPLOT command you should also summarize the data with the MINITAB command DESCRIBE to help you interpret the BOXPLOT.

MTB > DESCRIBE C1

	N	MEAN	MEDIAN	TRMEAN	STDEV	SEMEAN
C1	20	42.75	34.00	41.11	24.81	5.55

	MIN	MAX	Q1	Q3
C1	15.00	100.00	24.25	56.00

From MINITAB's DESCRIBE command you should notice that the 5-Number Summary values are 15, 24.25, 34, 56, and 100.

MTB > DESCRIBE C1

	N	MEAN	MEDIAN	TRMEAN	STDEV	SEMEAN
C1	20	42.75	34.00	41.11	24.81	5.55

	MIN	MAX	Q1	Q3
C1	15.00	100.00	24.25	56.00

For MINITAB's DESCRIBE command you should notice that the 5-Number Summary values are 15, 24.25, 34, 56, and 100.

A Comment On MINITAB's Procedure To Compute The First and Third Quartiles

At this time, we would like to mention that MINITAB's procedure to compute the first quartile (Q_1) and the third quartile (Q_3) is slightly different from the procedure discussed in the text. MINITAB's procedure involves the use of interpolation. We present the following discussion which explains MINITAB's procedure for students who may have noticed a slightly different result using the procedure within the text.

MINITAB Procedure to Calculate The First and Third Quartile

Step 1. Place the data values in numerical order ranging from the smallest to the largest value.

Step 2. The first quartile is defined to be the data value at the position $\frac{n+1}{4}$, where n represents the number of data values. If the value of $\frac{n+1}{4}$ is not an integer, then MINITAB uses interpolation to define the value of the first quartile.

Step 3. The third quartile is defined to be the data value at the position $\frac{3(n+1)}{4}$, where n represents the number of data values. If the value of $\frac{3(n+1)}{4}$ is not an integer, then MINITAB uses interpolation to define the value of the third quartile.

Let's consider an example using this procedure.

Using the data values of Example 5.14 and arranging the values in numerical order, we would obtain:

15 16 19 22 24 25 28 30 33 34 34 34 44 45 56 56 75 80 85 100

The position of the first quartile is determined by the formula: $\frac{n+1}{4}$, where n is equal to 20 data values. Therefore, the position of first quartile = $\frac{20+1}{4}$ = 5.25. This means that the position of the first quartile is 0.25 or $\frac{1}{4}$ the distance between the 5th and 6th data values. To determine the value of the first quartile, MINITAB uses interpolation. In this case, the formula to find the first quartile, (Q_1), is:

$$Q_1 = \text{5th data value} + \frac{1}{4} \text{(6th data value - 5th data value)}$$

The first quartile is:

$$Q_1 = 24 + \frac{1}{4}(25 - 24) = 24.25.$$

The position of the third quartile is determined by the formula: $\frac{3(n+1)}{4}$, where n is equal to 20 data values. Therefore, the position of third quartile = $\frac{3(20+1)}{4}$ = 15.75. This means that the position of the third quartile is 0.75 or $\frac{3}{4}$ the distance between the 15th and 16th data values. To determine the value of the third quartile, MINITAB uses interpolation. In this case, the formula to find the third quartile, (Q_3), is:

$$Q_3 = \text{15th data value} + \frac{1}{4} \text{(16th value - 15th data value)}$$

The third quartile is:

$$Q_3 = 56 + \frac{1}{4}(56 - 56) = 56.$$

An interesting interpretation of a Boxplot using MINITAB can be demonstrated if we change **two** of the data values in C1. For example, let's change the data value 30 to 200, and the data value 100 to 130. The resulting Boxplot using MINITAB is:
MTB > BOXPLOT C1

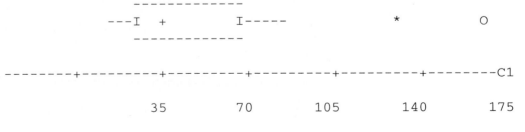

Figure 5.15

MTB > DESCRIBE C1

	N	MEAN	MEDIAN	TRMEAN	STDEV	SEMEAN
C1	20	52.8	34.0	46.7	45.1	10.1

	MIN	MAX	Q1	Q3
C1	15.0	200.0	24.2	70.3

Notice the letter "O" appearing to the extreme right side of Figure 5.15. MINITAB **identifies** data values that are more than 3(IQR) as probable outliers and symbolizes these probable outliers by an "O". Furthermore, notice the symbol "*" appearing to the left of the "O".

If there are data values that are between 1.5(IQR) and 3(IQR), MINITAB will identify these with an asterisk "*". These data values are considered to be potential outliers.

Using the DESCRIBE command makes it easy to determine the 5-Number Summary of the distribution.

MTB > DESCRIBE C1

	N	MEAN	MEDIAN	TRMEAN	STDEV	SEMEAN
C1	20	51.25	34.00	45.00	42.81	9.57

	MIN	MAX	Q1	Q3
C1	15.00	200.00	24.25	70.25

If you examine Figures 5.14 and 5.15, you will notice that the Boxplot for Figure 5.14 is more spread out. Why do you think this is the case?

Glossary

TERM	SECTION
z Score, z	5.2
Data Value, X	5.2
Raw Score, X	5.3
Percentile Rank, PR	5.4
Percentile	5.4
Decile	5.4
Quartile	5.4
Box-and-Whisker Plot	5.5
5-Number Summary	5.5
Interquartile Range(IQR)	5.5
Outlier	5.5

Exercises

PART I Fill in the blanks.

1. One method of ranking an individual data value within a distribution is to convert each data value into a _____.
2. The _____ indicates the number of standard deviations a data value is greater than or less than the mean.
3. The z score of a data value is calculated by the formula: z = _____.
4. The percentile rank of a data value indicates the percent of values within the distribution that are _____ than the given data value.
5. A data value that is one standard deviation greater than the mean has a z score of _____.
6. A data value that is one and a half standard deviations less than the mean has a z score of _____.
7. A data value that has a z score of zero is the _____.
8. If the z score of a data value is negative, then the data value is _____ than the mean.
9. If the z score of a data value is positive, then the data value is _____ than the mean.
10. For a distribution, $\mu = 100$ and $\sigma = 4$.
    a) if $X = \mu$, then z = _____.
    b) if $X = 90$, then z = _____.
    c) if z = 2, then X = _____.
    d) if z = 1.5, then X = _____.
11. The formula used to convert a z score to a raw score, X, is: X = _____.
12. If a raw score has a z score of 3, then the raw score is _____ standard deviations _____ than the mean.
13. If a raw score has a z score of -2, then the raw score is _____ standard deviations _____ than the mean.
14. In a distribution, if $X = 545$, $\mu = 525$ and $z = 5$, then σ = _____.
15. In a distribution, $\mu = 75$, and $\sigma = 5$. If $z = 3$ is the maximum z score value of a data value in the distribution, and $z = -2$ is the minimum z score value of a data value in the distribution, then, the range of the distribution is equal to _____.
16. In a distribution, if $\mu = 135$ and $\sigma = 0$, then the z score of any data value in this distribution is equal to _____.
17. Lorenzo's IQ score of 120 has a percentile rank of 90. This means that _____ percent of the individuals have IQ scores less than 120 and _____ percent have IQ scores greater than 120.
18. The percentile rank formula for a data value, x, is calculated by:

    PR of the data value $X = \dfrac{[B + (\frac{1}{2})E]}{N} (100)$

    where: B = the number of data values _____ than X.
    E = the number of data values _____ to X.
    N = the number of data values in the _____.

19. Percentile ranks are rounded off to _____ numbers.
20. The pth percentile is the data value in the distribution which has _____ percent of the data values less than this data value and the rest of the data values greater than it.
21. If $P_{20} = 75$, then we state that 75 is the 20th _____.
22. If $P_{25} = 80$, then we state that 80 is the 1st _____.
23. If $P_{50} = 100$, then we state that 100 is the 2nd _____.
24. If $P_{50} = 110$, then the median is _____.
25. If the IQ score 70 has a percentile rank of 10, then 70 is the 1st _____.
26. The percentile is a _____ of the distribution, while the percentile rank of a data value is a percentage value ranging from _____ to _____.
27. The Box-and-Whisker Plot is an _____ Data Analysis technique used to visually display several summary statistics to describe the location of the _____ of the data, the _____ and _____ of a distribution.
28. On a box-and-whisker plot, the middle 50% of the data values fall between the _____ and _____ quartiles.
29. The line connecting the 3rd quartile to the largest value of a box-and-whisker plot is called the _____ of the boxplot.
30. The Interquartile Range (IQR) is the difference between the _____ and the _____ quartile. On a boxplot, the IQR is represented by the length of the _____. The percent of data values that are represented by the IQR is _____.

PART II Multiple choice questions.

1. Which measure of variability is used in calculating the z score of a data value?
   a) range   b) variance   c) standard deviation
2. The z score of a data value indicates how many standard deviations the data value is above or below the _____.
   a) mean   b) mode   c) median
3. If a data value has a z score of zero, this indicates that the data value is the _____.

a) standard deviation b) median
c) mean d) none of these

4. A data value which is greater than its mean always has a positive _____ .
a) mode b) z score c) median d) range

5. Which of the following indicates the relative position of a data value within its distribution?
a) mean b) raw score
c) z score d) standard deviation

6. If we know the z score of a data value and wish to calculate its raw score, we multiply the z score by the standard deviation and add the _____ .
a) mean b) variance c) median d) range

For questions 7 thru 10, use the following information:

At Jefferson High, the final average with a percentile rank of 50 was 82. Mike's average has a percentile rank of 74. Pete's average is equal to the 9th decile. Maria's average has a percentile rank of 94. Joe's average is between the first and second quartile. Susan's average was less than 82. Jeff's average has a percentile rank which is lower than Maria's but greater than Mike's.

7. Of those students mentioned above, who had the highest average?
a) Joe b) Maria c) Mike d) Susan

8. What is the median average?
a) 74 b) 94 c) 82 d) 50

9. Of those students mentioned above, how many had an average less than 82?
a) 0 b) 3 c) 1 d) 2

10. Whose average was equal to the 90th percentile?
a) Maria b) Mike c) Pete d) Susan

PART III Answer each statement True or False.

1. A z score is used to summarize a distribution's variability.
2. In every distribution the mean has a z score equal to zero.
3. If a data value is two standard deviations less than the mean, it has a z score of +2.
4. If a standard deviation is equal to 10 units then the z score of a data value one standard deviation greater than the mean is equal to 10 units.
5. If a data value is 3 standard deviations greater than the mean, the z score of the data value is equal to +3.
6. Z scores are always whole numbers.
7. Negative raw scores can never have positive z scores.
8. If two distributions have equal means, then the data values that have z scores equal to 1 will always be the same.
9. If two distributions have equal standard deviations, then the data value that has a z score of -2 in the first distribution will always be equal to the data value that has a z score of -2 in the second distribution.
10. The percentile rank of a data value is the number of data values within the distribution that are less than the given data value.
11. A percentile is a data value of the distribution.
12. The 40th percentile is the data value, X, in a distribution that equals 40.
13. If the median mathematics SAT score is 460, then the score that represents the 50th percentile is 460.
14. The 7th decile is the data value, X, within a distribution that has the property that 70 percent of the data values are less than X and 30 percent of the data values are greater than X.
15. The data value that represents the 1st quartile is greater than the data value that represents the 3rd quartile.
16. A Box-and-Whisker Plot is an exploratory data analysis technique that uses a 5 number summary consisting of the smallest data value, the mean, the median, and the largest data value.
17. The Interquartile Range (IQR) is a measurement that indicates the middle 50% of the distribution.
18. The Whiskers in a Box-and-Whisker Plot can be used to indicate the presence of data values that are outliers.

PART IV Problems.

1. Using the z score formula, calculate the z scores for each raw score given in the table below:

	raw score(X)	mean(μ)	standard deviation(σ)
a)	24	16	8
b)	80	100	10
c)	440	460	15
d)	630	500	100
e)	1150	970	120

2. Using the raw score formula, calculate the raw score for each z score given in the table below:

	mean(μ)	standard deviation(σ)	z score(z)
a)	20	5	+3
b)	50	6	-2
c)	110	15	+2.5
d)	500	100	-2.33
e)	90	20	+1.96

3. Given the distribution of temperatures:
   80, 90, 69, 60, 85, 75
   a) Find the mean temperature, μ.
   b) Find the standard deviation, σ.
   c) Convert each temperature to a z score.

4. An instructor assigns letter grades in the following manner:

To achieve a letter grade of:	z score of test grade must be:
A	Greater than 1.5
B	From 0.5 to 1.5
C	Between -0.5 and 0.5
D	From -1.5 to -0.5
F	Less than -1.5

   a) If the mean is 83 and the standard deviation is 8, find the test scores that determine the letter grades.
   b) If the mean is 70 and the standard deviation is 18, find the test scores that determine the letter grades.

5. Watergate Price Inc., a consumer group, surveys a Washington, D.C. urban area to compare the basic service charges of the three most frequently called service people to determine the highest repair charge relative to their respective occupation.
   According to the survey, the 3 most frequently called service people were: a plumber, an auto mechanic and an electrician. In this area the survey indicates that the basic service charge per occupation is:
   plumber - $35/hour,
   auto mechanic - $40/hour,
   electrician - $47.50/hour.
   The following table represents the union mean and standard deviation service charges per occupation:

Occupation	Mean(μ)	Standard Deviation(σ)
plumber	$26.50/hour	$3.50/hour
auto mechanic	$23.75/hour	$5.75/hour
electrician	$35.90/hour	$4.25/hour

   Determine which of the three service people charge the highest repair charge relative to their respective occupation.

6. An Atlantic City casino credit manager claims that the mean loss suffered by a "high roller" is $10,000 a night with a standard deviation of $1800, while the mean loss for an average gambling patron is $500 a night with a standard deviation of $160. Recently, Elliot, a known high roller, claims he lost $11,000 and Lou, an average gambling patron, lost $700. Which person, relative to the type of gambler he is, has suffered the greater loss?

7. Michael and Lorenzo both took the same calculus test. Michael scored an 87 on the test. Lorenzo's test paper only had the z score of his grade which was +1.2. The instructor gave Lorenzo the following information about the distribution of test grades: the mean test grade is 76 and the standard deviation is 15. Who had the higher grade, Michael or Lorenzo?

8. Two track athletes, a sprinter and a long jumper, break their school's record in their respective events. To determine which feat is more outstanding relative to his respective event, the school's statistician collects the best performances for each event within the United States and list the necessary data below:

athlete	event	record	mean	standard deviation
sprinter	100m run	9.2 sec	9.6 sec	0.20 sec
long jumper	long jump	7.82 m	7.62 m	0.13 m

   Which of the two athletic records is more outstanding?

9. In certain manufacturing processes, parts must be machined to within specified tolerances. The parts must have a mean length of six inches plus or minus 0.26 inches. A sampling procedure has been established to determine whether the parts are within the specified tolerances. If for a sample of ten parts selected at random, the mean plus 1.5 standard deviations and the mean minus 1.5 standard deviations are within the specified limits of 5.74 inches to 6.26 inches, then the parts are meeting the specified tolerances. This means that a z score of 1.5 and a z score of -1.5 must be within the limits of 5.74 inches and 6.26 inches. The lengths (in inches) representing the ten parts are as follows:
   5.94, 6.20, 5.86, 6.08, 6.14, 6.12, 6.01, 5.92, 5.90, 6.23
   Procedure:
   a) calculate the mean, $\overline{x}$, and standard deviation, s.
   b) convert z = -1.5 and z = 1.5 into inches (data values).
   c) compare these values with the specified limits of 5.74 inches to 6.26 inches and determine if this sample meets the stated conditions.

10. Two amateur athletes, a bowler and golfer, are being considered for the Amateur Athlete of the Year Award in their country club. The award will be given to the athlete who has had the most outstanding season relative to their sport. The bowler's summer league average was 201, while the golfer averaged 74 for her summer league play. Based on these facts and the additional data given in the following table, determine which athlete should win the Amateur Athlete Award.

	mean score	standard deviation
Bowling League	170	21
Golf League	85	7

11. DEB Homes, Inc., a Long Beach real estate agency, sold three condominiums during the past week. Condo M, a waterfront condo, sold for $305,000 while Condo A, on the canal side of town, sold for $275,000 and Condo T, in the heart of town, sold for $215,000.
    In Long Beach, the purchase price of a condo is dependent upon its location. City Hall calculates the

mean price of a waterfront condo to be $245,000 (with σ = $50,000), while the mean price of a canal side condo is $230,000 (with σ = $23,000) and a town condo to have a mean price of $190,000 (with σ = $15,000).

Based upon its geographic location, which of the three condos was sold for the relatively:
a) higher purchase price?
b) lowest purchase price?

12. A New York Realty Association surveys local banks and mortgage lenders to determine the trends in 30 year fixed and adjustable mortgage rates. Based on their survey, the mean rate for a 30 year fixed mortgage is 10.97% (with σ = 0.14%) and the mean rate for a one year adjustable mortgage is 7.79% (with σ = 0.17%).
If Matcraig, a mortgage lender, is offering a 30 year fixed mortgage rate of 10.75% (& no points) and a one year adjustable mortgage rate of 7.5% (& no points), which rate would be considered the best deal relative to its type of mortgage?

13. Using the following collection of 50 IQ scores, compute the percentile rank of the IQ scores:
a) 87 b) 91 c) 105 d) 115 e) 119 f) 121

IQ Scores
81, 85, 85, 86, 87, 87, 88, 89, 89, 90, 90, 90, 91, 93, 93, 94, 95, 97, 98, 99, 100, 101, 102, 104, 105, 105, 107, 107, 110, 112, 112, 113, 114, 115, 115, 115, 115, 119, 120, 121, 121, 121, 122, 123, 125, 125, 126, 126, 127, 135

14. Using the results of problem #13, determine the IQ score representing the :
a) 7th decile   b) 3rd quartile   c) 2nd quartile
d) 1st decile   e) median   f) 1st quartile

15. Some results of the Graduate Record Examination (GRE) Mathematics part are recorded in the following table:

GRE Score	372	433	475	500	525	567	604	628	733
Percentile Rank	10	25	40	50	60	75	85	90	99

Using the table, determine the:
a) 1st decile score.
b) 1st quartile score.
c) median score.
d) 6th decile score.
e) 3rd quartile score.
f) percentage of students who scored below 604.
g) percentage of students who scored higher than 525.
h) percentage of students who scored between 433 and 567.
i) difference between the 3rd quartile score and the 1st quartile score.

16. The reaction time of 250 people is measured after a new drug is administered. Some results of this test are given in the following table.

reaction time (sec)	8	12	16	20	28	36	40
z score	-1.5	-1	-0.5	0	1	2	2.5
percentile rank	17	25	30	42	50	75	85

By inspection of the table, determine:
a) the mean reaction time of all the 250 people.
b) the median reaction time of all the 250 people.
c) the percent of all the 250 people who had reaction times 40 secs or below.
d) the reaction time representing the 3rd decile.
e) the reaction time representing the 1st quartile.
f) the reaction time representing the 3rd quartile.

17. The following table indicates the performance of six students on the law boards. Using this information, compute the percentile rank of each student if 10,000 students took the law boards.

student I.D.	number of students who scored less than this student	number of students who scored the same as this student (including student)
L	2,250	500
A	2,900	200
W	4,800	400
Y	7,475	50
E	8,000	100
R	9,000	25

18. To qualify for a scholarship at Richman's College, a student must score at least 1.5 standard deviations above the mean on the entrance exam. Which of the following students qualify if the mean and standard deviation for the entrance examination is 60 and 12 respectively?

student	score on exam
George	75
Alice	78
Barbara	81
Mauro	77
Mike	79

19. The mean annual family income in the state of West Virginia is $17,500 with a standard deviation of $1800. If a family's income is at least 2 standard deviations below the mean income, then the family is eligible for food stamps.
Would a family with an annual income of:
a) $13,345

b) $13,990
c) $14,300
be entitled to food stamps?

20. The following table lists the mean and standard deviation of a final exam given to three classes.

class	mean	standard deviation
I	80	2
II	75	8
III	80	15

a) If Chad scored an 83 on the final exam, in which class would his grade be considered more outstanding? less outstanding?
b) If Lauren scored a 65 on the final exam, in which class would his grade be considered more outstanding? less outstanding?
c) If Georgia scored an 81 on the final exam, in which class would her grade be considered more outstanding? less outstanding?

21. What percentage of a population of psychology test scores are:
a) above the third quartile?
b) below the first quartile?
c) greater than the 23rd percentile?
d) above the 6th decile?
e) between the first and third quartiles?
f) between the 2nd and 6th decile?

22. Some results of the verbal and mathematics college board scores along with their respective percentile ranks are given the following tables.

verbal score	490	510	570	650	740
percentile rank	40	50	75	80	92

mathematics score	450	500	590	610	650	750
percentile rank	45	50	65	75	85	90

Determine:
a) the 4th decile on the verbal college boards.
b) the median score on the verbal college boards.
c) the 3rd quartile on the verbal college boards.
d) the 2nd quartile on the mathematics college boards.
e) the median score on the mathematics college boards.
f) the 9th decile on the mathematics college boards.

23. On a tropical island, two natives, Luanna and Cousin Don, grow coconut and banana trees respectively. Each year they make a friendly wager over who will grow the largest produce relative to the type they grow. At harvest time each year, Luanna and Cousin Don compare their largest produce. This year, Cousin Don's largest banana was 8.3 inches and Luanna's largest coconut had a diameter of 5.9 inches. On this island for the past year, the mean banana length was 7.9 inches with $\sigma = 0.6$ inches and the mean diameter of the coconuts was 5.5 inches with $\sigma = 0.2$ inches.
Using this information, determine who won this year's contest and explain why?

24. Corinne, an Investment analyst, has been plotting the cyclic pattern of the stock of a fast food company. Based upon her analysis, she decides to purchase the stock when it is selling at least one and a half standard deviations below its mean cyclic price. Her plan is to sell the stock when its price is at least 1.65 standard deviations above the mean price. Determine her maximum buying price and her minimum selling price for this stock if its mean cyclic price is $27.65 with a standard deviation of $9.95.

25. If the mean cholesterol count of two hundred nutritionists is 176 with a standard deviation of 12, determine the cholesterol limits that cutoff all the cholesterol counts that are within:
a) one standard deviation of the mean
b) 1.96 standard deviations of the mean
c) two standard deviations of the mean
d) 2.58 standard deviations of the mean
e) three standard deviations of the mean.

26. A male is considered to be hypertensive for his age group, if he has a systolic blood pressure of more than two standard deviations greater than the mean. The following table contains the mean and standard deviation for systolic blood pressures for some male age groups.

Male systolic blood pressure parameters

age	mean	standard deviation
18	121	8
25	125	8
35	127	9
45	130	13
60	142	14

The following readings of systolic blood pressure were taken for a sample of males.

male	age	systolic blood pressure
Craig	18	134
Jim	25	142
Mike	35	142
Lou	60	150

Determine which (if any) of the males are classified as hypertensive.

27. Some test results of an economics test are listed in the following table. Using the information in the table, find:

a) the mean test grade of all the grades
b) the median test grade of all the grades
c) the standard deviation of all the grades
d) the percent of all the grades that are between 68 and 132.
e) what test grade corresponds to a z score of 1.5.
f) the percent of all the test grades within 1σ of the mean.
g) the percent of all the test grades that are at least 2σ above the mean.

test grade	z score	percentile rank
68	-2	2
84	-1	16
100	0	42
116	1	50
132	2	96
148	3	99

28. The following distributions represent the number of homeruns that Roger Maris and Babe Ruth hit during the designated years.

The number of home runs hit by Roger Maris during each baseball season for the years 1957-1968
14, 28, 16, 39, 61, 33, 23, 26, 8, 13, 9, 5

The number of home runs hit by Babe Ruth during each baseball season for the years 1919-1934
29, 54, 59, 35, 41, 46, 25, 47,
60, 54, 46, 49, 46, 41, 34, 22

For each home run distribution, determine:
a) the 1st quartile, median, and 3rd quartile.
b) Construct a Box-and-Whisker Plot for each distribution on the same scale.
c) Compare the Boxplots, and discuss the differences between the two distributions.
d) Use the Boxplots to compare the variability of each distribution. Which distribution has the smaller variability? the larger variability? Now, calculate the Interquartile Range and the standard deviation of each distribution. Compare the IQR and the standard deviation of each distribution. Are these results similar? With respect to variability, what is the relationship between the value of the IQR and the standard deviation of each distribution?
e) Identify any potential outliers for each distribution, and give an explanation for the outlier. Find the z score of any outlier and interpret the meaning of the z score.
f) Describe the shape of each distribution.

29. The following two Box-and-Whisker Plots represent the test results of a statistics final for two different instructors.

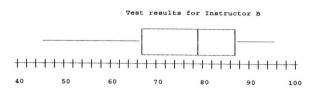

From each of the previous two Boxplots representing the test results of instructors A and B, approximate the:
a) 1st quartile test score.
b) 3rd quartile test score.
c) median test score.
d) smallest and largest test score.
e) Interquartile range(IQR).
f) shape of the distribution.
g) percent of test results to the left of the 3rd quartile.
h) percent of test results to the right of the median.
i) percent of test results between the 1st and 3rd quartile.
j) Compare the two test distributions and explain the differences about the test results.
k) Assume each instructor had different objectives in mind when constructing the final exam. One instructor was trying to determine the top students in the class, while the other instructor was trying to determine the minimum knowledge of the poorest students. Based on the Boxplots, explain which instructor you believe was trying to determine the top students in her class? Which test do you believe was the easiest test? Explain.

30. The quality control inspector at a nut and bolt company takes a random sample at the start of the day to check on the production process. One bolt producing machine at this company is set to produce a bolt with a length between 6.25 to 6.75 centimeters.

The following measurements represent 40 bolts that were selected from the production process at the start of the day.

length of the 40 bolts
(in centimeters)
6.72, 6.82, 6.74, 6.61, 6.64, 6.50, 6.65, 6.65,
6.57, 6.63, 6.75, 6.49, 6.41, 6.55, 6.63, 6.69,
6.46, 6.74, 6.55, 6.46, 6.50, 6.34, 6.54, 6.59,
6.60, 6.52, 6.59, 6.42, 6.57, 6.52, 6.66, 6.33,
6.24. 6.51, 6.63, 6.50, 6.45, 6.65, 6.57, 6.63
Using these measurements, determine the:

a) 1st quartile.
b) 3rd quartile.
c) median.
d) smallest and largest measurement.
e) Interquartile range(IQR).
f) Construct a box-and-whisker plot.
g) Describe the shape of the distribution of bolt lengths.
h) What does the boxplot indicate about the operation of this machine?
i) Are there any outliers? Explain.

31. An instructor from the school of business surveys the students in his Introduction to Business class to learn about the amount of dollars students spend per semester on textbooks. The distribution that follows indicates the results of his survey.

    Textbook Costs per Student per semester (in dollars)
    147 160 142 153 187 195 106 127
    178 178 272 175 178 133 186

    a) Determine the 5-Number Summary for the class' textbook spending data.
    b) Use the 5-Number Summary to construct a Box-and-Whisker Plot.
    c) Describe the shape of the Box-and-Whisker Plot.
    d) Calculate the interquartile range and determine the data values that are greater than
       1) 1.5(IQR).   2) 3(IQR).
    e) Are any of the data values considered to be outliers? Why? If there are any possible outliers discuss reasons that might contribute to these data values being an outlier.

32. The rate in which red blood cells settle in a tall narrow tube (known as the erythrocyte sedimentation rate) is an indication of a person's health. A sample of blood from 20 randomly selected elderly males was measured for this rate. The data values are as follows:

    red blood cell sedimentation rate
    (millimeters per 1 hour)
    13 9 9 105 8 4 5 10 8 120 8 5 12 5 7 10 9 8 5 10

    a) Determine the 5-number summary for the blood sedimentation rate data.
    b) Use the 5-number summary to construct a Box-and-Whisker Plot.
    c) Describe the shape of the Box-and-Whisker Plot.
    d) Calculate the interquartile range and determine the data values that are greater than
       1) 1.5(IQR).
       2) 3(IQR).
    e) Are any of the data values considered to be outliers? Why? If there are any possible outliers, discuss reasons that might contribute to these data values being an outlier.

33. The five number summaries of the distributions of the number of calories from fat per type of cookie are:

    calories from fat per cookie

	min	Q1	median	Q3	max
Chocolate Chip	20	25	36	40	55
Pecan	30	32	45	60	68
Oatmeal	20	24	27	32	40

    a) Construct a Boxplot for 5-number summary.
    b) What can be said about the calories from fat per type of cookie in terms of IQR and distribution shapes?

PART V What Do You Think?

1. **What Do You Think?**
   Examine the graphic entitled *Annual pay, by state* which represents a state-by-state estimate of the average salaries of classroom teachers in 1992-93, and the percentage increase over 1991-92.

Annual pay, by state

A state-by-state estimate of the average salaries of classroom teachers in 1992-93, and the percentage increase over 1991-92:

	Salary	Inc.		Salary	Inc.
Ala.	$27,490	2.0%	Mont.	$28,514	3.3%
Alaska	$46,373	3.7%	Neb.	$28,718	5.5%
Ariz.	$32,403	3.9%	Nev.	$34,119	0.8%
Ark.	$27,598	2.0%	N.Y.	$44,600	2.9%
Calif.	$41,400	3.0%	N.J.	$43,997	7.2%
Colo.	$33,541	1.4%	N.M.	$26,355	-0.1%
Conn.	$48,850	4.0%	N.H.	$33,931	2.3%
Del.	$36,217	4.8%	N.C.	$29,367	2.0%
D.C.	$38,168	-1.6%	N.D.	$25,211	2.9%
Fla.	$31,153	0.3%	Ohio	$34,600	4.1%
Ga.	$30,626	3.8%	Okla.	$26,051	2.8%
Hawaii	$36,470	5.6%	Ore.	$35,435	3.9%
Idaho	$27,156	3.1%	Pa.	$41,580	7.4%
Ill.	$38,576	5.8%	R.I.	$37,510	3.0%
Ind.	$37,446	7.6%	S.D.	$24,125	3.6%
Iowa	$30,124	3.2%	S.C.	$29,151	3.9%
Kan.	$33,133	7.8%	Tenn.	$29,313	2.4%
Ky.	$31,487	2.0%	Texas	$29,935	3.1%
La.	$26,074	0.5%	Utah	$26,997	2.5%
Maine	$30,258	0.5%	Vt.	$34,824	3.5%
Md.	$39,141	1.1%	Va.	$32,356	2.2%
Mass.	$39,245	5.3%	Wash.	$35,870	3.0%
Mich.	$43,331	5.3%	W.Va.	$30,301	10.7%
Minn.	$35,656	3.5%	Wis.	$36,477	3.5%
Miss.	$24,369	0.0%	Wyo.	$30,850	1.4%
Mo.	$29,410	1.8%	**USA**	**$35,334**	**3.6%**

Source: *1992-93 Estimates of School Statistics*, National Education Association

a) Using this information, and omitting the salary

for the D.C. area, find the mean and the standard deviation of the salaries.
b) For each of the following salaries, compute the z score and percentile rank:
1) $24,125
2) $29,151
3) $32,356
4) $32,403
5) $36,477
6) $48,850
c) Interpret the z score and percentile rank for each salary of part(b).
d) What is the median salary? What is the z score of the median salary?
e) What is the lowest salary? Is the percentile rank of the lowest salary zero? Why not?
f) What is the highest salary? Is the percentile rank of the highest salary 100? Why not?
g) Find the interquartile range(IQR), and determine the number of salaries that fall within this range. What percent of the salaries lie within the IQR?
h) Would you consider any of these salaries to be outliers? Explain.

2. **What Do You Think?**
Examine the graphic entitled **Bank Money Accounts** which represents annual effective yields, in percents, on money market accounts, three-month CDs, and six-month CDs.

Bank Money Accounts

Annual effective yields, in percent, on money market accounts, three-month certificates of deposit and six-month C.D.'s expected at selected institutions on Wednesday, April 21, 1993. Some C.D. yields assume reinvestment after maturity. Yields are based on the method of compounding and the rate stated for the lowest minimum to open an account, but minimum vary. Higher yields may be offered for larger deposits. C.D. figures are for fixed rates only.

New York Area

	Money Mkt.	3-mo. C.D.	6-mo. C.D.
COMMERCIAL BANKS			
Banco Popular	2.63	2.78	2.99
Bank of N.Y.	2.37	2.58	2.69
Bank of Tokyo Trust	2.17	2.19	2.25
Chase Manhattan	2.12	2.28	2.33
Chemical	2.07	2.65	2.70
Citibank	1.91	2.35	2.40
EAB	2.28	2.58	2.69
First Fidelity (N.J.)	2.50	2.60	2.85
Fleet Bank (Conn)	2.35	2.73	2.94
Key Bank of N.Y.	2.13	2.32	2.58
Marine Midland	2.15	2.42	2.53
Midlantic Bank (N.J.)	2.50	2.60	2.80
NatWest	2.53	2.43	2.48
Republic Nat'l	2.48	2.63	2.79
Shawmut Bank (Conn)	2.38	2.74	2.94
SAVINGS BANKS			
Anchor Savings	2.65	2.80	3.05
Apple Bank Savings	2.61	2.66	2.95
Astoria Fed Savings	2.94	2.79	2.89
Carteret Savings (N.J.)	2.53	2.80	3.00
Crossland Savings	2.65	2.75	3.08
Dime Savings Bank	2.64	2.64	3.00
Emigrant Savings	2.60	2.75	3.00
First Fed (Rochester)	3.05	2.95	3.05
First Nationwide	2.20	2.65	2.75
Greater NY Savings	2.60	2.80	3.00
Green Point Savings	3.00	N.O.	3.20
Home Savings Amer	2.63	2.69	3.05
Hudson City SB (N.J.)	3.00	3.30	3.40
People's Bank (Conn.)	2.53	2.74	3.05
River Bank America	2.80	2.85	3.10

Major Banks Outside N.Y.

	Money Mkt.	Wkly. Chg.	6 Mo. C.D.	Wkly. Chg.
Bank One Texas (Dallas)	2.43	0.00	2.95	0.00
Bank of Boston	2.63	0.00	2.78	0.00
Citibank, S.D.	2.76	0.00	3.51	0.00
Comerica (Detroit)	2.78	0.00	2.78	0.00
Fidelity Bank (Philadelphia)	2.50	0.00	2.85	0.00
First Bank (Minnesota)	2.22	-0.05	2.80	0.00
First Interstate (L.A.)	2.34	0.00	2.48	0.00
First National (Chicago)	2.53	0.00	2.73	0.00
First Union (Charlotte, N.C.)	2.55	-0.10	2.80	0.00
Fleet Bank (Boston)	2.68	0.00	2.94	0.00
Maryland Natl	2.75	0.00	3.10	-0.10
Mellon (Pittsburgh)	2.53	0.00	2.55	0.00
Meridian Bank (Reading Pa)	2.19	0.00	2.90	-0.05
National Bank (Detroit)	2.78	0.00	2.75	-0.05
NationsBank (Tampa)	2.35	0.00	3.10	0.00
Seafirst (Seattle)	2.50	0.00	3.00	0.00
Shawmut Bank (Boston)	2.48	0.00	2.80	0.00
Texas Commerce (Houston)	2.27	0.00	2.75	-0.03
Union Bank (S.F.)	2.43	0.00	2.53	0.00
Wachovia (Winston Salem, N.C.)	2.55	0.00	2.80	0.00

a) Find the z scores of all the money market rates for the commercial banks in the New York Area.
b) Without rounding off the results of part a, determine the mean and standard deviation of all the z scores.
c) Repeat steps a and b for the three-month CD rates. Do you get approximately the same results? What conclusions can you draw from these results?
d) Identify the mean rate and median rate for the money market rate and six-month CD rates for the commercial, and savings banks in the New York Area, and the Major Banks outside N.Y. Interpret these results.
e) Compare the money market rate of 2.53 for the New York commercial and savings banks, and the major banks outside the N.Y. area to determine in which category 2.53 is a better rate relative to the rates within each category. Explain.
f) Compare the six-month rate of 2.80 for the commercial and the major banks outside the N.Y. area to determine in which category 2.80 is a better rate relative to the rates within each category. Explain.
g) Compare the six-month rate of 2.95 for the savings banks, and the six-month rate of 2.94 for the major banks outside the N.Y. area to determine which rate is considered better relative to the rates within the respective categories. Explain.

3. **What Do You Think?**
Examine the graphic entitled *Salary Survey of Biostatistics and Biomedical Statistics Programs* which presents the results of a Fall 1992 survey of Biostatistics and other Biomedical Statistics Departments and Units. Previous salary survey results for Fall 1986 to Fall 1991 have been included for comparative purposes.

Salary Survey Results: Biostatistics and Other Biomedical Statistics Departments and Units (12 Months)

Rank/ Years in Rank	Percentile	Fall 1986 (Sample Size)	Fall 1987 (Sample Size)	Fall 1988 (Sample Size)	Fall 1989 (Sample Size)	Fall 1990 (Sample Size)	Fall 1991 (Sample Size)	Fall 1992 (Sample Size)
Assistant 1-3	25th	$33,188	$34,000	$35,438	$37,500	$43,224	$46,125	$47,380
	50th	36,000	37,392	40,000	40,000	47,500	47,858	50,050
	75th	40,000	42,350	43,392	44,034	52,700	53,000	56,015
		(53)	(64)	(53)	(45)	(60)	(62)	(46)
4 or more	25th	$34,949	$36,000	$32,635	$41,628	$41,939	$44,498	$48,271
	50th	39,530	39,550	40,320	45,236	45,166	54,300	50,157
	75th	46,000	42,400	46,500	49,880	50,325	60,925	59,007
		(38)	(28)	(29)	(30)	(32)	(26)	(33)
Associate 1-2	25th	$40,136	$39,143	$41,100	$45,732	$49,860	$52,373	$53,547
	50th	43,140	44,214	48,435	52,000	57,382	55,663	59,000
	75th	47,500	49,777	56,000	54,322	66,200	68,272	69,624
		(26)	(32)	(34)	(23)	(24)	(29)	(38)
3 or more	25th	$42,082	$40,185	$45,816	$48,098	$50,784	$50,270	$54,456
	50th	43,880	47,949	50,139	54,000	56,718	55,248	59,042
	75th	54,667	52,608	55,000	59,000	65,000	62,700	67,000
		(52)	(54)	(54)	(59)	(61)	(61)	(54)
Full 1-6	25th	$54,343	$51,372	$54,394	$56,583	$65,706	$51,195	$71,132
	50th	60,322	62,200	64,900	66,180	72,608	75,556	78,594
	75th	67,000	68,784	71,975	73,812	83,493	88,533	90,720
		(43)	(44)	(45)	(51)	(53)	(38)	(61)
7 or more	25th	$57,122	$58,793	$63,033	$63,803	$67,510	$71,381	$76,687
	50th	67,706	71,334	75,390	78,260	84,000	83,896	85,755
	75th	80,000	82,167	86,965	90,785	96,454	98,956	106,000
		(61)	(71)	(79)	(78)	(77)	(74)	(69)
Starting Assistant Professors	25th	$31,250	$34,960	$35,783	$37,250	$42,958	$41,000	$47,889
	50th	35,835	37,756,	37,009	42,000	45,000	45,000	49,440
	75th	40,000	41,350	45,150	44,250	47,500	57,000	54,600
		(26)	(37)	(32)	(29)	(21)	(10)	(5)

Using the Fall 1992 survey results, determine
a) what salary represents the 1st quartile for assistant professors with 1-3 years in rank?
b) what salary represents the 3rd quartile for assistant professors with 1-3 years in rank?
c) what percent of the assistant professors with 1-3 years in rank fall within Interquartile Range?
d) what is the Interquartile Range of associate professors with 1-2 years in rank?
e) Compare the Interquartile Range of the assistant professors with 1-3 years in rank with the Interquartile Range of the associative professors with 1-2 years in rank.
f) On the same scale, construct a box representing the 1st quartile, 3rd quartile and the median for the assistant professors with 1-3 years in rank and the associative professors with 1-2 years in rank. What can you conclude about the shape of each distribution?
g) Using the Fall 1991 and 1992 survey results, construct a box representing the 1st quartile, 3rd quartile and the median for the full professors with 1-6 years in rank on the same scale. Compare these two boxes, and comment on the differences. What is the Interquartile Range for each distribution, and why do you believe there is such a discrepancy? What do you think is the shape of each distribution?
h) For full professors with 7 or more years in rank:
   1) what salary represents the 3rd quartile?
   2) the 1st quartile?
   3) the median?
   4) What percent of the salaries fall within the interquartile range? How many professors have salaries within the interquartile range?
   5) Approximately how many of these professors have salaries between the first quartile and the median? between the median and the 3rd quartile?
   6) Calculate the range in salaries between the first quartile and the median? between the median and the 3rd quartile? Which has the smaller range?
   7) Associate with each range of part 6, the number and percent of professors that fall within that range. Comment on these results. What does the smaller range indicate regarding the variability of the salaries?

PART VI **Exploring DATA With MINITAB.**

1. For the distribution:
   Textbook Costs per Student per semester (in dollars)
   147 160 142 153 187 195 106 127
   178 178 272 175 178 133 186
   use MINITAB to:
   a) construct a Stem-and-Leaf Display and identify its shape.
   b) construct a Box-and-Whisker Plot and identify its shape.
   c) describe the data.
   d) find the 5-Number Summary.
   e) determine the Interquartile Range.
   f) identify potential or probable outliers.

2. Use MINITAB for the data values in the table to compare the amount of dietary fiber for three different cereal types.

   Amount of Dietary Fiber for Cereal Type
   (measured in grams)

Cereal	Whole-Grain Cereal with Bran	All-Bran Cereal
3.0	5.0	13
2.1	4.0	12
2.0	4.0	10
2.0	4.0	9
2.0	4.0	8
1.6	2.9	5

   For each cereal type use MINITAB to:

   a) construct a Box-and-Whisker Plot and identify its shape.
   b) describe the data.
   c) find the 5-Number Summary.
   d) identify potential or probable outliers.
   e) write a description comparing the similarities and/or differences for the cereal data.

PART VII **Projects.**

1. Look up the home run records of Babe Ruth, Mickey Mantle, Roger Maris, Mel Ott, Hank Greenberg, and Hank Aaron.

a) Calculate μ and σ for the number of home runs each player hit during his career.
b) For each player, take the maximum number of home runs he hit during any one season of his career and determine which player's maximum one year production is most outstanding relative to his career.

2. a) Calculate the mean and standard deviation of the number of electoral votes each Democratic and Republican presidential candidate received in each presidential election since 1900 inclusive.
b) From your data, choose the Democratic and Republican candidates who received the maximum number of electoral votes and calculate the z score of their electoral votes.
c) Compare the two z scores and decide which presidential candidate was the higher electoral vote getter relative to his party.

PART VIII Database.

The following exercises refer to the file DATABASE listed in Appendix A. We have indicated the appropriate MINITAB commands that are necessary to answer each exercise.

1. Using the MINITAB commands:
   RETRIEVE, DESCRIBE and BOXPLOT.
   a) Retrieve the file DATABASE.MTW and determine the measures of central tendency and variability, and the first and third quartiles for the variables:
   i) AGE
   ii) AVE
   iii) STY
   iv) GPA
   v) GFT
   b) For each of the variables of part (a), use the BOXPLOT command to construct a box-and-whisker plot. Are there any outliers?

2. Using the MINITAB commands:
   RETRIEVE AND DESCRIBE with the subcommand ALL
   Retrieve the file DATABASE.MTW and describe the variables:
   a) AGE
   What percent of the student ages are between 17 and 20?
   b) AVE
   What percent of the students have high school averages greater than 80?
   c) STY
   What percent of the students study more than 20 hours per week?
   d) GPA
   What percent of the students have grade point averages less than 3.0?
   e) GFT
   Determine the data value representing P_{90}.

3. Using the MINITAB commands:
   RETRIEVE, LET and PRINT
   Retrieve the file DATABASE.MTW, and compute and display a table representing, the data value, the z score for the data value, and the percentile rank for the data value for each of the variables:
   a) AGE
   b) AVE
   c) STY
   d) GPA
   e) GFT

CHAPTER 6
PROBABILITY

❄ 6.1 INTRODUCTION

The Birthday Problem.
A long long time ago
B.C.

By permission of Johnny Hart and Creators Syndicate, Inc.

Figure 6.1

200,000 years later
A. D.

Is Charlie Brown a loser? Lucy thinks so. In every wager Lucy makes with Charlie Brown, Charlie Brown loses. At tomorrow's Christmas Party Lucy plans to wager Charlie Brown a brand new bicycle that "at least two kids have the same birthday" (i.e. born on the same month and day).

Charlie Brown reasons that since there are going to be 50 kids at the Christmas Party and there are 365 possible birthdays, he believes that he is going to win his first bet ever with Lucy.

Do you think Charlie Brown's **chances** are good for winning this bet with Lucy?

Chance! Chance! Chance!

People have always been intrigued with **chance**. In fact an individual's own life involves chance. For example a person's sex is determined by the chromosomes of the parents. Also the purchase of a "lucky" lottery ticket could make one an instant millionaire.

As early as medieval times, man has been concerned with games of chance. In the 16th century, an Italian mathematician, Jerome Cordan (1501-1576) wrote the first book on this subject entitled: *The Book on Games of Chance*.

During the mid 17th century the French mathematician Pierre de Fermat (1601-1660) and Blaise Pascal (1623-1662) were inspired by a gambling friend to study the problem of how one should split the stakes in an interrupted dice game. Their study of this game of chance led to the development of the modern theory of **probability**.

In today's world probability plays an important role. The French mathematician Pierre Simon de Laplace (1749-1827) in his work, *Theorie Analytique des Probabilities* did a good job defining this role when he stated:

"...it is remarkable that a science which began with the consideration of games of chance should have become the most important object of human knowledge ...the most important questions of life are, for the most part, really only problems of probability."

Before we investigate problems of probability, let's develop some elementary probability concepts. By the way, Charlie Brown lost the bet! We will examine this problem in detail in Example 6.51.

6.2 SOME TERMS USED IN PROBABILITY

To begin the study of probability, several important terms used in probability need to be defined. The first term we need to examine is what we is meant by an experiment.

> **Definition: Experiment.** An experiment is the process by which an observation is made or obtained. An outcome of an experiment is any result that is obtained when performing the experiment.

For example rolling a fair die[1], tossing a fair coin, selecting a card from a deck of cards or selecting the winning numbers of a lottery ticket are all examples of an experiment. For the experiment: selecting a card from a regular deck of 52 playing cards, one possible outcome for this experiment is selecting the three of diamonds. Once an experiment is defined, it is necessary to determine how many outcomes and what outcomes are possible when performing an experiment. The collection of all the possible outcomes of an experiment is referred to as a sample space.

> **Definition: Sample Space.** A sample space is a representation of all the possible outcomes of an experiment.

---

[1] A fair die is a die that has the same chance of landing with a 1, 2, 3, 4, 5 or a 6 face up. Similarly, a fair coin is a coin that has the same chance of landing with heads face up as it does with tails face up.

In the next few examples, we will illustrate this idea of a sample space.

Example 6.1

A single toss of a fair coin is an experiment having **two** possible outcomes. Construct a sample space for this experiment.

Solution:

The sample space consists of the two possible outcomes: heads and tails. These outcomes are represented in Figure 6.2.

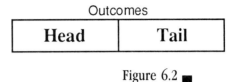

Figure 6.2 ∎

Example 6.2

A single roll of a fair die is an experiment having six possible outcomes. Construct a sample space for this experiment.

Solution:

The sample space for a single roll of a die is represented in Figure 6.3.

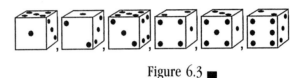

Figure 6.3 ∎

Example 6.3

Construct a sample space for the experiment of tossing a fair coin two times.

Solution:

Two tosses of a fair coin represent an experiment having four possible outcomes. The sample space consists of the following outcomes:

Outcome No. 1: A head on the 1st toss and a head on the 2nd toss, represented by (H,H).
Outcome No. 2: A head on the 1st toss and a tail on the 2nd toss, represented by (H,T).
Outcome No. 3: A tail on the 1st toss and a head on the 2nd toss, represented by (T,H).
Outcome No. 4: A tail on the 1st toss and a tail on the 2nd toss, represented by (T,T).

The sample space for two tosses of a fair coin is represented in Figure 6.4.

Outcome No. 1	Outcome No. 2
HH	**HT**
Outcome No. 3	Outcome No. 4
TH	**TT**

Figure 6.4 ∎

Example 6.4

An experiment consists of a single roll of a fair die and a single toss of a fair coin. Construct a sample space for this experiment.

Solution:

A single roll of a fair die has **6 possible outcomes** and a single toss of a fair coin has **2 possible outcomes**. To determine the total number of possible outcomes for this compound experiment, multiply the number of outcomes for rolling a single fair die by the number of outcomes for tossing a fair coin. This can be expressed using the following formula:

$$\begin{bmatrix} \text{number of outcomes} \\ \text{for rolling a die} \end{bmatrix} \times \begin{bmatrix} \text{number of outcomes} \\ \text{for tossing a coin} \end{bmatrix} = \begin{bmatrix} \text{total number of} \\ \text{possible outcomes} \end{bmatrix}$$

Therefore, there are 6 x 2 = 12 possible outcomes for this compound experiment. The 12 outcomes for this experiment are listed in Table 6.1.

Table 6.1

OUTCOME OF DIE	OUTCOME OF COIN
1	HEAD
1	TAIL
2	HEAD
2	TAIL
3	HEAD
3	TAIL
4	HEAD
4	TAIL
5	HEAD
5	TAIL
6	HEAD
6	TAIL

∎

Let's generalize the counting technique illustrated in Example 6.4. This counting principle is called the **Fundamental Counting Principle**. The Fundamental Counting Principle is used to determine the number of possible outcomes for a compound experiment or multi-stage experiment.

Fundamental Counting Principle
The total number of possible outcomes for a compound experiment is calculated by taking the product of the number of possible outcomes for each stage of the experiment.

When performing an experiment, we may be interested in a particular collection of outcomes. Such a collection of interested outcomes is called an **event**.

Definition: Event. An event is a particular collection of outcomes in the sample space of an experiment.

Let's examine this idea of an event in Example 6.5.

Example 6.5

Consider the experiment of tossing a fair coin twice.

a) Use the Fundamental Counting Principle to determine the number of possible outcomes in the sample space for this experiment.
b) Construct a sample space for this experiment.
Using the sample space, determine which outcomes satisfy the following events:
c) both tosses are heads
d) only one head
e) at least one head
f) at most one tail

Solution:

a) Using the **Fundamental Counting Principle** to determine the number of possible outcomes in the sample space, we need to multiply the number of possible outcomes for each toss.
That is,

$$\begin{bmatrix} \text{total number of} \\ \text{possible outcomes} \end{bmatrix} = \begin{bmatrix} \text{number of possible} \\ \text{outcomes for} \\ \text{the first toss} \end{bmatrix} \times \begin{bmatrix} \text{number of possible} \\ \text{outcomes for} \\ \text{the second toss} \end{bmatrix}$$

$$= [2] \times [2]$$

$$= 4$$

Thus, there are four possible outcomes for the experiment of tossing a fair coin twice.

b) The four outcomes in the sample space of this experiment are illustrated in Figure 6.5.

Outcome No. 1	Outcome No. 2
HH	HT
Outcome No. 3	Outcome No. 4
TH	TT

Figure 6.5

c) In examining the sample space of Figure 6.5, notice that there is **only one outcome** that satisfies **the event that both tosses of the coin are heads**. The outcome is: **HH**.

d) In the sample space of Figure 6.5, there are **two outcomes** that satisfy **the event that only one toss is a head**. The outcomes are: **HT and TH**.

e) **At least one head** *means* **one or more heads**. Therefore, we are looking for all the outcomes of this experiment that contain one or more heads. There are 3 outcomes satisfying **the event of at least one head**. As illustrated in Figure 6.5, these outcomes are: **HH, HT and TH**.

f) **At most one tail** *means* **one or less tails**. Therefore, we are looking for all the outcomes of this experiment that contain one or less tails. There are 3 outcomes satisfying **the event of at most one tail**. These 3 outcomes are: **HH, HT and TH**, as shown in Figure 6.5. ∎

The phrases "**at least**" and "**at most**" used to define events in Example 6.5 are frequently used in probability. It is important that you become familiar with the meaning of these expressions. Let's consider the general interpretation of these phrases.

Interpretation of the expressions:
"At Least" and "At Most"

"At least N" means N or more.

"At most N" means N or less.

Example 6.6

Consider the experiment of rolling a fair die once. The sample space consists of the following outcomes:

$$1, 2, 3, 4, 5, 6$$

Determine which outcomes satisfy the following events:
a) the outcome is an even number
b) the outcome is greater than 4
c) the outcome is less than 2
d) the outcome is a 7

Solution:

a) In the sample space, there are 3 outcomes that satisfy the event that the outcome is an even number. These outcomes are 2, 4, and 6.
b) In the sample space, there are 2 outcomes that satisfy the event that the outcome is greater than 4. These outcomes are 5 and 6.
c) In the sample space, there is 1 outcome that satisfies the event that the outcome is less than 2. The outcome is 1.
d) In the sample space, there is no outcome that satisfies the event that the outcome is a 7 since the highest number on a die is 6. ∎

❄ 6.3 Permutations and Combinations

When the total number of possible outcomes in a sample space is extremely large, there are techniques that can be used to count the outcomes of a sample space without having to list every possible outcome. The counting techniques are called **permutations** and **combinations**.

Definition: COUNTING RULE 3: Permutation Rule of N objects with k alike objects

Given N objects where n_1 are alike, n_2 are alike, ... , n_k are alike, then the number of permutations of these N objects is:

$$\frac{N!}{(n_1!)(n_2!) \cdots (n_k!)}$$

Example 6.15

Use counting rule 3 to determine the number of different arrangements for the letters in each of the following words:
a) OPINION
b) BANANA
c) MISSISSIPPI

Solution:

a) The word OPINION has seven letters with the letter O repeated twice, the letter I repeated twice and the letter N repeated twice, therefore by using counting rule 3, the number of different arrangements is:

$$\frac{7!}{(2!)(2!)(2!)} = \frac{(7)(6)(5)(4)(3)(2)(1)}{(2)(1)(2)(1)(2)(1)} = 630$$

b) The word BANANA has six letters with the letter A repeated three times and the letter N repeated twice, therefore the number of different arrangements is:

$$\frac{6!}{3!2!} = \frac{(6)(5)(4)(3)(2)(1)}{(3)(2)(1)(2)(1)} = 60$$

c) The word MISSISSIPPI has eleven letters with the letter I repeated four times, the letter S repeated four times and the letter P repeated twice, therefore the number of different arrangements is:

$$\frac{11!}{4!4!2!} = \frac{(11)(10)(9)(8)(7)(6)(5)(4)(3)(2)(1)}{(4)(3)(2)(1)(4)(3)(2)(1)(2)(1)} = 34,650 \quad \blacksquare$$

In our discussion of permutations we were concerned with the number of ways of arranging s objects selected from n objects in a *particular order*. Now we will consider a selection process where the *order is not important*. This selection process is called a combination.

Definition: Combination. A combination is a selection of objects **without** regard to order.

Let's examine this idea of a combination in Example 6.16.

Example 6.16

The chief Systems Analyst at H.A.L. Computer Systems, Inc. has to choose two programmers to work overtime on Sunday. If the company employs four programmers, then how many different ways can the Systems Analyst make her selection?

Solution:

We will use the letters A, B, C and D to represent the four programmers. The analyst would like all possible arrangements of the four programmers where the ordering of the programmers within each arrangement is *not important*. For example, the selection of programmers A and C where programmer A is chosen first and C second *is the same* as the selection of C and A where programmer C is chosen first and A second. Thus, the **order** *of selecting the programmers is* **not** *important*.

If **order were important**, then we could use Counting Rule 2 since this rule determines the number of ways to make a selection of n different objects taken s at a time. Thus, if order were important where n = 4 and s = 2, Counting Rule 2 would yield:

$$_4P_2 = (4)(3) = 12$$

However, the twelve different ways determined by counting rule 2 counts the number of permutations or arrangements of the four programmers if order were important. Within these 12 different arrangements, an arrangement like AC would be counted as a different arrangement than CA. However, the analyst would still end up with the *same* combination of programmers. Therefore to determine the number of different combinations of two programmers, we need to adjust these 12 arrangements for the fact that each pair of programmers is *counted twice*. Thus, we must divide the formula $_4P_2$ by 2 to determine the actual number of **different combinations** of 2 programmers.

Thus, there are:

$$\frac{_4P_2}{2} = \frac{(4)(3)}{2} = 6 \text{ different ways}$$

that the systems analyst can make her selection.

We can now generalize the counting technique developed in Example 6.16 to determine the number of ways that s objects can be selected from n objects where *order is not important*. We will call this technique counting rule 4.

Definition: COUNTING RULE 4: Number of Combinations of n objects taken s at a time

The number of combinations of n objects taken s at a time, symbolized as $\binom{n}{s}$, is:

$$\binom{n}{s} = \frac{_nP_s}{s!}$$

The formula for $\binom{n}{s}$ in factorial notation is written as:

$$\binom{n}{s} = \frac{n!}{s!(n-s)!}$$

Example 6.17

Use counting rule 4 to determine the number of combinations of n objects taken s at a time if:

a) n = 5 and s = 3.
b) n = 6 and s = 2.

For part a, use formula: $\binom{n}{s} = \dfrac{nP_s}{s!}$

For part b, use formula: $\binom{n}{s} = \dfrac{n!}{s!(n-s)!}$

Solution:

To determine the number of combinations of n objects taken s at a time, we will use counting rule 4.

a) If n = 5 and s = 3 then $\binom{5}{3} = \dfrac{{}_5P_3}{3!} = \dfrac{(5)(4)(3)}{(3)(2)(1)} = 10$

b) If n = 6 and s = 2 then $\binom{6}{2} = \dfrac{6!}{2!4!} = \dfrac{(6)(5)(4)(3)(2)(1)}{(2)(1)(4)(3)(2)(1)} = 15$

Example 6.18

Use counting rule 4 to determine the number of combinations for each of the following parts. For parts a and c, use the formula: $\binom{n}{s} = \dfrac{nP_s}{s!}$

For parts b and d, use the formula: $\binom{n}{s} = \dfrac{n!}{s!(n-s)!}$

a) the number of combinations of 6 objects taken 4 at a time.
b) the number of ways two books can be chosen from a group of five.
c) the number of ways a college student can choose 3 college courses out of 10 possible choices.
d) the number of ways a student can choose to answer 7 of the 10 questions on an examination.

Solution:

a) To determine the number of combinations of 6 objects taken 4 at a time, use the formula:
$\binom{n}{s} = \dfrac{nP_s}{s!}$

For n = 6 and s = 4, we have $\binom{6}{4} = \dfrac{{}_6P_4}{4!}$

Thus, the number of combinations of 6 objects taken 4 at a time is 15.

b) The number of ways to choose 2 books from a group of five is determined by using the formula:

$$\binom{5}{2} = \frac{5!}{10} = 10$$

Therefore, there are 10 ways to choose 2 books from five books.

c) The number of ways a student can select 3 college courses from 10 choices is:

$$\binom{10}{3} = \frac{_{10}P_3}{3!}$$

Thus, there are 120 ways to select 3 courses from 10 choices.

d) The number of ways a student can choose to answer 7 out of 10 questions is:

$$\binom{10}{7} = \frac{10!}{7!(10-7)!} = 120$$

Therefore, there are 120 ways to answer 7 out of 10 questions. ∎

Example 6.19

An offer from the Record of the Month Club to any new member is a choice of 5 records for 99 cents from a list of 25 records. How many combinations of 5 records can be chosen from this list?

Solution:

Since we are choosing from 25 records, then n = 25.

The fact that we are selecting 5 records makes s = 5. Therefore the number of combinations of 25 records taken 5 at a time is:

$$\binom{25}{5} = \frac{25!}{5!(25-5)!} = 53{,}130 \quad \blacksquare$$

Example 6.20

The Holiday Basketball Festival selection committee is considering 20 teams for their 12 team tournament. In how many ways can the 12 teams be selected?

Solution:

Because the committee is selecting from 20 teams, n = 20. Since they are selecting 12 teams, s = 12.

Therefore, the number of combinations of 20 teams taken 12 at a time is:

$$\binom{20}{12} = \frac{20!}{12!(20-12)!} = 125{,}970 \quad \blacksquare$$

Remark: The values of $\binom{n}{s}$ can be obtained using **TABLE I: The Success Occurrence in Appendix D: Statistical Tables** for values of n as large as 20. The use of TABLE I is discussed in Section 6.7.

6.4 PROBABILITY

In the previous sections, we developed techniques to help us count the total number of possible outcomes of a sample space. Now that we can determine the possible outcomes of an experiment we would like to investigate the chance or likelihood that a particular outcome or outcomes will happen. That is, we would like to calculate the probability of a particular outcome(s) of a sample space.

> **Definition: Probability.** The probability of an outcome is the chance or likelihood that the outcome will occur or happen.

The probability or chance of the outcome occurring is expressed as a number ranging from 0 to 1, or as a percentage ranging from 0 to 100%. Probability values can be expressed in the form of a: fraction, decimal or percent. For example, one could express the chance of getting a head on a single toss of a fair coin either as: ½ (fraction form) or 0.5 (decimal form) or 50% (percentage form).

If a particular outcome has a probability value of zero, then this is interpreted to mean that the outcome can never happen. While a probability value of one (or 100%) for an outcome indicates that the outcome will always happen. Figure 6.8 shows a scale representing the probability of an outcome and some interpretations of these probabilites.

Probability of an Outcome

fraction or decimal		percent	
0	——	0% :	outcome will never happen
½ or 0.50	——	50% :	outcome can go either way
1	——	100% :	outcome will definitely happen

Figure 6.8

There are different approaches to calculating the probability of an outcome. The approach we will use to assign an outcome a probability value is called the **Classical Approach to Defining Probability**. This classical approach can only be used if each outcome of all the possible outcomes of the experiment is equally likely.

> **Definition: Equally Likely.** All the possible outcomes of an experiment are equally likely if each outcome has the **same chance** of happening or occurring.

Consider the experiment of rolling a fair die once. The six outcomes: 1, 2, 3, 4, 5, 6 are equally likely, since the die is fair and each outcome is assumed to have an equal chance of happening.

If each possible outcome is equally likely, then the probability of an event E can be defined using the Classical Approach.

Classical Probability Definition for Equally Likely Outcomes

> **Definition: Probability of an Event E.** The probability of an Event E, written P(Event E), is equal to the number of outcomes satisfying the event E divided by the total number of outcomes in the sample space, provided each outcome within the sample space is equally likely.

This formula can be written as:

$$P(\text{Event E}) = \frac{\text{number of outcomes satisfying event E}}{\text{total number of outcomes in the sample space}}$$

Let's apply this classical approach to defining probability in Example 6.21.

Example 6.21

Consider once again the experiment of tossing a fair coin twice. The sample space for this experiment consists of the four outcomes illustrated in Figure 6.9.

| HH | HT | TH | TT |

Figure 6.9

Since the coin is *fair*, each outcome has the same chance of occurring. Find the probability of the event that when the fair coin is tossed two times it will land:
a) heads on both tosses, denoted P(HH).
b) tails on both tosses, denoted P(TT).
c) tails at least once, denoted P(at least one tail).

Solution:

a) The number of outcomes satisfying the event that the coin will land heads on both tosses is only 1. It is HH.
   Since the total number of outcomes in the sample space is 4 as illustrated in Figure 6.9, the probability of getting two heads is:

$$P(HH) = \frac{\text{number of outcomes satisfying two heads}}{\text{total number of outcomes}} = \frac{1}{4}$$

The probability of two heads is 1/4. This means that if the experiment of tossing a fair coin twice is repeated a large number of times, we would *expect* one-fourth of the outcomes to be two heads.

b) The number of outcomes satisfying the event two tails is one. From Figure 6.9, it is TT.
   Since the total number of outcomes in the sample space is 4, the probability of getting two tails is:

$$P(TT) = \frac{\text{number of outcomes satisfying two tails}}{\text{total number of outcomes}} = \frac{1}{4}$$

The probability of two tails is 1/4. Therefore, if this experiment were repeated a large number of times, we would expect 1/4 of the outcomes to be two tails.

c) At least one tail *means one or more tails*. The number of outcomes satisfying the event of at least one tail is 3. They are HT, TH and TT.

The total number of outcomes in the sample space is 4. Thus, the probability of at least one tail is:

$$P(\text{at least 1 tail}) = \frac{\text{number of outcomes satisfying at least 1 tail}}{\text{total number of outcomes}} = \frac{3}{4}$$

■

Example 6.22

An experiment consists of selecting one marble from an urn containing 4 red and 5 black marbles. The sample space consists of nine outcomes since there are nine marbles in the urn. The outcomes are listed in Figure 6.10.

Red	Red	Red	Red	Black	Black	Black	Black	Black

Figure 6.10

Since each marble has an equal chance of being selected the 9 outcomes are equally likely. Find the probability of selecting a:
a) red marble
b) black marble
c) green marble

Solution:

a) The number of outcomes satisfying the event of selecting a red marble from the urn is 4. The number of outcomes in the sample space is 9. Therefore,

$$P(\text{red marble}) = \frac{4}{9}$$

b) The number of outcomes satisfying the event of selecting a black marble from the urn is 5. The number of outcomes in the sample space is 9. Therefore,

$$P(\text{black marble}) = \frac{5}{9}$$

c) The number of outcomes satisfying the event of selecting a green marble from the urn is 0. Therefore,

$$P(\text{green marble}) = 0$$

■

Example 6.23

An experiment consists of tossing one fair die twice. Find the probability of the following events:
a) the sum of the two tosses is seven, denoted as P(sum is 7).
b) the sum of the two tosses is an odd number, denoted as P(sum is an odd number).
c) the sum of the two tosses is greater than six, denoted as P(sum > 6).

Solution:

To enable us to calculate these probabilities, we will use a sample space. But before we construct the sample space we should determine the *total number* of possible outcomes in the sample space by using the Fundamental Counting Principle. Thus, we have:

$$\begin{bmatrix} \text{total number} \\ \text{of} \\ \text{possible outcomes} \end{bmatrix} = \begin{bmatrix} \text{number of} \\ \text{possible outcomes} \\ \text{for first toss} \end{bmatrix} \times \begin{bmatrix} \text{number of} \\ \text{possible outcomes} \\ \text{for second toss} \end{bmatrix}$$

$$= [6] \times [6]$$

$$= 36$$

The 36 outcomes for this experiment are shown in Table 6.2.

Table 6.2

Outcome of First Toss	Outcome of Second Toss					
	1	2	3	4	5	6
1	1,1	1,2	1,3	1,4	1,5	1,6
2	2,1	2,2	2,3	2,4	2,5	2,6
3	3,1	3,2	3,3	3,4	3,5	3,6
4	4,1	4,2	4,3	4,4	4,5	4,6
5	5,1	5,2	5,3	5,4	5,5	5,6
6	6,1	6,2	6,3	6,4	6,5	6,6

a) There are six outcomes that satisfy the event that the sum of the tosses is seven. They are (6,1), (5,2), (4,3), (3,4), (2,5) and (1,6). The number of outcomes in the sample space is 36.

Therefore,

$$P(\text{sum is 7}) = \frac{6}{36}$$

b) The number of outcomes satisfying the event that the sum of the two tosses is odd is 18. Therefore,

$$P(\text{sum is an odd number}) = \frac{18}{36}$$

c) The number of outcomes satisfying the event the sum of the two tosses is greater than six is 21. Therefore,

$$P(\text{sum} > 6) = \frac{21}{36}$$ ■

Example 6.24

Five billiard balls numbered from 1 to 5 are placed in an urn. Two balls are selected in succession from the urn where the first ball selected is returned to the urn before the second ball is selected. Find the probability of selecting:
a) two even numbered balls
b) two odd numbered balls
c) one even and one odd numbered ball

Solution:

To help compute these probabilities, construct the sample space. To determine the total number of possible outcomes within the sample space, use the Fundamental Counting Principle.

$$\begin{bmatrix} \text{total number} \\ \text{of} \\ \text{possible outcomes} \end{bmatrix} = \begin{bmatrix} \text{number of} \\ \text{possible outcomes} \\ \text{for first selection} \end{bmatrix} \times \begin{bmatrix} \text{number of} \\ \text{possible outcomes} \\ \text{for second selection} \end{bmatrix}$$

$$= [5] \times [5]$$

$$= 25$$

The sample space for this experiment is listed in Table 6.3.

Table 6.3

Outcome of First Ball	Outcome of Second Ball				
	1	2	3	4	5
1	1,1	1,2	1,3	1,4	1,5
2	2,1	2,2	2,3	2,4	2,5
3	3,1	3,2	3,3	3,4	3,5
4	4,1	4,2	4,3	4,4	4,5
5	5,1	5,2	5,3	5,4	5,5

a) The number of outcomes satisfying the event of selecting two even numbered balls is 4. They are: (4,2), (2,4), (2,2) and (4,4). The number of outcomes in the sample space is 25. Therefore,

P(two even numbered balls) = $\frac{4}{25}$

b) The number of outcomes satisfying the event of selecting two odd numbered balls is 9. Therefore,

P(two odd numbered balls) = $\frac{9}{25}$

c) The number of outcomes satisfying the event of selecting one even and one odd numbered ball is 12. Therefore,

P(one odd and one even numbered ball) = $\frac{12}{25}$ ∎

CASE STUDY 6.1

Using the information contained in the three USA Snapshots of Figure 6.11, answer the following probability questions.

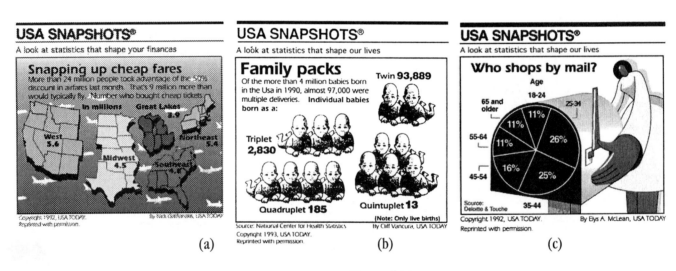

Figure 6.11

a) What is the probability that a person who purchased a cheap ticket last month was from the MidWest?
b) What is the probability that a person who purchased a cheap ticket last month was not from the Great Lakes region?
c) In 1990, what was the probability that a mother who had a multiple birth had twins?
d) What was the probability of a mother having 4 or more children during a multiple birth in 1990?
e) Approximate the probability of a mother having a multiple delivery during 1990.
f) What is the probability of a person who shops by mail is 18 to 24 years old?
g) What is the probability of a person who shops by mail is 65 or older?
h) What is the probability of a person who shops by mail is at least 35 years old? ∎

6.5 FUNDAMENTAL RULES AND RELATIONSHIPS OF PROBABILITY

Now that we have discussed the concept of probability, we are going to study some of the rules which probabilities must obey. The first basic rule states that the probability of an event must range from zero to one.

Probability Rule 1: Probability of an Event E ranges from 0 to 1

The probability of an event E is a number between 0 and 1 inclusive. Symbolically, this is expressed as:

$$0 \leq P(\text{Event E}) \leq 1$$

This rule is a consequence of the basic definition of the probability of an event.

The definition states:

$$P(\text{Event E}) = \frac{\text{number of outcomes satisfying event E}}{\text{total number of outcomes in the sample space}}$$

Since the number of outcomes satisfying event E and the total number of outcomes cannot be negative numbers, the number P(E) must be greater than or equal to zero. Furthermore since the number of outcomes satisfying event E would always be less than or equal to the total number of outcomes, the number P(E) would have to be less than or equal to one.

Probability Rule 1 indicates that the smallest probability of an event E is zero. Thus, if an event E cannot occur, then the probability of Event E is zero. That is, P(Event E) = 0.

Example 6.25

Determine the probability of rolling a fair die and obtaining a nine.

Solution:

Of the six possible outcomes there are zero outcomes corresponding to a nine. Therefore:

$$P(\text{rolling a 9}) = \frac{0}{6} = 0$$

Probability Rule 1 also indicates that the largest probability of an event E is one. An event with a probability of one is an event which is certain to occur.

Example 6.26

Determine the probability of rolling a fair die and obtaining a number less than seven.

Solution:

All six possible outcomes for this experiment are less than seven. Therefore:

$$P(\text{rolling a number less than seven}) = \frac{6}{6} = 1$$

The second probability rule explains the relationship about the sum of the probabilities of all the outcomes within a sample space.

**Probability Rule 2:
Sum of the Probabilities of All the outcomes of a Sample Space**

The sum of the probabilities of all the outcomes in the sample space of an experiment always equals 1.

Example 6.27

In the experiment of rolling a fair die, the sample space consists of the six outcomes illustrated in Figure 6.12.

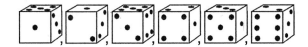

Figure 6.12

Examine the sample space. Notice that the probability for *each* of the six outcomes are as follows.

$$P(\text{rolling a 1}) = P(1) = \frac{1}{6}$$
$$P(\text{rolling a 2}) = P(2) = \frac{1}{6}$$
$$P(\text{rolling a 3}) = P(3) = \frac{1}{6}$$
$$P(\text{rolling a 4}) = P(4) = \frac{1}{6}$$
$$P(\text{rolling a 5}) = P(5) = \frac{1}{6}$$
$$P(\text{rolling a 6}) = P(6) = \frac{1}{6}$$

Observe that the sum of all these probabilities is one. That is,
$$P(1) + P(2) + P(3) + P(4) + P(5) + P(6) = 1.$$
This illustrates probability rule 2. ■

CASE STUDY 6.2

The American Cancer Society announced in January, 1991 that a woman's odds of getting breast cancer had risen to *1 in 9*.
The Cancer Society has been heavily publicizing this *1 in 9* statistic to persuade woman to have more regular mammograms and to examine their own breasts. However, this probability figure has terrified many women and made them feel doomed. In fact, many young woman have interpreted this *1 in 9* statistic to mean that one of their nine girl friends will get cancer this

year. However, this is not a correct conclusion based upon the calculated risk of getting breast cancer. Examine the chart shown in Figure 6.13 which indicates a woman's risk of getting breast cancer according to the American Cancer Society.

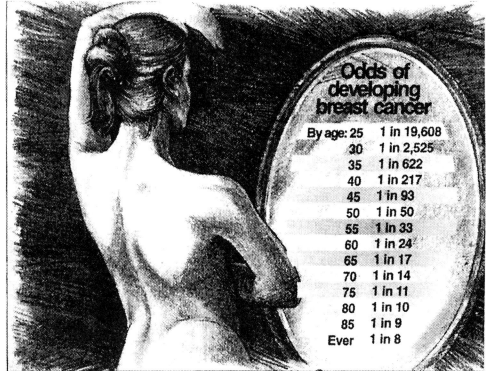

Figure 6.13

Using the information in Figure 6.13, what is the probability of a woman getting breast cancer:
a) by age 30?
b) by age 40?
c) by age 50?
d) by age 60?
e) by age 70?
f) by age 80?
g) by age 85?
h) By what age does a woman's risk rise to the *1 in 9* chance of getting breast cancer?
i) Would you interpret this *1 in 9* chance as a woman's risk of getting breast cancer for any particular age or the cumulative chance of a woman's risk of getting breast cancer during her lifetime? Explain.
j) Explain what you believe the American Cancer Society means when they say the *1 in 9* chance is meant to be "more of a metaphor than a hard figure." ■

The third probability rule explains how to calculate the probability of the event (**A or B**). This probability rule is called the **addition rule**.

Probability Rule 3: The Addition Rule

The probability of satisfying the event A or the event B is equal to the probability of event A plus the probability of event B minus the probability that both A and B occur at the same time.

The addition rule is written as:

$$P(A \text{ or } B) = P(A) + P(B) - P(A \text{ and } B)$$

Let's apply the addition rule to the probability problem in Example 6.28.

Example 6.28

In the experiment of tossing a fair die twice, what is the probability that the sum of the two tosses is greater than 9, or an odd sum?

Solution:

The sample space is illustrated in Table 6.4.

Table 6.4

Outcome of First Toss	Outcome of Second Toss					
	1	2	3	4	5	6
1	1,1	1,2	1,3	1,4	1,5	1,6
2	2,1	2,2	2,3	2,4	2,5	2,6
3	3,1	3,2	3,3	3,4	3,5	3,6
4	4,1	4,2	4,3	4,4	4,5	4,6
5	5,1	5,2	5,3	5,4	5,5	5,6
6	6,1	6,2	6,3	6,4	6,5	6,6

Using the addition rule, we write:

P(sum > 9 or odd sum) = P(sum > 9) + P(odd sum) − P(sum > 9 and odd).

Calculating each of the 3 probabilities (on the right of the equal sign), we have:

$P(\text{sum} > 9) = \dfrac{6}{36}$

$P(\text{odd sum}) = \dfrac{18}{36}$

$P(\text{sum} > 9 \text{ and odd}) = \dfrac{2}{36}$

Therefore, $P(\text{sum} > 9 \text{ or odd}) = \dfrac{6}{36} + \dfrac{18}{36} - \dfrac{2}{36} = \dfrac{22}{36}$

Example 6.29

Consider the experiment of selecting one card from an ordinary deck of 52 playing cards. The sample space for this experiment is shown in Figure 6.14.

Figure 6.14

The first row shows spades and clubs, while the second row shows hearts and diamonds. The club and spade suits are black and the diamond and heart suits are red. Each suit has 13 cards and contains the ranks: Ace, 2, 3, 4, 5, 6, 7, 8, 9, 10, jack, queen and king.

Find the probability of selecting:
a) a spade or a diamond
b) a king or a club
c) a picture card or a heart

Solution:

a) Using the addition rule, we have:

P(spade or diamond) = P(spade) + P(diamond) - P(spade and diamond)

Since:

$P(\text{spade}) = \frac{13}{52}$, $\quad P(\text{diamond}) = \frac{13}{52}$ and

since no *one* card is a spade *and* a diamond, then:

P(spade and diamond) = 0.

Therefore,

P(spade or diamond) = P(spade) + P(diamond) - P(spade and diamond)

$P(\text{spade or diamond}) = \frac{13}{52} + \frac{13}{52} - 0 = \frac{26}{52}$

b) Using the addition rule, we can write:

P(king or club) = P(king) + P(club) - P(king and club)

$P(\text{king}) = \frac{4}{52}$, $\quad P(\text{club}) = \frac{13}{52}$, and $\quad P(\text{king and club}) = \frac{1}{52}$

Therefore,
$$P(\text{king or club}) = \frac{4}{52} + \frac{13}{52} - \frac{1}{52} = \frac{16}{52}$$

c) Using the addition rule, we have:

P(picture card or heart) = P(picture card) + P(heart) − P(picture card and heart)
Since a picture card is either a king, a queen or a jack, then
$$P(\text{picture card}) = \frac{12}{52},$$
$$P(\text{a heart}) = \frac{13}{52},$$
$$P(\text{picture card and a heart}) = \frac{3}{52}$$
Therefore,
$$P(\text{picture card or heart}) = \frac{12}{52} + \frac{13}{52} - \frac{3}{52} = \frac{22}{52} \blacksquare$$

If two events A and B cannot occur at the same time, then the addition rule can be simplified. These events are said to be mutually exclusive. This leads us to the following definition.

Definition: Mutually Exclusive Events. Two events A and B are mutually exclusive if both events A and B cannot occur at the same time.

Since for mutually exclusive events P(A and B) = 0, then the addition rule for two mutually exclusive events A and B is simply the probability of event A plus the probability of event B. This leads us to the following probability rule.

Probability Rule 3A: The Addition Rule for Mutually Exclusive Events.

If two events A and B are mutually exclusive, then the probability that event A or event B will occur is equal to the sum of their probabilities.

This rule is written as:

$$P(A \text{ or } B) = P(A) + P(B)$$

The next example illustrates the concept of mutually exclusive events.

Example 6.30

Which of the following events are mutually exclusive?
a) Rolling a single die once and getting a 2 or an even number.
b) Rolling a single die once and getting a 2 or an odd number.
c) Rolling a single die once and getting a 3 or a 4.
d) Tossing a coin twice and getting two heads or at least one head.
e) Tossing a coin twice and getting at least two heads or at most one head.

Solution:

a) The sample space for rolling a single die once is given in Figure 6.13.

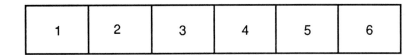

Figure 6.15

In Figure 6.16, we circled the outcomes that satisfy the event of getting a 2 and placed an X through the outcomes that satisfied the event of getting an even number.

Figure 6.16

Notice in Figure 6.16 the outcome 2 has a circle around it and an "X" through it. The events of getting a 2 or an even number can occur at the same time. Therefore, these two events are *not* mutually exclusive.

b) The sample space for rolling a single die once is given in Figure 6.17.

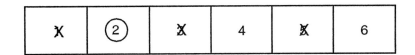

Figure 6.17

Circling the outcome(s) that satisfy the event of getting a 2 and placing an "X" through the outcome(s) that satisfy the event of getting an odd number we have:

| X | ② | X | 4 | X | 6 |

Figure 6.18

In Figure 6.18 we notice that there are no outcomes that satisfy both events at the same time. Therefore, the two events *are* mutually exclusive.

c) It is impossible to roll a single die once and get the two outcomes 3 and 4 at the same time. Therefore, the two events *are* mutually exclusive.

d) The sample space for tossing a coin twice is given in Figure 6.19.

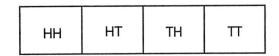

Figure 6.19

Circling the outcome(s) that satisfy the event of getting two heads and placing an "X" through the outcome(s) that satisfy the event of getting at least one head, we have:

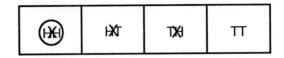

Figure 6.20

In Figure 6.20, we notice that there is one outcome that satisfies both events at the same time. Therefore, the two events are *not* mutually exclusive.

e) The sample space for tossing a coin twice is given in Figure 6.21.

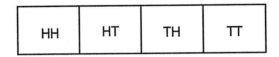

Figure 6.21

Circling the outcome(s) that satisfy the event of getting at least two heads and placing an "X" through the outcome(s) that satisfy the event of getting at most one head, we have:

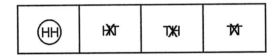

Figure 6.22

In Figure 6.22 we notice that there is no outcome that satisfies both events at the same time. Therefore, the two events *are* mutually exclusive. ∎

The fourth probability rule explains the relationship between the probability of an event E and the probability of the complement of event E. Let's first define the complement of event E.

> **Definition: Complement of event E.** Given an event E, the complement of event E, symbolized as E' and pronounced as E-complement, represents all the outcomes of an experiment that are not in event E.

Therefore, if event E occurs then the complement of event E cannot occur and vice versa. This leads us to the Complement Rule.

Probability Rule 4: The Complement Rule

The sum of the probabilities of event E and the complement of event E always equals one.

That is,

P(Event E) + P(complement of Event E) = 1

The complement rule can be symbolized as:

P(E) + P(E') = 1

The next few examples illustrate the application of the complement rule.

Example 6.31

Consider the experiment of rolling a fair die once. If event E represents the statement that the outcome is a 3, then compute the following probabilities:

a) P(E), that is P(rolling a 3)
b) P(E'), that is P(*not* rolling a 3)
c) P(E) + P(E')

Solution:

a) Of the six possible outcomes for a single roll of a fair die, only one is a 3. Therefore,

$$P(E) = \frac{1}{6}$$

b) Since there are five outcomes that are *not* a 3, then

$$P(E') = \frac{5}{6}$$

c) Using parts (a) and (b), we have:

$$P(E) + P(E') = \frac{1}{6} + \frac{5}{6}$$

$$P(E) + P(E') = 1 \quad \blacksquare$$

Example 6.32

A young couple planning to have 3 children have eight family possibilities which are illustrated in Table 6.5.

Table 6.5
Sample space for a 3 child family

First Child	Second Child	Third Child
boy	boy	boy
boy	boy	girl
boy	girl	boy
boy	girl	girl
girl	boy	boy
girl	boy	girl
girl	girl	boy
girl	girl	girl

Assuming each outcome is equally likely, compute the following probabilities.

a) P(couple will have 3 boys)
b) P(couple will not have 3 boys)
c) 1 − P(couple will have 3 boys)

Solution:

a) Of the eight possible outcomes listed in Table 6.5, only one outcome corresponds to the event the couple will have 3 boys. Therefore,

$$P(\text{couple will have 3 boys}) = \frac{1}{8}$$

b) Of the eight possible outcomes listed in Table 6.5, seven outcomes correspond to the event the couple will not have 3 boys. Therefore,

$$P(\text{couple will not have 3 boys}) = \frac{7}{8}$$

c) From part a, the $P(\text{couple will have 3 boys}) = \frac{1}{8}$

Therefore,

$$1 - P(\text{couple will have 3 boys}) = 1 - \frac{1}{8}$$

$$= \frac{7}{8} \quad \blacksquare$$

From Example 6.32, you should notice that the results of part b and part c are equal. That is,

P(couple will not have 3 boys) = 1 − P(couple will have 3 boys).

This is a direct result from the complement rule since it uses the probability of event E to compute the probability of the complement of event E. In general, this can be written as follows.

> **Computing the probability of the complement of event E using event E**
>
> $$P(\text{complement of event E}) = 1 - P(\text{event E})$$
>
> This can be further symbolized as:
>
> $$P(E') = 1 - P(E)$$
>
> Furthermore, we can also compute the probability of event E using the complement of event E. This is expressed as follows:
>
> $$P(\text{event E}) = 1 - P(\text{complement of event E})$$
>
> That is,
>
> $$P(E) = 1 - P(E')$$

The next example illustrates the use of these rules.

Example 6.33

If a young couple is planning on having three children then, the P(couple will have 3 girls) = $\frac{1}{8}$.

Using this information, compute the P(couple will not have 3 girls).

Solution:

Using: $P(E') = 1 - P(E)$, we have:
P(couple will not have 3 girls) = 1 - P(couple will have 3 girls)

$$= 1 - \frac{1}{8}$$
$$= \frac{7}{8}$$ ∎

Example 6.34

If a young couple is planning on having three children, then the P(couple will not have any girls) = $\frac{1}{8}$.

Using this information, compute the P(couple will have at least one girl).

Solution:

Using: $P(E) = 1 - P(E')$, we have:
P(couple will have at least one girl) = 1 - P(couple will not have any girls)

$$= 1 - \frac{1}{8}$$
$$= \frac{7}{8}$$ ∎

The next probability concept we will examine is the idea of independent events.

> **Definition: Independent Events.** Two events A and B are independent if the occurrence or nonoccurrence of event A has no influence on the occurrence or nonoccurrence of event B.

The following example illustrates the concept of independent events.

Example 6.35

For each of the following experiments determine which events are independent events.
a) Consider the experiment of rolling a single die twice. Let event A be the outcome that the 1st roll is a 5, while event B represents the outcome that the 2nd roll is a 2.
b) Consider the experiment of tossing a coin twice.
  Let event A be the outcome that the 1st toss is a head.
  Let event B be the outcome that the 2nd toss is a head.
c) Consider the experiment of selecting two marbles **without** replacement from an urn containing 5 blue marbles and 1 red marble. Let event A be the outcome that the first marble chosen is red while event B is the outcome that the 2nd marble chosen is red.

Solution:

a) When rolling a die twice, does the outcome of the first roll influence the outcome of the second roll? No.
  Thus, getting a 5 on the first roll has no influence on the second event B, getting a 2 on the second roll. Therefore, events A and B are independent.
b) When tossing a coin twice, does the outcome of the first toss influence the outcome of the second toss? No.
  Thus, getting a head on the first toss has no influence on getting a head on the second toss. Therefore, events A and B are independent.
c) If the red marble is selected on the first pick, then it cannot be selected on the second pick. Therefore, the two events are **not** independent. ■

Whenever two events, A and B, are independent, then the probability of A and B is very simple to calculate. This is illustrated in the next probability rule.

> **Probability Rule 5: Multiplication Rule for Independent Events.**
>
> If two events, A and B, are **independent**, then the probability of A and B equals the probability of A multiplied by the probability of B.
> This rule can be symbolized as:
>
> $$P(A \text{ and } B) = P(A) \cdot P(B)$$

The next few examples illustrate the use of the multiplication rule for independent events.

Example 6.36

Consider the experiment of tossing a fair coin twice. Find the probability of the following events:
a) both outcomes are heads
b) both outcomes are tails

Solution:

a) Since the successive tosses of a coin are independent, use the multiplication rule for independent events.

$$P(\text{head and head}) = P(\text{head}) \cdot P(\text{head}) = \left(\frac{1}{2}\right)\left(\frac{1}{2}\right) = \frac{1}{4}$$

Therefore, the probability of getting two heads in two tosses is $\frac{1}{4}$.

b) Using the multiplication rule for independent events, we have:
$P(\text{tail and tail}) = P(\text{tail}) \cdot P(\text{tail})$

$$= \frac{1}{2} \cdot \frac{1}{2} = \frac{1}{4}$$

Therefore, the probability of getting two tails in two tosses is $\frac{1}{4}$. ∎

Example 6.37

Two cards are selected **with replacement** from an ordinary deck of 52 playing cards. Find the probability of the following events:
a) both cards are clubs.
b) both cards are aces.
c) the first card is a picture card and the second card is an ace.

Solution:

a) First notice that in a deck of cards, two successive selections, with replacement of the card after each selection, are independent events. Therefore, we can apply the multiplication rule for independent events.

Therefore, $P(\text{club and club}) = P(\text{club}) \cdot P(\text{club}) = \frac{1}{4} \cdot \frac{1}{4} = \frac{1}{16}$

b) Similarly, for the event ace followed by ace with replacement, we can apply the multiplication rule for independent events.

Thus, $P(\text{ace and ace}) = \frac{1}{13} \cdot \frac{1}{13} = \frac{1}{169}$

c) Using the multiplication rule for independent events, we have:

$P(\text{picture card and ace}) = \frac{12}{52} \cdot \frac{1}{13} = \frac{12}{676}$

$$= \frac{3}{169}$$ ∎

Example 6.38

Two balls are selected **with replacement** from an urn containing 4 red, 6 green and 2 yellow balls. Find the probability of:
a) selecting 2 yellow balls.
b) selecting first a red ball and then a yellow ball.
c) selecting first a green ball and then a blue ball.

Solution:

a) Using the multiplication rule for independent events, we can write:
P(yellow and yellow) = P(yellow) · P(yellow)
$$= \frac{1}{6} \cdot \frac{1}{6} = \frac{1}{36}$$

b) Using the multiplication rule for independent events, we have:
P(red and then yellow) = P(red) · P(yellow)
$$= \frac{1}{3} \cdot \frac{1}{6} = \frac{1}{18}$$

c) Using the multiplication for independent events, we can write:
P(green and then blue) = P(green) · P(blue)
$$= \frac{1}{2} \cdot \frac{0}{12} = 0$$

PROBABILITY PROBLEMS USING PERMUTATIONS AND COMBINATIONS

In many instances when the outcomes of a sample space are too large to list, it becomes necessary to use the counting rules to calculate the probability. The next few examples illustrate the use of the counting rules of permutations and combinations in computing probability.

Example 6.39

If the four letters A, B, C, K are randomly arranged, what is the probability that the resulting arrangement is the word BACK?

Solution:

Since the order of the letters is important this is a permutation problem. First determine the number of ways the letters can be arranged using $_nP_n$, where n = 4.
Thus,
$$_4P_4 = 4!$$
$$= (4)(3)(2)(1)$$
$$= 24$$

Since only one of these arrangements is the word BACK, the probability that the resulting arrangement is the word BACK is:

$$P(\text{the arrangement spells BACK}) = \frac{1}{24}$$ ∎

Example 6.40

a) If the five letters A, I, C, R, G are randomly arranged what is the probability that the resulting arrangement is CRAIG?
b) If three of the previous five letters are selected and randomly arranged what is the probability that the result is CAR?

Solution:

a) Since the order of the letters is important, this is a permutation problem. There are $_5P_5$ ways to arrange all 5 letters. CRAIG is 1 of these arrangements, therefore:

$$P(\text{the arrangement spells CRAIG}) = \frac{1}{_5P_5} = \frac{1}{120}$$

b) Since we are concerned with the order of selecting and arranging only 3 of the 5 letters, there are $_5P_3$ or 60 arrangements. Therefore:

$$P(\text{the arrangement is CAR}) = \frac{1}{_5P_3} = \frac{1}{60}$$

■

Example 6.41

At a conference 9 executives consisting of 5 men and 4 women are required to sit along one side of a long table as illustrated in Figure 6.23. If they randomly take a seat what is the probability that the women occupy the even places?

Figure 6.23

Solution:

If we consider the seats to be numbered from 1 to 9, this arrangement requires that the four women occupy seats 2, 4, 6 and 8. There are $_4P_4$ ways that the women can be arranged in these even numbered seats. The 5 men must then occupy seats 1, 3, 5, 7 and 9. There are $_5P_5$ ways to arrange them in the odd numbered seats.

Hence using the **fundamental counting principle** we get:

$(_4P_4)(_5P_5)$ ways that the women can end up in the even numbered seats.

The total number of possible seating arrangements for all nine people is $_9P_9$. Thus, the probability that the women occupy the even places is:

$$\frac{(_4P_4)(_5P_5)}{_9P_9} = \frac{(24)(120)}{362{,}880}$$
$$= \frac{2880}{362{,}880}$$

■

Example 6.42

If 5 people are waiting for the bus and as it pulls up they get in line, what is the probability that Joe and Sean are *not* next to each other on line?

Solution:

There are $_5P_5$ or 120 different ways for the five people to get in line. How many of these *don't* have Joe and Sean next to each other? It's easier to determine the number of ways the people can get in line with Joe and Sean *next* to each and then subtract from 120 (the total number of ways) to get the number we need. For Joe and Sean to be next to each other they can be arranged either as Joe and Sean or Sean and Joe. If we treat Joe and Sean as **one item**, then there are now $_4P_4$ ways to arrange the other three people with the duo of Joe and Sean. Remembering that Joe and Sean can be arranged in 2 ways we get: $2(_4P_4) = 2(4!) = 48$ ways to form a line with Joe and Sean next to each other. Hence, the number of ways of Joe and Sean *not* being next to each other is:

$$_5P_5 - 2(_4P_4) = 120 - 48 = 72$$

Therefore, the probability of Joe and Sean *not* being next to each other on the bus line is:

$$P(\text{Joe and Sean are not next to each other}) = \frac{_5P_5 - 2(_4P_4)}{_5P_5} = \frac{72}{120}$$ ∎

Example 6.43

Using a regular deck of 52 playing cards:
a) how many different five card poker hands are possible?
b) what is the probability of being dealt five hearts in a five card poker hand?
c) what is the probability of being dealt a flush (all five cards of the same suit) in a five card poker hand?
d) what is the probability of being dealt 4 kings in a 5 card poker hand?
e) what is the probability of being dealt 4 of a kind in a 5 card poker hand?

Solution:

a) Since there are 52 cards in a deck, n = 52. A poker hand contains 5 cards, therefore: s = 5. Using the formula,

$$\binom{n}{s} = \frac{n!}{(n-s)!s!}$$

we have:

$$\binom{52}{5} = \frac{52!}{47!5!} = 2{,}598{,}960 \text{ different five card poker hands}$$

b) Since a deck contains thirteen hearts and any five would make a five heart hand, there are $\binom{13}{5}$ five card heart hands. Therefore,

$$P(\text{five heart poker hand}) = \frac{\binom{13}{5}}{\binom{52}{5}} = \frac{1287}{2{,}598{,}960} = 0.0005$$

c) Since there are four suits (diamonds, hearts, clubs and spades) and a flush is possible in any suit, then the probability of being dealt a flush is *four* times the probability of being dealt a flush in any one particular suit. From part (b) we know the probability of a flush in hearts is 0.0005. Thus the probability of a flush in any one suit is also 0.0005.

Therefore,

$$P(\text{being dealt a flush}) = 4(\text{probability of a flush in any one particular suit})$$

$$= 4(0.0005)$$

$$= 0.002$$

d) Since there are only 4 kings, you must be dealt all of them. This can only happen in one way. The fifth card can be any one of the remaining 48 cards. Therefore, there are 48 ways to get a fifth card or 48 five card hands which contain 4 kings. Thus,

$$P(\text{4 kings in a 5 card poker hand}) = \frac{48}{2{,}598{,}960} = 0.0000185$$

Remark: The probability of 4 of any one kind (4 aces, 4 sevens, etc) is the same as P(4 kings in a 5 card poker hand).

e) Since there are thirteen ranks within any suit, ace through king, and 4 of a kind is possible for any rank (i.e. 4 aces, 4 two, etc.) thus the
$$P(\text{being dealt 4 of a kind for any rank}) = 13(\text{probability of four of a kind for any particular rank})$$

$$= 13 \, (0.0000185)$$

$$= 0.0002405 \qquad \blacksquare$$

Example 6.44

If you're a member of an office staff of 25 clerks, what is the probability that you'll be selected as a member of a three person committee which is randomly chosen from the 25 clerks?

Solution:

Since there are 25 clerks in the office and 3 clerks are to be chosen where order is not important, then there are $\binom{25}{3}$ ways to select the committee of 3 clerks. If you're to be included on this committee then only two other clerks from the remaining 24 clerks can be chosen. This is done in $\binom{24}{2}$ ways.

Therefore,

$$P(\text{you are selected to committee of 3 clerks}) = \frac{\binom{24}{2}}{\binom{25}{3}} = \frac{276}{2300} = 0.12 \qquad \blacksquare$$

CASE STUDY 6.3

The USA Snapshot in Figure 6.24 indicates the number of $1 tickets necessary to cover every 6-number combination in selected lotto games.

Figure 6.24

a) What counting technique is used to determine the total number of possible 6-number choices?

b) What is the total number of ways to choose every possible 6-number combination out of 51 numbers? How does this result compare to the value listed in Figure 6.2 for the state of California?

c) What is the total number of ways to choose every possible 6-number combination out of 47 numbers? How does this result compare to the value listed in Figure 6.2 for the states of Michigan and Ohio?

d) For the state of California, what is the probability that a person who buys a $1 lotto ticket will win the lottery?
For the state of Michigan, what is the probability that a person who buys a $1 lotto ticket will win the lottery?

e) For the states New York and Illinois, what is the number of ways to choose every possible 6-number combination out of 54 numbers? What is the probability that a person who buys a $1 lotto ticket in New York will win the lottery?

f) In which state does a person have the greatest number of possible ways to select every 6-number combination?

h) For a $1 ticket, is the probability of winning the lottery greater for the state of California or the state of New York? Does this result agree with the result of part (f)? Explain.

CASE STUDY 6.4

The article entitled *1-in-a-Trillion Coincidence, You Say? Not Really, Experts Find* appeared in the New York Times on February 27, 1990. The article explains why seemingly unlikely events are almost certain to happen.

Science Times

1-in-a-trillion Coincidence, You Say? Not Really, Experts Find

Statisticians show that events that look unlikely are almost to be expected.

By GINA KOLATA

COINCIDENCES, those surprising and often eerie events that add spice to everyday life, may not be so unusual after all, researchers say.

After spending 10 years collecting thousands of stories of coincidences and analyzing them, two Harvard statisticians report that virtually all coincidences can be explained by some simple rules.

Some of the analyses performed by them or other statisticians showed that events that looked extremely unlikely were almost to be expected. When a woman won the New Jersey Lottery twice in four months the event was widely reported as an amazing coincidence that beat odds of one in 17 trillion. But when carefully analyzed, it turned out that the chance that such an event could happen to someone somewhere in the United States was more like one in 30.

It was an example of what the authors, Dr. Persi Diaconis, a professor of mathematics at Harvard University, and Dr. Frederick Mosteller, an emeritus mathematics professor at Harvard, call "the law of very large numbers." That long-understood law of statistics states, in their formulation: "With a large enough sample, any outrageous thing is apt to happen."

Some of the findings are published in the December issue of The Journal of the American Statistic Association; Others are now appearing in order professional journals, Dr. Diaconis, whose work also led to the recent discovery that seven shuffles are needed to mix a 52-card deck randomly, said the findings on coincidence were meeting mixed reactions. "Some people are enormously relieved, but others are furious," he said. Not everyone who supplied an amazing coincidence to the researchers wanted to hear that a cherished dramatic story was really nothing special.

"I think the whole subject is fascinating," said Dr. Erich Lehmann, a statistician at the University of California at Berkeley. Although it can be a difficult statistical problem to decide just how unlikely an event is, Dr, Lehmann said, there is no dispute about the validity of the findings of Dr. Diaconis and Dr. Mosteller, some of which have been discussed at statistical meetings in recent years.

The two Harvard statisticians reviewed a large body of calculations and analyses of coincidences performed by other researchers, and they devised new techniques and approaches for studying the phenomenon in a wide range of circumstances. The research results, Dr. Diaconis said, "are aimed at very basic problems of inference that arise in messy, real statistical problems."

Dr. Bradley Efron, a statistician at Stanford University, said coincidences arise "all the time" in statistical work. When researchers find clusters of odd cancers or birth defects or other diseases, statisticians are asked "to decide which events are the luck of the draw" and which may reflect some underlying cause, Dr. Efron said. "That's what Persi and Fred are trying to unravel," he added. "I think it's a very interesting enterprise."

Dr. Diaconis said one application of the new analyses is in scrutinizing data from clinical trials of new drugs. "as you are looking around in that mass of data," he said, "you find that in a certain subgroup, there are twice as many deaths in people taking drug A as there are in people taking drug B." Is it a coincidence or an indication that drug A is so dangerous to some patients that the trial must be stopped?

Dr. Diaconis and Dr. Mosteller said they decided to study coincidences because they were fascinated by the role these odd events play in everyone's lives." "All of us feel that our lives are driven by coincidences," Dr. Diaconis said. "Who we live with and where we work, why we do the things we do often rest on slim coincidences." These chance events "touch us very deeply," he said.

The two statisticians defined a coincidence as "a surprising concurrence of events, perceived as meaningfully related, with no apparent causal connection."

Dr. Diaconis and Dr. Mosteller began with the presumption that there are no extraordinary forces outside the realm of science that are acting to produce coincidences. But they also recognized that seeming coincidences are an important source of insight in science and so should not be dismissed out of hand. What looks like a coincidence may in fact have hidden cause, which can lead to a new understanding of a phenomenon. A sequence of odd blips on a chart, a clustering of cases of a rare disease can tell researchers that a new event is occurring.

Anthologies of Happenstance,

A decade ago, Dr. Diaconis and Dr. Mosteller started asking their colleagues, friends and friends of friends to send them examples of surprising coincidences. The collection quickly mushroomed. Dr. Mosteller said he had 13 notebooks, each three and a half inches thick, full of coincidences. "These notebooks are eating up the shelves in my den," he said. Dr. Diaconis said he had 200 file folders full of coincidences.

When they began to study these coincidences, they learned that they fell into several distinct groups. Some coincidences have hidden causes and are thus not really coincidences at all. Others arise from psychological factors, like selective memory or sensitivities, that make people think particular events are unusual whether they are or not. But many coincidences are simply chance events that turn out to be far more likely statistically than most people imagine. The analysis often required the researchers to develop new statistical methods, but in the end almost all coincidences could be analyzed.

The law of truly large numbers, which explains the double winner of the New Jersey Lottery, says that even if there is only a one-in-a-million chance that something will happen, it will happen eventually given enough time of enough people. "It's the blade-of-grass paradox," Dr. Diaconis said. "Suppose I'm standing in a large field and I put my finger on a blade of grass. The chance that I would choose that particular blade may be one in a million. But it is certain that I will choose a blade." So if something happens to only one in a million people per day and the population of the United States is 250 million, "you expect 250 amazing coincidences every day."

"If one-in-a-million thing happens to you, you start telling people about it," Dr. Diaconis went on. "You might say to me, 'so what do you think of that, wise guy?' And I say, 'It's an example of the law of truly large numbers.'"

Right Answer, Wrong Question

When a New Jersey woman won the lottery twice in a four-month period, it was reported as a one-in-17-trillion long shot. Narrowly speaking, that is correct. But as Dr. Diaconis and Dr. Mosteller reported, one in 17 trillion is the odds that a given person who buys a single ticket for exactly two New Jersey lotteries will win both times. The true question, they say, is, "What is the chance that some person, out of all the millions and millions of people who buy lottery tickets in the United States, hits a lottery twice in a lifetime?"

That event was called "practically a sure thing" by Dr. Stephen Samuels and Dr,. George McCabe, two statisticians at Purdue University. Over a seven-year period, they concluded, the odds are better than even that there will be a double lottery winner somewhere in the United States. Even over a four-month period, the odds of a double winner somewhere in the country are better than one in 30.

Another principle that demystifies many coincidences is what the researchers call "multiple end points"–occasions when what might qualify as a coincidence is not spelled out ahead of time, and when many chance events would qualify. This could apply, for example, at a party, where two people might discover that they come from the same town. this may seem a surprising coincidence. But the truth is that almost anything two strangers have in common would count as a coincidence – same first or last name, the same birthday, the same item of clothing.

"Clearly, the chances of getting a match in any of several things is bigger than if you look at just one thing," Dr. Diaconis said. He and Dr. Mosteller have developed a formula that can evaluate such problems.

Multiple end-point coincidences often sound amazing on the surface. For example, Dr. Diaconis said, "I had a friend who said, 'Gee, I was watching a James Bond movie and there was a four-digit code on a bomb that was exactly the same as the code on my Israeli bank account.'" It sounded extraordinary, since there are 10,000 possible four-digit numbers. But Dr. Diaconis went on: "If you know 120 numbers–Social Security

Researchers say a two-time lottery winner is 'practically a sure thing.'

numbers, bank codes, telephone numbers of friends–there are even odds that four digits of two of them will match," he said.

A third category of coincidences is those that are close but not exact. The odds of a coincidence then go up enormously. Once again, the two harvard statisticians have developed a formula to analyze these problems.

Birthdays and Near Birthdays

For example, with 23 people in a room, the chances are even that two of them will have the same birthday. But with only 14, there is an even chance of finding two people born within a day of each other, and if you ask instead that the birthdays match within a week, you need only seven people. "I have a friend who said, 'My daughter, myself, and my husband all have a birthday on the 11th of a month,'" Dr. Diaconis recalled. "O.K., there are 30 categories, 30 days of the month. How many birthdays do we have to know so that three are on the same day of the month?" The answer, he calculated, is 18. "So if you know 18 people, it's even odds that three will be born on the same day of the month," he explained. So the friend's birthday coincidence, he said, "is not so unusual."

By analyzing coincidences with these three principles in mind, Dr. Diaconis and Dr. Mosteller find that what looks unexpected usually turns out to be expected. No strange forces outside the realm of science are needed to explain coincidences.

"Why does an educated person think there might really be something in coincidences?" Dr. Diaconis asked, and answered his own question: "No one story holds up on its own. But taken together, they mean something." If you put all these near coincidences together, doesn't it indicate that something strange is going on?

It does not, Dr. Diaconis replied, and quoted another fundamental law of logic: "A lot of flawed arguments don't produce a sound conclusion."

a) Read the article and explain what is meant by the law of very large numbers.
b) To win the New York State lottery, an individual must select the six correct numbers out of 54 possible numbers.
   How many possible ways can a person select six such numbers?
c) If an individual gets to play two games for each $1 lottery ticket, then what is the probability that the individual will win when he/she buys a $1 lottery ticket?
d) As the lottery prize money increases, the number of people that play the lottery also increases. Keeping this in mind explain why when the lottery pot hits $20 million you might expect to have multiple winners even though the probability is extremely small as shown in part (c).
e) Explain what is meant by a coincidence?
f) Explain the different distinct groups that the coincidences fall into as categorized according to the article.
g) What is meant by the blade of grass paradox? How does this relate to an individual winning the lottery?

birthday, the same item of clothing.

"Clearly, the chances of getting a match in any of several things is bigger than if you look at just one thing," Dr. Diaconis said. He and Dr. Mosteller have developed a formula that can evaluate such problems.

Multiple end-point coincidences often sound amazing on the surface. For example, Dr. Diaconis said, "I had a friend who said, 'Gee, I was watching a James Bond movie and there was a four-digit code on a bomb that was exactly the same as the code on my Israeli bank account.'" It sounded extraordinary, since there are 10,000 possible four-digit numbers. But Dr. Diaconis went on: "If you know 120 numbers–Social Security

Researchers say a two-time lottery winner is 'practically a sure thing.'

numbers, bank codes, telephone numbers of friends–there are even odds that four digits of two of them will match," he said.

A third category of coincidences is those that are close but not exact. The odds of a coincidence then go up enormously. Once again, the two Harvard statisticians have developed a formula to analyze these problems.

Birthdays and Near Birthdays

For example, with 23 people in a room, the chances are even that two of them will have the same birthday. But with only 14, there is an even chance of finding two people born within a day of each other, and if you ask instead that the birthdays match within a week, you need only seven people. "I have a friend who said, 'My daughter, myself, and my husband all have a birthday on the 11th of a month,'" Dr. Diaconis recalled. "O.K., there are 30 categories, 30 days of the month. How many birthdays do we have to know so that three are on the same day of the month?" The answer, he calculated, is 18. "So if you know 18 people, it's even odds that three will be born on the same day of the month," he explained. So the friend's birthday coincidence, he said, "is not so unusual."

By analyzing coincidences with these three principles in mind, Dr. Diaconis and Dr. Mosteller find that what looks unexpected usually turns out to be expected. No strange forces outside the realm of science are needed to explain coincidences.

"Why does an educated person think there might really be something in coincidences?" Dr. Diaconis asked, and answered his own question: "No one story holds up on its own. But taken together, they mean something." If you put all these near coincidences together, doesn't it indicate that something strange is going on?

It does not, Dr. Diaconis replied, and quoted another fundamental law of logic: "A lot of flawed arguments don't produce a sound conclusion."

a) Read the article and explain what is meant by the law of very large numbers.
b) To win the New York State lottery, an individual must select the six correct numbers out of 54 possible numbers.
   How many possible ways can a person select six such numbers?
c) If an individual gets to play two games for each $1 lottery ticket, then what is the probability that the individual will win when he/she buys a $1 lottery ticket?
d) As the lottery prize money increases, the number of people that play the lottery also increases. Keeping this in mind explain why when the lottery pot hits $20 million you might expect to have multiple winners even though the probability is extremely small as shown in part (c).
e) Explain what is meant by a coincidence?
f) Explain the different distinct groups that the coincidences fall into as categorized according to the article.
g) What is meant by the blade of grass paradox? How does this relate to an individual winning the lottery?

Table 6.7
Selection of two balls *without replacement*

Outcome for First Ball	Outcome for Second Ball		
	1	2	3
1	X	1,2	1,3
2	2,1	X	2,3
3	3,1	3,2	X

Notice that the sample space does not have the outcomes 1,1; 2,2; and 3,3 since it is impossible to select the same numbered ball twice.

Using the sample space in Table 6.7, the
P(selecting two odd numbered balls without replacement) = 2/6 or 1/3

Comparing the answers to parts (a) and (b) of Example 6.45, notice that the results are different. This is attributed to the fact that in part (a), the two selections were *independent*. However in part (b), the selection of the second ball **was affected by the outcome of the first selection**. Therefore, we refer to this type of selection process as being *dependent*. Let's now formulate a multiplication rule for **dependent events**.

Probability Rule 6:   MULTIPLICATION RULE FOR DEPENDENT EVENTS: CONDITIONAL PROBABILITY

If two events A and B are dependent, then the probability of A *and* B equals the probability of A multiplied by the probability of B given that A has occurred.

This can be symbolized as:

P(A and B) = P(A)·P(B, given A has occurred)

This can be further symbolized as:

P(A and B) = P(A)·P(B | A)

where: P(B | A) means P(B, given A has occurred).

We will use the multiplication rule for dependent events to recalculate the P(selecting two odd numbered balls without replacement) as described in Example 6.45 part (b). Using Probability Rule 6, the P(selecting two odd numbered balls without replacement) can be written:

P(selecting 2 odd balls without replacement) = P(1st ball is odd) · P(2nd ball is odd | 1st ball was odd)

To compute the P(1st ball is odd), examine the urn pictured in Figure 6.25, where the three numbered balls have now been described as either odd or even.

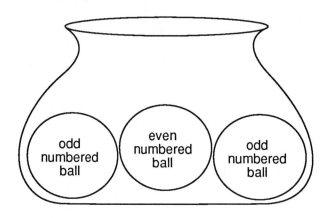

Figure 6.25

Using Figure 6.25,

P(1st ball selected is odd) = $\frac{2}{3}$

To calculate:

the P(2nd ball selected is odd | 1st ball selected was odd), the urn in Figure 6.25 must be **modified** to reflect the condition that the 1st ball selected was odd. This modified urn is illustrated in Figure 6.26.

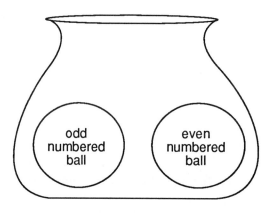

Figure 6.26

Using Figure 6.26, the:

P(2nd ball selected is odd | 1st ball selected was odd) = $\frac{1}{2}$

Therefore,

P(selecting 2 odd balls without replacement) = P(1st ball is odd) · P(2nd ball is odd | 1st ball was odd)

$$= \frac{2}{3} \cdot \frac{1}{2}$$
$$= \frac{2}{6} \text{ or } \frac{1}{3}$$

Notice the previous result obtained using probability rule 6 is the same as the result computed in Example 6.45 part b.

Example 6.46

If two balls are selected *without* replacement from an urn containing two red balls and three white balls, then what is the probability of choosing:
a) 2 white balls?
b) 2 red balls?
c) a red ball and a white ball?

Solution:

a) To compute the P(of selecting 2 white balls) without replacement we must use the multiplication rule for **dependent events** since the selection of the 2nd white ball is dependent upon the selection of the 1st white ball.

Using Probability Rule 6, we have:

P(selecting 2 white balls) = P(1st ball is white) · P(2nd ball is white | 1st ball was white)

To compute these probabilities, examine the urns shown in Figure 6.27.

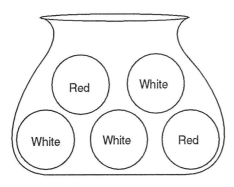
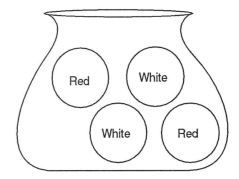

Urn prior to first selection   Urn prior to second selection

Figure 6.27a   Figure 6.27b

To determine the P(1st ball selected is white), examine the urn pictured in Figure 6.27a. Therefore,

$$P(\text{1st ball selected is white}) = \frac{3}{5}$$

Examine Figure 6.27b to help calculate the:

P(2nd ball selected is white | 1st ball was white).

Using Figure 6.27b, we have:

$$P(\text{2nd ball selected is white} \mid \text{1st ball was white}) = \frac{2}{4}$$

Now substitute these two results into Probability Rule 6, we get the following result:

P(selecting 2 white balls) = P(1st ball is white)·P(2nd ball is white | 1st ball was white)

$$= \frac{3}{5} \cdot \frac{2}{4}$$

$$= \frac{6}{20} \text{ or } \frac{3}{10}$$

b) To compute the P(of selecting 2 red balls) without replacement, use the multiplication rule for dependent events. Thus,

P(selecting 2 red balls) = P(1st ball is red)·P(2nd ball is red | 1st ball was red)

$$= \frac{2}{5} \cdot \frac{1}{4}$$

$$= \frac{2}{20} \text{ or } \frac{1}{10}$$

c) In order to compute the P(of selecting a red ball and a white ball), we must consider the fact that there are two distinct ways of selecting a red and a white ball from the urn. They are:

1st way	or	**2nd way**
selecting a red ball first then a white ball		*selecting a white ball first then a red ball*

That is,

a red ball then white   or   *a white ball then red*

Therefore, we can express

P(of selecting a red ball and a white ball) as

P(a red ball first then a white ball or a white ball first then a red ball).

Using the addition rule for mutually exclusive events, Probability Rule 3A, we have:

P(a red and a white ball) = P(first a red ball then a white ball or first a white ball then a red ball)

= P(first a red ball then a white ball) + P(first a white ball then a red ball)

First, we will compute the P(of selecting a red ball first and then a white ball), using Probability Rule 6.

Thus, we have:

P(1st a red ball then a white ball) = P(1st ball is red) · P(a white ball is selected | 1st ball was red)

$$= \frac{2}{5} \cdot \frac{3}{4}$$

$$= \frac{6}{20} \text{ or } \frac{3}{10}$$

Second, we will compute the:

P(of selecting a white ball first and then a red ball) using Probability Rule 6.

Thus,

P(1st a white ball then a red ball) = P(1st ball is white) · P(red ball is selected | the 1st ball was white)

$$= \frac{3}{5} \cdot \frac{2}{4}$$

$$= \frac{6}{20} \text{ or } \frac{3}{10}$$

Adding the last two results together, we have:

P(selecting a red and a white ball) =

= P(selecting a red ball first then a white ball) + P(selecting a white ball first then a red ball)

$$= \frac{3}{10} + \frac{3}{10}$$

$$= \frac{6}{10} \text{ or } \frac{3}{5}$$

In the previous example, you might have noticed that the P(of selecting a red ball first and then a white ball) and P(of selecting a white first and then a red ball) were equal. This is always true, and is referred to as the commutative probability rule. In general, we can state the commutative probability rule as:

Commutative Probability Rule

For two events A and B, the following probabilities are equal:
P(A and B) = P(B and A).

Using Probability Rule 6, we can write:

P(A and B) = P(A) · P(B | A)

and also

P(B and A) = P(B) · P(A | B)

These two results can be combined to form a new probability rule. This probability rule is called the **general multiplication rule**.

Probability Rule 7: General Multiplication Rule

For any two events A and B, the probability of A and B can be determined by either one of the following formulas:

$$P(A \text{ and } B) = P(A) \cdot P(B \mid A)$$

In words, the probability of A and B is equal to the probability of A multiplied by the probability of B, given that event A has occurred.

OR

$$P(A \text{ and } B) = P(B) \cdot P(A \mid B)$$

In words, the probability of A and B is equal to the probability of B multiplied by the probability of A, given that event B has occurred.

Example 6.47

Consider the hopper in Figure 6.28 containing 6 balls where two balls are numbered 1, two balls are numbered 2 and two balls are numbered 3.

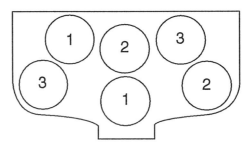

Figure 6.28

If this hopper is used to create a two digit numeral for a lottery game by selecting two balls in succession without replacement, find the probability that:
a) the lottery number 32 is selected.
b) the lottery number 33 is selected.

Solution:

a) To compute the P(the number 32 is selected) use Probability Rule 6 for dependent events since the second numbered ball selected is dependent upon the first selected ball.

P(the number 32 is selected) = P(first ball selected is a 3 and the second ball is a 2)
$$= P(\text{first ball is a 3}) \cdot P(\text{second ball is a 2} \mid \text{first ball is a 3})$$

$$= \frac{2}{6} \cdot \frac{2}{5}$$

$$= \frac{4}{30} \text{ or } \frac{2}{15}$$

b) To compute P(the number 33 is selected), we will use the Probability Rule 6 for dependent events.

P(the number 33 is selected) = P(first ball selected is a 3 and the second ball is a 3)

= P(first ball is a 3) · P(first ball is a 3 | second ball is a 3)

$$= \frac{2}{6} \cdot \frac{1}{5}$$

$$= \frac{2}{30} \text{ or } \frac{1}{15}$$

∎

Example 6.48

Five keys, of which only one works, are tried one after another to unlock a door. What is the probability that:
a) the door does not open on either of the first two tries?
b) the door does open within the first two tries?
(HINT: use Probability Rule 4: the complement rule and the result from part a).

Solution:

a) The statement that the door does not open on either of the first two tries means that the door did not open on the first try and the door did not open on the second try.

Therefore,

P(door doesn't open on either of the first two tries) =

= P(door doesn't open on 1st try) · P(door doesn't open on 2nd try | door didn't open on 1st try)

Since these two events are dependent, we will use Probability Rule 6 to calculate this probability.
Thus,

P(door doesn't open on either of the first two tries) =

= P(door doesn't open on 1st try) · P(door doesn't open on 2nd try | door didn't open on 1st try)

$$= \frac{4}{5} \cdot \frac{3}{4}$$

$$= \frac{12}{20} \text{ or } \frac{3}{5}$$

b) Use the complement rule:
P(E) = 1 − P(E')
to compute the P(door opens within the first two tries).

Thus,

P(door opens within the first two tries) = 1− P(door doesn't open on either of the first two tries)

From part a, we have:

P(door doesn't open on either of the first two tries) = $\frac{3}{5}$

Therefore,

P(door opens within the first two tries) = $1 - \frac{3}{5}$

$= \frac{2}{5}$ ∎

Using Probability Rule 6: the multiplication rule for dependent events, we can algebraically obtain a formula for the conditional probability of event A, given that event B has occurred. This result is called **Probability Rule 8: Conditional Probability Formula, denoted P(A | B)**, and is stated as follows:

Probability Rule 8: Conditional Probability Formula

For two events A and B, the probability of A given B has occurred is equal to the probability of A and B divided by the probability of B. This is rule symbolized as:

$$P(A \mid B) = \frac{P(A \text{ and } B)}{P(B)}$$

where P(B) > 0

Example 6.49

In a single toss of a fair die, what is the probability of getting a 1, given the occurrence of an odd number?

Solution:

Let's define event A to be the outcome is a 1 and event B to be the outcome is an odd number. Symbolically we want to determine P(A | B), i.e. the probability of getting a one, given that the outcome is odd. To compute this probability, we must first determine P(B) and P(A and B).

Since there are only six possible outcomes of a fair die and three are odd, then:

P(B) = P(outcome is odd)

$= \frac{3}{6}$

Since there is only one outcome satisfying both events A and B, then:
P(A and B) = P(outcome is a 1 and the outcome is odd)

$= \frac{1}{6}$

Therefore,

$$P(A \mid B) = \frac{P(A \text{ and } B)}{P(B)}$$

$$= \frac{\frac{1}{6}}{\frac{3}{6}} = \frac{1}{3}$$

Example 6.50

Two boxes identical in appearance are on a table.

Box 1 contains a nickel and a penny while box 2 contains two nickels. A person randomly picks a box and then randomly selects a coin from the box. If the coin selected is a nickel, what is the probability that the other coin in the box is also a nickel?

Solution:

Let's define event A to be the coin remaining in the selected box is a nickel and event B is the selected coin is a nickel. Symbolically we want to determine $P(A \mid B)$, i.e. the probability the coin remaining in the selected box is a nickel given the selected coin is a nickel. To compute this probability we must first determine $P(A \text{ and } B)$ and $P(B)$.

Thus,
$P(A \text{ and } B) = P(\text{coin remaining in box is a nickel and selected coin is a nickel})$

This is equivalent to writing:

$= P(\text{both coins in the selected box are nickels})$.

Since there are two boxes and only one box has two nickels, then:

$$P(A \text{ and } B) = \frac{1}{2}$$

Since a nickel can be chosen from either of the two boxes, we have to consider both these possibilities in computing $P(B)$.

That is,

$P(B) = P(\text{coin selected is a nickel})$

$= P(\text{box 1 is chosen}) \cdot P(\text{select a nickel} \mid \text{box 1 is chosen}) + P(\text{box 2 is chosen}) \cdot P(\text{select a nickel} \mid \text{box 2 is chosen})$

$$= \frac{1}{2} \cdot \frac{1}{2} + \frac{1}{2} \cdot 1$$

$$= \frac{3}{4}$$

Therefore, the probability the coin remaining in the box is a nickel given that the coin selected was a nickel can be computed using Probability Rule 8:

$$P(A \mid B) = \frac{P(A \text{ and } B)}{P(B)}$$

$$= \frac{\frac{1}{2}}{\frac{3}{4}}$$

$$= \frac{2}{3} \quad \blacksquare$$

Example 6.51

THE BIRTHDAY PROBLEM

In Section 6.1, Lucy wagered Charlie Brown that at least 2 children at their Christmas Party would have the same birthday (month and day).

Let's compute this probability assuming there are 365 days in a year and there were:
a) 3 children at the Christmas Party.
b) 4 children at the Christmas Party.
c) 10 children at the Christmas Party.

Solution:

The easiest way to calculate the:
P(at least 2 children have the same birthday) is to compute the P(no two children have the same birthday) and use the complement rule: $P(E) = 1 - P(E')$.

Using the complement rule, we can write:
P(at least 2 children have the same birthday) = 1 − P(no two children have the same birthday)
We will use this probability rule to solve the birthday problem.

a) For 3 children at the Christmas Party, we have:

P(no 2 children have the same birthday) = P(all 3 children have different birthdays)

= P(1st child has a birthday)

multiplied by

P(2nd child has a birthday | 2nd child's birthday is different from first child's birthday)

multiplied by

P(3rd child has a birthday | 3rd child's birthday is different from first 2 children's birthdays)

$$= \frac{365}{365} \cdot \frac{364}{365} \cdot \frac{363}{365}$$

= 0.992 (expressed to the nearest thousandths)

Thus,

P(at least 2 children have the same birthday) = 1 - 0.992

= 0.008

This probability is rather small for a group of three children.

b) For 4 children at the Christmas Party, we have:

P(no two children have the same birthday) = P(all 4 children have different birthdays)

$$= \frac{365}{365} \cdot \frac{364}{365} \cdot \frac{363}{365} \cdot \frac{362}{365}$$

= 0.984

Thus,

P(at least 2 children have the same birthday) = 1 - 0.984

= 0.016

c) For 10 children at the Christmas Party, we have:

P(no two children have same birthday) = P(all 10 children will have different birthdays)

$$= \frac{365}{365} \cdot \frac{364}{365} \cdot \frac{363}{365} \cdot \frac{362}{365} \cdot \frac{361}{365} \cdot \frac{360}{365} \cdot \frac{359}{365} \cdot \frac{358}{365} \cdot \frac{357}{365} \cdot \frac{356}{365}$$

= 0.883

Thus,

P(at least 2 children have the same birthday) = 1 - 0.883

= 0.117

Table 6.8 contains the probability that at least 2 children will have the same birthday for various sized groups. Notice the probability that at least 2 children will have the same birthday for a group of 50 children is 0.970 or 97%. Consequently, Lucy's chances of winning her wager with Charlie Brown (as mentioned in the introduction to this chapter) are extremely good.

Table 6.8

Number of People in the Group	Probability of at least two people having the same birthday
3	0.008 OR 0.8%
10	0.117 OR 11.7%
20	0.411 OR 41.1%
23	0.507 OR 50.7%
30	0.706 OR 70.6%
40	0.898 OR 89.8%
50	0.970 OR 97%
60	0.994 OR 99.4%
70	0.999 OR 99.9%

Example 6.52

Five people are each asked to select a number from 1 to 10 inclusive. What's the probability that at least two of them will select the same number?

Solution:

Using the complement rule we get:

P(at least 2 people select the same number) = 1 − P(no two people select the same number)

$$= 1 - \frac{10}{10} \cdot \frac{9}{10} \cdot \frac{8}{10} \cdot \frac{7}{10} \cdot \frac{6}{10}$$

$$= 1 - 0.302$$

$$= 0.698$$

CASE STUDY 6.5

The following article entitled *The Laws of Probability* appeared in Time magazine on January 8, 1965.

THE LAW

TRIALS
The Laws of Probability

Around noon one day last June, an elderly woman was mugged in an alley in San Pedro, Calif. Shortly afterward, a witness saw a blonde girl, her pony tail flying, run out of the alley, get into a yellow car driven by a bearded Negro, and speed away. Police eventually arrested Janet and Malcolm Collins, a married couple who not only fitted the witness's physical description of the fugitive man and woman but also owned a yellow Lincoln. The evidence, though strong, was circumstantial. Was it enough to prove the Collinses guilty beyond a reasonable doubt?

Confidently answering yes, a jury has convicted the couple of second–degree robbery because Prosecutor Ray Sinetar, 30, cannily invoked a totally new test of circumstantial evidence–the laws of statistical probability.

In presenting his case, Prosecutor Sinetar stressed what he felt sure was already in the jurors' minds: the improbability that at any one time there could be two couples as distinctive as the Collinses in San Pedro in a yellow car. To "refine the jurors' thinking," Sinetar then explained how mathematicians calculate the probability that a whole set of possibilities will occur at once. Take three abstract possibilities (A, B, C) and assign to each a hypothetical probability factor. A, for example, may have a probability of 1 out of 3: B, 1 out of 10: C, 1 out of 100. The odds against A, B and C occurring together are the product of their total probabilities (1 out of 3 X 10 X 100), or 3,000 to 1.

After an expert witness approved Sinetar's technique, the young prosecutor asked the jury to consider the six known factors in the Collins case: a blonde white woman, a pony–tail hairdo, a bearded man, a Negro man, a yellow car, an interracial couple.

Then he suggested probability factors ranging from 1–to–4 odds that a girl in San Pedro would be blonde to 1–to–1,000 odds that the couple would be negro–white. Multiplied together, the factors produced odds of 1 to 12 million that the Collinses could have been duplicated in San Pedro on the morning of the crime.

Public Defender Donald Ellertson strenuously objected on grounds that the mathematics of probability were irrelevant, and that Sinetar's probability factors were inadmissible as assumptions rather than facts. Sinetar, however, merely estimated the factors before inviting the jurors to substitute their own. And the public defender will not appeal because he found no trial errors strong enough to outweigh the strong circumstantial evidence. Convicted by math, Malcolm Collins received a sentence of one year to life. Janet Collins got "not less than one year."

a) Identify each of the six factors that the prosecutor used as evidence to convince the jury that the defendants committed the robbery.

b) What was the probability that the prosecutor assigned to the chance of a girl in San Pedro would have blonde hair? How do you think the prosecutor came up with this probability value? Are you absolutely convinced that this is the correct probability value?

c) What was the probability that the prosecutor assigned to the chance that an interracial couple would be black-white?
How do you think the prosecutor came up with this probability value? Are you absolutely convinced that this is the correct probability value?

d) To what event did the prosecutor assign the very convincing probability value of 1 in 12 million?

e) How did the prosecutor arrive at this 1 in 12 million probability value? What assumption did the prosecutor make about the relationship between the six factors before he could multiply each of the respective probability values?

f) Do you agree with the prosecutor's assumption about the relationship among the six factors? Explain.

g) If you were a lawyer who had to defend this couple, what would you say to dispute the prosecutor's mathematical argument?

6.7 Binomial Probability

There are many experiments in probability where we may be interested in only two possibilities. For example:

- the toss of a coin results in either a head or a tail.
- the toss of a die results in an outcome which is either a 5 or not a 5.
- the selection of a single card from a deck of cards results in either an ace or not an ace.
- a new drug will either be effective or not effective when administered to a single patient.
- in an ESP experiment using five objects, the subject will either guess the object or not guess the object correctly.
- in a quality control procedure an item tested will either be defective or non-defective.

These experiments can all be classified as **Binomial Experiments**.
Let's define a binomial experiment and consider some examples of a binomial experiment.

> **Definition:** **Binomial Experiment**. An experiment is binomial, if the experiment satisfies the following four conditions:
>
> 1) The number of trials is fixed.
> 2) Each trial is independent of the previous trials.
> 3) The outcome for each trial can be classified into one of two categories called a success or a failure.
> 4) The probability of a success is the same for each trial.

Within this definition of a binomial experiment, we use the terminology: **trial**, **success** and **failure**. Let's discuss these terms before we consider any examples. **The term trial refers to a repetition of an experiment**. Thus, if an experiment consists of a single roll of a fair die, and the die is rolled 20 times, then each roll is considered a trial. Therefore, there are a total of 20 trials for this experiment. For an experiment to be considered to be binomial, each trial can only be considered to have two outcomes called a success and a failure. The term **success is usually defined to be the outcome(s) which refer to the question being asked**, while the word **failure refers to the outcome(s) that is not referred to by the question**. For example, if the binomial probability question is stated as: "What is the probability of guessing 4 wrong answers on a 10 question multiple choice test?", then success would be defined to be a wrong answer, while a failure is defined to be a correct answer. Notice, the outcome **wrong answer** is being called a **success** because this is the outcome to which the question refers. Remember, the word success is simply a term used to refer to one of the two outcomes of a binomial experiment, and should not be confused with the outcome(s) that you may find to be desirable.

Let's now consider an example to explain how to determine whether an experiment is binomial.

Example 6.53

Consider the experiment of tossing a fair coin two times, where we are interested in getting a head.
Is this experiment a binomial experiment?

Solution:

Let's determine if the four conditions for a binomial experiment are satisfied.
1) Is the number of trials fixed? Yes, the number of trials is 2, because the coin is being tossed two times.
2) Is each trial independent of the previous trials? Yes, the outcome for the first toss does not influence the outcome for the second toss.
3) Can the outcome of each trial be classified into one of two categories: a success or a failure? Yes, there are only two outcomes: heads and tails. Thus, we define the outcome for a head to be a success, and the outcome for a tail to be a failure.
4) Is the probability of a success the same for each trial? Yes, on each toss of the fair coin the probability of a success equals the probability of a head which equals 1/2.

Therefore, this experiment is a binomial experiment. ∎

Example 6.54

Is the experiment of rolling a fair die once a binomial experiment, where we are interested in getting a four?

Solution:

1) The number of trials is fixed at 1, since the die is rolled only once.
2) Is this one trial independent? Yes.
3) Can the outcome be classified into one of two categories, success or failure? Yes, we can define the outcome 4 to be a success, then the outcomes 1, 2, 3, 5, and 6 would be considered failures.
4) There is only 1 trial and the probability of a success equals the probability of getting a 4 which equals 1/6.

Therefore, this experiment is a binomial experiment. ∎

Example 6.55

Consider spinning the spinner shown in Figure 6.29 two times and recording the two outcomes.

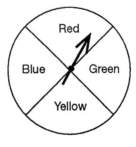

Figure 6.29

Can this experiment be considered a binomial, if we are interested in the outcome red?

Solution:

Let's check the four conditions of a binomial experiment.
1) Is the number of trials fixed? Yes, the number of trials is 2, since the spinner is spun twice.
2) Is each trial independent of the previous trials? Yes the outcome of the first spin has no influence on the outcome of the second spin.

3) Can the outcome of each trial be classified into one of two categories? Yes, the outcome red is defined to be a success, and the outcomes: green, blue and yellow are failures.
4) Is the probability of a success the same for each trial? Yes, the probability of red on the first spin is 1/4 and the probability of red on the second spin is also 1/4. ∎

Now that we've illustrated the four conditions of a binomial experiment, let's examine the Binomial Probability Formula which can be used to calculate the probabilities for a binomial experiment.

Probability Rule 9: Binomial Probability Formula.

For a binomial experiment, the probability of getting **s** successes in **n** trials is computed using the binomial probability formula. This formula is written:

$$P(s \text{ successes in n trials}) = \binom{n}{s} p^s q^{(n-s)}$$

where:
- n = **number of independent trials**
- s = **number of successes**
- (n–s) = **number of failures**

$\binom{n}{s}$ = the number of ways "s" successes can occur in "n" trials

- p = **probability of a success for one trial**
- q = **probability of a failure for one trial** = 1–p

A Discussion of $\binom{n}{s}$[2]: To evaluate $\binom{n}{s}$, we must determine the number of ways **s** successes can occur in **n** trials. The following example illustrates this calculation.

Example 6.56

For the binomial experiment of tossing a fair coin three times, where a success is getting a head on a single toss of the coin, determine the number of ways to get:

a) three successes in three tosses of the coin, written as $\binom{3}{3}$

b) two successes in three tosses of the coin, written as $\binom{3}{2}$

c) one success in three tosses of the coin, written as $\binom{3}{1}$

---

[2]. The evaluation of $\binom{n}{s}$ using a formula is discussed in Section 6.3.

d) no successes in three tosses of the coin, written as $\binom{3}{0}$

Solution:

a) There is only one way to get three successes in three tosses of a coin. It is a success on the first toss, a success on the second toss and a success on the third toss. This occurrence is symbolized as: sss.

Therefore, $\binom{3}{3}$, the number of ways of getting 3 successes in 3 trials, equals 1.

b) There are three ways to get two successes in three tosses of a coin. They are: ssf, sfs, and fss. Therefore, $\binom{3}{2}$, the number of ways of getting 2 success in 3 trials, is equal to 3.

c) There are three ways to get one success in three tosses of a coin. They are: sff, fsf, and ffs.

Therefore, $\binom{3}{1}$, the number of ways of getting 1 success in 3 trials, equals 3.

d) There is only one way to get no successes in three tosses of a coin. It is: fff. Therefore, $\binom{3}{0} = 1$. ■

Example 6.57

Consider the binomial experiment of rolling a fair die four times where a success is rolling a one. Determine:

a) the number of ways to get 4 successes in four rolls of the die.

b) $\binom{4}{2}$

c) $\binom{4}{0}$

Solution:

a) There is only one way to get 4 successes in four rolls of the die. It is: ssss. Therefore $\binom{4}{4} = 1$.

b) $\binom{4}{2}$ means the number of ways to get 2 successes in 4 trials which is 6, since we can get 2 successes in 4 trials six different ways. They are: ssff, sfsf, sffs, fsfs, ffss, and fssf.

c) $\binom{4}{0}$ means the number of ways to get no successes in 4 trials which is 1. Since there is only one way to get no successes in 4 trials. It is: ffff. ■

In general the number of ways **s** successes can occur in **n** independent trials, written as $\binom{n}{s}$, can be determined by using **TABLE I: The Success Occurrence Table** found in **Appendix D: Statistical Tables**.

EVALUATING $\binom{n}{s}$ USING TABLE I: THE SUCCESS OCCURRENCE TABLE

The expression: $\binom{n}{s}$ can be evaluated using TABLE I. To use TABLE I, you first find the row that corresponds to the value of n under the column labelled n. Second, you move across this row until you come to the column that corresponds to the value of s under the portion of the table labelled s. The number that you find in TABLE I is the value of $\binom{n}{s}$.

Let's consider evaluating the expression: $\binom{5}{3}$. For this expression, the value of n is 5 and the value of s is 3. To evaluate this expression, we will use TABLE I. First, locate the row corresponding to the number 5 under the column labelled n. Second, move across this row until you come to the column corresponding to the value of s, which is 3. You should be at the table value: 10. This is illustrated in Table 6.9, which displays a portion of TABLE I.

Table 6.9
TABLE I: The Success Occurrence Table, $\binom{n}{s}$

n	s=0	1	2	3	4	5	6	7	8	9	10	11	12	13	14	15
0	1			*												
1	1	1		*												
2	1	2	1	*												
3	1	3	3	*												
4	1	4	6	*												
5	*	*	*	10												
6	1	6	15													

Thus, from TABLE I, the expression: $\binom{5}{3} = 10$.

Example 6.58

Using TABLE I, determine the number of ways **s** successes can occur in **n** trials if:
a) n = 10, s = 7
b) n = 3, s = 2
c) n = 7, s = 0
d) n = 12, s = 4

Solution:

a) Since n = 10, and s = 7, then $\binom{n}{s} = \binom{10}{7}$

Using TABLE I, $\binom{10}{7} = 120$

b) For n = 3, and s = 2, then $\binom{n}{s} = \binom{3}{2}$

From TABLE I, $\binom{3}{2} = 3$

c) Since n = 7, and s = 0, then $\binom{n}{s} = \binom{7}{0}$

Using TABLE I, $\binom{7}{0} = 1$

d) For n = 12, and s = 4, then $\binom{n}{s} = \binom{12}{4}$

From TABLE I, $\binom{12}{4} = 495$ ∎

Before we work with binomial probability experiments using the binomial probability formula, it is important to review some rules dealing with exponents.

Rules For a Fraction Raised to an Exponent

The values for a, b, and n represent positive whole numbers.

I. $\left(\dfrac{a}{b}\right)^n = \dfrac{a^n}{b^n}$

II. $\left(\dfrac{a}{b}\right)^0 = 1$

Rule II states that any non-zero number raised to an exponent of zero is one.

Let's consider an example using these rules for a fraction raised to an exponent.

Example 6.59

Evaluate the expressions:

a) $(\frac{3}{5})^4$

b) $(\frac{4}{7})^0$

Solution:

a) The expression: $(\frac{3}{5})^4 = \frac{3^4}{5^4}$

$$= \frac{3}{5} \cdot \frac{3}{5} \cdot \frac{3}{5} \cdot \frac{3}{5}$$

$$= \frac{81}{625}$$

b) The expression: $(\frac{4}{7})^0 = 1$, since any non-zero number raised to an exponent of zero is one. ∎

Example 6.60

Use the binomial probability formula: P(**s** successes in **n** trials) = $\binom{n}{s} p^s q^{(n-s)}$ to determine the probability of getting 3 successes in 5 trials if p = 1/2.

Solution:

1) n = 5 (number of trials)
2) s = 3 (number of successes)
3) n−3 = 2
4) $\binom{n}{s} = \binom{5}{3}$ = 10 (from TABLE I)
5) p = 1/2
6) q = 1−p = 1/2

Substituting this information into Probability Rule 9: the Binomial Probability Formula, we get:

P(**3** successes in **5** trials) = $\binom{5}{3} (\frac{1}{2})^3 (\frac{1}{2})^2$

$$= 10\underline{(1)}\underline{(1)}$$
$$8\ \ 4$$
$$= \frac{10}{32}\qquad\blacksquare$$

USING A CALCULATOR TO EVALUATE: p^s AND $q^{(n-s)}$

To evaluate the quantities: p^s and $q^{(n-s)}$ on a calculator, you must use a special key that raises a number to an exponent. This key is usually displayed in one of two different ways on a calculator. It is written either as: y^x or x^y.

Let's consider the expression: $(\frac{2}{3})^4$

Mathematically, the expression: $(\frac{2}{3})^4$ means $(\frac{2}{3})(\frac{2}{3})(\frac{2}{3})(\frac{2}{3})$

To calculate the expression: $(\frac{2}{3})^4$ on a calculator, you can use the following procedure.

Step 1. Clear the display.
Step 2. Key in the number which appears in the numerator: 2.
Step 3. Press the special key: y^x or x^y. (whichever one appears on your calculator)
Step 4. Key in the power: 4.
Step 5. Press the = key. This result, which is 16, represents the numerator's value.
Step 6. Key in the number which appears in the denominator: 3.
Step 7. Key in the power: 4.
Step 8. Press the = key. This result, which is 81, represents the denominator's value.

Thus, the expression: $(\frac{2}{3})^4 = \frac{16}{81}$

The general calculator procedure for an expression of the form: $(\frac{n}{d})^s$ is:

Step 1. Clear the display.
Step 2. Key in the number which appears in the numerator: n.
Step 3. Press the special key: y^x or x^y. (whichever one appears on your calculator)
Step 4. Key in the power: s.
Step 5. Press the = key. This result represents the numerator's value.
Step 6. Key in the number which appears in the denominator: d.
Step 7. Key in the power: s.
Step 8. Press the = key. This result represents the denominator's value.

Example 6.61

For a binomial experiment where p = 2/3 determine the probability of 3 successes in 4 trials. This is written as P(s=3).

Solution:

1) $n = 4$ (number of trials)
2) $s = 3$ (number of successes)
3) $n-s = 1$
4) $\binom{n}{s} = \binom{4}{3} = 4$ (from TABLE I)
5) $p = 2/3$
6) $q = 1-p = 1/3$

Substituting this information into Probability Rule 9: the Binomial Probability Formula, we have:

$$P(3 \text{ successes in 4 trials}) = P(s=3)$$

$$= \binom{4}{3} \left(\frac{2}{3}\right)^3 \left(\frac{1}{3}\right)^1$$

$$= 4\left(\frac{8}{27}\right)\left(\frac{1}{3}\right)$$

$$= \frac{32}{81} \quad \blacksquare$$

Example 6.62

A student is going to guess at the answers to all the questions on a five question multiple choice test where there are four choices for each question. Calculate the probability of:
a) guessing three correct answers.
b) guessing five correct answers.
c) guessing at most one correct answer.
d) guessing at least four correct answers.

Solution:

a) Since the question pertains to a correct answer, we will define success to be guessing the correct answer to a question. Using this definition of success, and considering a trial to be a question, then we have the following values:

1. $n = 5$ (since there are 5 questions)
2. $s = 3$ (we are interested in 3 correct answers)
3. $n-s = 2$
4. $\binom{n}{s} = \binom{5}{3} = 10$ (from TABLE I)
5. $p = 1/4$ (since the probability of a correct answer is one chance out of four possible choices)
6. $q = 1-p = 3/4$
7. Substituting these values into Probability Rule 9, we have:

$$P(s=3) = \binom{5}{3} \left(\frac{1}{4}\right)^3 \left(\frac{3}{4}\right)^2$$

$$= 10(\frac{1}{64})(\frac{9}{16})$$

$$= \frac{90}{1024}$$

b) Again, we will define success to be guessing the correct answer to a question, since this is the outcome referred to within the question.
  1. n = 5 (because there are five questions)
  2. s = 5 (since we are interested in 5 correct answers)
  3. n-s=0
  4. $\binom{n}{s} = \binom{5}{5} = 1$ (from TABLE I)
  5. p = 1/4 (since the probability of getting a correct answer for one question is one chance out of four choices)
  6. q = 3/4 (1 - p)
  7. Substituting these values into Probability Rule 9, and remembering that **any non-zero number raised to an exponent of zero is one**, we get:

$$P(s=5) = \binom{5}{5}(\frac{1}{4})^5(\frac{3}{4})^0$$

$$= \frac{1}{1024}$$

c) Success is defined to be guessing the correct answer to a question, since this is the outcome referred to within the question.
   To determine the probability of guessing **at most one correct answer**, you must interpret the statement: **at most one correct answer. At most one correct means guessing one or less correct answers**. Thus, to calculate this probability, we need to determine two probabilities: the probability of getting one correct answer and the probability of zero correct answers.
That is:
P(at most 1 correct answer) = P(1 correct) + P(zero correct)

To determine the probability of one correct, we have the following values:
  1. n = 5 (number of questions)
  2. s = 1 (success is one correct answer)
  3. n-s = 4
  4. $\binom{n}{s} = \binom{5}{1} = 5$ (from TABLE I)
  5. p = 1/4 (probability of a correct answer)
  6. q = 3/4 (1 - p)
  7. Substituting these values into Probability Rule 9, we have:
     $$P(1 \text{ correct}) = \frac{405}{1024}$$

To calculate the probability of zero correct, we have the following values:
1. n = 5 (number of questions)
2. s = 0 (zero correct answers)
3. n-s = 5
4. $\binom{n}{s} = \binom{5}{0} = 1$ (from TABLE I)
5. p = 1/4 (probability of a correct answer)
6. q = 3/4 (1 - p)
7. Substituting these values into Probability Rule 9, and using the fact that **any non-zero number raised to an exponent of zero is one**, we have:

$$P(\text{zero correct}) = \frac{243}{1024}$$

Therefore, to calculate the probability of at most one correct, we add these two probabilities together. Thus, we have:

P(at most 1 correct answer) = P(**1 correct**) + P(**zero correct**)
$$= \frac{405}{1024} + \frac{243}{1024}$$
$$= \frac{648}{1024}$$

d) Success is defined to be guessing the correct answer to a question, since this is the outcome to which the question refers.

To determine the probability of guessing **at least four correct answers**, you must interpret the statement: **at least four correct answers**. **At least four correct means guessing four or more correct answers**. Since there are only five questions, the most correct answers that one can guess is five. Thus, to calculate this probability, we need to determine the two probabilities: the probability of getting four correct answers and the probability of five correct answers.

That is,

P(at least 4 correct answers) = P(**4 correct**) + P(**5 correct**)

To determine the probability of four correct, we have the following values:
1. n = 5 (number of questions)
2. s = 4 (four correct answers)
3. n-s = 1
4. $\binom{n}{s} = \binom{5}{4} = 5$ (from TABLE I)
5. p = 1/4 (probability of a correct answer)
6. q = 3/4 (1 - p)
7. Substituting these values into Probability Rule 9, we have:
$$P(\text{four correct}) = \frac{15}{1024}$$

To calculate the probability of five correct, we have the following values:
1. n = 5 (number of questions)
2. s = 5 (five correct answers)
3. n-s = 0
4. $\binom{n}{s} = \binom{5}{5} = 1$ (from TABLE I)
5. p = 1/4 (probability of a correct answer)
6. q = 3/4 (1 - p)
7. Substituting these values into Probability Rule 9, and using the fact that **any non-zero number raised to an exponent of zero is one**, we have:

$$P(\text{five correct}) = \frac{1}{1024}$$

Therefore, to calculate the probability of at least four correct, we add these two probabilities together. Thus, we have:

$$P(\text{at least 4 correct}) = P(4 \text{ correct}) + P(5 \text{ correct})$$
$$= \frac{15}{1024} + \frac{1}{1024}$$
$$= \frac{16}{1024}$$
∎

Example 6.63

From an urn containing 8 marbles, four of which are colored red, two are colored white, and two blue. Four marbles are selected with replacement. Calculate the probability of:
a) selecting four red marbles.
b) selecting at least two blue marbles.

Solution:

a) A success is defined to be the selection of a red marble, since this is the outcome referred to within the question.
1. n = 4 (since four marbles are being selected)
2. s = 4 (because the question refers to four red marbles)
3. n-s = 0
4. $\binom{n}{s} = \binom{4}{4} = 1$ (from TABLE I)
5. p = 1/2 (since half the marbles are red)
6. q = 1/2 (1 - p)
7. Using these values and Probability Rule 9, we have:
P(selecting 4 red marbles) = 1/16

b) A success is defined to be selecting a blue marble, since we are trying to determine the probability of at least two blue marbles.

The probability of selecting **at least two blue marbles** is equal to the probability of selecting **two or more blue marbles**. Since four marbles are being selected, then the maximum number of blue marbles that can be selected is four. Thus, the probability of at least 2 blue marbles is determined by using the following probability statement:

P(at least 2 blue marbles) = P(2 blue marbles) + P(3 blue marbles) + P(4 blue marbles)

Since: P(2 blue marbles) = $\frac{54}{256}$

P(3 blue marbles) = $\frac{12}{256}$

P(4 blue marbles) = $\frac{1}{256}$

then: P(at least 2 blue marbles) = $\frac{54}{256} + \frac{12}{256} + \frac{1}{256}$
= $\frac{67}{256}$ ∎

Notice that in the above example success was defined differently for each part of the example. Our definition of success was determined by the outcome referred to within each part of the question. For part a, we defined success to be obtaining a red marble, while for part b, we defined success to be obtaining a blue marble. It is very important before using the binomial probability formula, that you clearly define what you mean by a success.

Example 6.64

Imagine spinning the **fruit wheel** pictured in Figure 6.30 three times.

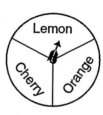

Figure 6.30

a) Find the probability of getting exactly 2 lemons.
b) Find the probability of getting exactly 1 cherry.

Solution:

a) Success is defined to be getting a lemon, since the question refers to getting 2 lemons.
1. n = 3 (since the wheel is being spun three times)
2. s = 2 (because the question refers to getting two lemons)
3. n−s = 1
4. $\binom{n}{s} = \binom{3}{2} = 3$ (from TABLE I)
5. p = 1/3 (since there is one lemon out of three fruits on the wheel)
6. q = 2/3 (1 − p)
7. Using Probability Rule 9, and the previous information we get:
P(exactly 2 lemons) = $\frac{6}{27}$

b) Success is now defined to be getting a cherry, since the question refers to getting a cherry.
   1. n = 3 (because the wheel is spun three times)
   2. s = 1 (since the question refers to getting one cherry)
   3. n−s = 2
   4. $\binom{n}{s} = \binom{3}{1} = 3$ (from TABLE I)
   5. p = 1/3 (because there is only one cherry on the fruit wheel)
   6. q = 2/3 (1 − p)
   7. Using Probability Rule 9, and the previous information we have:
      P(exactly 1 cherry) = $\frac{12}{27}$ ∎

CASE STUDY 6.6

The following article entitled *Trial By Mathematics* appeared in Time magazine on April 26, 1968.

THE LAW

DECISIONS
Trial by Mathematics

After an elderly woman was mugged in an alley in San Pedro, Calif., a witness saw a blonde girl with a ponytail run from the alley and jump into a yellow car driven by a bearded Negro. Eventually tried for the crime, Janet and Malcolm Collins were faced with the circumstantial evidence that she was white, blonde and wore a ponytail while her Negro husband owned a yellow car and wore a beard. The prosecution, impressed by the unusual nature and number of matching details, sought to persuade the jury by invoking a law rarely used in a courtroom – the mathematical law of statistical probability.

The jury was indeed persuaded, and ultimately convicted the Collinses (TIME, Jan. 8, 1965). Small wonder. With the help of an expert witness from the mathematics department of a nearby college, the prosecutor explained the probability of a set of events actually occurring is determined by multiplying together the probabilities of each of the events. Using what he considered "conservative" estimates (for example, that the chances of a car's being yellow were 1 in 10, the chances of a couple in a car being interracial 1 in 1,000), the prosecutor multiplied all the factors together and concluded that the odds were 1 in 12 million that any other couple shared the characteristics of the defendants.

Only One Couple. The logic of it all seemed overwhelming, and few disciplines pay as much homage to logic as do the law and math. But neither works right with the wrong premises. Hearing an appeal of Malcolm Collins' conviction, the California Supreme Court recently turned up some serious defects, including the fact that not even the odds were all they seemed.

To begin with, the prosecution failed to supply evidence that "any of the individual probability factors listed were even roughly accurate." Moreover, the factors were not shown to be fully independent of one another as they must be to satisfy the mathematical law: the factor of a Negro with a beard, for instance, overlaps the possibility that the bearded Negro may be part of an interracial couple. The 12 million to 1 figure, therefore, was just "wild conjecture." In addition, there was not complete agreement among the witnesses about the characteristics in question. "No mathematical equation," added the court, "can prove beyond a reasonable doubt (1) that the guilty couple *in fact* possessed the characteristics described by the witnesses or even (2) that only *one* couple possessing those distinctive characteristics could be found in the entire Los Angeles area."

Improbable Probability. To explain why, Judge Raymond Sullivan attached a four-page appendix to his opinion that carried the necessary math far beyond the relatively simple formula of probability. Judge Sullivan was willing to assume it was unlikely that such a couple as the one described existed. But since such a couple did exist–and the Collinses demonstrably did exist–there was a perfectly acceptable mathematical formula for determining the probability that another such couple existed. Using the formula and the prosecution's figure of 12 million, the judge demonstrated to his own satisfaction and that of five concurring justices that there was a 41% chance that at least one other couple in the area might satisfy the requirements.

"Undoubtedly," said Sullivan, "the jurors were unduly impressed by the mystique of the mathematical demonstration but were unable to assess its relevancy or value." Neither could the defense attorney have been expected to know of the sophisticated rebuttal available to them. Janet Collins is already out of jail, has broken parole and lit out for parts unknown. But Judge Sullivan concluded that Malcolm Collins, who is still in prison at the California Conservation Center, had been subjected to "trial by mathematics" and was entitled to a reversal of his conviction. He could be tried again, but the odds are against it.

This robbery was discussed in an earlier Case Study within this chapter. Recall, that the interracial couple was convicted of the crime, because they fit the characteristics of the robbers which had been established as a 1-in-12 million chance. The couple however appealed and were exonerated because the Judge used a mathematical argument of his own. The judge's argument assumed that the probability p that a couple has all the characteristics of the robbers was $\frac{1}{12,000,000}$ and that there does exist one such couple in a population of 12,000,000 (the prosecutor's figure) couples. The judge proceeded to calculate the chance that another such couple could exist. The judge's mathematical reasoning used binomial probability and conditional probability. He was trying to arrive at a value for the following probability:
P(2 or more such couples exist *given* at least one such couple exists).
This is equal to:

$$\frac{P(\text{2 or more such couples})}{P(\text{at least one such couple})}.$$

a) What probability rule was used to write the above statement?
This in turn equals:

$$\frac{1 - P(\text{one or less such couples})}{1 - P(\text{no such couple})}.$$

b) What probability rule allowed us to write the above probability statement?

If we use p to represent the probability that a couple has all the characteristics of the robbers, and let N be the number of couples in the population, then this probability results in the following formula:

$$\frac{1 - (1-p)^N - Np(1-p)^{N-1}}{1 - (1-p)^N}.$$

c) What probability formula did we use to arrive at the above formula?

If $\frac{1}{12,000,000}$ is substituted in for the value of p, and N equals 12,000,000 (the prosecutor's figure) couples, then the:
P(2 or more such couples exist *given* at least one such couple exists) is equal to approximately 41%.

d) Is the judge's argument based upon an unlikely or a likely chance that such a couple has the stated characteristics? What is this probability value that the judge is using in his argument?

Table 6.10 contains the probabilities for several values of N couples where p is equal to $\frac{1}{12,000,000}$.

Table 6.10

Number of Couples N (in millions)	Probability
1	4.02%
2	7.86%
3	11.60%
4	15.22%
5	18.75%
6	22.16%
7	25.47%
8	28.68%
9	31.79%
10	34.79%
15	48.35%
20	59.59%
25	68.75%
30	76.10%
40	86.44%
50	92.56%
75	98.52%
100	99.73%

e) Examining this probability table, what conclusion can you draw about the existence of another such couple as the population of N couples increases?

f) Was it to the judge's advantage or disadvantage to use the prosecutor's 12,000,000 population figure for N couples, if the judge was trying to demonstrate that there was a decent chance that another such couple could exist with the same characteristics? ∎

Case Study 6.7

The IVDS Probability Chart in Figure 6.31 was included within an advertising package to entice people to invest within a group that was going to file applications within many different cities across the USA hoping to win the operating license to Interactive Video and Data Services for that particular city.

IVDS Probability Chart

PROBABILITY OF WINNING 1 OR MORE, 2 OR MORE OR 3 OR MORE MARKETS FOR VARIOUS COMBINATIONS OF NUMBERS OF APPLICANTS AND MARKETS ENTERED WHEN TWO WINNERS ARE SELECTED FROM EACH MARKET

NUMBER OF APPLICANTS	NUMBER OF MARKETS FILED	PROBABILITY OF WINNING 1 OR MORE	WINNING 2 OR MORE	WINNING 3 OR MORE
250	100	55%	19%	5%
250	200	80%	48%	22%
250	300	91%	69%	43%
300	100	49%	14%	0%
300	200	74%	39%	1%
300	300	87%	59%	3%
400	100	39%	9%	0%
400	200	63%	26%	1%
400	300	78%	44%	2%
500	100	33%	6%	0%
500	200	55%	19%	0%
500	300	70%	34%	1%

Figure 6.31

Certain assumptions were made in constructing this probability chart:

1. Two winners will be selected from each market or city by a random selection process.
2. The number of applicants for each market or city is assumed to be the same as indicated in column one of the chart.

a) What happens to the probability of winning a license for a market or city as the number of applicants increases?
b) What happens to the probability of winning a license for a market or city as the number of applicants stays the same and the number of markets filed increases?
c) Can the binomial probability formula be used to calculate the probabilities shown within the last three columns of the chart? Explain.
d) If the binomial probability formula can be used, then what would p, the probability of a success, represent? What is the value of p for the third row where the number of applicants is 250 and the number of markets filed is 300?
What are the values of n and q for this row? What is the simplest approach to calculating the probability of winning one or more markets? That is, how could you rewrite the probability statement of winning one or more markets using the complement rule to make the calculations easier?
Calculate the probability of winning one or more markets using the complement rule and the values of n, p and q for the information pertaining to the third row of the chart in Figure 6.31?
e) How could you rewrite the probability statement of winning two or more markets using the complement rule to make the calculations easier? Calculate the probability of winning two or more markets using the complement rule and the values of n, p and q for the information pertaining to the third row of the chart in Figure 6.31? ■

❄ 6.8 Using MINITAB

To determine evaluations of binomial probabilities using the binomial probability formula:

$$P(\text{s successes in n trials}) = \binom{n}{s} p^s q^{(n-s)},$$

MINITAB uses the PDF command along with the BINOMIAL subcommand. In example 6.61 we determined by using the binomial probability formula that for a binomial experiment where $p = 2/3$ the probability of 3 successes in 4 trials to be $\frac{32}{81}$ which is approximately 0.3951.

Using the MINITAB commands we get:

MTB > PDF 3;
SUBC> BINOMIAL 4 .6667.

```
    K        P( X = K)
   3.00       0.3951
```

Along with the PDF command is the CDF command. The CDF gives the cummulative probability calculation for a binomial experiment. For example, in a binomial experiment where p = 0.25, the probability of at most 2 successes in 5 trials is determined by the MINITAB commands:

```
MTB >   CDF 2;
SUBC >  BINOMIAL 5 .25.
```

```
     K      P( X LESS OR = K)
   2.00          0.8965
```

To change the experiment from at most 2 successes in 5 trials to at most 20 successes in 50 trials the MINITAB commands become

```
MTB >   CDF 20;
SUBC >  BINOMIAL 50  .25.
```

```
     K      P( X LESS OR = K)
  20.00          0.9937
```

Notice that the change in the MINITAB commands are not much different! Consider the calculations you would have had to do to determine the probability in a binomial experiment for each of previous examples.

MINITAB also is able to generate a list of the probabilities for a binomial experiment.

```
MTB >   NAME C1='K'
MTB >   SET 'K'
DATA >  1(0:5/1)1
DATA >  END.
MTB >   NAME C2='P(X=K)'
MTB >   PDF 'K' 'P(X=K)'
MTB >   PDF 'K' 'P(X=K)';
SUBC >   BINOMIAL 5 .25.
MTB >   PRINT C1 C2
```

```
    ROW    K    P(X=K)

     1     0    0.237305
     2     1    0.395508
     3     2    0.263672
     4     3    0.087891
     5     4    0.014648
     6     5    0.000977
```

Using the graph command PLOT and the subcommand SYMBOL 'x' a graph of the binomial distribution can be constructed.

```
MTB >   PLOT C2 C1;
SUBC >    SYMBOL 'x'.
```

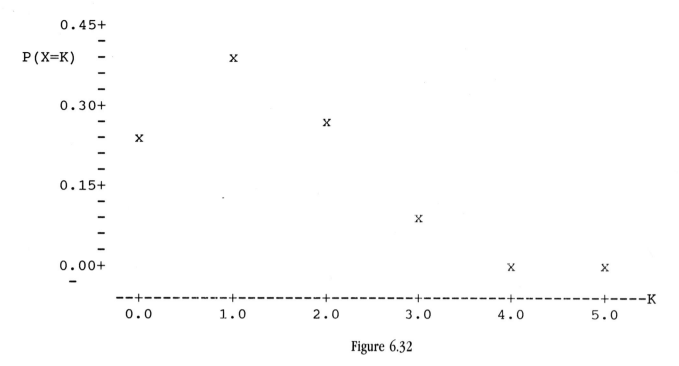

Figure 6.32

What shape is the graph? What would you think would happen to the shape of the graph if the experiment were repeated 50 times where the probability of success remains .25?

Let's use MINITAB to help explore this question.

```
MTB >   NAME C1='K'
MTB >   SET 'K'
DATA >  1(0:50/1)1
DATA >  END.
MTB >   NAME C2 = 'P(X = K)'
MTB >   PDF 'K' 'P(X = K)';
MTB >   BINOMIAL 50  .25.
MTB >   PRINT C1 C2
```

ROW	K	P(X=K)
1	0	0.000001
2	1	0.000009
3	2	0.000077
4	3	0.000411
5	4	0.001610
6	5	0.004938
7	6	0.012345
8	7	0.025865
9	8	0.046341
10	9	0.072087
11	10	0.098518
12	11	0.119416
13	12	0.129368
14	13	0.126050
15	14	0.111044
16	15	0.088836
17	16	0.064776
18	17	0.043184
19	18	0.026390
20	19	0.014816

21	20	0.007655
22	21	0.003645
23	22	0.001602
24	23	0.000650
25	24	0.000244
26	25	0.000084
27	26	0.000027
28	27	0.000008
29	28	0.000002
30	29	0.000001
31	30	0.000000
32	31	0.000000
33	32	0.000000
34	33	0.000000
35	34	0.000000
36	35	0.000000
37	36	0.000000
38	37	0.000000
39	38	0.000000
40	39	0.000000
41	40	0.000000
42	41	0.000000
43	42	0.000000
44	43	0.000000
45	44	0.000000
46	45	0.000000
47	46	0.000000
48	47	0.000000
49	48	0.000000
50	49	0.000000
51	50	0.000000

```
MTB >   Plot C2 C1;
SUBC >  Symbol 'x'.
```

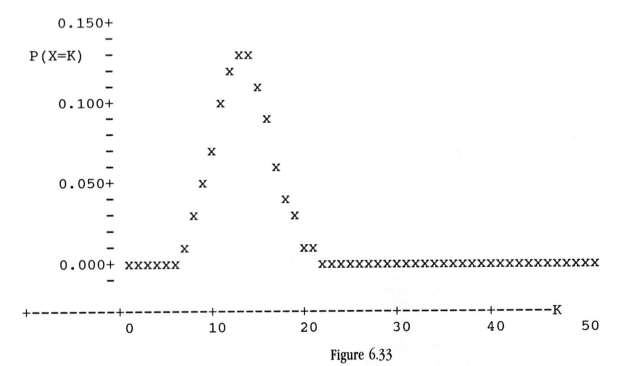

Figure 6.33

Is the shape of the graph in Figure 6.32 much different from the shape of the graph in Figure 6.33?

By examining the graph in Figure 6.33, estimate the probability of having more than 25 successes for a binomial experiment where the number of trials is 50 and the probability of successes for each trial is .25.

Glossary

TERM	SECTION
Chance	6.1
Experiment	6.2
Sample Space	6.2
Fundamental Counting Principle	6.2
Event	6.2
Permutation	6.3
n factorial, n!	6.3
Counting Rule 1: Permutation Rule, $_nP_n$	6.3
Counting Rule 2: Permutation Rule for n objects taken s at a time, $_nP_s$	6.3
Counting Rule 3: Permutation Rule of N objects with k alike objects	6.3
Combination	6.3
Counting Rule 4: The number of combinations of n objects taken s at a time, $\binom{n}{s}$	6.3
Probability	6.4
Probability Rule 1	6.5
Probability Rule 2	6.5
Probability Rule 3: The Addition Rule	6.5
Mutually Exclusive Events	6.5
Probability Rule 3A: The Addition Rule for Mutually Exclusive Events	6.5
Complement of an event	6.5
Probability Rule 4: Complement Rule	6.5
Independent Events	6.5
Probability Rule 5: Multiplication Rule for Independent Events	6.5
Conditional Probability	6.6
Selection With Replacement	6.6
Selection Without Replacement	6.6
Dependent Events	6.6
Probability Rule 6: Multiplication Rule for Dependent Events	6.6
Probability Rule 7: General Multiplication Rule	6.6
Probability Rule 8: Conditional Probability Formula	6.6
The Birthday Problem	6.7
Binomial Probability	6.7
Binomial Experiment	6.7
Binomial Trial	6.7
Binomial Success	6.7
Binomial Failure	6.7
Probability Rule 9: Binomial Probability Formula	6.7

Exercises

PART I Fill in the blanks.

1. An _____ is the process by which an observation is made or obtained.
2. A _____ is a representation of all possible outcomes of an experiment.
3. Using the Fundamental Counting Principle, the total number of possible outcomes for the experiment of tossing a fair die three times is _____.
4. An _____ is a particular collection of outcomes in the sample space of an experiment.
5. For the experiment of tossing a fair coin four times the event of obtaining at least 3 heads is interpreted as meaning the event of obtaining 3 or _____ heads. The outcomes satisfying this event are _____, _____, _____, and _____.
6. A permutation is an arrangement of objects in a _____ order.
7. 4! is read as four 4 _____ and is a shorthand notation for _____ x _____ x _____ x _____ which has a value of _____.
8. The Permutation Rule or Counting Rule #1 is used to determine the number of *permutations* (or arrangements) of n different objects taken altogether, denoted _____. The formula used to calculate this number of permutations is: $_nP_n$ = _____.
9. $_6P_6$ is equal to _____.
10. The number of different ways four photos can be matched with a list of four names is _____.
11. Counting Rule 2 is used to determine the number of permutations of n objects taken s at a time. This is denoted by: _____. The formula used to calculate this number of permutations is $_nP_s$ = _____.
12. The number of different ways to select two numbered balls from an urn containing 5 numbered balls is _____.
13. Counting Rule 3 is used to determine the number of permutations of N objects, where: n_1 are alike, n_2 are alike, ... , n_k are alike. The formula to calculate this number of permutations is : _____.
14. The number of different arrangements for the letters in the word **success** are _____.
15. A combination is a selection of objects in which _____ is not important.
16. Counting Rule 4 is used to determine the number of combinations of n objects taken s at a time which is symbolized as _____. The formulas used to calculate this number of combinations are _____ or _____.

17. $\binom{7}{4}$ is equal to _____.
18. The number of ways to select 3 video movies from a group of 8 videos is _____.
19. For equally likely events, the probability of an event E, denoted as P(E), is: P(E) = _____.
20. If we toss a fair coin and roll a fair die at the same time, then:
    a) the probability of getting a 6 on the die and a head on the coin is _____.
    b) the probability of getting an even number on the die and a head on the coin is _____.
    c) the probability of getting a 7 on the die and a tail on the coin is _____.
21. Probability Rule 1 states that the probability of an event E as a number between ____ and _____ inclusive.
22. Probability Rule 2 states that the sum of the probabilities of all the outcomes in the sample space of an experiment equals _____ .
23. Probability Rule 3 is the Addition Rule which states that the probability of satisfying the event A or the event B is equal to the probability of event _____ plus the probability of event B _____ the probability that both events A and B occur at the _____ time. This is symbolized as:
    P(A or B) = _____ + _____ − _____ .
24. One card is selected at random from an ordinary deck of 52 playing cards. The probability of selecting a queen or a red card is equal to the probability of selecting a _____ plus the probability of selecting a _____ minus the probability of selecting a _____ . In symbols, we have:
    P(queen or red card) = P(____) + P(____) − P(____).
    Thus, P(queen or red card) = ____ + ____ − ____ = ____.
25. If the probability a person owns a Chevy is 0.60, the probability a person owns a BMW is 0.40, and the probability a person owns both a Chevy and a BMW is 0.30, then the probability a person owns a Chevy or a BMW is equal to: ____+____−____ = _____.
26. Two events A and B are _____ if both events A and B cannot occur at the same time.
27. The Addition Rule for mutually exclusive events A and B is:
    P(A or B) = _____ + _____ .
28. If two events C and D are mutually exclusive, then:
    a) P(C and D) = _____.
    b) P(C or D) = _____.
29. Consider the experiment of selecting one card at random from an ordinary deck of 52 playing cards.

Let event A represent the event of selecting an ace and event B represent the event of selecting a king. Then, events A and B are _____ since a card cannot be both an ace and a king at the _____ time.
  Therefore, P(ace or king) = P(___) + P(___) = ___ + ___ = ___.

30. In a single roll of a fair die, the event of getting a 4 and the event of getting a 5 are _____ events, because a die cannot show a _____ and a _____ at the same time.

31. In a single roll of a fair die, the event of getting a 4 and the event of getting an even number are _____ events, because a die can show a ____ and an _____ at the same time. An outcome that satisfies both events is the outcome _____.

32. Probability Rule 4: The Complement Rule states that an event E can either occur or not occur. Thus, the sum of these events always equals _____ . This rule is symbolized as:
P(____) + P(_____) = _____.

33. For the experiment of tossing a fair coin four times, the probability of not getting 4 heads can be rewritten using the complement rule as:
P(not getting 4 heads) = 1 − P(_____).

34. Two events A and B are _____ if the occurrence or nonoccurrence of event A has no influence on the occurrence or nonoccurrence of event B.

35. Probability Rule 5 is the Multiplication Rule for _____ events. This rule states that if two events A and B are independent, then the probability of event A ____ B equals the probability of A _____ the probability of B.
This is symbolized as : P(___) = P(___)·P(___).

36. An urn contains 4 white marbles and 3 red marbles. If two marbles are selected with replacement from this urn, then the probability of selecting two white marbles equals the probability of selecting _____ on the first selection _____ the probability of selecting _____ on the second selection since both events are _____. Therefore,
P(two white marbles) = P(white marble) · P(white marble) = _____ · _____ = _____.

37. Consider the experiment of selecting two balls without replacement from an urn containing 6 numbered balls. If event A is the first ball selected and event B is the second ball selected, then these two events are said to be _____ .
  This type of probability problem is classified as _____ probability.

38. Probability Rule 6 is the multiplication rule for _____ events. This rule states that if two events A and B are dependent, then the probability of A and B is equal to the probability of A times the probability of _____ given that event A has _____. This is symbolized as:
P(_____) = P(A) · P(_____).

39. An urn contains 4 coins: a penny, two nickels and a dime. If two coins are selected without replacement, and event A is first coin selected is a nickel, while event B is second coin selected is a nickel, then events A and B are _____ events.
  To calculate P(A and B), one can use Probability Rule ___.
This is written as: P(A and B) = P(A) · P(____).
Calculating these probabilities, we have:
P(A) = P(first coin is a nickel) = _____.
P(B | A) = P(second coin is nickel | first coin is nickel)= _____ .
  Thus, P(first coin is a nickel and second coin is a nickel)
= P(A and B) = P(A) · P(B | A) = _____ .

40. Probability Rule 7: General Multiplication Rule states that for any two events A and B, the probability of A and B can be determined by either one of the following formulas:
P(A and B) = P(A) · _____. This formula is read as: the probability of A and B is equal to the probability of A multiplied by the _____. Or the formula:
P(A and B) = P(B) · _____. This formula is read as: the probability of A and B is equal to the probability of B multiplied by the _____.

41. Probability Rule 8: The Conditional Probability Formula states the probability of event A, given that event B has occurred, equals the probability of event A ___ B divided by the probability of event _____.
This is symbolized as: P(_____) = P(A and B) .
                                          P(_____)

42. The six chambers of a handgun randomly were filled with 4 blanks and 2 bullets. If the chambers are spun once, and the gun is fired twice in succession, then to calculate the probability a bullet is fired on the second shot given that the first shot was a blank is determined by using Probability Rule 8: The _____ Probability Formula. Thus,
P(second shot is a bullet | first shot is a blank)
= P(first shot is a blank and second shot is a bullet) = 8/30
   ───
                P(_____)   _____
= _____.

43. For the Birthday Problem, the complement rule is used to write:
P(at least 2 people have the same birthday)= 1 − P(_____).
If there are 35 people in a class, then the probability that at least 2 people have the same birthday is _____.

44. For an experiment to be binomial, it must satisfy the following conditions:
    a) the number of trials is _____.
    b) each trial is _____ of the previous trials.
    c) the outcome of each trial can be classified into one of two categories called _____ or a _____.
    d) the probability of a _____ is the same for each trial.

45. To calculate the probabilities for a binomial experiment, we use Probability Rule 9: The Binomial Probability Formula. This is symbolized as:

    $$P(\text{s successes in n trials}) = \binom{n}{s} p^s q^{(n-s)}$$

    where: n = number of _____ trials
    s = number of _____
    n-s = number of _____

    $\binom{n}{s}$ = number of ____ s successes can occur in n trials

    p = probability of a _____
    q = 1 - p = probability of a _____

46. For a binomial experiment, n = 6 and s = 3, then:
    $\binom{n}{s} = \binom{6}{3} = $ _____.
    Therefore, the number of ways to get ____ successes in ____ independent trials is _____.

47. For the binomial experiment of tossing a fair coin 4 times the number of ways to get 3 successes (consider a head as a success) in 4 independent trials is:
    $\binom{4}{3} = $ _____.
    Those four successes would be ____, ____, ____, and ____.

48. For a binomial experiment, if $\binom{7}{6} = 7$, then the number of ways to get ____ successes in ____ independent trials is _____. The seven successes are: sssssssf, ____, ____, ____, ____, ____, and ____.

49. For a binomial experiment, where n = 5, s = 3, and p = 1/3, calculate:
    a) P(3 successes in 5 trials) = P(3) = $\binom{5}{3} (\frac{1}{3})^3 (\frac{2}{3})^2$
    = ___ ___ ___
    = _____.

    b) P(5 successes in 5 trials) = $\binom{5}{5} (\frac{1}{3})^5 (\frac{2}{3})^0$
    P(5) = ___ ___ ___
    = _____.

50. For a binomial experiment, where p = 1/2, calculate: P(4 successes in 5 trials) = _____.

51. For a binomial experiment, if: $P(s=5) = \binom{7}{5} (\frac{1}{4})^5 (\frac{3}{4})^2$
    then: n = ____, s = ____, p = ____, and q = ____.
    The number of ways to get 5 successes in 7 trials is _____.
    The number of successes is _____.
    $\binom{7}{5} = $ _____.

52. For a binomial experiment, if p = 4/5, then q = _____.

PART II **Multiple choice questions.**

1. If three dice are tossed, how many different outcomes are in the sample space?
   a) 72 b) 42 c) 216 d) 18
2. The letters in the word MOM can be put into how many different arrangements?
   a) 6 b) 3 c) 27 d) 9
3. If a spinner with four different outcomes is spun and a die is tossed, how many different outcomes are in the sample space?
   a) 40 b) 10 c) 24 d) 1296
4. How many different teams containing five members each can be formed from a group of 9 players?
   a) 126 b) 15,120 c) 45 d) 180
5. If two cards are selected from a deck of 52 cards without replacement, how many different outcomes are in the sample space?
   a) 2704 b) 103 c) 1326 d) 2652
6. You've just arrived in Las Vegas for a vacation. You plan to see four dinner shows during your visit. If there are eight shows to select from, then how many ways can you select the four shows you'll see?
   a) 32 b) 8 c) 1680 d) 70
7. The coach of the basketball team would like to determine how many different five man teams he can put on the court if each team must have a center, two forwards and two guards. His team has two centers, five forwards and four guards.
   a) 120 b) 64 c) 462 d) 40
8. The chairperson of the scholarship committee wants to select a subcommittee of 2 men and 2 women to evaluate applications. If there are 7 women and 4

men on the committee, then how many different subcommittees can be selected?
a) 28   b) 333   c) 126   d) 27

9. Two actors are to be chosen to read the lines of a new play. If there are 17 actors applying, what is the probability Nick and Jack are randomly chosen?
a) 1/68   b) 1/136   c) 1/17   d) 1/272

PART III Answer each statement True or False.

1. A sample space is a representation of some of the outcomes for an experiment.
2. The Fundamental Counting Principle is used to determine the number of outcomes in a sample space.
3. If two events A and B are mutually exclusive then P(A and B) does not equal zero.
4. Permutations are used as a counting technique when the objects are represented in a definite order.
5. To determine the number of ways six CDs can be chosen from a selection of 140 one must use the combination formula $\binom{140}{6}$.
6. The number $\binom{n}{s}$, determined by use of TABLE I, gives us the number of ways "s" successes can occur in "n" independent trials.
7. It is impossible for P(E) to equal zero (i.e. P(E) = 0).
8. It is impossible for P(E) to equal one (i.e. P(E) = 1).
9. It is impossible for P(E) to be greater than one (i.e. P(E) > 1).
10. Two events that are mutually exclusive must also be independent.
11. Two events that are independent must also be mutually exclusive.

PART IV Problems.

The problems in this part are separated according to concepts. Problems 1-20 deal with elementary probability concepts. Problems 21-43 deal with counting concepts including permutations. Problems 44-50 are probability problems dealing with counting concepts including permutations. Problems 51-70 deal with combinations. Problems 71-84 are complex problems dealing with a mixture of all elementary concepts and conditional probability. Finally, problems 85-105 deal with binomial probability.

Problems preceded with "*" are to be completed using a calculator.

1. Construct the sample space for the following probability experiments:

   a) selecting one ball from an urn containing 4 red, 3 yellow and 3 blue balls.
   b) spinning the following spinner 2 times.

   c) tossing a fair coin three times.
   d) tossing a biased coin three times.
   e) arranging the letters in the word BYE.
   f) arranging the letters in the word GOOD.

2. Construct the sample space for the following probability experiments.
   a) selecting two balls without replacement from an urn containing 1 red, 1 yellow and 2 blue balls.
   b) selecting two balls with replacement of the first ball before the second is selected from an urn containing 1 red, 1 yellow and 2 blue balls.
   c) selecting two cards without replacement from the four cards: jack, queen, king and ace.
   d) selecting two cards with replacement from the four cards: jack, queen, king and ace.

3. Find the probability of each of the following events:
   a) selecting the ace of spades by randomly choosing one card from a well shuffled deck containing 52 cards.
   b) randomly selecting the lottery ticket numbered 0007 from a bin containing lottery tickets numbered 0001 to 9999.
   c) guessing the month in which your teacher was born.
   d) randomly selecting an incorrect response to a five item multiple choice question.
   e) randomly selecting a loser in an eight horse race.

4. Find the probability of each of the following events:
   a) selecting an eight by randomly choosing one card from a well shuffled deck containing 52 cards.
   b) randomly selecting a lottery ticket beginning with the digit 5 from a bin containing lottery tickets numbered 00001 to 99999.
   c) selecting the ace of hearts when randomly selecting one card from a deck of 52 cards.
   d) randomly selecting your best friend's birthday from a bin containing all 365 days.
   e) randomly selecting the correct answer to a five choice multiple choice question.
   f) randomly selecting the winner of an eight horse race.

5. From a regular deck of 52 cards, what's the probability of
   a) selecting two picture cards with replacement.
   b) selecting two picture cards without replacement.
   c) selecting two cards with replacement whose face values are even numbers.
   d) selecting two cards without replacement whose face values are even numbers.
6. Construct a sample space for the sum of two dice.
   a) What is the probability of each outcome? (Hint: look at the sample space for Example 6.23.)
   b) What is the probability that one rolls a sum of 7 or 11?
   c) What is the probability that one doesn't roll a sum of 7 or 11?
   d) What is the probability that the sum is less than 9?
   e) What is the probability that the sum is odd?
7. In the game of chuck-a-luck a participant wages on the sum of two dice to be above or below seven.
   a) What's the probability of the sum being below seven?
   b) What's the probability of the sum not being below seven?
8. A wine tasting contest is being conducted at the Bowery Hotel. There are 4 varieties of wine labeled: B, M, S, and U to be rated. If one of the contestants has a hangover and has no ability to distinguish a difference in taste between the wines, then what is the probability that he will:
   a) rank wine B as the least desirable tasting wine?
   b) rank wine S as the best tasting wine and M as the least desirable tasting wine?
   c) rank the four wines in the following order: B, U, M, S?
9. a) Construct a sample space for a four child family. Assuming each outcome is equally likely, what is the probability of:
   b) exactly one boy?
   c) at least one boy?
   d) at least two girls?
   e) at most one boy?
   f) no boy?
10. A young couple plans to have three children. Assuming that the probability of having a boy is 1/2 determine:
    a) the probability of having 1 boy and 2 girls.
    b) the probability of having only 1 girl.
    c) the probability of having at least 2 boys.
11. A sociology class consists of 12 boys (5 tall, 6 average height, 1 short) and 18 girls (3 tall, 10 average height, 5 short). Suppose the name of each student is written on a slip of paper and placed in a box. After the slips are mixed vigorously, one slip is selected at random. What is the probability that the name on the slip is:
    a) a girl?
    b) a boy?
    c) a tall boy?
    d) a student of average height?
    e) a short girl or a tall boy?
    f) not a student of average height?
12. Two Italians, Lorenzo and Luigi, play a game where they simultaneously exhibit their right hands with one, two, three or four fingers extended.
    a) List all the possible outcomes in the sample space.
    Determine the probability that:
    b) both players extend the same number of fingers.
    c) both players together extend an even number of fingers.
    d) Lorenzo shows an odd number of fingers.
13. In New York State's numbers game, an individual chooses a 3 digit number to play the game. This 3 digit number can range from 000 to 999. What is the probability that Tony will win the game, if he selects:
    a) the number 123?
    b) the digits 4, 2, 7 in all possible arrangements?
    c) all numbers beginning with a 7?
    d) all numbers beginning with 46?
    e) all possible numbers using the digits 1, 2, 6?
14. Derek has the following four coins: A penny, a nickel, a dime and a quarter. Construct a sample space listing all the possible sums of money Derek can form with one or more of these coins. Determine the probability that the sum of money:
    a) is greater than 10 cents?
    b) is less than 35 cents?
    c) is an even amount of money?
    d) contains a penny?
    e) contains at least one silver colored coin (nickel, dime, or quarter)?
15. A letter of the English alphabet is selected at random. Determine the probability that the letter selected:
    a) is a consonant.
    b) is a vowel (ie. a, e, i, o, u).
    c) follows the letter L.
    d) follows the letter L *and* is a vowel.
    e) follows the Letter L *or* is a vowel.
    f) is one of the letters in the word "bushes".
    g) is not in the word "bushes".
    h) is one of the letters in the word, "bushes" or the word "tushes".

i) is one of the letters in both the words "tushes" and "bushes".
j) is one of the letters in the word "bushes" but not in the word "tushes".

16. Four balls numbered 1 to 4 inclusive are placed in an urn. Randomly choose the four balls from the urn without replacement. Record the number appearing on each ball as it is selected so as to form a four digit number.
What is the probability that:
a) the number 1234 is obtained?
b) the number obtained ends in a 4?
c) the number obtained is greater than 3200?
d) the number obtained is not greater than 4000?
e) the number obtained is either even or less than 3500?
f) the number obtained is odd and greater than 5000?

17. Four defective flashlight batteries are mistakenly mixed with two non-defective flashlight batteries which all look identical. If you randomly choose 2 batteries without replacement, what is the probability that:
a) exactly one is defective?
b) both are defective?
c) both are non-defective?
d) at most one is non-defective?

18. An eight cylinder automobile engine has two defective spark plugs. If two plugs are removed from the engine without replacement, determine the probability of selecting:
a) no defective plug.
b) one defective plug.
c) two defective plugs.

19. Millie, a hat check girl, mistakenly mixes up the tickets on three hats. If she randomly places the mixed up tickets on the three hats, what is the probability that:
a) exactly one guest will receive the correct hat?
b) exactly all 3 guests will receive the correct hats?
c) exactly two guests will receive the correct hats?
d) no one will receive the correct hat?

20. If Toni, a craps player, rolls a five on her first roll, what is the probability that she makes her point? (In this case, making the point means rolling a five, before rolling a seven).

21. In a given semester, a computer science major must select one of five science courses, one of three English courses and one of four psychology courses. How many different programs are available if there are no time conflicts?

22. When ordering a new car the buyer has the choice of four body styles, five different engines and twelve colors.

a) In how many different ways can a buyer order one of these cars?
b) If the buyer also has the option of ordering the car with or without air conditioning, with one of three transmissions, with one of four radio options and with or without a sun roof, in how many ways can the buyer order one of these cars.

23. Read the following famous nursery rhyme and answer the counting questions pertaining to this rhyme.

> As I was going to St. Ives,
> I met a man with seven wives,
> Every wife had seven sacks,
> Every sack has seven cats,
> Every cat has seven kits,
> Kits, cats, sacks and wives,

a) How many are going to St. Ives? Determine the number of:
b) sacks.
c) cats.
d) kits.

24. Calculate each of the following
a) $_2P_2$ e) $_7P_3$
b) $_4P_4$ *f) $_{11}P_7$
c) $_4P_2$ *g) $_{12}P_5$
d) $_4P_3$

25. a) Write the letters of the word DOG in all possible three letter arrangements.
b) Use the formula $_nP_n = n!$ to compute the number of arrangements.

26. a) Write the letters of the word PIG in all possible three letter arrangements.
b) Use the formula: $_nP_n = n!$ to compute the number of arrangements.

*27. In how many ways can eight teachers be assigned to eight classrooms?

28. Five textbooks are to be arranged on a shelf. In how many ways can they be arranged?

29. On the way to work Allison must pass through five traffic lights. How many different sequences of green, red could Allison observe as she drives to work?

30. If seven cars are entered in an auto race and only four cars can be placed in the front row, in how many ways can four cars be arranged in the front row?

31. In how many ways can the officers for President, Treasurer and Secretary be filled from the 12 members of the ski club?

*32. Joe, the manager of a California baseball team, decides to change the batting order of his present starting lineup to help his team get out of their slump. If he places the names of his present starting team in a hat and randomly draws out the lineup, how many different batting orders can he select? If he decides to place the names of all sixteen non-pitching players in the hat and randomly draw a group of 9 players, how many different teams of 9 players can he choose?

*33. How many four letter arrangements can be made from ten different letters if:
   a) repetitions are allowed?
   b) repetitions are *not* allowed?
   c) a letter can be used at most twice?

34. A *byte* is a computer "word" which consists of a sequence of 0's and 1's. If a particular manufacturer uses a byte of length 8, how many different bytes can be formed?

*35. Four married couples have purchased eight seats in a row for a Broadway show.
   a) In how many different ways can these eight people be seated?
   b) In how many different ways can they be seated if the husband must be seated immediately to the right of his wife?
   c) In how many different ways can they be seated if each couple must sit together?
   d) In how many different ways can they be seated if all the men sit together and all the women sit together?

36. a) Write the letters in the word MISS into all possible four letter arrangements.
   b) Use Counting Rule 3 to determine the number of arrangements.

37. Use Counting Rule 3 to determine the number of 5 letter arrangements for the letters in the word FUNNY.

38. In how many different ways can the letters in the word TOOTH be arranged?

*39. In how many different ways can the letters in the following words be arranged?
   a) calculator
   b) Tennessee
   c) scissors
   d) infinity

40. A company wishes to issue ID plates to it's employees.
   a) If they use a two letter code for each plate, how many different plates can be made?
   b) If no letter can be repeated in the same code, how many different plates can be made?
   c) If the code consists of two letters and one digit in the last position, now many different plates can be made?
   d) If the code consists of two letters and one digit in any order, how many different plates can be made?

41. How many different license plates can be made:
   a) If the first 3 characters are letters and the last 3 characters are numerals?
   b) If all six characters are letters?
   c) If all six characters are numerals?
   d) If all six characters are letters with none repeated?
   e) If all six characters are numerals with none repeated

*42. How many different signals (hanging on a vertical line) can be made with 4 identical green flags, 3 identical blue flags and 3 identical red flags?

*43. In a new version of *Master Blaster Brain* one of the players must construct a sequence of 15 colored pegs. There are 2 red, 4 green, 1 blue, 3 white, 2 yellow and 3 black pegs. How many different sequences can be made?

44. If you remember the first five digits of a telephone number but you cannot recall the last two digits, what is the probability that you select the correct missing digits on your first guess?

45. If a random drawing is performed and five letters are selected without replacement from the alphabet, determine the probability that:
   a) the first three letters in order spell YES.
   b) the first five letters in order spell GREAT.

46. Given five *Scrabble* blocks, each having one of the letters a, c, e, p, l:
   a) how many 5 letter arrangements are possible?
   b) how many 3 letter arrangements are possible?
   c) If the 5 blocks are randomly arranged on a table, what is the probability that the word *place* is spelled?

47. Using six different *Scrabble* blocks with the letters a, e, o, n, r, s on them:
   a) how many six letter arrangements are possible?
   b) how many three letter arrangements are possible?
   c) If the 6 blocks are randomly arranged on a table, what is the probability that the word *reason* is spelled?

48. Of seven trees in a line, two are diseased. If any of the seven are equally likely to be affected, what is the probability that the diseased trees are side by side?

49. Amy has four cassettes to be played at a party.
Cassette 1 contains songs from the early 60's
Cassette 2 contains songs from the late 60's
Cassette 3 contains songs from the 70's
Cassette 4 contains songs from present day rock.
   a) How many ways can Amy play the cassettes?

(Assuming once started a cassette is played to its completion).

b) What is the probability that if the cassettes were randomly selected from a bag, the order would be cassette 1, cassette 2, cassette 3 and cassette 4?

c) How many ways can Amy play the cassettes if cassette 1 must be played first?

d) What is the probability that after playing cassette 1, cassette 2, 3 and 4 are played in that order?

e) In how many ways can Amy pick two of the four cassettes?

f) What is the probability if Amy selects only 2 cassettes, that she chooses cassette 1 and 2 in that order?

*50. Matthew, a playboy, is returning to his penthouse on the thirteenth floor late one night. He is concerned that his expected date is already waiting for him in his penthouse. If seven other people get on the elevator with him and no two are together, and he knows that no one in the elevator besides himself is going to the penthouse, what is the probability that:

a) they all get off at different floors, thus delaying Matthew as long as possible?

b) all the other passengers get off at the same floor, thus delaying Matthew the least?

51. a) Find the number of permutations that are possible if 5 distinct objects are taken:
1. one at a time.
2. two at a time.
3. three at a time.
4. four at a time.
5. five at a time.

b) Find the number of combinations that are possible if 5 distinct objects are taken:
1. one at a time.
2. two at a time.
3. three at a time.
4. four at a time.
5. five at a time.

c) Why are the answers to part (a) larger than the answers to part (b)?

*52. Calculate:

a) $\binom{6}{3}$ b) $\binom{5}{2}$ c) $\binom{10}{4}$ d) $\binom{25}{13}$ e) $\binom{52}{7}$

*53. Calculate:

a) $\binom{4}{3}$ b) $\binom{7}{5}$ c) $\binom{12}{4}$ d) $\binom{52}{2}$ e) $\binom{52}{13}$

*54. In how many different ways can the I.R.S. select 7 tax returns to audit out of 50 tax returns?

55. Ten people meet at a pub. Find the number of handshakes that take place if each person shakes hands with everyone else in the group.

56. Using the seven points pictured in the following diagram, how many straight lines can be constructed using only two points for each line?

*57. Doug recently was sent an advertisement to join a record club. As an introductory offer he could choose *any* five selections from the one hundred listed. In how many different ways can Doug make his selections?

*58. As an incentive to join a book-of-the-month club, Grace can purchase any six selections from a list of 40 for only 99 cents. In how many different ways can Grace make her selections?

59. How many different grade grievance committees can be selected from 15 Professors if the committee has four members?

60. How many different three member bowling teams can be selected from the eight people trying out for the team?

*61. Two attorneys must select nine members for a jury from a group of 15 individuals.

a) How many different ways can the nine members be selected?

b) Of the 15 individuals, if 9 are women and 6 are men and the attorneys would like a jury of 5 women and 4 men, how many different ways can the jury be selected?

62. Twelve college graduates, of which four are female, have applied for seven vacancies in a computer company. If the company has decided to hire three females, how many ways could the company make the seven job offers?

63. On a history test part I has five questions of which three must be answered, where part II has six questions of which four must be answered. In how many ways can a student choose the questions they must answer on this test?

*64. A five card poker hand is dealt to you. Determine the probability of the following:

a) the hand contains exactly 1 club?
b) the hand contains at least 1 diamond?

65. A set of ten cards contains two jokers. There are two players, Bob and Audrey. Bob chooses six cards at random and Audrey gets the remaining four cards. What is the probability that Bob has at least one joker?

*66. Find the probability that a poker hand will contain:
   a) full house (1 pair and 1 triple)
   b) 4 of one kind
   c) one pair (hint: more complicated than you may think)
   d) two pair
   e) a flush (all cards of the same suit)

67. From a regular deck of 52 cards, what's the probability of selecting two cards without replacement which form *blackjack*? A blackjack has one ace and either a jack, queen, king or ten.

*68. In the game of bridge four hands of 13 cards each are dealt from a regular deck of 52 cards.
   a) How many different bridge hands are possible?
   b) How many different ways can one be dealt a bridge hand of all hearts?
   c) What is the probability of being dealt a bridge hand of ten clubs?

69. An urn contains 10 marbles. If there are 6 red marbles and 4 blue marbles, in how many ways can one choose 3 of these marbles (one at a time) without replacement so that:
   a) none of the red marbles are selected?
   b) exactly one of the blue marbles is selected?
   c) 1 red and 2 blue marbles are selected?
   d) 3 marbles, regardless of color, are selected?

70. A jar contains 5 red, 3 green, 4 blue and 2 yellow marbles. Four marbles are selected without replacement. Determine the probability of the following:
   a) one marble of each color is selected.
   b exactly three red marbles are selected.
   c) exactly two blue marbles are selected.

71. a) If two people are randomly selected and it is known that they are both born in October, what is the probability that they have the same birthday (day only) ?
   b) If three people are randomly selected and it is known that they are all born in April, what is the probability that their birthdays are all different?
   c) What is the probability that at least two of these three people have the same April birthday?

72. Mary, an intoxicated individual, has five keys on her key ring, one of which opens the front door to her house. If she randomly selected one key at a time and is unable to remember which keys she has already tried, what is the probability that:
   a) she selects the correct key for the front door on the first try?
   b) she selects the correct key for the front door on the second try?
   c) she selects the correct key for the front door on the third try?
   d) she selects the correct key for the front door within the first four tries?

73. Matt The Magician in his "three card Monte" act uses three colored cards. One is red on both sides, one is blue on both sides and the third is red on one side and blue on the other side. If one card is selected at random and the top side is red, what is the probability that the other side is also red?

74. Suppose a two child family is moving onto the block where you live.
   a) If you learn that not both children are girls, what is the probability that both children are boys?
   b) If you know that one child is a boy, what is the probability that the oldest child is a boy?
   c) If you learn that both children are of the same sex, what is the probability that both children are girls?

75. The following table gives a breakdown of the students that attended the graduation picnic.

	BOYS	GIRLS	
5th Grade	24	29	53
6th Grade	18	29	47
	42	58	100

If a student is selected at random to be in charge of coordinating the games, what's the probability of:
   a) the student selected being a girl?
   b) the student selected is a girl if the student is a 5th grader?
   c) the student selected is a 6th grader if the student selected is a boy?

76. On the planet Krypton the calendar was almost identical to ours with the exception that every one of the twelve months had 30 days. What's the probability that the day the planet exploded was:
   a) the thirteenth of the month?
   b) a Friday?
   c) a Friday given that it was the thirteenth?
   d) the thirteenth given that it was a Friday?

77. Craig, an investment advisor, has a portfolio of 90 stocks: 60 blue-chip and 30 growth stocks. Of the 60 blue-chip stocks, 40 have increased in price during the past month, and 25 of the 30 growth stocks have increased in price. If a stock is selected at random from the portfolio, what is the probability:
   a) it will be a growth stock that has not increased in price?
   b) If the stock has not increased in price, then it is a blue-chip stock?

78. The DPT Petroleum Enterprise exploring for oil decides to drill two wells, one after the other. Based upon their research of the area, the probability of striking oil with the first well is 0.31. Given the first attempt is successful, the probability of striking oil with the second well is 0.85. What is the probability of striking oil with both wells?

79. The past weather records of a New England Airport indicate that the probability it will snow is 0.2. Given that it snows, the probability is 0.5 that the airplane will depart on time, and given it does not snow, the probability is 0.85 that the airplane will depart on time. Find the probability that:
   a) it will snow and the plane will depart on time.
   b) it will not snow and the plane will depart on time.
   c) the plane will not depart on time.

80. Vinny, the manager of a Baltimore baseball team, uses probability to help make crucial decisions during the game. In one such instance he estimated that the present pitcher had a 70% chance of getting the next batter out. However, his relief pitcher has a 95% chance of getting the batter out if he is at his best that day, but only a 45% chance if he is not. His pitching coach states that the relief pitcher has a 60% chance of being at his best.
   a) What is the probability of the relief pitcher getting the batter out?
   b) What should the manager's decision be if he uses probability to make his decisions?

81. In the casino game of CRAPS various wagers are made regarding the sum of the dice. A typical game begins when the "stickman" pushes two dice to a gambler and announces "new 'shooter' coming out." Then the shooter will start rolling the dice. On the first roll of the dice, three possible things can happen:
   a) A natural (a sum of 7 or 11) may be thrown. This is an outright win. Determine the probability of getting a natural.
   b) A craps (a sum of 2, 3, or 12) may be thrown. This is an outright loss. Determine the probability of getting a craps.
   c) A sum of 4, 5, 6, 8, 9 or 10 may be thrown. In this case, the number you have rolled is your "point", and you then continue to roll the dice until either one of two things occur:
      1) your point is thrown again, in which case you win;
      2) a 7 is thrown, in which case you lose.
      If your point is a 4, what is the probability of making your point (i.e. rolling a 4 before rolling a 7)?
   d) If your point is a 5, what is the probability of making your point?
   e) If your point is an 8, what is the probability of making your point?
   *f) One possible wager in the game of CRAPS is called the "pass line bet". This wager is won by either rolling a natural on the first roll or rolling a 4, 5, 6, 8, 9 or 10 and making the point on the subsequent rolls of the dice. Determine the possibility of winning the pass line bet. (Calculate your answer to 5 decimal places)

82. **A classic probability and logic problem:**
   A missionary is captured by a tribe of cannibals who decide to have her for dinner. However, this tribe, being a more sporting group, decide to give the missionary a chance to save her life. They place before her two empty coconuts and twenty berries, ten red berries and ten blue berries. She must distribute *all* the berries as she pleases between the two coconuts, putting at least one berry in each coconut. The head cannibal will randomly choose one berry from the selected coconut. If the berry is blue, the missionary will be set free. However, if it is red, the missionary will be eaten for dinner.
   a) How should the missionary distribute the berries to maximize her chances for freedom?
   b) Using this distribution, what is the probability that she will be set free?

83. **The following legendary problem from India prophesies the *end of the world*.**
   In a great temple located at the center of the world, rests a brass plate in which are fixed three ivory rods. At the creation of the world, God placed sixty-four discs of pure gold on one of these rods with the largest disc resting on the brass plate and the remaining discs getting smaller and smaller up to the smallest disc at the top. This is referred to as the *Tower of Brahma*. Day and night, the priest on duty transfers the disc from one ivory rod to another adhering to the ancient laws of Brahma. These laws stipulate that the priest must move only one disc at a time, and he must place these discs on the rods so that there never is a smaller disc beneath a larger disc. When all the sixty-four discs have been transferred from the initial ivory rod, on which God placed them, to one of the other rods, the tower and temple will crumble into dust and with a thunderclap the world will vanish.
   a) If there were only two discs on the initial rod, what is the *least* number of moves needed to transfer the discs from the initial rod to one of the other rods?

b) Determine the *least* number of moves using:
   1) three discs.
   2) four discs.
   3) five discs.

c) Complete the following table and deduce a general rule to fit the pattern established in the table.

number of discs	least number of moves
1	1
2	
3	
4	
5	
.	
.	
.	
N	

d) Determine the least number of moves using 64 discs.

e) If the priests worked day and night, without stopping, making one move every second, how much time would it take for the priests to accomplish their task?

84. **The following dialogue took place between the TV personality Monte Hall of *Let's Make a Deal* and a contestant.**[1]

Monte Hall: One of the three boxes labeled A, B, and C contains the keys to that new 1975 Lincoln Continental. The other two are empty. If you choose the box containing the keys, you win the car.
Contestant: Gasp!
Monte Hall: Select one of the boxes.
Contestant: I'll take box B.
Monte Hall: Now box A and box C are on the table and here is box B (contestant grips box B tightly). Is it possible the car keys are in that box? I'll give you $100 for the box.
Contestant: No, thank you.
Monte Hall: How about $200?
Contestant: NO!
Audience: NO!
Monte Hall: Remember that the probability of your box containing the keys to the car is one-third and the probability of your box being empty is two-thirds. I'll give you $500.
Audience: NO!
Contestant: NO, I think I'll keep this box.
Monte Hall: I'll do you a favor and open one of the remaining boxes on the table (he opens box A).
Monte Hall: It's empty! (Audience:applause). Now either box C or your box B contains the car keys. Since there are two boxes left, the probability of your box containing the keys is now one-half. I'll give you $1,000 cash for your box.

WAIT!!!!

Is Monte right? The contestant knows that at least one of the boxes on the table is empty. He now knows it was box A. Does this knowledge change his probability of having the box containing the keys from 1/3 to 1/2? One of the boxes on the table has to be empty. Has Monte done the contestant a favor by showing him which of the two boxes was empty?

a) Is the probability of winning the car 1/2 or 1/3?
Contestant: I'll trade you my box B for the box C on the table.
Monte Hall: That's weird!!
Do you agree with Monte that the contestant's proposal is "weird," or is the contestant making a logical choice?

b) What is the probability that box C contains the keys?

85. A *biased* coin is flipped 5 times. If the probability is 1/3 that it will land heads on any toss, calculate the probability that the coin:
a) lands heads all five times.
b) lands heads at least three times.
c) lands tails at most one time.

86. If you guess at the answers to five multiple choice questions where each question has four choices what is the probability that:
a) you guess at least 4 correct?
b) you guess less than 3 correct?
c) you guess them all wrong?

87. Candy and Sam, a young couple, plan to have five children. Assuming that the probability of having a boy is 1/2 determine:
a) the probability of having 2 boys and 3 girls.
b) the probability of having only 1 boy.
c) the probability of having at least 4 boys.

88. The probability that a softball player named Casey gets a base hit during a single time at bat is 4/10. If he goes to bat 4 times in his next game (which is played in Mudville), determine:
a) the probability that he gets 4 hits.

b) the probability that he gets no hits, causing no joy in Mudville.
c) the probability that he gets 3 or more hits.

*89. Craig "The Magician" claims his hands are quicker than your eyes. To illustrate this he places a red ball under one of three cups, does some quick manipulations and asks you to tell which cup is covering the red ball. Since his hands are indeed faster than your eyes, calculate the probability that you do not guess the correct cup ten times in succession. (Calculate probability to three decimal places).

90. If you roll a pair of fair dice, the probability that the sum is seven is 1/6. If the dice are rolled four times, what is the probability that:
a) none of the outcomes are seven?
b) all of the outcomes are seven?
c) at most 1 of the outcomes is a seven?

91. If five cards are selected with replacement from a regular deck of 52 playing cards, what's the probability that:
a) at least three of the cards are clubs?
b) none of the cards are clubs?
c) one of the five cards is a club?

92. Each week during the football season, a New York newspaper has a contest that involves predicting the 10 winners of 10 specified football games. If Daniel randomly guesses the winner of each of these games, and if there are no ties, what is the probability that Daniel will guess all ten correct?

93. From past experience, Caryl, a golfer, knows that she will hit her drive into a sand trap from the green one-fourth of the time. Using this figure, what is the probability that she will hit the ball into a sand trap on exactly four of the first nine holes?

94. In the United States the probability that a person has chapped lips is 30%. On Charlene's next five dates, what is the probability that:
a) at least four of the dates have chapped lips?
b) less than two have chapped lips?
c) none have chapped lips?

95. Suppose that 70% of the voters are in favor of the repeal of a county tax assessment law. If a random sample of nine voters were selected, what is the probability that the majority of the sample would be against the repeal?

96. The probability that a stock analyst of the Wall Street firm DPT will convince one of his customers to take a "position" in a recommended stock is 40%. The analyst presents his recommendations to five customers. Assuming independence, what is the probability that:
a) all five will take a "position"?
b) none will take a "position"?
c) at most one will take a "position"?

97. A national bank has determined that the probability of a default on one of their loans is 5%. If they approve seven loans during the next business day, what is the probability that:
a) there are no defaults?
b) there are at least six defaults?

98. Medical records show that 40% of all persons affected by a certain viral illness recover. A pharmaceutical company has developed a new vaccine for this illness. Ten people with this illness were selected at random and injected with the vaccine. Eight of the ten people recovered shortly after receiving the vaccine. Suppose that the vaccine was absolutely worthless:
a) what is the probability that at least eight of the ten people injected with the vaccine recover?
b) Based on your answer to part (a), do you think that the vaccine was helpful in the recovery of these people?

99. In Denmark the probability that a carton of a dozen eggs contains one rotten egg is 1/20. If a boy named Hamlet buys 5 cartons of eggs, what is the probability that at least 1 carton contains a rotten egg (causing Hamlet to mumble: "Something is rotten in Denmark!")?

100. A slot machine has four windows. The wheel behind each window has the same five identical objects. They are: cherry, grape, peach, star and jackpot. After the handle is pulled, the four wheels revolve independently several times before coming to a stop. If the slot machine is played once, what is the probability that:
a) exactly three cherries appear in the windows?
b) at least three peaches appear in the windows?
c) four jackpots appear in the windows?
d) at most 1 grape appears in the windows?
e) less than 3 stars appear in the windows?
f) only fruits appear in the windows?
g) only 1 fruit appears in the windows?

101. Four out of five patients who have an artery bypass heart operation are known to survive at least three years. Of five patients who recently had the operation, what is the probability that:
a) all five will survive at least 3 years?
b) at most two will *not* survive at least 3 years?
c) at least four will survive at least 3 years?

102. A psychologist claims that he has trained a rat named Algernon to select the correct path of a maze that leads to food 90% of the time. Based upon this assumption, what is the probability that Algernon will select the correct path six times in eight previously unknown mazes?

103. Suppose an urn contains 10,000 balls, of which 4,000 are red and 6,000 are blue. If five balls are randomly selected one at a time, with replacement, from this urn find the probability that:
   a) exactly two blue balls were selected.
   b) more than four red balls were selected.
   c) at most two blue balls were selected.
   d) no red balls were selected.

104. Anne Marie, a botanist, is studying a new hybrid Kentucky blue grass seed. It is known that these grass seeds have a 95% probability of germinating. Anne Marie plants six seeds. What is the probability that:
   a) exactly four seeds will germinate?
   b) at least five will germinate?
   c) at most one will not germinate?

105. Medical records indicate that 40% of the people have type O+ blood. What is the probability that of the next seven people donating blood that:
   a) at least five have type O+ blood?
   b) at most one has type O+ blood?
   c) more than five do *not* have type O+ blood?

PART V What Do You Think?

1. **What Do You Think?**

   An article entitled *Behind Monty Hall's Doors: Puzzle, Debate and Answer?* appeared in the Sunday New York Times on July 21, 1991. The article discussed the debate that was raging among mathematicians, readers of the "Ask Marilyn" column of Parade Magazine and the fans of the TV game show "Let's Make a Deal". The argument began in September, 1990 when Ms. vos Savant, who is listed in the Guinness Book of World Records Hall of Fame for the highest IQ, posed the following question in her "Ask Marilyn" column of Parade Magazine:

   "Suppose you're on a game show, and you're given the choice of three doors: Behind one door is a car; behind the others, goats. You pick a door, say No. 1, and the host, who knows what's behind the other doors, opens another door, say No. 3, which has a goat. He then says to you, 'Do you want to pick door No. 2?' Is it to your advantage to take the switch?"

   Since Ms. vos Savant gave her answer, she has received approximately 10,000 letters with the great majority disagreeing with her answer. The most vehement from mathematicians and scientists who have called her the "goat". Her answer has been debated from the halls of the CIA to the mathematicians of MIT. Even Monty Hall conducted a simulation of the problem in his dining room. After which Monty came to the conclusion that Ms. vos Savant's critics were dead wrong provided that the host is required to open a door all the time and offer you a switch. However, Monty suggested since he has a choice on the show as to whether he allows the contestant the choice to switch or not, then the question cannot be answered without knowing the motivation of the host. As Monty Hall said: "If he (the host) has the choice whether to allow the switch or not, beware. *Caveat emptor*. It all depends on his mood. My only advice is, if you can get me to offer you $5,000 not to open the door, take the money and go home."

   a) Monty Hall states in the article that the odds on the car being behind door No. 1 are still only 1 in 3 even after he opened another door to reveal a goat. Explain why this is true.
   b) Some people argued with Ms. vos Savant that if one door is shown to be a loser, that information changes the probability of either remaining choice - neither of which has any reason to be more likely - to ½. Explain why this is not correct reasoning.
   c) Perform a simulation of this game with one of your friends by turning over three playing cards instead of opening doors and using the ace of diamonds as the prize.

   After doing this simulation 90 times, determine the chance of winning the car when the strategy is to switch doors. What did you get as the chance of winning the car when the strategy was not to switch doors? Are these results close to the actual probabilities?

2. **What Do You Think?**
   Tests for AIDS and Drugs: How Accurate Are They? appeared in the "Ask Marilyn" column of Parade Magazine in the March 28, 1993 issue. The question posed was:

   "Suppose we assume 5% of the people are drug users. A test is 95% accurate, which we'll say means if a person is a user, the result is positive 95% of the time; and if she or he isn't, it's negative 95% of the time. A randomly chosen person tests positive. What is the probability the individual is a drug user"?

To help answer this question, let's assume the population consists of 10,000 people.
a) How many people are drug-users?
b) How many people are non-users?
c) How many of the non-users will test negative?
d) How many of the non-users will test positive?
e) How many of the drug-users will test negative?
f) How many of the drug-users will test positive?
g) How many are positive results?
h) How many are negative results?
i) How many of the positive results are drug-users?
j) How many of the positive results are non-users?
k) What is the probability that given a randomly chosen person tested positive, then the person is a drug-user?
l) What is the probability that given a randomly chosen person tested positive, then the person is a non-user?
m) What would happen to the results of parts (k) and (l) if the percent of drug users in the population was 2%, 1%, 0.5% or 0.25%? What happens as the percentage of drug-users gets closer to 0%?
n) What would happen to the results of parts (k) and (l) if the percent of accuracy for the test is increased from 95% to 98% or 99%? What effect does this have on the results?

PART VI Exploring DATA With MINITAB.

1. Use MINITAB to calculate the binomial probabilities for the binomial experiment where:
   a) $n = 4$, $p = 0.5$ and $s = 2$
   b) $n = 40$ $p = 0.5$ and $s = 20$
   c) $n = 80$ $p = 0.5$ and $s = 40$
2. Use MINITAB to calculate the binomial probabilities for the binomial experiment where:
   a) $n = 10$, $p = 0.3$ and $s = 5$
   b) $n = 100$, $p = 0.3$ and $s = 5$
   c) $n = 200$, $p = 0.3$ and $s = 100$
3. Use MINITAB to generate a table successes along with the probability of success for a binomial experiment where
   a) $n = 4$ and $p = 0.5$
   b) $n = 40$ and $p = 0.5$
   c) $n = 80$ and $p = 0.5$
   d) For parts a,b and c construct a graph for each binomial experiment.
   e) Describe the shape of each graph generated in part d.
   f) For parts a, b and c, determine the values of s for which the $P(s) = 0$.
4. Use MINITAB to generate a table successes along with the probability of success for a binomial experiment where
   a) $n = 10$ and $p = 0.3$
   b) $n = 100$ and $p = 0.3$
   c) $n = 200$ and $p = 0.3$
   d) For parts a,b and c construct a graph for each binomial experiment.
   e) Describe the shape of each graph generated in part d.
   f) For parts a, b and c, determine the values of s for which the $P(s) = 0$.
5. Use MINITAB to generate a table successes along with the probability of success for a binomial experiment where
   a) $n = 10$ and $p = 0.7$
   b) $n = 100$ and $p = 0.7$
   c) $n = 200$ and $p = 0.7$
   d) For parts a,b and c construct a graph for each binomial experiment.
   e) Describe the shape of each graph generated in part d.
   f) For parts a, b and c, determine the values of s for which the $P(s) = 0$.
6. Examine the graphs produced in problems 3d, 4d, and 5d.
   a) What is the basic shape for each graph?
   b) Where is the peak located on each of these graphs?
   c) Is the length of each tail the same or different for each of the graphs?
   d) Try to generate a rule between the length of the tail on the left side of the graph and the value of p?
   e) Based on your rule which of the following binomial experiments would have the shorter left tail?
      1) $n = 100$ and $p = 0.2$
      2) $n = 100$ and $p = 0.8$

PART VII Projects.

1. Reread Example 6.51 *The Birthday Problem*. Ask 20 people to write a whole number between 1 and 100 inclusive onto a piece of paper. Fold the paper in half and put the folded paper into a large envelope. Before examining the results calculate the probability that at least two of the people have the same number.
2. Administer the following test to 32 students who have no previous knowledge about the material within each question.

a) Complete the table given below with your test results.
b) Construct a histogram using the data of the test results.
c) Compare the test results and the histogram of the test results to the binomial experiments table and histogram.

Test: Answer each question True or False.
1) If $f(x) = e^x$, x is a real number, then:
$$\int f(x)\, dx = f(x) + C,$$ where C is a constant.

2) Set A is an open set of a metric space, M, if for every x belonging to A, there exists a $\delta > 0$ such that the interval $(x-\delta, x+\delta) \, \varepsilon \, A$.

3) In 1733, DeMoivre developed the equation for the T distribution.

4) For all values of x,
$$e^x = 1 + x + \frac{x^2}{2!} + \ldots + \frac{x^n}{n!} + \ldots .$$

5) If a function $f(x)$ is differentiable at x_0, then it is continuous at x_0.

Answers to test: (1)T, (2) T, (3) F, (4) T, (5) T

Test results:

Number of correct answers	Number of Students per correct answer
0	
1	
2	
3	
4	
5	

The binomial experiment with $p = 1/2$ and $n = 5$ has the following probabilities:

$$P(s=0) = \binom{5}{0} \left(\frac{1}{2}\right)^0 \left(\frac{1}{2}\right)^5 = \frac{1}{32}$$

$$P(s=1) = \binom{5}{1} \left(\frac{1}{2}\right)^1 \left(\frac{1}{2}\right)^4 = \frac{5}{32}$$

$$P(s=2) = \binom{5}{2} \left(\frac{1}{2}\right)^2 \left(\frac{1}{2}\right)^3 = \frac{10}{32}$$

$$P(s=3) = \binom{5}{3} \left(\frac{1}{2}\right)^3 \left(\frac{1}{2}\right)^2 = \frac{10}{32}$$

$$P(s=4) = \binom{5}{4} \left(\frac{1}{2}\right)^4 \left(\frac{1}{2}\right)^1 = \frac{5}{32}$$

$$P(s=5) = \binom{5}{5} \left(\frac{1}{2}\right)^5 \left(\frac{1}{2}\right)^0 = \frac{1}{32}$$

These probabilities can be written in tabular form as follows:

The Binomial Experiment Table

Number of successes	Probability of S successes
0	1/32
1	5/32
2	10/32
3	10/32
4	5/32
5	1/32

Histogram for the Binomial Experiment

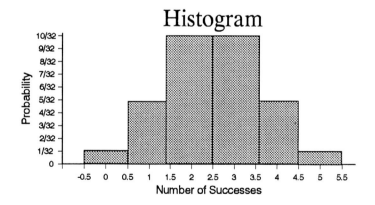

3. Randomly survey (a) 10, (b) 40, (c) 70, and (d) 100 people within the election district that you live and ask them:

   (1) Whether they are eligible voters within your election district?

   (2a) If they answer yes to question number one, ask them to select the appropriate following statement.

   I am registered in the following political party:

   Democratic
   Republican
   Conservative
   Liberal
   Other

   or

   I am *not* registered for a political party.

   (2b) If they answer no to question number one, exclude them from your survey.
   Use the probability formula:

$$P(\text{Event E}) = \frac{\text{number of outcomes satisfying Event E}}{\text{total number of outcomes}}$$

   and your survey results to calculate the probability that a person chosen at random from your election district is a:
   (a) Democrat, (b) Republican, (c) Conservative, (d) Liberal, (e) not registered for a political party.

   Contact the Board of Elections to determine the number of eligible voters in your election district and the actual number of Democrats, Republicans, Conservatives, Liberals and voters which are not registered for a political party within your election district.

   Using the Board of Election information, recalculate the above probabilities.

   Compare the survey probability results to the Board of Election probability results.

   Is there any relationship between the two as the survey size increases?

   Explain this relationship, if any exists.

PART VIII **Database.**

The following exercises refer to the file DATABASE listed in Appendix A. We have indicated the appropriate MINITAB commands that are necessary to answer each exercise.

1. Using the MINITAB commands:
   RETRIEVE, TALLY and LET.
   Retrieve the file DATABASE.MTW and count the number of *each type* of categorical value for the variables:
   a) SEX
   b) MAJ
   c) EYE
   d) HR
   e) MOB
   What is the probability of selecting each type of categorical value?

2. Using the MINITAB commands:
   RETRIEVE and HISTOGRAM with subcommands START and INCREMENT
   Retrieve the file DATABASE.MTW and produce a histogram for the variables:
   a) AGE
   b) GPA
   c) GFT
   Use the histogram to determine the probability that a student selected at random:
   (1) will be at least 19 years of age.
   (2) will have a grade point average of 2.5.
   (3) will spend at most $100 for a gift.

1. This problem was reprinted from The American Statistician, vol. 29. no. 1. February, 1975.

CHAPTER 7
THE NORMAL DISTRIBUTION

❄ 7.1 INTRODUCTION

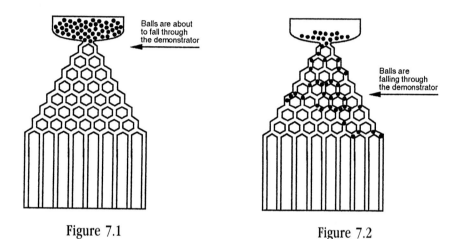

Figure 7.1   Figure 7.2

If one uses the pinball demonstrator pictured in Figure 7.1, a collection of balls is released. As the balls fall through the different paths as shown in Figure 7.2, they are constantly hitting other balls and consequently the paths they follow are random. Because the paths are random one might expect to get a *radically different* figure each time the balls are released. However, this is not the case. In fact, a bell-shaped figure similar to the one pictured in Figure 7.3 usually occurs all the time.

Machine diagram of bell-shaped results of experiment

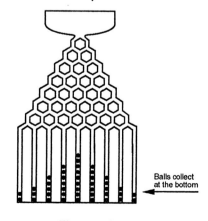

Figure 7.3

This occurrence was first observed by Sir Frances Galton when he used a similar pinball machine to demonstrate how many unknown factors acting together yield a phenomena whose distribution is bell-shaped. The demonstration led to the discovery that the **normal distribution**, pictured in Figure 7.4, could be used as a model for this type of phenomena.

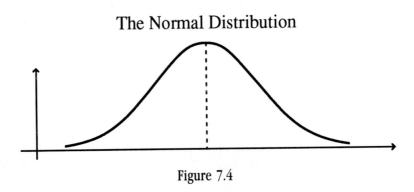

Figure 7.4

The significance of this finding is that it provides an explanation of why many observable phenomena (common occurrences) in nature have *approximately **normal distributions***.

For example, the distribution of weights for adult males is approximated by a normal distribution since each adult weight is influenced by many random factors acting together. These factors include: hereditary, environment, physiological and diet.

Other examples of distributions that can be approximated by the normal distribution include IQ scores of individuals, heights and weights of people, the diameters of tree trunks, blood pressures of men and women, scores on standardized exams, tire wear, the size of red blood cells, the time required to get to work, and the actual amount of soda dispensed into a 2 liter soda bottle. These variables that have been observed to have an approximately normal distribution are continuous variables. Thus, a normal distribution is a distribution that represents the values of a continuous variable. The importance of a normal distribution is that it provides as a good model in approximating many distributions of real world phenomena. When a continuous variable is said to be approximately normal, then the normal distribution, also referred to as a normal curve, is a distribution that has a symmetric bell-shaped curve with a single peak at the center as illustrated in Figure 7.4. However, you should understand that not all bell-shaped curves represent normal distributions. We will discuss another bell-shaped distribution in Chapter 10 that is not a normal distribution. Therefore, a normal distribution is a specific kind of bell-shaped curve with certain important properties. Let's now discuss the properties of a normal distribution.

7.2 Properties of the Normal Distribution

In order to analyze distributions which are approximately normal, or that can be approximated by the normal distribution, we must discuss the properties that a normal distribution possesses.

their graphs will differ as pictured in Figures 7.8, and 7.9. *Examine carefully* the percentages corresponding to the z scores. Notice that, although the data values are more dispersed for distribution B as compared to distribution A, the percent of data values between the z scores **are the same.**

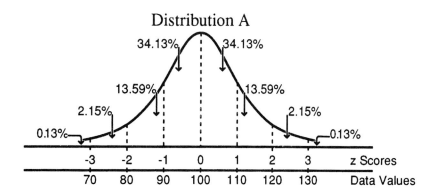

Figure 7.8

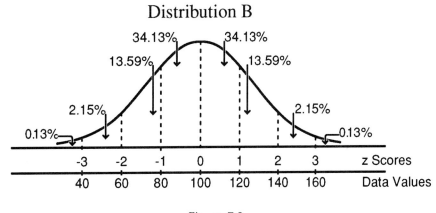

Figure 7.9

Even though the *shapes* of the two normal distributions A and B can be different, the *percent* of data values between any two z scores in Distribution A is equal to the percent of data values between the same two z scores in Distribution B. For example, in Figures 7.8 and 7.9, the percent of data values between the z scores of -1 and 0 is the same for *both* distributions. ∎

We have just illustrated that the percent of area between any two z scores is the same for all normal distributions. These percents can be found by using **TABLE II: The Normal Curve Area Table located in Appendix D: Statistical Tables.** Let's now explain how TABLE II is used to find the percent of area under the normal curve.

❄ 7.3 Using the Normal Curve Area Table: Table II

The **Normal Curve Area Table, TABLE II** located in **Appendix D: Statistical Tables** at the end of the text, gives the relationship between a z score and the percent of area under the normal curve. In particular, when a z score is looked up in TABLE II, the percent of area under the normal curve to the **left of this z score** is given. In Example 7.2, we will examine how to use TABLE II to find different areas under the normal curve.

Example 7.2

Find the percent of area under the normal curve:
a) to the left of z= 1.28.
b) to the left of z= -0.53.
c) to the right of z= +0.67.
d) to the right of z= -1.28.
e) between z= 0 and z= +1.5.
f) between z= -1.96 and z= +1.96.
g) between z= -1.25 and z= +1.0.

Solution:

a) It's helpful to make a sketch of the normal curve and to shade in the area one wants to determine. In Figure 7.10, we shade the area under the normal to the left of z= 1.28.

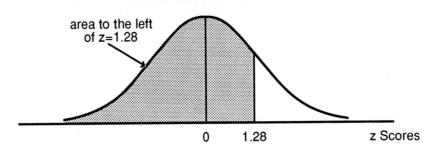

Figure 7.10

To determine this area, locate z= 1.28 in TABLE II. To do this, we look up 1.2 in the column labeled z (in general, look up the integer value and first decimal place of the z score under the column labeled z). The second decimal place (i.e the number 8) is found by moving across the row labeled 1.2 of the table until you reach the column labeled 8. The entry in TABLE II corresponding to z= 1.28 is 89.97. Thus, 89.97 represents the percent of area to the **left** of z= 1.28. This is illustrated in Figure 7.11.

TABLE II: The Normal Curve Area Table

z	0	1	2	3	4	5	6	7	8	9
.									*	
.									*	
.									*	
0.9									*	
1.0									*	
1.1									*	
1.2	**									**89.97**
1.3										
1.4										
.										
.										
.										

Figure 7.11

b) In Figure 7.12, the area to the left of z = -0.53 is shaded.

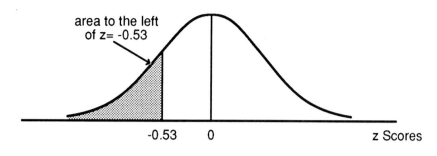

Figure 7.12

Reading directly from TABLE II, 29.81% of the area under the normal curve is to the *left* of z= -0.53. This result is illustrated in Figure 7.13.

TABLE II: The Normal Curve Area Table

z	0	1	2	3	4	5	6	7	8	9
				*						
.				*						
.				*						
.				*						
-1.0				*						
-0.9				*						
-0.8				*						
-0.7				*						
-0.6				*						
-0.5**************				**29.81**						
.										
.										
.										

Figure 7.13

Therefore, the percent of area to the left of z= -0.53 is 29.81%.

We can generalize the procedure to find the percent of area to the left of a z score.

Procedure to find the percent of area to the left of a z score

Percent of Area = Entry in TABLE II
to the left of z   corresponding to z

---

c) The area to the right of z= +0.67 is shaded in Figure 7.14.

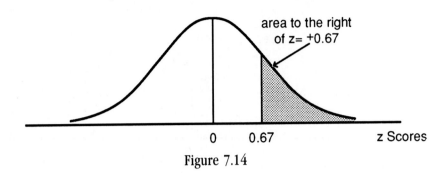

Figure 7.14

The entry in TABLE II corresponding to z= +0.67 is 74.86%. Consequently, 74.86% of the **total** area is to the left of z= 0.67. To find the area to the right of z= 0.67 subtract 74.86% from 100%, since the total percent of area under the normal curve is 100%. The percent of area to the *right* of z= 0.67 is 25.14%.

d) The area to the right of z= -1.28 is shaded in Figure 7.15.

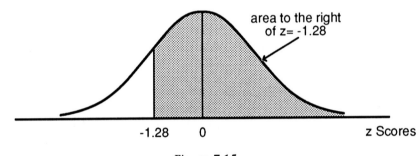

Figure 7.15

The entry in TABLE II corresponding to z= -1.28 is 10.03% which represents the percent of area to the left of z= -1.28. To find the percent of area to the right of z= -1.28, we subtract 10.03% from 100%. The percent of area to the right of z= -1.28 is 89.97%.

We can now generalize the procedure to find the area to the right of a z score.

**Procedure to find the percent of area
to the right of a z score**

Percent of Area to the right of z = 100% − Percent of Area to the left of z

---

Another procedure which can be used to determine the percent of area to the right of a z score is to utilize the symmetric property of the normal curve (property 2).

Alternate Procedure to find the percent of area to the right of a z score
Percent of Area to the right of z = Percent of Area to the left of −z

For example, in Part (c), the percent of area to the right of z= +0.67 can be found by looking up the percent of area to the left of z= −0.67, which is 25.14%. In Part (d), the percent of area to the right of z= −1.28 can be found by looking up the percent of area to the left of z= +1.28, which is 89.97%.

e) In Figure 7.16, the area between z= 0 and z= 1.5 is shaded. The procedure to determine this shaded region is outlined in the following two steps.

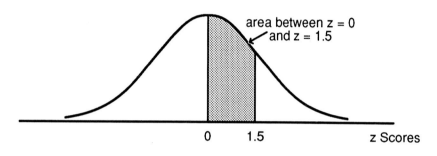

Figure 7.16

1. To find the percent of area of the shaded region in Figure 7.16, first determine the percent of area to the left of z= 0 and the percent of area to the left of z= 1.5. From TABLE II you will find that the percent of area to the left of z= 0 is 50% and the percent of area to the left of z= 1.5 is 93.32%.
2. The percent of area between the two z scores can be found by subtracting the **smaller area** from the **larger area**. Thus, the percent of area between z= 0 and z= 1.5 is:

$$93.32\% - 50\% = 43.32\%$$

f) In Figure 7.17, the region between z= −1.96 and z= +1.96 is shaded. To calculate this area, use the following two steps.

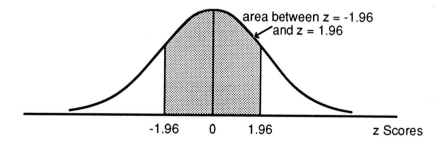

Figure 7.17

1. Find the percent of area to the left of each z score.

z score	% of area to the left
-1.96	2.50
1.96	97.50

2. Subtracting the smaller area from the larger, we find the percent of area between the two z scores is 95%.

g) In Figure 7.18, the region between z= -1.25 and z= 1.0 is shaded.

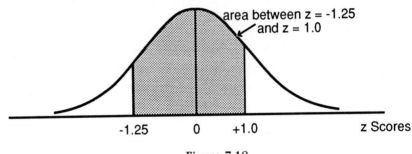

Figure 7.18

To calculate the shaded region, perform the following two steps.
1. Find the percent of area to the left of z= -1.25. The percent of area is 10.56. The percent of area to the left of z= 1.0 is 84.13.

2. Subtract the smaller area from the larger area. The percent of area *between* the two z scores is 73.57%. ∎

We can now generalize the procedure to find the percent of area between two z scores.

Procedure to find the percent of area between two z scores

Percent of Area between two z scores = Percent of Area to the left of larger z score *minus* Percent of Area to the left of smaller z score

---

Since the normal curve represents a continuous variable, then the area associated with a particular z score is zero. This is true because there is no area under a continuous curve and exactly over a z score. Therefore, we have the following rule.

Percent of area associated to a single z score

Percent of area for a particular z score equals zero.

❄ 7.4 APPLICATIONS OF THE NORMAL DISTRIBUTION

Let's now consider some applied problems in which it will be assumed that the distributions under consideration can be approximated by a normal distribution.

Assume the distribution of tire wear for a radial brand of auto tire is approximately normal with $\mu = 48{,}000$ miles and $\sigma = 2{,}000$ miles. Find the percent of tires which would be expected to wear:
a) less than 46,000 miles.
b) greater than 49,000 miles
c) between 47,000 miles and 51,000 miles.

Solution:

Since this distribution is approximately normal, we will assume all of the properties of the normal distribution and use the normal curve as a sketch of the distribution of tire wear. This is shown in Figure 7.19.

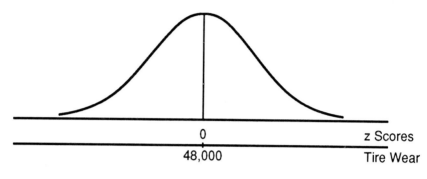

Figure 7.19

In your examination of Figure 7.19, notice that the mean is placed at the center of the normal curve below the z score line and corresponding to its z score of zero.

a) In Figure 7.20, the shaded area represents the percent of tires which would be expected to wear less than 46,000 miles.

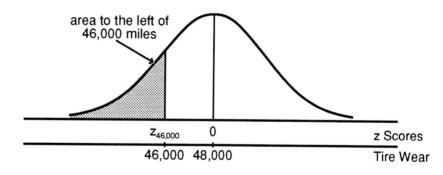

Figure 7.20

To determine the percent of tires which would be expected to wear less than 46,000 miles we must find the percent of area to the left of 46,000 miles. To determine this, we first calculate the z score for 46,000 miles, using:

$$z = \frac{X - \mu}{\sigma}$$

Thus,

$$z = \frac{46{,}000 - 48{,}000}{2{,}000}$$

$$z = \frac{-2000}{2000} = -1$$

The percent of area to the left of $z = -1$ is 15.87%. Therefore the percent of tires which would be expected to wear less than 46,000 miles is 15.87%.

b) In Figure 7.21, the shaded area represents the percent of tires which would expect to wear greater than 49,000 miles.

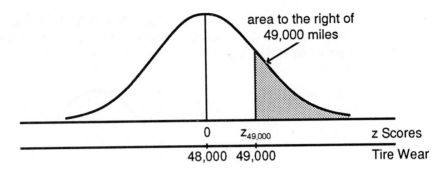

Figure 7.21

To find the percent of tires which would be expected to wear greater than 49,000 miles, find the percent of area to the right of 49,000 miles. To determine this, first calculate the z score of 49,000 miles.

$$z = \frac{49,000 - 48,000}{2,000} = 0.50$$

Remember that TABLE II gives the percent of area to the left of 49,000 miles or $z = 0.50$, which is 69.15%. To find the percent of area to the *right* of 49,000 miles, subtract 69.15% from 100%. Therefore the percent of tires which would expect to wear greater than 49,000 miles is 30.85%.

c) In Figure 7.22, the shaded area represents the percent of tires which would expect to wear between 47,000 miles and 51,000 miles.

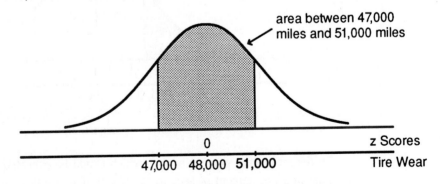

Figure 7.22

To find the area between 47,000 miles and 51,000 miles, first calculate their z scores.

For z of 47,000

$$z = \frac{47,000 - 48,000}{2,000} = -0.50$$

For z of 51,000

$$z = \frac{51{,}000 - 48{,}000}{2{,}000} = 1.50$$

To find the percent of area *between* the two z scores, determine each of the corresponding areas and then subtract the smaller area from the larger area.

tire wear(in miles)	z score	percent of area to the left
47,000	-0.5	30.85
51,000	1.5	93.32

shaded area = larger area - smaller area
= 93.32% - 30.85%
= 62.47%

Therefore, the percent of tires that last between 47,000 miles and 51,000 miles is 62.47%.

■

Example 7.4

A distribution of IQ scores is normally distributed with a mean of 100 and a standard deviation of 15. Find the percent of IQ scores that are:
a) greater than 127.
b) less than 88.
c) greater than 145 or less than 70.
d) less than one standard deviation from the mean.
e) within two standard deviations of the mean.

Solution:

a) In Figure 7.23, the shaded area represents the percent of IQ scores greater than 127.

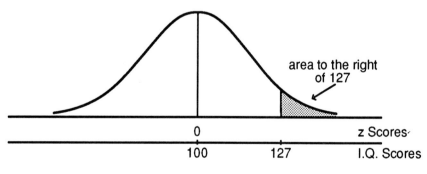

Figure 7.23

To determine this area calculate the z of 127.

$$z = \frac{127 - 100}{15} = 1.80$$

The percent of area to the left 127 is 96.41%. Therefore the percent of IQ scores greater than 127 is equal to:

$$100\% - 96.41\% = 3.59\%$$

b) In Figure 7.24 the shaded area represents the percent of IQ scores less than 88.

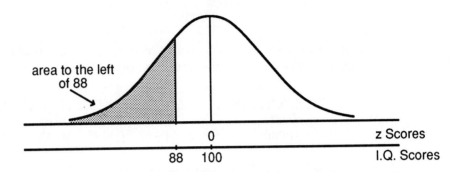

Figure 7.24

To determine this area calculate z of 88.

$$z = \frac{88 - 100}{15} = -0.80$$

The percent of IQ scores less than 88 is 21.19%.

c) In Figure 7.25, the shaded area represents the percent of IQ scores less than 70 or greater than 145.

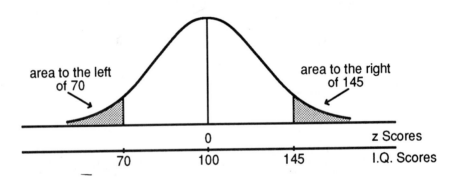

Figure 7.25

As can be seen in Figure 7.25 there are two separate areas to be determined. To calculate the *total* shaded area, calculate each area separately and add the two areas together.

For the percent of area less than 70, calculate the z of 70.

$$z = \frac{70 - 100}{15} = -2.00$$

Using TABLE II, the percent of area less than 70 is 2.28. For the percent of area greater than 145, calculate z of 145.

$$z = \frac{145 - 100}{15} = 3.00$$

From TABLE II, the percent of area to the right of 145 is 0.13. Therefore the total shaded area is the sum of these two areas: 0.13% + 2.28% = 2.41%. Thus the percent of IQ scores less than 70 or greater than 145 is 2.41%.

d) In Figure 7.26, the shaded area represents the percent of IQ scores within one standard deviation of the mean.

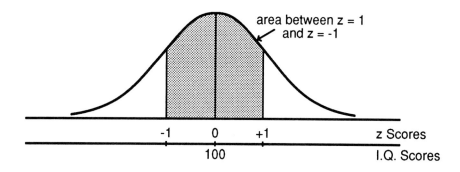

Figure 7.26

Less than one standard deviation from the mean would include all IQ scores between $z = -1$ and $z = +1$ as shown in Figure 7.26. We need to determine the area between the z scores of -1 and $+1$.

Using TABLE II, we have:

z score	percent of area to the left
-1	15.87
+1	84.13

To find the area between the z scores of -1 and $+1$ subtract the smaller area from the larger area.

shaded area = larger area - smaller area
= 84.13% - 15.87%
= 68.26%.

Therefore the percent of IQ scores less than one standard deviation from the mean is 68.26%.

e) In Figure 7.27, the shaded area represents the percent of IQ scores within two standard deviations of the mean.

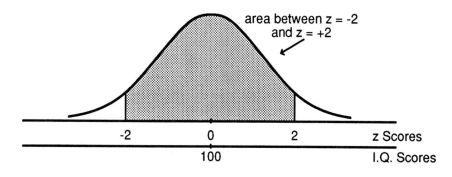

Figure 7.27

The IQ scores within two standard deviations from the mean would include all the IQ scores between $z = -2$ and $z = +2$ as shown in Figure 7.27. Let's determine the area between the z scores of −2 and +2. Using TABLE II, we find the corresponding areas to the left of each z score:

z score	percent of area to the left
−2	2.28
+2	97.72

Therefore the percent of IQ scores within two standard deviations of the mean is 95.44%. ∎

Example 7.5

One thousand students took a standardized psychology exam. The results were approximately normal with $\mu = 83$ and $\sigma = 8$. Find the *number* of students who scored:
a) greater than 87.
b) less than 67.

Solution:

a) In Figure 7.28, the shaded area represents the percent of students who scored greater than 87.

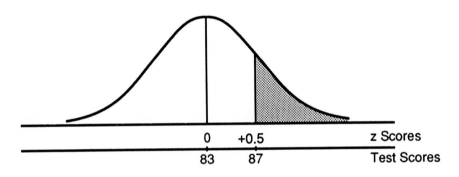

Figure 7.28

The z of 87 is 0.50. From TABLE II, the percent of the area to the left of z= 0.50 is 69.15. Therefore the percent of area to the right of 87 equals 30.85.

To find the *number* of students who scored greater than 87 multiply the proportion of area by the total number of students who took the exam. (Since the *area* to the right of 87 represents the proportion of students in the distribution who scored greater than 87.) Thus, the number of students who scored greater than 87 is equal to:

$$30.85\% \text{ of } 1000 = (0.3085)(1000)$$
$$= 308.5.$$

We round this off to 309 students.

Remark: To convert a percent to a proportion move the decimal point two places to the left. For example:

percent ⟶ **proportion**
30.85%   (move decimal point two places to left)   0.3085

b) In Figure 7.29, the shaded area represents the percent of students who scored less than 67.

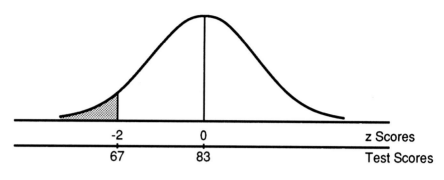

Figure 7.29

The z score of 67 is: z = -2 . Using TABLE II, the percent of area to the left of 67 is 2.28. Thus, the number of students who scored less than 67 is:

$$2.28\% \text{ of } 1000 = (0.0228)(1000)$$
$$= 22.8$$

This is rounded off to 23 students. ∎

7.5 PERCENTILES

Percentiles are widely used to describe the position of a data value within a distribution. Let's examine the definition of a percentile rank for a data value within a normal distribution.

> **Definition: Percentile Rank of a Data Value within a Normal Distribution.** The percentile rank of a data value X within a normal distribution is equal to the percent of area under the normal curve to the left of the data value X. Percentile ranks are always expressed as a whole number.

Example 7.6

Matthew earned a grade of 87 on his history exam. If the grades in his class were normally distributed with μ = 80 and σ = 7, find Matthew's percentile rank on this exam.

Solution:

The shaded area in Figure 7.30 represents the percentile rank of the grade 87.

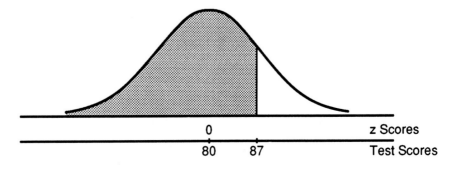

Figure 7.30

To find the area to the left of 87 we must first calculate z of 87.

$$z = \frac{87 - 80}{7} = 1$$

From TABLE II, the area to the left of 87 is 84.13%. Since percentile ranks are always expressed as whole numbers, 84.13% is rounded off to the nearest whole percent which is 84%. Therefore, the **percentile rank of 87 is 84**. Using Percentile rank notation, we can express this result as: $87 = P_{84}$.

Matthew's grade of 87 has a percentile rank of 84 which means that his grade of 87 was better than 84% of the test grades. ∎

Percentile rank is an important statistical measure that describes the position of a data value within a distribution. It can also be used to compare data values from different distributions. This is illustrated in Example 7.7.

Example 7.7

Mary Ann took two exams last week. Her grades and class results were as follows:

Exam	Her Grade	Class Mean (μ)	Class Standard Deviation (σ)
English	83	80	6
Math	77	74	5

Assuming the results of both exams to be normally distributed, use percentile ranks to determine on which exam Mary Ann did better relative to her class.

Solution:

a) The shaded area in Figure 7.31 represents the percentile rank of Mary Ann's english grade of 83.

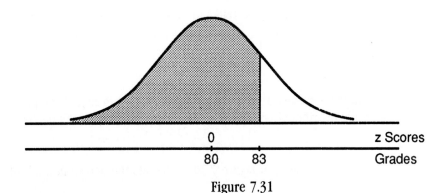

Figure 7.31

To determine this area, calculate the z score of 83.

$$z = \frac{83 - 80}{6} = 0.5$$

From TABLE II, the area to the left of 83 is 69.15%. Since percentile ranks are expressed in whole numbers, 69.15% rounded off to the nearest whole percent is 69. Therefore, the percentile rank of Mary Ann's english grade is 69, which can be expressed as: $P_{69} = 83$.

b) The shaded area in Figure 7.32 represents the percentile rank of Mary Ann's math grade of 77.

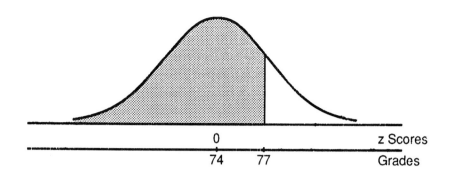

Figure 7.32

To determine this area, compute the z score of 77.

$$z = \frac{77 - 74}{5} = 0.60$$

The area corresponding to the z = 0.6 is 72.57%. Therefore, the percentile rank of her math exam grade is 73, which can be expressed as: $P_{73} = 77$. The percentile rank of her math grade is larger than the percentile rank of her english grade. Thus, *relative* to her class, Mary Ann did better on her math exam. ∎

Example 7.8

One of the entrance requirements at a local college is a percentile rank of at least 45 on the verbal SAT exam. Bill scored 470 on his verbal SAT exam. Does he meet the minimum requirement of this college, if the verbal SAT distribution is normally distributed with $\mu = 500$ and $\sigma = 100$?

Solution:

The shaded area in Figure 7.33 represents the percentile rank of the SAT score 470.

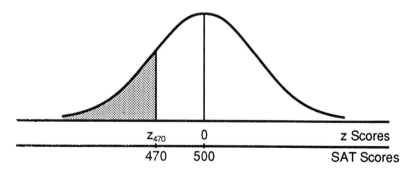

Figure 7.33

Find z of 470.

$$z = \frac{470 - 500}{100} = -0.30$$

In TABLE II, the entry corresponding to the z = -0.30 is 38.21%. Thus, the percentile rank of Bill's SAT score is 38, which is expressed as: $P_{38} = 470$. Therefore, Bill does not meet the entrance requirement. ∎

Example 7.9

An entrance requirement at an eastern university is a high school average in the top 25% of one's high school class. Craig would like to know if he qualifies for entrance into this college. His high school average is 85. If the high school averages of his class are normally distributed with a mean of 75 and a standard deviation of 10 does he meet the requirement?

Solution:

Being in the top 25% of the graduating class means that Craig's average must be better than **at least 75%** of the graduating class. Therefore the percentile rank of Craig's average must be at least 75%. To determine if Craig qualifies for entrance into this college, find the percentile rank of Craig's high school average and compare it to the minimum requirement of a percentile rank of 75%.

The shaded area in Figure 7.34 represents the percentile rank of Craig's high school average of 85.

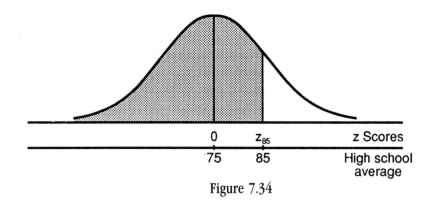

Figure 7.34

To determine this area, find the z score of 85.

$$z = \frac{85 - 75}{10} = 1$$

From TABLE II, the percent of area corresponding to the z = 1 is 84.13. Therefore, the percentile rank of Craig's average is 84, which can be expressed as $P_{84} = 85$. Since Craig's percentile rank is greater than 75, he meets this entrance requirement. ∎

CASE STUDY 7.1

The information contained in the table of Figure CS7.1 represents the percentile ranks of Achievement Tests administered by the College Board to high school seniors who are college-bound. An Achievement Test measures the specific knowledge of a subject.

Percentile ranks of Achievement Test Scores (for college-bound seniors)

Score	English		Mathematics			History		Sciences		
	Eng. Comp. EN-ES*	Literature LR	Math. Level I M1	Math. Level II M2	Math. Level IIC 2C	Amer. Hist. & Social Studies AH	Euro. Hist. & World Cult. EH	Biology BY	Chemistry CH	Physics PH
800	99+	99+	99+	92	92	99+	99+	99+	99	97
750	99	99+	99	79	79	98	98	97	94	90
700	96	95	93	60	59	93	93	89	86	80
650	87	85	83	39	41	84	83	76	73	66
600	75	70	67	20	19	71	70	61	57	50
550	60	54	48	9	8	54	53	43	40	32
500	41	38	30	4	4	36	34	27	24	16
450	25	23	15	2	1	20	18	15	12	6
400	13	12	5	1	..	8	6	7	4	1
350	5	5	1	1–	..	2	1	3	1	1–
300	1	1	1–	1–	..	1–	1–	1	1–	..
250	1–	1–	1–	..	..	1–	..	1–	..	..
200	..	..	..	..	..	..	..	..	..	..
Average Score	519	529	549	666	668	537	541	561	575	601

Figure CS7.1

a) Do you think that the test results of a standardized achievement test would be approximately normally distributed when administered to a large group of students across the country?

b) If the Mathematics Achievement Test results for Level I were approximately normal, explain what this information would mean to you?

c) Examine the table in Figure CS7.1 and determine the mean test score for the Mathematics Level I (M1), American History and Social Sciences (AH), and the Physics (PH) tests.

d) Approximate the median test score for each of these tests. If these test results are approximately normal, then what value would you expect the median test score to equal?

e) Let's assume that the standard deviation for each of the achievement tests in part (c) is approximately 100.
For the mathematics Level I test, the American and Social Studies test, and the Physics test, approximate the percent of test scores that are within one, and two standard deviations of the mean.

f) For a normal distribution, what percentage of the data values are within one, two and three standard deviations of the mean.

g) Compare the results of parts (e) and (f).

h) Which exam(s) seem to approximate the percentages of a normal distribution? Do these results have any affect on the way you answered part (a)? ∎

Example 7.10

Consider a normal distribution with a mean of 75 and $\sigma = 5$. Find the data value(s) in the distribution that:
a) cut(s) off the top 25% of the data values.
b) cut(s) off the bottom 30% of the data values.

c) has a percentile rank of 55.
d) cut(s) off the middle 40% of the data values.

Solution:

a) The shaded area in Figure 7.35 represents the top 25% of the data values in a normal distribution.

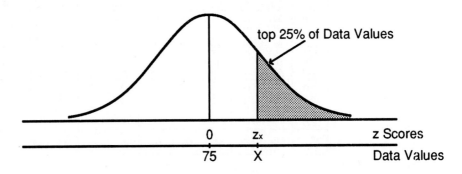

Figure 7.35

To determine the data value that cuts off the top 25% of the data values, we must find the data value that has 25% of the data values greater than this value and 75% of the data value less than this value. Since TABLE II gives the relationship between z scores and the percent of area to the left of a z score, then knowing the area to the left of this unknown z score will enable us to determine this z score.

We can now begin to calculate this data value by the using the following procedure.

Step 1. Determine the percent of area to the left of the data value, X.

Since 25% is to the right of the data value, X, then 75% is to the left of the data value, X.

Step 2. Find the area in Table II which is closest to the area that is to the left of the data value and determine the z score corresponding to this area.

In TABLE II find the area closest to 75%. This is illustrated in Figure 7.36.

TABLE II: The Normal Curve Area Table

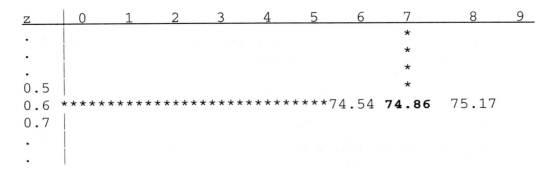

Figure 7.36

Thus, the closest area to 75% which can be found in the table is 74.86%. Thus, the corresponding z score for the data value X is 0.67.
Therefore the data value X which cuts off the top 25% has a z score of 0.67.

Step 3. Using the raw score formula, $X = \mu + (z)\sigma$, convert the z score to a data value.

In order to convert the z score of 0.67 to a data value, we need to use the raw score formula:
$$X = \mu + (z)\sigma$$

The data value for $z = 0.67$ is:
$$\begin{aligned}X &= \mu + (z)\sigma \\ &= 75 + (0.67)(5) \\ &= 78.35\end{aligned}$$

Therefore the data value which cuts off the top 25% in this normal distribution is 78.35.

b) To determine the data value that cuts off the bottom 30% of the data values within a normal distribution. We will follow the procedure outlined in part (a) of this question.

Step 1. Determine the percent of area to the left of the data value, X.

We are looking for the data value X that has 30% of the area under the normal curve to the left of this data value.

Step 2. Find the area in TABLE II which is closest to the area that is to the left of the data value and determine the z score corresponding to this area.

We must find the z score that corresponds to the bottom 30% of area under the normal curve as illustrated in Figure 7.37.

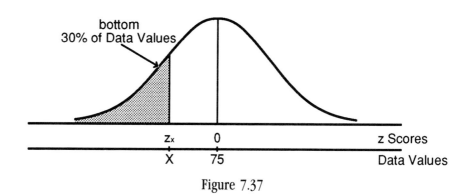

Figure 7.37

In TABLE II find the area closest to 30%. This is illustrated in Figure 7.38.

TABLE II: The Normal Curve Area Table

```
z     |  0       1       2       3       4       5       6       7       8       9
      |                  *
  .   |                  *
  .   |                  *
-0.6  |                  *
-0.5  ****30.50       30.15    29.81
-0.4  |
  .   |
```

Figure 7.38

Thus, the closest area to 30% is 30.15%. The corresponding z score is -0.52.

Step 3. Using the raw score formula, $X = \mu + (z)\sigma$, convert the z score to a data value.

Using the formula: $X = \mu + (z)\sigma$, we can transform this z score to the data value X.

$$X = 75 + (-0.52)(5)$$
$$X = 72.4.$$

Therefore the data value X which cuts off the bottom 30% in this distribution is 72.4.

c) To determine the data value that has a percentile rank of 55, we will follow the procedure outlined in part (a) of this problem.

Step 1. Determine the percent of area to the left of the data value, X.

The data value that has a percentile rank of 55 cuts off the bottom 55% of the data values in the distribution.
This is illustrated in Figure 7.39.

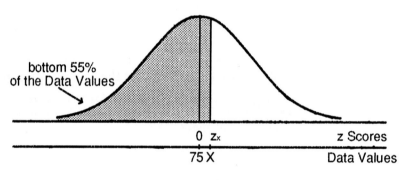

Figure 7.39

Step 2. Find the area in TABLE II which is closest to the area that is to the left of the data value and determine the z score corresponding to this area.

From TABLE II, the area closest to 55% is 55.17%, which corresponds to a z score of 0.13.

Step 3. Using the raw score formula, $X = \mu + (z)\sigma$, convert the z score to a data value.

Using the formula: $X = \mu + (z)(\sigma)$ we can transform the z score to a data value.

$$X = 75 + (0.13)(5)$$
$$X = 75.65$$

Therefore, the data value that has a percentile rank of 55 is 75.65.

d) In a normal distribution, the mean is located at the center. Therefore the middle 40% of the data values will be centered about the mean, that is 20% of the data values will be less than the mean and 20% of the data values will be greater than the mean. This is illustrated in Figure 7.40.

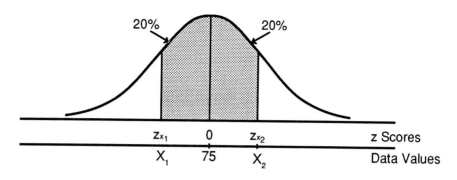

Figure 7.40

Notice in Figure 7.40 that we are looking for two data values, which we will symbolize as X_1 and X_2.

To determine these data values, we will follow the procedure outlined in part (a) of this problem.

Step 1. Determine the percent of area to the left of the data value, X.

To determine the percent of area to the left of X_1, observe that 20% of the area is between the data value X_1 and the mean. Since the total percent of area to the left of the mean is 50%, then the remaining area to the left of X_1 is 50% - 20% = 30%.

Step 2. Find the area in TABLE II which is closest to the area that is to the left of the data value and determine the z score corresponding to this area.

In TABLE II the area closest to 30% is 30.15%. Its corresponding z score is -0.52.

Step 3. Using the raw score formula, $X = \mu + (z)\sigma$, convert the z score to a data value.

Using the formula $X = \mu + (z)\sigma$ we can transform z of X_1 = -0.52 to a data value as follows:

$$X_1 = 75 + (-0.52)(5)$$

$$X_1 = 72.4$$

Using the same procedure, we can determine the value of X_2.

Step 1. Determine the percent of area to the left of the data value, X.

To find X_2 first determine the area to the left of X_2. Since the area to the left of the mean is 50% and the area between the mean and X_2 is 20%, then the area to the left of X_2 is 50% + 20% = 70%.

Step 2. Find the area in Table II which is closest to the area that is to the left of the data value and determine the z score corresponding to this area.

Using TABLE II, the area closest to 70% is 69.85% which has a z score of 0.52.

Step 3. Using the raw score formula, $X = \mu + (z)\sigma$, convert the z score to a data value.

Using the formula $X = \mu + (z)\sigma$ we have:

$$X_2 = 75 + (0.52)(5)$$

$$X_2 = 77.6$$

Therefore the middle 40% of the distribution is located between the two data values 72.4 and 77.6. ■

We can now generalize the procedures to determine the data value that cuts off the bottom p% in a normal distribution and the data value that cuts off the top q% in a normal distribution.

**Procedure to Find Data Value (or Raw Score)
That Cuts Off Bottom p % Within a Normal Distribution**

1. Using TABLE II, find the z score that has the closest area to the percentage p.

2. Substitute this z score into the raw score formula to determine the data value that corresponds to this z score.
$$X = \mu + (z)\sigma$$

**Procedure to Find Data Value (or Raw Score)
That Cuts Off Top q % Within a Normal Distribution**

1. Calculate the percentage of data values below the raw score by subtracting q from 100: (100 - q).

2. Using TABLE II, find the z score that has the closest area to the percentage to: 100 - q.

3. Substitute this z score into the raw score formula to determine the data value that corresponds to this z score:
$$X = \mu + (z)\sigma$$

Example 7.11

In a large physical education class the grades on the final exam were normally distributed with $\mu = 80$ and $\sigma = 8$. Professor Stevenson assigns letter grades in the following way: the top 10% receive A's, the next 15% receive B's, the lowest 4% receive F's, the middle 50% receive C's, and the rest of the grades are D's.

Solution:

The information is interpreted in Figure 7.41.

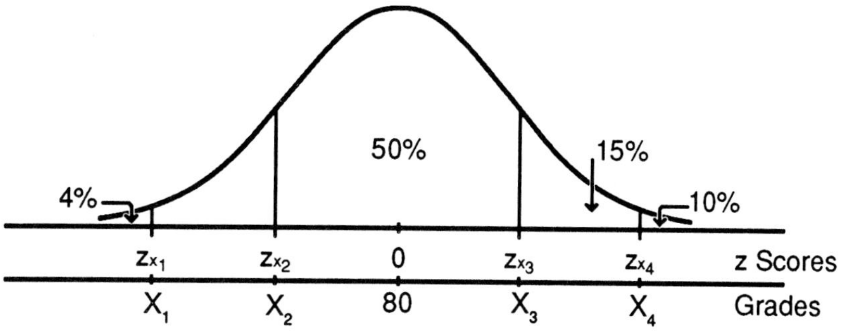

Figure 7.41

Using Figure 7.41, let us determine the test grades X_1, X_2, X_3, and X_4 that separate the letter grades.

To determine X_1:
1. Find the percent of area to the left of X_1. This area is 4%.
2. Find the closest entry to 4% in TABLE II. The closest entry to 4% is 4.01%. The z score that corresponds to 4.01% is -1.75.

To find the value of X_1, we can use the raw score formula:
$$X_1 = \mu + (z)\sigma$$
$$X_1 = 80 + (-1.75)(8)$$
$$X_1 = 66$$

Therefore, the grade which separates the F's from the D's is 66.

To determine X_2:
1. Find the percent of area to the left of X_2. This area is 25%.
2. The closest entry in TABLE II to 25% is 25.14%. The z score that corresponds to 25.14% is -0.67. The formula to determine X_2 is:
$$X_2 = \mu + (z)\sigma$$
$$X_2 = 80 + (-0.67)(8)$$
$$X_2 = 74.64$$

Therefore, the test grade which separate grades C and D is 74.64.

To determine X_3:
1. The percent of area to the left of X_3 is 75%.
2. The closest entry in TABLE II is 74.86%. It has a z score of 0.67. Using the raw score formula, we have:
$$X_3 = \mu + (z)\sigma$$
$$X_3 = 80 + (0.67)(8)$$
$$X_3 = 85.36$$

The test grade which separate grades B and C is 85.36.

Finally, to determine X_4:
1. The percent of area to the left of X_4 is 90%.
2. The closest entry in TABLE II is 89.97, which has a z score of 1.28. Using the raw score formula, we have:
$$X_4 = \mu + (z)\sigma$$
$$X_4 = 80 + (1.28)(8)$$
$$X_4 = 90.24$$

Thus the test grade which separate grades A and B is 90.24. ■

Example 7.12

To determine the size of the annual Christmas bonus, Grace, the personnel and budget director of a telemarketing firm, decides to use annual sales per salesperson as the criterion. The salespeople in the top 20% of annual sales will receive $3000 in bonus, those in the next 30% will receive $2000 in bonus, those in the next 30% will receive $1000 and the rest will receive $500 in bonus. If the annual sales per salesperson is normally distributed with $\mu = \$120{,}000$ and $\sigma = \$15{,}000$ determine the annual sales required to obtain:

a) a bonus of $3000.
b) a bonus of $2000.

Solution:

The relevant information is summarized in Figure 7.42.

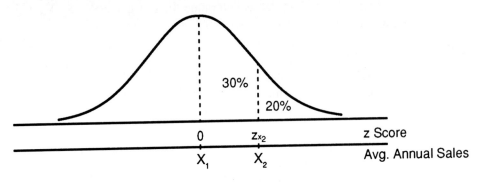

Figure 7.42

a) To determine X_2:

1. The percent of area to the left of X_2 is 80%.
2. The closest entry in TABLE II to 80% is 79.95% which has the z score of 0.84. Converting this z score to a raw score, we get $X_2 = \$132{,}600$.

Thus, annual sales of at least $132,600 are required to get a $3000 bonus.

b) To determine X_1:

1. Notice that X_1 has 50% of the area to the left of it. Thus, X_1 is the mean.
2. Therefore, $X_1 = \$120{,}000$.

Thus, annual sales of at least $120,000 are required to get at least a $2000 bonus. ∎

Example 7.13

A local community college decides to award scholarships based upon the results of a standardized achievement test. If this year's 5,000 test results are approximately normal with a $\mu = 137$ and a $\sigma = 15$, then determine the minimum test grade needed to win a scholarship if the top:

a) 20 students receive scholarships.
b) 30 students receive scholarships.

Solution:

a) To find the minimum test grade needed to win a scholarship if the top 20 students receive scholarships, we must first determine the *percent* of the students who will win scholarships. Since there are 20 students receiving scholarships out of 5,000 students, the percent of scholarship winners equals:

$$\frac{20}{5000} \times 100\% = (0.0040)(100\%)$$
$$= 0.40\%$$

This is illustrated in Figure 7.43.

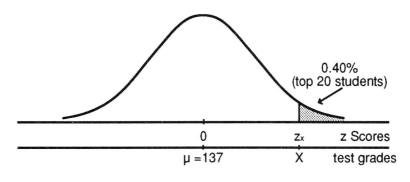

Figure 7.43

To determine the test grade X which cuts off the top 0.40% we need to determine the percent of area to the left of the test grade X. Since 0.40% is to the right of the test grade X, then 99.60% is to the left of X. Using **TABLE II**, the area 99.60% corresponds to a z score of +2.65.

In order to convert this z score into a test grade, we need to use the formula:

$$X = \mu + (z)\sigma$$
$$X = 137 + (2.65)(15)$$
$$X = 176.75$$

Therefore, the minimum test grade needed to win a scholarship if the top 20 students will receive scholarships is 176.75.

b) To find the minimum test grade needed to win a scholarship if the top 30 students will receive scholarships, we must determine the *percent* of the students who will win scholarships. Since 30 students out of 5,000 students who will win scholarships, the percent of scholarship winners equals:

$$\frac{30}{5000} \times 100\% = (0.0060)(100\%)$$
$$= 0.60\%$$

This is illustrated in Figure 7.44.

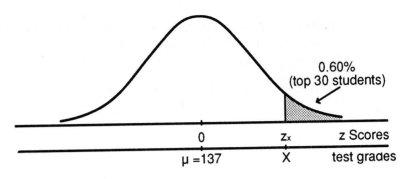

Figure 7.44

To find the value of the test grade X we need to determine the percent of area to the left of the test grade X. Since 0.60% is to the right of X, then the percent to the left is 99.40%. Using TABLE II, the area 99.40% corresponds to a z score of 2.51.

To convert this z score into a test grade, we use the formula: $X = \mu + (z)\sigma$. Thus, the test grade X for z = 2.51 is:

$$X = \mu + (z)\sigma$$
$$X = 137 + (2.51)(15)$$
$$X = 174.65$$

Therefore, the minimum test grade needed to win a scholarship if the top 30 students will receive scholarships is 174.65. ∎

7.6 PROBABILITY

In a normal distribution, the area under the normal curve can be interpreted to represent probability. This follows from the definition of probability given in Chapter 6, which is:

$$P(event\ E) = \frac{Number\ of\ outcomes\ satisfying\ event\ E}{Total\ number\ of\ outcomes\ in\ the\ sample\ space}$$

Since we are working with a normal distribution, we need to redefine this definition in terms of the area under the normal curve. Thus, we have:

$$P(event\ E) = \frac{Area\ under\ the\ normal\ curve\ satisfying\ Event\ E}{Total\ area\ under\ the\ normal\ curve}$$

In a normal distribution, the *total area* under the normal curve is 1.[1]

Thus, for a normal distribution the Probability of event E is:

$$P(event\ E) = \frac{Area\ under\ the\ normal\ curve\ satisfying\ Event\ E}{1}$$

This leads us to the following definition.

---

1 This is a mathematical property of the normal curve.

> **Definition: Probability of an Event E in a Normal Distribution.** The probability of an Event E in a normal distribution is defined to be:
>
> **P(Event E) = Percent of area under the normal curve satisfying Event E.**

This is illustrated in Figure 7.45.

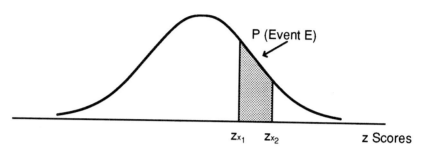

Figure 7.45

This leads to the following definition for the probability that a data value falls between two data values X_1 and X_2 within a normal distribution.

> **Definition: Probability of a data value X falling between two data values, X_1 and X_2.** The probability that a data value X selected at random from a normal distribution will fall between the two data values X_1 and X_2 is equal to the **percent of area** between X_1 and X_2. This can be expressed as:
>
> $P(X_1 \leq X \leq X_2)$ = Area under the normal curve between X_1 and X_2

Remember that a normal curve represents the distribution of a continuous variable, and the percent of area under the normal curve that is associated with a particular data value is zero! Therefore, the probability of randomly selecting a data value from a normal distribution that is exactly equal to the value X is zero. This can be expressed as:

> **Probability of selecting a data value equal to the value X within a normal distribution**
>
> **P(Data Value is X) = 0**

Example 7.14

If the distribution of the lengths of major league baseball players playing careers is approximately normal, with a mean of 8 years and a standard deviation of 4 years, find the probability that a player selected at random will have a career that lasts:
a) longer that 14 years
b) less than 6 years
c) between 4 and 10 years.

Solution:

a) 1. In Figure 7.46, the shaded region represents the probability that a player's career lasts longer than 14 years.

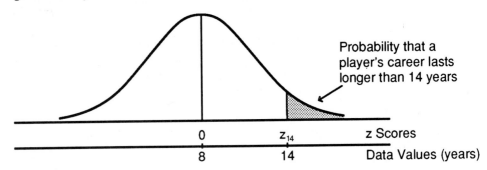

Figure 7.46

2. To determine the probability that a player selected at random has a career that lasts longer than 14 years, calculate the percent of area to the right of 14 years. Since, z of 14 = 1.5, then the percent of area to the left of z = 1.5 is 93.32%. Thus the percent of area to the right of z = 1.5 is 6.68%

Therefore, the probability that a player's career lasts longer than 14 years is 6.68%.

b) 1. The shaded region in Figure 7.47 represents the probability that a player's career lasts less than 6 years.

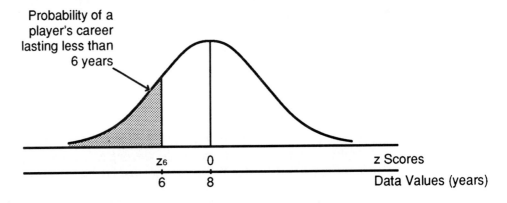

Figure 7.47

2. To determine the probability that a player selected at random has a playing career that lasts less than 6 years, calculate the area to the left of 6 years.

The percent of area to the left of z = -0.5 is 30.85%. Therefore the probability that a player's career lasts less than 6 years is 30.85%.

c) 1. The shaded region in Figure 7.48 represents the probability that a player's career lasts between 4 and 10 years.

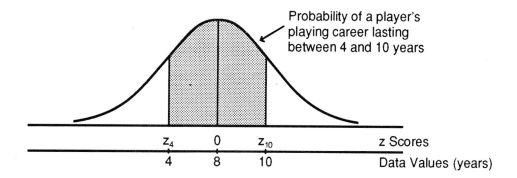

Figure 7.48

2. To determine the probability that a player selected at random has a career that lasts between 4 and 10 years, calculate the percent of area between 4 and 10 years.

Determine the z scores of 4 and 10 years:

z of 4 = -1 and   z of 10 = 0.5

Notice that the percent of area between z = -1 and z = +0.5 is 69.15% - 15.87% = 53.28%. Therefore the probability that a player's career lasts between 4 and 10 years is 53.28%. ∎

Example 7.15

The amount of soda that a filling machine dispenses into a 32 ounce bottle is normally distributed with $\mu = 32$ oz. and $\sigma = 0.5$ oz. Find the probability that a bottle selected at random will contain:
a) more than 33 oz.
b) less than 31.6 oz.
c) between 31 oz. and 32.7 oz.

Solution:

a) The shaded area in Figure 7.49 represents the probability that a bottle selected at random will contain more than 33 oz.

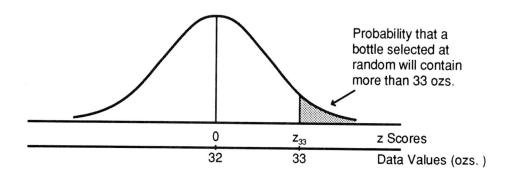

Figure 7.49

To determine the probability that a bottle selected at random will contain more than 33 oz., calculate the percent of area to the right of 33 oz.

Calculating the z score of 33, we get: z = 2. The percent of area to the left of z = 2 is 97.72%. Thus the probability that a bottle selected at random will contain more than 33 oz. is 2.28%.

b) The shaded area in Figure 7.50 represents the probability that a bottle selected at random will contain less than 31.6 oz.

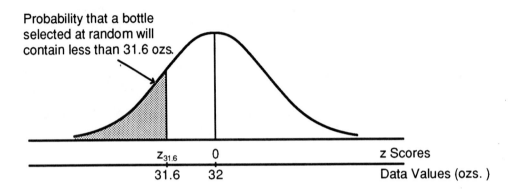

Figure 7.50

To determine the probability that a bottle selected at random will contain less than 31.6 oz., we must calculate the percent of area to the left of 31.6 ozs.

Calculating the z score of 31.6, we find: z = -0.8. The percent of area to the left of z = -0.8 is 21.19%. Therefore the probability that a bottle selected at random will contain less than 31.6 oz. is 21.19%.

c) The shaded area in Figure 7.51 represents the probability that a bottle selected at random will contain between 31 and 32.7 oz.

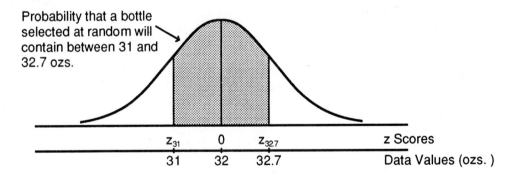

Figure 7.51

To determine the probability that a bottle selected at random will contain between 31 and 32.7 oz., calculate the percent of area between 31 oz. and 32.7 oz.
Determine the z scores of 31 and 32.7 oz:

z of 31 = -2   and   z of 32.7 = +1.4

The percent of area between z = -2 and z = 1.4 is equal to:
91.92% - 2.28% = 89.64%

Therefore, the probability that a bottle selected at random will contain between 31 and 32.7 oz. is 89.64%. ■

7.7 THE NORMAL APPROXIMATION TO THE BINOMIAL DISTRIBUTION

An important use of the normal distribution is to approximate binomial probabilities. However, before we consider this binomial approximation let us re-examine the binomial experiment first introduced in Chapter 6. To compute the probabilities for a binomial experiment we used the formula:

$$P(s \text{ successes in n trials}) = \binom{n}{s} p^s q^{(n-s)}, \text{ where: } q = 1-p.$$

For the binomial experiment where n = 2 and p = 1/2 let us use the binomial probability formula to compute the probability for all possible number of successes. This is illustrated in Table 7.1.

Table 7.1

Binomial Distribution for n = 2 and p = 1/2	
number of successes "s"	Probability of "s" successes $P(s) = \binom{n}{s} p^s q^{(n-s)}$ where: q = 1-p
2	$\binom{2}{2} (\frac{1}{2})^2 (\frac{1}{2})^0 = \frac{1}{4}$
1	$\binom{2}{1} (\frac{1}{2})^1 (\frac{1}{2})^1 = \frac{2}{4}$
0	$\binom{2}{0} (\frac{1}{2})^0 (\frac{1}{2})^2 = \frac{1}{4}$

Table 7.1 represents the binomial distribution for n = 2 and p = 1/2. A histogram and frequency polygon of this binomial distribution is illustrated in Figure 7.52, where the vertical axis represents the probability of **s** successes and the horizontal axis represents the number of **s** successes.

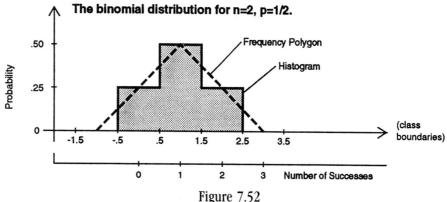

Figure 7.52

Notice that in Figure 7.52 the probability for each success is represented by the *area* of the rectangle associated with it. Notice that the probability of getting zero successes is represented by the rectangle bounded by the boundary values of zero: -0.5 and 0.5. Since the width of this rectangle is 1 and the height is 1/4 then using the formula for the area of a rectangle we have:

$$\text{Area of rectangle} = (\text{height})(\text{width})$$
$$= (1/4)(1)$$
$$= 1/4.$$

Therefore, the area of the rectangle representing the probability of getting zero successes is 1/4.

Like all distributions we have studied the binomial distribution has a mean and standard deviation. Symbolically, the **mean of a binomial distribution** is expressed μ_s and is pronounced *mu sub s*. The **standard deviation of a binomial distribution** is expressed σ_s and is pronounced *sigma sub s*. To calculate the mean and standard deviation of a binomial, we use the following definitions:

Definition: Mean of the Binomial Distribution. For a binomial distribution, the mean number of successes, denoted μ_s, is determined by the formula:

$$\mu_s = np$$

where: n = the number of trials
p = the probability of a success in one trial

Definition: Standard Deviation of the Binomial Distribution. In a binomial distribution, the standard deviation, denoted σ_s, is determined by the formula:

$$\sigma_s = \sqrt{np(1-p)}$$

where: n = the number of trials
p = the probability of a success in one trial
1-p = the probability of a failure in one trial

For the binomial distribution illustrated in Table 7.1 where n = 2 and p = 1/2, the mean is:

$$\mu_s = np$$
$$= (2)(½)$$
$$= 1$$

And the standard deviation of this binomial distribution is:

$$\sigma_s = \sqrt{np(1-p)}$$
$$= \sqrt{(2)\left(\frac{1}{2}\right)\left(\frac{1}{2}\right)}$$
$$= \sqrt{0.5}$$
$$\sigma_s \approx 0.71$$

Example 7.16

Examine Table 7.2 for the binomial distribution where $n = 4$ and $p = 1/2$ and compute the mean, μ_s, and the standard deviation, σ_s, of this binomial distribution.

Table 7.2

Binomial Distribution for $n = 4$ and $p = 1/2$	
number of successes "s"	Probability of "s" successes $P(s) = \binom{n}{s} p^s q^{(n-s)}$ where: $q = 1-p$
4	$\binom{4}{4} (\frac{1}{2})^4 (\frac{1}{2})^0 = \frac{1}{16}$
3	$\binom{4}{3} (\frac{1}{2})^3 (\frac{1}{2})^1 = \frac{4}{16}$
2	$\binom{4}{2} (\frac{1}{2})^2 (\frac{1}{2})^2 = \frac{6}{16}$
1	$\binom{4}{1} (\frac{1}{2})^1 (\frac{1}{2})^3 = \frac{4}{16}$
0	$\binom{4}{0} (\frac{1}{2})^0 (\frac{1}{2})^4 = \frac{1}{16}$

Solution:

To compute mean, μ_s, use the formula:
$$\mu_s = np$$

If $n = 4$ and $p = 1/2$, then: $\mu_s = 2$.

To compute standard deviation, σ_s, use the formula:
$$\sigma_s = \sqrt{np(1-p)}$$

For $n = 4$, $p = 1/2$ and $1-p = 1/2$,

$$\sigma_s = \sqrt{4(\frac{1}{2})(\frac{1}{2})}$$

$$\sigma_s = 1$$

The graph of this binomial distribution is illustrated in Figure 7.53. ∎

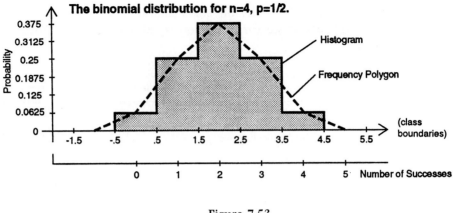

Figure 7.53

In many applications of the binomial distribution the number of trials, n, is large. Figure 7.54 illustrates the effect of *increasing the number of trials, n*, on the shape of the binomial distribution with p = 1/2.

Binomial distributions with p=1/2

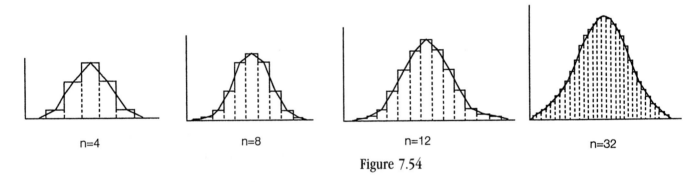

Figure 7.54

In Figure 7.54, notice that **as n *increases*, the shape of the binomial distribution gets *closer* to the bell-shape of a normal distribution**. In fact, the binomial distribution *can* be approximated by a normal distribution with a mean $\mu_s = np$ and a standard deviation $\sigma_s = \sqrt{np(1-p)}$. However, this approximation is reasonable only when the following conditions are *both* true:

$$np > 5 \text{ and } n(1-p) > 5.$$

Let's summarize these conditions.

Conditions of Normal Approximation of a Binomial Distribution

A binomial distribution with a probability of a success p and n trials can be approximated by a normal distribution with a mean $\mu_s = np$ and a standard deviation $\sigma_s = \sqrt{np(1-p)}$ only when both:

np and n(1-p) are greater than 5.

Example 7.17 illustrates the use of a normal approximation to a binomial distribution.

Example 7.17

Consider the following binomial experiment: a fair coin is tossed 20 times. Determine the probability of getting at least 14 heads.

Solution:

The probability of getting at least 14 heads can be determined using the binomial probability formula:

$$P(s \text{ successes in n trials}) = \binom{n}{s} p^s q^{(n-s)}$$

Using this formula, we would obtain:[2]

P(at least 14 heads) = P(14 or more heads)
    = P(14 heads) + P(15 heads) + P(16 heads) +
      P(17 heads) + P(18 heads) + P(19 heads) +
      P(20 heads)
P(at least 14 heads) = 0.0577.

The calculations associated with using the binomial probability formula to determine the result of 0.0577 are both long and tedious.

Let's consider an easier approach which uses ***the normal distribution to approximate this binomial probability***. This approximation is reasonable since both np and n(1-p) are greater than 5. That is,

$$np = (20)(1/2) \quad \text{and} \quad n(1-p) = (20)(1/2)$$
$$= 10 \qquad \qquad \qquad = 10$$

To use the normal approximation to the binomial distribution, we need to calculate the mean and standard deviation for this binomial distribution where:

$$n = 20, \quad p = 1/2 \quad \text{and} \quad 1-p = 1/2$$

The mean of the binomial distribution is:

$$\mu_s = np$$
$$= (20)(1/2)$$
$$= 10$$

The standard deviation of the binomial distribution is:

$$\sigma_s = \sqrt{np(1-p)}$$

$$\sigma_s = \sqrt{20(\tfrac{1}{2})(\tfrac{1}{2})}$$
$$= \sqrt{5}$$

---

[2] The actual specifics of the calculation of this binomial formula can be found in Chapter 6, in the section on binomial probability.

$\sigma_s \approx 2.24$

Thus, the normal distribution with $\mu_s = 10$ and $\sigma_s = 2.24$ will be used to obtain an approximation for the binomial probability of getting at least 14 heads in 20 tosses of a fair coin. This is illustrated in Figure 7.55.

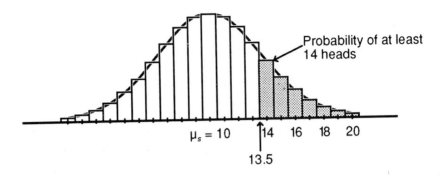

Figure 7.55

In Figure 7.55, the histogram illustrates the binomial distribution where **the shaded region represents the probability of getting at least 14 heads**, since this region represents **the probability of 14 or more heads**. This area can be approximated by **the area under the normal curve to the *right of 13.5*** which is the lower class boundary of 14.

Notice that we are finding the area to the right of 13.5, the boundary value of 14, and not to the right of 14. We are using the boundary value (13.5) rather than the actual value (14) because the Binomial Distribution is a discrete distribution and the Normal Distribution is a continuous distribution. We refer to the use of 13.5 rather than 14 as a **correction for continuity**. This **continuity correction** is due to the fact that we cannot find the probability of a specific value, like 14 heads, within a normal distribution. Why?

Therefore, the probability of exactly 14 heads corresponds to the area under the normal curve between the boundary values of 13.5 and 14.5. Thus, any time we use the normal distribution to approximate a binomial probability, we must make a slight modification which is to either add or subtract 0.5 from the data value.

Therefore, the probability of at least 14 heads corresponds to the area under the normal curve which is to the right of 13.5. To determine the area to the right of 13.5, we must first compute the z score for X = 13.5. Using the z score formula,

$$z = \frac{X - \mu_s}{\sigma_s}$$

and substituting 13.5 for X, we have:

$$z = \frac{13.5 - 10}{2.24} = 1.56$$

From TABLE II, the percent of area to the left of z = 1.56 is 94.06%. The percent of area to the right of z = 1.56 is 5.94%.

This percentage of 5.94% represents the normal approximation of the binomial probability of getting at least 14 heads in 20 tosses of a fair coin.

Let's compare the actual binomial probability of at least 14 heads to the normal approximation:

Binomial Calculation: $0.0577 = (0.0577)(100\%) = \mathbf{5.77\%}$
Normal Approximation: $\mathbf{5.94\%}$

Notice that the normal approximation (**5.94%**) is *very* **close** to the actual value (**5.77%**), since the difference between these two values is only **0.17% or 0.0017.** ∎

When using the normal distribution to approximate binomial probabilites, the *class boundaries* of a data value are used to get a better approximation of the binomial probability. This is referred to as a continuity correction, since the binomial distribution, which is a discrete distribution, is being corrected so that is can be approximated by the normal distribution, a continuous distribution.

Continuity Correction When Using the Normal Distribution to Approximate a Binomial Probability

When using the normal distribution to approximate a binomial probability, you must either add or subtract 0.5 from the data value depending upon the binomial probability question.

Let's summarize the procedure to approximate the binomial distribution using the normal distribution.

Normal Approximation Procedure to the Binomial Distribution

1. Determine the values of n, number of trials, and p, probability of a success, needed to solve the binomial probability problem.
2. Calculate the mean and standard deviation of the binomial distribution using the formulas:

$$\mu_s = np$$

$$\sigma_s = \sqrt{np(1-p)}$$

3. Identify the class boundary or boundaries required for the solution to the binomial probability. The boundary value of the data value is found by either adding or subtracting 0.5 from the data value depending upon the binomial probability question.
4. Shade the area under the normal curve representing the solution to the binomial probability problem.
5. Calculate the z score for each class boundary and use TABLE II to determine the percent of area corresponding to the shaded area identified in step 4. This result represents the normal approximation to the binomial probability.

Example 7.18

Consider the binomial experiment of tossing a fair coin 100 times. Use the normal approximation to calculate the probability of getting:
a) at least 55 heads.
b) between 40 and 60 heads.
c) exactly 54 heads.

d) at most 45 heads.

Solution:

To use the normal approximation to the binomial distribution, calculate the mean and standard deviation for the binomial distribution where n = 100, p = 1/2, and 1 - p = 1/2.

The mean is: $\mu_s = np$
$= 50$

The standard deviation is:

$$\sigma_s = \sqrt{np(1-p)}$$
$$= \sqrt{(100)(\tfrac{1}{2})(\tfrac{1}{2})}$$
$$= \sqrt{25}$$

$\sigma_s = 5$

Thus, the normal distribution with a $\mu_s = 50$ and a $\sigma_s = 5$ will be used to approximate these binomial probabilities.

a) To calculate the probability of getting at least 55 heads, we must determine the probability of getting 55 or more heads. This is illustrated in Figure 7.56.

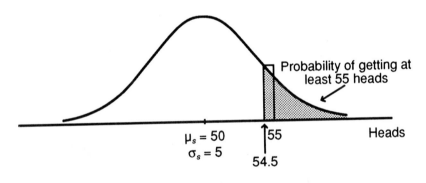

Figure 7.56

To approximate the probability of getting 55 or more heads use the lower boundary of 55 which is 54.5 and compute z of 54.5.

$$z = \frac{54.5 - 50}{5} = \frac{4.5}{5} = 0.90$$

Using TABLE II the area to the left of z = 0.90 is 81.59%. Thus, the percent of area to the right of z = 0.90 is 18.41. Therefore, the probability of at least 55 heads is approximately 18.41%.

b) To calculate the probability of getting between 40 and 60 heads, compute the probability of getting more than 40 heads and less than 60 heads. This is illustrated in Figure 7.57.

SECTION 7.7 THE NORMAL APPROXIMATION TO THE BINOMIAL DISTRIBUTION • **373**

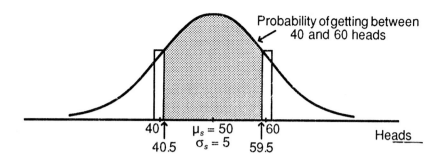

Figure 7.57

To approximate the probability of getting between 40 and 60 heads, use the upper boundary of 40, which is 40.5, and the lower boundary of 60, which is 59.5.

To determine the shaded area in Figure 7.57 compute z of 40.5 and z of 59.5. We obtain:

z of 40.5 is − 1.90
z of 59.5 is 1.90

From TABLE II the percent of area to the left of z = −1.9 is 2.87%, and the percent of area to the left of z = 1.9 is 97.13%. Therefore, the percent of area between z = −1.9 and z = 1.9 is 94.26% Hence, the probability of getting between 40 and 60 heads is approximately 94.26%.

c) The probability of getting exactly 54 heads is approximated by the shaded area in Figure 7.58.

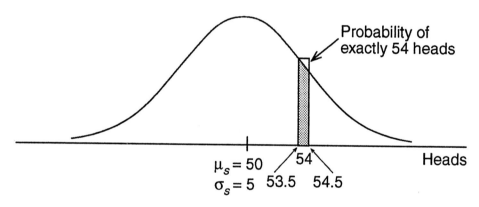

Figure 7.58

To approximate the probability of exactly 54 heads, use the lower boundary of 54, which is 53.5, and the upper boundary of 54, which is 54.5.

To determine the shaded area in Figure 7.58 compute z of 53.5 and z of 54.5. We obtain:

z of 53.5 = 0.70
z of 54.5 = 0.90

Using TABLE II the percent of area to the left of z = 0.70 is 75.80%, and the percent of area to the left of z = 0.90 is 81.59%. Thus, the percent of area between z = 0.70 and z = 0.90 is 5.79%. Therefore, the probability of getting exactly 54 heads is approximately 5.79%.

d) To calculate the probability of getting at most 45 heads, we must determine the probability of getting 45 or less heads. This is illustrated in Figure 7.59.

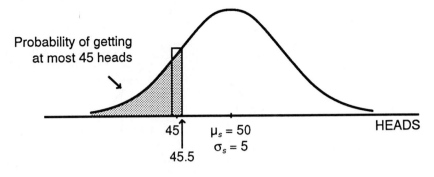

Figure 7.59

To approximate the probability of getting 45 or less heads use the upper boundary of 45 which is 45.5 and compute z of 45.5.

$$z = \frac{45.5 - 50}{5} = \frac{-4.5}{5} = -0.90$$

Using TABLE II the area to the left of z = -0.90 is 18.41%. Therefore, the probability of at most 45 heads is approximately 18.41%. ■

In Example 7.18, we examined four different types of probability questions. For each question, we determined the boundary value required to find the appropriate area under the normal curve to approximate the binomial probability. Let's examine the different probability questions and the boundary values that where used to approximate the probability. Table 7.3 shows the boundary needed to find the normal curve area to answer the different types of binomial probability problem.

Table 7.3

Binomial Probability Problem	Boundary Value Used to Solve Problem	Area Under Normal Curve corresponding to Binomial Probability
AT LEAST 55 HEADS	BOUNDARY VALUE is 54.5	AREA TO THE RIGHT OF 54.5 (see Figure 7.56)
BETWEEN 40 AND 60 HEADS	LOWER BOUNDARY is 40.5 AND UPPER BOUNDARY is 59.5	AREA BETWEEN 40.5 AND 59.5 (see Figure 7.57)
EXACTLY 55 HEADS	LOWER BOUNDARY is 54.5 AND UPPER BOUNDARY is 55.5	AREA BETWEEN 54.5 AND 55.5 (see Figure 7.58)
AT MOST 45 HEADS	BOUNDARY VALUE is 45.5	AREA TO THE LEFT OF 45.5 (see Figure 7.59)

Example 7.19

An automobile salesperson is successful in selling new cars to 40% of the customers that come into the showroom. What is the probability that the salesperson will sell new cars to at least 30 of the next 50 customers?

Solution:

This problem can be considered to be a binomial problem where the probability that the salesperson successfully sells a new car to the next customer is 0.40, and the probability the salesperson does not sell a new car to the next customer is: 1 - 0.4 = 0.60.

To determine if the normal approximation to the binomial distribution can be used to solve this problem, calculate np and n(1-p).

$$np = (50)(0.4) \quad \text{and} \quad n(1-p) = (50)(0.6)$$
$$= 20 \quad\quad\quad\quad\quad\quad = 30$$

Since np and n(1-p) are both greater than 5 the normal approximation to the binomial distribution can be used to solve this problem. To use this normal approximation, we must first calculate the mean and standard deviation of the binomial distribution.

The mean of the binomial distribution is:

$$\mu_s = np = 20$$

The standard deviation of the binomial distribution is:

$$\sigma_s = \sqrt{np(1-p)}$$
$$\sigma_s = \sqrt{12}$$
$$\sigma_s \approx 3.5$$

Thus, a normal distribution with $\mu_s = 20$ and $\sigma_s \approx 3.5$ will be used to obtain an approximation for the binomial probability of the salesperson selling new cars to at least 30 of the next 50 customers. This is illustrated in Figure 7.60.

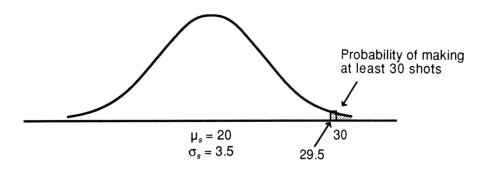

Figure 7.60

To approximate the probability of selling at least 30 new cars, we need to calculate the probability of selling 30 or more cars. Thus, we will use the lower boundary of 30 which is 29.5 and find the area to the right. This is illustrated by the shaded area in Figure 7.60.

To calculate the percent of area to the right of 29.5, find the z of 29.5. Looking up z of 29.5 = 2.71 in TABLE II, we find the area to the left of 2.71 is 99.66%.

Thus, the area to the right of z = 2.71 is 0.34%.

Therefore, the probability that the salesperson successfully sells at least 30 new cars to his next 50 customers is approximately 0.34%. ∎

Example 7.20

Tom, a daily commuter to Wall Street, is late to work 20% of the time. Calculate the probability that in his next 40 trips to work Tom will be late at most 15 times.

Solution:

Let's consider using the normal approximation to the binomial distribution for this problem, where the probability that Tom gets to work late is 0.20 and the probability that Tom does not get to work late is 0.80.

Calculate np and n(1-p) to determine if both expressions are greater than 5.

$$np = (40)(0.20) \quad \text{and} \quad n(1-p) = (40)(0.80)$$
$$= 8 \qquad\qquad\qquad\qquad = 32$$

Since np and n(1-p) are both greater than 5, the *normal* approximation to the binomial distribution can be used to solve this problem.

To use the normal approximation, calculate the mean and standard deviation of the *binomial* distribution.

The mean of the binomial distribution is:

$$\mu_s = 8$$

The standard deviation of the binomial distribution is:

$$\sigma_s \approx 2.53$$

Thus, the normal distribution with $\mu_s = 8$ and $\sigma_s \approx 2.53$ will be used to obtain an approximation for the binomial probability that Tom gets to work late at most 15 times in his next 40 trips. This is illustrated in Figure 7.61.

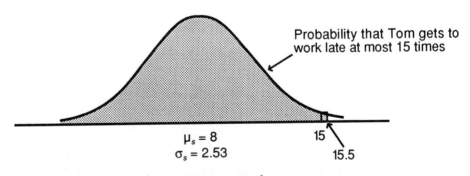

Figure 7.61

To approximate this binomial probability, use the *upper* boundary of 15 which is 15.5 since we are trying to find the probability of 15 or less. Thus, this probability is determined by finding the percent of area to the left of 15.5. Finding the z of 15.5 = 2.96 in TABLE II, the area to the left of z = 2.96 is 99.85%. Therefore, the probability that Tom gets to work late at most 15 times is approximately 99.85%. ∎

Example 7.21

A single fair die is rolled 36 times. Calculate the probability of getting:
a) exactly 7 sixes
b) at least 10 fours

Solution:

a) For this binomial probability problem, we can use the normal approximation since np and n(1-p) are both greater than 5. Let's verify this.

Since the fair die is rolled 36 times, n = 36. For a fair die, the probability of getting a six in one roll is 1/6.

Therefore, np = 6

and

$$n(1-p) = (36)(5/6) = 30.$$

To use the normal approximation, calculate the mean and standard deviation of the binomial distribution.

The mean is: $\mu_s = 6$

The standard deviation is: $\sigma_s \approx 2.24$

Thus, the normal distribution with $\mu_s = 6$ and $\sigma_s \approx 2.24$ can be used to obtain an approximation for the binomial probability of getting 7 sixes in 36 rolls of a fair die. This is illustrated in Figure 7.62.

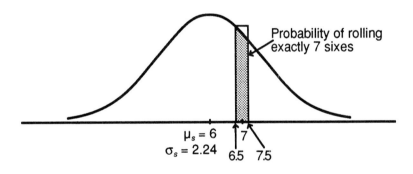

Figure 7.62

To approximate this binomial probability, use the lower boundary of 7 which is 6.5 and the upper boundary of 7 which is 7.5 since we are trying to find the area of exactly 7 sixes.

To determine the shaded area represented in Figure 7.62, we must compute the z scores for 6.5 and 7.5.

$$z \text{ of } 6.5 = 0.22 \text{ and } z \text{ of } 7.5 = 0.67$$

Using TABLE II the percent of area to the left of z = 0.22 is 58.71% and the percent of area to the left of z = 0.67 is 74.86%. Therefore, the percent of area between z = 0.22 and z = 0.67 is 16.15%. Thus, the probability of getting exactly 7 sixes in 36 rolls of a fair die is approximately 16.15%.

b) As in part (a), this binomial probability problem has n = 36 trials or rolls. For a fair die, the probability of getting a four in one roll is: p = 1/6. Notice, these values for n and p are the same as those computed in Part a of this example, hence we can use the normal approximation to the binomial distribution where $\mu_s = 6$ and $\sigma_s \approx 2.24$. Thus, the normal distribution illustrated in Figure 7.63 is used to obtain an approximation for the binomial probability of getting at least 10 fours in 36 rolls of a fair die.

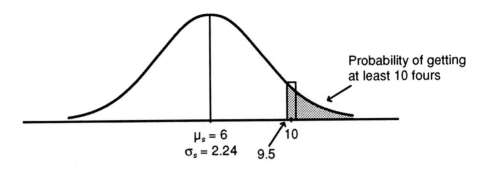

Figure 7.63

This binomial probability can be approximated by determining the percent of area to the right of the lower boundary 9.5 since we are trying to find the probability of 10 or more fours. To determine this area, compute the z score for 9.5.

$$z \text{ of } 9.5 = 1.56.$$

Using TABLE II, the area to the left of the $z = 1.56$ is 94.06%. The area to the right of $z = 1.56$ is 5.94%. Therefore, the probability of getting at least 10 fours in 36 rolls of fair a die is approximately 5.94%. ∎

CASE STUDY 7.2

In Case Study 6.6, the article entitled *Trial By Mathematics* which appeared in Time magazine on April 26, 1968 discussed a robbery committed by an interracial couple. The couple accused of the crime had been convicted because the likelihood of a couple having the known characteristics of the robbers had been established at 1 in 12 million. In the couples' appeal, the judge assumed the probability p of a couple having all the characteristics of the robbers to be: $p = \dfrac{1}{12,000,000}$. Assume that there does exist another such couple where each of the N couples in the population serve as a trial of a binomial experiment. Using this assumption, the judge then calculated the probability that another couple has the same characteristics of the convicted couple for the value of N = 12,000,000.

a) Would the normal distribution be a good approximation to the binomial distribution? What values of N, p, and (1 - p) would you use, if you could use the normal approximation?
b) What is the mean and standard deviation of this binomial distribution?
c) If you used the normal approximation to the binomial distribution, what boundary value would you use to calculate the probability that two or more such couples within the population of N couples satisfy the characteristics of the robbers?
d) Using the normal distribution, what did you get for the probability that two or more such couples satisfy the characteristics of the robbers? How does this compare to the judges' calculation of approximately 41%?
e) Calculate this probability again for the following different values of N couples (in millions): 15, 20, 25, 30, 40, 50, 75, and 100. Compare your results using the normal approximation to the results of Table 6.10 which represent the probabilities calculated using the binomial

distribution. Are the results of the normal approximation close to the binomial distribution? What happens to the difference between the normal approximation results and the binomial distribution results as the value of N gets larger? ∎

7.8 USING MINITAB

In this chapter we have explored the characteristics of the normal distribution and some of its applications. MINITAB has many commands relating to the normal distribution. For example, there are a set of commands that can *simulate* a sample of data values such that a graph of the distribution of these data values approximates the graph of the normal distribution.

Using the MINITAB commands RANDOM with subcommand NORMAL, and HISTOGRAM with subcommand START, a collection of data values can be assigned to a column from which a histogram of the column entries will produce a distribution that is **approximately** normal.

For example,

```
MTB >   RANDOM 100 C1;
SUBC>   NORMAL 0 1.
MTB >   HISTOGRAM C1;
SUBC>   START -3.5.
```

Histogram of C1   N = 100

Midpoint	Count	
-3.5	0	
-3.0	0	
-2.5	1	*
-2.0	3	***
-1.5	9	*********
-1.0	12	************
-0.5	19	*******************
0.0	18	******************
0.5	18	******************
1.0	12	************
1.5	4	****
2.0	3	***
2.5	1	*

Figure 7.64

The shape of the histogram pictured in Figure 7.64 appears approximately normal. Most of the data values are within two standard deviations of the mean. In fact for this sample, 98% of the data values are within two standard deviations of the mean. You have learned that in a normal distribution 95% of the data values are within two standard deviations of the mean. Can you find any other differences between the histogram pictured in Figure 7.64 and the properties of the normal distribution? What do you think will happen if we used MINITAB to produce graphs for samples of data values where the number of values is larger than 100? What do you think contributes to the differences between the characteristics of the distribution of data values and the characteristics of the normal distribution?

In Chapter 6 it was shown how MINITAB can produce a table of probability values and a distribution for a binomial experiment. In Section 7.7, we examined how the normal distribution is used to approximate a binomial distribution. Let's use MINITAB to produce a binomial distribution along with a table of binomial probability values.

```
MTB >   NAME C1='K'
MTB >   SET 'K'
DATA>   1(0:32/1)1
DATA>   END.
MTB >   NAME C2='P(X=K)'
MTB >   PDF 'K' 'P(X=K)';
SUBC>   BINOMIAL 32 .5.
MTB >   PLOT C2 C1;
SUBC>   SYMBOL 'x'.
```

The graph in Figure 7.65 and the subsequent table of probability values of the binomial experiment is generated for which n = 32 and p = 0.5.

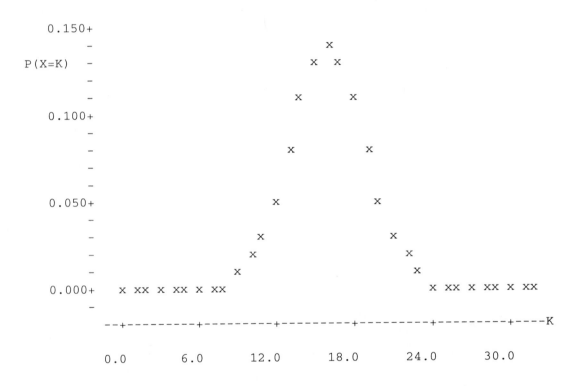

Figure 7.65

```
MTB >   PRINT C1 C2

        ROW    K    P(X=K)
         1     0    0.000000
         2     1    0.000000
         3     2    0.000000
         4     3    0.000001
         5     4    0.000008
         6     5    0.000047
         7     6    0.000211
         8     7    0.000784
         9     8    0.002449
        10     9    0.006531
        11    10    0.015020
        12    11    0.030041
        13    12    0.052571
        14    13    0.080879
        15    14    0.109765
        16    15    0.131718
        17    16    0.139950
        18    17    0.131718
        19    18    0.109765
        20    19    0.080879
        21    20    0.052571
        22    21    0.030041
        23    22    0.015020
        24    23    0.006531
        25    24    0.002449
        26    25    0.000784
        27    26    0.000211
        28    27    0.000047
        29    28    0.000008
        30    29    0.000001
        31    30    0.000000
        32    31    0.000000
        33    32    0.000000
```

Notice that the shape of the graph in Figure 7.65 approximates the shape of the normal distribution for this binomial experiment where n = 32 and p = 0.5.

The MINITAB command CDF and the subcommand BINOMIAL yields the cumulative probability less than or equal to 's'. Using the MINITAB command to calculate the probability of s ≤ 10 we get

```
MTB >   CDF 10;
SUBC >  BINOMIAL N = 32 P = .5.

     K   P( X LESS OR = K)
   10.00     0.0251
```

You should verify this cumulative probability from the above binomial probability table. In Section 7.7, we learned that the normal distribution can be used to approximate a binomial distribution whenever both, np and n(1-p) is greater than five. For this distribution both np and n(1-p) are greater than five. Thus, the normal distribution can be used to model the binomial distribution. The normal approximation to the binomial probability is approximately 0.0261. Compare the normal approximation result to the actual binomial result. Are these results very close? Did you expect this to happen? Explain.

Glossary

TERM	SECTION
Normal Distribution	7.1
Bell-Shaped	7.1, 7.2
Symmetry	7.2
Percentiles	7.5
Percentile Rank	7.5
Probability	7.6
Binomial Distribution	7.7
Mean of a binomial distribution, μ_s	7.7
Standard Deviation of a binomial distribution, σ_s	7.7
Normal Approximation	7.7
Class Boundaries	7.7
Continuity Correction	7.7

Exercises

PART I Fill in the blanks.

1. The data values in a normal distribution tend to cluster about the _____.
2. In a normal distribution, the mean is located at the _____ of the distribution.
3. The percent of area under the normal curve which is to the left of the mean is equal to _____.
4. In a normal distribution if the median is 100, then the mean is _____.
5. The percent of data values in a normal distribution that are located within:
   a) one standard deviation of the mean (that is, between z = +1 and z = -1) is approximately _____.
   b) two standard deviations of the mean (that is, between z = 2 and z = -2) is approximately _____.
   c) three standard deviations of the mean (that is, between z = +3 and z = -3) is approximately _____.
6. For a normal distribution the percent of area between z = 1 and z = -1 is approximately 68.26%. Therefore the percent of data values between z = -1 and z = 1 is approximately _____.
7. Consider the following information for two normal distributions A and B.

A	B
$\mu = 50$	$\mu = 75$
$\sigma = 10$	$\sigma = 20$

   For distribution A, the percent of data values between 60 and 50 is _____. For distribution B the percent of data values between 75 and _____ is 34.13%.
8. In a normal distribution, the percent of data values below z = -0.53 is the same as the percent of data values above z = _____.
9. In a normal distribution, the percent of data values between z = 0 and z = -2 is the same as the percent of data values between z = _____ and z = _____.
10. In a normal distribution, the percent of area to the left of z = -1 is _____. Therefore, the percent of area to the right of z = -1 is _____.
11. In a normal distribution, if the percent of data values greater than the data value 60 is 84.13% then, the percentile rank of 60 is _____.
12. In a normal distribution if the percentile rank of a data value is P_{50}, then the data value is equal to the _____, and its z score is equal to _____.
13. In a normal distribution with $\mu = 80$ and $\sigma = 10$, if a data value has a percentile rank of P_{70} then the data value has a value of _____.
14. The frequency polygon of a binomial distribution looks more like a _____ distribution as the number of trials, n, gets larger.
15. The normal distribution is a good approximation of the binomial distribution when both np and n(1-p) are greater than _____.
16. For a binomial distribution, the formula for:
    a) the mean of the binomial distribution, μ_s, is _____.
    b) the standard deviation of the binomial distribution, σ_s, is _____.
17. For a binomial distribution, if n = 100 and p = 1/2, then:
    a) $\mu_s = np = $ _____.
    b) $\sigma_s = \sqrt{np(1-p)} = $ _____.
18. a) For a binomial distribution with n = 10 and p = 1/5, would the normal distribution be a good approximation to the binomial distribution? _____.
    b) For a binomial distribution with n = 8 and p = 3/4, would the normal distribution be a good approximation to the binomial distribution? _____
    c) For the binomial distribution with n = 30 and p = 1/5, would the normal distribution be a good approximation to the binomial distribution? _____.

PART II Multiple choice questions.

1. In a normal distribution approximately two-thirds of the data values are within _____ standard deviation(s) of the mean.
   a) 3   b) 1   c) 2   d) 1 and 1/2
2. In a normal distribution if the mean is 75 then the median is:
   a) 0   b) 50   c) 75   d) insufficient information
3. In a normal distribution with $\mu = 50$ and $\sigma = 10$ the percent of data values greater than 65 is:
   a) 6.68%   b) 69.15%   c) 93.32%   d) 15.87%
4. In a normal distribution with $\mu = 75$ and $\sigma = 15$ the data value with a percentile rank of 77 is:
   a) 77   b) 75   c) 80   d) 86.1

5. For a binomial distribution with n = 120 and p = 0.4 what is the value of σ_s?
   a) 0.24   b) 28.8   c) 5.37   d) 14.4
6. For a binomial distribution with n = 200 and p = 0.37 what is the value of μ_s?
   a) 74   b) 200   c) 37   d) 100
7. Assuming that the probability that someone will respond positively to a survey question is 40% what is the chance that more than half of the 150 people surveyed will respond positively?
   a) 40%   b) 20%   c) 50%   d) 0.49%
8. The normal distribution can be used as a good approximation to the binomial distribution if np and n(1-p) are both greater than:
   a) 10   b) 5   c) 0.5   d) 25
9. If a fair coin is tossed 100 times what is the probability that the coin lands heads more than 60 times?
   a) 98.21%   b) 5%   c) 1.79%   d) 95%
10. If a pitcher has a 25% probability of striking out a batter what's the probability that this pitcher strikes out 18 or more of the next 40 batters he faces?
    a) 0.31%   b) 5%   c) 1%   d) 99.69%

PART III Answer each statement True or False.

1. The shaded areas (A1 and A2) are equal.

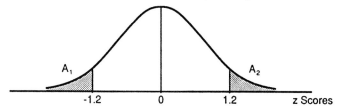

2. The shaded area to the right of z = 0 is equal to the shaded area to the left of z = 0.

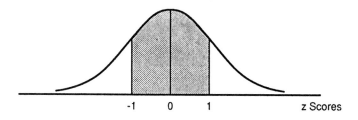

3. The shaded area A1 is equal to the shaded are A2.

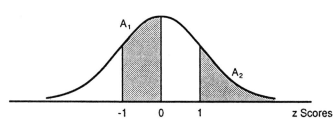

4. The z score of 0 divides the normal distribution into two equal parts.

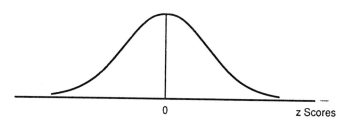

5. The area to the left of a z score of 1 equals the area to the right of a z score of 1.

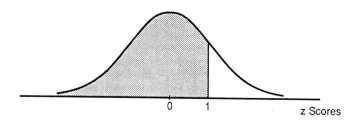

6. The area to the right of a z score of -1 is greater than the area to the left of z score of -1.

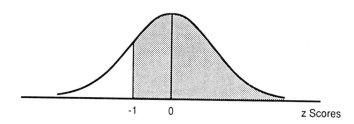

7. The area between the z scores -1 and -2 is equal to the area between the z scores +1 and +2.

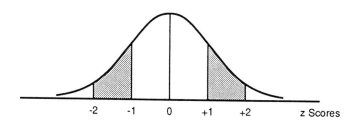

8. The total area under the normal curve is infinite.
9. The binomial distribution can be approximated by a normal distribution if either np or n(1-p) is greater than 5.
10. To get a good approximation of a binomial probability problem by using the normal approximation, real boundaries should be applied.

PART IV Problems.

1. In a normal distribution find the percent of area:
   a) less than a z score of +2.50.
   b) between the z scores of −2.33 and +0.5.
   c) greater than the z score of −1.30.
2. In a normal distribution find the probability that a data value picked at random has z score:
   a) greater than $z = 1.75$.
   b) less than $z = -1.16$.
   c) between $z = 1.0$ and $z = +0.5$.
3. In a normal distribution find the z score(s) that:
   a) cuts off the top 5% of the z scores.
   b) cuts off the bottom 12% of the z scores.
   c) cuts off the middle 60% of the z scores.
   d) determines the third decile.
   e) determines the first quartile.
4. In a normal distribution what is the percentile rank of the data value which is:
   a) two standard deviations greater than the mean?
   b) one standard deviation less than the mean?
   c) one-half a standard deviation greater than the mean?
5. In a normal distribution with $\mu = 90$ and $\sigma = 15$ find the percent of data values which are:
   a) less than 106.
   b) between 78 and 107.
   c) greater than 100.
6. In a normal distribution with $\mu = 900$ and $\sigma = 350$ find the probability that a data value picked at random is:
   a) greater than 1100.
   b) less than 500.
   c) between 800 and 1200.
7. In a normal distribution with $\mu = 250$ and $\sigma = 20$ find the percentile rank of the following data values:
   a) 286.
   b) 265.
   c) 225.
8. In a normal distribution with $\mu = 300$ and $\sigma = 40$, approximate the data value which is:
   a) P_{85}.
   b) P_{35}.
   c) P_{70}.
   d) the first decile.
   e) the third quartile.
   f) the sixth decile.
9. During the past season the number of points scored per game by the local High School football team was approximately normally distributed with $\mu = 18$ and $\sigma = 10$.
   Use the normal distribution to answer the following questions:
   a) If one assumes this season's play is indicative of next season's play then estimate the percent of next season's games that the team will score more than 31 points?
   b) If the team plays 30 games, in approximately how many would they score more than 6 points?
   c) What's the probability that the team will score at least 22 points in any given game?
10. If two hundred thousand families in a local community have annual incomes which are normally distributed with a mean of $32,500 and a standard deviation of $5,000, then determine the following:
    a) the percent of families with incomes greater than $38,000.
    b) the percent of families with incomes between $30,000 and $39,000.
    c) the number of families who have incomes less than $29,000.
    d) the family income that represents the 8th decile.
    e) the family income that represents the 1st quartile.
    f) the family income that represents the 65th percentile.
    g) the family income that cuts off the lowest 200 incomes.
    h) the family income that cuts off the highest 100 incomes.
    i) the maximum family income eligible for food stamps if the five hundred families with the lowest incomes in this community are eligible for food stamps.
11. Suppose the systolic blood pressure for an adult male above age 55 is normally distributed with a mean of 145 and a standard deviation of 20. If a sixty-year old male is selected at random, what is the probability that his blood pressure is:
    a) greater than 170?
    b) between 140 and 160?
    c) If 4000 adult males above age 55 have their blood pressure checked, then how many would you expect to have a blood pressure less than 130?
    d) Volunteers from a Red Cross mobile unit that is checking blood pressures decide to recommend for additional medical diagnosis only those males above age 55 whose blood pressures are at the third quartile. What would be the minimum systolic blood pressure required to be recommended for additional medical diagnosis?
12. The length of time required to complete a college placement exam is normally distributed with a mean of 90 minutes and a standard deviation of 10 minutes. How much time should the college proctor allow the students for the exam if he would like:
    a) 95% of the students to complete the exam?
    b) 90% of the students to complete the exam?
    c) 99% of the students to complete the exam?
13. Three hundred first-year law students at a local law school had a mean of 165 on the LSAT, the law boards, with a standard deviation of 5. Assuming these results are approximately normally distributed, determine:
    a) the percentile rank of the LSAT score 176.
    b) the LSAT score that cuts off the top 15% of the LSAT scores (to the nearest whole number).

c) the LSAT score that cuts off the top 80 LSAT scores (to the nearest whole number).
d) the LSAT score that represents the 6th decile (to the nearest whole number).
e) the LSAT score that represents the first quartile (to the nearest whole number).
f) the LSAT score that represents the third quartile (to the nearest whole number).
g) the LSAT score that represents the 35th percentile (to the nearest whole number).

Suppose one of these first-year law student's is selected at random, and you have to guess the student's LSAT score. If you guess it correctly to within 7 points, then you will be given ten dollars.

h) what LSAT score should you guess and what is your chance of winning the ten dollars?

14. A marketing research agency states that the number of responses received from 2,000 mailed questionnaires is normally distributed with a mean of 1400 and a standard deviation of 80. If Craig, a market researcher, plans to mail out 2,000 questionnaires for a consumer product study, what is the probability that Craig will receive at least 1500 responses?

15. Recently, Ms. Piper bought a tire for her car. The manufacturer claimed that the tire wear is normally distributed with $\mu = 56{,}000$ miles and $\sigma = 4{,}000$ miles. What is the probability:
a) that Ms. Piper's tire lasts at least 60,000 miles?
b) that Ms. Piper's tire lasts less than 50,000 miles?
c) that Ms. Piper's tire lasts between 60,000 and 70,000 miles?

16. The Motor Vehicle Commissioner reports that the weight of passenger cars is normally distributed with $\mu = 2700$ lbs. and $\sigma = 300$ lbs. Determine the following:
a) the percent of cars weighing more than 2200 lbs.
b) the percent of cars weighing less than 3500 lbs.
c) the percent of cars weighing either less than 2100 lbs. or more than 3800 lbs.

17. At a local high school 5000 juniors and seniors recently took an aptitude test. The results of the exam were normally distributed with $\mu = 450$ and $\sigma = 50$. Calculate the following:
a) the percent of students that scored over 575.
b) the number of students that scored less than 425.
c) the probability of a student selected at random having scored between 400 and 510.

18. An automatic machine process produces ball bearings whose diameters are normally distributed with $\mu = 0.075"$ and $\sigma = 0.004"$. Find the probability that a ball bearing selected at random will have a diameter:
a) less than 0.071".
b) greater than 0.083".
c) between 0.077" and 0.087".

19. According to bank records, the amount of daily cash withdrawals from a Long Beach bank is normally distributed with a mean of $75,000 and a standard deviation of $10,000. How much cash should the bank have on hand in order to cover the daily withdrawals 90% of the time?

20. Debbie, an architect, is designing the interior doors to a new squash club. She wants the doors to be high enough so that 90% of the men using the doors will have at least one foot clearance. Assuming the heights to be normally distributed with a mean of 69 inches and a standard deviation of 4 inches, how high must Debbie design the doors to satisfy her requirements?

21. Matthew, an experimental psychologist, indicates that the age at which a child learns to walk is normally distributed with a mean age of 10.9 months with a standard deviation of 0.5 months. If Matthew classifies an early walker as a child who is among the youngest 10% to walk, then by what age must a child walk to be identified as an early walker?

22. A new drug developed to immobilize wild life is administered to a herd of four hundred buffaloes which must be transported to another grazing area for environmental reasons. The buffaloes reaction time to this drug is approximately normal with a mean of 17.4 seconds and a standard deviation of 3.4 seconds. What is the probability that a buffalo selected at random will have a reaction time to the drug:
a) greater than 20 seconds?
b) less than 15 seconds?
c) between 13 and 23 seconds?
Determine:
d) What is the maximum time required for the ten most *susceptible* buffaloes to react to this drug?

23. Bruce, an accountant, determines that the mean time required to prepare a tax return using the computer package "Tax Pro" is normally distributed with a mean of 2.8 hours and a standard deviation of 0.7 hours. If Bruce will prepare 800 tax returns during the year, then approximately how many will Bruce complete in less than 3 hours and 45 minutes?

24. It is estimated that on a given night at a harness racing track Aaron, a typical gambler, will lose on the average $65.00 with a standard deviation of $25.00. If the distribution of losses is normally distributed, what is the probability that a typical harness racing bettor will:
a) lose more than $100?
b) *break even or make money*?
c) If two friends that bet independently spend an evening at the track what is the probability that they both lose more than $80? (Hint: use the multiplication rule for independent probability events in Chapter 6.)

25. The production process at an automobile manufacturing company uses ball bearings whose service life is

approximately normally distributed with a mean of 820 hours and a standard deviation of 40 hours. If the maintenance supervisor decides to replace the ball bearings before no more than 5% have worn out, what is the maximum time the ball bearings can be in operation before being replaced?

26. The lifetime of a car battery is found to be normally distributed with $\mu = 5$ years and $\sigma = 3/4$ year. If the company which manufactures this battery is only willing to replace 5% of these batteries under its guarantee how many months should they guarantee their batteries?

27. If the amount of soda which a dispensing machine puts into a two liter bottle is normally distributed with $\sigma = 0.5$ oz. find the mean amount of soda which the machine will put into a two liter bottle if only 65.54% of the bottles the machine filled will have less than 67.6 oz. of soda.

28. If the grades of a math exam are normally distributed with $\sigma = 5$ find the mean grade on this exam, if the percentile rank of the grade 90 is P_{92}.

29. The amount of time required for Jain to travel to and from home and the office can be approximated by a normal distribution with $\mu = 55$ minutes and $\sigma = 10$ minutes.
    a) If Jain leaves home at 7:45 a.m., what is the probability that she will arrive at the office no later than 9:00 a.m.?
    b) If coffee is served at the office between 8:45 - 9:00 a.m., what is the probability that Jain will miss the coffee break (arrives after 9 a.m.) if she leaves home at 8:05 a.m.?
    c) If Jain has an appointment with a client at the office promptly at 9:05 a.m., what is the probability she will make the appointment if she leaves home promptly at 8:15 a.m.?

30. A machine manufactures a special nut whose diameter is found to be normally distributed with $\mu = 1.5$ inches and $\sigma = 0.10$ inches. The specifications, for this special nut, call for the diameters to be within the limits of 1.4 plus or minus 0.30 inches. If a nut does not meet this specification, it is considered to be defective. What percentage of the nuts this machine manufactures will be considered to be defective?

31. An entrance requirement for a certain University is a percentile rank of at least 85 on the Math SAT exam. Ten applicants are chosen at random and their Math SAT scores are listed below. The Math SAT scores are normally distributed with $\mu = 500$ and $\sigma = 100$. Determine which of the ten applicants fail to meet the entrance requirement.

Applicants	Math SAT Score
1. Tina	590
2. Sean	640
3. Jaime	625
4. Michael	575
5. Craig	700
6. Allison	520
7. Nick	730
8. Mary Ann	595
9. Matthew	650
10. Ryan	615

32. Use the normal distribution to approximate the following binomial probabilities:
    a) If $n = 100$ and $p = 0.5$, then find the probability that the number of successes "s" is:
       1. greater than 60, i.e. $P(s > 60)$.
       2. equal to 60, i.e. $P(s = 60)$.
       3. greater than or equal to 60, i.e. $P(s \geq 60)$.
       4. less than 60 or greater than 60, i.e. $P(s < 60 \text{ or } s > 60)$.
    b) If $n = 100$ and $p = 0.1$, then find the probability that the number of successes "s" is:
       1. equal to 10, i.e. $P(s = 10)$.
       2. less than 7, i.e. $P(s < 7)$.
       3. greater than or equal to 7, i.e. $P(s \geq 7)$.
       4. less than 7 or greater than or equal to 7, i.e. $P(s < 7 \text{ or } s > 7)$.
    c) If $n = 10,000$ and $p = 0.8$, then find the probability that the number of successes "s" is:
       1. greater than 8060, i.e $P(s > 8060)$.
       2. greater than or equal to 8060, i.e. $P(s \geq 8060)$.
       3. less than 7960, i.e. $P(s < 7960)$.
       4. equal to 7960, i.e. $P(s = 7960)$.
       5. between 7960 and 8060, i.e. $P(7960 < s < 8060)$.

33. For the binomial experiment of tossing a fair coin 400 times, use the normal distribution to approximate the probability that:
    a) the coin lands heads at least 217 times.
    b) the coin lands tails between 180 and 195 inclusive.
    c) the coin lands heads exactly 200 times.
    d) the coin lands heads less than 185 or more than 210.

34. The probability that a camcorder battery lasts three years is 80%. What is the probability that between 330 and 340 of the next 400 batteries sold last three years?

35. A multiple choice test consists of 100 questions, each with five possible answers. If a student must answer at least 33 questions correctly in order to pass, what is the probability that the student passes the test? (Assume the student guesses at all the answers.).

36. An airline company knows that 90% of the people making flight reservations for a certain flight will

show up for the flight. What is the probability that out of the next 10,000 people making flight reservations:
a) at least 9030 of them will show up for the flight?
b) 9075 or less will show up for the flight?
c) exactly 1000 will not show up for the flight?

37. Over the past year, 60% of the patients waiting to see Dr. Hawkeye, an ophthalmologist, for their scheduled appointment have had to wait at least 45 minutes. What is the probability that out of the next 600 patients:
a) more than 390
b) less than 340
will have to wait at least 45 minutes for their scheduled appointments?

38. According to a study by Dow Jones & Co., the best time to buy stocks is on a Monday afternoon since the chance of the market rising is 43% while the best time to sell is on a Friday since the chance of the market rising is 58%.

Use the results of this study to determine the probability that over the next 100 trading weeks, the market will:
a) rise on a Monday afternoon at least 55 times.
b) rise on a Friday at most 45 times.
c) not rise on Friday more than 40 times.
d) not rise on a Monday afternoon less than 60 times.

39. It has been established that 10 percent of the customers that are willing to listen to a sales promotion from DPT marketing will sign a contract. If 10,000 customers listen to the sales promotion determine the following probabilities:
a) that less than 950 clients will sign the contract.
b) that more than 920 will sign the contract.
c) that between 1070 and 1100 will sign the contract.

40. A southwest regional labor official stated that only 80 percent of this year's graduating college class will have found jobs by Sept. 1. From the graduating classes of all the colleges in the southwest region 400 college graduates are selected. Determine the following probabilities that by Sept 1:
a) 310 or less have jobs.
b) more than 340 have jobs.
c) between 300 and 320 have jobs.

41. If the probability of winning a game of chance is 1/10 and the game is played 100 times find the probability of:
a) winning more than 17 times.
b) winning less than 7 or more than 12 times.
c) between 8 and 15 wins inclusive.

42. A gambler is trying to determine which method of play will yield the best profit. In the casino game roulette, a wager on red has a payoff of $1 for each $1 wagered, and the probability that red occurs is $18/38 \approx 0.47$. Calculate the probability that the gambler will make a profit:
a) if he bets $1,000 one time on red.
b) if he makes 100 bets at $10 each on red.
c) if he makes 1000 bets at $1 each on red.
d) Determine which of these betting strategies has the highest chance of yielding a profit?

43. A baseball slugger hits a home run once out of every fifteen times he bats, that is, the probability that he will hit a home run at any given time at bat is 1/15. If he bats 450 times in a season what is the probability that he hits:
a) at least 40 home runs?
b) between 15 and 30 home runs (inclusive)?
c) more than 49 home runs?

44. A recent survey showed that only 17% of the married women in the labor force have husbands who were earning over $25,000 annually. If a random sample of 100 married women is selected from the labor force, what is the probability that:
a) more than 25 have husbands earning over $25,000 annually?
b) less than 76 have husbands earning $25,000 or less annually?

45. According to a survey conducted by Pulltone, an antifreeze manufacturer, 7 out of 10 cars on the road have rust in their radiators. Assuming this statement is accurate, and if 100 cars are randomly inspected for rust in their radiators, what is the probability that:
a) at most 20 cars do *not* have rust in their radiators?
b) at most 55 cars have rust in their radiators?

46. A New York municipal court cites "incompatibility" as the legal cause for 80 percent of all divorce cases. If one hundred divorce cases are randomly selected from the court files, what is the probability that:
a) at least 85 cite incompatibility as the grounds for divorce?
b) not more than 72 cite incompatibility as the grounds for divorce?
c) less than 15 do *not* cite incompatibility as the grounds for divorce?

47. Jacqueline, a sales representative of Day News, a Nassau County newspaper, makes 200 calls a day trying to persuade potential customers to subscribe to Day News. If she has a 30% chance of selling a new subscription, what is the probability that Jacqueline will sell:

a) at least 78 new subscriptions for the day?
b) fewer than 40 new subscriptions for the day?

48. Last year, the records of a Brooklyn college show that 80% of their graduates who were recommended to U.S. medical schools were admitted. If the records of 100 of these Brooklyn college graduates who were recommended to U.S. medical schools were randomly selected from the school's files, determine the probability that:
   a) at least 90 of these graduates were admitted to a U.S. medical school.
   b) at most 15 of these graduates were not admitted to a U.S. medical school.

49. Medical researchers at a Boston university report that 20% of babies born to American mothers this year will be delivered by cesarean sections. What is the probability that out of the next 300 deliveries:
   a) not more than 85 will be cesarean sections?
   b) at least 260 will not be cesarean sections?

50. Medical records show that 25% of all people suffering from "hardening of the arteries" have side effects from a certain type of medication. If 200 people suffering from "hardening of the arteries" are given this medication, what is the probability that:
   a) no more than 35 will have side effects?
   b) at least 160 will not have side effects?

51. Clinical records indicate that 20% of the outpatients in a facility for emotional disorders have been treated in this facility for at least 3 years. What is the probability that a sample of 400 outpatients from this facility will include:
   a) not more than 60 of the patients who have been treated at least 3 years?
   b) no fewer than 90 patients who have been treated at least 3years?

52. The accountant firm of a major credit card company reports that 30% of all the company's delinquent accounts will eventually require legal action to collect the payment due. If this statement is accurate, what is the probability that of the 10,000 delinquent accounts due:
   a) at least 3100 accounts will require legal action?
   b) at most 6940 accounts will not require legal action?

53. The Best Fit Screw Company institutes the following quality control procedure to improve the quality of the screws it uses. Before a shipment of screws is accepted a sample 100 screws is selected and tested for defects. If more than six of the screws are found to be defective, the shipment is rejected.
   a) If a shipment contains 20 percent defective, what is the probability that it will be accepted?
   b) If a shipment contains 10 percent defective, what is the probability that it will be accepted?

54. A money market fund company reports that next to individual retirement accounts (IRA's), the tax-shelter most frequently used by their shareholders is the uniform gifts to minors account. If 10% of their shareholders are custodians of an account for a minor, then what is the probability that out of the next 40,000 shareholders opening new accounts no more than 4100 will be custodian accounts for a minor?

55. A study released by the Bureau of Justice Statistics found that eight out of 10 young adults paroled from prison are rearrested for serious crimes within six years. If 1600 cases of young parolees are reviewed after six years, then what is the probability that:
   a) more than 1300 were rearrested for a serious crime?
   b) less than 1320 were rearrested for a serious crime?

56. From past experience a restaurant owner knows that 80% of the drinks ordered will be alcoholic. For the next 400 drinks ordered:
   a) how many would be expected to be alcoholic?
   b) what is the probability that at least 300 drinks will be alcoholic?

   On a *good* weekend the owner sells at least $990 worth of alcoholic drinks at $3.00 per drink.

   c) if 400 drinks are ordered next weekend what is the probability that next weekend is good?
   d) The owner estimates that 25,600 drinks will be sold during the next year. He knows that for every 30 alcoholic drinks sold he uses one bottle of liquor. If each bottle of liquor costs him $9.00 how much money should he budget to be 95% confident his liquor costs are covered?

PART V What Do You Think?

1. **What Do You Think?**

   The histogram in the following figure represents the ages of the *Forbes* magazine's 400 Richest People in the United States for 1992. To be ranked on this list, you had to be worth at least $265 million.

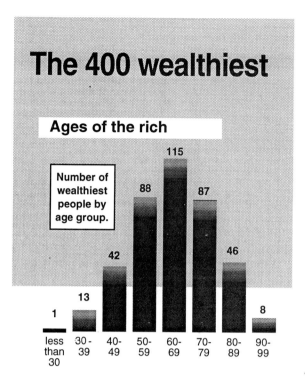

a) What variable is being displayed in this histogram?
b) Is this variable: discrete or continuous?
c) What shape does this histogram have?
d) What do you think the shape of the histogram representing the ages of all the people living in the USA would look like? Explain.
e) Would you expect the mean age or median age of all the people living in the USA to be larger? Explain.
f) In a normal distribution, what is the relationship between the values of the mean, median and the mode?
g) Using the information contained in the histogram, what age would you use to represent the typical age of each class if you wanted to calculate the mean age?
h) Omitting the class "less than 30", calculate the mean, median and modal age (to the nearest whole number). How do these results compare to part (f)?

Using the histogram and omitting the "less than 30" class, approximate the percentage of rich people that are:
i) from 50 to 65 years old and from 65 to 79 years old?
j) from 50 to 79 years old?
k) from 40 to 89 years old?
l) from 30 to 65 years old?
m) from 65 to 99 years old?

n) Omitting the class "less than 30", calculate the standard deviation of the ages of the richest 399 people using part (g). Using this information and assuming the distribution of ages of the richest people is approximately normal, answer the questions of parts (i) to (m).
o) Compare your answers from parts (i) to (m) to the results of part (n). Do you think that it is appropriate to make use of the normal distribution to answer questions regarding the distribution of the wealthiest people? Explain.

What is the probability of selecting a person from the distribution of wealthiest people who is:
p) from 70 to 79 years old?
q) at least 50 years old?
r) at most 69 years old?
s) Calculate the same probabilities using the normal distribution as your model for the distribution of the ages and compare your results to parts (p) to (r).

2. **What Do You Think?**

Examine the two graphs on the following page representing the annual distribution of Mean Monthly Precipitation and Temperature for the years 1921 - 1970 for Dodge City, Kansas and Portland, Oregon.

a) Describe the shape of the bar graph within each graphic?
b) What does the bar graph represent?
c) What does the height of each bar represent?
d) Compare the monthly precipitation for Dodge City, Kansas and Portland, Oregon during the months from May to August. Explain.
e) Describe the shape of the mean monthly temperature graph for Dodge City, Kansas and Portland, Oregon.
f) What was the mean monthly temperature for the month of May for both Dodge City and Portland?
g) What was the median monthly temperature for both Dodge City and Portland?
h) In which month was the greatest mean monthly temperature for both Dodge City and Portland?
i) Which city would be described by wet winters and dry summers? Explain.
j) Which city would have over 60% of the annual precipitation during the months May, June and July?
k) Which city has the driest months in late summer and early fall?

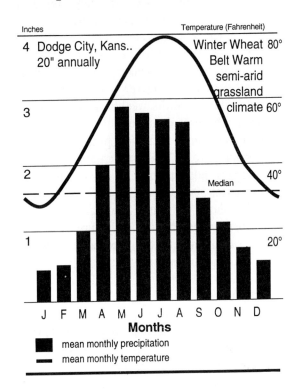

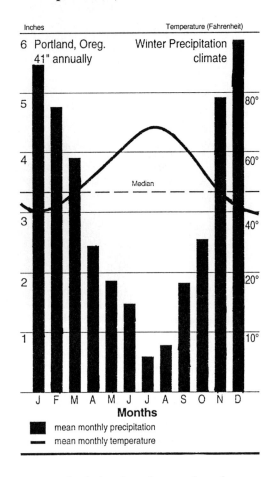

3. **What Do You Think?**

An article entitled *Student Finds Penny IS Tail-Heavy* appeared in The Washington Post on November 27, 1965.

The article gives an account of a high school senior who flipped a penny 17,950 times and determined that the penny was tail-heavy. According to the story, a student flipped a penny 17,950 times and he recorded 464 more heads than tails. From his experiment, he proclaimed that the United States Mint produces tail-heavy pennies.

a) Is this a binomial experiment? If it is, then what are the values of n, p, and (1 − p)?

b) Can you use the normal approximation to the binomial distribution to determine probabilities associated with this experiment?

c) How many heads and tails did Edward get during his experiment?

d) If we define p to be the *probability of getting a head*, then what is the expected mean number of heads for this experiment? How does the mean number of heads compare to the result in part (c)?

e) If a coin is fair, would you expect the number of times a coin lands heads to be equal to the mean number of heads found in part (d)? If not, then how many heads might you expect the coin to deviate from this expected mean number?

f) What is the probability that a coin will land at least 9,207 heads in 17,950 tosses, using the normal approximation?

g) Is this probability small or large? If the event is rare, then would you expect the probability of the event to be small or large? What can you say about this event?

h) Do you think that the coin is fair or biased?

i) The number of heads can be 9,207 or 9,208, but it can't be any value in between. Is the number of heads a discrete or continuous variable? Since the normal distribution represents values between whole numbers, does it represent a discrete or continuous variable? Why are you able to use the normal distribution to approximate the binomial distribution? Do any adjustments have to be made to perform this approximation? If so, what are these adjustments?

j) What did you use as the boundary value to calculate the probability in part (f)? Explain why it is necessary to use a boundary value rather than the actual value of 9,207, when using the normal approximation to the binomial?

PART VI Exploring DATA with MINITAB

1. Use MINITAB to vary the sample size, for n = 10, 100, 200 and 500, to produce samples that are approximately normal with a $\mu = 0$ and $\sigma = 1$. For each sample construct a histogram.
   a) Describe the shape of each histogram.
   b) Does the shape of the histogram change as the sample size increases? Explain.
   c) Repeat parts (a) and (b). Are the results the same as before? Explain.
   d) **Without** using MINITAB, what do you believe would be the shape of the resulting histogram if the sample size was 100,000?

2. Use MINITAB to vary the sample size, for n = 10, 100, 200 and 500, to produce samples that are approximately normal with a $\mu = 500$ and $\sigma = 100$. For each sample construct a histogram.
   a) Describe the shape of each histogram.
   b) Does the shape of the histogram change as the sample size increases? Explain.
   c) Repeat parts (a) and (b). Are the results the same as before? Explain.
   d) For each histogram calculate the percent of data values that are within two standard deviations of the mean.
   e) Compare the results obtained in part (d) to the actual percent of data values that are within two standard deviations of the mean in a normal distribution.
   f) As the sample size is increased, did the percent of data values within two standard deviations of the mean for the samples, get closer to or farther from the percent of data values that are within two standard deviations of the mean for a normal distribution?

3. a) Use MINITAB to produce a binomial distribution along with a table of binomial probability values for the binomial experiment where n = 100 and p = 0.1.
   b) Use the CDF command to calculate $P(s \leq 45)$.
   c) Compare this probability value to the one obtained by using the normal approximation to the binomial distribution. Are these results very close? Did you expect this to happen? Explain.

4. a) Repeat problem 3(a) for n = 200, 300 and 400.
   b) 1) For n = 200 use the CDF command to calculate $P(s \leq 90)$. Compare this probability value to one obtained by using the normal approximation to the binomial distribution.
   2) For n = 300 use the CDF command to calculate $P(s \leq 135)$. Compare this probability value to one obtained by using the normal approximation to the binomial distribution.
   3) For n = 400 use the CDF command to calculate $P(s \leq 180)$. Compare this probability value to one obtained by using the normal approximation to the binomial distribution.
   c) By analyzing the results of parts (a) and (b), write a conclusion regarding the normal approximation to the binomial distribution as n increases?

5. In the casino game roulette the probability of success for the outcome "black" is approximately 0.474. Use MINITAB to generate a table of successes along with the probability of success for choosing "black" and to construct a graph of each distribution for the game being played
   a) n = 10 times b) n = 100 times c) n = 1000 times
   What happens to the probability of success in the game as the number of times you play the game increases?

PART VII Projects.

1. a) Randomly select 36 people and record their heights (or weights).
   b) Construct a frequency table, a histogram and a frequency polygon using your data.
   c) Calculate the mean and standard deviation of your data.
   d) Find the percent of heights (or weights) that are within 1 standard deviation of the mean.
   e) For a normal distribution which has the same mean and standard deviation obtained in part (c) calculate the percent of data values within one standard deviation of the mean.
   f) Compare the results of parts (d) and (e).
   g) Repeat parts (a) through (f), using samples of size 64, 100, and 200.

2. Consider the uniform distribution pictured in the histogram on the following page.
   a) On twenty-one small pieces of paper write the numbers 0 through 20, one number to a piece of paper.
   b) Place these pieces of paper into a container.
   c) Select ten numbers from the container with replacement and record the result of each selection.

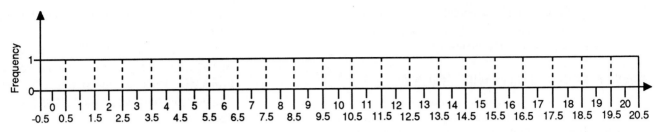

d) Calculate the sum of the ten numbers selected.
e) Repeat steps (c) and (d) until 100 sums have been calculated.
f) Construct a histogram and frequency polygon using an interval width of 5 for the 100 sums.
g) Compare their general shape to that of the normal distribution.

PART VIII Database.

The following exercises refer to the file DATABASE listed in Appendix A. We have indicated the appropriate MINITAB commands that are necessary to answer each exercise.

1. Using the MINITAB commands:
   RETRIEVE, MEAN, STDEV, NORMAL MU= SIGMA=, CDF, SORT, PRINT
   retrieve the file DATABASE.MTW and determine the mean and standard deviation for the variable AVE.
   a) Use the Minitab commands CDF and NORMAL MU = SIGMA = to obtain the probability of getting an average less than 78.
   b) Sort and display the data values for the variable AVE and find the portion of data values below 78.
   c) Compare the results of parts (a) and (b).

2. Using the MINITAB commands:
   RETRIEVE, MEAN, STDEV,
   NORMAL MU= SIGMA= , CDF, SORT, PRINT
   retrieve the file DATABASE.MTW and determine the mean and standard deviation for the variable GPA.
   a) Use the Minitab commands CDF and NORMAL MU = SIGMA = to obtain the probability of getting an average less than 3.00.
   b) Sort and display the data values for the variable GPA and find the portion of data values less than 3.00.
   c) Compare the results of parts (a) and (b).

CHAPTER 8
SAMPLING DISTRIBUTIONS

❄ 8.1 INTRODUCTION

Populations are usually quite large (or infinite) and obtaining all the data values necessary to calculate their mean and standard deviation is impractical. In practice, samples from populations are used to obtain estimates of the population parameters. The population mean, population median and population standard deviation are examples of a **parameter** since they are computed on the basis of the **entire data in the population. Parameters are constant, since they are fixed for a population.** That is, there is only one value for the mean of a given population. However, a sample mean, sample median or sample standard deviation are **statistics** since they are computed on the basis of **sample data** and they may be expected to vary from sample to sample. The distribution of a statistic, like a sample mean, is called a **Sampling Distribution**.

> **Definition: Sampling Distribution.** A sampling distribution is a distribution of all the values of a sample statistic that would occur if all possible random samples of a fixed sample size, n, are selected from a population.

Sampling Distributions are of prime importance in statistical inference. In fact, all statistical inferences are made on the basis of sampling distributions. In this chapter, we will only consider the sampling distribution of the mean. First, let's consider the concept of sampling.

Suppose a consumer group is interested in determining the mean life of a 60 watt light bulb. Obviously it is impractical to test the entire population of 60 watt light bulbs to calculate the *true* population mean life. Therefore a sample from the population is **randomly** selected to obtain an estimate of the population mean life.

The goal of sampling is to obtain a *representative* sample. This is critical, for only then can we expect to make reasonable statistical inferences about a population using information obtained from a sample. ***Random sampling*** is a technique used to try to obtain a representative sample.

> **Definition: Random Sampling.** Random sampling is a technique used to select a sample of size n from a population where *every possible sample of size n has an equal chance of being selected.*

Random sampling plays an important role in formulating statistical inferences about population parameters.

Example 8.1

Determine if each of the following samples have been selected randomly.
a) Consider the population to be females between the ages of twenty and thirty-five living in New York State. From the population we select a sample consisting of 100 college seniors.
b) Consider the population to be all the employees of the H.A.L. microcomputer software company. From the population we select a sample of 50 managerial employees.
c) Consider the population to be the faculty members at a mid-eastern university. From this population we select a sample of 25 faculty members using the following procedure:
   The name of each faculty member is placed in a large drum. The drum is rotated several times before each of the 25 faculty names are selected.

Solution:

a) The sample of 100 college seniors is *not* random because it excluded all the females who are not *college* seniors.
b) The sample of 50 managerial employees is *not* random because it excluded all employees who are not in *managerial* positions.
c) The sample of 25 faculty members *is* random because every possible sample of size 25 had an *equal* chance of being selected. ■

CASE STUDY 8.1

The article entitled *Statisticians sample universe* shown in Figure CS8.1, on the following page, appeared in the magazine *Purchasing* on December 12, 1991. This article discusses the importance of sampling in different applications.

a) Explain some reasons of how sampling is necessary and discuss an application for each reason.
b) Explain two problem areas that one should be aware of when using a sample to make an inference from a population.
c) Discuss an example on how a sample is used to determine whether to accept or reject a sample lot of a product.
d) Why do you think that a 1% chance is used in determining whether to reject the sample lot? Do you think that a 15% chance is practical to use in determining whether to reject the sample lot? Explain.
e) What does the statement "99% confidence that we took the correct action" mean to you?
f) What does the phrase "degree of confidence or accuracy" mean to you? ■

NEWS

FORECAST BASICS

Statisticians sample universe

Here's the fifth in a series titled "Forecast Basics" exploring the mathematical building blocks for strategic planning

Without sampling, statistical analysis and forecasting would be impossible. Quality control people, who, on the basis of limited testing, decide whether to accept or reject a production lot are essentially sampling. So are government price collectors who study a limited number of transactions to come up with the nation's overall price trend. Sampling sometimes yields less-than perfect results, but the practice is indispensable for a variety of reasons:

• Cost: If sampling 5 or 10% of the data can give a reliable estimate of the population or universe, a tremendous saving can be achieved.

• Practicality: It is frequently impossible to collect the universe of data, even if money is available. For example, counting the total number of unemployed or every single price transaction each month is not physically feasible.

• Waste: it doesn't make sense to use data that are more precise than needed. It might make little difference to a soap producer whether 84, 85 or 86% of its customers like the product smell.

• Destruction: A sample is often necessary because the product is destroyed in the process of being tested. Case in point: light bulbs.

• Spurious accuracy: Count the number of marbles in a glass urn several times; the chances are that no two counts will agree. The same accuracy can be attained by counting the number of marbles in one-tenth of the urn (a sample) and then extrapolating the results to the entire urn.

Sampling can offer tremendous advantages, but it is not foolproof. There is no guarantee of 100% accuracy. Two challenges await anyone who samples.

The first is a matter of common sense: **A sample must represent the attributes of a given population or universe.**

A classic example of how a sample can go awry involves a poll conducted more then 50 years ago in the presidential contest between Franklin D. Roosevelt and Alfred M. Landon. The *Literary Digest* randomly samples telephone numbers. Their prediction: Landon by a landslide. The magazine was woefully wrong because its sample didn't accurately represent the electorate; at that time only the relatively wealthy had telephones. Inferring total votes on the basis of the wealthy was unjustified and the magazine was folded soon after the fiasco.

Beyond the problem of building a representative sample — no matter how much time and effort are devoted — dash the possibility remains **that the sample will vary from the true population or universe on the basis of chance.** So sampling results should always be couched in terms of probabilty or odds (see box).

This "chance" concept can be applied to a myriad of industrial sampling problems, for example, statistical quality control. The rationale is straightforward as described in the following steps:

• Measure a sample lot of some product (say, steel sheet) for some attribute like weight, width, or tensile strength.

• Calculate the difference between the sample result and the specifications for the given attribute.

• Calculate the probability of getting such a difference on the basis of pure chance.

Set acceptance-rejection limits. Example: If the probability of obtaining such a difference due to chance is (say 1%) then reject the sample lot, dubbing it "significantly different" from the population spec for the given attribute.

Such a rejection might not be warranted 1% of the time, but we could say with 99% confidence that we took the correct action. The business, in most cases, is willing to accept the 1% risk of making the wrong decision in return for the tremendous potential savings derived from being able to differentiate most of the time between good and bad lots.

Not every sampling procedure requires the same degree of accuracy. A parachute maker would want 100% precision. A bulb maker may settle for considerably less. And risk is not the only determinant of required accuracy. The amount of money riding on a decision is another key consideration.

Probability: A simple sample

Problem: What is the most likely result if you toss two coins into the air? What are the odds of obtaining that result?

Answer. One head and one tail is the most likely result. There is a 50% chance of obtaining that result.

Rationale. The probability of obtaining one head when tossing the first coin is one-half. The odds of getting a head on the second coin is also one-half. The odds of two heads is then calculated by multiplying the odds together 50% x 50% = 25%. If there is a 25% chance of getting two heads, then there is an equal 25% chance of getting two tails. This leaves a 50% probability for one head and one tail.

Figure CS8.1

In the previous chapters we were able to compare an individual data value of a population to the entire population. For example, if a high school principal wants to determine the percentile ranking of *a student* whose SAT score is 650, the principal compares this *individual* score to the population of SAT scores. However suppose the principal wants to determine the percentile rank of the mean SAT score of 480 that represents the 400 senior class members, can she compare this sample mean to the population of individual test scores? No, because the *sample mean* of 480 represents the SAT scores of *four hundred* students while *each score* in the population of SAT scores represents *one student's* SAT score. This is illustrated in Figure 8.1.

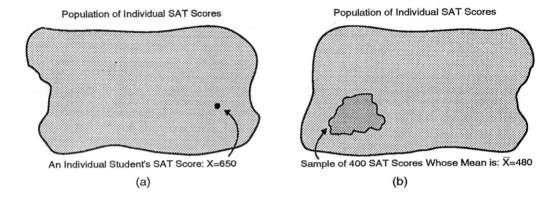

Figure 8.1

Therefore, in order for the principal to make a *valid* comparison, she would have to compare the **sample mean** of the 400 students to a distribution comprised of all possible sample means of size 400. This sampling distribution is referred to as the **Sampling Distribution of The Mean** or **The Distribution of Sample Means**.

❋ 8.2 The Sampling Distribution of The Mean

In this section we are going to discuss the Sampling Distribution of The Mean. This distribution will enable us to make inferences about the population from which the samples were selected.

> **Definition: The Sampling Distribution of The Mean**
> **or**
> **The Distribution of Sample Means.**
> This distribution is made up of the means of all possible random samples of size n selected from a population.

Referring to the SAT problem discussed in Section 8.1 we can see that in order for the principal to determine the percentile rank of the mean SAT score of 480 that represents the 400 senior class members, the principal would have to consider the Sampling Distribution of The Mean. *Each* score of this distribution represents the mean of a sample of 400 SAT scores. This concept is illustrated in Figure 8.2.

In practice the Sampling Distribution of the Mean is seldom constructed because it is extremely difficult or impossible to enumerate all the possible random samples. However, we can explain the essential characteristics of the Sampling Distribution of the Mean through the use of Theorems 8.1 and 8.2.

The first theorem states the relationship between the mean and standard deviation of the *population of individual data values* to the mean and standard deviation of the *Sampling Distribution of the Mean*.

THEOREM 8.1 Mean and Standard Deviation of the Sampling Distribution of The Mean

If all possible random samples of size n are selected from an infinite population with a mean, μ, and a standard deviation, σ, then:

a) The **mean of the Sampling Distribution of the Mean**, denoted $\mu_{\bar{x}}$, is equal to the mean of the population, denoted μ, from which the samples were selected.
This is expressed symbolically as: $\mu_{\bar{x}} = \mu$

b) The **standard deviation of the Sampling Distribution of the Mean**, denoted $\sigma_{\bar{x}}$, is equal to the population standard deviation, σ, divided by the square root of the sample size, n. The formula for $\sigma_{\bar{x}}$ is expressed as[1]:

$$\sigma_{\bar{x}} = \frac{\sigma}{\sqrt{n}}$$

These results are illustrated in Figure 8.2.

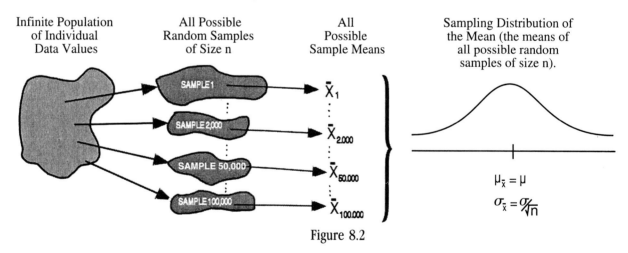

Figure 8.2

---

[1] If we are sampling without replacement from a finite population then the exact formula for $\sigma_{\bar{x}}$ is:

$$\sigma_{\bar{x}} = \frac{\sigma}{\sqrt{n}} \cdot \sqrt{\frac{N-n}{N-1}}$$

where N represents the population size, n represents the sample size and σ represents the population standard deviation. However, if the sample size, n, is less than 5% of the population size, N, then it is appropriate to use $\sigma_{\bar{x}} = \frac{\sigma}{\sqrt{n}}$.

Example 8.2

At a large community college the mean IQ score of the student body is 107 with a standard deviation of 20. Suppose a random sample of size n is selected from the student body, compute the mean, $\mu_{\bar{x}}$, and standard deviation, $\sigma_{\bar{x}}$, of the sampling distribution of the mean, if:
a) n = 4
b) n = 25
c) n = 400

Solution:

a) The mean of the sampling distribution of the mean is determined by the formula: $\mu_{\bar{x}} = \mu$. Since the mean of the population, μ, is 107, then $\mu_{\bar{x}} = 107$.
The standard deviation of the Sampling Distribution of the Mean is determined by the formula:

$$\sigma_{\bar{x}} = \frac{\sigma}{\sqrt{n}}$$

Since the standard deviation of the population, σ, is 20 and the sample size, n, is 4, then:

$$\sigma_{\bar{x}} = \frac{\sigma}{\sqrt{n}}$$

$$\sigma_{\bar{x}} = \frac{20}{\sqrt{4}}$$

$$\sigma_{\bar{x}} = 10$$

b) Since the mean of the population, μ, is 107, then $\mu_{\bar{x}} = 107$. To determine $\sigma_{\bar{x}}$ we use the standard deviation of the population, σ, and the sample size, n. For $\sigma = 20$ and n = 25 we get,

$$\sigma_{\bar{x}} = \frac{\sigma}{\sqrt{n}}$$

$$\sigma_{\bar{x}} = \frac{20}{\sqrt{25}}$$

$$\sigma_{\bar{x}} = 4$$

c) Since μ is 107 then $\mu_{\bar{x}} = 107$. For a sample size of 400 and a population standard deviation, σ, of 20 then:

$$\sigma_{\bar{x}} = \frac{\sigma}{\sqrt{n}}$$

$$\sigma_{\bar{x}} = \frac{20}{\sqrt{400}}$$

$$\sigma_{\bar{x}} = 1 \blacksquare$$

Notice that the mean of the sampling distribution of the mean, $\mu_{\bar{x}}$, is *always equal* to the population mean, μ, **regardless** *of the sample size*. However, the standard deviation of the sampling distribution of the mean, $\sigma_{\bar{x}}$, is *dependent* upon the sample size. That is, as the sample size **increases** the standard deviation of the sampling distribution of the mean, $\sigma_{\bar{x}}$, **decreases**.

Theorem 8.1 enabled us to determine the mean, $\mu_{\bar{x}}$, and standard deviation, $\sigma_{\bar{x}}$, for the sampling distribution of the mean. Additional information about the sampling distribution of the mean is obtained through the use of Theorem 8.2, ***The Central Limit Theorem***.

THEOREM 8.2 The Central Limit Theorem

The Sampling Distribution of the Mean can be closely approximated by a normal distribution if the sample size, n, is large (that is, n is greater than 30).

The Central Limit Theorem is a very powerful theorem because it guarantees that the Sampling Distribution of the Mean is approximately normal as long as the sample size is large enough (greater than 30) *regardless of the shape* of the population distribution from which the samples were selected. In essence, since the **population can have any shape** it is not even necessary to know the shape of the population **provided the sample size is large enough**. Since in practice the population is usually not known this emphasizes the importance of the Central Limit Theorem.

Combining the results of Theorems 8.1 and 8.2 we can state the characteristics of the sampling distribution in Theorem 8.3.

THEOREM 8.3 Characteristics of the Sampling Distribution of the Mean

If the following three conditions are satisfied:

- given **any infinite** population with mean, μ, and standard deviation, σ, and
- all possible random samples of size n are selected from the population to form a Sampling Distribution of the Mean, and
- the sample size, n, is large (greater than 30),

then:

1) the Sampling Distribution of the Mean is **approximately normal**.
2) the mean of the Sampling Distribution of the Mean is equal to the mean of the population.

That is, $\quad \mu_{\bar{x}} = \mu$

3) the standard deviation of the Sampling Distribution of The Mean is equal to standard deviation of the population divided by the square root of the sample size. That is,

$$\sigma_{\bar{x}} = \frac{\sigma}{\sqrt{n}}$$

In many practical situations, the population being sampled is *finite*. However, if the population size, N, is large *relative* to the size, n, of the sample, then it is appropriate to use the standard deviation formula: $\sigma_{\bar{x}} = \dfrac{\sigma}{\sqrt{n}}$.

The rule used to determine whether the population is large in relation to the sample is: **the sample size, n , is less than 5% of the population size, N** . That is, $\dfrac{n}{N} < 5\%$.

In this chapter, we will generally assume the population is either infinite or is large relative to the sample size.

To illustrate Theorem 8.3, consider selecting all possible random samples of *size n* = 100 from each of the populations shown in Figure 8.3. Thus, each of the resulting Sampling Distributions of the Mean will be approximately normal with their respective mean and standard deviation shown in Figure 8.3.

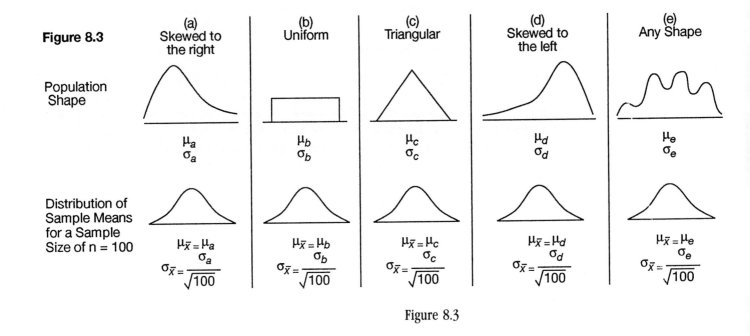

Figure 8.3

The Sampling Distribution of the Mean When the Original Population is Normal

If the original population is either a normal distribution or a large population that is bell-shaped, then the Sampling Distribution of the Mean is normal regardless of the sample size. That is, even samples of size n = 2 would produce a Sampling Distribution of the Mean which is normal.

In the introduction of this chapter, we considered the problem of a high school principal who wanted to determine the percentile rank of the senior class' mean SAT score. Let's use Theorem 8.3 to solve this problem.

Example 8.3

If the principal knows that the mean SAT score of the population of students that took the exam is 470 with a standard deviation of 100, then determine the percentile rank of the senior class of 400 students whose mean SAT score is 480.

Solution:

To solve this problem we must make use of Theorem 8.3 since we need to compare the mean SAT score for a high school class of 400 students to the mean SAT score of all other samples of the *same* size. We can use Theorem 8.3 because the conditions of this theorem have been satisfied. Since:

a) the population mean, μ, is 470.
b) the population standard deviation, σ, is 100.
c) the sample size, n = 400, is considered large since it is greater than 30, then, the Sampling Distribution of the Mean is **approximately normal** with:

$$\mu_{\bar{x}} = \mu$$
$$\mu_{\bar{x}} = 470,$$

and

$$\sigma_{\bar{x}} = \frac{\sigma}{\sqrt{n}}$$

$$\sigma_{\bar{x}} = \frac{100}{\sqrt{400}}$$

$$\sigma_{\bar{x}} = 5$$

Figure 8.4 represents this sampling distribution of the mean.

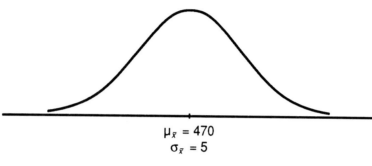

$\mu_{\bar{x}} = 470$
$\sigma_{\bar{x}} = 5$

Figure 8.4

Since the sample mean SAT score of 480 representing the 400 seniors is one score from the Sampling Distribution of the Mean shown in Figure 8.4 we can now use this sampling distribution of the mean which is a normal distribution to determine the percentile rank for the sample mean of 480.

To determine the percentile rank of a data value in a normal distribution we must find the percent of data values which are less than the data value. In particular the percentile rank of the sample mean of 480 is represented by the shaded area illustrated in Figure 8.5.

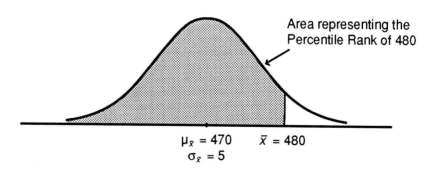

Area representing the Percentile Rank of 480

$\mu_{\bar{x}} = 470$ $\bar{x} = 480$
$\sigma_{\bar{x}} = 5$

Figure 8.5

To find the area to left of 480 we must first calculate z of 480. Since the data values of the distribution are sample means, $\bar{x}$, the z score formula is now symbolized as:

$$z \text{ of } \bar{x} = \frac{\bar{x} - \mu_{\bar{x}}}{\sigma_{\bar{x}}}$$

Thus, $$z \text{ of } 480 = \frac{480 - 470}{5}$$

$$z \text{ of } 480 = 2$$

Using TABLE II, the percent of area to the left of $z = 2$ is 97.72. Therefore, the percentile rank of the sample mean 480 is P_{98}. ∎

Example 8.4

The population mean weight of newborn babies in a western suburb is 7.4 lbs. with a standard deviation of 0.8 lbs. What is the probability that a sample of 64 newborns selected at random will have a mean weight greater than 7.5 lbs.?

Solution:

Since we are dealing with the mean of a sample of 64 newborns, we must compare this sample mean $\bar{x} = 7.5$ lbs. to the sampling distribution of the mean with samples of size 64 selected from the population of all newborn babies of this western suburb. Because the population mean, 7.4, and standard deviation, 0.8, are known and the sample size is greater than 30 we can apply Theorem 8.3. Thus, the sampling distribution of the mean is approximately normal with:

$$\mu_{\bar{x}} = 7.4$$

and

$$\sigma_{\bar{x}} = \frac{0.8}{\sqrt{64}}$$

$$\sigma_{\bar{x}} = 0.1$$

To find the probability that the sample mean will be greater than 7.5 lbs. we must find the percent of area to the *right* of 7.5 lbs. This is illustrated in Figure 8.6.

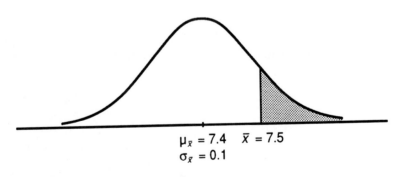

$\mu_{\bar{x}} = 7.4 \quad \bar{x} = 7.5$
$\sigma_{\bar{x}} = 0.1$

Figure 8.6

To determine the shaded area, compute the z score of 7.5:

$$z = \frac{7.5 - 7.4}{0.1} = 1$$

Using TABLE II, we must find the percent of area to the right of $z = 1$. Since the percent of area to the left of $z = 1$ is 84.13, then the percent of area to the right of $z = 1$ is 15.87. Therefore, the probability of randomly selecting a sample of 64 newborns whose mean weight is greater than 7.5 lbs. is 15.87%. ∎

Example 8.5

According to a consumer survey, the mean length of time that American families keep a T.V. set is 48 months with a standard deviation of 9 months. What is the probability that the 36 T.V. sets owned by families in a new condominium complex will be kept between 48 and 51 months?

Solution:

Since the sample size is 36, then the sampling distribution of the mean is approximately normal with:

$$\mu_{\bar{x}} = 48$$

and

$$\sigma_{\bar{x}} = \frac{9}{\sqrt{36}} = 1.5$$

To determine the probability that the sample mean will be between 48 and 51 months, we must find the shaded percent of area illustrated in Figure 8.7.

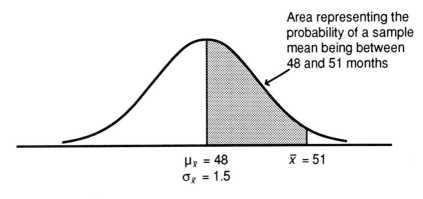

Figure 8.7

To determine this shaded area compute the z scores for 48 and 51 months. Thus,

$$z \text{ of } 48 = \frac{48 - 48}{1.5} \quad \text{and} \quad z \text{ of } 51 = \frac{51 - 48}{1.5}$$

$$z \text{ of } 48 = 0 \quad \text{and} \quad z \text{ of } 51 = 2$$

From Table II, the percent of area to the left of z = 2 is 97.72. While the percent of area to the left of z = 0 is 50.

Thus, the percent of area between z of 48 and z of 51 is found by subtracting the smaller area, 50%, from the larger area, 97.72%, which is 47.72%. Therefore, the probability that a sample mean of 36 T.V. sets will be kept between 48 and 51 months is 47.72%. ∎

CASE STUDY 8.2

Examine the information on the *Average Annual Income For Families With Children* in Figure CS8.2 which appeared in the New York Times on November 9, 1992.

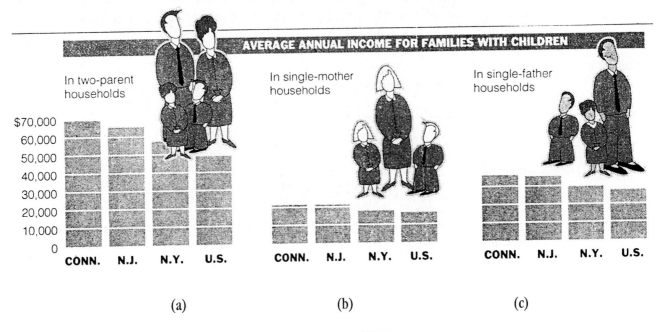

(a)   (b)   (c)

Figure CS8.2

a) Define the population for Figure CS8.2 (a). What do you believe is the shape of the population? Explain.

For the population stated in Figure CS8.1(a):

b) if all possible samples of size 100 are selected from the population of Figure CS8.1(a) and the means are computed for these samples, what distribution shape would you expect for the resulting distribution?

c) what is the name of the resulting distribution?

d) what value would you expect the mean of the resulting distribution to have?

e) if the population standard deviation is σ, then what would be the standard deviation of the distribution containing all the means of the possible samples?

For the population stated in Figure CS8.1(b):

f) define the population and what do you think is the shape of this population? Explain.

g) if all possible samples of size 64 are selected from the population of Figure CS8.1(b) and the means are computed for these samples, what distribution shape would you expect for the resulting distribution?

h) what is the name of the resulting distribution?

i) what value would you expect the mean of the resulting distribution to have?

j) if the population standard deviation is σ, then what would be the standard deviation of the distribution containing all the means of the possible samples?

For the population stated in Figure CS8.1(c):

k) define the population and what do you think is the shape of this population? Explain.

l) if all possible samples of size 36 are selected from the population of Figure CS8.1(c) and the means are computed for these samples, what distribution shape would you expect for the resulting distribution?

m) what is the name of the resulting distribution?

n) what value would you expect the mean of the resulting distribution to have?

o) if the population standard deviation is σ, then what would be the standard deviation of the distribution containing all the means of the possible samples? ■

❄ 8.3 THE EFFECT OF SAMPLE SIZE ON THE SAMPLING DISTRIBUTION OF THE MEAN

Let's examine the effect of an increase in the sample size on the standard deviation of the Sampling Distribution of the Mean.

Example 8.6

Consider the distribution of 5 feet 9 inch American males whose mean weight is 160 lbs. with a standard deviation of 10 lbs. Determine the mean, $\mu_{\bar{x}}$, and standard deviation, $\sigma_{\bar{x}}$, for the Sampling Distribution of the Mean for samples of size 36, 100, 400 and 1600.

Solution:

a) For n = 36, μ = 160, σ = 10, then:

$$\mu_{\bar{x}} = \mu \quad \text{and} \quad \sigma_{\bar{x}} = \frac{\sigma}{\sqrt{n}}$$

$$\mu_{\bar{x}} = 160 \quad \sigma_{\bar{x}} = \frac{10}{\sqrt{36}}$$

$$\sigma_{\bar{x}} = 1.67$$

b) For n = 100, μ = 160, σ = 10, then:

$$\mu_{\bar{x}} = 160 \quad \text{and} \quad \sigma_{\bar{x}} = \frac{10}{\sqrt{100}} = 1$$

c) For n = 400, μ = 160, σ = 10, then:

$$\mu_{\bar{x}} = 160 \quad \text{and} \quad \sigma_{\bar{x}} = \frac{10}{\sqrt{400}} = 0.5$$

d) For n = 1600, μ = 160, σ = 10 then

$$\mu_{\bar{x}} = 160 \quad \text{and} \quad \sigma_{\bar{x}} = \frac{10}{\sqrt{1600}} = 0.25 \quad \blacksquare$$

Using the results of Example 8.5, notice as the *sample size* increases from n = 36 to n = 1600, the *standard deviation* of the Sampling Distribution of the Mean *decreases* from $\sigma_{\bar{x}} = 1.67$ to $\sigma_{\bar{x}} = 0.25$. This relationship is illustrated in Figure 8.8.

(a) n = 36

(b) n = 100

(c) n = 400

(d) n = 1600

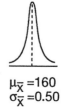

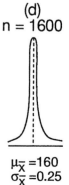

$\mu_{\bar{x}}$ =160
$\sigma_{\bar{x}}$ =1.67

$\mu_{\bar{x}}$ =160
$\sigma_{\bar{x}}$ =1

$\mu_{\bar{x}}$ =160
$\sigma_{\bar{x}}$ =0.50

$\mu_{\bar{x}}$ =160
$\sigma_{\bar{x}}$ =0.25

Figure 8.8

Also, notice that as the *sample size increases* the sample means become more clustered about the mean, $\mu_{\bar{x}}$. Figure 8.9 illustrates that the size of the interval of sample means containing the middle 99% of the sample means *decreases* as the sample size *increases*.

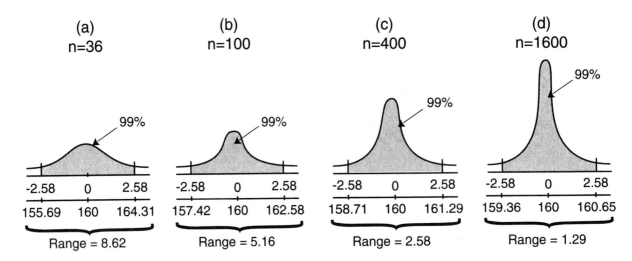

Figure 8.9

For instance when n = 36, 99% of the sample means lie within the interval 155.69 to 164.31. Whereas when n = 1600, 99% of the sample means lie within the *smaller* interval 159.36 to 160.65.

Suppose you didn't know the mean weight for 5 foot 9 inch American males discussed previously and you were going to estimate the population mean by selecting a random sample from the population of American males. Using Figure 8.9, which sample size would you use if you wanted to be 99% confident that the sample mean estimate would be within approximately one pound of the actual population mean?

You should choose a sample size of 1600 since 99% of the sample means are within approximately one pound of the actual population mean. In fact for this sample size, the sample mean is within 0.65 lbs. of the actual population mean. Thus, we can now summarize the effect of the sample size on the standard deviation of the sampling distribution of the mean and the accuracy of using a sample mean to estimate the population mean.

The Effect of the Sample Size on the Standard Deviation of the Sampling Distribution of the Mean

As the sample size increases, the standard deviation of the sampling distribution of the mean decreases. In general, when using a sample mean to estimate a population mean, **as the sample size increases the more confidence you have in the accuracy of the sample mean as an estimate of the population mean.**

❄ 8.4 Using MINITAB

The Central Limit Theorem and its consequences are perhaps some of the most important results in elementary statistics.

An advantage of using statistical software such as MINITAB is that it usually enables us to simulate concepts that would otherwise be difficult to understand. The following set of MINITAB commands will effectively simulate Theorems 8.2 and 8.3 that are stated within

Section 8.2. Let's explore a simulation on The Central Limit Theorem that illustrates its assumptions and consequences.

A set of MINITAB commands to Simulate The Central Limit Theorem

LINE NUMBER
1. STORE 'CLT'
2. RANDOM K1 C2;
3. INTEGER 1 7.
4. LET C3(K2)=MEAN(C2)
5. LET K2=K2+1
6. PRINT K2
7. END
8. LET K1=49
9. LET K2=1
10. NOECHO
11. EXECUTE 'CLT' 200
12. PRINT C3
13. HISTOGRAM C3
14. DESCRIBE C3

The commands have been numbered so that the line numbers can be explained.

Line 1 is used to save (STORE) a sequence of lines (i.e. numbered 2 through 7 in this case) until the END command is executed. Essentially, Lines 2 through 7 is called a *program, named CLT*.

Lines 8 and 9 are used to "initialize" variables within the program. For example, Line 8 sets the number of random data values that are chosen. In this case 49 random data values will be selected. (i.e. K1=49). These 49 values are listed in C2.

Each time the program is run the following will happen.

(1) An INTEGER from 1 to 7 inclusive is randomly selected from a UNIFORM distribution.
(2) Step (1) is repeated 49 times.
(3) Then the mean of the **sample** of size 49 is calculated and listed as the first entry of column C3.

Line 10 is optional. Line 11 enables the program to run this sequence 200 times. Thus, after Line 11 is executed, MINITAB has listed 200 means each representing a sample of size 49 in column C3.

The results of the simulation are then printed (i.e. Line 12). A Histogram is displayed (i.e. Line 13). Summary statistics are displayed using the DESCRIBE command (i.e. Line 14).

When the sequence of all fourteen commands are executed the first thing you see is:

MINITAB OUTPUT FOR THE SIMULATION

K2   2.00000
K2   3.00000
K2   4.00000
K2   5.00000

K2 6.00000
K2 7.00000
K2 8.00000
K2 9.00000
K2 10.0000

.
. [*The lines K2 continue until 200 sample*
. *means are* calculated and listed in column C3]

K2 197.000
K2 198.000
K2 199.000
K2 200.000
K2 201.000

Then the 200 sample means are printed.

MINITAB OUTPUT FOR THE SIMULATION

C3

3.65306	3.79592	3.75510	4.36735	3.81633	4.32653	4.04082
4.73469	4.59184	3.93878	4.02041	3.69388	3.91837	3.87755
4.12245	4.32653	3.26531	4.40816	3.97959	3.53061	4.69388
4.14286	3.85714	4.04082	3.97959	4.22449	3.38776	4.44898
3.79592	3.89796	4.06122	4.30612	3.65306	3.73469	4.44898
3.71429	3.93878	4.16327	3.93878	4.48980	4.48980	3.93878
3.55102	3.91837	3.87755	3.59184	4.38775	4.26531	4.14286
4.00000	4.24490	4.30612	4.16327	3.69388	4.22449	4.08163
4.04082	3.53061	3.75510	4.02041	4.26531	4.04082	3.81633
3.73469	4.48980	3.61224	3.87755	4.04082	4.79592	4.00000
4.00000	3.51020	4.14286	4.12245	4.04082	4.06122	4.10204
3.87755	4.02041	3.93878	4.28571	3.75510	4.10204	3.93878
4.14286	3.97959	4.04082	3.69388	4.34694	3.89796	3.95918
3.73469	3.89796	4.06122	4.26531	4.38775	4.30612	4.12245
3.63265	3.77551	3.65306	3.83673	4.08163	4.32653	3.97959
3.46939	4.12245	4.36735	3.75510	3.57143	3.91837	4.18367
3.73469	3.79592	4.28571	4.40816	4.16327	3.89796	3.59184
4.12245	3.69388	3.97959	4.18367	3.57143	3.42857	4.30612
4.42857	4.51020	4.42857	3.75510	3.79592	3.40816	4.46939
3.51020	3.81633	4.18367	3.77551	3.91837	3.63265	4.18367
4.08163	4.24490	3.67347	4.20408	3.97959	4.14286	3.73469
4.28571	4.14286	4.10204	3.69388	3.40816	4.16327	4.00000
4.00000	4.44898	3.79592	3.83673	4.44898	3.75510	3.77551
3.97959	4.57143	3.73469	3.65306	4.00000	3.93878	4.10204
3.85714	4.08163	3.87755	3.93878	3.97959	4.20408	3.83673
4.04082	3.91837	3.63265	3.67347	3.97959	3.95918	3.69388
3.97959	4.20408	3.85714	3.79592	3.83673	4.20408	4.44898
4.20408	3.97959	3.44898	4.32653	3.83673	4.34694	3.57143
4.34694	4.04082	4.28571	3.67347			

A Histogram of the sample means is displayed.

MINITAB OUTPUT FOR THE SIMULATION

Histogram of C3   N = 200
Each * represents 2 obs.

Midpoint	Count	
3.2	1	*
3.4	6	***
3.6	27	*************
3.8	42	*********************
4.0	51	*************************
4.2	39	********************
4.4	28	**************
4.6	4	**
4.8	2	*

And finally the DESCRIBE command lists the summary statistics for the 200 sample means.

MINITAB OUTPUT FOR THE SIMULATION

	N	MEAN	MEDIAN	TRMEAN	STDEV	SEMEAN
C3	200	3.9966	3.9796	3.9951	0.2895	0.0205

	MIN	MAX	Q1	Q3
C3	3.2653	4.7959	3.7806	4.1990

By examining the histogram you should see that:

(1) the shape of the resulting sampling distribution of the mean is approximately normal.
(2) the summary statistics indicate the mean of the sampling distribution is 3.9966 while the STDEV is 0.2895.

Are these results consistent to the results of Theorems 8.2 and 8.3? Explain.

The population from which the samples were selected is a uniform distribution of whole numbers from 1 to 7. For this population, $\mu=4$ and $\sigma=2$.

According to Theorems 8.2 and 8.3, the distribution of sample means constructed using random samples selected from this uniform population has the following characteristics:

(1) The sampling distribution of the mean is approximately normal.
(2) $\mu_{\bar{x}} = \mu$

$\mu_{\bar{x}} = 4$

(3) $\sigma_{\bar{x}} = \dfrac{\sigma}{\sqrt{n}}$

For samples of size n = 49,

$\sigma_{\bar{x}} = \dfrac{2}{\sqrt{49}}$

$\sigma_{\bar{x}} = 0.2857$

Therefore since the mean for this simulation of 200 sample means is 3.9966 and the estimate of the population standard deviation is 0.2895 you can see that the simulation is indeed very close to the results of the Central Limit Theorem. In practice such simulations are usually carried out many more thousands of times.

Glossary

TERM	SECTION
Sampling Distribution	8.1
Random Sampling	8.1
The Sampling Distribution of the Mean	8.2
The Distribution of Samples Means	8.2
Mean of the Sampling Distribution of the Mean, $\mu_{\bar{x}}$	8.2
Standard Deviation of the Sampling Distribution of The Mean, $\sigma_{\bar{x}}$	8.2
Theorem 8.2: Central Limit Theorem	8.2
Theorem 8.3: Characteristics of the Sampling Distribution of the Mean	8.2
Simulation	8.4

Exercises

PART I Fill in the blanks.

1. The distribution of a statistic is called a _____.
2. A _____ is a portion or part of a population.
3. The goal of sampling is to obtain a _____ sample.
4. Random sampling is a technique used to select a sample of size n from a population in which every possible sample of size n has an _____ chance of being selected.
5. The Sampling Distribution of the Mean is made up of the means of _____ possible random samples of size n selected from a population.
6. If all possible random samples of size n are selected from an infinite population with a mean, μ, and a standard deviation, σ, then:
   a) the mean of the Sampling Distribution of the Mean, denoted $\mu_{\bar{x}}$, is equal to the _____ of the population from which the samples were selected.
   This is symbolized: $\mu_{\bar{x}} = $ _____.
   b) the standard deviation of the Sampling Distribution of the Mean, denoted $\sigma_{\bar{x}}$, is equal to the population standard deviation divided by the square root of the sample _____.
   This is symbolized: $\sigma_{\bar{x}} = $ _____.
7. a) The mean of a sample is symbolized by _____.
   b) The mean of the population is symbolized by _____.
   c) The mean of the Sampling Distribution of the Mean is symbolized by _____.
   d) The standard deviation of the Sampling Distribution of the Mean is symbolized by _____.
   e) The standard deviation of the population is symbolized by _____.
8. For a Sampling Distribution of the Mean, the standard deviation of the Sampling Distribution of the Mean, $\sigma_{\bar{x}}$, is dependent upon the sample _____, n. That is, as the sample size increases, the standard deviation of the Sampling Distribution of the Mean, $\sigma_{\bar{x}}$, _____.
9. The Central Limit Theorem states that the Sampling Distribution of the Mean can be closely approximated by a _____ distribution if the sample size is _____. That is, usually greater than _____. The importance of the Central Limit Theorem is that it guarantees that the Sampling Distribution of the Mean is approximately normal as long as the sample size is large enough regardless of the _____ of the population from which the samples were selected.
10. If all possible random samples of size 100 are selected from an infinite population whose mean is 130 and standard deviation is 10, and we form a new distribution called the Sampling Distribution of the Mean, then we can conclude that:
    a) this Sampling Distribution of the Mean is approximately _____.
    b) the mean of the Sampling Distribution of the Mean is _____, that is, $\mu_{\bar{x}} = $ _____.
    c) the standard deviation of the Sampling Distribution of the Mean is _____, that is, $\sigma_{\bar{x}} = $ _____.
    Result "a" is a consequence of the _____ _____ Theorem.
11. The Sampling Distribution of the Mean is normal for all sample sizes, even as small as n = 2, if the original population, from which the samples were selected, is a _____ distribution.

PART II Multiple choice questions.

1. The symbol used to represent the mean of a sample is a) $\mu_{\bar{x}}$ b) μ c) X d) $\bar{x}$
2. If random samples of size 25 are selected from a population with a mean of 45 and a standard deviation of 15, then the mean of all the sample means will be:
   a) 9 b) 45 c) 60 d) 15
3. If random samples of size 16 are selected from a population with a mean of 80 and a standard deviation of 20, then the standard deviation of the Sampling Distribution of the Mean will be:
   a) 5 b) 16 c) 4 d) 20
4. If random samples of size 10 are selected from a normal distribution, then the Sampling Distribution of the Mean will be:
   a) unknown b) any shape
   c) uniform d) normal
5. If random samples of size 20 are selected from a distribution of unknown shape, then the Sampling Distribution of the Mean will be:
   a) unknown b) any shape
   c) uniform d) normal
6. If random samples of size 50 are selected from a uniform distribution, then the Sampling Distribution of the Mean will be:
   a) unknown b) any shape
   c) uniform d) normal
7. To be certain that the Sampling Distribution of the Mean will be normally distributed, the sample size should be at least:
   a) 25 b) 30 c) 50 d) 100
8. The standard deviation of the Sampling Distribution of the Mean gets smaller as the size of the sample
   a) increases b) remains the same
   c) decreases d) not enough information
9. If the sample size, n, is increased, then the mean of the Sampling Distribution of the Mean:

a) increases   b) remains the same
c) decreases   d) not enough information
10. The symbol used to represent the standard deviation of the Sampling Distribution of the Mean is
a) $\sigma_{\bar{x}}$ b) σ c) s   d) $\bar{x}$

PART III   Answer each statement True or False.

1. Random sampling techniques are very important because of their use in formulating valid generalizations about population parameters.
2. The Sampling Distribution of the Mean is made up of the means of *many* random samples of size n selected from an infinite population.
3. The Sampling Distribution of the Mean must be constructed in order to determine $\mu_{\bar{x}}$ and $\sigma_{\bar{x}}$.
4. In forming a Sampling Distribution of the Mean, the sample size n should vary, but it must be larger than 30.
5. If the population mean, μ, is known then $\mu_{\bar{x}}$ is equal to μ.
6. If all possible random samples of size n are selected from an infinite population with a standard deviation of σ then, the standard deviation of the Sampling Distribution of the Mean, $\sigma_{\bar{x}}$, will equal $\frac{\sigma}{\sqrt{n}}$ only if n is greater than 30.
7. The standard deviation of the Sampling Distribution of the Mean is dependent upon the sample size.
8. As the sample size increases the standard deviation of the Sampling Distribution of the Mean, $\sigma_{\bar{x}}$, also increases.
9. The Central Limit Theorem is important because it states that the Sampling Distribution of the Mean can be approximated by a normal distribution.
10. To apply the Central Limit Theorem the shape of the original population must be known.

PART IV   Problems.

1. Classify each of the following sampling techniques as either random or not random.
    a) A college student is trying to determine an estimate for the mean height of the entire student body. The student interviews and measures 200 students in the math and science building.
    b) A political pollster wants to determine her town's opinion on a tax proposition. She decides to interview 100 registered voters within her neighborhood.
    c) A social worker wants to study the 5000 senior citizens in a Florida retirement community. The social worker decides to select 100 senior citizens by first assigning an identification number, from 0000 to 4999, to each of the 5000 senior citizens. She then spins the number wheels pictured in Figure A to form a four digit number. This is continued until a sample of 100 unique identification numbers has been obtained. These 100 senior citizens will constitute the sample.

Spun for first digit

Spun for remaining digits

Figure A

2. Find the mean, $\mu_{\bar{x}}$, of the Sampling Distribution of the Mean and the standard deviation, $\sigma_{\bar{x}}$, of the Sampling Distribution of the Mean if the population from which the random samples are selected has $\mu = 100$ and $\sigma = 9$ where the sample size, n, is:
a) 9   b) 81   c) 100   d) 400

3. For each of the following parts, you are given the approximate *population* shape, the mean, μ, the standard deviation, σ, and you are required to form a Sampling Distribution of the Mean by selecting all possible random samples of size n.
Determine:
1) the approximate shape (normal or unknown) for the sampling distribution of the mean
2) the mean of the sampling distribution of the mean, $\mu_{\bar{x}}$
3) the standard deviation of the sampling distribution of the mean, $\sigma_{\bar{x}}$.

a) shape: uniform
mean, $\mu = 160$
standard deviation, $\sigma = 8$
sample size, n = 64

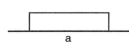
a

b) shape: triangular
mean, $\mu = 16$
standard deviation, $\sigma = 3$
sample size, n = 81

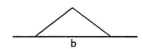
b

c) shape: any shape
mean, μ = 4000
standard deviation,
σ = 200
sample size, n = 25

d) shape: normal
mean, μ = 1200
standard deviation,
σ = 400
sample size, n = 16

e) shape: unknown
μ = 32
σ = 5
n = 100

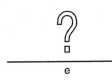

f) shape: bimodal
μ = 500
σ = 100
n = 400

g) shape: any shape
μ = 20,000
σ = 4500
n = 225

h) shape: unknown
μ = 8
σ = 1.6
n = 16

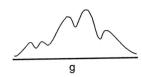

i) shape: normal
μ = 1000
σ = 100
n = 4

j) shaped: skewed
μ = 24,000
σ = 6000
n = 9

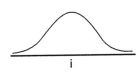

4. The population of IQ scores from a large community college has a mean of 110 with a standard deviation of 10. Find the probability that a random sample of size
   a) 36   b) 81   c) 100
   selected from this population will have a sample mean greater than 113.

5. The mean systolic blood pressure of young female adults is 120 with a standard deviation of 15. Find the probability that a random sample of 100 young female adults will have a mean systolic blood pressure:

a) less than 118.
b) between 116 and 123.
c) greater than 125.

6. The heights of 4000 male athletes have a mean of 72" and a standard deviation of 2". If all possible random samples of size 64 are taken from this population, determine:
   a) the mean and standard deviation of the sampling distribution of the mean.
   b) the percent of sample means that fall between 71.75" and 72.25".
   c) the probability a sample mean is below 71.35".
   d) Below what sample mean can we expect to find the lowest 20% of all the sample means?

7. A sample of size 49 is randomly selected from a population with a mean of 60 and standard deviation of 14. What is the probability that the sample mean is:
   a) less than 64?
   b) greater than 58?
   c) between 56 and 62?

8. The mean age of New York City legal secretaries is 24 years with σ = 3 years. If a sample of 36 secretaries is selected at random, what is the probability that the sample mean age is:
   a) less than 25?
   b) between 24 and 26?
   c) greater than 28?

9. Recently, it was reported that a baby born in 1987 is expected to live to a mean age of 74 years. If the standard deviation for this population is 10 years, use the Central Limit Theorem to answer questions about a Sampling Distribution of the Mean formed by taking from the population of babies born in 1987, all possible random samples of size 100.
   a) Find $\mu_{\bar{x}}$
   b) Determine $\sigma_{\bar{x}}$
   c) Can this Sampling Distribution of the Mean be approximated by a normal distribution?
   d) If a random sample of 100 babies born in 1987 were selected, what is the probability that these babies will live to at least a mean age 72? to at least a mean age 80?
   e) If the population of life expectancies for those born in 1987 is approximately normal, what is the probability that a baby born in 1987 will live to:
   at least age 72?
   at least age 80?
   f) Why are the answers to parts (d) and (e) different?

10. Air travel from JFK New York to Sarasota, Florida takes an average 150 minutes with σ = 8 minutes. Use the Central Limit Theorem to determine the characteristics and to answer questions about a Sampling Distribution of the Mean formed by selecting from this population random samples of size 64.
   a) Find $\mu_{\bar{x}}$.
   b) Determine $\sigma_{\bar{x}}$.
   c) Can this Sampling Distribution of the Mean be approximated by a normal distribution?
   d) What is the probability that a random sample of 64 flights from JFK to Sarasota will average (mean average) more than 10 minutes late?
   e) If the population of air travel time from JFK to Sarasota is normally distributed, what is the probability that a flight from JFK to Sarasota is more than 10 minutes late?
   f) Why are the answers to parts (d) and (e) different?

11. The mean tire life of a certain brand of radial tires is 40,000 miles with a standard deviation of 2000 miles. If the life of this brand of tires is normally distributed then:
   a) what is the probability a tire will last longer than 38,000 miles?
   b) what is the probability that the mean life of a random *sample* of 100 tires is greater than 40,400 miles?
   c) what is the probability that the mean life of a random *sample* of 400 tires is less than 40,200 miles?

12. Last year, the mean batting average of all major league baseball players was .275 with a standard deviation of .018. What is the probability that the mean of a random sample of 36 baseball players is at most .280?

13. The mean weight of newborn babies in a Long Island Community is 7.5 lbs. with a standard deviation of 1.4 lbs. What is the probability that a random sample of 49 babies has a sample mean weight of at least 7.2 lbs?

14. During the month of December, the mean telephone bill for families in a Tucson, Arizona community was $48 with a standard deviation of $9. If a random sample of 36 families is selected at random from this community, what is the probability that the telephone bills for the month of December for the sampled families would have a mean between $45.50 and $50.50?

15. A manufacturer of automobile batteries states that their "slow die" battery has a mean life of 50 months with a standard deviation of 6 months. If a consumer protection group randomly samples 49 of these batteries, what is the probability that the mean life of the consumer group's sample will be at most 52 months (assuming the manufacturer's claim is true)?

16. The Environmental Protection Agency's (E.P.A.) rating of an automobile manufacturer's Z car is 27 m.p.g. with a standard deviation of 6 m.p.g.. Suppose the Miami Police Department has randomly purchased a sample of 36 Z cars from this manufacturer. If the police officers have driven these cars exclusively in the city and have recorded a mean average of 24 m.p.g., then:
   a) what are the chances of getting a random sample mean of at most 24 m.p.g.?
   b) would you question the E.P.A.'s rating for this model car based upon this sample result?

17. A light bulb manufacturer states that its energy saver 60 watt light bulb has a mean life of 1850 hours with a standard deviation of 240 hours. If random samples of size 64 of these light bulbs are selected and tested, then within what interval (centered about the mean) would you expect 95% of these sample means to fall?

18. The mean IQ score of a tribe of American Indians is 110 with a standard deviation of 14. If the distribution of sample means is formed for random samples of size 49 between what two sample means centered at the mean will:
   a) 90% of all sample means lie?
   b) 95% of all sample means lie?
   c) 99% of all sample means lie?

19. The production characteristics for a small engine part are as follows: mean is 2.36", standard deviation is 0.04" and normally distributed.
   a) What percent of these parts fail to meet the engineering specifications of 2.35" plus or minus 0.1"?
   b) If sixteen of these parts are selected at random, what is the probability that the average length is less than 2.35"?
   c) A quality control procedure requires that a random sample of size 4 be selected and that the sample mean be in the interval 2.32" to 2.38". If not, then the case of parts is declared defective. What percent of these samples will fail to meet this test?

20. If the mean length of the parts made by an old cutting machine is 6.74 inches with a standard deviation of 0.12 inches then:
   a) what is the probability that a random sample of 36 of these parts will have a mean length greater than 6.75 inches?
   b) an acceptance sampling procedure requires that the mean length for a random sample of 144 of these parts be between 6.725 inches and 6.755 inches. What is the probability of the sample mean not being in this interval?

PART V What Do You Think?

1. **What Do You Think?**

   The following article entitled *Order of the Draft Drawing* which appeared in the New York Times shows the order in which birth dates were selected during the 1969 draft lottery. On December 1, 1969 the lottery was held in the Selective Service headquarters in Washington, D.C. to determine the draft status of all 19 year-old males for the Vietnam Conflict. Each of the 366 possible birth dates was inserted within a small capsule. The capsules were placed in a stationary drum beginning with January, then February and so on until all the birth dates of December were placed in the drum last. The first capsule drawn would have the highest draft priority, and so on.

   ### Order of the Draft Drawing

   Special to The New York Times

   WASHINGTON, Dec. 1—Following is the order in which birth dates were drawn tonight in the draft lottery:

#	Date	#	Date	#	Date	#	Date
1	Sept. 14	35	May 7	69	June 13	101	Jan. 5
2	April 24	36	Aug. 24	70	Dec. 22	102	Aug. 15
3	Dec. 30	37	May 11	71	Sept. 10	103	May 30
4	Feb. 14	38	Oct. 30	72	Oct. 12	104	June 19
5	Oct. 18	39	Dec. 11	73	June 17	105	Dec. 8
6	Sept. 6	40	May 3	74	April 27	106	Aug. 9
7	Oct. 26	41	Dec. 10	75	May 19	107	Nov. 16
8	Sept. 7	42	July 13	76	Nov. 6	108	March 1
9	Nov. 22	43	Dec. 9	77	Jan. 28	109	June 23
10	Dec. 6	44	Aug. 16	78	Dec. 27	110	June 6
11	Aug. 31	45	Aug. 2	79	Oct. 31	111	Aug. 1
12	Dec. 7	46	Nov. 11	80	Nov. 9	112	May 17
13	July 8	47	Nov. 27	81	April 4	113	Sept. 15
14	April 11	48	Aug. 8	82	Sept. 5	114	Aug. 6
15	July 12	49	Sept. 3	83	April 3	115	July 3
16	Dec. 29	50	July 7	84	Dec. 25	116	Aug. 23
17	Jan. 15	51	Nov. 7	85	June 7	117	Oct. 22
18	Sept. 26	52	Jan. 25	86	Feb. 1	118	Jan. 23
19	Nov. 1	53	Dec. 22	87	Oct. 6	119	Sept. 23
20	June 4	54	Aug. 5	88	July 28	120	July 16
21	Aug. 10	55	May 16	89	Feb. 15	121	Jan. 16
22	June 26	56	Dec. 5	90	April 18	122	Mar. 7
23	July 24	57	Feb. 23	91	Feb. 7	123	Dec. 28
24	Oct. 5	58	Jan. 19	92	Jan. 26	124	April 13
25	Feb. 19	59	Jan. 24	93	July 1	125	Oct. 2
26	Dec. 14	60	June 21	94	Oct. 28	126	Nov. 13
27	July 21	61	Aug. 29	95	Dec. 24	127	Nov. 14
28	June 5	62	April 21	96	Dec. 16	128	Dec. 18
29	Mar. 2	63	Sept. 20	97	Nov. 8	129	Dec. 1
30	Mar. 31	64	June 27	98	July 17	130	May 15
31	May 24	65	May 10	99	Nov. 29	131	Nov. 15
32	April 1	66	Nov. 12	100	Dec. 31	132	Nov. 25
33	Mar. 17	67	July 25				
34	Nov. 2	68	Feb. 12				

   Continued on Page 20, Column 4

#	Date	#	Date	#	Date	#	Date
133	May 12	195	Sept. 24	256	Mar. 23	313	May 31
134	June 11	196	Oct. 24	257	Sept. 28	314	Dec. 12
135	Dec. 20	197	May 9	258	Mar. 24	315	Sept. 30
136	Mar. 11	198	Aug. 14	259	Mar. 13	316	April 22
137	June 25	199	Jan. 8	260	April 17	317	Mar. 9
138	Oct. 13	200	Mar. 19	261	Aug. 3	318	Jan. 13
139	Mar. 6	201	Oct. 23	262	April 28	319	May 23
140	Jan. 18	202	Oct. 4	263	Sept. 9	320	Dec. 15
141	Aug. 18	203	Nov. 19	264	Oct. 27	321	May 8
142	Aug. 12	204	Sept. 21	265	Mar. 22	322	July 15
143	Nov. 17	205	Feb. 27	266	Nov. 4	323	Mar. 10
144	Feb. 2	206	June 10	267	Mar. 3	324	Aug. 11
145	Aug. 4	207	Sept. 16	268	Mar. 27	325	Jan. 10
146	Nov. 18	208	April 30	269	April 5	326	May 22
147	April 7	209	June 30	270	July 29	327	July 6
148	April 16	210	Feb. 4	271	April 2	328	Dec. 2
149	Sept. 25	211	Jan. 31	272	June 12	329	Jan. 11
150	Feb. 11	212	Feb. 16	273	April 15	330	May 1
151	Sept. 29	213	Mar. 8	274	June 16	331	July 14
152	Feb. 13	214	Feb. 5	275	Mar. 4	332	Mar. 18
153	July 22	215	Jan. 4	276	May 4	333	Aug. 30
154	Aug. 17	216	Feb. 10	277	July 9	334	Mar. 21
155	May 6	217	Mar. 30	278	May 18	335	June 9
156	Nov. 21	218	April 10	279	July 4	336	April 19
157	Dec. 3	219	April 9	280	Jan. 20	337	Jan. 22
158	Sept. 11	220	Oct. 10	281	Nov. 28	338	Feb. 9
159	Jan. 2	221	Jan. 12	282	Nov. 10	339	Aug. 22
160	Sept. 22	222	Jan. 28	283	Oct. 8	340	April 26
161	Sept. 2	223	Mar. 28	284	July 10	341	June 18
162	Dec. 23	224	Jan. 6	285	Feb. 29	342	Oct. 9
163	Dec. 13	225	Sept. 1	286	Aug. 25	343	Mar. 25
164	Jan. 30	226	May 29	287	July 30	344	Aug. 26
165	Dec. 4	227	July 19	288	Oct. 17	345	April 29
166	Mar. 16	228	June 2	289	July 27	346	April 12
167	Aug. 28	229	Oct. 29	290	Feb. 22	347	Feb. 6
168	Aug. 7	230	Nov. 24	291	Aug. 21	348	Nov. 3
169	Mar. 15	231	April 14	292	Feb. 18	349	Jan. 29
170	Mar. 26	232	Sept. 4	293	Mar. 5	350	July 2
171	Oct. 15	233	Sept. 27	294	Oct. 14	351	April 25
172	July 23	234	Oct. 7	295	May 13	352	Aug. 27
173	Dec. 26	235	Jan. 17	296	May 27	353	June 29
174	Nov. 30	236	Feb. 24	297	Feb. 3	354	Mar. 14
175	Sept. 13	237	Oct. 11	298	May 2	355	Jan. 27
176	Oct. 25	238	Jan. 14	299	Feb. 28	356	June 14
177	Sept. 19	239	Mar. 20	300	Mar. 12	357	May 26
178	May 14	240	Dec. 19	301	June 3	358	June 24
179	Feb. 25	241	Oct. 19	302	Feb. 20	359	Oct. 1
180	June 15	242	Sept. 12	303	July 26	360	June 20
181	Feb. 8	243	Oct. 21	304	Dec. 17	361	May 25
182	Nov. 23	244	Oct. 3	305	Jan. 1	362	Mar. 29
183	May 20	245	Aug. 26	306	Jan. 7	363	Feb. 21
184	Sept. 8	246	Sept. 18	307	Aug. 13	364	May 5
185	Nov. 20	247	June 22	308	May 28	365	Feb. 26
186	Jan. 21	248	July 11	309	Nov. 26	366	June 8
187	July 20	249	June 1	310	Nov. 5		
188	July 5	250	May 21	311	Aug. 19		
189	Feb. 17	251	Jan. 3	312	April 8		
190	July 18	252	April 23				
191	April 29	253	April 6				
192	Oct. 20	254	Oct. 16				
193	July 31	255	Sept. 17				
194	Jan. 9						

   Following is the order of the alphabet to be applied to the first letter of last names in determining the order of call for inductees with the same birth dates:
   J, G, D, X, N, O, Z, T, W, P, Q, Y, U, C, F, I, K, H, S, L, M, A, R, E, B, V

   Theoretically, if the capsules were adequately mixed, then the resulting draft order should have been random with respect to each month. For example, a birth date in December should have the same chance of being selected early in the draft as a birth date for any other month, such as January. Examine the list of birth dates within the previous article which represents the order of the 1969 draft drawing.

a) For the first 122 birth dates selected in the 1969 draft lottery, determine the frequency for each birth month.
b) If the selection process represented a random selection, then determine the expected frequency per month for the first 122 selections.
c) Repeat parts (a) and (b) for all the birth dates.
d) Based on the results of the first three parts, do you believe this draft lottery represented a random selection? Explain.

What do you think might have contributed to this type of result?

2. **What Do You Think?**

Each of the following statements represent an excerpt from different studies which indicate how the sample for the study was selected. For each sampling procedure, indicate whether you think the procedure to select the sample result is biased or unbiased. Explain the reason for your answer, in some cases you may wish to answer "not enough information given". In such cases, explain why.

a) "A market research survey was mailed to a random sample of 250 owners of new automobiles. A total of 85 completed surveys was received yielding a return rate of 34%."
b) "Forty-six rats, all coming from one of several litters, were separated from their mothers when they were three weeks old. One half (Group A) was put into individual cages; the other half (Group B) was left together to see what effect a rat's early environment has on its behavior later in life."
c) "A total of 356 persons were interviewed, 160 men and 196 women, to test whether men have nightmares as often as women."
d) "The subjects in this sample were 90 children, ranging in age from 6 to 12, who were selected from a list of students attending an elementary school. Three children were selected at random from each of the classes within the school."
e) "A telephone survey was conducted to determine current attitudes toward birth control education in the public high schools. The respondents were 1,345 men and women of voting age; all were white"
f) "A total of 199 persons, all complaining of frequent headaches, agreed to participate in a study to compare the effectiveness of four different headache remedies. The participants were divided into four groups (I, II, III, and IV)."

3. **What Do You Think?**

The article entitled *Poll Involved Queries to 1,540* describes the procedure used to select adults for a New York Times/CBS News Public Opinion Poll.

Poll Involved Queries to 1,540

The latest New York Times/CBS News Poll is based on telephone interviews conducted from Jan. 11 through Jan. 15 with 1,540 adults around the United States.

The sample of telephone exchanges called was selected by a computer from a complete list of exchanges in the country. The exchanges were chosen in such a way as to insure that each region of the country was represented in proportion to its population. For each exchange, the telephone numbers were formed by random digits, thus permitting access to both listed and unlisted residential numbers.

The results have been weighted to take account of household size and to adjust for variations in the sample relating to region, race, sex, age and education.

In theory, it can be said that in 95 cases out of 100 the results based on the entire sample differ by no more than 3 percentage points in either direction from what would have been obtained by interviewing all adult Americans. The error for smaller subgroups is larger, depending on the number of sample cases in the subgroup.

The theoretical errors do not take into account a margin of additional error resulting from the various practical difficulties in taking any survey of public opinion.

Read the procedure that was used to conduct this public opinion poll.
a) Define the population for this poll.
b) What do you think are some problems with selecting a random sample of adults for a national poll using a telephone interview?
c) What does the statement "the exchanges were chosen in such a way as to insure that each region of the country was represented in proportion to its population" mean? What is the purpose of doing this?
d) What does the statement "the results have been weighted to take account of household size and to adjust for variations in the sample mean relating to region, race, sex, age and education"? What do you believe is the purpose for doing this?
e) How would you translate the statement "95 cases out of 100 the results based on the entire sample differ by no more than 3 percentage points in either direction from what would be obtained by interviewing all adult Americans"?

PART VI Exploring DATA with MINITAB

1. Using the "SIMULATION OF THE CENTRAL LIMIT THEOREM" sequence of commands outlined in Section 8.4 replace line 8 with
    LET K1=9
   and execute the simulation.
   (a) What is the effect of this change?
   (b) For this simulation the population parameters are $\mu=4$ and $\sigma=2$. What should be the values of $\mu_{\bar{x}}$ and $\sigma_{\bar{x}}$ if the conditions of The Central Limit Theorem are satisfied?
   (c) Compare the results you obtain with those listed in Section 8.4. Explain.

2. Using the "SIMULATION OF THE CENTRAL LIMIT THEOREM" sequence of commands outlined in Section 8.4 replace line 8 with
    LET K1=81
   and execute the simulation.
   (a) What is the effect of this change.
   (b) For this simulation the population parameters are $\mu=4$ and $\sigma=2$. What should be the values of $\mu_{\bar{x}}$ and $\sigma_{\bar{x}}$ if the conditions of The Central Limit Theorem are satisfied?
   (c) Compare the results you obtain with those listed in Section 8.4. Explain.

3. The command XBARCHART will construct sample mean charts. The XBARCHART is a special chart known as a *statistical process control chart (SPC)*. It is used to graphically display a distribution of sample means for a given sample size n. The plot of the sample means yields (1) a center line, representing an estimate of the average value for the population mean, (2) an upper control limit (UCL) drawn three standard deviations above the center line, and (3) a lower control limit (LCL) drawn three standard deviations below the center line. Any sample mean falling outside the UCL or LCL would be considered very unusual. The sample corresponding to a sample mean falling outside the UCL or LCL should be examined.
   (a) The following sequence of MINITAB commands will construct XBARCHARTs for samples of size:
   MTB > RANDOM n C1;
   SUBC > INTEGER 0 10.
   MTB > DESCRIBE C1
   (1) n=4
   (2) n=16
   (3) n=25
   (4) n=100.
   (b) Examine the sample mean charts to determine if there are any samples that are considered very unusual.
   (c) Compare the sample mean charts to determine if there are relationships between the variability of the sample means and the sample size. Explain.

PART VII Projects.

1. a) Take a set of 50 blank plastic chips and write the following numbers on the chips:

Number on chips	Frequency
10	6
9	5
8	6
7	4
6	7
5	7
4	4
3	5
2	5
1	1

   b) Construct a histogram for the above distribution using a class width of 1.
   c) Calculate:
      1) the mean (μ) of the population.
      2) the standard deviation (σ) of the population.
   d) Place the 50 chips in a box. Shake the box rigorously and randomly choose 20 samples

of size 9 with replacement. (Be sure to shake the box each time before a new sample is chosen) For *each* sample chosen, calculate the sample mean, $\bar{x}$, and indicate the results in the table below.

Sample	$\bar{x}$
1	
2	
3	
.	
.	
.	
20	

e) Using the 20 sample means, construct a histogram and frequency polygon.

f) Calculate the mean of all the sample means, $\mu_{\bar{x}}$, and the standard deviation of the sample means, $\sigma_{\bar{x}}$. Place these values in the following table.

g) Calculate the expected mean of the sample means, and the expected standard deviation of the sample means using the formulas:

$$\mu_{\bar{x}} = \mu, \text{ and } \sigma_{\bar{x}} = \frac{\sigma}{\sqrt{n}}.$$

Place these values in the following table.

Computed Values	Expected Values
$\mu_{\bar{x}} =$	$\mu_{\bar{x}} = \mu$
$\sigma_{\bar{x}} =$	$\sigma_{\bar{x}} = \frac{\sigma}{\sqrt{n}}$

h) Compare the computed values to the expected values of $\mu_{\bar{x}}$ and $\sigma_{\bar{x}}$.

i) Repeat the above procedure for twenty random samples of size 25 with replacement. Compare and interpret the results of both experiments.

2. Theorem 8.1 deals with the formulation of $\mu_{\bar{x}}$ and $\sigma_{\bar{x}}$ given the mean, μ, and standard deviation, σ, of a population and the selection of all possible random samples of size n.
Given the population: 1,2,3,4,5.
a) Calculate μ and σ
b) Form a distribution by selecting *with replacement* all samples of size n = 2.
c) Calculate the mean of each sample found in part (b).
d) Form a Sampling Distribution of the Mean from the sample means found in part (c).
e) Find the mean and standard deviation of the sample means found in part (c). (i.e. find $\mu_{\bar{x}}$ and $\sigma_{\bar{x}}$)
f) Form a distribution by selecting *without replacement* all samples of size n = 2.
g) Calculate the mean of each sample found in part (f).
h) Form a Sampling Distribution of the Mean from the sample means found in part (g).
i) Find the mean and standard deviation of the sample means found in part (g).
j) Compute the theoretical values for $\mu_{\bar{x}}$ and $\sigma_{\bar{x}}$ using Theorem 8.1.
k) Compare the results of parts (e) and (j).
l) Compare the results of parts (i) and (j). Notice that the results of parts (i) and (j) will not be equal unless the finite correction factor given at the bottom of Theorem 8.1 is applied.

PART VIII Database.

The following exercises refer to the file DATABASE listed in Appendix A. We have indicated the appropriate MINITAB commands that are necessary to answer each exercise.

1. Using the MINITAB commands:
RETRIEVE, SAMPLE, and DESCRIBE
Retrieve the file DATABASE and use the SAMPLE command to select a random sample of 20 student records for the variables:
a) AGE
b) AVE
c) STY
d) CUM
e) GFT
For each variable use the DESCRIBE command to compare the selected random sample to the original data set.

2. Using the MINITAB commands:
RETRIEVE, SAMPLE, MEAN, SET, DESCRIBE and PRINT
Retrieve the file DATABASE and display a table representing 10 random samples of data, where each sample contains 20 data values, randomly selected from the data values for the variables:
a) AGE b) HS c) WK
For each variable the table should include columns representing the 10 different random samples of size 20.

Using the MEAN command determine and record the mean of each random sample. Form a distribution of the 10 sample means by using the SET command.

Determine the mean and standard deviation of the *distribution of sample means*.

Compare these results to the original mean and standard deviation for each of the variables by explaining the *differences* in the statistics in light of the theorems and properties discussed in this chapter.

CHAPTER 9
INTRODUCTION TO HYPOTHESIS TESTING

❄ 9.1 INTRODUCTION

Hypothesis testing is an important statistical decision-making tool that has applications in a variety of fields such as education, economics, medicine, psychology, marketing, sociology, management, etc. The following represent some typical situations in which a decision is required and where the hypothesis testing procedure can be used to help arrive at this decision:

- A new drug is being tested to determine its effectiveness and safety. These results will be compared to those of a drug currently in use. Based on the comparison of the test results, a decision will be made to either market the new drug or abandon it.
- The registrar at a local college is considering a new registration procedure. The effectiveness of the new procedure is to be determined by comparing data collected under the existing system with the data collected using the new registration system. The registrar's decision as to which procedure to use for their registration system will depend upon the analysis of the test results.
- A new assembly procedure is being tested to determine its effectiveness in reducing the number of defective finished products. These test results will be compared to the number of defective products under the present assembly procedure. Based on the comparison of the test results, a decision will be made to either implement the new assembly procedure or to retain the present procedure.
- The developer of a special reflective coating for windows claims that when it is applied to the windows of a house it will reduce air conditioning costs by at least forty percent. Before a local manufacturer will agree to purchase the manufacturing and marketing rights of this product a test will be conducted to determine if this product is as effective as claimed by its developer. Based on the test results the manufacturer will decide if it should market the product.
- Many companies use acceptance sampling to determine if the quality of parts received from suppliers meet their specifications. Before the company accepts the supplier's shipment of parts a sample of the parts is selected from the shipment and tested to determine if the parts meet the specifications. Based on the analysis of the test results, the company will either accept or reject the shipment.

9.2 Hypothesis Testing

Hypothesis testing is a statistical procedure. In many ways, the hypothesis testing procedure is similar to the U.S. judicial process. In our development of hypothesis testing we will draw an analogy to the judicial process to provide a rationale and better understanding of the procedure used in the statistical testing of a hypothesis.

Within the judicial system, once an individual has been accused of committing a crime, a trial is conducted to determine if there is enough evidence to convict the individual. Throughout the trial, the *fundamental assumption* is that the individual is innocent. However, it is the objective of the prosecutor to present enough evidence to convince the jury to reject the assumption of innocence and find the individual guilty as charged.

In hypothesis testing the statistician is confronted with a similar problem. The statistician begins by formulating a fundamental assumption referred to as the **null hypothesis**. Similar to the judicial system where it is the prosecutor's task to present evidence to cause the rejection of the assumption of innocence, in the hypothesis testing procedure it is the objective of the statistician to present data (evidence) to cause the rejection of the null hypothesis.

Thus, the only way the prosecutor can obtain the objective of a guilty verdict is by convincing the jury to *reject* the assumption of innocence. This last statement is the essence of hypothesis testing.

In the hypothesis testing procedure, the statistician formulates *two hypotheses:* the **null hypothesis** and an **alternative hypothesis**. The alternative hypothesis is stated in such a way that by *rejecting the null hypothesis* the statistician is able to support his belief, the alternative hypothesis. The null hypothesis is symbolized by H_o and the alternative hypothesis is symbolized by H_a. Let's now define these two hypotheses.

Definition: Null Hypothesis, symbolized by H_o. The null hypothesis is the **tested** hypothesis that is initially **assumed to be true**. The null hypothesis is a statement indicating **no change** in the status quo or **no difference** and is formulated for the sole purpose of trying to reject it.

Definition: Alternative Hypothesis, symbolized by H_a. The alternative hypothesis is the hypothesis that is suspected to be true and is supported by the rejection of the null hypothesis.

Example 9.1

For **each** of the following four problems state the null hypothesis and the alternative hypothesis.
- *A recent poll indicated that 60% of the U.S. population believe that the Space Shuttle program should be continued. A statistician believes that more than 60% of the U.S. population believe the Space Shuttle program should be continued.*

Solution:

Since the statistician wants to *show* that **more than 60 percent** believe the Space Shuttle program should be continued this is the *Alternative* Hypothesis, H_a. In order to support the alternative hypothesis, H_a, the statistician must *test* the hypothesis that **60 percent** believe the Space Shuttle program should be continued. This is the *Null* Hypothesis, H_o. ∎

- *The school newspaper claims that full-time college students work an average of 20 hours a week at a part-time job. A marketing professor feels that students work less than 20 hours per week.*

Solution:

Since the professor wants to show that full-time college students work **less than an average of 20 hours per week** this is the *Alternative* Hypothesis, H_a. The hypothesis to be tested is that full-time college students work **an average of 20 hours per week**. This is the *Null* hypothesis, H_o. ∎

- *In a recent issue of a sociology journal, it was stated that 50 percent of all marriages will end in divorce. A sociologist believes that this claim is too low.*

Solution:

Since the sociologist believes the claim is too low, the sociologist would like to show that the divorce rate is **greater than 50 percent**. This is the *Alternative* Hypothesis, H_a. The *Null* Hypothesis to be tested is that **50 percent of all marriages end in divorce**, since this is the hypothesis to be tested. ∎

- *Ten years ago, the average age of a college freshman was 18 years. Today a college admissions officer believes that the average age of a college freshman has increased.*

Solution:

Since the admissions officer believes that the average age of a college freshman has increased, she wants to show that the average is **greater than 18 years**. This is the *Alternative Hypothesis*, H_a. The *Null* Hypothesis, H_o, is the **average age is 18 years**, since this is the hypothesis to be tested. ∎

In the four problems of **Example 9.1**, each **alternative** hypothesis contained words as: more than, less than, too low, or has increased. Alternative hypotheses that contain such words are called *directional* alternative hypotheses.

> **Definition: Directional Alternative Hypothesis.** A directional alternative hypothesis is an alternative hypothesis that considers **only one specified direction of difference** away from the value specified in the null hypothesis.
> A directional alternative hypothesis is stated using words equivalent to: **less than** or **greater than**.

Example 9.2

For *each* of the following two problems state the null and alternative hypotheses.

- *A new surgical procedure used to reverse vasectomies is claimed to be 80 percent effective. A family planning group decides to test this claim.*

Solution:

The hypothesis to be tested is that *the new surgical procedure is 80 percent effective*. This is referred to as the Null Hypothesis, H_o.

Since the family planning group has **not** indicated a particular belief or feeling about the claim such as, it is too high or it is too low, *both* of these *alternatives* must be considered when stating the alternative hypothesis. This is accomplished by stating the alternative hypothesis, H_a, as: the new surgical procedure is *not* 80 percent effective.

When we need to consider **both** directions for the alternative hypothesis, that is, when phrases such as: too high, too low, more than, less than, etc. are not explicitly used or implied then the alternative hypothesis is a ***nondirectional*** alternative hypothesis. ∎

Definition: Nondirectional Alternative Hypothesis. A nondirectional alternative hypothesis is an alternative hypothesis that considers **both directions** away from the value specified in the null hypothesis.

A **nondirectional** alternative hypothesis is stated using phrases equivalent to: **is not**, or **is not equal to**.

• *A new environmentally safe organic fertilizer developed by a botanist is claimed to have no effect on the average yield of 50 vegetables per plant. The U.S. Agriculture Department wants to test the claim that the new fertilizer has no effect on the average yield of 50 vegetables per plant.*

Solution:

The hypothesis to be tested is: *the new fertilizer has no effect on the average yield of 50 vegetables per plant.* This is referred to as the *Null* Hypothesis, H_0. Since the Agriculture Department would like to test whether the average yield **has been** effected (i.e., decreased or increased), both possibilities must be considered when stating the alternative hypothesis. This is done by stating the alternative hypothesis as: **the average yield per plant is *not* 50 vegetables.** ∎

After the hypotheses are formulated the statistician's task is to design a procedure to test the null hypothesis. We will outline the statistician's procedure and show how it is similar to the procedure followed in the judicial system. This statistical procedure consists of five steps and is referred to as the **hypothesis testing procedure.** Let's now outline this procedure and compare it to the judicial system procedure.

Hypothesis Testing Procedure	Analogy of the Hypothesis Testing Procedure to the Judicial System
Step 1 Formulate the hypotheses. State the null hypothesis, H_o, and alternative hypothesis, H_a.	**Step 1** H_o: individual is innocent. H_a: individual is guilty.
Step 2 Design an experiment to test the null hypothesis. In this design the statistician identifies an appropriate distribution to serve as a model and calculates its mean and standard deviation. The mean and standard deviation values are referred to as the **expected results** of the experiment and are computed based upon the **assumption that the null hypothesis is true**.	**Step 2** The design is how the trial will be conducted. This includes the selection of a jury and the strategies formulated by both the defense attorney and the prosecutor. **This trial is based upon the assumption that the accused is presumed to be innocent** (i.e. the null hypothesis is assumed to be true).
Step 3 Formulate a decision rule. A decision rule is a statement formulated by the statistician which defines the criteria necessary for the rejection of the null hypothesis.	**Step 3** Each juror, in his/her own mind, will determine how much evidence is required to convict the individual. (i.e. reject the assumption of innocence)
Step 4 Conduct the experiment. Now that the experiment has been designed the statistician collects the necessary data to calculate the **experimental outcome**.	**Step 4** The trial is conducted. The prosecutor and the defense attorney present their side of the case.
Step 5 Determine the conclusion. The statistician compares the experimental outcome to the decision rule and determines the conclusion. The conclusion is either: a) **Reject the Null Hypothesis, H_o** OR b) **Fail to Reject the Null Hypothesis, H_o.**	**Step 5** The jury evaluates the evidence presented and renders their verdict. This is either: a) **Individual is guilty** (that is, **reject the null hypothesis**) OR b) **Individual is not guilty** (that is, **fail to reject the null hypothesis**)

When the statistician follows the hypothesis testing procedure an important question to be considered is: **will this procedure always lead to the correct conclusion?** Unfortunately, the answer is no. Let's examine why by looking at the judicial system. The possible outcomes of a trial are shown in Table 9.1

Table 9.1
POSSIBLE OUTCOMES OF A TRIAL

VERDICT:	TRUTH:	
	NOT GUILTY	*GUILTY*
GUILTY	WRONG DECISION	CORRECT DECISION
NOT GUILTY	CORRECT DECISION	WRONG DECISION

By examining Table 9.1 we see that there are four possible outcomes. In two of the outcomes the correct decision is rendered but in the other two outcomes the wrong decision is reached. These wrong decisions are: to convict an innocent individual, and to let a guilty individual go free. A good procedure tries to *minimize* the probabilities of making both types of wrong decisions.

In hypothesis testing the statistician can make similar incorrect decisions. These outcomes are summarized in Table 9.2.

Table 9.2
POSSIBLE OUTCOMES TO A STATISTICAL HYPOTHESIS TEST

CONCLUSION:	REALITY:	
	H_o IS TRUE	H_o IS FALSE
REJECT H_o	TYPE I ERROR	CORRECT DECISION
FAIL TO REJECT H_o	CORRECT DECISION	TYPE II ERROR

In examining Table 9.2 notice that there are two types of errors that a statistician can make. They are referred to as *TYPE I ERROR* and *TYPE II ERROR*.

> **Definition: TYPE I ERROR.** A TYPE I ERROR is the error that is made when a **true null hypothesis is incorrectly rejected**.

In the Judicial System a TYPE I ERROR is equivalent to convicting an innocent person.

> **Definition: TYPE II ERROR.** A TYPE II ERROR is the error that is made when a **false null hypothesis is not rejected**.

A TYPE II ERROR in the judicial system is equivalent to letting a guilty person go free. In the **Hypothesis Testing Procedure** the statistician will try to minimize the effects of TYPE I and TYPE II ERRORS. These techniques will be discussed in Section 9.3.

Now, we will illustrate the hypothesis testing technique through the following four examples.

Remark: **In order to concentrate on the concepts of hypothesis testing, the calculations for the mean, standard deviation and distribution model usually calculated in step 2 and the calculations for the decision rule(s) usually calculated in step 3 are deliberately omitted. You will NOT be responsible for these calculations in this chapter. However, these calculations will be discussed and included in subsequent chapters.**

Example 9.3

Let's use the hypothesis testing procedure to determine if a coin is biased.

Solution:

Step 1. *Formulate the hypotheses*
Null Hypothesis, H_o: coin is fair.
Alternative hypothesis, H_a: coin is biased.

Step 2. *Design an experiment to test the null hypothesis.*

To test the null hypothesis we will toss the coin 100 times. *Assuming the null hypothesis is true* we expect a fair coin to land heads 50 times out of 100 tosses with a standard deviation of 5. The distribution of all possible outcomes is approximately normal and these expected results are shown in Figure 9.1.

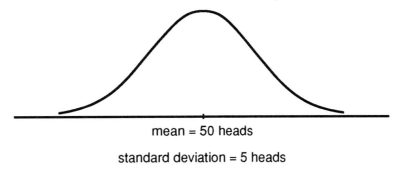

mean = 50 heads

standard deviation = 5 heads

Figure 9.1

Step 3. *Formulate the decision rule.*

Before one can formulate a decision rule, it is necessary to define a criteria for the rejection of the null hypothesis. This criteria will **always** be based upon the assumption that the null hypothesis is **true**.

Based upon the assumption that the coin is fair, one would expect the coin to land heads about *50* times out of 100 tosses. On the other hand, if the coin were to land heads, say, more than 90 times or fewer than 10 times then this might be considered an *unusual* outcome and could lead one to conclude that the coin is biased.

Suppose we then choose the decision rule to be: *reject the null hypothesis*, H_o, if, after 100 tosses of the coin, the coin lands heads either *more than 90 times or less than 10 times*.

This decision rule is illustrated in Figure 9.2.

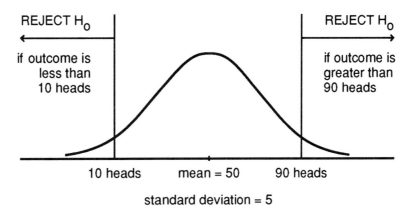

Figure 9.2

Step 4. *Conduct the experiment.*

To conduct the experiment, the coin is tossed 100 times and the number of heads is recorded. This result is referred to as the **experimental outcome**.

To illustrate the conclusions that are possible let's consider three different experimental outcomes. State a conclusion for each of the following experimental outcomes of tossing the coin:

a) *92 heads.*
b) *7 heads.*
c) *75 heads.*

Step 5. *Determine the conclusion.*

Compare the experimental outcome to the *decision rule* and determine the conclusion.

a) For an experimental outcome of 92 heads, the conclusion is to reject the null hypothesis since 92 heads is greater than 90 heads. This is illustrated in Figure 9.3.

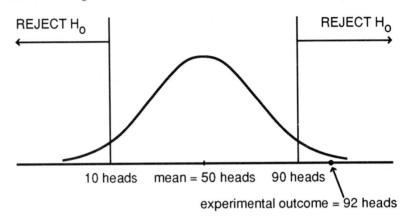

Figure 9.3

b) For an experimental outcome of 7 heads, the conclusion is to *reject* the null hypothesis since 7 heads is fewer than 10 heads. This is illustrated in Figure 9.4.

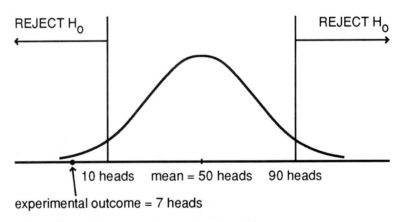

Figure 9.4

In parts (a) and (b) the rejection of the null hypothesis means that one does *not* agree with the assumption that the coin is fair. Therefore, this leads us to conclusion that we believe the coin is biased. Thus, in statistical terms, a *rejection* of the null hypothesis leads to the **acceptance** of the alternative hypothesis. Furthermore, a rejection of the null hypothesis means that the *difference* between the experimental outcome and the expected outcome (the mean) is *significant* enough to support the alternative hypothesis. We can then say that this difference is **statistically significant**.

c) For an experimental outcome of 75 heads, the conclusion is *not* to reject the null hypothesis or, in other words, to *fail to reject* the null hypothesis since 75 heads is not greater than 90 or less than 10 heads. This is illustrated in Figure 9.5.

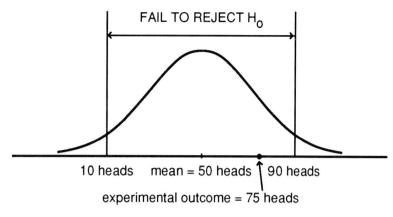

Figure 9.5

When the experimental outcome does *not* fall within the rejection region of the decision rule, the conclusion is stated as: *fail to reject* the null hypothesis.

In part (c), the interpretation of the conclusion (i.e. fail to reject the null hypothesis) means that, according to the decision rule, the experimental outcome of 75 heads is *not* significantly different from the expected mean number of 50 heads. Therefore, although there is a difference, this difference is **not statistically significant** (not large enough) to conclude that the coin is biased.

In general, when the statistician *fails* to reject the null hypothesis, this means that the difference between the experimental outcome and the expected outcome (the mean) is *not statistically significant.* ∎

Let us use the hypothesis testing procedure for the following example.

Example 9.4

Use the 5-Step hypothesis testing procedure to outline a hypothesis test for the following problem.

An automotive engineer claims to have developed a new computerized fuel injection system which will increase gas mileage. Before an auto manufacturer will invest in the computerized fuel injection system the engineer must show that this fuel injection system will **significantly increase** gas mileage over the gas mileage obtained using the standard fuel injection system. Using the standard fuel injection system the L model cars get an average MPG of 31.6 with a standard deviation of 3.

Solution:

Step 1. *Formulate the hypotheses.*

The hypothesis to be tested is that the new computerized fuel injection system is only as effective as the standard fuel injection system which yields an average MPG of 31.6. Therefore, the null hypothesis is:

H_o: Mean MPG for the new computerized fuel system **is** 31.6

Since the engineer wants to show that this computerized fuel injection system will *increase* gas mileage, the alternative hypothesis is:

H_a: Mean MPG for the new computerized fuel system **is greater than** 31.6

Step 2. *Design an experiment to test the null hypothesis.*
(Note: **As indicated by the remark prior to Example 9.3 all calculations necessary to obtain the mean, standard deviation, model distribution and decision rule(s) are deliberately omitted to simplify the presentation).**

To test the null hypothesis the engineer equips 36 of the Model L cars with the new computerized fuel injection systems. Assuming the null hypothesis is true implies that the new computerized fuel injection system is only as good as the standard fuel injection system.

Therefore the expected mean mpg for the distribution of samples of 36 cars is 31.6 with a standard deviation of 0.5. The distribution for all possible outcomes is approximately normal and these expected results are shown in Figure 9.6.

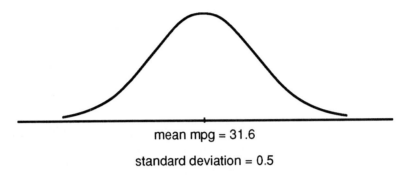

Figure 9.6

Step 3. *Formulate the decision rule.*

If the auto manufacturer stipulates that in order to accept the new computerized fuel injection system the sample of 36 cars must average more than 32.8 mpg, then, based upon the manufacturer's criterion, our **decision rule is stated as: reject the null hypothesis (H_o) if the sample mean is greater than 32.8 mpg.**

This decision rule is illustrated in Figure 9.7.

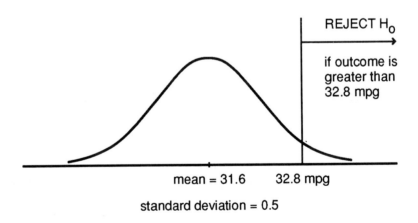

Figure 9.7

Step 4. *Conduct the experiment.*

The 36 cars equipped with the computerized fuel injection system are tested and their sample mean mpg is computed.

State a conclusion for each of the following experimental outcomes:
a) 34 mpg
b) 32 mpg
c) 33.4 mpg

Step 5. *Determine the conclusion.*

For each part (a), (b) and (c), compare the experimental outcome to the decision rule and determine the conclusion.

   a) For an experimental outcome of 34 mpg the conclusion is to reject the null hypothesis since the outcome of 34 mpg is greater than our decision rule of 32.8 mpg. This is illustrated in Figure 9.8.

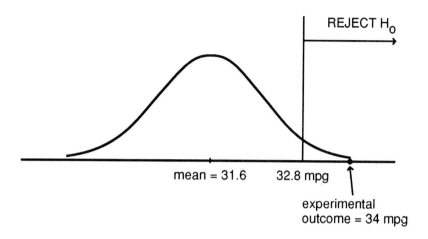

Figure 9.8

Our conclusion is to **reject H_o**, and **accept the alternative hypothesis, H_a,** that the mean mpg is greater than 31.6. Because the experimental outcome of 34 mpg is significantly greater than the expected mean of 31.6 mpg, the auto manufacturer will invest in the computerized fuel injection system.

   b) For an experimental outcome of 32 mpg the conclusion is to **fail to reject H_o** since the outcome of 32 mpg is less than the decision rule of 32.8 mpg. This illustrated in Figure 9.9.

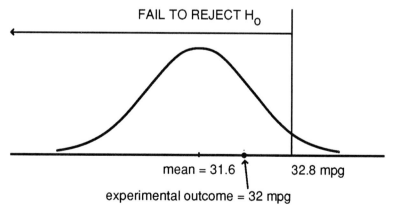

Figure 9.9

Our conclusion is to **fail to reject H_o since the experimental outcome of 32 mpg is not significantly greater than the expected mean of 31.6 mpg**. Therefore the auto manufacturer *will not* invest in the computerized fuel injection system.

   c) For the experimental outcome of 33.4 mpg the conclusion is to *reject H_o* since the outcome of 33.4 is greater than the decision rule of 32.8 mpg. This illustrated in Figure 9.10.

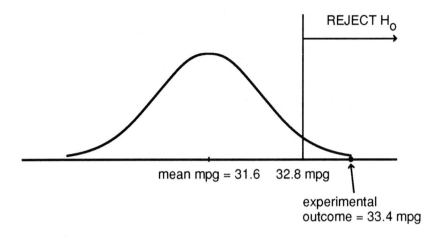

Figure 9.10

Our conclusion is to reject H_o, which means we *accept* the alternative hypothesis that the mean mpg is greater than 31.6. Since the **experimental outcome of 33.4 mpg is significantly greater than the expected mean of 31.6 mpg** the auto manufacturer *will* invest in the computerized fuel injection system. ■

Example 9.5

Use the 5-Step hypothesis testing procedure to outline a hypothesis test for the following problem.

A video cassette recorder manufacturer claims that 90% of its video cassette recorders (VCR's) operate three years before repairs are necessary. A consumer group feels that the 90% figure is too high and decides to test the claim.

Solution:

Step 1. *Formulate the hypotheses.*

The claim that 90% of the manufacturer's VCR's operate three years before needing repair represents the status quo, therefore the null hypothesis is:

H_o: 90% of the VCR's operate three years before needing repair.

The consumer group wants to show that the 90% figure is too high, therefore the alternative hypothesis is:

H_a: *less than 90%* of the VCR's will operate three years before needing repair.

Step 2. *Design an experiment to test the null hypothesis.*

To test the null hypothesis the consumer group randomly selects 100 people who have owned one of the manufacturer's video recorders for at least three years. Assuming the *null* hypothesis to be true we expect that 90 of the 100 video recorders should have lasted at least three years before needing repair with a standard deviation of 3.
(**Remember all calculations necessary to obtain the mean, standard deviation and decision rule are omitted. The student should concentrate on the concepts of the hypothesis testing procedure.**)

The distribution of all possible outcomes is approximately normal and these expected results are shown in Figure 9.11.

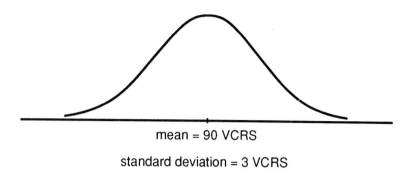

mean = 90 VCRS

standard deviation = 3 VCRS

Figure 9.11

Step 3. *Formulate the decision rule.*

Based upon the assumption that the manufacturer's claim is correct one would expect 90 of the video recorders to have lasted at least three years before needing repair. Taking this into consideration the consumer group formulates its decision rule to be:

Reject the null hypothesis if less than 85 of the 100 video recorders have operated at least three years before needing repair. This is illustrated in Figure 9.12.

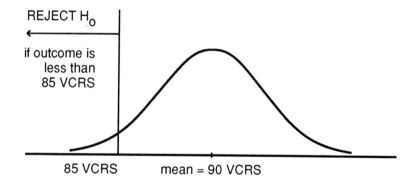

Figure 9.12

Step 4. *Conduct the experiment.*

The 100 selected people are interviewed to determine how long their VCR's operated before needing repair.

State your conclusion if the outcome of the experiment is:

a) *84 VCR's operated at least three years before needing repair.*
b) *87 VCR's operated at least three years before needing repair.*
c) *92 VCR's operated at least three years before needing repair.*

Step 5. *Determine the conclusion.*

For experimental outcomes (a), (b) and (c) compare the experimental outcome to the decision rule and determine the conclusion.

a) For an experimental outcome of 84 VCR's the conclusion is to *reject* the null hypothesis since the outcome of 84 VCR's is less than our decision rule of 85 VCR's. This is illustrated in Figure 9.13

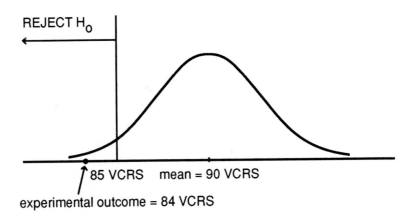

Figure 9.13

Rejecting the null hypothesis means that we feel that the manufacturer's claim of 90% is *incorrect*. Therefore we accept the alternative hypothesis since the experimental outcome of 84 VCR's is *significantly* less than the expected result of 90 VCR's.

b) For an experimental outcome of 87 VCR's the conclusion is *fail to reject H_o* since the outcome of 87 VCR's is greater than our decision rule of 85 VCR's. This is illustrated in Figure 9.14.

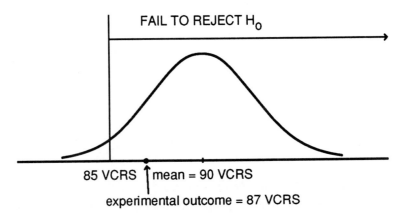

Figure 9.14

Our conclusion is *fail to reject H_o* since the experimental outcome of 87 VCR's is *not* significantly less than the expected mean of 90 VCR's.

c) For an experimental outcome of 92 VCR's the conclusion is *fail to reject H_o* since the outcome of 92 VCR's is greater than our decision rule of 85 VCR's. This is illustrated in Figure 9.15.

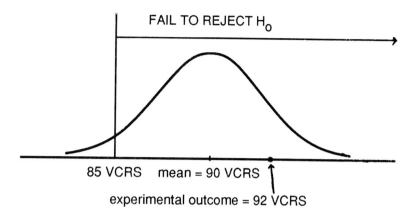

Figure 9.15

In this case the experimental outcome is not only greater than the decision rule (85 VCR's) but it is also **greater than** the expected mean of 90 VCR's. Consequently our conclusion is *fail to reject* H_o which means that the manufacturer's claim is supported. ■

Example 9.6

Use the 5-Step hypothesis testing procedure to outline a hypothesis test for the following problem.

The Clean Chemical Company has been granted permission from the Environmental Protection Agency (EPA) to return its waste water to a nearby river if the pollutant level of a certain toxic chemical is 17 parts per million (ppm) or less with a standard deviation of 2ppm. The EPA periodically tests the waste water to determine if the pollutant level is exceeding the allowable level. If the pollutant level is found to exceed the allowable level, permission to dispose of their waste water will be suspended until the company can adjust its purification equipment. The EPA takes thirty samples of the waste water over a two day period to determine if the company is meeting the allowable toxic level in waste water.

Solution:

Step 1. *Formulate the Hypotheses.*

Since the company can dispose of its waste water in the river only if the pollutant level is 17 ppm or less the null hypothesis is:

H_o: mean pollutant level is 17 ppm.

If the pollutant level goes above 17 ppm, permission to dispose of the waste water in the river will be suspended. Therefore the alternative hypothesis is:

H_a: mean pollutant level is *greater than* 17 ppm.

Step 2. *Design an experiment to test the null hypothesis.*

To test the null hypothesis the EPA takes thirty samples of the waste water over a two day period. Assuming the null hypothesis is true implies that the expected mean pollutant level for the distribution of samples of size 30 is 17 ppm with a standard deviation of 0.37 ppm. The distribution for all possible outcomes is approximately normal.

These expected results are shown in Figure 9.16.

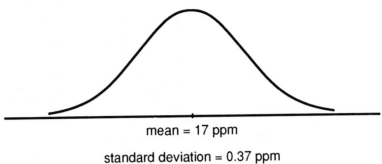

mean = 17 ppm

standard deviation = 0.37 ppm

Figure 9.16

Step 3. *Formulate the decision rule.*

If the EPA's test results exceed 17.86 ppm, then the waste water disposal permit will be suspended until such time as the purification equipment is functioning properly. Based on this criterion the decision rule will be to reject the null hypothesis if the sample mean is greater than 17.86 ppm. This decision rule is illustrated in Figure 9.17.

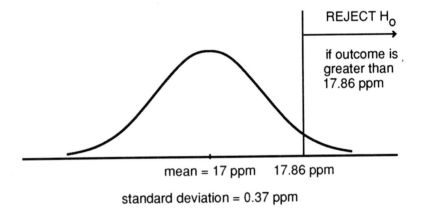

Figure 9.17

Step 4. *Conduct the experiment.*

The thirty samples of waste water are collected over a two day period and tested to determine their toxic level.

State a conclusion for each of the following experimental outcomes:
a) *17.91 ppm.*
b) *17.63 ppm.*
c) *16.94 ppm.*

Step 5. *Determine the conclusion.*

Compare the experimental outcomes for parts (a), (b) and (c) to the decision rule, and determine the conclusion for each part.

a) For an experimental outcome of 17.91 ppm the conclusion is to *reject the null hypothesis* since the experimental outcome (17.91 ppm) is greater than the decision rule of 17.86 ppm.

This is illustrated in Figure 9.18.

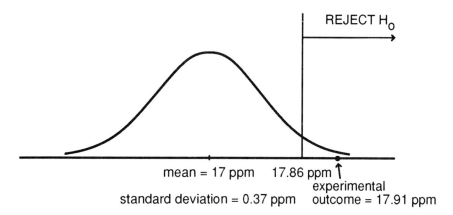

Figure 9.18

The conclusion is reject H_o which means we accept the alternative hypothesis that the mean ppm is greater than 17. This means that the *EPA will suspend the chemical company's permit for disposing of waste water into the river until the purification equipment is adjusted.*

b) For an experimental outcome of 17.63 ppm the conclusion is *fail to reject H_o* since the outcome of 17.63 ppm is less than the decision rule of 17.86 ppm. This is illustrated in Figure 9.19.

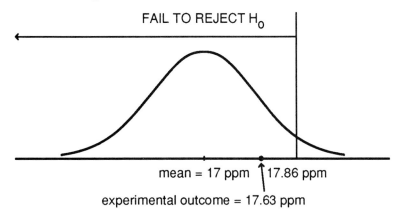

Figure 9.19

The conclusion is fail to reject H_o since the experimental outcome of 17.63 ppm is *not significantly* greater than the expected mean of 17 ppm. Therefore *the EPA will allow the Clean Chemical Company to continue to dispose of its waste water into the river.*

c) For an experimental outcome of 16.94 ppm the conclusion is *fail to reject H_o* since the outcome of 16.94 ppm is *less* than the decision rule of 17.86 ppm. This is illustrated in Figure 9.20.

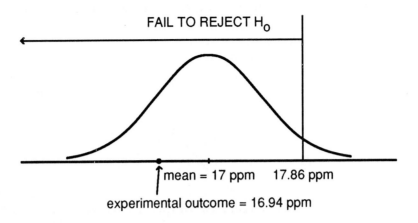

Figure 9.20

In this case the experimental outcome is not only less than the decision rule (17.86 ppm) but it is also *less* than the expected mean of 17 ppm. Consequently the conclusion is fail to reject H_o which means that *the Clean Chemical Company is meeting the EPA's allowable toxicity in its waste water.* ∎

❄ 9.3 THE DEVELOPMENT OF A DECISION RULE

The objective of a hypothesis test is to use sample data, referred to as the experimental outcome, to decide if one should reject or fail to reject the null hypothesis. Statisticians often use a predetermined decision rule to decide whether the sample data supports or refutes the null hypothesis. That is, the statistician compares the sample data to the value(s) of the decision rule, referred to as the ***critical value(s)***, to decide whether to reject H_o or fail to reject H_o. Let's examine this idea of the decision rule further through the use of Figure 9.21.

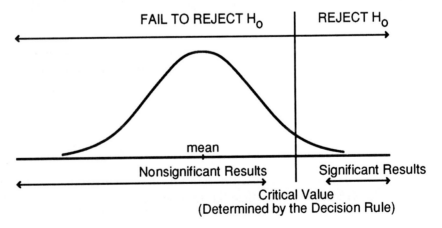

Figure 9.21

According to the decision rule pictured in Figure 9.21 one *rejects* the null hypothesis if the experimental outcome is *greater than* the critical value. One *fails to reject* the null hypothesis if the experimental outcome is *less than* the critical value. Any experimental outcome falling *beyond* the critical value is referred to as a ***significant*** result because it leads to the *rejection* of the null hypothesis. For the decision rule illustrated in Figure 9.21, any experimental outcome less than the critical value is referred to as a nonsignificant result since it leads one to fail to reject the *null hypothesis*.

The area to the right of the critical value is referred to as the ***level of significance***, since any experimental outcome falling within this area is considered *significantly* greater than the expected outcome (the mean) and will lead to a rejection of the null hypothesis. Let's now define the level of significance.

Definition: Level of Significance *(definition one)*
The level of significance is the probability of an experimental outcome falling beyond the critical value, assuming that the null hypothesis is true. The level of significance is represented by the lower-case Greek letter α, pronounced alpha.

This is illustrated in Figure 9.22.

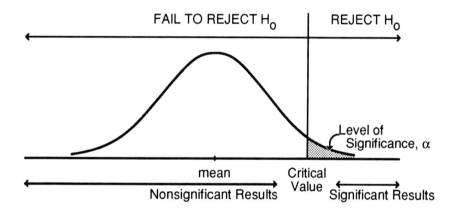

Figure 9.22

Example 9.7

Using the decision rule from Example 9.6, which is illustrated in Figure 9.23, determine the level of significance, α.

Solution:

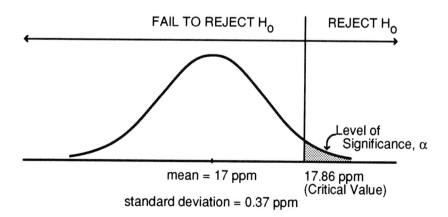

Figure 9.23

By computing the **z score** of the critical value, 17.86, and using **TABLE II: The Normal Curve Area Table in Appendix D: Statistical Tables**, we have:

$$z = \frac{17.86 - 17}{0.37} = 2.32$$

From TABLE II, the area to the left of z = 2.32 is 98.98%.

Thus, the area to the right of z = 2.32 is 1.02%.

Therefore, **the level of significance is approximately 1%, symbolized** $\alpha \approx 1\%$. ∎

Example 9.8

Using the decision rule from Example 9.5, which is illustrated in Figure 9.24, determine the level of significance, α.

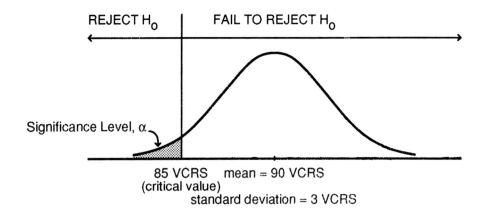

Figure 9.24

Solution:

For this decision rule, the level of significance, α, is the area to the left of the critical value of 85 VCR's because any experimental outcome which falls below 85 VCR's is considered a significant result. Since this distribution can be approximated by the normal distribution we can determine the level of significance by computing the z score of 84.5 (the lower boundary of 85 is used for binomial outcomes, see Section 7.7) and by using TABLE II: The Normal Curve Area Table.

$$z = \frac{84.5 - 90}{3} = -1.83$$

Using TABLE II the area to the left of z = -1.83 is 3.36%. Therefore the level of significance is 3.36%, written $\alpha \approx 3.36\%$. ∎

Example 9.9

Matthew decides to test his lucky $5.00 gold coin to determine if it is biased. He decides to test the null hypothesis, the coin is fair, by tossing the coin 100 times. His decision rule is to reject the null hypothesis if he obtains more than 60 heads or less than 40 heads in the 100 tosses of the coin. This decision rule is illustrated in Figure 9.25. What is the significance level for this decision rule?

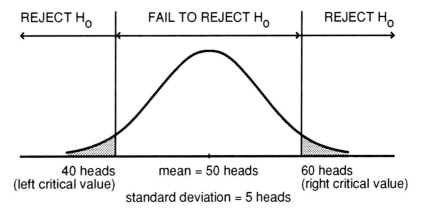

Figure 9.25

Solution:

From the diagram in Figure 9.25, the decision rule has two areas which must be considered in the evaluation of the significance level. These areas are: the area that exceeds the right critical value of 60 heads, and the area that is less than the left critical value of 40 heads. If the experimental outcome falls into **either** one of the areas then the experimental result is **considered significant enough to reject the null hypothesis**.

The level of significance, α, is the *sum* of the two shaded areas in Figure 9.25. Using the normal approximation to the binomial distribution (Section 7.7) we can determine the level of significance by computing the z scores of 39.5 and 60.5 (the boundary values of 40 and 60 are used for binomial outcomes, see Section 7.7) and by using TABLE II: The Normal Curve Area Table.

The z score of 39.5 is:

$$z = \frac{39.5 - 50}{5} = -2.10$$

From TABLE II, the area to the left of z = -2.10 is 1.79%. Calculating the z score of 60.5, we get:

$$z = \frac{60.5 - 50}{5} = 2.10$$

From TABLE II, the area to the left of z = 2.10 is 98.21%, therefore the area to the right of z = 2.10 is 1.79%. (see Figure 9.26)

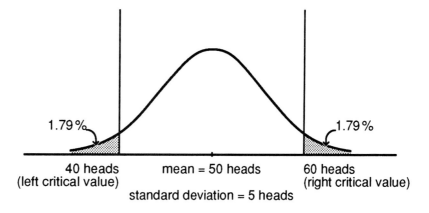

Figure 9.26

Since the significance level, α, represents the two shaded areas illustrated in Figure 9.26, then α is equal to the sum of these two shaded areas. Thus, we have:

α = the area to the left of 40 heads plus the area to the right of 60 heads

α = 1.79% + 1.79%

α = 3.58% ∎

Let's take another look at TYPE I and TYPE II errors and their relationship to the significance level, α. Recall that a TYPE I error is made when a true null hypothesis is incorrectly rejected. Therefore *whenever* the null hypothesis is rejected the possibility of a TYPE I error occurs. Consequently, a TYPE I error will occur whenever the null hypothesis is true *and* an experimental outcome falls beyond the critical value, leading to the rejection of the true null hypothesis. This is illustrated in Figure 9.27.

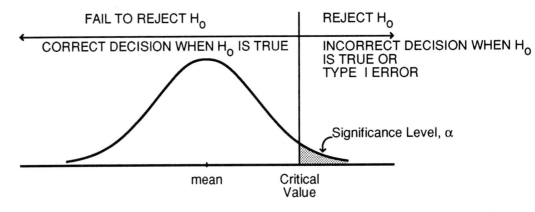

Figure 9.27

As can be seen from the diagram in Figure 9.27, when the null hypothesis is true the probability of committing a TYPE I error equals the level of significance, α.

Thus an alternative definition of the level of significance can be defined as follows:

Definition: Level of significance, α. *(definition two)*
The level of significance is the probability of committing a TYPE I error or, the probability of rejecting a true null hypothesis.

The decision rule determines the level of significance.

Thus, by choosing different decision rules the level of significance can be increased or decreased. Since the level of significance is the probability of committing a TYPE I error one might ask: "why not choose a decision rule that will have the *smallest* possible significance level?" Example 9.10 addresses this question.

Example 9.10

Suppose an experiment is conducted to determine if a coin is biased. If the coin is tossed 100 times consider the following three decision rules that are illustrated in Figures 9.28, 9.29 and 9.30, and its effect on TYPE I and TYPE II errors.

Solution:

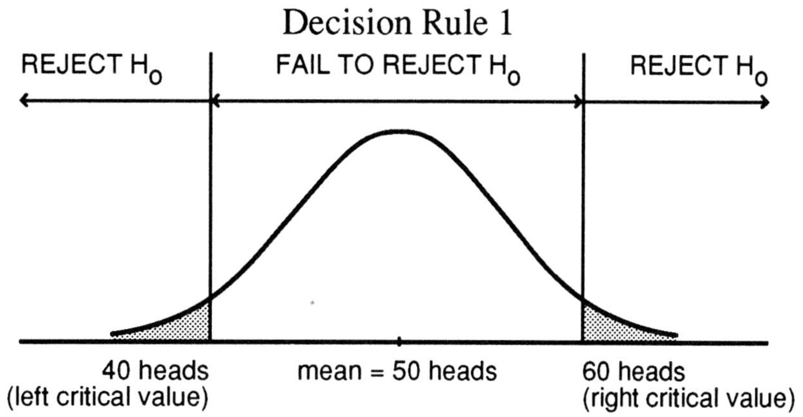

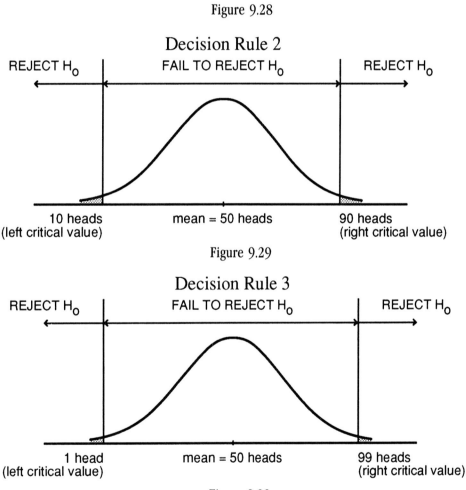

Figure 9.28

Figure 9.29

Figure 9.30

Notice in Decision Rule 1 illustrated in Figure 9.28, a coin will be considered biased if the number of heads in 100 tosses is either *fewer* than 40 heads or *greater than* 60 heads. In Decision Rule 2 illustrated in Figure 9.29, a coin will be considered biased if the number of heads in 100 tosses is either fewer than 10 heads or greater than 90 heads. And finally, using Decision Rule 3 illustrated in Figure 9.30, a coin is considered biased if the number of heads is fewer than 1 head or greater than 99 heads.

Thus, as the critical values for the decision rules are chosen *farther* from the mean of 50 heads (i.e. the expected outcome) the significance level, represented by the shaded areas, *decreases*. In reality one never knows whether the coin being tossed is fair or biased. Let's consider how choosing the critical values of our decision rule further from the mean value affects the probabilities of a TYPE I and TYPE II error for a coin that is in *reality* fair (possibility #1) verses biased (possibility #2).

Possibility #1: Coin is Fair

If the coin being tossed is *fair*, then the effect of choosing the critical values farther from the mean (or expected value) is to decrease the chance of concluding that a fair coin is biased. Thus the chance of committing a TYPE I is *decreased*.

Possibility #2: Coin is Biased

If the coin being tossed is *biased*, then the effect of choosing the critical values farther from the mean (or expected value) is to *decrease* the chance of concluding that the coin is biased and the chance of concluding that this biased coin is fair is increased. Thus, the chance of committing a TYPE II error is *increased*.

To illustrate the relationship between choosing the critical values farther from the mean and the chance of committing a TYPE II error let's consider tossing a **biased** coin (possibility #2) 100 times. If the coin lands heads 85 times out of 100 tosses let's examine the possible conclusions for each of the three previously stated decision rules shown in Figures 9.28, 9.29 and 9.30.

Using the decision rule 1 (Figure 9.28) we would conclude that the coin is biased since 85 heads falls beyond the right critical value of 60 heads. This happens to be the correct decision for *this* coin.

However using decision rules 2 or 3 (Figures 9.29 or 9.30 respectively), which have the critical values farther from the mean, would lead to the *incorrect* conclusion that the biased coin is fair, since 85 heads does not fall beyond the right critical values of 90 or 99 heads, respectively.

■

Now let's reconsider the question posed prior to Example 9.10: *"why not choose a decision rule that will have the smallest possible significance level?"*

We can conclude from the previous discussion that as the **significance level decreases the chance of committing a TYPE II error increases**. Figure 9.31 illustrates this relationship. **The probability of committing a TYPE II error will be denoted by the lower-case Greek letter ß, pronounced beta.**

(NOTE: Although 70 is used as the mean in Figure 9.31b, *any* value other than 50 could have been used.)

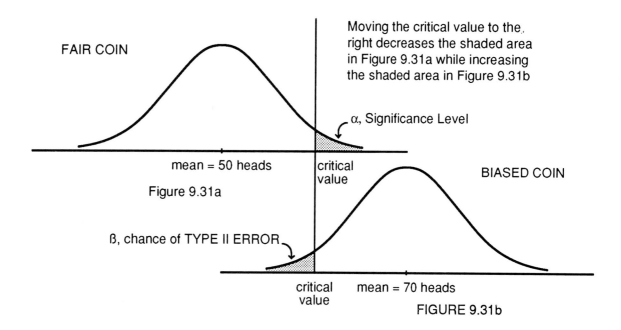

Figure 9.31

By carefully examining Figure 9.31, we observe that as the critical value is moved to the right, **the significance level, α, decreases *but* the chance of committing a TYPE II error, ß, *increases*.** On the other hand, if the critical value is moved to the left, **the significance level, α, increases while the chance of committing a TYPE II error, ß, *decreases*.**

Now that we have examined the relationship between the significance level, α, and the probability of committing a TYPE II error, ß, one might ask: "**what *is* an appropriate value for the significance level?**"

As can be seen from the previous discussion, if the significance level, α, is chosen too small then this increases the risk of committing a TYPE II error. Usually a significance level, α, of less than 1% is generally not used. On the other hand, values greater than 10% are generally not chosen for the significance level, α, since this would allow for a greater chance of committing a TYPE I error. Consequently in practice **the two most commonly used values for the significance level, α, are 1% and 5%.**

After the statistician has selected a suitable significance level, the critical value(s) of the decision rule can be determined using the raw score formula ($X = \mu + z\sigma$) and the appropriate z score found in TABLE II.[1]

In the following examples we will illustrate the technique used to determine the critical value(s) for the decision rule.

Example 9.11

A consumer testing agency is conducting a hypothesis test to determine the validity of an advertised claim made by the Eye Saver Light Bulb Co. that the mean life of its new 60 watt bulb is 1800 hours with a standard deviation of 100 hours. To test this claim the agency randomly selects 400 bulbs. Determine the critical value(s) for the decision rule if:

---

[1]Assuming the Distribution is Normal. Other Distributions will be discussed in subsequent chapters.

a) the consumer agency believes the claim is too high and selects a significance level of 5%.
b) the consumer agency wants to test the claim using a nondirectional test and selects a significance level of 5%.

Solution to part (a)

a) In order to determine the decision rule we must follow the first three steps in the hypothesis testing procedure.

Step 1. *Formulate the hypotheses.*
The null hypothesis is:

H_o: The mean life of the new 60 watt bulb is 1800 hrs.
Since the consumer agency believes that the claim is *too high*, the alternative hypothesis is:

H_a: The mean life of the new 60 watt bulb is *less than* 1800 hrs.

Step 2. *Design an experiment to test the null hypothesis.*
(Note: all calculations necessary to obtain the mean and standard deviation are deliberately omitted to simplify the presentation).

To test this claim, the consumer agency randomly selects 400 bulbs.

Assuming the null hypothesis is true implies that the mean life for samples of 400 bulbs is 1800 hrs with a standard deviation of 5 hrs.

The distribution of all possible outcomes is approximately normal and the expected results are shown in Figure 9.32.

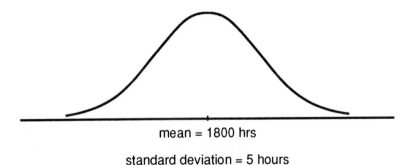

mean = 1800 hrs

standard deviation = 5 hours

Figure 9.32

Step 3. *Formulate the decision rule.*
First, we must determine whether we have a directional or nondirectional alternative hypothesis. Since the alternative hypothesis is the mean life is *less than* 1800 hours, we have a *directional alternative hypothesis*.

Therefore the critical value of the decision rule will be located to the *left* of the mean. This is illustrated in Figure 9.33.

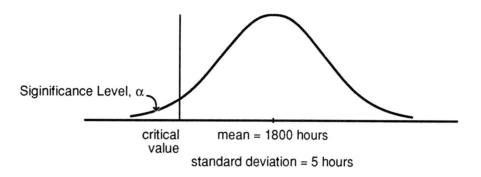

Figure 9.33

In this example the consumer agency selected a significance level of 5%. The significance level of 5% is represented by the shaded region in Figure 9.33. Using this information we can determine the z score of the critical value. This z score is called the ***critical z score***, and is written as z_c. Since we have a normal distribution and the area to the left of the critical z score, z_c, is 5% we can find the value of z_c using TABLE II. The value for z_c is -1.65. This is illustrated in Figure 9.34.

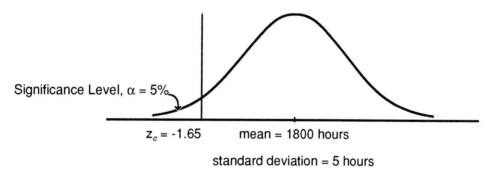

Figure 9.34

Now that the critical z score, has been determined it can be substituted into the raw score formula, $X = \mu + z\sigma$, to determine the critical value, denoted X_c. This leads us to the formula for the critical value, X_c.

Formula for the Critical Value, X_c

The formula for the critical value is:

$$X_c = \text{mean} + (z_c)(\text{standard deviation})$$

where: z_c is the critical z score

Using this formula to calculate the critical value, X_c, where:
mean = 1800 hours
standard deviation = 5
$z_c = -1.65$

We have: X_c = mean + (z_c)(standard deviation)
          X_c = 1800 hours + (-1.65)(5 hours)
          X_c = 1800 hours - 8.25 hours
          X_c = 1791.75 hours

For the critical value, X_c = 1791.75, we formulate the following decision rule: **reject the null hypothesis if the mean life of the sample of 400 bulbs is less than the critical value of 1791.75 hours.** This is illustrated in Figure 9.35.

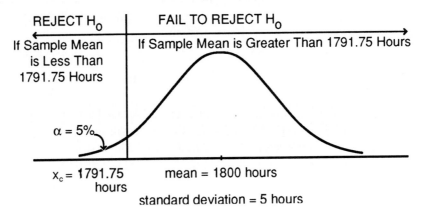

Figure 9.35

Since the region for rejecting the null hypothesis is only on the left side (or in the left tail) of the distribution this type of hypothesis test is referred to as a *one-tailed test*, abbreviated **1TT**.

One-Tailed Test, 1TT

A one-tailed test is conducted when the alternative hypothesis is directional.

A one-tailed test is illustrated in Figure 9.36.

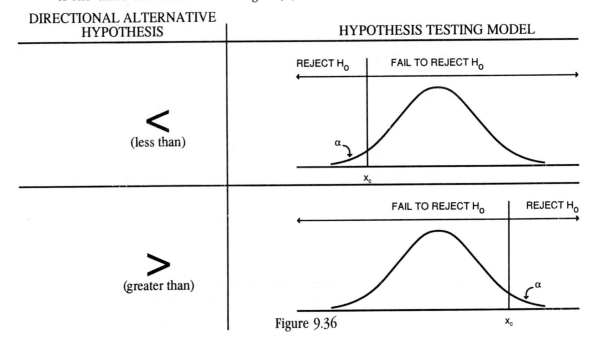

Figure 9.36

Solution to part (b)

b) As in part (a) in order to determine the decision rule, we must follow the first three steps in the hypothesis testing procedure.

Step 1. *Formulate the hypotheses.*
The null hypothesis is:
H_o: The mean life of the new 60 watt bulb is 1800 hrs.

Since the consumer agency wants to test the null hypothesis using a nondirectional test, the alternative hypothesis is:
H_a: The mean life of the new 60 watt bulb is *not* 1800 hrs.

Step 2. *Design an experiment to test the null hypothesis.*
(**Note all calculations necessary to obtain the mean and standard deviation are deliberately omitted to simplify the presentation**). To test the null hypothesis the consumer agency randomly selects 400 bulbs. Assuming the null hypothesis is true implies that the mean life for samples of 400 bulbs is 1800 hrs with a standard deviation of 5 hrs. The distribution for all possible outcomes is approximately normal and the expected results are shown in Figure 9.37.

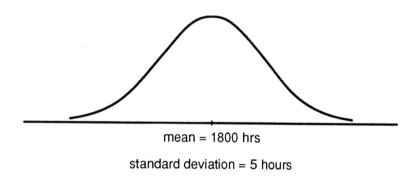

Figure 9.37

Step 3. *Formulate the decision rule.*
First we must determine whether we have a directional or nondirectional alternative hypothesis. Since the alternative hypothesis is **nondirectional** we must determine *two* critical values, one located to the left of the mean and one located to the right of the mean. The critical value located to the left of the mean is referred to as the *left critical value*, denoted X_{LC}, and the critical value located to the right of the mean is referred to as the *right critical value*, denoted X_{RC}. This is illustrated in Figure 9.38.

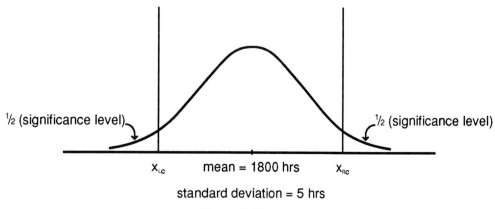

Figure 9.38

Since the significance level for this example is 5%, then the sum of the two shaded regions in Figure 9.38 must total 5%. The conventional procedure is to **divide the significance level equally** between these two shaded regions. Therefore the area of each shaded region is half the significance level (i.e. ½ α). Since the significance level is 5% the area in each shaded region (or tail) is:

$$\tfrac{1}{2}\alpha = \tfrac{1}{2}(5\%)$$
$$= 2.5\%$$

This is illustrated in Figure 9.39.

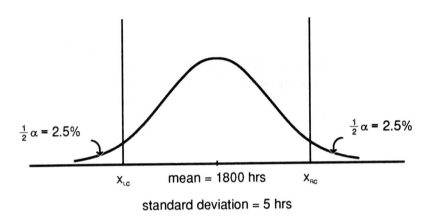

Figure 9.39

Since the decision rule requires two critical values we need two critical z scores, z_c, one for the left critical value (referred to as the *left critical z score*, denoted z_{LC}) and one for the right critical value (referred to as the *right critical z score*, denoted z_{RC}). Using the information in Figure 9.39 and TABLE II, we can find the value of the two critical z scores, z_{LC} and z_{RC}.

To find z_{LC} we must determine the z score which has 2.5% of the area to the left. From TABLE II, $z_{LC} = -1.96$. Similarly to find z_{RC} we must determine the z score which has 97.5% of the area to the left. From TABLE II, $z_{RC} = +1.96$. This is illustrated in Figure 9.40.

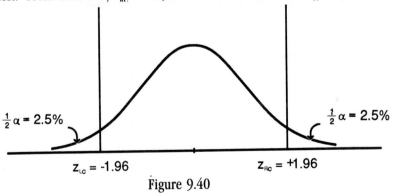

Figure 9.40

Now that the critical z scores, z_{LC} and z_{RC}, have been determined they can be **substituted into the raw score formula**, $X = \mu + z\sigma$, to determine the critical values, X_{LC} and X_{RC}, respectively. Therefore the formula for the left critical value, X_{LC}, is:

$$X_{LC} = \text{mean} + (z_{LC})(\text{standard deviation})$$
$$= 1800 \text{ hrs} + (-1.96)(5 \text{ hrs})$$
$$= 1790.2 \text{ hours}$$

Similarly the formula for the right critical value, X_{RC}, is:
$$X_{RC} = \text{mean} + (z_{RC})(\text{standard deviation})$$
$$= 1800 \text{ hrs} + (+1.96)(5 \text{ hrs})$$
$$= 1809.8 \text{ hours}$$

Using these critical values we formulate the decision rule: **reject the null hypothesis if the mean life of the sample of 400 bulbs is either less than the left critical value of 1790.2 hrs or more than the right critical value of 1809.8 hours.** This is illustrated in Figure 9.41. ■

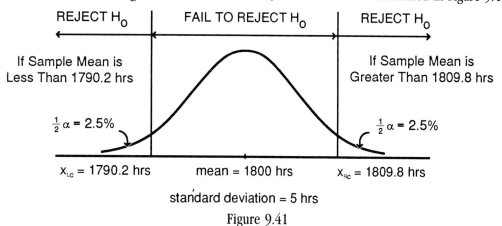

Figure 9.41

In the previous example there were two regions on the normal curve for rejecting the null hypothesis. One rejection region was in the left tail and the other rejection region was in the right tail of the curve. This type of hypothesis test is referred to as a ***two-tailed test***, abbreviated **2TT**.

Two-Tailed Test, 2TT

A two-tailed test is conducted when the alternative hypothesis is nondirectional. In a two-tailed test, the level of significance, α, is divided equally between the two tails. Therefore, the area in each tail is half the significance level, ½ α.

A two-tailed test is illustrated in Figure 9.42.

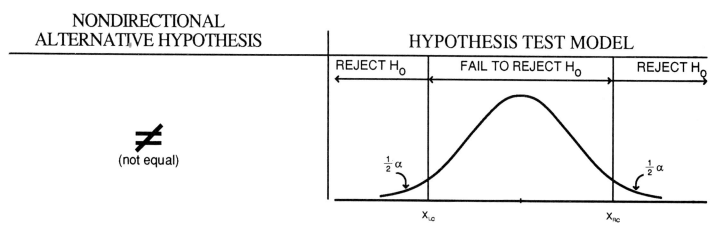

Figure 9.42

Example 9.12

A national department store chain has determined that in previous years 40% of its customers used the department store's special Holiday credit card to make Christmas purchases. This year because of changing economic conditions the department store management would like to determine if the 40% figure is still accurate. To test the 40% figure, 150 customers are randomly selected during the week of December 10th and asked whether they were using the store's special Holiday credit card. Determine the decision rule for the following hypothesis tests. Use a significance level of 1%.

a) Assume the distribution of outcomes for samples of 150 customers is normal with a mean of 60 customers and a standard deviation of 6 customers.
Test the null hypothesis:

H_o: 40% of the customers use the special Holiday credit card.

Against the alternative hypothesis:

H_a: *More than* 40% of the customers use the special Holiday credit card.

b) Assume the distribution of outcomes for samples of 150 customers is normal with a mean of 60 customers and a standard deviation of 6 customers.

Test the null hypothesis:

H_o: 40% of the customers use the special Holiday credit card.

Against the alternative hypothesis:

H_a: The percent of customers using the special Holiday credit card is *not* 40%.

Solution to part (a)

a) Since the first two steps of the hypothesis testing procedure are given we can begin with Step 3.

Step 3. *Formulate the decision rule.*
We have a directional alternative hypothesis since it is written as *more than* 40% of the customers use the special Holiday credit card. Therefore the critical value will be located to the *right* of the mean. This is illustrated in Figure 9.43.

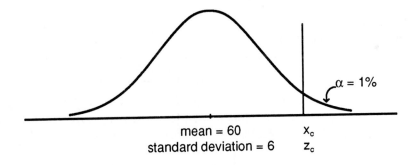

Figure 9.43

Using the information in Figure 9.43 and TABLE II: the Normal Curve Area Table, we can determine the critical z score, z_c. Since the percent of the area to the left of z_c is 99% then $z_c = 2.33$.

Now that the critical z score has been determined it can be substituted into the formula for the critical value, X_c.

X_c = mean + (z_c)(standard deviation)

$X_c = 60 + (2.33)(6)$

$X_c = 73.98$

For the critical value, $X_c = 73.98$, we formulate the decision rule: reject the null hypothesis if the number of customers in the sample of 150 using the special Holiday credit card is *more* than the critical value of 73.98.

This is illustrated in Figure 9.44.

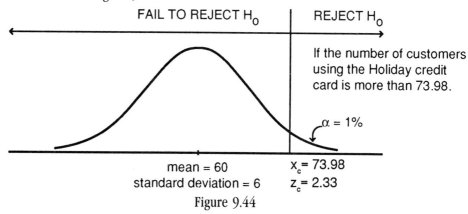

Figure 9.44

Solution to part (b)

b) Since the first two steps of the hypothesis testing procedure are given we can begin with Step 3.

Step 3. *Formulate the decision rule.*

We have a *nondirectional* alternative hypothesis since it is stated as: the percent of customers using the special holiday credit card is *not* 40%. Therefore we will have two critical values, one located to the *left* of the mean, denoted X_{LC}, and one located to the *right* of the mean, denoted X_{RC}. Thus, the 1% significance level is **divided equally between the two tails**. This is illustrated in Figure 9.45.

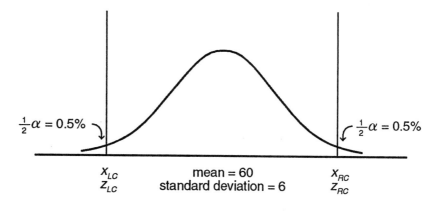

Figure 9.45

Using the information in Figure 9.45 and the TABLE II: the Normal Curve Area Table, we can determine the two critical z scores, z_{LC} and z_{RC}.

To find z_{LC} we must determine the z score which has *0.5%* of the area to the left. Using TABLE II, $z_{LC} = -2.58$. Similarly to find z_{RC} we must determine the z score which has *99.5%* of the area to the left. Using TABLE II, $z_{RC} = 2.58$.

Now that the critical z scores, z_{LC} and z_{RC}, have been determined they can be substituted into the formula for the critical value to determine X_{LC} and X_{RC} respectively.

$$X_{LC} = \text{mean} + (z_{LC})(\text{standard deviation})$$
$$X_{LC} = 60 + (-2.58)(6)$$
$$X_{LC} = 44.52$$

Similarly, $X_{RC} = \text{mean} + (z_{RC})(\text{standard deviation})$

$$X_{RC} = 60 + (2.58)(6)$$
$$X_{RC} = 75.48$$

Using these critical values we formulate the decision rule: **reject the null hypothesis if the number of customers using the special Holiday credit card is either less than 44.52 or more than 75.48 customers.** This is illustrated in Figure 9.46. ■

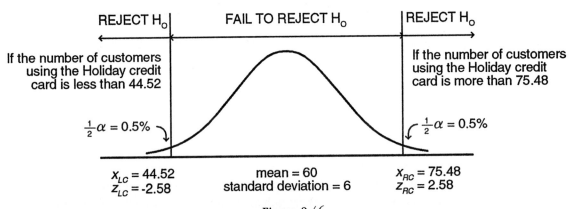

Figure 9.46

From the previous two examples, we can now generalize the procedure to formulate the decision rule.

Procedure to Formulate the Decision Rule

a) Determine the type of alternative hypothesis (directional or nondirectional).

b) Determine the type of test (1TT or 2TT).

c) Identify the significance level ($\alpha = 5\%$ or $\alpha = 1\%$).

d) Construct the appropriate hypothesis test model.

e) Calculate the critical value(s) using the formula:
X_c = mean + (z_c)(standard deviation).

f) State the decision rule.

Tables 9.3 and 9.4 summarize the procedure for formulating the decision rule for the significance levels of 5% and 1%.

Table 9.3

Formulating the Decision Rule Using 5% Significance Level

Type of Alternative Hypothesis, H_a	Directional (less than) <	Directional (greater than) >	Nondirectional (not equal to) ≠
Type of Test	One-tailed (1TT)	One-tailed (1TT)	Two-tailed (2TT)
Hypothesis Test Model	Reject H_o at $z_c = -1.65$, $\alpha = 5\%$; Fail to Reject H_o at mean	Fail to Reject H_o at mean; Reject H_o at $z_c = 1.65$, $\alpha = 5\%$	Reject H_o at $z_{LC} = -1.96$, $\frac{1}{2}\alpha = 2.5\%$; Fail to Reject H_o at mean; Reject H_o at $z_{RC} = 1.96$, $\frac{1}{2}\alpha = 2.5\%$
Formula for Critical Values	$x_c = \text{mean} + (-1.65)(sd)$	$x_c = \text{mean} + (+1.65)(sd)$	$x_{LC} = \text{mean} + (-1.96)(sd)$ and $x_{RC} = \text{mean} + (+1.96)(sd)$
Decision Rule	Reject H_o, if experimental outcome is: less than X_c	Reject H_o, if experimental outcome is: greater than X_c	Reject H_o, if experimental outcome is either: less than X_{LC} or greater than X_{RC}

Formulating the Decision Rule Using 1% Significance Level

Table 9.4

	Directional (less than) $<$	Directional (greater than) $>$	Nondirectional (not equal to) $\neq$
Type of Alternative Hypothesis, H_a	Directional (less than) $<$	Directional (greater than) $>$	Nondirectional (not equal to) $\neq$
Type of Test	One-tailed (1TT)	One-tailed (1TT)	Two-tailed (2TT)
Hypothesis Test Model	REJECT H_o | FAIL TO REJECT H_o; $\alpha = 1\%$; $z_C = -2.33$; mean	FAIL TO REJECT H_o | REJECT H_o; $\alpha = 1\%$; $z_C = 2.33$; mean	REJECT H_o | FAIL TO REJECT H_o | REJECT H_o; $\frac{1}{2}\alpha = 0.5\%$; $z_{LC} = -2.58$; $z_{RC} = 2.58$; mean
Formula for Critical Values	$x_c = \text{mean} + (-2.33)(\text{sd})$	$x_c = \text{mean} + (+2.33)(\text{sd})$	$x_{LC} = \text{mean} + (-2.58)(\text{sd})$ and $x_{RC} = \text{mean} + (+2.58)(\text{sd})$
Decision Rule	Reject H_o, if experimental outcome is: less than X_c	Reject H_o, if experimental outcome is: greater than X_c	Reject H_o, if experimental outcome is either: less than X_{LC} or greater than X_{RC}

460 • CHAPTER 9 INTRODUCTION TO HYPOTHESIS TESTING

Now that we have completed the introduction to hypothesis testing, let us summarize the hypothesis testing procedure.

HYPOTHESIS TESTING PROCEDURE

Step 1. Formulate the two hypotheses, H_o and H_a.

Step 2. Design an experiment to test the null hypothesis, H_o.

Under the assumption H_o is true, calculate the expected results: (**These results will be discussed in the subsequent chapters.**)

a) Identify the appropriate distribution used as the hypothesis testing model.
b) Mean of the distribution
c) Standard deviation of the distribution

Step 3. Formulate the Decision Rule:

a) Determine type of alternative hypothesis, directional or nondirectional.
b) Determine type of test, 1TT or 2TT.
c) Identify the significance level, $\alpha = 1\%$ or $\alpha = 5\%$.
d) Construct the appropriate hypothesis test model.
e) Calculate the critical value(s) using the formula:

$$X_c = \text{mean} + (z_c)(\text{standard deviation})$$

f) State the decision rule.

Step 4. Conduct the experiment.

Collect the necessary sample data to calculate the experimental outcome.

Step 5. Determine the conclusion.

Compare the experimental outcome to the decision rule and either:

a) **Reject H_o and accept H_a at α,**
or
b) **Fail to reject H_o at α.**

Example 9.13 illustrates the use of the hypothesis testing procedure to perform a hypothesis test.

Example 9.13

Last year, 75% of adults within the USA stated that they believed a college degree was very important. This year a national poll of 1,020 American adults who were randomly selected found that 794 adults believed that a college degree is very important. Can you conclude that the percent of adults within the USA who believe a college degree is very important has significantly increased this year?

Use mean = 765 and standard deviation = 13.83 to perform a hypothesis test at $\alpha = 5\%$ to answer this question.

Solution:

We will use the hypothesis testing procedure to perform this hypothesis test. The first step is to formulate the hypotheses.

Step 1. Formulate the two hypotheses, H_o and H_a.
The hypothesis being tested is that 75% of the adults within the USA believe a college degree is important. Therefore, the null hypothesis is:

H_o: 75% of adults within the USA believe a college degree is important.

Since we are trying to show that the percent of adults who believe a college degree is important *has increased*, therefore, the alternative hypothesis is:

H_a: *more than* 75% of adults within the USA believe a college degree is important.

The second step involves designing an experiment to test the null hypothesis.

Step 2. Design an experiment to test the null hypothesis, H_o.
To test the null hypothesis, a national poll of 1,020 American adults were randomly selected to determine their belief about the importance of a college degree. Under the assumption H_o is true, we need to calculate the expected results. As we have mentioned repeatedly throughout Chapter 9, the expected results are given so we can concentrate on the concepts of the hypothesis testing procedure. Thus, the distribution of all possible outcomes for samples of 1,020 American adults is approximately normal with a mean = 765 and a standard deviation = 13.83. These expected results are shown in Figure 9.47.

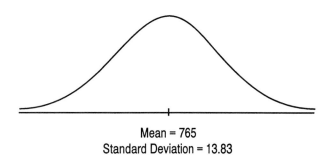

Mean = 765
Standard Deviation = 13.83

Figure 9.47

Now that we have calculated the expected results based upon the assumption that the null hypothesis is true, we need to formulate a decision rule.

Step 3. Formulate the decision rule.
a) **Determine type of alternative hypothesis.**
Since the alternative hypothesis is in the form of a *more than statement*, it is a directional hypothesis.
b) **Determine the type of test.**
A directional alternative hypothesis is a one tail test (1TT). Since it is in the form of a *more than statement*, the critical value will be located to the *right* of the mean.

c) Identify the significance level.
The significance level is $\alpha = 5\%$.

d) Construct the appropriate hypothesis test model.
This is illustrated in Figure 9.48.

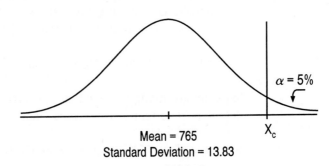

Figure 9.48

e) Calculate the critical value using the formula:

$$X_c = \text{mean} + (z_c)(\text{standard deviation})$$

Using the information in Figure 9.48 and Table II: the Normal Curve Area Table, we can determine the critical z score, z_c. Since the percent of the area to the left of z_c is 95%, then $z_c = 1.65$.

Now that the critical z score has been determined it can be substituted into the formula for the critical value, X_c.

$X_c = \text{mean} + (z_c)(\text{standard deviation})$

$X_c = 765 + (1.65)(13.83)$

$X_c \approx 787.82$

For the critical value, $X_c \approx 787.82$, we formulate the decision rule: reject the null hypothesis if the number of American adults in the sample of 1,020 who believe that a college degree is very important is *more* than the critical value of 787.82. This is illustrated in Figure 9.49.

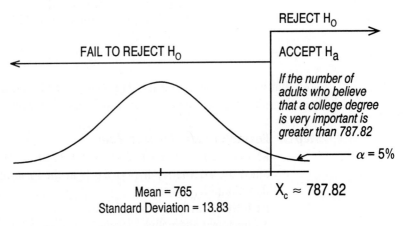

Figure 9.49

Now that the decision rule has been formulated, we need to conduct the experiment.

Step 4. Conduct the experiment.
The experiment is the process of collecting the sample information from the national poll of 1,020 randomly selected American adults. The 794 adults of the sample who said that a college degree is very important represents the experimental outcome.

If we compare the experimental outcome of 794 adults to the decision rule, we can draw a conclusion.

Step 5. Determine the conclusion.
Since the experimental outcome of 794 adults is greater than the critical value of 787.82, then the conclusion is *reject H_o and accept H_a at $\alpha = 5\%$*. This is illustrated in Figure 9.50.

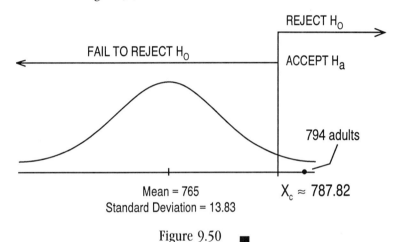

Figure 9.50

CASE STUDY 9.1

Using the information contained in the USA Snapshot entitled *Trying to kick the habit* shown in Figure CS9.1, answer the following questions.

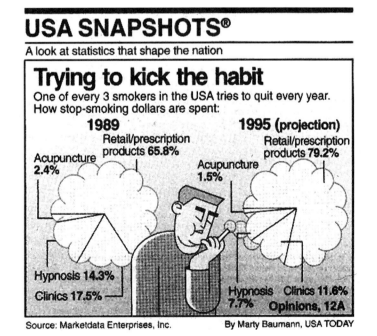

Figure CS9.1

a) Define the population of people being illustrated in Figure CS9.1.
b) Define the variable(s) being discussed in the snapshot in Figure CS9.1.

If you use the information pertaining to the year 1989 as representing the status quo, and the information regarding the projected year 1995 as the direction you believe the percentages for each category are headed, then:

c) formulate a null hypothesis and an alternative hypothesis for each of the four categories shown in Figure CS9.1.
d) indicate the type of alternative hypothesis.
e) determine the type of test (i.e. 1TT or 2TT), and on which side of the curve the critical outcome would be positioned.
f) describe how you would select an appropriate sample to test the null hypothesis.
g) When calculating the expected results in step 2 of the hypothesis testing procedure, what assumption are you making? Explain why it is necessary to make this assumption.
h) For the hypothesis test you outlined regarding Acupuncture, what would be the conclusion to your test if the experimental outcome were to fall to the left of your critical outcome?
i) For the hypothesis test you outlined regarding Hypnosis, what would be the conclusion to your test if the experimental outcome were to fall to the right of your critical outcome? ■

CASE STUDY 9.2

Using the information contained in the USA Today caption entitled *Skyrocketing cable rates* shown in Figure CS9.2, answer the following questions assuming you are a statistician who feels that cable rates are increasing.

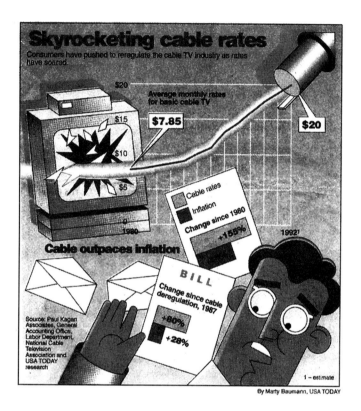

Figure CS9.2

a) What variable is being shown by the graph in Figure CS9.2?

If you use the information regarding cable rates for the year 1992 as representing the status quo, and the implied direction of the graph as your belief about what cable rates are doing now then:

b) formulate a null hypothesis and an alternative hypothesis.
c) indicate the type of alternative hypothesis.
d) determine the type of test (i.e. 1TT or 2TT), and on which side of the curve the critical outcome would be positioned.
e) describe how you would select an appropriate sample to test the null hypothesis.
f) When calculating the expected results in step 2 of the hypothesis testing procedure, what assumption are you making? Explain why it is necessary to make this assumption.
g) For your hypothesis test, what would be the conclusion to your test if the experimental outcome were to fall to the right of your critical outcome?

9.4 p-VALUES FOR HYPOTHESIS TESTING

When performing a hypothesis test, the last step or step 5 of the hypothesis testing procedure requires that the experimental outcome (or the sample outcome) be compared to the critical value X_c of the decision rule to determine whether to either reject the null hypothesis or fail to reject the null hypothesis. In Example 9.13, a one-tailed test(1TT) on the right side was performed to determine if we could reject the null hypothesis at a 5% level of significance. In step 5 of the hypothesis test, the experimental outcome of 794 adults was compared to the critical value $X_c = 787.82$. Since the experimental outcome was *greater than* the critical value, the *null hypothesis H_o was rejected at the 5% level of significance* (that is, $\alpha = 5\%$). This is illustrated in Figure 9.51.

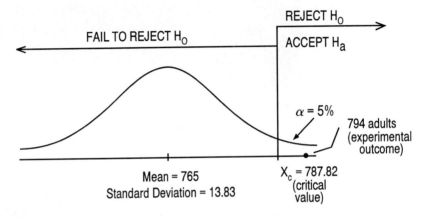

Figure 9.51

A question that might be asked is: *could we have rejected the null hypothesis for a different significance level, such as $\alpha = 4\%$ or $\alpha = 3\%$ or $\alpha = 2\%$ or even $\alpha = 1\%$?*

For example, if the significance level of Example 9.13 was $\alpha = 4\%$, would the null hypothesis still have been rejected? To answer this question, we need to determine the critical value X_c for a significance level of $\alpha = 4\%$, and a 1TT on the right side. Since the percent of area to the right of the critical value X_c is 4%, then the percent of area to the left of the critical z score, z_c, is 96%. Using Table II: the Normal Curve Area Table, the nearest percentage to 96% is 95.99%, thus the critical z score is: $z_c = 1.75$. Substituting the critical z score into the formula for the critical value, X_c, we obtain:

$$X_c = \text{mean} + (z_c)(\text{standard deviation})$$

$$X_c = 765 + (1.75)(13.83)$$

$$X_c \approx 789.20$$

Since the experimental outcome of 794 adults **is still greater than** this critical value of $X_c \approx 789.20$ for a significance level of 4%, the null hypothesis is rejected at the significance level of $\alpha = 4\%$. This is illustrated in Figure 9.52.

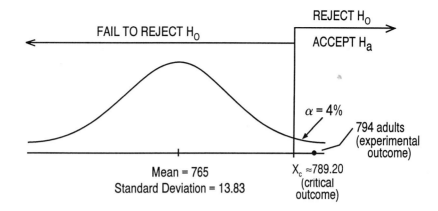

Figure 9.52

Using this same procedure to determine the critical value X_c for each of the significance levels of $\alpha = 3\%$ and $\alpha = 2\%$, we obtain a critical value of $X_c \approx 791$ for $\alpha = 3\%$, and a critical value of $X_c \approx 793.35$ for $\alpha = 2\%$. Since the experimental outcome of 794 adults is still greater than both of the critical values: $X_c \approx 791$ (for $\alpha = 3\%$) and $X_c \approx 793.35$ (for $\alpha = 2\%$), then the null hypothesis is rejected at each of these significance levels. This is illustrated in Figure 9.53(a) and Figure 9.53(b) respectively.

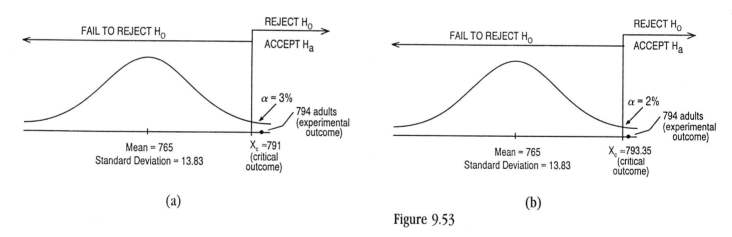

Figure 9.53

However, at the significance level of $\alpha = \mathbf{1\%}$, the null hypothesis **is not rejected**, since the experimental outcome of 794 adults is now **less than** the critical value of $X_c \approx 797.22$. This is illustrated in Figure 9.54.

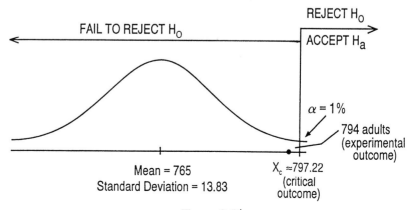

Figure 9.54

Examining the results for the different significance levels, we can begin to obtain an estimate to the answer of the question: *how large would the significance level, α, have to be to reject the null hypothesis?* The null hypothesis of Example 9.13 can be rejected using a significance level of $\alpha = 2\%$ or greater. This is illustrated in Figure 9.55.

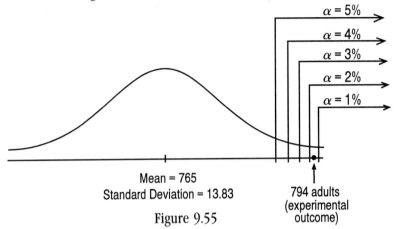

Figure 9.55

From Figure 9.55, we should notice it is possible to use a significance level smaller than 2% but greater than 1% that would still allow us to reject the null hypothesis. In essence, we are looking for the *true level of significance* for this hypothesis test or the **smallest value of** α for which the null hypothesis can be rejected? The smallest value of α is called the **probability-value** or **p-value of a hypothesis test**. The p-value of a hypothesis test is also referred to as the **observed significance level of a hypothesis test**. The p-value of a hypothesis test is usually expressed in decimal form rather than a percent. We will express the p-value as both a decimal and a percent.

> **Definition: p-Value of a Hypothesis Test.** The p-value, or observed significance level, of a hypothesis test is the **smallest level of significance(α)** at which the null hypothesis H_o can be rejected using the experimental outcome (or sample outcome).

To find the smallest value of α at which the null hypothesis of Example 9.13 can be rejected, that is, the p-value, it is necessary to find the percent of area to the right of the experimental outcome of 794 adults. This is illustrated in Figure 9.56.

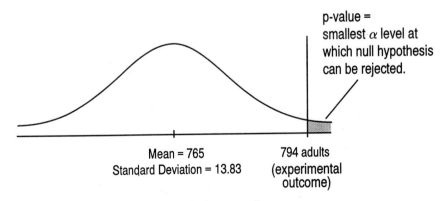

Figure 9.56

To determine the area to the right of 794, we need to calculate the z score of 794. Using the z score formula, we obtain:

$$z \text{ of } 794 = \frac{794 - 765}{13.83} \approx 2.10.$$

Using Table II: The Normal Curve Area Table, the area to the left of $z \approx 2.10$ is 98.21%, thus the area to the right which represents the p-value is:

$$\text{p-value} = 100\% - 98.21\% = 1.79\%.$$

Thus, the p-value of this hypothesis test is 0.0179 or 1.79%. Consequently, the smallest value of α at which the null hypothesis of Example 9.13 can be rejected is 1.79%. This is illustrated in Figure 9.57.

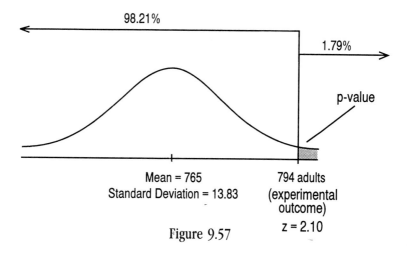

Figure 9.57

Therefore, we say that this experimental outcome, or sample outcome, of 794 adults is "extremely significant," since the p-value or probability of observing an experimental outcome of 794 is only 0.0179 or 1.79%. **Reporting the p-value of a hypothesis test is more informative than stating the conclusion of a hypothesis test.** Suppose instead of stating the conclusion of the hypothesis test in Example 9.13 as reject the null hypothesis at $\alpha = 5\%$, the p-value of 0.0179 or 1.79% was reported. This knowledge about the p-value of the hypothesis conveys more information to the reader. It tells the reader that the null hypothesis can be rejected at $\alpha = 4\%$ and that it can be rejected even at smaller values of α such as $\alpha = 2\%$. In fact, this p-value of 0.0179 or 1.79% indicates that the null hypothesis can be rejected at values of α as small as $\alpha = 1.79\%$. Compare the content information indicated by the statement "the p-value is 0.0179 or 1.79%" with the very limited information contained by the statement "reject the null hypothesis at $\alpha = 5\%$".

Computer software programs, like MINITAB, carry out hypothesis tests by calculating the p-value for you. In fact, many researcher articles containing statistical tests of hypothesis publish p-values. Thus, instead of selecting a significance level, α, before conducting the experiment as outlined in the hypothesis testing procedure, the researcher or the computer software reports the p-value of the hypothesis test along with the experimental outcome (or sample outcome). It is left to the reader to interpret the significance of the hypothesis test. That is, the reader must determine whether to reject the null hypothesis based upon the reported p-value. Usually, the null hypothesis is rejected if the p-value is less than the significance level, α, that the reader decides to choose. Thus, the smaller the p-value, the stronger the evidence against the null hypothesis provided by the experimental outcome. To decide whether to reject the null hypothesis when the p-value of a hypothesis test is reported, you can use the following guideline.

Deciding whether to Reject the null hypothesis when a p-value is reported.

Compare the p-value of the hypothesis test to the chosen significance level, α, that you are willing to tolerate. If the p-value is less than or equal to α, then the hypothesis test is significant at the selected significance level, α. That is,

if p-value $\leq \alpha$,
then reject H_o at significance level, α.

Remember, however, that a p-value conveys more information than than just reporting whether or not to reject the null hypothesis at a given significance level, α. Thus, the p-value indicates how unusual the experimental outcome (or sample outcome) assuming that the null hypothesis is true. Therefore, the p-value will provide an indication as to whether or not the experimental outcome is very significant to not significant. The following rules of thumb established by statisticians will help you to interpret p-values regarding the significance of the experimental outcome (or sample outcome).

Interpreting p-values

1. If the **p-value is less than 0.01 (or 1%)**, the experimental outcome(or sample outcome) is *very significant*.
2. If the **p-value is between 0.01 (or 1%) and 0.05 (or 5%)**, the experimental outcome (or sample outcome) is *significant*.
3. If the **p-value is between 0.05 (or 5%) and 0.10 (or 10%)**, the experimental outcome (or sample outcome) is *marginally or not significant*.
4. If the **p-value is greater than 0.10 (or 10%)**, the experimental outcome (or sample outcome) is *not significant*.

Example 9.14

A statistical hypothesis test was conducted using a significance level of $\alpha = 5\%$. If the p-value of this test is equal to 0.06 (or 6%), can the null hypothesis be rejected?

Solution:

Since the p-value of a hypothesis test is the smallest level of significance(α) at which the null hypothesis H_o can be rejected, then for this test the null hypothesis cannot be rejected because the p-value of 0.06 (or 6%) is greater than the significance level of $\alpha = 5\%$. ■

Example 9.15

The null hypothesis of a statistical hypothesis test was rejected at a significance level of $\alpha = 5\%$. If the p-value of this test is equal to 0.025 (or 2.5%), then what is the smallest value of α for which the null hypothesis can be rejected?

Solution:

Since the p-value of a hypothesis test is the smallest level of significance(α) at which the null hypothesis H_o can be rejected, then the smallest value of α is equal to 2.5%. ■

PROCEDURE TO CALCULATE THE p–VALUE OF A HYPOTHESIS TEST

Let's discuss how to compute the p-value of a hypothesis test for the following three possible cases.

Case One: a one-tailed test on the left side

For a one-tailed test on the left side, the p-value of a hypothesis test is equal to the percent of area in the tail to the left of the experimental outcome (or sample outcome) of the test. This is illustrated in Figure 9.58.

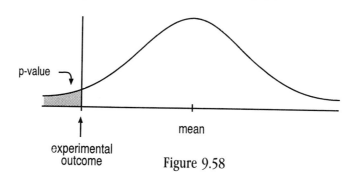

Figure 9.58

Case Two: a one-tailed test on the right side

For a one-tailed test on the right side, the p-value of a hypothesis test is equal to the percent of area in the tail to the right of the experimental outcome (or sample outcome) of the test. This is illustrated in Figure 9.59.

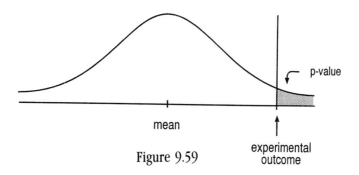

Figure 9.59

Case Three: a two-tailed test

For a two-tailed test, the p-value of a hypothesis is equal to twice the percent of area in the tail beyond the direction of the experimental outcome (or sample outcome) of the test. This is illustrated in Figure 9.60.

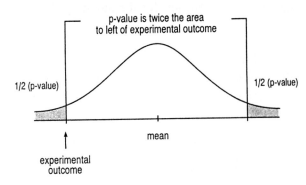
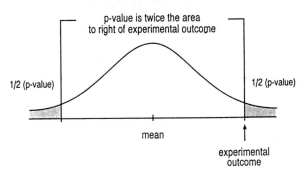

Figure 9.60

Example 9.16

Use the information contained within the given hypothesis test to answer the following questions. The first two steps of the hypothesis testing procedure are given.

A consumer testing agency is conducting a hypothesis test to determine the validity of an advertised claim made by the Eye Saver Light Bulb Co. that the mean life of its new 60 watt bulb is 1800 hours with a standard deviation of 100 hours. The consumer agency believes the claim is too high and randomly selects 400 bulbs to test the company's claim and determines the mean life of these 400 bulbs to be 1788 hours.

Step 1. *Formulate the hypotheses.*
H_o: The mean life of the new 60 watt bulb is 1800 hrs.
H_a: The mean life of the new 60 watt bulb is *less than* 1800 hrs.

Step 2. *Design an experiment to test the null hypothesis.*
To test the null hypothesis, the consumer agency randomly selects 400 bulbs. Assuming the null hypothesis is true implies that the mean life for all possible samples of 400 bulbs is 1800 hrs with a standard deviation of 5 hrs.

The distribution of all possible outcomes is approximately normal and the expected results are shown in Figure 9.61.

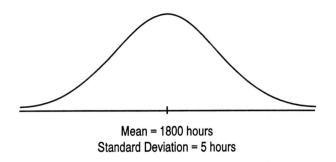

Figure 9.61

a) Calculate the p-value of this hypothesis test.
b) Can the consumer agency reject the null hypothesis at $\alpha = 5\%$? at $\alpha = 1\%$?
c) What is the smallest level of significance at which the null hypothesis H_o can be rejected?

Solution:

a) To calculate the p-value, we need to find the area to the left of the experiment outcome, 1788 hours, since this is a 1TT on the left side. This is illustrated in Figure 9.62.

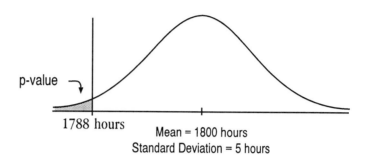

Figure 9.62

To determine this area find the z score of 1788. Using the z score formula, we obtain:

$$z \text{ of } 1788 = \frac{1788 - 1800}{5} = -2.40.$$

Using Table II: The Normal Curve Area Table, the area to the left of $z = -2.40$ is 00.82%. Thus the p-value is 0.82%. In decimal form the p-value is 0.0082.

b) Yes, it can be rejected at both significance levels, since the p-value of 0.0082 or 0.82% is less than both 5% and 1%.
c) The smallest level of significance at which the null hypothesis can be rejected is equal to the p-value which is 0.0082 (or 0.82%) ∎

❄ 9.5 USING MINITAB

Using statistical software is generally the way a researcher analyzes hypotheses tests. MINITAB has many different models of hypothesis tests. Although each model is specific to the distribution from which a sample is found, they have one common thread - each reports the **significance** of the test using a p-value.

Let's use MINITAB to simulate an experiment of selecting 100 random numbers from a normal distribution whose mean is 480 and standard deviation is 100 and placing the random sample's data values into a column within the MINITAB worksheet. Furthermore, let's do this twenty times and place the results in the columns labelled C1 to C20.

To accomplish this use the MINITAB command RANDOM with its subcommand NORMAL in the following way:

```
MTB >   RANDOM 100 C1-C10;
SUBC >  NORMAL MU=480 SIGMA=100.
```

How many of these random samples would you expect to have sample means that are significantly different from 480, using an $\alpha=5\%$? [answer: 5% of 20 = 1 random sample]

The ZTEST command of MINITAB use the p-value to indicate the probability that a random sample is from a distribution whose mean is 480 and standard deviation is 100.

For the twenty random samples that were simulated let's use the ZTEST command for each one to determine the probability of each one coming from a normal distribution whose mean is 480 and whose standard deviation is 100. This is accomplished by using the ZTEST command for each of the column data values.

MTB > ZTEST MU=480 SIGMA 100, C1-C10

An output for the sequence of the previous three commands is:
TEST OF MU = 480.000 VS MU N.E. 480.000
THE ASSUMED SIGMA = 100

	N	MEAN	STDEV	SE MEAN	Z	P VALUE
C1	100	481.667	96.773	10.000	0.17	0.87
C2	100	481.107	100.927	10.000	0.11	0.91
C3	100	480.061	95.140	10.000	0.01	1.00
C4	100	486.809	104.383	10.000	0.68	0.50
C5	100	471.296	101.909	10.000	-0.87	0.38
C6	**100**	**460.075**	**100.314**	**10.000**	**-1.99**	**0.047**
C7	100	485.719	80.881	10.000	0.57	0.57
C8	100	487.935	99.079	10.000	0.79	0.43
C9	100	467.486	99.981	10.000	-1.25	0.21
C10	100	482.801	99.469	10.000	0.28	0.78
C11	100	482.216	88.678	10.000	0.22	0.82
C12	100	483.884	105.896	10.000	0.39	0.70
C13	100	487.653	95.341	10.000	0.77	0.44
C14	100	465.011	99.123	10.000	-1.50	0.13
C15	100	483.081	101.363	10.000	0.31	0.76
C16	100	486.568	106.316	10.000	0.66	0.51
C17	100	481.208	104.617	10.000	0.12	0.90
C18	100	474.395	95.205	10.000	-0.56	0.58
C19	100	466.774	98.412	10.000	-1.32	0.19
C20	100	488.163	99.748	10.000	0.82	0.41

There are many things to notice about the MINITAB output.

(1) A test of MU=480 verses MU N.E. 480 is listed first.
This is MINITAB's way of describing that the ZTEST is a hypothesis test about a population mean, where the alternative hypothesis, H_a, is that the population mean is NOT EQUAL to 480.

(2) The ASSUMED value of the SIGMA is 100.
The SIGMA value was assigned using the ZTEST command.

(3) Descriptive statistics are given for each of the twenty random samples. These include:
    a) the number, N, within each random sample.
    b) the sample mean, MEAN for each random sample.
    c) the sample standard deviation, STDEV.
    d) a statistic labelled S.E. MEAN, referred to as the standard error of the mean.
    e) the z-score, Z, for the sample mean.
    f) the **p-value**, p, for the sample mean.

(4) The descriptive statistics for column C6 is the **only line whose p-value is less than 0.05.**

The MINITAB output indicates that if twenty samples were randomly selected from a population whose mean is 480 and whose standard deviation is 100 than the probability that the random sample in column C6 is from that population is p=0.047.

Although it is possible that random sample C6 is from a population whose mean is 480 and standard deviation is 100 it is **not likely**.

On the other hand, if you were conducting a hypothesis test about the population mean = 480 and sample C6 were selected, you would conclude that the **random sample probably did not come from a distribution whose mean is 480 and standard deviation is 100. Therefore, the hypothesis that the mean of the population is 480 would be rejected at an α level of 5%.**

Let's use MINITAB to simulate an experiment where random samples are selected from a population whose mean is 460 and whose standard deviation is 100. Furthermore let's put the data values for these random samples in columns C1 to C20.

MTB > RANDOM 100 C1-C20;
SUBC > NORMAL MU=460 SIGMA=100.

Use each of the twenty randomly selected samples to test the hypothesis that the random sample was selected from a population whose mean is 480. What do you expect to happen?

ANSWER: If we assume that these samples come from a population whose mean is 480, then many of the random samples would have a p-value less than 0.05.

The ZTEST command could test the hypothesis that the population mean is 480.

MTB > ZTEST MU=480 SIGMA=100, C1-C20

The MINITAB output for this experiment is:
TEST OF MU = 480.000 VS MU N.E. 480.000
THE ASSUMED SIGMA = 100

	N	MEAN	STDEV	SE MEAN	Z	P VALUE
C1	100	457.922	109.011	10.000	-2.21	0.027
C2	100	458.434	93.907	10.000	-2.16	0.031
C3	100	459.207	102.387	10.000	-2.08	0.038
C4	100	469.864	98.914	10.000	-1.01	0.31
C5	100	476.835	100.023	10.000	-0.32	0.75
C6	100	455.691	104.313	10.000	-2.43	0.015
C7	100	466.296	94.863	10.000	-1.37	0.17
C8	100	453.546	101.436	10.000	-2.65	0.0083
C9	100	464.107	101.043	10.000	-1.59	0.11
C10	100	450.517	99.219	10.000	-2.95	0.0033
C11	100	473.914	97.206	10.000	-0.61	0.54
C12	100	445.526	104.296	10.000	-3.45	0.0006
C13	100	449.224	96.955	10.000	-3.08	0.0021
C14	100	453.015	86.942	10.000	-2.70	0.0071
C15	100	453.316	95.321	10.000	-2.67	0.0077
C16	100	452.035	88.727	10.000	-2.80	0.0053
C17	100	453.487	102.948	10.000	-2.65	0.0081
C18	100	471.239	100.821	10.000	-0.88	0.38
C19	100	455.464	99.577	10.000	-2.45	0.014
C20	100	447.833	107.117	10.000	-3.22	0.0013

As you can readily see, **most of the random samples have p-values less than 0.05.** (i.e. 14 out the 20 randomly simulated samples had p-values less than 0.05).

If you were conducting a hypothesis test that the population is 480 and a random sample was selected that had a p-value less than 0.05 than your conclusion would be that the random sample probably did not come from a population whose mean is 480.

It turns out that when a hypothesis test is conducted a random sample is collected and compared to the null hypothesis, H_o. If it is significantly different, based upon some predesigned significance level, then you reject the null hypothesis and consequently accept the alternative hypothesis, H_a.

In this chapter you have been introduced to the hypothesis testing model. You have learned that the first step in doing a hypothesis test is to state the null and alternative hypotheses.

If the alternative hypothesis, H_a is *two-tailed nondirectional*, the MINITAB ZTEST is written as:

MTB > ZTEST MU=(null hypothesis value) SIGMA=(assumed value), C1

If the alternative hypothesis, H_a is *one-tailed directional*,
the MINITAB ZTEST is written as:

MTB > ZTEST MU=(null hypothesis value) SIGMA=(assumed value), C1;
SUBC > ALTERNATIVE=-1.

for a one-tail directional to the *left*,
or

MTB > ZTEST MU=(null hypothesis value) SIGMA=(assumed value), C1;
SUBC > ALTERNATIVE= 1.

for a one-tail directional to the *right*.

For example, if an hypothesis test was conducted where the alternative hypothesis is that the population mean is less than 80 with a standard deviation of 5, then the MINITAB command would be:

      MTB > ZTEST MU=80 SIGMA=5 C1;
      SUBC > ALTERNATIVE=-1.

In this use of the command it is assumed that a random sample of data values have been selected from a specified population and the data values are entered into C1.

An output for this could be:

TEST OF MU = 80.000 VS MU L.T. 80.000
THE ASSUMED SIGMA = 5.00

	N	MEAN	STDEV	SE MEAN	Z	P VALUE
C1	49	78.160	4.988	0.714	-2.58	0.0051

Such an output would indicate that the null hypothesis is rejected and the alternative hypothesis is accepted. This is indicated because of the p-value, 0.0051. In fact, such a low p-value gives strength to the rejection of the null hypothesis!

Suppose that you believe that an alternative hypothesis is that the mean of a population is greater than 2.8 where the population standard deviation is 0.4. This directional alternative hypothesis would have the MINITAB form:

MTB > ZTEST MU=2.8 SIGMA=.4 C1;
SUBC > ALTERNATIVE=1.

An output for this could be:

TEST OF MU = 2.8000 VS MU G.T. 2.8000
THE ASSUMED SIGMA = 0.400

	N	MEAN	STDEV	SE MEAN	Z	P VALUE
C1	81	2.8286	0.3749	0.0444	0.64	0.26

Since the p-value (p=0.26) for this result is greater than 0.05, there would be no rejection of the null hypothesis, that is, you would fail to reject the null hypothesis.

Glossary

TERM	SECTION
Null hypothesis, H_o	9.2
Alternative hypothesis, H_a	9.2
Directional alternative hypothesis	9.2
Nondirectional alternative hypothesis	9.2
Hypothesis Testing Procedure	9.2
Type I error	9.2
Type II error	9.2
Statistically significant	9.2
Critical z score, z_c	9.3
Critical value, X_c	9.3
Level of significance, α	9.3
Chance of Type II error, β	9.3
Left critical z score, z_{LC}	9.3
Right critical z score, z_{RC}	9.3
One-tailed test, 1TT	9.3
Two-tailed test, 2TT	9.3
Left critical value, X_{LC}	9.3
Right critical value, X_{RC}	9.3
Probability value	9.4
p-Value	9.4
Observed significance level	9.4
Smallest level of significance	9.4
True level of significance	9.4

Exercises

PART I Fill in the blanks.

1. The null hypothesis is the hypothesis to be tested that is initially assumed to be _____ and formulated for the sole purpose of trying to _____. The null hypothesis is symbolized as _____.
2. The alternative hypothesis is the hypothesis which is supported by the rejection of the _____ hypothesis. The alternative is symbolized as _____.
3. When a true null hypothesis is rejected, one has committed a type _____ error.
4. When a false null hypothesis is accepted, one has committed a type _____ error.
5. A rejection of the null hypothesis means that the difference between the experimental outcome and the _____ outcome, the mean, is statistically _____.
6. If an experimental outcome statistically significant, then the _____ hypothesis is accepted.
7. The level of significance is the probability of committing a type _____ error.
8. The Greek letter used to symbolize the level of significance is _____.
9. As the significance level decreases the chance of committing a Type I error _____, while the chance of committing a Type II error _____.
10. The Greek letter used to symbolize a Type II error is _____.
11. Whenever the alternative hypothesis is directional we have a _____ tailed test.
12. Whenever the alternative hypothesis is nondirectional we have a _____ tailed test.
13. In a two-tailed test, the significance level is divided _____ between the two tails. Therefore, for $\alpha = 1\%$, the percent of the area in each tail is _____.
14. The general formula used to calculate a critical value, X_c, is:
    X_c = _____ + (_____) (_____).
15. The critical z score for a one-tailed test at $\alpha = 5\%$ is either _____ or _____.
16. The two critical z scores for a two-tailed test at $\alpha = 5\%$ are _____ and _____.
17. The critical z score for a one-tailed test at $\alpha = 1\%$ is either _____ or _____.
18. The two critical z scores for a two-tailed test at $\alpha = 1\%$ are _____ and _____.
19. In a hypothesis test, if the experimental outcome falls beyond the critical outcome, then the conclusion is stated as _____ H_o and _____ H_a.
20. In a hypothesis test, if the experimental outcome does not fall beyond the critical outcome, then the conclusion is stated as _____ H_o.
21. The p-value, or observed significance level, of a hypothesis test is the _____ level of significance at which the null hypothesis H_o can be rejected using the experimental outcome (or sample outcome).
22. Reporting a p-value of 0.89% indicates that the null hypothesis can be rejected at values of α as small as $\alpha =$ _____ %.
23. In a hypothesis test, the null hypothesis is rejected if the p-value is _____ or _____ the significance level, α. Thus, the smaller the p-value, the stronger the evidence against the _____ hypothesis provided by the experimental outcome.
24. When interpreting p-values, a rule of thumb that can be followed is:
    a) if the p-value is less than 1%, the experimental outcome (or sample outcome) is ____ significant.
    b) if the p-value is between 1% and 5%, the experimental outcome (or sample outcome) is ____ .
    c) if the p-value is between 5% and 10%, the experimental outcome (or sample outcome) is ____ significant.
    d) if the p-value is greater than 10%, the experimental outcome (or sample outcome) is ____ significant.
25. If the p-value of a statistical hypothesis test is equal to 3%, can the null hypothesis be rejected at $\alpha = 5\%$? For this test, the smallest significance level of α in which the null hypothesis can be rejected is _____.
26. In a two-tailed test, the p-value of a hypothesis is equal to ____ the percent of area in the tail beyond the direction of the experimental outcome (or sample outcome) of the test.

PART II Multiple choice questions.

1. If a false null hypothesis is rejected then
   a) a Type I error has been made
   b) a Type II error has been made
   c) no error has been made
2. If a false alternative hypothesis is accepted, then:
   a) a Type I error has been made
   b) a Type II error has been made
   c) no error has been made
3. If the experimental outcome is beyond the critical value, X_c, then the outcome is considered:

a) to be a Type I error
b) statistically significant
c) to be a Type II error
4. If a nondirectional alternative hypothesis is being tested then:
a) a one-tailed test is performed
b) a two-tailed test is performed
c) not enough information is given
5. If the null hypothesis is rejected then the alternative hypothesis is:
a) accepted
b) rejected
c) neither accepted nor rejected
6. The significance level, α, is the probability of:
a) rejecting a false null hypothesis
b) making a Type II error
c) rejecting a true null hypothesis
7. When the experimental outcome falls beyond the critical outcome(s) then the decision is to:
a) reject H_o, and accept H_a
b) fail to reject H_o
c) seek more information
8. When the experimental outcome does not fall beyond the critical outcome(s) then the decision is to "fail to reject H_o." What type of error may have been made?
a) no error
b) Type I error
c) Type II error
9. When a directional hypothesis is being tested and the critical value is to the right of the mean then the alternative hypothesis must be of the form:
a) less than
b) more than
c) not equal to
10. When a nondirectional hypothesis is being tested then and the significance level is 5% then the critical z scores are:
a) ± 1.96
b) ± 2.58
c) ± 1.65

PART III Answer each statement True or False.

1. In the hypothesis testing procedure, the main objective of the statistician is to present enough data to show that the null hypothesis is not correct.
2. If the null hypothesis is rejected then the alternative hypothesis is accepted.
3. A Type I error is an error that is made when a true null hypothesis is incorrectly rejected.
4. "Statistically significant" means that the experimental outcome is different enough from the expected outcome to support the alternative hypothesis.
5. When the difference between the experimental outcome and the expected outcome is *not* statistically significant the statistician concludes that the null hypothesis is true.
6. The level of significance is the probability of an experimental outcome supporting the null hypothesis.
7. As the significance level decreases, the chance of committing a Type I error decreases.
8. As the significance level increases, the chance of committing a Type II error decreases.
9. The decision rule is usually determined after a suitable significance level is chosen.
10. A directional alternative hypothesis results from a two-tailed test.
11. The p-value, or observed significance level is the largest level of significance, α, at which the null hypothesis H_o can be rejected.
12. P-values can be represented as either percents or decimals.
13. If a null hypothesis is rejected at an $\alpha=1\%$ then the p-value for the experimental outcome must be less than 0.01.
14. If for an experimental outcome, the p-value is greater than 10%, then the experimental outcome is not significant.

PART IV Problems.

For problems 1 - 10 state the null and alternative hypotheses assuming you are the individual or agency performing the hypothesis test. Also indicate whether these tests are directional or nondirectional.

1. The Department of Consumer Affairs is going to conduct a study to determine if the following claim made by Height Examiners Inc. is an exaggeration. "The mean weight loss during the first 10 days is 12 pounds!"
2. The Environmental Protection Agency is going to conduct a series of tests to determine if V & D Inc. is exceeding the allowable pollutant levels. The allowable pollutant level is a mean of 13 ppm.
3. According to a research study linking maternal smoking and childhood asthma, 10% of the children whose mothers smoked during pregnancy are asthmatic. A local official from a public health agency has doubts about this statement and decides to test this claim against her suspicion that it is not 10%.
4. The National Safety Council has recently stated that only 11% of all licensed drivers wear seat belts. A research team believes this claim is too low and decides to test it.

5. In a recent issue of *Psychology for Tomorrow* researchers report that 20% of the wives who hold full-time jobs now earn more than their husbands. A sociologist from the midwest believes this claim is too high and decides to test the claim.
6. The National Education Association (NEA) states that the national mean salary of public school teachers in the United States is $25,000. A teacher's union wants to show the their mean salary is less than the national mean.
7. A consumer protection group is interested in testing a new energy-saving refrigerator because of their suspicion that the new refrigerator uses more than the advertised mean consumption of 4.9 kilowatt hours of electricity per day.
8. A national magazine recently claimed that 3 out of every 10 women between the ages of 16 and 40 years have a form of herpes virus. Researchers representing the New World Medical Society believe the claim is not true and decide to test the claim.
9. The U.S. Navy is going to conduct a series of tests to determine if an automatic opening device for parachutes meets the manufacturer's claim of a mean opening time of 6 seconds.
10. According to the National Center for Education Statistics, the national average daily rate of attendance for kindergarten through 12th grade is 94%. The school superintendent of a city school district believes her district has a daily average attendance rate that is higher than the national average.

For problems 11 - 15 you are given the null hypothesis (H_o), the alternative hypothesis (H_a), a diagram indicating the distribution model, the mean, standard deviation, critical outcome(s) and an experimental outcome.

11. H_o: mean = 100

    H_a: mean ≠ 100

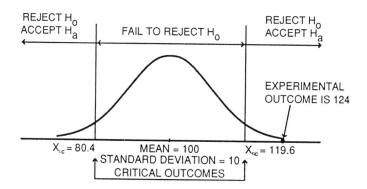

a) State the decision rule.
b) Do you reject the null hypothesis?
c) Do you accept the alternative hypothesis?

12. H_o: percent = 50

    H_a: percent > 50

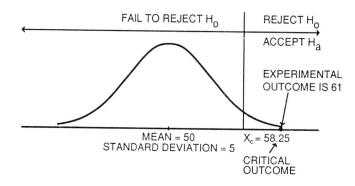

a) State the decision rule.
b) Do you reject the null hypothesis?
c) Do you reject the alternative hypothesis?

13. H_o: mean = 114

    H_a: mean < 114

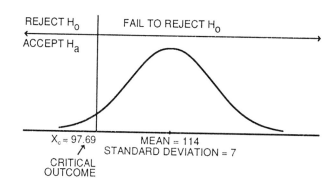

a) State the decision rule.
If the experimental outcome is 100, then:
b) Do you reject the null hypothesis?
c) Do you accept the alternative hypothesis?
If the experimental outcome is 93, then:
d) Do you reject the null hypothesis?
e) Do you accept the alternative hypothesis?

14. H_o: mean = 12
    H_a: mean < 12

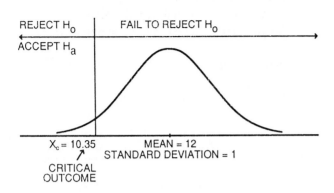

a) State decision rule.
If the experimental outcome is 10, then:
b) Do you reject the null hypothesis?
c) Do you accept the alternative hypothesis?
If the experimental outcome is 15, then:
d) Do you reject the null hypothesis?
e) Do you accept the alternative hypothesis?

15. H_o: percent = 20%
    H_a: percent ≠ 20%

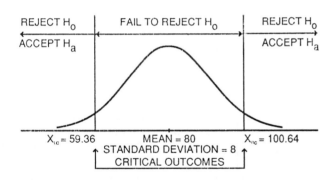

a) State the decision rule.
b) If the experimental outcome is 98, state your conclusion.
c) If the experimental outcome is 55, state your conclusion.
d) If the experimental outcome is 103, state your conclusion.

For the problems 16–20 use the given information to:
a) Label each diagram with critical outcome(s), reject and fail to reject areas and the experimental outcome.
b) Determine the conclusion.
c) Is the difference between the experimental outcome and the expected outcome, the mean statistically significant?

16. H_o: percent = 90
    H_a: percent > 90
    Decision Rule: Reject H_o if experimental outcome is greater than 96.99. The experimental outcome is 96.

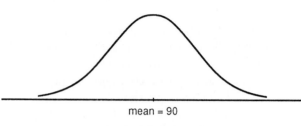

17. H_o: mean = 190
    H_a: mean ≠ 190
    Decision Rule: Reject H_o if the experimental outcome is either less than 170.4 or greater than 209.6. The experimental outcome is 212.

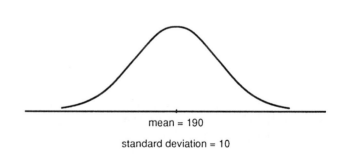

18. H_o: mean = 14.5
    H_a: mean < 14.5
    Decision Rule: Reject H_o if the experimental outcome is less than 11.2. The experimental outcome is 10.7.

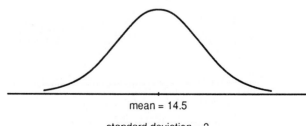

19. H_o: percent = 80
H_a: percent < 80
Decision Rule: Reject H_o if experimental outcome is less than 306.8. The experimental outcome is 307.

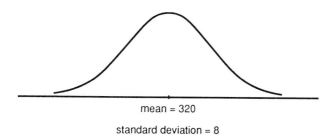

mean = 320
standard deviation = 8

20. H_o: mean = 520
H_a: mean > 520
Decision Rule: Reject H_o if experimental outcome is greater than 543.3. The experimental outcome is 549.

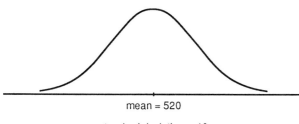

mean = 520
standard deviation = 10

For problems 21-30 go back to problems 11-20 and:
a) Change each critical outcome to a z score.
b) Use TABLE II to determine the chance of a Type I error.

31. An unemployment agency claims that the mean age of recipients of unemployment benefits is 37 years with a standard deviation of 5 years. A trade union association believes this claim is not correct. The association randomly interviews 400 recipients of unemployment benefits and obtains a mean age of 35 years. The statistician computes the decision rule for this test to be: reject H_o if the sample mean age is greater than 37.49 or less than 36.51 years.
Use the following model to perform this hypothesis, and:
a) State the null and alternative hypotheses.
b) Label the diagram with the critical outcome(s), reject and fail to reject areas, and the experimental outcomes.

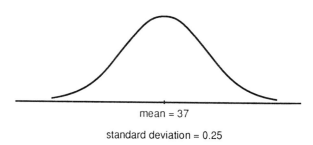

mean = 37
standard deviation = 0.25

c) Determine the p-value for the experimental outcome.
d) State your conclusion.

32. According to a recent insurance company claim an American family of four spends a mean of $2000 per year for medical care with a standard deviation of $600. The medical association believes this claim is too high. They randomly survey 3600 families and find the mean expenditure to be $1990. Their decision rule is to reject H_o if the sample mean is less than $1980.
Use the following model to perform this hypothesis, and:
a) State the null and alternative hypothesis.
b) Label the diagram with the critical outcome(s), reject and fail to reject areas, and the experimental outcome.

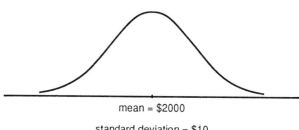

mean = $2000
standard deviation = $10

c) Determine the p-value for the experimental outcome.
d) State your conclusion.

33. A manufacturer wants to advertise a product in *Playperson* magazine. The product is specifically designed for a person aged 21 through 35 years. The magazine has stated that 70% of its subscribers are in this age group. The marketing supervisor for the manufacturer believes the claim is too high and randomly surveys 100 of the magazine's subscribers and determines that 64 are in this age group. The decision rule is reject H_o if the sample outcome is less than 62.
a) State the null and alternative hypothesis.

b) Label the diagram with the critical outcome(s), reject and fail to reject areas, and the experimental outcome.

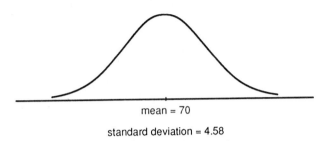

mean = 70
standard deviation = 4.58

c) Determine the p-value for the experimental outcome.
d) State your conclusion.

34. According to a recent census report 50% of U.S. families earn more than $20,000 per year. A sociologist from a northeastern university believes this percent is too low. He randomly selects 100 families and determines that 64 have incomes of more than $20,000. The decision rule is to reject H_0 if more than 62 have incomes greater than $20,000.
   a) State the null and alternative hypothesis.
   b) Label the diagram with the critical outcome(s), reject and fail to reject areas, and the experimental outcome.

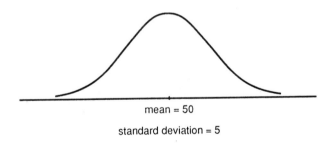

mean = 50
standard deviation = 5

c) Determine the p-value for the experimental outcome.
d) State your conclusion.

For problems 35 - 40 the null and alternative hypotheses, the mean and standard deviation, the significance level, α, and a diagram will be given. Calculate the critical outcome(s) and formulate the decision rule. Label the diagram with the mean, standard deviation, significance level, critical outcome(s) and reject and fail to reject areas. Answer the appropriate questions.

35. H_0: mean = 2500
    H_a: mean > 2500
    mean = 2500
    standard deviation = 100
    significance level, α = 5%

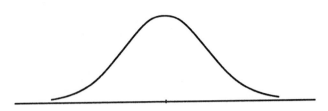

a) If the experimental outcome is:
   1) 2715   2) 2600   3) 2400   4) 1200
   then, determine the conclusion.
b) Which of these outcomes are statistically significant?

36. H_0: mean = 18
    H_a: mean < 18
    mean = 18
    standard deviation = 3
    significance level, α = 1%

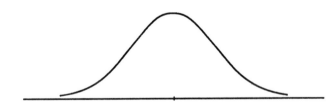

a) If the experimental outcome is:
   1) 13   2) 21   3) 11   4) 28
   then, determine the conclusion.
b) Which of these outcomes are statistically significant?

37. H_0: mean = 90
    H_a: mean ≠ 90
    mean = 90
    standard deviation = 10
    significance level, α = 5%

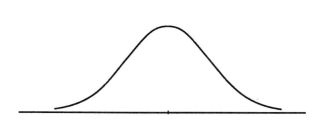

a) If the experimental outcome is:
   1) 83   2) 79   3) 110   4) 103
   then, determine the conclusion.
b) Which of these outcomes are statistically significant?

38. H_o: percent = 30
    H_a: percent > 30
    mean = 300
    standard deviation = 14.5
    significance level, α = 5%

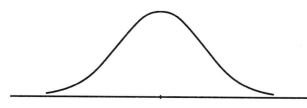

a) If the experimental outcome is:
   1) 314   2) 330   3) 320   4) 298
   then, determine the conclusion.
b) Which of these outcomes are statistically significant?

39. H_o: percent = 80
    H_a: percent < 80
    mean = 2000
    standard deviation = 20
    significance level, α = 1%

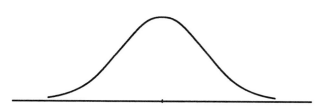

a) If the experimental outcome is:
   1) 1874   2) 1884   3) 2094   4) 1984
   then, determine the conclusion.
b) Which of these outcomes are statistically significant?

40. H_o: percent = 40
    H_a: percent $\neq$ 40
    mean = 16
    standard deviation = 3.1
    significance level, α = 1%

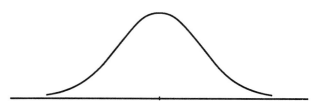

a) If the experimental outcome is:
   1) 25   2) 13   3) 15   4) 2
   then, determine the conclusion.
b) Which of these outcomes are statistically significant?

For problems 41 - 46:

a) Formulate the null and alternative hypotheses.
b) Use the mean, standard deviation, and significance level, α, given in the problem to calculate the critical value(s) and formulate the decision rule.
c) Label the diagram with the mean, standard deviation, significance level, reject and fail to reject areas.
d) Determine the p-value for the experimental outcome.
e) Use the experimental outcome to determine the conclusion.
f) State whether the experimental outcome is statistically significant.
g) Answer the questions in each problem.

41. An independent research group is interested in showing that the percent of babies delivered by Cesarean Section is increasing. Last year, 20% of the babies born were delivered by Cesarean Section. The research group randomly inspects the medical records of 100 recent births and finds that 25 of the births were by Cesarean Section. Can the research group conclude that the percent of births by Cesarean section has increased? Use mean = 20, standard deviation = 4 and α = 5%.

42. According to the latest figures unemployment is at 8.7%. The Congressman from the third district believes that the unemployment rate is lower in his district. To test his belief he interviews 200 residents of his district and finds 9 of them to be unem-

ployed. Is the Congressman's belief correct? Use mean = 17.4, standard deviation = 3.99 and α = 5%.

43. According to the norms established for a mathematics achievement test fifth graders should have a mean score of 88. If 64 randomly selected fifth graders from a local school district average 89.4, can the district superintendent conclude that her fifth graders are above the norm? Use mean = 88, standard deviation = 1.2 and α = 1%.

44. A beverage company has a machine that is supposed to fill soda bottles with two liters of soda. To check that the machine is operating correctly, (i.e., not underfilling or overfilling the bottles), a sample of 81 bottles was selected and the contents of each bottle was determined.

    The sample mean was 1.93 liters of soda. Does this indicate that the machine is improperly filling the bottles? Use mean = 2.0 liters, standard deviation = 0.02 liter and α = 1%.

45. A pharmaceutical company claims to have developed a new antibiotic which is more effective against type A bacteria than its current product. The current product is 90% effective against type A bacteria. To test its claim a random sample of 400 individuals who were infected with this bacteria are treated with the new antibiotic and 380 recover. Do these results support the pharmaceutical company's claim? Use mean = 360, standard deviation = 6 and α = 1%.

46. A psychiatrist wants to determine if a small amount of alcohol decreases the reaction time in adults. The mean reaction time for a specified test is 0.20 seconds. A sample of 64 people were given a small amount of alcohol and their reaction time averaged 0.17 seconds. Does this sample result support the psychiatrist's hypothesis? Use mean = 0.20 sec., standard deviation = 0.01 sec. and α = 1%.

47. The Federal Aviation Agency(FAA) claims that 10% of all U. S. airline flights are delayed at least 15 minutes. The Air Transport Association(ATA) who believes the claim is too high decides to test this claim. The ATA selects a random sample of 1000 U. S. airline flights and determines that 98 flights were delayed at least 15 minutes. Use mean = 100 and standard deviation = 9.49.
    a) What is the p-value for this hypothesis test?
    b) Can the hypothesis be rejected at α = 1%?
    c) What is the smallest level of significance that the test can be rejected?

48. The college newspaper at Rich Man's College states that the average time that students study per week is 8.5 hours. A skeptical student decides to test the claim to determine if the claim is true. The student selects a random of 400 students from the student body and determines the average study time of the sample to be 9.2 hours. Use mean = 8.5 and standard deviation = 0.12.
    a) What is the p-value for this hypothesis test?
    b) Can the hypothesis be rejected at α = 5%?
    c) What is the smallest level of significance that the test can be rejected?

PART V **What Do You Think?**

1. **What Do You Think?**

   Examine the graphic within the USA Snapshot entitled *Back to School*.

   a) What different populations are being represented within the USA snapshot?

   If we assume the stated figures within the graphic represent the population parameters for each population, then determine:
   b) the null hypothesis for each percentage figure within the graphic.
   c) an alternative hypothesis if you believe that the figure is too high for the general population, but too low for the African American population, and the Hispanic population.
   d) indicate the type of alternative hypothesis for each population.
   e) determine the type of test (i.e. 1TT or 2TT), and

on which side of the curve the critical outcome would be positioned.

f) describe how you would select an appropriate sample to test the null hypothesis.

g) When calculating the expected results in step 2 of the hypothesis testing procedure, what assumption are you making? Explain why it is necessary to make this assumption.

h) Is there an overlap between the members of the three populations shown in the USA snapshot? If so, then explain why you think that the percentage for the general population can be smaller than the percentages of the other populations that are represented within the general population?

2. **What Do You Think?**

Examine the graphic within the USA Snapshot entitled *Spending for the Holidays*.

a) What different populations are being represented within the USA snapshot?
b) Does the average amount of money spent by each age group represent both males and females?
c) What claim do the men make regarding their spending during the holidays?
d) What claim do the women make regarding their spending during the holidays?
e) How many people were surveyed to determine their spending plans for the holiday? Do you have any information regarding their breakdown with respect to their age and sex? Why do think that information might be helpful in interpreting the results stated in the snapshot?

If you are planning to test the statement about what men spend for the holidays and your belief is that the figure within the graphic is too high, what would:

f) the null and alternative hypothesis be for your test?
g) indicate the type of alternative hypothesis for each population.
h) determine the type of test (i.e. 1TT or 2TT), and on which side of the curve the critical outcome would be positioned?
i) describe how you would select an appropriate sample to test the null hypothesis.
j) When calculating the expected results in step 2 of the hypothesis testing procedure, what assumption are you making? Explain why it is necessary to make this assumption.
k) For the hypothesis test you outlined regarding the amount men spend for the holidays, what would be the conclusion to your test if the experimental outcome were to fall to the left of your critical outcome?

If you are planning to test the statement about what women spend for the holidays and your belief is that the figure within the graphic is too low, what would:

l) the null and alternative hypothesis be for your test?
m) indicate the type of alternative hypothesis for each population.
n) determine the type of test (i.e. 1TT or 2TT), and on which side of the curve the critical outcome would be positioned?
o) describe how you would select an appropriate sample to test the null hypothesis.
p) When calculating the expected results in step 2 of the hypothesis testing procedure, what assumption are you making? Explain why it is necessary to make this assumption.
q) For the hypothesis test you outlined regarding the amount women spend for the holidays, what would be the conclusion to your test if the experimental outcome were to fall to the right of your critical outcome?

If you are planning to test the statement about what people within the age group 25 - 34 spend for the holidays and your belief is that the figure within the graphic is too low, what would:

r) the null and alternative hypothesis be for your test?
s) indicate the type of alternative hypothesis for each population.
t) determine the type of test (i.e. 1TT or 2TT), and on which side of the curve the critical outcome would be positioned?
u) describe how you would select an appropriate sample to test the null hypothesis.
v) When calculating the expected results in step 2 of the hypothesis testing procedure, what assumption are you making? Explain why it is necessary to make this assumption.
w) For the hypothesis test you outlined regarding the amount people within the age group 25-34 spend for the holidays, what would be the conclusion to your test if the experimental outcome were to fall to the right of your critical outcome?

PART VI Exploring DATA With MINITAB

1. Use MINITAB to simulate the experiment of randomly selecting 50 data values from a normal population whose mean=75 and standard deviation is 10. Repeat the simulation 20 times. For each random sample use the ZTEST command to test the hypothesis that MU=78 and SIGMA=10 against the alternative hypothesis MU N.E. 78.
   (a) Before you actually use MINITAB estimate how many of the twenty random samples you expect to have a p-value < 0.05.
   (b) How many random sample samples actually had p-values less than 0.05?
   (c) Repeat the exercise, and compare your results.
   (d) If the results are different explain the differences.

2. Use MINITAB to simulate the experiment of randomly selecting 50 data values from a normal population whose mean=75 and standard deviation is 10. Repeat the simulation 20 times. For each random sample use the ZTEST command to test the hypothesis that MU=78 and SIGMA=10 against the alternative hypothesis the MU < 78.
   (a) Before you actually use MINITAB estimate how many of the twenty random samples you expect to have a p-value < 0.05.
   (b) How many random samples actually had p-values less than 0.05?
   (c) Repeat the exercise, and compare your results.
   (d) If the results are different explain the differences.

3. Use MINITAB to simulate the experiment of randomly selecting 40 data values from a normal population whose mean=16 and standard deviation is 2. Repeat the simulation 20 times. For each random sample use the ZTEST command to test the hypothesis that MU=17 and SIGMA=2 against the alternative hypothesis MU > 17.
   (a) Before you actually use MINITAB estimate how many of the twenty random samples you expect to have a p-value < 0.05.
   (b) How many random sample samples actually had p-values less than 0.05?
   (c) Repeat the exercise, and compare your results.
   (d) If the results are different explain the differences.

4. (a) Use the SET command to place the following data values into a MINITAB worksheet in column C1:
   44 66 84 33 22 99 97 66 58 23 19
   95 86 66 27 38 76 85 91 39 43 51
   88 21 67 27 32 10 64 29 45 75 57
   (b) Use the ZTEST command to test the claim that the sample of data values in C1 is from a population whose mean is less than 60. Test at $\alpha=5\%$.

PART VII Projects.

1. Read a research article from a professional journal in a field such as medicine, education, marketing, science, social science, etc. Write a summary of the research article which includes the following:

   a) Name of article, author and journal (include volume, number and date).
   b) The statement of the research problem.
   c) The null and alternative hypotheses.
   d) How the experiment was conducted (include the sample size and the sampling technique).
   e) Determine the experimental outcome and the level of significance.
   f) State the conclusion in terms of the null and alternative hypotheses.

2. Select a claim found in a magazine, newspaper or stated on television or radio. Formulate an opinion regarding this claim, and devise a hypothesis testing procedure to test the claim. Be sure to include the following:

   a) The source from which you found this claim.
   b) The null and alternative hypotheses.
   c) Define the population and how you would go about selecting a sample from this population.
   d) State the method you would use to obtain your data from the sample.

PART VIII **Database.**

The following exercises refer to the file DATABASE listed in Appendix A. We have indicated the appropriate MINITAB commands that are necessary to answer each exercise.

1. Using the MINITAB commands:
    RETRIEVE and TTEST
    Retrieve the file DATABASE.MTW and identify the column number:
    for each of the variable names:
    a) AGE
    b) AVE
    c) STY
    d) GPA
    e) GFT
    For each variable use the TTEST command to test the null hypothesis that that the population mean, mu = k
    a) k= 18 for the variable AGE.
    b) k= 85 for the variable AVE.
    c) k= 8 for the variable STY.
    d) k= 2.75 for the variable GPA.
    e) k= 100 for the variable GFT.
    Write the TTEST command if the alternative hypothesis for each variable is *nondirectional*. Have your instructor help you interpret the output.

2. Using the MINITAB commands:
    RETRIEVE and TTEST
    Retrieve the file DATABASE.MTW and identify the column number:
    for each of the variable names:
    a) AGE
    b) AVE
    c) STY
    d) GPA
    e) GFT
    For each variable use the TTEST command to test the null hypothesis that the population mean, mu = k
    a) k= 18 for the variable AGE.
    b) k= 85 for the variable AVE.
    c) k= 8 for the variable STY.
    d) k= 2.75 for the variable GPA.
    e) k= 100 for the variable GFT.
    Write the TTEST command if the alternative hypothesis for each variable is *directional* by adding the following additional subcommand:
    a) ALTERNATIVE = +1 for the variable AGE.
    b) ALTERNATIVE = -1 for the variable AVE.
    c) ALTERNATIVE = +1 for the variable STY.
    d) ALTERNATIVE = +1 for the variable GPA.
    e) ALTERNATIVE = -1 for the variable GFT.
    Have your instructor help you interpret the output.

CHAPTER 10
HYPOTHESIS TESTING INVOLVING A POPULATION MEAN

❋ 10.1 Introduction

The general concept of hypothesis testing was introduced in Chapter 9. In this chapter we will examine in detail the procedure required to test hypotheses (or claims) made about a population mean **using information obtained from a sample drawn from the population**.

To illustrate the use of this procedure, consider the following story about the Precision Tool and Die Company, which was the principal employer in a southern Nevada town. This company has been credited with building up the town not only from an economic point of view, but also in a physical sense, since they supplied precision manufactured screws used to assemble many of the town's building projects. In fact, today the company still boasts: "we're the company that screwed up this town."

The success of this company can be attributed to the quality control department whose motto is: "we manufacture only precision screws." The quality control manager is responsible for testing every batch of screws before they leave the plant to insure they meet the manufacturers specifications.

To determine if the **entire batch** of 4" screws have been manufactured to the required length specification, the manager **randomly selects a sample** of the screws and **computes the mean length of the sample. Her decision to reject the entire batch is based upon a comparison of the sample mean to the length specification. If the difference between the sample mean length and the manufacturer's length specification is statistically significant, then the entire batch will be rejected**. Otherwise, the manufacturer is **confident** that the entire batch meets the required length specification based upon the sample tested.

The underlying theorems which support the statistical comparison made by the quality control manager in the previous story are the **Central Limit Theorem and the Theorem on the Properties of Sampling Distributions** (discussed in Chapter 8).

Before we can illustrate the use of these theorems in the statistical calculations required to perform the **testing of hypotheses involving a population mean**, we need to develop a procedure for testing a hypothesis involving a population mean.

The development of the hypothesis testing procedure concerning a population mean will incorporate the previous two theorems into the general hypothesis testing procedure discussed in Chapter 9. Let's now develop the hypothesis testing procedure involving a population mean.

❋ 10.2 Hypothesis Testing Involving a Population Mean

We are now going to outline the hypothesis testing procedure for testing the mean of a population by modifying the general five step hypothesis testing procedure developed in Chapter 9. The first step of the hypothesis testing procedure is to formulate the null and alternative hypotheses. Let's examine how to state these hypotheses for a test involving a population mean.

Step 1. *Formulate the two hypotheses, H_o and H_a.*

Null Hypothesis:
In general, the null hypothesis has the following form.

Null Hypothesis Form:

H_o: The mean of the population, μ, is claimed to be equal to a numerical value which will be symbolized as: μ_o.

Thus, the null hypothesis can be symbolized as:

$$H_o: \mu = \mu_o$$

Alternative Hypothesis:
Since the alternative hypothesis, H_a, can be stated as either *greater than*, *less than*, or *not equal* to the numerical value, μ_o, then the alternative hypothesis can have one of three forms:

Form (a): Greater Than Form For H_a

H_a: The mean of the population, μ, is claimed to be **greater than** the numerical value, μ_o.

This is symbolized as:

$$H_a: \mu > \mu_o$$

Form (b): Less Than Form For H_a

H_a: The mean of the population, μ, is claimed to be **less than** the numerical value, μ_o.
This is symbolized as:

$$H_a: \mu < \mu_o$$

Form (c): Not Equal Form For H_a

H_a: The mean of the population, μ, is claimed to be **not equal** to the numerical value, μ_o.
This is symbolized as:

$$H_a: \mu \neq \mu_o$$

Step 2. *Design an experiment to test the null hypothesis, H_o.*

Two important design considerations are choosing a sample size, n, and determining the sampling distribution. Since we are testing a hypothesis about a population mean using a sample mean, $\bar{x}$, we use as the hypothesis test model the sampling distribution of the mean and its properties, which were discussed in Chapter 8. Using this information, we can draw the following conclusions.

Under the following assumptions that:
a) the null hypothesis, H_o, is true, i.e. $\mu = \mu_o$, and,
b) the standard deviation of the population, σ, is **known**, and,
c) the sample size, n, is greater than 30,

then we can conclude that:
a) the sampling distribution of the mean is approximately **normal** and serves as the hypothesis test model,
b) the mean of the sampling distribution of the mean is:

$$\mu_{\bar{x}} = \mu_o$$

(i.e. the mean, $\mu_{\bar{x}}$, is equal to the claimed population mean, μ_o).

c) the standard deviation of the sampling distribution of the mean is:

$$\sigma_{\bar{x}} = \frac{\sigma}{\sqrt{n}}$$

This is illustrated in Figure 10.1.

HYPOTHESIS TESTING MODEL

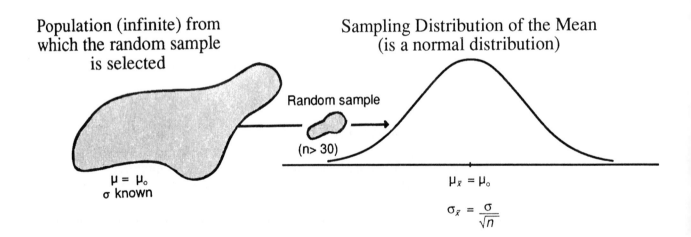

Figure 10.1

In theory, the population being sampled must be infinite to use the standard deviation formula:

$$\sigma_{\bar{x}} = \frac{\sigma}{\sqrt{n}}$$

However it is appropriate to use this formula when the population is *finite* and the population size, N, is large relative to the sample size, n. The rule used to determine whether the population is large relative to the sample is: the sample size, n, is less than 5% of the population size, N. That is, $\frac{n}{N} < 5\%$.

Step 3. *Formulate the decision rule.*

a) Determine the type of alternative hypothesis (i.e., is the alternative hypothesis **directional** or **nondirectional**?)
b) Determine the type of hypothesis test (i.e., is the test **1TT** or **2TT**?)
c) Identify the significance level (i.e., is $\alpha = 1\%$ or $\alpha = 5\%$?)
d) Calculate the critical value(s) using the formula:
$$X_c = \mu_{\bar{x}} + (z_c)(\sigma_{\bar{x}})$$
e) Construct the appropriate *hypothesis testing model*.
f) State the decision rule.

Step 4. *Conduct the experiment.*

Randomly select the necessary sample data from the population to calculate the experimental outcome, the sample mean, symbolized as $\bar{x}$.

Step 5. *Determine the conclusion.*

Compare the sample mean, $\bar{x}$, (referred to as the **experimental outcome**) to the **critical value**, X_c, of the decision rule and draw one of the following conclusions:

(a) **Reject H_0 and Accept H_a at α**

or

(b) **Fail to reject H_0 at α**

In summary, the procedure just outlined to perform a hypothesis test involving a population mean, μ, where the **population standard deviation, σ, is known** and **n is greater than 30**, contains three refinements to the general hypothesis testing procedure developed in Chapter 9. They are:

I The **sampling distribution of the mean** is used as the appropriate hypothesis testing model.
II The properties of the sampling distribution of the mean are used to calculate the expected values.
III The formula: $X_c = \mu_{\bar{x}} + (z_c)(\sigma_{\bar{x}})$ is used to calculate the critical value(s) for the decision rule.

Example 10.1 demonstrates the use of the hypothesis testing procedure to test a population mean.

Example 10.1

An automatic opening device for parachutes has a stated mean release time of 10 seconds with a standard deviation of 3 seconds. A local parachute club decides to test this claim against the alternative hypothesis that the release time is not 10 seconds. The club purchases 36 of these devices and finds that the mean release time is 8.5 seconds. Does this sample result indicate that the opening device does not have a mean release time of 10 seconds? Use $\alpha = 1\%$.

Solution:

Step 1. *Formulate the hypotheses.*

H_o: The population mean release time is 10 seconds.
This is symbolized as: $H_o: \mu = 10$ secs.

H_a: The population mean release time *is not* 10 seconds.
This is symbolized as: $H_a: \mu \neq 10$ secs.

Step 2. *Design an experiment to test the null hypothesis, H_o.*

To test the null hypothesis, the club decides to purchase 36 parachutes.

Since the sample size, n, is greater than 30, we can use the Central Limit Theorem. Using the results of this theorem, the sampling distribution of the mean is a normal distribution and *serves as the hypothesis testing model.*

Assuming the null hypothesis is true, the *expected results* for the sampling distribution of the mean are:

$$\mu_{\bar{x}} = \mu$$

$$\mu_{\bar{x}} = 10 \text{ seconds}$$

and

$$\sigma_{\bar{x}} = \frac{\sigma}{\sqrt{n}}$$

$$\sigma_{\bar{x}} = \frac{3}{\sqrt{36}}$$

$$\sigma_{\bar{x}} = 0.50 \text{ seconds}$$

Figure 10.2 illustrates the sampling distribution of the mean hypothesis testing model.

The sampling distribution of the mean
(is a normal distribution)

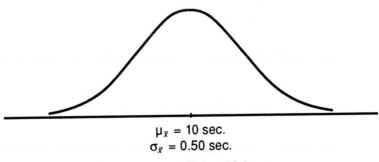

$\mu_{\bar{x}} = 10$ sec.
$\sigma_{\bar{x}} = 0.50$ sec.

Figure 10.2

Step 3. *Formulate the decision rule.*

    a)   the alternative hypothesis is nondirectional.
    b)   the type of test is two-tailed, 2TT.
    c)   the significance level is: $\alpha = 1\%$.
    d)   To calculate the critical values for a two-tailed test at $\alpha = 1\%$, first find the two critical z scores, z_{LC} and z_{RC}.

Using **TABLE II: The Normal Curve Area Table, found in Appendix D: Statistical Tables**, we find:

$$z_{LC} = -2.58 \text{ and } z_{RC} = 2.58$$

The critical values, X_{LC} and X_{RC}, are computed using the following formulas:

$$X_{LC} = \mu_{\bar{x}} + (z_{LC})(\sigma_{\bar{x}})$$
$$X_{LC} = 10 + (-2.58)(0.5)$$
$$X_{LC} = 8.71 \text{ seconds}$$

and

$$X_{RC} = \mu_{\bar{x}} + (z_{RC})(\sigma_{\bar{x}})$$
$$X_{RC} = 10 + (2.58)(0.5)$$
$$X_{RC} = 11.29 \text{ seconds}$$

    e)   Figure 10.3 illustrates the decision rule for this hypothesis test.

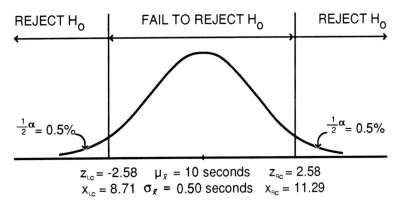

Figure 10.3

The decision rule is:
    f)   Reject the null hypothesis if the experimental outcome (the sample mean) is less than 8.71, or more than 11.29 seconds.

Step 4. *Conduct the experiment.*

The **sample mean** that represents the experimental outcome is:

$$\bar{x} = 8.5 \text{ seconds.}$$

Step 5. *Determine the conclusion.*

Since the experimental outcome is less than 8.71 seconds; we *reject* H_o and *accept* the alternative hypothesis, H_a, at $\alpha = 1\%$. Therefore, the mean release time for the automatic opening device is **significantly different** than 10 seconds. ■

Example 10.2

A tire manufacturer advertises that its brand of radial tires has a mean life of 40,000 miles with a standard deviation of 1,000 miles.

A consumer's research team decides to investigate this claim after receiving several complaints from people who believe this advertisement is false (i.e. the mean life of 40,000 miles is too high).

If the research team tests 100 of these radial tires and obtains a sample mean tire life of 39,750, is the advertisement legitimate? Use $\alpha = 1\%$.

Solution:

Step 1. *Formulate the hypotheses.*

H_o: The population mean tire life is 40,000 miles.
This is symbolized as:

$$H_o: \mu = 40{,}000 \text{ miles}$$

H_a: The population mean tire life is less than 40,000 miles.
This is symbolized as:

$$H_a: \mu < 40{,}000 \text{ miles}$$

Step 2. *Design an experiment to test the null hypothesis, H_o.*

Under the assumption that the null hypothesis is true and the sample size, n, is 100, then the *expected results* are:

$$\mu_{\bar{x}} = \mu$$

$$\mu_{\bar{x}} = 40{,}000 \text{ miles}$$

and

$$\sigma_{\bar{x}} = \frac{\sigma}{\sqrt{n}}$$

$$\sigma_{\bar{x}} = \frac{1000}{\sqrt{100}}$$

$$\sigma_{\bar{x}} = 100 \text{ miles}$$

From the Central Limit Theorem, the sampling distribution of the mean is approximately normal and serves as the hypothesis testing model. This is illustrated in Figure 10.4.

The sampling distribution of the mean
(is a normal distribution)

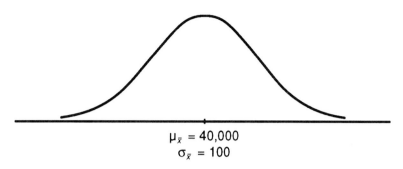

$\mu_{\bar{x}} = 40,000$
$\sigma_{\bar{x}} = 100$

Figure 10.4

Step 3. *Formulate the Decision Rule.*

a) The alternative hypothesis is directional, since the research team is trying to show that the mean tire life is too high.
b) The type of test is one-tailed on the left side.
c) The significance level is: $\alpha = 1\%$.
d) To calculate the critical value, X_c, for a one-tailed test on the left side at $\alpha = 1\%$, first find the critical z score, z_c. Using TABLE II, (the Normal Curve Area Table), we find:
$z_c = -2.33$.

Thus, the critical value, X_c, is:

$$X_c = \mu_{\bar{x}} + (z_c)(\sigma_{\bar{x}})$$
$$X_c = 40,000 + (-2.33)(100)$$
$$X_c = 39,767 \text{ miles}$$

e) Figure 10.5 illustrates the decision rule for this hypothesis test.

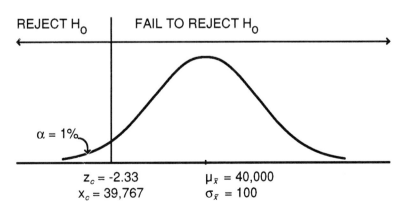

Figure 10.5

The decision rule is:

f) Reject the null hypothesis if the experimental outcome (the sample mean) is less than 39,767 miles.

Step 4. *Conduct the experiment.*

The experimental outcome is: $\bar{x}$ = 39,750 miles.

Step 5. *Determine the conclusion.*

Since the experimental outcome of 39,750 miles is less than the critical outcome of 39,767 miles we *reject H_o and accept H_a* at α = 1%. Thus, the research team can conclude that the advertisement was *not* legitimate. ∎

10.3 Introduction to the *t* Distribution

There are many practical instances when one performs a hypothesis test involving a population mean, μ, where **the population standard deviation, σ, is unknown.**

For example:
- An automobile manufacturer claims that its new subcompact averages 50 m.p.g. for highway driving.
- A cereal company claims that their family-sized cereal box has an average net weight of 15 oz.
- An archaeologist claims that the dinosaur species *Tyrannosaurus Rex* has an average length of 50 feet.

In such instances, when the **population standard deviation is unknown**, it becomes necessary to obtain a **good estimate of the population standard deviation**. A good estimate of the population standard deviation is determined by using the standard deviation of a **random sample** selected from the population. The following formulas are used to estimate the population standard deviation. They are referred to as the **definition formula and the computational formula for the sample standard deviation, s**.

FORMULAS TO ESTIMATE THE POPULATION STANDARD DEVIATION

Definition Formula for the Sample Standard Deviation:

$$s = \sqrt{\frac{\Sigma(x-\bar{x})^2}{n-1}}$$

Computational Formula for the Sample Standard Deviation:

$$s = \sqrt{\frac{\Sigma x^2 - \frac{(\Sigma x)^2}{n}}{n-1}}$$

where: s = sample standard deviation
$\bar{x}$ = sample mean
n = sample size

When the sample size, n, is **large**, then s is considered to be a **good estimator for** σ because the bias in s is very small. That is, s is just as likely to be larger than the population standard deviation as it is to be smaller. This concept is similar to an unbiased coin which is just as likely to come up heads as it is tails on any toss.

Consequently, **whenever the population standard deviation, σ, is unknown, then the standard deviation of the sampling distribution of the mean must also be estimated**.

Since s serves as a good estimate of the population standard deviation, σ, then the **estimate of the standard deviation of the sampling distribution of the mean** will make use of s. This estimate of the standard deviation of the sampling distribution of the mean is symbolized by $s_{\bar{x}}$ and is computed by the following formula.

Formula to Estimate the Standard Deviation of the Sampling Distribution of the Mean

$$s_{\bar{x}} = \frac{s}{\sqrt{n}}$$

where:

$s_{\bar{x}}$ = **estimate** of the standard deviation of the sampling distribution of the mean, $\sigma_{\bar{x}}$

s = **estimate** of the population standard deviation, σ

n = sample size

Thus, when s is used to estimate σ, then $s_{\bar{x}}$ is used to estimate $\sigma_{\bar{x}}$.

When we are confronted with having to estimate the population standard deviation when performing a hypothesis test involving a population mean, we need to use a distribution developed by a statistician named William S. Gosset called the **Student's *t* distribution** or simply, the *t* distribution rather than the normal distribution as our model for the sampling distribution of the mean. Gosset's research work was done while he was employed by the famous Guinness Brewery in Ireland and was published under his pen-name "Student." Gosset's development of the *t* distribution allows us to test hypotheses involving a population mean when the population standard deviation, σ, is unknown. Let's examine this *t* distribution.

The *t* Distribution

If samples of size n are randomly selected from a **normal or approximately normal population** with an unknown standard deviation, then the sampling distribution of the mean is a *t* distribution where the values of the following formula are called *t scores*.

$$t = \frac{\bar{x} - \mu}{s_{\bar{x}}}$$

where:

$\bar{x}$ = sample mean

μ = population mean

$s_{\bar{x}}$ = estimate of the standard deviation of the sampling distribution of the mean

It is important to realize that when you are performing a hypothesis test involving a population mean where the population is normal and s is used as an estimator of the unknown population standard deviation, then the sampling distribution of the mean is a *t* distribution. Let's now examine the properties of a *t* distribution.

In general the *t* distribution is similar to the normal distribution in that it is **symmetric about the mean and bell-shaped**. However, the *t* distribution is more dispersed than a normal distribution, that is, it has more area in the tails and less in the center than the normal distribution. Thus, in appearance, a *t* curve is lower and flatter than the normal curve. The *t* distribution is composed of a family of *t* curves, where there is a different *t* curve for each sample size. Thus, the area under a particular *t* curve is **dependent upon the size of the sample, n,** or the value of: **n − 1**, known as the concept of **degrees of freedom**.

Definition: Degrees of Freedom. The degrees of freedom, denoted by *df*, for each sample of size n, is one less than the sample size. Thus, for a sample of size n, the degrees of freedom are given by formula:

$$df = n-1$$

Figure 10.6 shows the relationship between the normal distribution and two particular *t* distributions of sample sizes n = 5 and n = 12, or degrees of freedom $df = 4$ and $df = 11$, respectively.

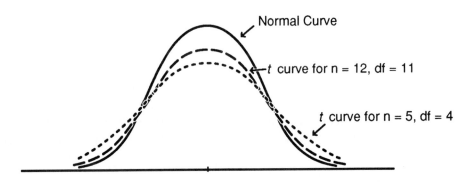

Figure 10.6

Notice, from Figure 10.6, the *t* distributions are **flatter and more dispersed** than the normal distribution, and, as the sample size or the degrees of freedom *increase* the *t* distribution is getting closer to the normal distribution. In fact, as the number of degrees of freedom (or the sample size) increases, the *t* distribution begins to approach the normal distribution until the two distributions are virtually identical. For most practical purposes, whenever the degrees of freedom are greater than 30, the *t* and normal distributions are considered to be sufficiently close.

Let's summarize the properties of the *t* distribution.

Properties of the *t* Distribution

- The *t* distribution is bell-shaped.
- The *t* distribution is symmetric about the mean.
- There is a different *t* distribution for each sample size, n or degrees of freedom, **df =n-1**.
- As the number of degrees of freedom increases, the *t* distribution approaches the normal distribution. They are sufficiently close when the degrees of freedom are at least 30.

Since the *t* distribution is composed of a family of *t* curves that are dependent upon degrees of freedom, the table representing the area associated with the *t* distribution is given with degrees of freedom. The *t* distribution table is different than the normal distribution table in that it lists the critical *t* scores, denoted t_c, for significance levels of 5% and 1% which are needed to perform a hypothesis test. These critical *t* scores, t_c, are found in **TABLE III: Critical Values for the *t* Distribution in Appendix D: Statistical Tables**. Let's examine TABLE III and discuss how to use it.

10.4 USING TABLE III: CRITICAL VALUES FOR THE t DISTRIBUTION

To find the critical *t* score, t_c, in TABLE III needed to perform a hypothesis test, you need the following information:

- The significance level, α.
- Type of hypothesis test. (i.e. 1TT or 2TT)
- Degrees of freedom, *df*.

Example 10.3 illustrates how to use TABLE III, given the previous information.

Example 10.3

Use TABLE III to determine the critical *t* score, t_c, for a hypothesis test where:

1) $\alpha = 1\%$.
2) one-tailed test on the right side.
3) *df* = 16.

Solution:

To determine the critical *t* score on the right side, t_c, examine TABLE III. A portion of TABLE III is shown in Table 10.1.

Table 10.1

TABLE III: Critical Values for the t Distribution

α = 1%

df	one tail critical t left (t_c)	critical t right (t_c)	two-tail critical t left (t_{LC})	critical t right (t_{RC})
		$t_{99\%}$		$t_{99.5\%}$
1		*		
2		*		
3		*		
.		*		
.		*		
.		*		
15		*		
16	*************	2.58		
17				
.				
.				
.				
normal distribution				

From Table 10.1, the critical *t* score on the right side, t_c, is 2.58, because it is found under the column for one-tail test at α = 1%, critical *t* right and in the row for *df* = 16. ∎

Example 10.4

Find the critical *t* score, t_c, for the *t* distribution in Figure 10.7 if:
1) α = 5%
2) one-tailed test on the left side.
3) *df* = 30.

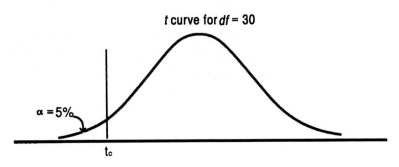

Figure 10.7

Solution:

From TABLE III, the t_c = −1.70 since it is in the column for one-tail test at α = 5%, critical *t* left and in the row for *df* = 30. ∎

Example 10.5

Find the critical *t* scores, t_{LC} and t_{RC}, for a two-tailed test, α = 5% and *df* = 9 as illustrated in Figure 10.8

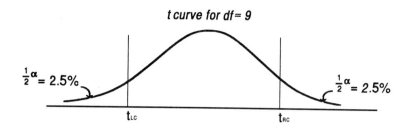

Figure 10.8

Solution:

From TABLE III, the critical *t* scores are $t_{LC} = -2.26$ and $t_{RC} = 2.26$. ∎

Example 10.6

Using the information in Figure 10.9, find the critical *t* score, t_c.

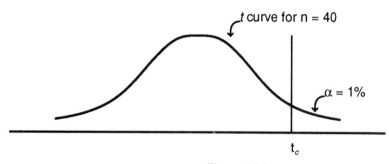

Figure 10.9

Solution:

Before TABLE III can be used, the degrees of freedom must be calculated using $df = n-1$. For $n = 40$, $df = 39$. Since there is no row for 39 degrees of freedom in TABLE III, we use the row for 40 degrees of freedom because it is *closest* to 39 df. Therefore, $t_c = 2.42$. ∎

Example 10.7

Using the information in Figure 10.10, find the critical *t* scores, t_{LC} and t_{RC}.

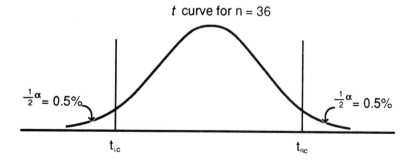

Figure 10.10

Solution:

Before TABLE III can be used, the degrees of freedom *and* the significance level must be determined. The degrees of freedom for n = 36 is: *df* = 35. The level of significance, α, is determined by adding together the two shaded areas in Figure 10.10. Thus, α = 1%. In TABLE III there is **no** row corresponding to *df* = 35. Since *df* = 35 falls exactly between *df* = 30 and *df* = 40 we will use the t_c score that is *farthest* from *t* = 0. Thus we use the *t* scores corresponding to *df* = 30. Therefore, t_{LC} = -2.75 and t_{RC} = 2.75.

Procedure to Locate a Critical *t* Score in TABLE III

To locate a critical *t* score in TABLE III, use the following procedure.
1. Select the appropriate α value column.
2. Identify the type of hypothesis test: one or two-tailed test.
3. Select the appropriate *df* row using the following guidelines:

   (i) If the *df* value is found in the table, use this row to locate the critical *t* score.
   (ii) If the exact *df* value does not appear in the table, then choose the row corresponding to the closest *df* value to locate the critical t score.
   (iii) If the *df* value does not appear in the table but falls exactly halfway between two *df* values, then choose the row that corresponds to the critical *t* score that is farthest from the *t* score: *t* = 0.

❄ 10.5 Hypothesis Testing: Population Standard Deviation Unknown

The *t* Distribution is important in hypothesis tests involving a population mean where the population is approximately normal and the population standard deviation is *unknown*. In these instances the population standard deviation, σ, is estimated by s, the sample standard deviation, and the sampling distribution of the mean is a *t* distribution. Therefore, the *t* distribution is used in hypothesis testing under the following conditions.

When the Sampling Distribution of the Mean is a *t* Distribution

The *t* distribution is used in performing a hypothesis test involving the population mean, when the following two conditions are true:

1. the population is approximately normal
2. the population standard deviation is unknown

The five step hypothesis testing procedure has to be modified when performing a hypothesis test involving a population mean when the *t* distribution serves as the model for the Sampling Distribution of the Mean. Let's discuss these minor changes in the hypothesis testing procedure.

Hypothesis Testing Procedure Changes When the Sampling Distribution of the Mean is a *t* Distribution

In performing a hypothesis test where the sampling distribution of the mean is a *t* distribution, the hypothesis testing procedure using a *t* distribution is similar to the hypothesis testing procedure using a normal distribution. The same five step procedure is used with the only differences being:

1. the use of the *t* distribution in place of the normal distribution

2. $\sigma_{\bar{x}}$ has been estimated by $s_{\bar{x}}$, where $s_{\bar{x}} = \dfrac{s}{\sqrt{n}}$

3. the critical value, X_c, becomes: $X_c = \mu_{\bar{x}} + (t_c)(s_{\bar{x}})$. where the critical *t* score, t_c, has replaced the critical z score, z_c, and $\sigma_{\bar{x}}$ has been replaced by $s_{\bar{x}}$, where $s_{\bar{x}} = \dfrac{s}{\sqrt{n}}$.

Let's now consider a few examples of hypothesis testing involving a population mean where the *t* distribution serves as the model for the sampling distribution of the mean.

Example 10.8

An automobile manufacturer claims that their economy car averages 55 mpg for highway driving. An engineer representing a leading automotive magazine believes this claim is too high. To test the manufacturer's claim, the engineer randomly selects sixteen of the economy cars and finds that they averaged 53 mpg with s = 1.6 mpg. Does this sample result indicate that the manufacturer's claim is too high at $\alpha = 1\%$?

Solution:

Step 1. *Formulate the hypotheses.*

H_o: The population mean mpg for the economy car is 55.
In symbols, H_o: $\mu = 55$ mpg.
H_a: The population mean mpg for the economy car is less than 55.
In symbols, H_a: $\mu < 55$ mpg.

Step 2. *Design an experiment to test the null hypothesis, H_o.*

To test the null hypothesis, a random sample of n = 16 cars was selected and tested. **Since σ is unknown, it is necessary to estimate σ using s, thus, the sampling distribution of the mean is a *t* distribution** and serves as the hypothesis testing model.
For this sampling distribution of the mean, the expected results are:

$$\mu_{\bar{x}} = \mu$$
$$\mu_{\bar{x}} = 55 \text{ mpg}$$

and, **since s is an estimate for σ, it is necessary to estimate** $\sigma_{\bar{x}}$ using the formula:

$$s_{\bar{x}} = \dfrac{s}{\sqrt{n}}$$

$$s_{\bar{x}} = \frac{1.6}{\sqrt{16}}$$
$$s_{\bar{x}} = 0.4 \text{ mpg}$$

The hypothesis testing model is illustrated in Figure 10.11.

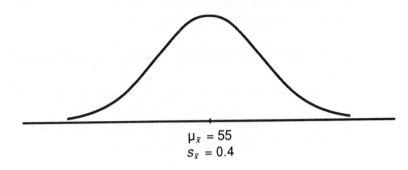

The sampling distribution of the mean
(is a t distribution)

$\mu_{\bar{x}} = 55$
$s_{\bar{x}} = 0.4$

Figure 10.11

Step 3. *Formulate the decision rule.*

a) The alternative hypothesis is directional since the engineer believes that the manufacturer's claim is *too high*.
b) The type of hypothesis test is one-tailed (1TT) on the left side.
c) The significance level is: $\alpha = 1\%$.
d) To calculate the critical value, X_c, for a *t* distribution, we use the following formula:
$$X_c = \mu_{\bar{x}} + (t_c)(s_{\bar{x}}).$$
Notice that the critical value formula for a *t* distribution has two changes from the critical value formula for a normal distribution. The critical *t* score, t_c, has replaced the critical z score, z_c, and $\sigma_{\bar{x}}$ has been replaced by $s_{\bar{x}}$.

Thus, to calculate the critical value, X_c, for a one-tailed test on the left side with $\alpha = 1\%$ and $df = 15$, we must first find the critical *t* score on the left, t_c, using TABLE III. From TABLE III, $t_c = -2.60$. Thus, the critical value on the left is:
$$X_c = \mu_{\bar{x}} + (t_c)(s_{\bar{x}})$$
$$X_c = 53.96 \text{ mpg}$$

e) Figure 10.12 illustrates the decision rule for this hypothesis test.

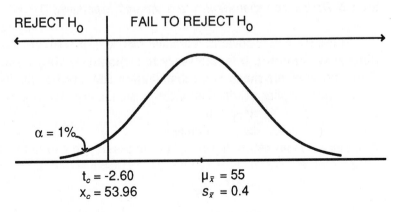

REJECT H_0 | FAIL TO REJECT H_0

$\alpha = 1\%$

$t_c = -2.60$
$X_c = 53.96$

$\mu_{\bar{x}} = 55$
$s_{\bar{x}} = 0.4$

Figure 10.12

The decision rule is:

f) reject the null hypothesis if the experimental outcome (the sample mean) is less than 53.96 mpg.

Step 4. *Conduct the experiment.*

The experimental outcome (sample mean) is:

$$\overline{x} = 53 \text{ mpg}$$

Step 5. *Determine the conclusion.*

Since the experimental outcome (53) is less than the critical value of 53.96, we reject H_o and accept H_a at $\alpha = 1\%$. Thus, the experimental result is *significantly different* than the manufacturer's claim of 55 mpg. Therefore, the engineer can conclude that the manufacturer's claim is too high at $\alpha = 1\%$.

Example 10.9

A medical research team decides to study whether pressure exerted on a person's upper arm will increase bleeding time. The average bleeding time for a pricked finger where no pressure is applied to the upper arm is 1.6 minutes. To test their claim the research team randomly selects 64 subjects. They find that the average bleeding time is 1.9 minutes with s = 0.80 minutes when 50 mm of pressure is applied to the upper arm.

Does this sample result support the claim that pressure applied to the upper arm will increase bleeding time? (use $\alpha = 5\%$)

Solution:

Step 1. *Formulate the hypotheses.*

H_o: The population mean bleeding time when 50 mm of pressure is applied to the upper arm is 1.6 minutes.
In symbols, H_o: $\mu = 1.6$ minutes.
H_a: The population mean bleeding time when 50 mm of pressure is applied to the upper arm is greater than 1.6 minutes.
In symbols, H_a: $\mu > 1.6$ minutes.

Step 2. *Design an experiment to test the null hypothesis, H_o.*

To test the null hypothesis, a random sample of n = 64 subjects was selected and tested. **Since σ is unknown, s was used to estimate σ, and s was found to be 0.8 minutes**.

Under the assumption H_o is true, the sampling distribution of the mean has the expected results:

$$\mu_{\overline{x}} = \mu$$
$$\mu_{\overline{x}} = 1.6 \text{ minutes}$$

and

$$s_{\overline{x}} = \frac{s}{\sqrt{n}}$$

$$s_{\overline{x}} = 0.1 \text{ minutes}$$

The sampling distribution of the mean is a *t* distribution with *df* = 63 and serves as the hypothesis testing model. This is illustrated in Figure 10.13.

The sampling distribution of the mean
(is a *t* distribution)

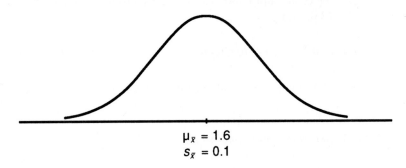

$\mu_{\bar{x}} = 1.6$
$s_{\bar{x}} = 0.1$

Figure 10.13

Step 3. *Formulate the decision rule.*

a) The alternative hypothesis is directional since the research team believes that pressure applied to the upper arm will *increase* bleeding time.
b) The type of hypothesis test is one-tailed on the right side.
c) The significance level is: $\alpha = 5\%$.
d) To calculate the critical value, X_c, for a *t* distribution, we use the following formula:

$$X_c = \mu_{\bar{x}} + (t_c)(s_{\bar{x}}).$$

Notice that the critical value formula for a *t* distribution has two changes from the critical value formula for a normal distribution. The critical *t* score, t_c, has replaced the critical z score, z_c, and $\sigma_{\bar{x}}$ has been replaced by $s_{\bar{x}}$.

Thus, to calculate the critical value for a one-tailed test on the right side with $\alpha = 5\%$ and $df = 63$, find the critical *t* score on the right, t_c, in TABLE III using the closest *df* table entry.
From TABLE III,
$t_c = 1.67$
The critical value on the right is:
$X_c = \mu_{\bar{x}} + (t_c)(s_{\bar{x}})$
$X_c = 1.77$ minutes.

e) Figure 10.14 illustrates the decision rule for this hypothesis test.

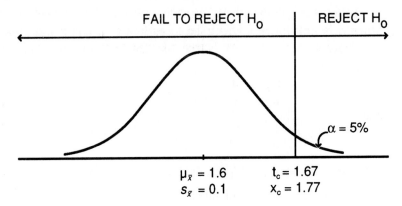

Figure 10.14

The decision rule is:
f) reject the null hypothesis if the experimental outcome (the sample mean) is greater than 1.77 minutes.

Step 4. *Conduct the experiment.*

The experimental outcome (sample mean) is:

$$\overline{x} = 1.9 \text{ minutes.}$$

Step 5. *Determine the conclusion.*

Since the experimental outcome (1.9) is *greater than* the critical value of 1.77, we reject H_o and accept H_a at $\alpha = 5\%$.

Therefore, the experimental result is statistically significant and the research team can conclude that pressure applied to the upper arm will increase bleeding time at $\alpha = 5\%$.

Example 10.10

American Computer Machines, a large computer firm, has branch offices in several major cities of the world. From past experience, ACM knows that its employees relocate on the average once every ten years. Due to recent population trends, the company wants to determine if the average relocation time has changed (that is, is the average relocation time *different* from ten years?). To determine if a change has occurred, a random sample of twenty-five employees were interviewed and it was found that their mean relocation time was 9.5 years with s = 4.5 years.

Does this sample result indicate a change has occurred in the mean relocation time? (use $\alpha = 5\%$)

Solution:

Step 1. *Formulate the hypotheses.*

H_o: The population mean relocation time is 10 years.
   In symbols, H_o: $\mu = 10$ years.

H_a: The population mean relocation is not 10 years.
   In symbols, H_a: $\mu \neq 10$ years.

Step 2. *Design an experiment to test the null hypothesis, H_o.*

To test the null hypothesis, a random sample of n = 25 employees were interviewed. **Since σ is unknown, s was used to estimate σ and determined to be 4.5 years.** Under the assumption H_o is true, the expected results for the sampling distribution of the mean are:

$$\mu_{\overline{x}} = \mu$$
$$\mu_{\overline{x}} = 10 \text{ years}$$

and

$$s_{\overline{x}} = \frac{s}{\sqrt{n}}$$

$$s_{\overline{x}} = 0.9 \text{ years}$$

The hypothesis testing model is the sampling distribution of the mean which is a *t* distribution. This is illustrated in Figure 10.15.

The sampling distribution of the mean
(is a t distribution)

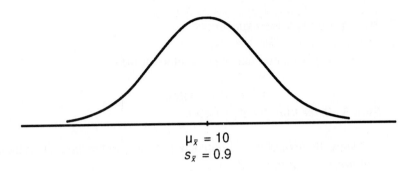

$\mu_{\bar{x}} = 10$
$s_{\bar{x}} = 0.9$

Figure 10.15

Step 3. *Formulate the Decision Rule.*

a) The alternative hypothesis is *nondirectional* since the company is testing whether the mean relocation time has decreased or increased.
b) The type of hypothesis test is two-tailed.
c) The significance level is: $\alpha = 5\%$.
d) To calculate the critical value, X_c, for a *t* distribution, we use the following formula:
$$X_c = \mu_{\bar{x}} + (t_c)(s_{\bar{x}})$$
Use TABLE III to calculate the critical values for a two-tailed test with $\alpha = 5\%$ and $df = 24$. Using TABLE III, find the two critical *t* scores, t_{LC} and t_{RC}. From TABLE III:

$t_{LC} = -2.06$ and $t_{RC} = 2.06$

The critical values are:

$$X_{LC} = \mu_{\bar{x}} + (t_{LC})(s_{\bar{x}})$$

$X_{LC} = 8.15$ years

and

$$X_{RC} = \mu_{\bar{x}} + (t_{RC})(s_{\bar{x}})$$

$X_{RC} = 11.85$ years

e) Figure 10.16 illustrates the decision rule for this hypothesis test.

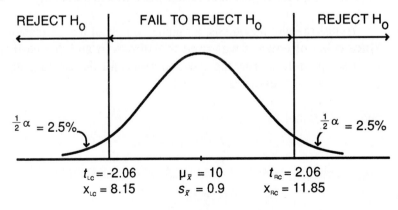

Figure 10.16

The decision rule is:

f) reject the null hypothesis if the experimental outcome (the sample mean) is either less than 8.15 or greater than 11.85 years.

Step 4. *Conduct the experiment.*

The experimental outcome (sample mean) is: $\bar{x}$ = 9.5 years.

Step 5. *Determine the conclusion.*

Since the experimental outcome falls between the critical outcomes of 8.15 and 11.85 years, we *fail to reject* H_o at α = 5%.

Although there is a *difference* between the sample mean of 9.5 years and the hypothesized mean of 10 years, this difference is **not statistically significant**. Therefore, the company cannot conclude that the mean relocation time of 10 years has changed. ■

Example 10.11

The IRS claims that the mean amount of time it takes to transfer tax information from a tax return to a computer record is 35 minutes. Before they will invest into a new system, they decide to do a test run to determine if the new system is significantly better than the present system.

Four hundred tax returns are randomly selected and the data is transferred using the new system. The mean amount of time it takes to transfer the data from a tax return to a computer record is 34.3 minutes with s = 4.2 minutes. Based upon this sample result should the IRS invest in the new system? (use α= 5%).

Solution:

Step 1. *Formulate the hypotheses.*

In hypothesis testing problems dealing with a comparison between a present (or existing) system and a new system, the null hypothesis is *always* stated as: the mean of the new system is *equal* to the mean of the present (or existing) system. The null hypothesis is stated in this form because if the null hypothesis is rejected, then this would *indicate* the new system is significantly better than the present system.

On the other hand, if the null hypothesis *cannot* be rejected, then the new system is *not* considered to be significantly better. Thus, in this instance, the present system would *not* be replaced by the new system.

Therefore, the null hypothesis is:

H_o: Under the new transfer system, the population mean time to transfer a tax return to a computer record is 35 minutes.

In symbols, H_o: μ = 35 minutes.

H_a: Under the new transfer system, the population mean time to transfer a tax return to a computer record is less than 35 minutes.

In symbols, H_a: μ < 35 minutes.

Step 2. *Design an experiment to test the null hypothesis, H_o.*

To test the null hypothesis, a random sample of n= 400 tax returns were transferred to computer records using the new transfer system. Since the population standard deviation, σ, is unknown, s was used to estimate σ. The value of s was determined to be 4.2 minutes. Under the assumption H_o is true, the expected results for the Sampling distribution of the mean are:

$$\mu_{\bar{x}} = \mu$$

$$\mu_{\bar{x}} = 35 \text{ minutes}$$

and

$$s_{\bar{x}} = \frac{s}{\sqrt{n}}$$

$$s_{\bar{x}} = 0.21 \text{ minutes}$$

The hypothesis testing model is the sampling distribution of the mean which is a **t** distribution. This is illustrated in Figure 10.17.

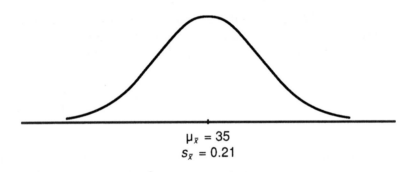

Figure 10.17

Step 3. *Formulate the decision rule.*

a) The alternative hypothesis, H_a, is directional.
b) The type of hypothesis test is one-tailed on the left side.
c) The significance level is: $\alpha = 5\%$.
d) The critical *t* score for a one-tailed test on the left side with $\alpha = 5\%$ and $df = 399$ is found in TABLE III. It is $t_c = -1.65$.
e) The left critical value is:

$$X_c = \mu_{\bar{x}} + (t_c)(s_{\bar{x}})$$
$$= 35 + -0.35$$
$$X_c = 34.65 \text{ minutes}$$

f) Figure 10.18 illustrates the decision rule for this hypothesis test.

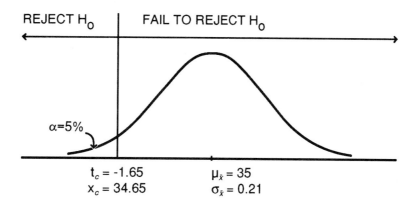

Figure 10.18

The decision rule is:

f) reject the null hypothesis if the experimental outcome (the sample mean) is less than 34.65 minutes.

Step 4. *Conduct the experiment.*

The experimental outcome (sample mean) is: $\overline{x}$ = 34.3 minutes.

Step 5. *Determine the conclusion.*

Since the experimental outcome falls below the critical outcome of 34.65 minutes, we reject H_o, and accept H_a at α = 5%. Thus, there is a significant difference between the sample mean time of 34.3 minutes and the expected mean time of 35 minutes. Therefore, the IRS should consider investing into the new data transferring system. ■

CASE STUDY 10.1

The USA Snapshot entitled *Cost to visit doc skyrockets* shown in Figure CS10.1 shows the average cost of a visit to the doctor for the years 1983 and 1992.

Figure CS10.1

For each year shown in Figure CS10.1,
a) define the population.
b) define the variable being displayed.
c) is the variable - discrete or continuous?
d) For the year 1982, what does the value $25.11 indicate about the population?
e) For the year 1992, what does the value $46.43 indicate about the population?

*Assuming you are an individual who believes the value for the year 1992 is **too low**, and decides to perform a hypothesis test to test this value of $46.43.*

f) What is the null and alternative hypothesis for this test? Is the alternative hypothesis directional or nondirectional? Explain.
g) Which population parameter are you assuming you know? What is the value of this parameter?
h) Which population parameter are you missing? How would you go about obtaining an estimate of this parameter?
i) What is the name of the sampling distribution that will serve as the model for this test? Is this sampling distribution a normal or a t distribution, if we assume that the distribution of doctor costs is a normal distribution? Explain.
j) What is the mean of this sampling distribution?
What formula would you use to estimate the standard deviation of the sampling distribution?
k) Which tail of the curve would the critical outcome of the test be placed?
l) If the experimental outcome (or sample outcome) falls to the right of the critical outcome, what is the conclusion to this test? Would this result indicate the average cost for a visit to the doctor is more or less than $46.43? Explain.
m) If the p-value of the hypothesis test is 3.96% (or 0.0396), then can you reject the null hypothesis at $\alpha = 1\%$? What would you say about the average cost for a visit to the doctor? Explain.
n) If the p-value of the hypothesis is 0.98% (or 0.0098) and $\alpha = 1\%$, could you conclude that the average cost for a visit to the doctor is more than $46.43? Explain. ∎

CASE STUDY 10.2

A study of sick days conducted by the National Center for Health Statistics was based on a survey of 117,000 households containing a total of 303,000 persons.

Figure CS10.2, on the following page, displays a table containing some of findings of the study.

a) The statement "Women averaged 5.5 lost work days per year" pertains to what population?
b) What parameter is being mentioned within the statement?
Define the variable being displayed. Is the variable - discrete or continuous?
c) According to the study the men averaged 4.3 lost work days per year. Can you speculate on some reasons as to why women might average more sick days per year than men?
d) If you were to conduct a hypothesis test to determine if the stated claim is too high, what would be the null and alternative hypothesis? Is the alternative hypothesis directional or nondirectional? Explain.
e) Which population parameter are you assuming you know? What is the value of this parameter?
f) Which population parameter are you missing? How would you go about obtaining an estimate of this parameter?
g) What is the name of the sampling distribution that will serve as the model to this test? Is this sampling distribution a normal or a *t* distribution, if we assume that the distribution of sick days is a normal distribution? Explain.

Women Use More Sick Days
Men take one fewer day per year, study says

A Sick Survey

The National Center for Health Statistics released a study yesterday comparing the amount of sick days taken by working men and women. Among its findings:

- Women averaged 5.5 lost work days per year, compared with 4.3 for men.
- 11.2 percent of women took eight or more sick days per year, while 7 percent of men had that many.
- By job, the fewest lost work days were 3.0 for men in the professions and 3.3 for women working in farming, forestry and fishing.
- 78.5 percent of working women reported visiting a doctor during one year compared with 61.8 percent of male workers.
- 9.4 percent of women reported a hospital stay, compared with 7.5 percent of men. But the average hospital stay was 6.1 days for men and 5.4 days for women.

SOURCE: The Associated Press

Figure CS10.2

h) What is the mean of this sampling distribution? What formula would you use to estimate the standard deviation of the sampling distribution?

i) Which tail of the curve would the critical outcome of the test be placed?

j) If the experimental outcome (or sample outcome) falls to the left of the critical outcome, what is the conclusion to this test? Would this result indicate the average number of sick days for women is more or less than 5.5? Explain.

k) If the p-value of the hypothesis test is 4.50% (or 0.0450), then can you reject the null hypothesis at $\alpha = 5\%$? What would you say about the average number of sick days for women? Explain.

l) If the p-value of the hypothesis is 1.76% (or 0.0176) and $\alpha = 1\%$, could you conclude that the average number of sick days for women s more than 5.5? Explain. ∎

Summary of Hypothesis Tests Involving a Population Mean

The following outline is a summary of the 5 step hypothesis testing procedure for a hypothesis test involving one population mean.

Step 1. *Formulate the two hypotheses, H_o and H_a.*
Null Hypothesis, H_o: $\mu = \mu_o$
Alternative Hypothesis, H_a: has one of the following three forms:

$H_a: \mu < \mu_o$ or $H_a: \mu > \mu_o$ or $H_a: \mu \neq \mu_o$

Step 2. *Design an experiment to test the null hypothesis, H_o*
Under the assumption H_o is true, the expected results are:
a) the Sampling Distribution of The Mean is either:
   (i) a normal distribution, if the standard deviation of the population is known and the sample size is greater than 30,
   or
   (ii) a *t* Distribution with degrees of freedom: $df = n-1$ if the standard deviation of the population is unknown.
b) the mean of the Sampling Distribution of The Mean is given by the formula: $\mu_{\bar{x}} = \mu_o$
   (since H_o is true :i.e. $\mu = \mu_o$)
c) *for a normal distribution: the **standard deviation** of the Sampling Distribution of The Mean, denoted by $\sigma_{\bar{x}}$, is:* $\sigma_{\bar{x}} = \dfrac{\sigma}{\sqrt{n}}$
   or

for a t distribution: the estimate of the standard deviation of the Sampling Distribution of The Mean, written $s_{\bar{x}}$, is given by the formula:

$$s_{\bar{x}} = \dfrac{s}{\sqrt{n}}$$

Step 3. *Formulate the decision rule.*
a) alternative hypothesis: directional or nondirectional.
b) type of test: 1TT or 2TT.
c) significance level: $\alpha = 1\%$ or $\alpha = 5\%$.
d) critical value formula:

for a normal distribution: $X_c = \mu_{\bar{x}} + (z_c)(\sigma_{\bar{x}})$

for a t distribution: $X_c = \mu_{\bar{x}} + (t_c)(s_{\bar{x}})$.

e) draw hypothesis testing model.
f) State the decision rule.

Step 4. *Conduct the experiment.*
Compute the experimental outcome (or sample mean): $\bar{x}$.

Step 5. *Determine the conclusion.*
Compare the experimental outcome, $\bar{x}$, to the critical value(s), X_c, of the decision rule and state the conclusion:
either: (a) **Reject H_o and Accept H_a** at α
or (b) **Fail to Reject H_o** at α

Table 10.2 contains a condensed summary of the hypothesis tests involving a population mean

Table 10.2

SUMMARY OF HYPOTHESIS TESTS FOR TESTING THE VALUE OF A POPULATION MEAN			
FORM OF THE NULL HYPOTHESIS	CONDITIONS OF TEST	CRITICAL VALUE FORMULA	SAMPLING DISTRIBUTION IS A:
$\mu = \mu_0$	KNOWN σ and $n > 30$	$X_c = \mu_{\bar{x}} + (z_c)(\sigma_{\bar{x}})$ where: $\sigma_{\bar{x}} = \dfrac{\sigma}{\sqrt{n}}$	Normal Distribution
$\mu = \mu_0$	UNKNOWN σ	$X_c = \mu_{\bar{x}} + (t_c)(s_{\bar{x}})$ where: $s_{\bar{x}} = \dfrac{s}{\sqrt{n}}$	t Distribution where: $df = n - 1$

10.6 USING MINITAB

In this chapter we have developed hypotheses tests involving a population mean. MINITAB has two types of commands that can be used for these types of hypotheses tests:

ZTEST and the TTEST command

To use the ZTEST command, it is assumed that the sampling distribution of the mean is modelled by a normal distribution.

Depending on the type of alternative hypothesis, (i.e. whether it is nondirectional or directional) the ZTEST command takes on one of three forms:

for a **two-tailed hypothesis test**:

ZTEST MU=(numerical value) SIGMA=(numerical value) C1

for a **one-tailed** hypothesis test, where the alternative hypothesis, H_a, is **on the left** is:

ZTEST MU=(numerical value) SIGMA=(numerical value) C1;
ALTERNATIVE= -1.

for a **one-tailed** hypothesis test, where the alternative hypothesis, H_a, is **on the right** is:

ZTEST MU=(numerical value) SIGMA=(numerical value) C1;
ALTERNATIVE= 1.

In each form the notation assumes that the population mean, μ, is MU, the population standard deviation, σ, is SIGMA, and, that the sample used to test the population mean is in a MINITAB worksheet in column C1.

The ZTEST for a nondirectional alternative hypothesis test, where MU=20 and SIGMA=4 and where the sample data values are in C1 is:

MTB> ZTEST MU=20 SIGMA=4 C1

This command would produce a MINITAB output like:
TEST OF MU = 20.000 VS MU N.E. 20.000
THE ASSUMED SIGMA = 4.00

	N	MEAN	STDEV	SE MEAN	Z	P VALUE
C1	36	18.413	4.546	0.667	-2.38	0.018

Notice that the test of the population mean, $\mu = 20$ versus μ not equal to 20 is significant for an $\alpha=2\%$ or higher since the p-value for the sample in C1 is 0.018. In most social science applications, p-values less than 0.05 are considered significant.

When the sampling distribution of the mean is modelled by a t-distribution then the ZTEST command is replaced with the TTEST command. Unlike the ZTEST command, the TTEST command does not assume knowledge of the population standard deviation. However, similar to the ZTEST command, the TTEST command has three forms:

for a **two-tailed hypothesis test**:

TTEST MU=(numerical value) C1

for a **one-tailed** hypothesis test, where the alternative hypothesis, H_a, is **on the left** is:

TTEST MU=(numerical value) C1;
ALTERNATIVE= -1.

for a **one-tailed** hypothesis test, where the alternative hypothesis, H_a, is **on the right** is:

TTEST MU=(numerical value) C1;
ALTERNATIVE= 1.

The TTEST command for a one-tailed test on the right about a population mean, MU=70 where the sample data is contained in C1 is:

MTB> TTEST MU=70 C1;
MTB> ALTERNATIVE=1.

A possible MINITAB output for this test could be:
TEST OF MU = 70.000 VS MU G.T. 70.000

	N	MEAN	STDEV	SE MEAN	T	P VALUE
C1	25	71.470	4.540	0.908	1.62	0.059

In this example notice that the output indicates that the alternative hypothesis, H_a, is that the population mean is greater than 70. By examining the p-value for this test you should notice that it is 0.059. Since this p-value is greater than 0.05, the sample data would **not** indicate a rejection of the null hypothesis. Thus, a conclusion from this output would be that indicate a rejection of the null hypothesis. Thus, a conclusion from this output would be that the we fail to reject H_o, although there is a difference between the sample mean (i.e. sample mean, $\bar{x}$ =71.470) and the population mean (i.e. population mean, μ = 70) the difference is **not statistically significant** at an alpha, α = 5%.

Glossary

TERM	SECTION
Hypothesis Testing Procedure Involving a Population Mean	10.2
Hypothesis testing model	10.2
Sample mean, $\bar{x}$	10.2
$\mu_{\bar{x}}$: mean of the sampling distribution of the mean	10.2
$\sigma_{\bar{x}}$: standard deviation of the sampling distribution of the mean	10.2
t distribution	10.3
s: estimate of the population standard deviation	10.3
$s_{\bar{x}}$: estimate of the standard deviation of the sampling distribution of the mean	10.3
t score	10.3
degrees of freedom, df	10.3
critical t score, t_c	10.3

EXERCISES

PART I Fill in the blanks.

For questions 1 - 3, complete the following statements with the word: **directional or nondirectional**.

1. If the null hypothesis has the form: $H_o: \mu=\mu_o$, and the alternative hypothesis has the form: $H_a: \mu < \mu_o$, then this is a _____ hypothesis test.
2. If the null hypothesis has the form: $H_o: \mu=\mu_o$, and the alternative hypothesis has the form: $H_a: \mu \neq \mu_o$, then this is a _____ hypothesis test.
3. If the null hypothesis has the form: $H_o: \mu=\mu_o$, and the alternative hypothesis has the form: $H_a: \mu > \mu_o$, then this is a _____ hypothesis test.
4. The expected results of a hypothesis test are calculated under the assumption that the null hypothesis is _____.
5. The statistic $s = \sqrt{\dfrac{\sum(x-\overline{x})^2}{n-1}}$ is a good _____ of the population standard deviation, σ, when the sample size is large.
6. The estimate of the standard deviation of the sampling distribution of the mean is symbolized as _____. The formula is _____.
7. Given the null hypothesis is $H_o: \mu=\mu_o$, the standard deviation of the population is σ, and the size of the random sample selected from the population is greater than 30, then:
   a) the sampling distribution of the mean is _____.
   b) the mean of the sampling distribution of the mean is given by the formula:
      $\mu_{\overline{x}}$ = _____ , and
   c) the standard deviation of the sampling distribution of the mean is given by the formula:
      $\sigma_{\overline{x}}$ = _____.
8. To perform a hypothesis test involving a population mean, the sampling distribution of _____ is used as the hypothesis testing model.
9. The sampling distribution of the mean is a t distribution if the standard deviation of the _____ is unknown and the population from which the sample is selected is a _____ distribution.
10. The properties of a t distribution are:
    a) the graph of a t distribution is _____ .
    b) the t distribution is symmetric about the _____.
    c) the percent of area to the left of a t score is dependent upon the degrees of _____ .
11. For a sample of size n, the degrees of freedom is equal to the sample size minus _____. In symbols, $df =$ _____.
12. The null hypothesis for a hypothesis test involving a population mean is stated as: H_o: the population mean, μ, is _____ to the numerical value, μ_o. In symbols this would be, $H_o: \mu$____μ_o.
13. The alternative hypothesis of a hypothesis test involving a population mean can have one of the following three forms. They are:
    a) the population mean, μ, is *greater than* the numerical value, _____. This is symbolized as: $H_a: \mu$___μ_o.
    b) the population _____, μ, is *less than* the numerical value,_____. This is symbolized as: $H_a: \mu$___μ_o.
    c) the _____ mean, μ, is *not equal* to the numerical value, μ_o. This is symbolized as: $H_a: \mu$___μ_o.
14. When performing a hypothesis test about a population mean where H_o is $\mu = 75$, the expected result $\mu_{\overline{x}} = 75$ is determined under the assumption that _____ is true.
15. The formula for the critical value, X_c, for a hypothesis test involving a population mean is:
    a) when the sampling distribution of the mean is approximately normal, and the standard deviation of the population, σ, is known:
       $X_c = \mu_{\overline{x}} + ($____$)($____$)$.
    b) when the sampling distribution of the mean is a t distribution:
       $X_c = \mu_{\overline{x}} + ($____$)($____$)$.
16. The hypothesis testing model for a hypothesis test involving a population mean where n=25 and the population standard deviation, σ, is unknown, is a t distribution with _____ degrees of freedom.

PART II   Multiple choice questions.

1. Which of the following statements could be used for the null hypothesis, H_o?
   a) $\mu < \mu_o$ b) $\mu > \mu_o$ c) $\mu \neq \mu_o$ d) $\mu = \mu_o$
2. If the population is normal with $\sigma=8$, and samples of size 16 are selected, then the standard deviation of the sampling distribution of the mean is:
   a) 8 b) 2 c) 0.5 d) 1
3. If the alternative hypothesis, H_a, is $\mu < 27$, then the hypothesis test uses a:

a) 1TT on right   b) 2TT
c) 1TT on left   d) 2TT on left
4. The statistic, s, is a good estimate for:
a) μ b) $\sigma_{\bar{x}}$ c) $\bar{x}$ d) σ
5. Which of the following is *not* a property of the *t* distribution?
a) less dispersed than the normal distribution
b) symmetric with respect to the mean
c) bell-shaped
d) different *t* distribution for each sample size
6. If samples of size 36 are selected, then the degrees of freedom for the *t* distribution is:
a) 30   b) 37   c) 6   d) 35
7. The formula used to determine the *df* with one sample is:
a) $df = n$ b) $df = n+1$ c) $df = n-1$ d) $df = n$
8. The critical *t* score for a 1TT on the right with $\alpha = 5\%$ and $df = 17$ is:
a) 1.74   b) 2.11   c) 2.57   d) 2.90
9. Which of the following could be an alternative hypothesis?
a) $\mu \neq 17$ b) $\mu = 26$ c) $\mu = 4.3$ d) $\mu = 0.10$
10. Which of the following could be the alternative hypothesis for a 1TT on the left?
a) $\mu < 56$ b) $\mu > 173$ c) $\mu \neq 93$ d) $\mu = 63$

PART III   Answer each statement True or False.

1. When the mean of a sample, $\bar{x}$, is used to test the null hypothesis, $\mu = \mu_o$, one must use the sampling distribution of the mean and its properties.
2. To insure that the sampling distribution of the mean can be approximated by the normal distribution, the sample size must be less than 30.
3. When the standard deviation of the population is known then $\sigma_{\bar{x}}$ is equal to the standard deviation of the population.
4. The sample mean is used as the mean for the sampling distribution of the mean.
5. When doing a hypothesis test about a population mean, μ, and the population standard deviation, σ, is known, then the sampling distribution of the mean is always normal, regardless of the size, n, of the random sample.
6. Whenever the population standard deviation, σ, is unknown, a good estimate of the population standard deviation is obtained from a random sample taken from the population.
7. When σ is unknown and the formula $s = \sqrt{\dfrac{\sum(x-\bar{x})^2}{n-1}}$ is used as an estimate for σ, then s is said to be a good estimator.

8. The *t* distribution is used as a model for the normal distribution whenever the sample size is less than 30.
9. Degrees of freedom is a concept used to determine critical *t* scores.
10. The *t* distribution is generally less dispersed than the normal distribution.

PART IV   Problems.

1. Use TABLE III to determine the critical *t* score(s) t_c, for a hypothesis test where:
a) $\alpha = 5\%$, 1TT on left side, and $df = 25$.
b) $\alpha = 1\%$, one-tail test on left side, and $df = 11$.
c) $\alpha = 5\%$, one-tail test on right side, and $df = 22$.
d) $\alpha = 1\%$, 1TT on right side, and $df = 17$.
e) $\alpha = 5\%$, 2TT, and $df = 43$.
f) $\alpha = 1\%$, 2TT, and $df = 56$.
2. Use the information in each of the Figures a through f to determine the critical *t* score(s), t_c, from TABLE III.

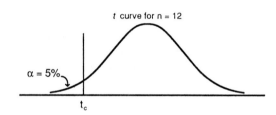

Figure (a)

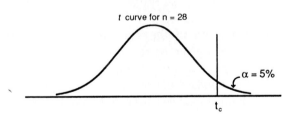

Figure (b)

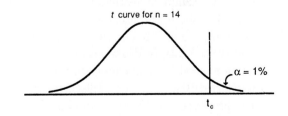

Figure (c)

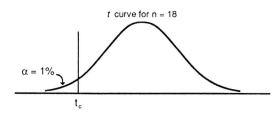

Figure (d)

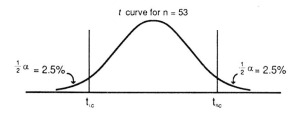

Figure (e)

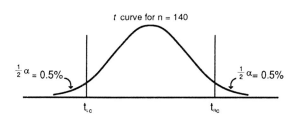

Figure (f)

3. a) Find the critical t score, t_c, for $\alpha = 5\%$, and a one-tailed test on left side for each value of n given in the following table.

n	t_c
5	
10	
50	
100	
500	

b) From TABLE II: The Normal Distribution Table, find the critical z score, z_c, for $\alpha = 5\%$ and a one-tailed test on the left side.

c) Compare the values of the critical t scores, t_c, to the critical z scores, z_c, as the sample size, n, increases.

For problems 4 through 32:
If the population standard deviation is unknown, assume that the population is approximately normal.

For each of the hypothesis testing problems, find the following:
a) State the null and alternative hypotheses.
b) Calculate the expected results for the hypothesis test assuming the null hypothesis is true and determine the hypothesis test model.
c) Formulate the decision rule.
d) Determine the experimental outcome.
e) Determine the conclusion and answer the question(s) posed in the problem, if any.

4. The mean systolic blood pressure for twenty-five year old females is given to be 120 with a standard deviation of 18. A sociologist believes that twenty-five year old females who live in the southwestern section of the United States will have significantly lower blood pressures. To test her claim, the sociologist randomly selects 36 of these females and determines their mean systolic blood pressure to be 114. Do these sample results support the sociologist's claim at $\alpha = 5\%$?

5. A sports equipment company claims to have developed a new manufacturing process which will increase the mean breaking strength of their new tennis racquet. The present manufacturing process produces a tennis racquet with a mean breaking strength of 90 lbs. with a standard deviation of 6 lbs. Larry, a wealthy tennis perfectionist who is known for his intense play that usually results in broken racquets, employs a product testing company to test the manufacturer's claim. The testing company randomly purchases sixty-four of these new racquets and tests them. If the testing company determines the mean breaking strength to be 93 lbs., can Larry agree with the manufacturer's claim? (use $\alpha = 1\%$)

6. An unemployment agency claims that the average age of recipients of unemployment benefits are 37 years old with a standard deviation of 5 years. A trade union association wants to test this claim. They randomly interview 400 recipients of unemployment benefits and obtain an average age of 35 years. Using a 1% significance level, can the trade union reject the unemployment agency's claim?

7. The population norms (i.e. parameters μ, and σ) for a mathematics anxiety test are known to be $\mu = 6$ and $\sigma = 1.6$. A teacher wants to determine if the students at her school will yield significantly different test results from the population norm. She

administers the test to a random sample of 100 students and determines the sample mean test score to be 5.73. Can the teacher conclude that these results are different from the population norm at $\alpha = 5\%$?

8. The personnel director of a Wall Street brokerage house believes that the executives working in the "Big Apple" tend to have higher IQ's than the national executive mean IQ. The national mean IQ score for executives is 112 with a population standard deviation of 18. If a random sample of 81 "Big Apple" executives yields a mean IQ score of 116, does these sample results support the personnel director's claim at $\alpha = 1\%$?

9. A large computer corporation states that their new computer would significantly lower a university's registration time. Before investing in this new computer, the university decides to test this claim at $\alpha = 1\%$. Under the present system, the mean registration time had been 90 minutes with a standard deviation of 20 minutes. With the new computer system, 100 students selected at random had an average registration time of 84 minutes. Should the university invest in this new computer?

10. The LIRR carries a mean average of 125,000 passengers daily with a standard deviation of 4000. The railroad has averaged 126,000 passengers daily for the last 36 snow storms. Does this represent a significant increase over the mean average passenger load at $\alpha = 5\%$?

11. A St. Louis meteorologist claims that the number of cloudy days per month in the St. Louis metropolitan area is increasing over what it was 50 years ago. To support his claim, the meteorologist examines the number of cloudy days for the past 40 months and determines the mean number of cloudy days to be 12. If 50 years ago the mean number of cloudy days per month was 10, with a standard deviation of 3, does the data support the meteorologist at $\alpha = 5\%$?

12. A Florida citrus grower claims to have devised a new method for harvesting oranges which is faster and just as efficient as the method currently used. The current method takes a mean average of 5 hours to harvest 1 acre of orange trees with a standard deviation of 45 minutes. One hundred acres harvested by the new method averaged 4 hours and 45 minutes. The average yield for the new method was equal to that of the current method. Do the sample results support the citrus grower's claim? (Use a significance level of 1%).

13. A traffic engineer believes that the time needed to drive from the downtown area out to the suburbs has significantly decreased due to increased use of public transportation. A study done five years ago determined that it took a mean of 45 minutes for this trip with a standard deviation of 7 minutes. Recently, the mean average time for a sample of 49 trips was 42 minutes. Do these data support the traffic engineer? Use a significance level of 5%.

14. The Fort Knox Federal Savings Bank claims that their regular passbook savings accounts have an average balance of $1236.45 with a standard deviation of $317.68. A group of federal bank auditors believes this claim is too high. To test the bank's claim the auditors randomly selected 400 accounts and obtained a mean balance of $1217.69. Do these sample results indicate that the bank's claim was too high? (use $\alpha = 5\%$)

15. An automobile manufacturer claims that their leading compact car averages 36 mpg in the city. The standard deviation is 4 mpg for city driving. Suppose a city police department purchases 64 cars from this auto manufacturer. If these cars were driven exclusively under city conditions and averaged 33 mpg, can one argue that the manufacturer's claim is too high at $\alpha = 5\%$?

16. A drug manufacturer specifies that their brand of cough suppressant contains a mean average of 25 milligrams of the expectorant drug DOKE for every milliliter of the suppressant. If 36 randomly selected one milliliter samples of this expectorant were found to have a mean of 25.4 mg of the expectorant drug DOKE with a standard deviation of 0.6, does this indicate that the product is not meeting the manufacturer's specifications? ($\alpha = 5\%$)

17. A certain type of Oak tree has a mean growth of 14.3 inches in 4 years. A forestry biologist believes that a new variety will have a greater mean growth during this same length of time. A sample of 36 trees of this new variety are studied. If it is found that the mean growth is 15.4 inches with s = 1.8 inches, can one conclude the biologist is correct? (use $\alpha = 5\%$)

18. Presently, the mean life expectancy of a rare strain of bacteria is 12 hours. A scientist claims she has developed a medium that will increase the mean life of the bacteria. The scientist tests 16 cultures of the newly treated bacteria and finds that they have a mean life of 13 hours with s = 1 hour. Do these results show that the medium is effective in increasing the bacteria's life expectancy? (Use $\alpha = 1\%$)

19. The Federal Aviation Agency (FAA) claims that on the average an air passenger's luggage weighs 35 pounds. An air carrier believes that the FAA is incorrect and randomly selects 25 passengers and

determines that the average weight of the luggage is 36 pounds with s=4 pounds. Would these results indicate that the FAA's claim is incorrect? (Use α = 1%)

20. An automobile club claims that the average time for a non-stop trip from New York City to Washington, D.C. by automobile is 4.5 hours. A disbelieving driver feels the claim is incorrect. The driver randomly questions 49 people who recently made the trip and finds that their average non-stop time was 4.3 hours with s = 0.35. Do these sample results refute the automobile club's claim at α = 1%?

21. A construction company specifies in their ordering of rivets that the rivets must have a mean sheer strength of 80,000 lbs. or more to be acceptable for their next contract. To test the quality of their shipment, the construction company takes a random sample of 16 rivets and finds the mean sheering strength to be 79,880 lbs. with s = 200 lbs. Should the company accept this shipment? (Test at α = 1%)

22. A manufacturer of space vehicle components specifies that a particular part must be able to withstand a mean heat of 5,000 degrees Fahrenheit or more. To determine if the parts meet the manufacturer's specifications, a random sample of 9 parts is chosen from the supplier's shipment and found to have a mean heat resistance of 4,930 F degrees with s = 150 degrees. Should the manufacturer accept the shipment? (Use α = 5%)

23. An ecologist claims that because of environmental influences, the mean life of Louisiana shellfish is 27 months. Feeling the claim is too low, a congressman decides to test the ecologist's claim at α = 5%. If the mean life of a random sample of 16 shellfish is 28.05 months with s = 2.8, do the sample results indicate that the ecologist's claim is too low?

24. An advertisement for an alkaline battery claims the batteries have a mean life of 100 hours. Forty-nine batteries are tested and found to have a mean life of 97.45 hours with s = 7. Based on the sample results would you feel that a mean life of 100 hours is too high? (Use α = 1%)

25. A new training procedure for athletes was developed and has been used during the past season. The trainer claims that this new training procedure has reduced the number of injuries to his players. Using the present training method, there were, on the average, 9 injuries per week. For a random sample of 12 weeks during the period in which the new training procedure was used, the mean number of injuries per week was only 7 with s = 2. Do the sample results support the trainer's claim at α=5%?

26. A pharmaceutical company claims that their muscle relaxing tablets have a mean effective period of 4.4 hours. Researchers representing the Food and Drug Administration suspect the company's claim is too high. The researchers administer the muscle relaxing tablets to a random sample of 121 patients and find that the mean effective period is 4.2 hours with s = 1.1 hours. Do the sample results support the researcher's suspicions? (Use α = 1%)

27. A National Research Group reports that children between the ages of 6 and 10 years watch television an average of 21 hours per week. A sociologist believes that the children within her upper-middle class neighborhood watch television less than an average of 21 hours per week. She randomly selects 25 children aged 6 to 10 years from her neighborhood and determines the mean number of hours that the children watch television is 20 hours with s = 3 hours. Do these sample results support the sociologist's claim at α = 5%?

28. A new chemical process has been developed for producing gasoline. The company claims that this new process will increase the octane rating of the gasoline. Sixteen samples of the gasoline produced with the new process are selected at random and their octane ratings were: 94, 93, 97, 92, 96, 94, 95, 91, 98, 95, 92, 91, 95, 96, 97, 93. If the mean octane using the existing process is 93, is the company's claim correct? (Use α = 5 %)

29. The commerce department in a California County claims that the mean rental for a 3 room apartment is $850. A real estate company believes the commerce department's claim is not correct and randomly surveys 144 apartment dwellers and finds that their average rental is $872, with a standard deviation of $90. Do the sample data indicate that the commerce department's claim is *not* correct at α = 1%?

30. A pharmaceutical company manufacturers a blood pressure drug which they claim will lower blood pressure an average of 20 point over a three week period. Dr. Zorba administers this drug to 36 patients for three weeks and records the following drops in their blood pressure: 22, 18, 19, 17, 24, 22, 15, 17, 21, 18, 15, 19, 24, 20, 14, 16, 19, 23, 24, 13, 18, 22, 21, 17, 24, 19, 15, 17, 21, 19, 22, 17, 19, 21, 17, 20. Do these sample data indicate that the pharmaceutical company's claim is incorrect at α = 5%?

31. A law student decides to check her professor's claim that the average time to disposition is 90 days. To test this claim, she randomly selects 15 cases of a

similar type from county court records and determines the number of days to disposition for each case. The results of her sample are: 85, 97, 93, 88, 93, 92, 89, 88, 94, 87, 96, 92, 94, 89, 95. Can the law student reject her professor's claim at $\alpha = 5\%$?

32. A prominent psychologist at a New York hospital claims that a new form of psychotherapy for a certain type of mental disorder will help a patient to be ready for outpatient treatment in an average of twelve weeks. A resident psychologist decides to test this claim under the suspicion that the claim is too low. The resident psychologist uses this new form of psychotherapy on a random sample of 40 patients. The time required for these patients to reach out patient status were (in days): 91, 88, 86, 95, 90, 93, 87, 89, 83, 91, 93, 86, 92, 91, 89, 83, 92, 88, 87, 90, 93, 87, 94, 86, 84, 92, 87, 85, 90, 88, 84, 92, 94, 86, 91, 89, 84, 89, 85, 95. Do these sample data support the claim made by the resident psychologist at $\alpha = 1\%$?

33. **Why state the conclusion as "Fail to Reject H_o" rather than Accept H_o?**

    In the hypothesis testing procedure, when the null hypothesis cannot be rejected we state the conclusion as "fail to reject H_o" rather than "accept H_o". This problem examines the reason why we prefer to state the conclusion as "fail to reject H_o."

    Two researchers, Maria and Larry, (who often don't agree on what the null hypothesis should be) decide to perform a hypothesis test on the same population of students to determine if the population mean IQ score has increased at the 5% level of significance.

    Maria states her null hypothesis as: "The population mean is 120." Larry states his null hypothesis as: "The population mean is 122." Both researchers agree that the population standard deviation is 12. Their hypothesis tests are shown side by side in the following table.

          Maria's Test   Larry's Test

    1. **Formulate the hypotheses:**

    H_o: population mean is 120 H_o: population mean is 122
          $\mu = 120$ $\mu = 122$
    H_a: population mean is H_a: population mean is
         greater than 120   greater than 122
         $\mu > 120$ $\mu > 122$

    2. **Design an experiment to test the null hypothesis, H_o.**

    To test their null hypotheses, Maria and Larry select a random sample of 36 students. The expected results for their respective tests are:

Sampling distribution of the mean
(is a Normal Distribution)

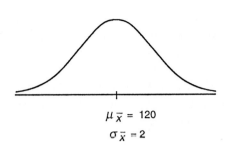

Sampling distribution of the mean
(is a Normal Distribution)

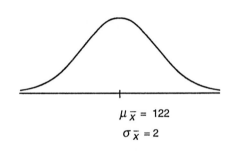

3. **Formulate the decision rule.**

   Maria's decision rule: Larry's decision rule:

   Reject H_o if the   Reject H_o if the
   experimental outcome   experimental outcome is
   is greater than 123.3.   greater than 125.3.

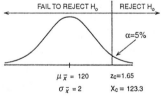 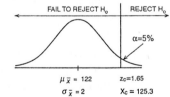

4. **Conduct the experiment.**
   Maria and Larry randomly selected 100 students and found the sample mean IQ (the experimental outcome) to be: $\bar{x} = 123$.
   Using this sample result, determine:
   a) the conclusion to Maria's hypothesis test.
   b) the conclusion to Larry's hypothesis test.

c) Was the experimental outcome statistically significant for either Maria's or Larry's hypothesis test?
d) If Maria had decided to state her conclusion as **accept H_o** rather than **fail to reject H_o**, what would she be accepting as the population mean IQ score?
e) If Larry had decided to state his conclusion as **accept H_o** rather than **fail to reject H_o**, what would he be accepting as the population mean IQ score?
f) Using the ideas of this problem, explain what is wrong with the stating the conclusion as:
accept H_o.

PART V What Do You Think?

1. **What Do You Think?**

   Examine the two graphs shown in the figure entitled *Student aid falters* which represents the ten year trend for the student-aid programs: Stafford Loans and Pell Grants.

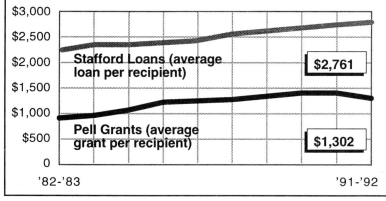

Source: The College Board   By Marty Baumann, USA TODAY

a) What parameter is represented by both graphs? Define the variable being displayed in the graphs. Is the variable - discrete or continuous?

For each student-aid program, design a hypothesis test that includes:

b) a null hypothesis. Use the information contained in the graph for the '91 - 92 year as the status quo.
c) an alternative hypothesis. Use the direction that the line graph is heading as to your belief regarding the true value of the population parameter for this year.
d) Is the alternative hypothesis directional or nondirectional? Is this a 1TT or a 2TT? Explain.
e) Which population parameter are you assuming you know? What is the value of this parameter?
f) Which population parameter are you missing? How would you go about obtaining an estimate of this parameter?
g) State the name of the sampling distribution that will serve as the model for this test? Is this sampling distribution a normal or a t distribution, if we assume that the distribution pertaining to the student-aid program is a normal distribution? Explain.
h) What is the mean of the sampling distribution? What formula would you use to estimate the standard deviation of the sampling distribution?
i) Which tail of the curve would the critical outcome of the test be placed?
j) If you decide to use a significance level of 5% for each hypothesis test, then write the critical value formula and substitute the values of the variable that you know into the formula.
k) If the experimental outcome (or sample outcome) falls to the right of the critical outcome for the Stafford Loan hypothesis test, what is the conclusion to this test? Could you agree that the conclusion to this hypothesis test follows the trend shown in the graph? Explain.
l) If the p-value of the hypothesis test is 6% (or 0.06), then can you reject the null hypothesis at $\alpha = 5\%$? Does this indicate that the Stafford Loan average is following the trend shown in the graph? Explain.
m) If the p-value of the hypothesis is 2.90% (or 0.0290) and $\alpha = 2\%$, could you conclude that the Stafford Loan average is following the trend shown in the graph? Explain.

2. **What Do You Think?**

The USA Snapshot entitled *Staying longer on the road* shown in the following figure indicates the average length of trip for the years 1989 - 1991.

USA SNAPSHOTS®
A look at statistics that shape your finances

Staying longer on the road
To cut travel costs, companies are sending employees on fewer — but longer — trips.

Average length of trip

Source: U.S. Travel Data Center
By Bob Laird, USA TODAY

a) Define the population.
b) Define the variable being displayed.
c) Is the variable - discrete or continuous?
d) What does the graph indicate about the average length of business trips over the years shown?
e) What does the number 4.4 days indicate about the population?

Assuming you are an individual who believes that the average length number of business trips days is following the trend shown in the USA snapshot:

f) what is the null and alternative hypothesis for this test? Is the alternative hypothesis directional or nondirectional? Explain.
g) Which population parameter are you assuming you know? What value does this parameter have?
h) Which population parameter are you missing that you need to perform this test? How would you go about obtaining an estimate of this parameter?
i) What is the name of the sampling distribution that will serve as the model for this test? Is this sampling distribution a normal or a t distribution, if we assume that the distribution representing the length of business trips is a normal distribution? Explain.
j) What is the mean of this sampling distribution? Which formula would you use to estimate the standard deviation of the sampling distribution?
k) Which tail of the curve would the critical outcome of the test be placed?
l) If the experimental outcome (or sample outcome) falls to the right of the critical outcome, what is the conclusion to this test? Would this result indicate the average length of business trips is more or less than 4.4? Explain.
m) If the p-value of the hypothesis test is 2.47% (or 0.0247), then can you reject the null hypothesis at $\alpha = 3\%$? What would you say about the average length of business trips? Explain.
n) If the p-value of the hypothesis is 0.05% (or 0.0005) and $\alpha = 5\%$, could you conclude that the average length of business trips is more than 4.4? Explain.

PART VI Exploring DATA With MINITAB

1. a) Use the SET command of MINITAB to enter the data values into column C1 of a MINITAB worksheet.
   68 73 50 67 69 56 76 70 56 69 60 60
   70 60 51 54 63 69 70 76 69 61 80 58
   77 63 71 66 65 79 64 85 61 81 43 74
   b) Using the data in C1 and the ZTEST command where MU=70 and SIGMA=10, test the hypothesis that the mean of the population is not equal to 70.
   c) State the null and alternative hypothesis.
   d) Is this a one or a two-tailed test? Explain.
   e) Why is the ZTEST command appropriate. Explain.
   f) Where is the data values for the random sample?
   g) Name and explain which values on the output are used to calculate the p-values.
   h) Report the p-value for this sample.
   i) Is the p-value significant at $\alpha=5\%$? Explain.
   j) Is the p-value significant at $\alpha=1\%$? Explain.
   k) State the conclusion in terms of the null and alternative hypothesis.

2. a) Use the SET command of MINITAB to enter the data values into column C1 of a MINITAB worksheet.
   116 119 104 115 119 148 120 132 140
   104 138 134 143 94 158 117 89 110
   119 127 130 113 119 117 122 128 138
   117 97 162 119 110 135 120 112 142
   b) Using the data in C1 and the ZTEST command where MU=120 and SIGMA=20, test the hypothesis that the mean of the population is greater than 120.
   c) State the null and alternative hypothesis.
   d) Is this a one or a two-tailed test? Explain.
   e) Why is the ZTEST command appropriate. Explain.
   f) Where are the data values for the random sample?

g) Name and explain which values on the output are used to calculate the p-value.
h) Report the p-value for this sample.
i) Is the p-value significant at $\alpha=5\%$? Explain.
j) Is the p-value significant at $\alpha=1\%$? Explain.
k) State the conclusion in terms of the null and alternative hypothesis.

3. a) Use the SET command of MINITAB to enter the data values into column C1 of a MINITAB worksheet.
   2.257 2.592 1.723 2.310 3.085 2.708 1.941
   2.867 2.727 2.393 2.012 2.413 2.476 2.133
   2.561 2.291 2.452 2.587 3.053 2.457 2.127
   2.218 2.708 2.707 3.425
   b) Using the data in C1 and the TTEST command where MU=2.8, test the hypothesis that the mean of the population is less than 2.8.
   c) State the null and alternative hypothesis.
   d) Is this a one or a two-tailed test? Explain.
   e) Why is the TTEST command appropriate. Explain.
   f) Where are the data values for the random sample?
   g) Name and explain which values on the output are used to calculate the p-value.
   h) Report the p-value for this sample.
   i) Is the p-value significant at $\alpha=5\%$? Explain.
   j) Is the p-value significant at $\alpha=1\%$? Explain.
   k) State the conclusion in terms of the null and alternative hypothesis.

4. a) Use the SET command of MINITAB to enter the data values into column C1 of a MINITAB worksheet.
   23 12 14 20 13 23 26 20 18 24 22 16
   16 23 12 21
   b) Using the data in C1 and the TTEST command where MU=18, test the hypothesis that the mean of the population is greater than 18.
   c) State the null and alternative hypothesis.
   d) Is this a one or a two-tailed test? Explain.
   e) Why is the TTEST command appropriate. Explain.
   f) Where are the data values for the random sample?
   g) Name and explain which values on the output are used to calculate the p-value.
   h) Report the p-value for this sample.
   i) Is the p-value significant at $\alpha=5\%$? Explain.
   j) Is the p-value significant at $\alpha=1\%$? Explain.
   k) State the conclusion in terms of the null and alternative hypothesis.

PART VII **Projects.**

1. Hypothesis Testing Project
A. Selection of a topic
   1. Select a reported fact or claim about a population mean made in a local newspaper, magazine or other source.
   2. Suggested topics:
      a) Test the claim that the average number of hours worked per week by a student is 20 hours.
      b) Test the claim that the cumulative grade point average is higher today then it was ten years ago. The cumulative grade point average ten years ago was 2.64.
      c) Test the claim that the average SAT score on the mathematics section is greater than 480.
      d) Test the claim that the average pulse rate at rest for a woman between the ages of 16 and 23 inclusive is lower than 80.
B. Use the following procedure to test the claim selected in part A.
   1. State the claim, identify the population referred to in the statement of the claim and indicate where the claim was found.
   2. State your opinion regarding the claim (i.e. do you feel it is too high, too low, or simply don't agree with the claim) and clearly identify the population you plan to sample to support your opinion.
   3. State the null and alternative hypotheses.
   4. Develop and state a procedure for selecting a random sample from your population. Indicate the technique you are going to use to obtain your sample data.
   5. Compute the expected results, indicate the hypothesis testing model, and choose an appropriate level of significance.
   6. Calculate the critical value(s) and state the decision rule.
   7. Perform the experiment and calculate the experimental outcome, $\bar{x}$.
   8. Construct a hypothesis testing model and place the experimental outcome on the model.
   9. Formulate the appropriate conclusion and interpret the conclusion with respect to your opinion (as stated in step 2).
2. Read a research article from a professional journal in a field such as: medicine, education, marketing, science, social science, etc. Write a summary of the research article that includes the following:
   a) Name of article, author, and journal (include volume number and date).

b) Statement of the research problem.
c) The null and alternative hypotheses.
d) How the experiment was conducted. (include the sample size and sampling technique)
e) Determine the experimental outcome(s) and the level of significance.
f) State the conclusion(s) in terms of the null and alternative hypotheses.

PART VIII Database.

The following exercises refer to the file DATABASE listed in Appendix A. We have indicated the appropriate MINITAB commands that are necessary to answer each exercise.

1. Using the MINITAB commands: RETRIEVE and TTEST

    Retrieve the file DATABASE.MTW and identify the column number:
    for each of the variable names:
    a) SHT
    b) WRK
    For each variable use the TTEST command to test the null hypothesis that that the population mean, mu = k
    a) k = 68 for the variable SHT.
    b) k = 15 for the variable WRK.
    State the null hypothesis for each variable in words.
    Write and execute the TTEST command if the alternative hypothesis for each variable is *nondirectional*.
    State the alternative hypothesis for each variable in words.
    Is the null hypothesis rejected? Identify the p value and interpret.
    State the conclusion.

2. Using the MINITAB commands: RETRIEVE and TTEST
    Retrieve the file DATABASE.MTW and identify the column number:
    for each of the variable names:
    a) SHT
    b) WRK
    For each variable use the TTEST command to test the null hypothesis that the population mean, mu = k
    a) k = 68 for the variable SHT,
    b) k = 15 for the variable WRK,
    if the alternative hypothesis subcommand for each variable is *directional* with form:
    a) ALTERNATIVE = -1 for the variable SHT.
    b) ALTERNATIVE = +1 for the variable SHT.
    c) ALTERNATIVE = -1 for the variable WRK.
    d) ALTERNATIVE = +1 for the variable WRK.
    State the null and alternative hypotheses for each variable in words. Interpret the output in terms of the null and alternative hypotheses, and p value.

CHAPTER 11
HYPOTHESIS TESTING INVOLVING TWO POPULATION MEANS

11.1 INTRODUCTION

In Chapter 10, we discussed hypothesis tests involving one population mean. In this chapter we will study hypothesis tests that involve the comparison of two population means to determine if the population means are statistically different.

For example:

- Is there a difference in the mean physical fitness scores of third grade girls and boys? This involves the comparison of two populations: the population of physical fitness scores for the girls and the population of physical fitness scores for the boys.

- Will Goodday automobile tires give better average mileage wear than Goodpoor tires? (Neither company owns a blimp!) The mileage data for Goodday tires would comprise one of the populations and the mileage data for the Goodpoor tires would comprise the other population.

- Does a toothpaste that contains fluoride significantly lower the mean number of cavities as compared to the toothpaste without fluoride? One population represents those people that used the toothpaste with fluoride and the other population represents those people that used the tooth paste without fluoride.

To answer these questions about whether there is a significant difference between the means of two independent populations, we have to select a random sample from each population, to calculate the mean of each sample and to determine the difference between the sample means. We would then have to test whether the **difference between the two sample means** is significant enough to conclude that the two population means are different. To help us perform this test, we need to examine the **sampling distribution of the difference between two means.**

11.2 THE SAMPLING DISTRIBUTION OF THE DIFFERENCE BETWEEN TWO MEANS

To perform a hypothesis test involving two population means, an independent random sample[1] from each population will be selected and compared to determine whether **the difference between the two sample means** is significantly different. This is illustrated in Figure 11.1.

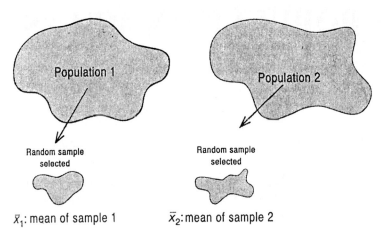

Figure 11.1

To determine if the difference between the sample means, denoted by $d = \bar{x}_1 - \bar{x}_2$, is statistically significant, we must compare this difference to a distribution formed by calculating the difference between the means of all possible random sample pairs, where the first random sample with size n_1 for the pair is selected from an infinite population 1 and the second random sample with size n_2 of the pair is selected from an infinite population 2. This sampling distribution is referred to as the **Sampling Distribution of The Difference Between Two Means** or the **Sampling Distribution of $\bar{x}_1 - \bar{x}_2$**. This is illustrated in Figure 11.2.

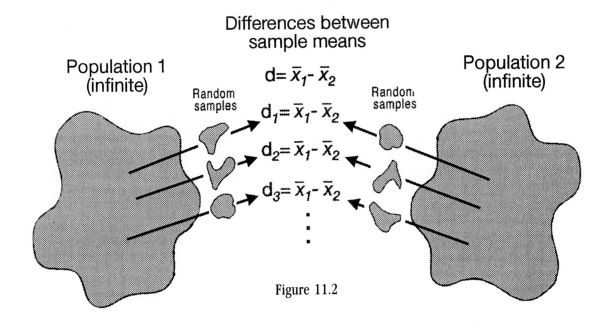

Figure 11.2

---

[1] By independent random samples, we mean that the selection of one random sample has no influence upon the selection of the other random sample.

The distribution of the differences between the means of all possible independent random sample pairs form the Sampling Distribution of The Difference Between Two Means.

Let's now examine Theorem 11.1 which explains the properties of the Sampling Distribution of the Differences Between Two Means where the population standard deviations are known and the sample sizes are greater than 30.

Theorem 11.1 Characteristics of the Sampling Distribution of The Difference Between Two Means with Population Standard Deviations Known.

Given two populations, referred to as population 1 and population 2, with means μ_1 and μ_2 respectively and standard deviations σ_1 and σ_2 respectively. If all possible independent random samples of size n_1 and n_2 (where **both sample sizes are greater than 30**) are selected from population 1 and population 2 respectively, the mean of all these pairs of samples are determined, and the differences between the pairs of sample means ($\overline{x}_1 - \overline{x}_2$) are computed, then the Sampling Distribution of The Difference Between Two Means is formed and has the following characteristics.

1. The **mean** of the Sampling Distribution of the Difference Between Two Means, denoted by $\mu_{\overline{x}_1 - \overline{x}_2}$, is:

$$\mu_{\overline{x}_1 - \overline{x}_2} = \mu_1 - \mu_2$$

2. The **standard deviation** of the Sampling Distribution of the Difference Between Two Means, denoted by $\sigma_{\overline{x}_1 - \overline{x}_2}$, is:

$$\sigma_{\overline{x}_1 - \overline{x}_2} = \sqrt{\frac{\sigma_1^2}{n_1} + \frac{\sigma_2^2}{n_2}}$$

3. The Sampling Distribution of the Difference Between Two Means is **approximately normal**.

In most instances the standard deviations of the two populations are *unknown*. Thus, the **estimates** of the two population standard deviations, s_1 and s_2 respectively, are used to approximate the two population standard deviations, σ_1 and σ_2 respectively, and the Central Limit Theorem may not apply. That is, the Sampling Distribution of the Difference Between Two Means may not be approximated by a normal distribution. However, if the two populations are approximately normal with equal standard deviations, and the sample standard deviations are used to estimate the population standard deviations, then the Sampling Distribution of the Difference Between Two Means can be approximated by a *t* distribution. This result and the characteristics of the Sampling Distribution of the Difference Between Two Means with population standard deviations unknown is stated in Theorem 11.2.

Theorem 11.2 Characteristics of the Sampling Distribution of the Difference Between Two Means with Population Standard Deviations Unknown.

Given two approximately normal populations, referred to as population 1 and population 2 with means, μ_1 and μ_2 and approximately equal standard deviations σ_1 and σ_2, respectively.

If all possible independent random samples (i.e. the selection of one random sample has no influence upon the selection of the other random sample) of size n_1 and n_2 are selected from population 1 and population 2 respectively, the means of all these pairs of samples are determined, and the differences between the pairs of sample means ($\bar{x}_1 - \bar{x}_2$) are computed, then the Sampling Distribution of the Difference Between Two Means has the following characteristics:

1. The mean of the Sampling Distribution of the Difference Between Two Means, denoted by $\mu_{\bar{x}_1 - \bar{x}_2}$, is: $\mu_{\bar{x}_1 - \bar{x}_2} = \mu_1 - \mu_2$

2. The standard deviation of the Sampling Distribution of the Difference Between Two Means cannot be determined since the population standard deviations, σ_1 and σ_2, are unknown.

   However, since the two sample standard deviations, s_1 and s_2, provide good estimates of the population standard deviations, σ_1 and σ_2, we can combine the sample standard deviations to obtain an estimate of the standard deviation of the Sampling Distribution of the Difference Between Two Means. The process of combining the two independent sample results, which provides an estimate of the standard deviation of the Sampling Distribution of the Difference Between Two Means, is referred to as **pooling**.

   The formula to calculate this pooled estimate of the standard deviation of the Sampling Distribution of the Difference Between Two Means, written $s_{\bar{x}_1 - \bar{x}_2}$, is:

$$s_{\bar{x}_1 - \bar{x}_2} = \sqrt{\frac{(n_1 - 1)s_1^2 + (n_2 - 1)s_2^2}{n_1 + n_2 - 2} \cdot \left(\frac{1}{n_1} + \frac{1}{n_2}\right)}$$

   where:
   s_1 is the estimate of the standard deviation of population 1.
   n_1 is the sample size for the sample selected from population 1.
   s_2 is the estimate of the standard deviation of population 2.
   n_2 is the sample size for the sample selected from population 2.

3. The Sampling Distribution of The Difference Between Two Means is approximated by a *t* distribution with degrees of freedom which is given by: $df = n_1 + n_2 - 2$.

11.3 HYPOTHESIS TESTING INVOLVING TWO POPULATION MEANS AND UNKNOWN POPULATION STANDARD DEVIATIONS

Applying the general five step hypothesis testing procedure discussed in Chapter 9 to hypotheses involving **two population means**, where the population standard deviations are unknown, we will develop a hypothesis testing procedure that will allow us to determine if there is a significant difference between the means of two independent population. This type of hypothesis test is often referred to as a **two sample t test**.

Two Sample t Test

Step 1. *Formulate the two hypotheses, H_o and H_a.*
  Null Hypothesis, H_o

The **Null Hypothesis**, H_o, in testing the difference between two population means, has the following form:

Null Hypothesis Form

H_o: There is **no difference** between the mean of population 1, written μ_1, and the mean of population 2, written μ_2.

This is symbolized as:
$$H_o: \mu_1 - \mu_2 = 0$$

By stating H_o as there is no difference between the population means, μ_1 and μ_2, we are making the assumption that μ_1 **equals** μ_2. Consequently, only by **rejecting the null hypothesis**, H_o, can we conclude that there is a *statistically significant difference* between μ_1 and μ_2. Therefore, by rejecting H_o, we can conclude that the alternative hypothesis is true.

Alternative Hypothesis, H_a

The **Alternative Hypothesis, H_a**, in testing the difference between two population means, has one of the following three forms:

Form (a): Less Than Form for H_a

H_a: the mean of population 1, written μ_1, is **less than** the mean of population 2, written μ_2. This can also be stated as: **the difference ($\mu_1 - \mu_2$) is significantly less than zero.**
This is symbolized as:
$$H_a: \mu_1 - \mu_2 < 0$$

Form (b): Greater Than Form for H_a

H_a: the mean of population 1, written μ_1, is **greater than** the mean of population 2, written μ_2. This can also be stated as: **the difference ($\mu_1 - \mu_2$) is significantly greater than zero.**
This is symbolized as:
$$H_a: \mu_1 - \mu_2 > 0$$

Form (c): Not Equal Form for H_a

H_a: the mean of population 1, written μ_1, is **not equal to** the mean of population 2, written μ_2. This can also be stated as: **difference ($\mu_1 - \mu_2$) is significantly different from zero.**
This is symbolized as:
$$H_a: \mu_1 - \mu_2 \neq 0$$

Step 2. *Design an experiment to test the null hypothesis, H_o. Under the assumption H_o is true, calculate the expected results.*

Since we are testing the hypothesis that the **difference between two population means ($\mu_1 - \mu_2$)** is significantly different from zero by **using the difference between two sample means, $\overline{x}_1 - \overline{x}_2$**, then we will use the **Sampling Distribution of The Difference Between Two Means as the hypothesis test model.**

Using Theorem 11.2, and **under the assumption H_o is true** (that is: $\mu_1 - \mu_2 = 0$ is true), then **the expected results for a two sample t test are:**

a) the Sampling Distribution of The Difference Between Two Means is approximately a t Distribution with degrees of freedom:

$$df = n_1 + n_2 - 2$$

b) the mean of the Sampling Distribution of The Difference Between Two Means is given by the formula:

$$\mu_{\bar{x}_1 - \bar{x}_2} = \mu_1 - \mu_2$$

Under the assumption H_o is true (i.e. $\mu_1 - \mu_2 = 0$ is true), then we can state that:

$$\mu_{\bar{x}_1 - \bar{x}_2} = 0$$

c) the estimate of the standard deviation of the Sampling Distribution of The Difference Between Two Means, written $s_{\bar{x}_1 - \bar{x}_2}$, is given by the formula:

$$s_{\bar{x}_1 - \bar{x}_2} = \sqrt{\frac{(n_1-1) s_1^2 + (n_2-1) s_2^2}{n_1 + n_2 - 2} \cdot \left(\frac{1}{n_1} + \frac{1}{n_2}\right)}$$

Step 3. *Formulate the decision rule.*

a) Determine the type of alternative hypothesis: directional or nondirectional.

b) Determine the type of test: 1TT or 2TT.

c) Identify the significance level: $\alpha = 1\%$ or $\alpha = 5\%$.

d) Calculate the critical value(s). The formula for the critical value:

$$X_c = \mu_{\bar{x}_1 - \bar{x}_2} + (t_c)(s_{\bar{x}_1 - \bar{x}_2})$$ can be simplified by using the assumption that H_o is true, that is: $\mu_1 - \mu_2 = 0$, and $\mu_{\bar{x}_1 - \bar{x}_2} = \mu_1 - \mu_2$, then we obtain: $\mu_{\bar{x}_1 - \bar{x}_2} = 0$. Thus, the critical value formula is simplified to:

$$X_c = t_c (s_{\bar{x}_1 - \bar{x}_2})$$

e) Construct the hypothesis testing model.

f) State the decision rule.

Step 4. *Conduct the experiment.*

> Randomly select an independent sample from each population and calculate the mean of each sample, denoted $\bar{x}_1$ and $\bar{x}_2$. **Determine the difference between the two sample means, the experimental outcome, using the formula:**
> $d = \bar{x}_1 - \bar{x}_2$.

Step 5. *Determine the conclusion.*

> Compare the experimental outcome, $d = \bar{x}_1 - \bar{x}_2$, to the critical value(s), X_c, of the decision rule and draw one of the following conclusions:
>
> (a) **Reject H_o and Accept H_a at α** or (b) **Fail to reject H_o at α**

In summary, the procedure just used outlined to perform a **two sample t test**, there were four refinements to the general hypothesis testing procedure developed in Chapter 9. They were:

1) The **Sampling Distribution of the Difference Between Two Means**, which is a t distribution, is used as the appropriate hypothesis testing model.
2) The properties of the Sampling Distribution of the Difference Between Two Means are used to calculate the expected values.
3) The formula: $X_c = (t_c)(s_{\bar{x}_1 - \bar{x}_2})$ is used to calculate the critical value(s) for the decision rule.
4) The experimental outcome represents the difference between the two sample means and is determined by the formula: $d = \bar{x}_1 - \bar{x}_2$

Example 11.1 illustrates the use of a two sample t test to perform hypothesis test involving two population means.

Example 11.1

A computer programming aptitude test was given to 50 men and 60 women. The women's sample mean test score was 84 with s = 7, while the men's mean test score was 82 with s = 5. Can one conclude that there is a significant difference between the mean test scores for the population of men and the population of women on this test? (Use a = 1%)

Solution:

Step 1. *Formulate the hypotheses.*

If we let population 1 represent the men's computer aptitude scores and population 2 represent the women's computer aptitude scores then the null hypotheses is:

H_o: There is **no difference** between the means of the population computer programming aptitude test scores for men and women.
In symbols, H_o: $\mu_1 - \mu_2 = 0$

Since we are trying to show that there is a significant difference between the population mean test scores of the men and women we state the null hypothesis in such a way that by *rejecting* H_o we can conclude that there *is* a difference between the population means. Therefore, we will always state the null hypothesis as "there is no difference between the mean of population 1 and the mean of population 2."
This is symbolized as: H_o: $\mu_1 - \mu_2 = 0$.

We are trying to show that there is a difference between the population mean for the men's and women's aptitude test scores. Thus, the alternative hypothesis is nondirectional and is stated as:

H_a: The difference between the mean computer aptitude test scores for the men and women **is significantly different** from *zero*. This is symbolized as:
H_a: $\mu_1 - \mu_2 \neq 0$

Step 2. *Design an experiment to test the null hypothesis, H_o*

To test the null hypothesis an independent random sample is selected from each population. The sample results are summarized in Table 11.1.

Table 11.1

Random Sample selected from Population 1 (men)	Random Sample selected from Population 2 (women)
$\overline{x}_1 = 82$	$\overline{x}_2 = 84$
$s_1 = 5$	$s_2 = 7$
$n_1 = 50$	$n_2 = 60$

Since the difference between the two sample means, $\overline{x}_1 - \overline{x}_2$, will be used to determine if $(\mu_1 - \mu_2)$ is **significantly different** from **zero, the Sampling Distribution of the Differences Between Sample Means** is used as the **hypothesis testing model**. To determine the characteristics of this model we use Theorem 11.2 and the assumption H_o is true. Thus, **the expected results** are:

a) The Sampling Distribution of the Difference Between Two Means is approximated by a *t* distribution with degrees of freedom:
$$df = n_1 + n_2 - 2 = 108.$$

b) The mean of the Sampling Distribution of the Difference Between Two Means is:

$$\mu_{\bar{x}_1 - \bar{x}_2} = \mu_1 - \mu_2$$

$$= 0$$

c) The estimate of the standard deviation of the Sampling Distribution of the Difference Between Two Means formula is:

$$s_{\bar{x}_1 - \bar{x}_2} = \sqrt{\frac{(n_1-1)s_1^2 + (n_2-1)s_2^2}{n_1+n_2-2} \cdot \left(\frac{1}{n_1} + \frac{1}{n_2}\right)}$$

Substituting $s_1 = 5$, $n_1 = 50$, $s_2 = 7$ and $n_2 = 60$ into the formula for $s_{\bar{x}_1 - \bar{x}_2}$, we have:

$$s_{\bar{x}_1 - \bar{x}_2} = \sqrt{\frac{(n_1-1)s_1^2 + (n_2-1)s_2^2}{n_1+n_2-2} \cdot \left(\frac{1}{n_1} + \frac{1}{n_2}\right)}$$

$$= \sqrt{\frac{(50-1)5^2 + (60-1)7^2}{50+60-2} \left(\frac{1}{50} + \frac{1}{60}\right)}$$

$$= \sqrt{\frac{(49)25 + (59)49}{108} \left(\frac{1}{50} + \frac{1}{60}\right)}$$

$$\approx \sqrt{(38.11111)(0.03666)}$$

$$s_{\bar{x}_1 - \bar{x}_2} \approx 1.18$$

Figure 11.3 illustrates the Sampling Distribution of the Difference Between Two Means hypothesis testing model.

Sampling Distribution of the Difference Between Two Means
*(is a **t** Distribution with **df**=108)*

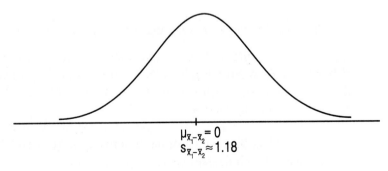

Figure 11.3

Step 3. *Formulate the decision rule.*

a) The alternative hypothesis is nondirectional since we are trying to show that there is a significant difference between the population mean scores of the men and women.
b) The type of hypothesis test is two-tailed (2TT).
c) The significance level is: $\alpha = 1\%$.
d) To calculate the critical values for a two-tailed test with $\alpha = 1\%$ and $df = 108$ first determine the critical t scores, t_{LC} and t_{RC}, using **TABLE III: Critical Values for the t Distribution found in Appendix D: Statistical Tables**.

From TABLE III, the critical t scores are:
$$t_{LC} = -2.63 \quad \text{and} \quad t_{RC} = 2.63$$

The critical values, X_{LC} and X_{RC}, are determined using the critical outcome formula:

$$X_c = t_c (s_{\bar{x}_1 - \bar{x}_2})$$

$$X_{LC} = (t_{LC})(s_{\bar{x}_1 - \bar{x}_2}) \quad \text{and} \quad X_{RC} = (t_{RC})(s_{\bar{x}_1 - \bar{x}_2})$$

$$X_{LC} = (-2.63)(1.18) \qquad X_{RC} = (2.63)(1.18)$$

$$X_{LC} = -3.10 \qquad\qquad X_{RC} = 3.10$$

e) Figure 11.4 illustrates the decision rule for this hypothesis test.

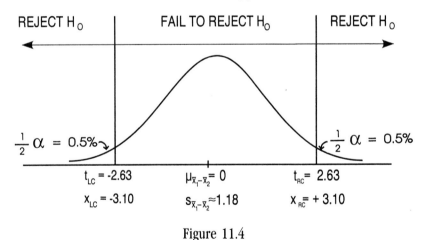

Figure 11.4

The decision rule is:

f) Reject H_0 if the difference between the two sample means, $\bar{x}_1 - \bar{x}_2$, is either less than -3.10 or more than 3.10.

Step 4. *Conduct the experiment.*

Calculate the experimental outcome using $d = \bar{x}_1 - \bar{x}_2$, which represents the difference between the two sample means.

$$d = \bar{x}_1 - \bar{x}_2$$
$$d = 82 - 84$$
$$d = -2$$

Step 5. *Determine the conclusion.*

Since the experimental outcome, $d = \bar{x}_1 - \bar{x}_2 = -2$, falls between the critical values, $X_{LC} = -3.10$ and $X_{RC} = 3.10$, we fail to reject H_o at $\alpha = 1\%$. Therefore, we can conclude that the difference in the population mean aptitude test scores for men and women is **not statistically significant** at $\alpha = 1\%$. ■

Example 11.2

A sociologist claims that women in the South marry at a younger age than women in the North. An independent random sample of 40 recently married southern women had a mean age of 21 years with s = 2 years. An independent random sample of 40 recently married northern women had a mean age of 23.5 years with s = 3 years. Do these sample results support the sociologist's claim at an $\alpha = 5\%$?

Solution:

Step 1. *Formulate the hypotheses.*

If we let population 1 represent the ages of the southern brides and population 2 represent the ages of the northern brides then the null hypothesis is:

H_o: there is **no difference** in the population mean age of southern and northern brides.
In symbols we have, H_o: $\mu_1 - \mu_2 = 0$.

H_a: the population mean age for southern brides is **less than** the population mean age of northern brides.

In symbols we have, H_a: $\mu_1 - \mu_2 < 0$.

Step 2. *Design an experiment to test the null hypothesis, H_o.*

To test the null hypothesis, an independent random sample is selected from each population and the sample results are summarized in Table 11.2.

Table 11.2
Summary of the data for the random samples
selected from each population.

Population 1 (South)	Population 2 (North)
$\bar{x}_1 = 21$	$\bar{x}_2 = 23.5$
$s_1 = 2$	$s_2 = 3$
$n_1 = 40$	$n_2 = 40$

Since the differences between the two sample means, $\overline{x}_1 - \overline{x}_2$, will be used to determine if $(\mu_1 - \mu_2)$ is significantly less than zero, the Sampling Distribution of the Difference Between Two Means is used as the hypothesis testing model. To determine the characteristics of this model we use Theorem 11.2 and the assumption H_o is true. Thus, **the expected results** are:

a) The Sampling Distribution of the Difference Between Two Means is approximated by a t distribution with $df = 78$.

b) The mean of the Sampling Distribution of the Difference Between Two Means is:

$$\mu_{\overline{x}_1 - \overline{x}_2} = \mu_1 - \mu_2 = 0$$

c) The estimate of the standard deviation of the Sampling Distribution of the Differences Between Sample Means formula is:

$$s_{\overline{x}_1 - \overline{x}_2} = \sqrt{\frac{(n_1 - 1) s_1^2 + (n_2 - 1) s_2^2}{n_1 + n_2 - 2} \cdot \left(\frac{1}{n_1} + \frac{1}{n_2} \right)}$$

Substituting $s_1 = 2$, $n_1 = 40$, $s_2 = 3$ and $n_2 = 40$ into $s_{\overline{x}_1 - \overline{x}_2}$, we have:

$$s_{\overline{x}_1 - \overline{x}_2} = \sqrt{\frac{(n_1 - 1) s_1^2 + (n_2 - 1) s_2^2}{n_1 + n_2 - 2} \cdot \left(\frac{1}{n_1} + \frac{1}{n_2} \right)}$$

$$= \sqrt{\frac{(40-1) 2^2 + (40-1) 3^2}{40 + 40 - 2} \left(\frac{1}{40} + \frac{1}{40} \right)}$$

$$= \sqrt{\frac{(39) 4 + (39) 9}{78} \left(\frac{1}{40} + \frac{1}{40} \right)}$$

$$= \sqrt{(6.5)(0.05)}$$

$$s_{\overline{x}_1 - \overline{x}_2} \approx 0.57$$

Figure 11.5 illustrates the Sampling Distribution of the Difference Between Two Means hypothesis testing model.

*Sampling Distribution of the Difference
Between Two Means*
(is a *t* Distribution with *df*=78)

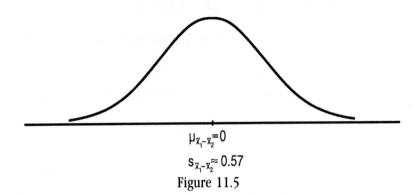

Figure 11.5

Step 3. *Formulate the decision rule.*

a) the alternative hypothesis is directional since we are trying to show that the population mean age of southern brides is significantly less than the population mean age of northern brides.
b) the type of hypothesis test is one-tailed (i.e. 1TT) on the **left side** since we are trying to show that $(\mu_1 - \mu_2)$ is **less than** zero.
c) the significance level is: $\alpha = 5\%$.
d) to calculate the critical value for a one-tailed test on the left side with $\alpha = 5\%$ and $df = 78$, find the critical t score, t_c, using TABLE III.
From TABLE III, we find: $t_c = -1.66$.

The critical value, X_c, is determined using:

$$X_c = (t_c)(s_{\bar{x}_1 - \bar{x}_2}), \quad s_{\bar{x}_1 - \bar{x}_2} = 0.57 \text{ and } t_c = -1.66.$$

Thus,
$$X_c = (-1.66)(0.57)$$
$$X_c = -0.95$$

e) Figure 11.6 illustrates the decision rule for this hypotheses test.

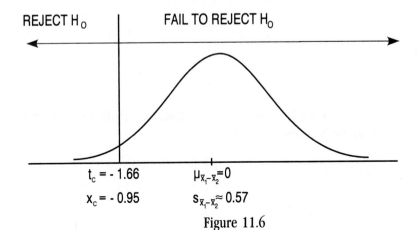

Figure 11.6

The decision rule is:

f) Reject H_o if the difference between the two sample means, $d = \bar{x}_1 - \bar{x}_2$, is less than -0.95.

Step 4. *Conduct the experiment.*

Calculate the experimental outcome using $d = \bar{x}_1 - \bar{x}_2$.
(i.e. the difference between the two sample means.)

$$d = \bar{x}_1 - \bar{x}_2$$
$$d = 21 - 23.5$$
$$d = -2.5.$$

Step 5. *Determine the conclusion.*

Since the experimental outcome, $d = \bar{x}_1 - \bar{x}_2 = -2.5$, is *less than* the critical value of -0.95 we *reject H_o and accept H_a* at $\alpha = 5\%$. Therefore, we can agree with the sociologist's claim that women in the south marry at a younger mean age than women in the north. ■

11.4 HYPOTHESIS TESTS COMPARING TREATMENT AND CONTROL GROUPS

Frequently, an experimenter wants to compare two populations or groups where one group receives a treatment called the ***treatment group***, while a second group called the ***control group*** does not receive the treatment. Ideally, the members of **both groups** are **similar** in every way **except** that the members of the **control** group are **not** given the **treatment**.

For example, a new medication is tested to determine if it is effective. A random sample of the population called the treatment group is selected and given this new medication. Another random sample of the population called the control group is selected and does not receive the medication. However, the control group is given a **placebo**, that is, a pill which looks like the real medication but does not contain the medication. This is done to guard against the psychological effect of the treatment group feeling better not because of the new medication but because they were picked for the special treatment. This type of experimental design is called a ***single-blind study***, because the *subjects* do **not** know who is receiving the *real* medication. Sometimes it has been shown that studies have been effected by the *doctors* who are administering the medication, since *they* know who is and who is not receiving the placebo. Consequently, it becomes necessary to conduct a study where *both* the subjects and doctors are *not* aware of who is receiving the medication and who is receiving the placebo. This type of experimental design is called a ***double-blind study***.

Example 11.3 illustrates the comparison of treatment verses control groups using a two sample *t* test.

Example 11.3

A pharmaceutical company has developed a diet pill called, Effective Anti-Hunger Tablet (E.A.T.), which the company believes will significantly reduce an individual's weight at the end of one week. To support their claim, the company selects 30 overweight individuals and randomly divides them into two independent groups: a **treatment** group of 15 subjects who are administered the diet pill E.A.T. and a **control** group of 15 subjects who are given a **placebo** tablet (i.e. **a pill that has *no effect* on weight**).

After one week, the treatment group had a sample mean weight loss of $\bar{x}$ = 7.1 lbs. with s = 3.6 lbs. while the control group had a sample mean weight loss of $\bar{x}$ = 4.2 lbs. with s = 2.3 lbs.

Can the company conclude that their diet pill E.A.T. will significantly reduce an individual's weight within one week at α = 5%?

Remark: In this study the administration of the diet pill E.A.T. to the treatment group is the only difference between the two groups. Therefore, if we can show a significant difference between the treatment and control groups then this ***difference can be attributed only to the treatment (the diet pill).***

Solution:

Step 1. *Formulate the hypotheses.*

If we let population 1 represent the treatment group and population 2 represent the control group then the null hypothesis is:

H_o: there is **no difference** in the population mean weight loss for the treatment group and the control group.

In symbols we have, H_o: $\mu_1 - \mu_2 = 0$

H_a: the population mean weight loss for the treatment group is **greater than** the population mean weight loss for the control group.

In symbols we have, H_a: $\mu_1 - \mu_2 > 0$.

Step 2. *Design an experiment to test the null hypothesis, H_o.*

To test the null hypothesis the company selects 30 overweight individuals and randomly divides them into two independent groups and the sample results are summarized in Table 11.3.

Table 11.3
Summary of the data for the random samples
selected from each population.

Population 1 (Treatment)	Population 2 (Control)
$\bar{x}_1$ = 7.1	$\bar{x}_2$ = 4.2
s_1 = 3.6	s_2 = 2.3
n_1 = 15	n_2 = 15

Since the differences between the two sample means, $\bar{x}_1 - \bar{x}_2$, will be used to determine if $(\mu_1 - \mu_2)$ is significantly greater than zero, the Sampling Distribution of the Difference Between Two Means is used as the hypothesis testing model. To determine the characteristics of this model we use Theorem 11.2 and the assumption H_0 is true. Thus, **the expected results** are:

a) The Sampling Distribution of the Difference Between Two Means is approximated by a *t* distribution with *df* = 28.

b) The mean of the Sampling Distribution of the Difference Between Two Means is:
$$\mu_{\bar{x}_1-\bar{x}_2} = \mu_1 - \mu_2 = 0$$

c) The estimate of the standard deviation of the Sampling Distribution of the Differences Between Sample Means formula is:

$$s_{\bar{x}_1-\bar{x}_2} = \sqrt{\frac{(n_1-1)s_1^2 + (n_2-1)s_2^2}{n_1+n_2-2} \cdot \left(\frac{1}{n_1} + \frac{1}{n_2}\right)}$$

Substituting $s_1=3.6$, $n_1=15$, $s_2=2.3$ and $n_2=15$ into $s_{\bar{x}_1-\bar{x}_2}$, we have:

$$s_{\bar{x}_1-\bar{x}_2} = \sqrt{\frac{(15-1)(3.6)^2 + (15-1)(2.3)^2}{15+15-2} \left(\frac{1}{15} + \frac{1}{15}\right)}$$

$$= \sqrt{\frac{181.44 + 74.06}{28}(0.13)}$$

$$\approx \sqrt{(9.1250)(0.1333)}$$

$$s_{\bar{x}_1-\bar{x}_2} \approx 1.10$$

Figure 11.7 illustrates the Sampling Distribution of the Difference Between Two Means hypothesis testing model.

Sampling Distribution of the Difference Between Two Means
(is a *t* Distribution with *df*=28)

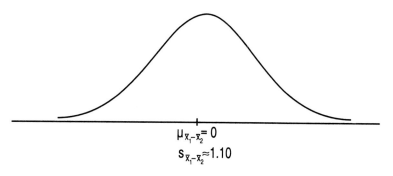

$\mu_{\bar{x}_1-\bar{x}_2} = 0$
$s_{\bar{x}_1-\bar{x}_2} \approx 1.10$

Figure 11.7

Step 3. *Formulate the decision rule.*

a) the alternative hypothesis is **directional** since we are trying to show that the population mean weight loss for the treatment group is **significantly greater** than the population mean weight loss for the control group.

b) the type of hypothesis test is one-tailed (1TT) on the **right side** since we are trying to show that $(\mu_1 - \mu_2)$ is **greater than** zero.

c) the significance level is: $\alpha = 5\%$.

d) to calculate the critical value for a one-tailed test on the right with $\alpha = 5\%$ and $df = 28$, first determine the critical t score, t_c, using TABLE III. From TABLE III, $t_c = 1.70$.

The critical value, X_c, is determined using:

$$X_c = (t_c)(s_{\bar{x}_1 - \bar{x}_2})$$

Thus,
$$X_c = (1.70)(1.10)$$
$$X_c = 1.87$$

e) Figure 11.8 illustrates the decision rule for this hypothesis test.

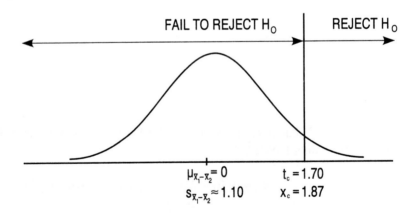

Figure 11.8

Step 4. *Conduct the experiment.*

Calculate the experimental outcome (i.e. the difference between the two sample means), using $d = \bar{x}_1 - \bar{x}_2$.

$$d = \bar{x}_1 - \bar{x}_2$$
$$d = 7.1 - 4.2$$
$$d = 2.9$$

Step 5. *Determine the conclusion.*

Since the experimental outcome, d = $\overline{x}_1 - \overline{x}_2$ = 2.9, is greater than the critical value of 1.87, *reject* H_o and *accept* H_a at α = 5%.

Therefore, the company can conclude that their diet pill (E.A.T.) will **significantly reduce** an individual's weight within the end of one week at α = 5%. ∎

CASE STUDY 11.1

A research article entitled *Differences in Extracurricular Activity Participation, Achievement, And Attitudes Toward School Between Ninth-grade Students Attending Junior High School And Those Attending Senior High School* appeared in Adolescence. The following table represents a comparison of the Mean Number of Total Extracurricular Activities Participated, Overall GPA, and Mean Attitude Toward Self and School for Ninth Graders in Junior High (n = 771) and Senior High Settings (n = 825).

GROUPS	MEAN	STANDARD DEVIATION	t value
EXTRACURRICULAR ACTIVITIES			
NINTH GRADERS IN JHS	2.68	2.30	6.20*
NINTH GRADERS IN SHS			
OVERALL GPA			
NINTH GRADERS IN JHS	2.59	0.89	6.91*
NINTH GRADERS IN SHS	2.24	1.11	
ATTITUDE TOWARD SELF AND SCHOOL			
NINTH GRADERS IN JHS	77.87	8.07	0.60
NINTH GRADERS IN SHS	78.12	8.48	

*p < 0.01

Let's define population 1 as: Ninth Graders in Junior High School (JHS) and population 2 as: Ninth Graders in Senior High School (SHS).

a) If you are trying to show that there is a significant difference in the mean number of extracurricular activities for these two populations, then state the null hypothesis and alternative hypothesis for this two-sample t test.
b) If you want to show that there is a significant difference in the mean overall GPA for these two populations, then state the null and alternative hypothesis for this two-sample t test.
c) If you would like to show that there is a significant difference in the mean attitude toward self and school for these two populations, then state the null and alternative hypothesis for this two-sample t test.
d) What is the name and mean of the sampling distribution which serves as a model for this test? Is this a normal or a *t* distribution? Explain.
e) For the Extracurricular Activities Study, which group had the greater variability? For the Overall GPA Study, which group had the less variability? For the Attitude Study, which group had the greater variability? Do these measures of variability represent a parameter or a statistic? Explain.
f) Was there a significant difference between the populations with respect to the mean number of extracurricular activities? Explain.

g) If the significance level for the extracurricular activities had been 0.5%, would the null hypothesis be rejected? Explain.
h) Was there a significant difference between the populations with respect to the mean overall GPA? Was the null hypothesis rejected? Explain.
i) Would the null hypothesis have been rejected for the overall GPA study if $\alpha = 1\%$? Explain.
j) Was there a significant difference between the populations with respect to the mean attitude? Was the null hypothesis rejected? Explain.
k) Would the null hypothesis have been rejected for the attitude study if $\alpha = 2\%$? Explain.

CASE STUDY 11.2

In Figure CS11.1 is an article entitled *Study says 'placebo effect' is potent* appeared in the Orange County Register, a California newspaper, on July 1, 1993.

Study says 'placebo effect' is potent

HEALTH: Research shows people with certain diseases get better 70 percent of the time with dummy treatments.

By LEE SIEGEL
The Associated Press

LOS ANGELES — People with certain diseases get better 70 percent of the time even when they receive dummy treatments, according to a study that suggests the "placebo effect" can be a powerful healer.

"Even if a treatment is not effective, large numbers of people will feel they've been helped if both the therapist and the patient believe in its effectiveness," said Alan H. Roberts, principal author of the study, being published Thursday in Clinical Psychology Review.

The research is "consistent with what we knew, but documents it better than before," said William Jarvis, a preventive medicine professor at Loma Linda University and president of the National Council Against Health Fraud.

The findings suggest doctors as well as patients may be too quick to use experimental treatments that seem promising but ultimately prove ineffective, said Roberts, chief psychologist at Scripps Clinic and Research Foundation in La Jolla.

He said the research also indicates people seek unconventional, unproven therapies because they often seem to be helped by them, but that they would do even better with scientifically proven treatments.

That would allow them to benefit from both real medicine and the "placebo effect" — the improvement patients get from inert drugs or ineffective treatments purely because of belief the treatments will work, he said.

Roberts and colleagues at San Diego State University and the University of Michigan, Ann Arbor, analyzed dozens of old studies on five treatments that were medically accepted in the 1960s or 1970s but later found ineffective.

The treatments were glomectomy, an asthma-relief surgery; briefly freezing the stomach lining in an attempt to treat peptic ulcers; and three drugs applied to herpes simplex sores.

Of the 6,931 patients treated in the old studies, the outcome was excellent for 40 percent, good for 30 percent and poor for only 30 percent.

That shows "a lot of healing from medicine — and probably from psychotherapy — is the result of what are generally called placebo effects," Roberts said.

The most frequently cited research on the subject has been a 1955 study that found the placebo effect accounted for one-third of the improvement in people who got better after being given either a medicine or a placebo for a variety of ailments, Roberts said.

Figure CS11.1

Read the article and comment on the following the questions.
a) What does the term 'dummy treatment' mean to you? Is there another name for this?
b) What does the author mean by the statement: "the 'placebo effect' can be a powerful healer"?
c) What does the "placebo effect" mean? Give examples from the article to illustrate this effect?
d) How did the researchers come to the conclusion of a placebo effect? What did they use to support their claim?

Summary of the Hypothesis Tests Involving Two Population Means

The following outline is a summary of the 5 step hypothesis testing procedure for a hypothesis test involving two population means.

Hypothesis Testing Involving Two Population Means

Step 1. *Formulate the two hypotheses, H_o and H_a.*
Null Hypothesis, H_o: $\mu_1 - \mu_2 = 0$
Alternative Hypothesis, H_a: has one of the following three forms:

H_a: $\mu_1 - \mu_2 < 0$ or
H_a: $\mu_1 - \mu_2 > 0$ or
H_a: $\mu_1 - \mu_2 \neq 0$

Step 2. *Design an experiment to test the null hypothesis, H_o.*
Under the assumption H_o is true, the expected results are:
a) the Sampling Distribution of The Difference Between Two Means is either:
   (i) a normal distribution, if the standard deviations of both populations are known and both sample sizes are greater than 30, or
   (ii) is a t Distribution with degrees of freedom: $df = n_1 + n_2 - 2$, if the standard deviations of both populations are unknown.

b) the mean of the Sampling Distribution of The Difference Between Two Means is given by the formula:

$\mu_{\bar{x}_1 - \bar{x}_2} = \mu_1 - \mu_2 = 0$ (since H_o is true: i.e. $\mu_1 - \mu_2 = 0$)

c) for a normal distribution: the **standard deviation** of the Sampling Distribution of the Difference Between Two Means, denoted by $\sigma_{\bar{x}_1 - \bar{x}_2}$, is:

$$\sigma_{\bar{x}_1 - \bar{x}_2} = \sqrt{\frac{\sigma_1^2}{n_1} + \frac{\sigma_2^2}{n_2}}$$

or for a t distribution: the estimate of the standard deviation of the Sampling Distribution of The Difference Between Two Means, written $s_{\bar{x}_1 - \bar{x}_2}$, is given by the formula:

$$s_{\bar{x}_1 - \bar{x}_2} = \sqrt{\frac{(n_1-1)s_1^2 + (n_2-1)s_2^2}{n_1+n_2-2} \cdot \left(\frac{1}{n_1} + \frac{1}{n_2}\right)}$$

Step 3. *Formulate the decision rule.*
a) alternative hypothesis: directional or nondirectional.
b) type of test: 1TT or 2TT.
c) significance level: $\alpha = 1\%$ or $\alpha = 5\%$.
d) critical value formula:
   for a normal distribution:

   $X_c = (z_c)(\sigma_{\bar{x}_1 - \bar{x}_2})$

   for a t distribution: $X_c = (t_c)(s_{\bar{x}_1 - \bar{x}_2})$

e) draw hypothesis testing model.
f) State the decision rule.

Step 4. *Conduct the experiment.*
Compute the difference between the two sample means:

$d = \bar{x}_1 - \bar{x}_2$.

Step 5. *Determine the conclusion.*
Compare the experimental outcome, $d = \bar{x}_1 - \bar{x}_2$, to the critical value(s), X_c, of the decision rule and state the conclusion: either:
(a) **Reject H_o and Accept H_a at α**
or
(b) **Fail to Reject H_o at α**

Table 11.4 contains a condensed version of the hypothesis tests involving two population means.

Table 11.4

SUMMARY OF HYPOTHESIS TESTS INVOLVING TWO POPULATION MEANS			
FORM OF THE NULL HYPOTHESIS	**CONDITIONS OF TEST**	**CRITICAL VALUE FORMULA**	**SAMPLING DISTRIBUTION IS A:**
$\mu_1 - \mu_2 = 0$	KNOWN: σ_1 and σ_2 and n_1 and n_2 are both greater than 30	$X_c = (z_c)(\sigma_{\bar{x}_1 - \bar{x}_2})$ where: $\sigma_{\bar{x}_1 - \bar{x}_2} = \sqrt{\dfrac{\sigma_1^2}{n_1} + \dfrac{\sigma_2^2}{n_2}}$	Normal Distribution
$\mu_1 - \mu_2 = 0$	UNKNOWN: σ_1 and σ_2 and $\sigma_1 \approx \sigma_2$	$X_c = (t_c)(s_{\bar{x}_1 - \bar{x}_2})$ where: $s_{\bar{x}_1 - \bar{x}_2} = \sqrt{\dfrac{(n_1-1)s_1^2 + (n_2-1)s_2^2}{n_1 + n_2 - 2} \cdot \left(\dfrac{1}{n_1} + \dfrac{1}{n_2}\right)}$	t Distribution where: $df = n_1 + n_2 - 2$

❄ 11.5 Using MINITAB

Using MINITAB commands to do hypothesis testing about two population means is similar in format to the commands we used to do hypothesis testing about one population mean (Chapter 10). However there are some differences also.

Suppose you would like to compare the means of two populations. The first step in the hypothesis testing procedure is to determine the null and alternative hypothesis. In this chapter we have learned that the null hypothesis is based on the assumption that there is no difference between the population means. This is symbolized as:

$$H_o: \mu_1 - \mu_2 = 0$$

On the other hand, we have also learned that the alternative hypothesis is expressed in one of three forms:

$$H_a: \mu_1 - \mu_2 < 0$$

or

$$H_a: \mu_1 - \mu_2 > 0$$

or

$$H_a: \mu_1 - \mu_2 \neq 0$$

When a hypothesis test involving two population means is conducted, random samples are collected from each of the populations and the random samples are compared using the hypothesis testing procedure outlined in this chapter.

To use MINITAB, each random sample's data values are first entered into a MINITAB worksheet. A comparison of the random samples is then developed using the MINITAB TWOSAMPLE command.

The TWOSAMPLE command has subcommands that can indicate that the sample standard deviations are to be *pooled*, if appropriate, and, to indicate the form of the alternative hypothesis.

To illustrate this let's assume data values for two independent random samples have been entered into columns C1 and C2 of a MINITAB worksheet. The MINITAB command that tests if the samples are significantly different using a **nondirectional** alternative hypothesis is:

MTB> TWOSAMPLE C1 C2

If we assume that the population standard deviations are equal for this test then the subcommand POOLED is added

MTB> TWOSAMPLE C1 C2;
SUBC> POOLED.

If, in addition, a **directional** alternative hypothesis is used then the subcommand added is either

SUBC> ALTERNATIVE=-1

for a 1TT on the left,

or the subcommand

SUBC> ALTERNATIVE=1

for a 1TT on the right.

For example, suppose a two tailed nondirectional hypothesis test is conducted using data values in C1 and C2, if the standard deviations from the population are assumed equal then

MTB> TWOSAMPLE C1 C2;
SUBC> POOLED.

would produce output similar to:

```
              TWOSAMPLE T FOR C1 VS C2
        N      MEAN    STDEV    SE MEAN
C1      36     105.4   13.8     2.3
C2      36     95.6    17.0     2.8
```

95 PCT CI FOR MU C1 - MU C2: (2.5, 17.1)

TTEST MU C1 = MU C2 (VS NE): T= 2.68 P=0.0093 DF= 70

POOLED STDEV =   15.5

We have seen that MINITAB indicates the results of a hypothesis test by reporting the p-value for the test statistic. Examine the output and find the p-value. You should see that it is contained on the output line

TTEST MU C1 = MU C2 (VS NE): T= 2.68 P=0.0093 DF= 70

Notice that the p-value for this test is p-value=0.0093. Find the experiment outcome (i.e. Test statistic) of T=2.68.
How is the p-value interpreted in terms of the significance of the experiment outcome? Does this p-value indicate that the null hypothesis is rejected? Why?

You should also notice that the output line indicates that the test is a 2TT from the portion of the statement

TTEST MU C1 = MU C2 (VS NE)

Finally, the output line indicates that for this hypothesis test involving two population means, the degrees of freedom, DF= 70.

The output contains other information. For example, the statement

POOLED STDEV =   15.5

indicates that the pooled estimate of the standard deviation of the difference between two means is used and calculated. The output also shows summary statistics about each of the samples used.

	N	MEAN	STDEV	SE MEAN
C1	36	105.4	13.8	2.3
C2	36	95.6	17.0	2.8

How many data values were in random sample C1? What was the sample standard deviation for the random sample contained in C2? What is the standard error of the mean for the random sample contained in C2? How was it calculated?

There is one last item included on the output.

95 PCT CI FOR MU C1 - MU C2: (2.5, 17.1)

This line has to do with estimating the mean of the differences between two populations. This concept is presented in Chapter 13.

Let's consider another illustration.

These MINITAB commands

MTB >   TWOSAMPLE C1 C2;
SUBC>   ALTERNATIVE=1;
SUBC>   POOLED.

produced the output:

TWOSAMPLE T FOR C1 VS C2

	N	MEAN	STDEV	SE MEAN
C1	36	100.5	16.6	2.8
C2	33	97.3	17.8	3.0

95 PCT CI FOR MU C1 - MU C2: (-4.9, 11.3)

TTEST MU C1 = MU C2 (VS GT): T= 0.79 P=0.22 DF= 67

POOLED STDEV =   17.2

For this output:
    Identify the p-value. What is the experimental outcome? How is the significance of the p-value interpreted. Is the null hypothesis rejected at $\alpha=5\%$? Why? Is the test a 1TT or a 2TT? How do you know? How is DF calculated?

For the MINITAB commands

MTB > TWOSAMPLE C1 C2;
SUBC> ALTERNATIVE=-1;
SUBC> POOLED.

Examine the output:

TWOSAMPLE T FOR C15 VS C19

	N	MEAN	STDEV	SE MEAN
C15	15	6.66	2.64	0.68
C19	15	8.14	2.04	0.53

95 PCT CI FOR MU C15 - MU C19: (-3.25, 0.28)

TTEST MU C15 = MU C19 (VS LT): T= -1.72 P=0.048 DF= 28
POOLED STDEV =   2.36

For this output:
    Identify the p-value. What is the experimental outcome? How is the significance of the p-value interpreted. Is the null hypothesis rejected at $\alpha=5\%$? Why? Is the test a 1TT or a 2TT? How do you know? How is DF calculated?

    It is quite common to collect alot of information for each subject within a study. For example, to learn about student spending habits, a researcher might collect background data such as, gender, number of hours worked per week, weekly salary, the amount of money spent on social functions, etc.
    As a result when these data values are placed within a MINITAB worksheet, a given column might contain information that needs to be separated into two or more sections. For example, the researcher might be interested in a comparison of the number of hours worked per week for males and females.
    The TWOT command is used to perform this type of comparison. However, since the data values within one column contain both male and female number of hours worked per

week, an **additional** column that identifies (i.e. codes) the gender for the number of hours worked per week is needed.

To illustrate this, suppose a researcher surveys sixteen students, and records, for each, **two data values**. The first data value is the amount of hours the student works per week. The second data value represents the student's gender. For this second data value the researcher codes the gender variable as 0 for female and 1 for male. The sixteen data values are entered in a MINITAB worksheet in columns C1 and C2 respectively. The researcher uses the TWOT command to determine if there is a significant difference between the number of hours worked per week between females and males. Thus, for the data in C1 and C2,

Student	hours C1	gender C2
1	22	1
2	18	0
3	25	1
4	21	1
5	24	0
6	24	1
7	35	1
8	45	0
9	24	0
10	23	1
11	35	1
12	12	1
13	23	0
14	22	0
15	24	0
16	30	0

a 2TT is conducted using the TWOT command.

MTB > TWOT C1 C2;
SUBC> pooled.

The MINITAB output is:

TWOSAMPLE T FOR C1

C2	N	MEAN	STDEV	SE MEAN
1	8	24.63	7.54	2.7
0	8	26.25	8.26	2.9

95 PCT CI FOR MU 1 - MU 0: (-10.1, 6.9)

TTEST MU 1 = MU 0 (VS NE): T= -0.41 P=0.69 DF= 14

POOLED STDEV =   7.91

For this output:

Identify the p-value. What is the experimental outcome? How is the significance of the p-value interpreted. Is the null hypothesis rejected at $\alpha=5\%$? Why? Is the test a 1TT or a 2TT?
How do you know? How is DF calculated?

GLOSSARY

TERM	SECTION
Difference between the sample means, $\bar{x}_1 - \bar{x}_2$	11.2
Sampling Distribution of the Difference Between Two Means	11.2
Independent random samples	11.2
$\mu_{\bar{x}_1-\bar{x}_2}$, mean of the Sampling Distribution of the Difference Between Two Means	11.2
$s_{\bar{x}_1-\bar{x}_2}$, estimate of the standard deviation of the Sampling Distribution of the Difference Between Two Means	11.2,5
Pooled standard deviation	11.2,5
Two sample t test	11.3
Treatment Group	11.4
Control Group	11.4
Placebo	11.4
Single-Blind Study	11.4
Double-Blind Study	11.4

Exercises

PART I Fill in the blanks.

1. To perform a hypothesis test involving two population means, the Sampling Distribution of the Difference Between _____ is used as the hypothesis testing model.

2. The notation $\bar{x}_1 - \bar{x}_2$ represents the difference between the _____ _____, where:
   $\bar{x}_1$ represents the _____ of sample 1 and
   $\bar{x}_2$ represents the _____ of sample 2.

3. The Sampling Distribution of the Difference Between Two Means has the following characteristics:
   a) The mean of the Sampling Distribution of the Difference Between Two Means, denoted _____, is determined by the formula:
   $$\mu_{\bar{x}_1 - \bar{x}_2} = \underline{} - \underline{},$$
   where: μ_1 represents the mean of _____ one and
   μ_2 represents the mean of _____ two.
   b) The formula for the estimate of the standard deviation of the Sampling Distribution of the Difference Between Two Means, denoted $s_{\bar{x}_1 - \bar{x}_2}$ is dependent upon the sample standard deviations and the sample sizes. The formula for $s_{\bar{x}_1 - \bar{x}_2}$ is _____.
   c) Under Theorem 11.2, the Sampling Distribution of the Difference Between Two Means is approximated by a _____ distribution with df, degrees of freedom, given by the formula:
   $df = $ _____.

4. The null hypothesis of a hypothesis test involving two population means, μ_1 and μ_2, is stated as: There is _____ difference between the _____ of population 1 and the _____ of population 2. In symbols, H_o is written as: _____ = 0.

5. The alternative hypothesis of a hypothesis test involving two population means, μ_1 and μ_2, has one of three possible forms. They are:
   a) The mean of population 1 is greater than the _____.
   In symbols, this H_a is written: _____.
   b) The _____ of population 1 is less than the mean of _____. In symbols, this H_a is written: _____.
   c) The mean of _____ 1 is not equal to the _____ of population _____. In symbols, this H_a is written: _____.

6. When performing a hypothesis test involving two population means, the expected result $\mu_{\bar{x}_1 - \bar{x}_2} = 0$ is determined under the assumption that _____ is true.

7. The formula for the critical value, X_c, for a hypothesis test involving two population means is:
   $X_c = \mu_{\bar{x}_1 - \bar{x}_2} + $ _____.
   Under the assumption H_o is true, $\mu_{\bar{x}_1 - \bar{x}_2} = $ _____, and thus, the formula for X_c is reduced to:
   $X_c = $ _____.

8. To determine the conclusion of a hypothesis test involving two population means, the experimental outcome, denoted _____, is compared to the critical outcome(s).

9. When using the t distribution with $n_1 = 16$ and $n_2 = 20$ to approximate the Sampling Distribution of the Difference Between Two Means, the degrees of freedom for the t distribution is _____.

10. To study whether a new pill will significantly reduce the blood pressure of patients, forty patients are randomly selected and divided randomly into two equal sized groups, A and B. Group A is administered the new blood pressure pill, while Group B is administered a placebo. Then, the control group is Group _____, while the treatment group is Group _____.

11. If $n_1 = 35$, $n_2 = 40$, $\sigma_1 = 4$ and $\sigma_2 = 6$, then
    $\sigma_{\bar{x}_1 - \bar{x}_2} = $ _____.

12. If $n_1 = 25$, $n_2 = 50$, $s_1 = 3$ and $s_2 = 5$, then
    $s_{\bar{x}_1 - \bar{x}_2} = $ _____.

PART II Multiple choice questions.

1. Symbolically, the difference between two sample means is expressed as:
   a) $\bar{x}_1 - \bar{x}_2$ b) $\mu_1 - \mu_2$ c) $\mu_{\bar{x}_1 - \bar{x}_2}$ d) $P_1 - P_2$

2. The formula for the degrees of freedom of a t Distribution for testing the difference between two population means using two independent samples is:
   a) $df = n - 1$ b) $df = n_1 + n_2 - 2$
   c) $df = n_1 - n_2$

3. The null hypothesis, H_o, for testing the difference between two population means can be represented as:
   a) $H_o: \mu_1 = 0$ b) $H_o: \mu_1 = \mu_2$ c) $H_o: \mu_1 - \mu_2 > 0$
   d) $H_o: \mu_1 - \mu_2 = 0$

4. The formula for the mean of the Sampling Distribution of The Difference Between Two Means is:
   a) $\mu_{\bar{x}_1 - \bar{x}_2} = \mu_1 + \mu_2$ b) $\mu_{\bar{x}_1 - \bar{x}_2} = \mu_1 - \mu_2$

   c) $\mu_{\bar{x}_1 - \bar{x}_2} = \bar{x}_1 - \bar{x}_2$ d) $\mu_{\bar{x}_1 - \bar{x}_2} > 0$

5. The formula $\sigma_{\bar{x}_1-\bar{x}_2} = \sqrt{\dfrac{\sigma_1^2}{n_1} + \dfrac{\sigma_2^2}{n_2}}$ can be used only if:
   a) both population standard deviations are known
   b) both population standard deviations are unknown
   c) both s_1 and s_2 are known
6. The Distribution used to approximate the Sampling Distribution of The Differences Between Sample Means, when the population standard deviations are unknown, is the:
   a) binomial distribution b) t distribution
   c) normal distribution
7. The treatment group is frequently compared to the:
   a) control group b) null group c) test group
   d) pseudo group
8. If the difference between the two sample means, $\bar{x}_1 - \bar{x}_2$, is less than the critical value, X_c, then the conclusion is:
   a) Reject H_o b) Reject H_a c) Fail to Reject H_o
   d) Accept H_a
9. If the sample sizes, n_1 and n_2, are increased and the sample standard deviations, s_1 and s_2, remain the same then the value of $s_{\bar{x}_1-\bar{x}_2}$ will:
   a) increase b) remain the same c) decrease
10. When performing a two-tailed hypothesis test, if the experimental outcome falls between the two critical values, X_{LC} and X_{RC}, then the conclusion is:
    a) Reject H_o b) Reject H_a c) Fail to Reject H_o
    d) Accept H_a

PART III Answer each statement True or False.

1. The Sampling Distribution of the Difference Between Two Means is formed by finding the differences between all possible independent sample means of the form $\bar{x}_1 - \bar{x}_2$, where $\bar{x}_1$ represents the mean of an independent random sample selected from population 1 and $\bar{x}_2$ represents the mean of an independent random sample selected from population 2.
2. The mean of The Sampling Distribution of the Difference Between Two Means is computed by the formula: $\bar{x}_1 - \bar{x}_2$.
3. The standard deviation of the Sampling Distribution of The Difference Between Two Means is estimated by the formula: $s_{\bar{x}_1-\bar{x}_2} = s_1 - s_2$.
4. To determine if the difference between the sample means is significantly different from zero it must be compared to a critical value, X_c, that is determined by:
   $$X_c = \mu_{\bar{x}_1-\bar{x}_2} + (t_c)(s_{\bar{x}_1-\bar{x}_2})$$
5. The null hypothesis used in most comparisons involving two population means is based upon the assumption that mean of population 1 equals the mean of population 2.
6. If the conclusion of a hypothesis test involving two population means is that there is no significant difference between the population means then the experimenter has failed to reject H_o.
7. When the experimenter shows no significant difference between the sample means, then the mean of population 1 is said to equal the mean of population 2.
8. Only by rejecting H_o can you conclude that there is a difference between the population means.
9. If the difference between the sample means, $\bar{x}_1 - \bar{x}_2$ is different from zero then there is a significant difference between the sample means.
10. If the difference between the sample means, $\bar{x}_1 - \bar{x}_2$ does not fall beyond X_c then the null hypothesis, H_o, is assumed true.

PART IV Problems.

For each of the following problems:
a) Define population 1 and population 2.
b) Formulate the null and alternative hypotheses.
c) Calculate the expected results for the hypothesis test assuming the null hypothesis is true and determine the hypothesis test model.
d) Formulate the decision rule.
e) Determine the experimental outcome, $d = \bar{x}_1 - \bar{x}_2$.
f) Determine the conclusion and answer the question(s) posed in the problem, if any.

   For problems 1 and 2, use the results of Theorem 11.1 and the hypothesis testing procedure.

1. An educational researcher wants to determine if the mean score of urban sixth grade students is significantly greater than the mean score of rural sixth grade students on a standardized reading comprehension test. From past experience, the standard deviation of the scores on this standardized test is 10. If a random sample of 64 urban students had a mean test score of 78.4, and a random sample of 36 rural students had a mean test score of 73.9, can the educational researcher conclude that the urban sixth grade students performed significantly better on the standardized test at $\alpha = 5\%$?
2. A psychologist who needs two groups of students for a learning experiment decides to select a random sample 100 of female college students and 81 male college students. Before the experiment, the psychologist administers an IQ test to both groups to determine if there is a significant difference in their mean IQ scores. The standard deviation for this IQ test is known to be 12. If the sample mean IQ score of the

100 female students is 116.9 and the sample mean IQ score of the 81 male students is 120.8, can the psychologist conclude that there is a significant difference in the mean IQ score of both groups at $\alpha = 1\%$?

3. A light bulb manufacturer has developed a new light bulb which the manufacturer claims has a longer mean life than that of their leading competitor. The mean life for an independent random sample of 50 of the manufacturer's bulbs is 2,500 hours with s = 100 hours. The mean life for an independent random sample of 50 of their competitor's bulbs is 2,450 hours with s = 90. Do these sample results support the manufacturer's claim at $\alpha = 5\%$?

4. A new insecticide was developed by a chemical company. The company claims that the new insecticide has a longer effective life than their present insecticide. An independent random sample of 40 cans of the new insecticide had a mean effective life of 45 minutes with s = 8 minutes. An independent random sample of 50 cans of the present insecticide had a mean effective life of 42 minutes with s = 6 minutes. Do these sample results support the manufacturer's claim, at $\alpha = 1\%$?

5. Members of the student government at Valley College in Sacramento believe that female students receive significantly lower final grades in the liberal arts math course than male students. From an independent random sample of 100 female students it was found that the mean final grade was 78 with s = 10. From an independent random sample of 100 male students the mean final grade was 80 with s = 6. Do these sample results support the student government's claim at $\alpha = 5\%$?

6. A school nutritionist believes that her special diet given along with a rigorous training routine will significantly increase stamina. She randomly selects 30 students and divides them into two equal independent groups. Group I, called the treatment group, is put on the special diet and the rigorous training routine. Group II, called the control group, is just given the rigorous training routine with no special diet. After six months the increase in stamina level of all subjects is measured. The results are given in the following table.

Treatment Group	Control Group
$n_1 = 15$	$n_2 = 15$
$\overline{x}_1 = 26$	$\overline{x}_2 = 18$
$s_1 = 4$	$s_2 = 10$

Do these sample results support the nutritionist's belief at $\alpha = 5\%$?

7. A new toothpaste containing an anti-cavity substance DPT claims to significantly reduce cavities in children. To test this claim, researchers from a dental school randomly select two groups of children. To one group, called the treatment group, the children are given the toothpaste containing DPT. To the other group, called the control group, the children are given the toothpaste <u>without</u> DPT. After one year the mean number of cavities for each group is determined.

The results for each group are summarized in the following table:

Treatment Group	Control Group
$n_1 = 64$	$n_2 = 100$
$\overline{x}_1 = 3$	$\overline{x}_2 = 3.2$
$s_1 = 0.4$	$s_2 = 0.6$

Do these sample results indicate that DPT significantly reduces cavities at $\alpha = 1\%$?

8. The personnel director of a large computer company has been hiring word processing students from community colleges and from secretarial schools. The director would like to determine if there is a significant difference in the mean typing speed of these graduates. Each graduate was administered a typing test and the sample results are summarized in the following table:

Community College Graduates	Secretarial School Graduates
$\overline{x}_1 = 65$ words per minute	$\overline{x}_2 = 61$ words per minute
$s_1 = 12$	$s_2 = 15$
$n_1 = 40$	$n_2 = 36$

Can the personnel director conclude that there is significant difference in the mean typing speed of the Community College graduates and the secretarial school graduates at $\alpha = 5\%$?

9. An insurance company claims that male drivers under age 25 have more accidents per year than male drivers 25 years of age or older. An independent sample of size 40 is randomly selected from each age group. The mean number of accidents per year for the male drivers under age 25 was found to be 1.2 with s = 0.1 while the mean number of accidents per year for the drivers 25 years of age or older was found to be 0.94 with s = 0.2. Do you believe the claim of the company is correct at $\alpha = 5\%$?

10. A farmer believes that his special fertilizer will significantly increase his tomato crop yield. To test his belief, he randomly divides 200 acres into two groups. In Group I, he administers his special fertilizer. In Group II, he administers the usual tomato

fertilizer. The table below summarizes the tomato yields for this experiment.

	Group I	Group II
Number of acres, n	100	100
sample average bushels per acre, $\bar{x}$	830	805
sample standard deviation, s	40	36

Do these sample results support the farmers claim at $\alpha = 1\%$?

11. A sociologist claims that marriage lowers the cumulative average of graduate students. If the cumulative average of 16 randomly selected married graduate students was 3.35 with s = 0.3 and the cumulative average of 9 unmarried graduate students was 3.56 with s = 0.5, do you agree with the sociologists claim at $\alpha = 1\%$?

12. A New York sportswriter claims the American League pitchers have lower earned run averages than National League pitchers. If 20 randomly selected American League pitchers have a mean E.R.A. of 2.50 with s = 0.15 while 16 randomly selected National League pitchers have a mean E.R.A. of 2.6 with s = 0.25. Is the New York sportswriter's claim justified at $\alpha = 1\%$?

13. The Affirmative Action Officer of a city newspaper decides to determine if the recently hired male journalists earn a higher mean salary than the recently hired female journalists. The Affirmative Action Officer randomly selected independent samples of 16 recently hired male and 16 recently hired female journalists. The male journalists had a mean salary of $20,750 per year with s = $700 and the female journalists had a mean salary of $19,650 per year with s = $1200. Can the Affirmative Action Officer of the city newspaper conclude there is a significant difference in their population mean salaries at $\alpha = 5\%$?

14. A statistics instructor wants to determine if the addition of statistics projects integrated into his statistics classes will result in significantly better performance by the students on the final exam. In one of his statistics classes, he integrates the statistics projects with his regular lessons. In another of his statistics classes he does not integrate the projects. The results in the following table summarize the results of his experiment.

Statistic Class (with projects)	Statistic Class (without projects)
$n_1 = 25$	$n_2 = 22$
$\bar{x}_1 = 82$	$\bar{x}_2 = 79$
$s_1 = 8$	$s_2 = 10$

Based upon these sample results, can the instructor conclude that the addition of statistics projects significantly increases student performance on the final exam at $\alpha = 5\%$?

15. A psychologist administered a perception test to two randomly selected groups. The first group of 10 subjects were asked to estimate the length of an object dangling from a rod at a distance of 20 feet. These subjects had both eyes open while observing the object. The second group of 12 subjects estimated the length of the object with one eye covered. The results are summarized in the following table.

Group I (using both eyes)	Group II (one eye covered)
$n_1 = 10$	$n_2 = 12$
$\bar{x}_1 = 12.5$ inches	$\bar{x}_2 = 14.25$ inches
$s_1 = 1.25$ inches	$s_2 = 2.5$ inches

Can the psychologist (with both eyes open) conclude that there is a significant difference in the perception of the two groups at $\alpha = 1\%$?

16. A medical researcher wants to study whether a diet deficient in Vitamin E affects the mean amount of Vitamin A stored in mice. She randomly selects two samples of fifteen mice each. Group I, called the control group, is fed a normal diet while the second group, called the treatment group, is fed a diet deficient in vitamin E. She collected the following data:

Control Group I.U. of Vitamin A	Treatment Group I.U. of Vitamin A
3850	3300
3700	3250
3650	2900
3500	2700
3450	2000
3450	2550
3900	2400
3700	2600
2900	2300
3400	2450
3200	1900
3150	2100
3850	2500
3000	2350
2800	2800

Can the researcher conclude that the mean amount of Vitamin A for mice on the deficient diet is less than the mean amount of Vitamin A for mice on a normal diet at $\alpha = 1\%$?

17. A botanist wants to test whether the new product "Miraculous Growth" would significantly increase the number of flowers per plant over an eight week period. Two trays of 12 plants were prepared. Tray

I received the "Miraculous Growth" during the test period while Tray II served as the control group. The plants were grown in a controlled environment for this eight week period. At the end of this period, the number of flowers per plant were recorded. The results are summarized in the following table:

Tray I (treatment group)		Tray II (control group)	
14	17	12	9
16	19	10	10
20	20	13	15
19	18	14	12
16	17	15	13
15	16	12	14

Can the botanist conclude that the plants treated with "Miraculous Growth" had greater number of flowers per plant than the plants in the control group at $\alpha = 5\%$?

18. An educational researcher wants to determine whether the Scholastic Aptitude Test (SAT) scores of students attending an Ivy League College is significantly different than the SAT scores of students attending a Military Academy. An independent random sample is chosen from each population and the SAT scores for the verbal, the mathematics and the combined sections are summarized in the following table.

	Verbal		Math		Combined	
	Ivy	Military	Ivy	Military	Ivy	Military
$\bar{x}$	680	640	675	710	1355	1350
s	50	60	70	50	86	78
n	50	50	50	50	50	50

Using this sample data, can the educational researcher conclude that the:
(I) population mean verbal score is higher for the Ivy League students at $\alpha = 1\%$?
(II) population mean math score is higher for the military academy students at $\alpha = 1\%$?
(III) population mean score of the combined SAT scores is significantly different from the Ivy League and military academy students at $\alpha = 1\%$?

PART V What Do You Think?

1. **What Do You Think?**

A research article entitled *Japanese And American Tendencies To Argue* appeared in Psychological Reports on June, 1990.

The article discussed argumentativeness as a personality trait composed of the tendencies to approach and avoid arguments. The authors defined people with approach tendencies as being predisposed to advocate positions on controversial issues and attack conflicting positions (argue) while people with avoidance tendencies try to keep arguments from happening.

The purpose of the study was to compare the argumentativeness in Japan and the United States using an argumentativeness scale called Infante–Rancer which measures the approach and avoidance tendencies as well as the respondents' argumentativeness.

The study involved 168 Japanese subjects representing three Tokyo Universities and 153 Americans attending West Virginia University. The authors indicated that there was no significant difference between the Japanese and Americans in age, proportion of men and women, and years of education. The following table represents the analysis of the study.

Tendency	Culture	Mean	Standard Deviation	t value	p
Approach	Japanese American	29.00 33.00	6.54 6.65	5.19	< 0.01
Avoidance	Japanese American	27.00 26.00		ns	
Arguementativeness	Japanese American	2.00 7.00	9.70 11.10	3.66	< 0.01

a) Define the two populations within the study. By examining the t value can you determine which culture has been defined as population 1? Explain.
b) What variables are being measured? How are they being measured?
c) What is the difference between people who have an approach tendency and people with an avoidance tendency?
d) If a larger mean score indicates a greater inclination to have that tendency, then how might you interpret the mean scores for each of these cultures on the approach tendency?
e) If the null hypothesis for the approach tendency aspect of the study is stated as: there is no significant in the difference between the mean scores of the Japanese and American subjects on the approach tendencies, then state the alternative hypothesis for a nondirectional test.
f) Can we accept this alternative hypothesis at a 1% level of significance? Explain.
g) Does the t test analysis indicate that there is a significant or nonsignificant difference between the Japanese and the Americans with respect to approach tendencies? Explain.
h) Which statement agrees with the conclusion of the results of the approach tendency? The Japanese are more inclined to argue or the Americans are more inclined to argue? Explain.
i) State the null and alternative hypothesis for the study pertaining to the avoidance tendency?
j) Can we reject the null hypothesis at a 1% level of significance? Explain.
k) Does the t test analysis indicate that there is a significant or nonsignificant difference between the Japanese and the Americans with respect to avoidance tendencies? Explain.
l) State the null and alternative hypothesis for the study pertaining to the arguementativeness tendency?
m) Can we reject the null hypothesis at a 5% level of significance? Explain.
n) Does the t test analysis indicate that there is a significant or nonsignificant difference between the Japanese and the Americans with respect to argumentativeness tendencies? Explain.
o) Indicate whether you agree or disagree with the following statements as they pertain to the results of this study.
  (1) The Japanese are more nonargumentative.
  (2) Americans feel that arguing offers intellectual challenges and is exciting.
  (3) Japanese view arguments positively.
  (4) Americans value group harmony and shun controversy.

2. **What Do You Think?**

A research article exploring the relations between the personality characteristics and concepts and skills necessary for leadership development in students. The study involved 53 girls and 42 boys attending a Leadership Studies Program for one week on a university campus. The following table presents the means and standard deviations for the girls and boys on the Leadership Skills Inventory subscale scores and each of the High School Personality Questionnaire factors along with the t values of the comparison of the mean scores. The purpose of administering these instruments was to assess the strengths and weaknesses of the students' leadership concepts and skills.

MEANS, STANDARD DEVIATIONS, AND t RATIOS FOR SUBSCALE SCORES ON LEADERSHIP SKILLS INVENTORY AND FACTORS (STEN SCORES) OF HIGH SCHOOL PERSONALITY QUESTIONNAIRE FOR BOYS AND GIRLS IN LEADERSHIP STUDIES PROGRAM.

	Variable	All Subjects		Boys		Girls		t^*
		M	SD	M	SD	M	SD	
1.	Fundamentals of Leadership	50.04	11.09	49.61	12.02	50.37	10.40	-.33
2.	Written Communication Skills	55.27	7.98	53.66	6.88	56.45	8.60	-1.77
3.	Speech Communication Skills	53.21	9.48	52.19	9.30	54.01	9.64	-.93
4.	Values Clarification	52.96	6.78	51.35	6.08	54.24	7.09	-2.10*
5.	Decision-making Skills	54.10	7.48	53.09	7.43	54.92	7.50	-1.18
6.	Group Dynamic Skills	53.58	7.53	51.97	6.72	54.86	7.97	-1.88
7.	Problem-solving Skills	55.08	9.17	54.11	10.38	55.84	8.10	-.91
8.	Personal Development Skills	54.55	6.39	53.26	6.51	55.58	6.16	-1.78
9.	Planning Skills	55.94	7.20	54.90	7.55	56.77	6.88	-1.26
10.	A Warmth	5.98	1.93	5.88	2.34	6.07	1.56	-.48
11.	B Intelligence	6.52	1.85	6.40	1.72	6.62	1.95	-.57
12.	C Emotional Stability	6.26	1.82	5.81	1.89	6.62	1.71	-2.20*
13.	D Excitability	5.69	1.89	5.85	2.07	5.56	1.74	.74
14.	E Dominance	5.89	2.16	5.33	1.85	6.34	2.30	-2.30*
15.	F Cheerfulness	5.41	2.10	5.54	2.23	5.30	2.00	.56
16.	G Conformity	6.35	1.89	6.38	1.88	6.34	1.91	.11
17.	H Boldness	6.07	2.16	5.95	2.23	6.17	2.11	-.49
18.	I Sensitivity	6.45	2.21	6.45	2.10	6.45	2.31	.00
19.	J Withdrawal	5.90	1.95	5.59	1.93	6.15	1.94	-1.38
20.	O Apprehension	4.26	2.14	4.40	2.16	4.15	2.13	.57
21.	Q_2 Self-sufficiency	5.36	1.95	5.00	1.87	5.66	1.99	-1.65
22.	Q_3 Self-discipline	6.11	1.86	5.88	1.86	6.30	1.86	-1.09
23.	Q_4 Tension	5.15	1.90	5.28	2.06	5.05	1.78	.58
24.	Extraversion	6.90	1.60	7.11	1.70	6.74	1.51	1.09
25.	Anxiety	4.71	1.86	5.13	1.85	4.38	1.83	1.97
26.	Tough Poise	5.13	2.19	5.09	2.20	5.16	2.20	-.16
27.	Independence	6.30	1.93	5.71	1.60	6.77	2.06	-2.71*
28.	Delinquency	6.49	2.04	6.22	1.84	6.70	2.17	-1.15
29.	Accident Proneness	4.55	1.91	4.73	1.73	4.41	2.05	.81
30.	Creativity	6.75	1.98	6.77	1.99	6.82	1.99	-.36
31.	Leadership Potential	6.52	1.98	6.74	2.09	6.34	1.89	.95
32.	School Achievement	6.51	2.10	6.25	1.95	6.70	2.20	-1.08
33.	Vocational Growth	6.62	2.26	6.22	2.21	6.93	2.28	-1.52
34.	Vocational Success	5.91	2.20	5.91	2.15	5.91	2.27	-.02

*$p = .05$.

a) Define the two populations within the study. By examining the t value can you determine which group has been defined as population 1? Explain.

b) What variables are being measured? How are they being measured?

c) On which of the variables are the results of the t test significant?

d) Would these results be significant at a $\alpha = 1\%$? Explain.

e) State the null and alternative hypotheses for each of these significant t tests if the alternative hypothesis was a nondirectional hypothesis.

f) What is the name and mean of the sampling distribution that serves as the model for this test?

g) Did the boys or girls score significantly higher on the Dominance Factor?

h) Who scored significantly higher on the Independence Factor?

i) For the Dominance and Independence Factor, was the null hypothesis rejected at $\alpha = 5\%$? at $\alpha = 1\%$? Explain.

j) For the Fundamentals of Leadership Skill, was the null hypothesis rejected at $\alpha = 5\%$? at $\alpha = 1\%$? Explain.

k) Does the t test analysis indicate that there is a significant or nonsignificant difference between the girls and boys with respect to Speech Communication Skills at $\alpha = 5\%$? Explain.

l) Does the t test analysis indicate that there is a significant or nonsignificant difference between the girls and boys with respect to Values Clarification at $\alpha = 5\%$? at $\alpha = 1\%$? Explain.

3. **What Do You Think?**

The following excerpt is from a recent research article from the New England Journal of Medicine regarding a study concerning the effects of oat bran on cholesterol.
Read this excerpt and answer the following questions.

A Harvard study from the New England Journal of Medicine found no significant difference on cholesterol levels between people who ate oat bran and those that ate a placebo supplement, which consisted of just plain flour. In both cases, the cholesterol levels decreased slightly, mainly because the volunteers ate grains instead of fatty foods. In the study, 20 healthy people from the staff of Brigham and Women's Hospital in Boston were put for six weeks on an oat bran high-fiber diet, two weeks on their normal diets, and then six weeks on a diet high in white flour, which is low in fiber. The study was double-blinded.

a) Define the two groups contained within the study.
b) State the purpose of the study, and a null and an alternative hypothesis for this study.
c) What was the result of the study? Was the null hypothesis rejected? Explain.
d) What does "placebo supplement" mean?
e) What is the purpose of giving a group a placebo? How does this help to conduct the study?
f) Explain what is meant by the statement: "the study was double-blinded"?
g) Comment on the following statement about this study:
"twenty volunteers are too few to be meaningful and that the participants had cholesterol levels of 186, which is less than the 235 cutoff point set by the American Heart Association for serious concern about heart disease"
h) Comment on the following statement about this study:
"Because the cholesterol levels dropped slightly, but equally, regardless of whether the volunteers ate high or low fiber foods, the oat bran hype is clearly overblown. This is a classic case of where the oat bran industry has run ahead of science."

PART VI Exploring DATA With MINITAB

1. Write a MINITAB command with the appropriate subcommand(s) for a hypothesis test involving two population means if you know that the test is: (assume population standard deviations are equal)
   a) a 2TT, random sample data values are in C1 and C2 respectively.
   b) a 1TT, H_a: $\mu_1 - \mu_2 <$ 0, C1 contains information on both males and females, C2 codes the information within C1 as to male and female.
   c) a 1TT, H_a: $\mu_1 - \mu_2 >$ 0, random sample data values are in C1 and C2 respectively.
   d) H_a: $\mu_1 - \mu_2 \neq$ 0, C1 contains information on both males and females, C2 codes the information within C1 as to male and female.

2. Given the MINITAB output:

TWOSAMPLE T FOR C1 VS C2

	N	MEAN	STDEV	SE MEAN
C1	15	33.04	4.93	1.3
C2	17	37.84	4.62	1.1

95 PCT CI FOR MU C1 - MU C2: (-8.2, -1.3)

TTEST MU C1 = MU C2 (VS NE): T= -2.84
P=0.0081 DF= 30
POOLED STDEV = 4.77

For this output answer the following:
a) Identify the p-value.
b) What is the experimental outcome?
c) How is the significance of the p-value interpreted.
d) Is the null hypothesis rejected at α=5%? Why?
e) Is the test a 1TT or a 2TT? How do you know?
f) How is DF calculated?

3. Given the MINITAB output:

TWOSAMPLE T FOR C5 VS C6

	N	MEAN	STDEV	SE MEAN
C5	20	92.99	9.64	2.2
C6	17	83.60	6.06	1.5

95 PCT CI FOR MU C5 - MU C6: (4.1, 14.7)

TTEST MU C5 = MU C6 (VS GT): T= 3.60
P=0.0005 DF= 32

For this output answer the following:
a) How many data values were in random sample C5?
b) What is the sample standard deviation for the random sample contained in C6?
c) What is the standard error of the mean for the random sample contained in C6? How was it calculated?
d) Identify the p-value.
e) What is the experimental outcome?
f) How is the significance of the p-value interpreted.
g) Is the null hypothesis rejected at $\alpha=5\%$? Why?
h) Is the test a 1TT or a 2TT? How do you know?

4. Given the MINITAB output:

TWOSAMPLE T FOR C5 VS C6

	N	MEAN	STDEV	SE MEAN
C5	15	33.97	6.55	1.7
C6	17	37.12	5.88	1.4

95 PCT CI FOR MU C5 - MU C6: (-7.6, 1.3)

TTEST MU C5 = MU C6 (VS NE): T= -1.44
P=0.16 DF= 30
POOLED STDEV =   6.20

For this output answer the following:
a) How many data values were in random sample C5?
b) What is the sample standard deviation for the random sample contained in C6?
c) What is the standard error of the mean for the random sample contained in C6? How was it calculated?
d) Identify the p-value.
e) What is the experimental outcome?
f) How is the significance of the p-value interpreted.
g) Is the null hypothesis rejected at $\alpha=5\%$? Why?
h) Is the test a 1TT or a 2TT? How do you know?

5. Given the MINITAB output:

TWOSAMPLE T FOR C9

C10	N	MEAN	STDEV	SE MEAN
2	54	36.45	6.01	0.82
1	74	34.87	4.46	0.52

95 PCT CI FOR MU 2 - MU 1: (-0.25, 3.42)

TTEST MU 2 = MU 1 (VS NE): T= 1.71   P=0.089
DF= 126
POOLED STDEV =   5.17

For this output answer the following:
a) How many data values were in random sample C10?
b) What is the sample standard deviation for the random sample coded 1?
c) What is the sample standard deviation for the random sample coded 2?
d) Identify the p-value.
e) What is the experimental outcome?
f) How is the significance of the p-value interpreted.
g) Is the null hypothesis rejected at $\alpha=5\%$? Why?
h) Is the test a 1TT or a 2TT? How do you know?

6. A medical researcher collects the following data for 10 nursing students:

sex	wt	ht	sys	dia	chl	ldl
0	124	64	123	78	210	145
1	150	65	140	95	240	160
1	170	68	125	85	190	140
0	160	67	123	78	238	160
0	115	63	110	68	196	135
1	145	70	115	75	205	168
1	195	71	125	85	230	180
1	230	72	140	90	240	190
0	145	70	120	80	205	155
0	105	63	115	60	195	135

(sex: 0 is female, 1 is male)

Use MINITAB to determine compare the population means for the variables:
a)
   1) sex and weight(wt). 1TT.
   2) sex and height(ht). 1TT.
   3) sex and diastolic(dia). 2TT
   4) sex and ldl. 2TT
b) For each output answer the following:
   1) Identify the p-value.
   2) What is the experimental outcome?
   3) How is the significance of the p-value interpreted.
   4) Is the null hypothesis rejected at $\alpha=5\%$? Why?
   5) Write out the conclusion using the variable names.

PART VII Projects.

1. Hypothesis Testing Project:
   A. Selection of a topic.
      1. Select a reported fact or claim involving two population means made in a local newspaper, magazine or other source.
      2. Suggested topics:
         a) Compare the cumulative averages of males and females at your school and test the hypothesis that there is no significant difference between their cumulative averages.
         b) Compare the mean number of traffic violations incurred during the past 3 years for drivers under 25 against the average for drivers over 25 to determine if a significant difference exists between these averages.
         c) Compare the average pulse rate of football players and cheerleaders at your school and test the hypothesis that there is no significant difference between their average pulse rates.
         d) Compare the average number of part-time hours worked by freshmen and sophomores at your school to determine if there is a significant difference between these averages.

   B. Use the following procedure to test the claim selected in Part A.
      1. State the claim, identify the two populations referred to in the statement of the claim, and indicate where the claim was found.
      2. State your opinion regarding the claim of the two population means, μ_1 and μ_2 (i.e. $\mu_1 > \mu_2$, $\mu_1 < \mu_2$ or $\mu_1 \neq \mu_2$).
      3. State the null and alternative hypotheses.
      4. Develop and state a procedure for selecting an independent random sample from each population. Indicate the technique you are going to use to obtain your sample data.
      5. Compute the expected results, indicate the hypothesis testing model, and choose an appropriate level of significance.
      6. Calculate the critical value(s) and state the decision rule.
      7. Perform the experiment and calculate the experimental outcome.
      8. Construct a hypothesis testing model and place the experimental outcome on the model.
      9. Formulate the appropriate conclusion and interpret the conclusion with respect to your opinion (as stated in Step 2).

2. Read a research article from a professional journal in a field such as medicine, education, marketing, science social science, etc. Write a summary of the research article that includes the following:
   a) Name of article, author, and journal (include volume number and date).
   b) Statement of the research problem.
   c) The null and alternative hypotheses.
   d) How the experiment was conducted (give the definition of the treatment group, treatment used and the sample size; give the definition of the control group and the sample size. Were placebos used? If so, what were they?)
   e) Determine the experimental outcome and the level of significance.
   f) State the conclusion in terms of the null and alternative hypotheses.

PART VIII Database.

The following exercises refer to the file DATABASE listed in Appendix A. We have indicated the appropriate MINITAB commands that are necessary to answer each exercise.

1. Using the MINITAB commands:
       RETRIEVE and TWOT
   Retrieve the file DATABASE.MTW and identify the column number for each of the variable names:
   a) AVE
   b) SEX
   Use the TWOT command to test the hypothesis that the mean high school average of females is not equal to the mean high school average of males.
   State the null hypothesis for this test.
   Write and execute the TWOT command if the alternative hypothesis is *nondirectional*.
   State the alternative hypothesis for this test.
   Is the null hypothesis rejected? Identify the p value and interpret.
   State the conclusion.

2. Using the MINITAB commands:
       RETRIEVE and TWOT
   Retrieve the file DATABASE.MTW and identify the column number for each of the variable names:
   a) GFT
   b) SEX
   Use the TWOT command to test the hypothesis that the mean amount of dollars spent on a gift by males is more than the mean amount of dollars spent on a gift by females.
   State the null and alternative hypotheses for this test. Interpret the output in terms of the null and alternative hypotheses, and p value.

CHAPTER 12
HYPOTHESIS TESTING INVOLVING A POPULATION PROPORTION

❄ 12.1 INTRODUCTION

ARE THE FOLLOWING CLAIMS STARTLING ??
WHAT ARE YOUR FEELINGS ABOUT THESE CLAIMS ??

Example 12.1

The Bureau of Statistics claims that 75% of new small businesses will go bankrupt within ten months.

Success

Failure

Example 12.2

A sociologist claims that two out of three couples that wed this year will divorce within five years.

Newlyweds

Divorce Court

Example 12.3

Educational Research Associates (ERA) claims that at least 40% of the male graduates of the State University have a reading level below 9th grade.

Entering College

Graduating College

In each of the previous claims, the population characteristic described is based upon **count data**. That is, individuals in the population are **categorized** and **counted** according to a particular characteristic like sex, race, religion, marital status, educational level, political affiliation, state of health, etc. When dealing with count data, the **proportion** of the **population** that possess each of these characteristics can be determined.

Definition: Population (or True) Proportion. The population (or true) proportion is the ratio of the number of occurrences (or observations) in the population that possess the particular characteristic to the population size.

Thus, the population proportion, denoted by p, can be expressed by the formula:

$$p = \frac{x}{N}$$

where: p = population proportion
x = number of occurrences (or observations) in the population possessing the particular characteristic
N = population size

The population proportion is often expressed in an equivalent percentage form.

For example, according to the registration records of the 26,000 students attending a large state community college, 6,240 students were enrolled in a psychology course. Thus, the proportion of community college students enrolled in a psychology course is:

$$p = \frac{\text{number of students enrolled in a psychology course}}{\text{total number of students}}$$

$$= \frac{6,240}{26,000}$$

$$= 0.24$$

The population proportion, p = 0.24, can also be expressed as 24%. Therefore, we can state that: 24% of the community college students are enrolled in a psychology course.

However, there are many instances where it is either impractical or impossible to calculate the population proportion.

For example, suppose you wanted to determine the proportion of red colored M & M's manufactured by the Mars Corporation. Since it is impractical to determine the population (or true) proportion of red colored M & M's, it becomes necessary to **estimate the population proportion.** To estimate this proportion, we would have to select a random sample of M & M candies and calculate the proportion of red colored M & M's within the sample. If there were 500 M & M's in the sample and 150 were red, then we would have $\frac{150}{500}$ or 30% red colored M & M's. This result is called a **sample proportion.**

Definition: Sample Proportion. The sample proportion is the ratio of the number of occurrences (or observations) in the sample that possess the particular characteristic to the sample size.

Thus, the sample proportion, denoted by $\hat{p}$ (read p-hat), can be expressed by the formula:

$$\hat{p} = \frac{x}{n}$$

where: $\hat{p}$ = sample proportion

x = number of occurrences (or observations) in the sample possessing the particular characteristic

n = sample size

The sample proportion of 30% red colored M & M's is a statistic that can be used to estimate the unknown population value p, population proportion of red colored M & M's. Keep in mind, however, if a second random sample of M & M's were selected and the proportion of red colored M & M's determined, we would most likely obtain a different sample proportion of red M & M's. Since different samples consist of different items from the population, we expect the different samples to provide different values for the sample proportion. Therefore, if we are interested in either estimating or testing claims about a population proportion, it becomes necessary to examine the distribution of all possible values of the sample proportion. This distribution is referred to as the Sampling Distribution of The Proportion.

In this chapter, we will be applying the general hypothesis testing procedure discussed in Chapter 9 to test a claim made about a population proportion. For instance, considering Example 12.3, suppose that you believe that the claim: 40% of the State University male graduates having a reading level below 9th grade **is too high**. How would you go about statistically testing this claim while trying to support your point of view?

To test this claim you would select a **random sample of size n** from the population and compute the **sample proportion** of State University male graduates that have a reading level below 9th grade. This **sample proportion, $\hat{p}$, is compared** to the **claimed value for the population proportion, p,** to determine if the **difference between the sample proportion and the population proportion is statistically significant.** If the difference between the sample proportion and the population proportion is statistically significant then your belief that the claim about the State University male graduates is too high would be statistically supported.

The statistical procedure needed to perform this hypothesis test involves the *general five step hypothesis testing procedure* developed in Chapter 9 where the Sampling Distribution of The Proportion will be used as the hypothesis testing model. In concept, the hypothesis test involving a population proportion is similar to the hypothesis test involving a population mean. The main difference is that the Sampling Distribution of The Proportion is the hypothesis testing model rather than the Sampling Distribution of The Mean. The sampling distribution of the proportion has the same properties as the Binomial Distribution which was discussed in Chapter 7. These properties will be discussed in Section 12.2.

CASE STUDY 12.1

Examine the pie chart information contained in the USA Snapshot entitled *Mutual funds bearing loads* shown in Figure CS12.1. The pie chart presents the different type of mutual funds along with the number of funds belonging to each type.

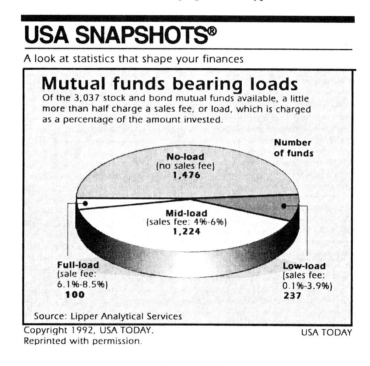

Figure CS12.1

    a) Define the population represented by the pie chart?
    b) What variable is represented by the pie chart?
    c) Is this variable numeric or categorical? Explain.
    d) What is the total number of funds represented within the pie chart?
    e) Is this variable numeric or categorical? Explain.
    f) Determine the proportion of no-load mutual funds represented within the pie chart.
    g) Determine the proportion of full-load mutual funds represented within the pie chart.
    h) Determine the proportion of low-load mutual funds represented within the pie chart.
    i) Determine the proportion of mid-load mutual funds represented within the pie chart.
    j) What percent of the mutual funds represented by the pie chart have no sales fee?
    k) What percent of the mutual funds represented by the pie chart have a sales fee from 0.1% to 6%?

12.2 THE SAMPLING DISTRIBUTION OF THE PROPORTION

Let us now discuss the characteristics of the distribution model needed to perform a hypothesis test involving a population proportion: **The Sampling Distribution of The Proportion.**

Theorem 12.1 The Characteristics of the Sampling Distribution of The Proportion

If all possible random samples of size n are selected from an infinite or large binomial population[1] having a population proportion p and, the sample proportion, $\hat{p}$, is computed for each of these samples, then:

a) the Sampling Distribution of The Proportion is approximately normal when **both** np and n(1 - p) are greater than 5.

b) the mean of the Sampling Distribution of The Proportion, denoted $\mu_{\hat{p}}$, is:

$$\mu_{\hat{p}} = p$$

c) the standard deviation of the Sampling Distribution of The Proportion, denoted $\sigma_{\hat{p}}$, is:[2]

$$\sigma_{\hat{p}} = \sqrt{\frac{p(1-p)}{n}}$$

Theorem 12.1 is illustrated in Figure 12.1.

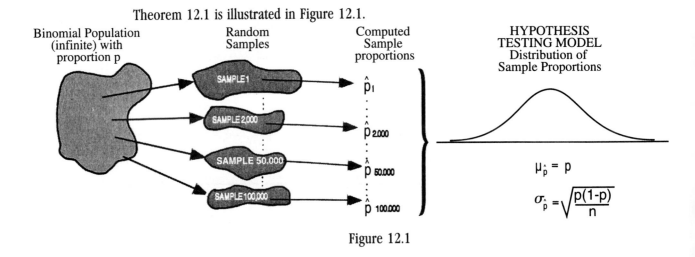

Figure 12.1

---

[1] A binomial population is a population where the scores included can be classified into one of two categories.(See Section 6.7 for more information)

[2] If we are sampling without replacement from a finite binomial population then the exact formula for $\sigma_{\hat{p}}$ is:

$$\sigma_{\hat{p}} = \sqrt{\frac{p(1-p)}{n}} \sqrt{\frac{N-n}{N-1}}$$

where: N is the population size and n is the sample size.
If the sample size, n, is less than 5% of the population size, N, then it is appropriate to use:

$$\sigma_{\hat{p}} = \sqrt{\frac{p(1-p)}{n}}$$

Let's apply the results of Theorem 12.1 to the following problem.

Example 12.4

If 75% of new small businesses fail within 10 months, what's the probability that a random sample of 100 newly formed small businesses will have a proportion of at least 80% failures within 10 months?

Solution:

From Theorem 12.1, the mean of the Sampling Distribution of The Proportion is: $\mu_{\hat{p}} = p$. Since the population proportion p is 75% or 0.75, then: $\mu_{\hat{p}} = 0.75$.

Using n = 100, p = 0.75 and 1 - p = 0.25, the standard deviation of the Sampling Distribution of The Proportion, $\sigma_{\hat{p}}$, determined by the formula:

$$\sigma_{\hat{p}} = \sqrt{\frac{p(1-p)}{n}} \text{, is:}$$

$$\sigma_{\hat{p}} = \sqrt{\frac{0.75(0.25)}{100}}$$

$$\sigma_{\hat{p}} = 0.0433$$

The Sampling Distribution of The Proportion is approximately a normal distribution since:

$$\begin{array}{ll} np = (100)(0.75) & \text{and} \quad n(1-p) = (100)(0.25) \\ = 75 & \phantom{\text{and} \quad n(1-p)} = 25 \end{array}$$

are both greater than 5. The Sampling Distribution of The Proportion is illustrated in Figure 12.2.

Sampling Distribution of The Proportion
(is approximately a normal distribution)

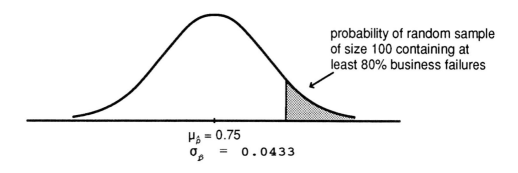

Figure 12.2

Since the Sampling Distribution of The Proportion is approximately a normal distribution, we need to calculate a z score to find the shaded area in Figure 12.2. The z score formula for the Sampling Distribution of The Proportion is given by:

$$z \text{ of } \hat{p} = \frac{\hat{p} - \mu_{\hat{p}}}{\sigma_{\hat{p}}}$$

Thus, the z score of 0.80 is

$$z = \frac{0.80 - 0.75}{0.0433}$$

$$z \text{ of } 0.80 = 1.15$$

From **TABLE II, the Normal Curve Area Table found in Appendix D: Statistical Tables**, the percent of the area to the left of z = 1.15 is 87.49%. Thus, the percent of area to the right of z = 1.15 is 12.51%. Therefore, the probability of a sample of 100 newly formed businesses will have at least 80% failures within 10 months is 12.51%. ∎

Example 12.5

E.R.A. claims that 40% of the male graduates of the State University System have a reading level below the 9th grade. What is the probability that less than 38% of a random sample of 270 male State University graduates will have a reading level below 9th grade?

Solution:

From Theorem 12.1, we obtain the mean of the Sampling Distribution of The Proportion using: $\mu_{\hat{p}} = p$.

Thus, $\mu_{\hat{p}} = 0.40$.

Since n = 270, p = 0.40 and 1 - p = 0.60, the standard deviation of the Sampling Distribution of The Proportion is determined by the formula:

$$\sigma_{\hat{p}} = \sqrt{\frac{p(1-p)}{n}}$$

Thus, $$\sigma_{\hat{p}} = \sqrt{\frac{0.40(0.60)}{270}}$$

$$\sigma_{\hat{p}} = 0.0298$$

Because np and n(1-p) are both greater than 5, then the Sampling Distribution of The Proportion is approximately a normal distribution. This is illustrated in Figure 12.3.

Sampling Distribution of The Proportion
(is approximately a normal distribution)

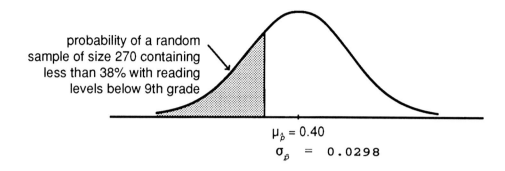

Figure 12.3

Since we are dealing with a normal distribution, we need to calculate a z score of 0.38 to determine the shaded area illustrated in Figure 12.3.

$$z = \frac{0.38 - 0.40}{0.0298} = -0.67$$

From TABLE II, the percent of area to the left of $z = -0.67$ is 25.14%. Therefore, the probability that less than 38% of the State University's male graduates have a reading level below 9th grade is 25.14%. ∎

❊ 12.3 HYPOTHESIS TESTING INVOLVING A POPULATION PROPORTION

Now that we have discussed the Sampling Distribution of The Proportion model, let's develop the hypothesis testing procedure for hypotheses involving a population proportion using the five step general hypothesis testing procedure discussed in Chapter 9.

Throughout this chapter, we will be assuming that the Sampling Distribution of The Proportion satisfies the conditions of Theorem 12.1.

Hypothesis Testing Procedure Involving a Population Proportion

Step 1. *Formulate the two hypotheses, H_o and H_a.*

Null Hypothesis:

The null hypothesis has the following form:

Null Hypothesis Form:

H_o: The population proportion, p, is claimed to be **equal to** the proportion p_o.

This is symbolized as:

$$H_o: p = p_o$$

Alternative Hypothesis:

Since the alternative hypothesis, H_a, can be stated as either *greater than*, *less than*, or *not equal to* the numerical value, μ_o, then the alternative hypothesis can have one of three forms:

Form (a): Greater Than Form For H_a

H_a: The population proportion, p, is claimed to be **greater than** the numerical value, p_o.

This is symbolized as:
$$H_a: p > p_o$$

Form (b): Less Than Form For H_a

H_a: The population proportion, p, is claimed to be **less than** the numerical value, p_o.

This is symbolized as:
$$H_a: p < p_o$$

Form (c): Not Equal Form For H_a

H_a: The population proportion, p, is claimed to be **not equal** to the numerical value, p_o.

This is symbolized as:
$$H_a: p \neq p_o$$

Step 2. *Design an experiment to test the null hypothesis, H_o.*

Under the assumption H_o is true, calculate the expected results. Since we are testing a hypothesis about a population proportion using a sample proportion, $\hat{p}$, we use as our model the Sampling Distribution of The Proportion and should choose a sample size which will make both np and n(1-p) greater than 5.

Using Theorem 12.1 and under the assumption H_o is true (i.e. $p = p_o$), then:

---

the expected results are:
a) the Sampling Distribution of The Proportion is *approximately normal* and serves as the hypothesis testing model.
b) the mean of the Sampling Distribution of The Proportion is: $\mu_{\hat{p}} = p_o$,
(i.e. the mean $\mu_{\hat{p}}$ equals the claimed population proportion, p_o).
c) the standard deviation of the Distribution of Sample Proportions is:
$$\sigma_{\hat{p}} = \sqrt{\frac{p_o(1-p_o)}{n}}$$

This is illustrated in Figure 12.4

HYPOTHESIS TESTING MODEL

Binomial Population from which the random sample is selected

Sampling Distribution of The Proportion (is approximately a normal distribution)

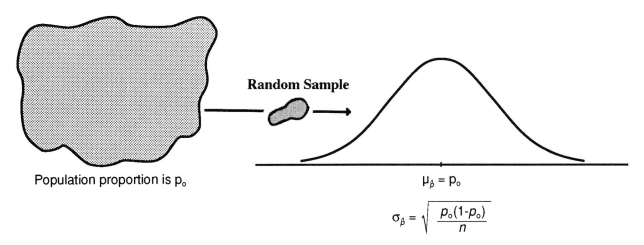

Population proportion is p_o

$\mu_{\hat{p}} = p_o$

$\sigma_{\hat{p}} = \sqrt{\dfrac{p_o(1-p_o)}{n}}$

Figure 12.4

In theory, the binomial population being sampled must be infinite to use the standard deviation formula: $\sigma_{\hat{p}} = \sqrt{\dfrac{p(1-p)}{n}}$

However, it is appropriate to use this formula when the population is *finite* and the population size, N, is large relative to the sample size, n. The rule used to determine whether the population is large relative to the sample size is:

the sample size, n, is less than 5% of the population size, N.

That is, $\dfrac{n}{N} < 5\%$.

Step 3. *Formulate the decision rule.*

---

a) Determine the type of alternative hypothesis: directional or nondirectional.
b) Determine the type of test: 1TT or 2TT.
c) Identify the significance level: $\alpha = 1\%$ or $\alpha = 5\%$.
d) Calculate the critical value(s) using the formula:

$$X_c = \mu_{\hat{p}} + (z_c)(\sigma_{\hat{p}})$$

e) Construct the ***hypothesis testing model***.
f) State the decision rule.

Step 4. *Conduct the experiment.*

> Randomly select the necessary data from the binomial population and calculate the experimental outcome, (i.e the sample proportion), $\hat{p}$, using the formula:
>
> $$\hat{p} = \frac{\text{number of occurrences possessing particular characteristic}}{\text{sample size}}$$

Step 5. *Determine the conclusion.*

> Compare the sample proportion, $\hat{p}$, (the experimental outcome) to the critical value(s), X_c, of the decision rule and draw one of the following conclusions:
>
> a) **Reject H_o and Accept H_a at α**
>
> or
>
> b) **Fail to reject H_o at α**

Example 12.6

An advertising agency developed the theme for the commercials for the daytime soap: *The Nights of Our Lives* on the assumption that 80% of its viewers are women. The ad agency would consider changing the theme of the commercials if the proportion of women viewers is **significantly lower** than 80%.

To test this population proportion, the ad agency conducted a random survey of 400 viewers of this show and determined that 311 of these viewers were women. Based upon this sample proportion, should the ad agency change the theme of the commercials at $\alpha = 5\%$?

Solution:

Step 1. *Formulate the hypotheses.*

H_o: The population proportion of women viewers who watch the day time soap *The Nights of Our Lives* is 0.80.

In symbols, we have:
$$H_o: p = 0.80$$

Since the advertising agency is interested in determining if the population proportion is *lower than* 0.80, then the alternative hypothesis is directional and is stated as:

H_a: The population proportion of women viewers who watch the day time soap *The Nights of Our Lives* is *less than* 0.80.

In symbols, we have:
$$H_a: p < 0.80$$

Step 2. *Design an experiment to test the null hypothesis, H_o.*

Under the assumption H_o is true, calculate the expected results. Using Theorem 12.1 and

the assumption H_o is true, the expected results are:
a) The Sampling Distribution of The Proportion is approximately a normal distribution since both np and n(1-p) are greater than 5.
b) For p = 0.80, then the mean of the Sampling Distribution of The Proportion is:

$$\mu_{\hat{p}} = 0.80$$

c) Since n = 400, p = 0.80 and 1 - p = 0.20, then the standard deviation of the Sampling Distribution of The Proportion is:

$$\sigma_{\hat{p}} = \sqrt{\frac{p(1-p)}{n}}$$

$$\sigma_{\hat{p}} = \sqrt{\frac{0.80(0.20)}{400}}$$

$$\sigma_{\hat{p}} = 0.02$$

Figure 12.5 illustrates the Sampling Distribution of The Proportion Hypothesis Testing model.

Sampling Distribution of The Proportion
(is approximately a normal distribution)

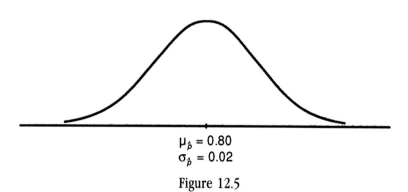

$\mu_{\hat{p}} = 0.80$
$\sigma_{\hat{p}} = 0.02$

Figure 12.5

Step 3. *Formulate the decision rule.*

a) The alternative hypothesis is *directional* since the ad agency wants to determine if the proportion of women viewers is significantly *lower* than 0.80.
b) The type of hypothesis test is *one-tailed* on the left side.
c) The significance level is: $\alpha = 5\%$.
d) The critical z score, z_c, for a one-tailed test on the left side with $\alpha = 5\%$ is -1.65.

The critical value, X_c, is:

$$X_c = \mu_{\hat{p}} + (z_c)(\sigma_{\hat{p}})$$
$$X_c = 0.80 + (-1.65)(0.02)$$
$$X_c = 0.767$$

e) Figure 12.6 illustrates the decision rule for this hypothesis test.

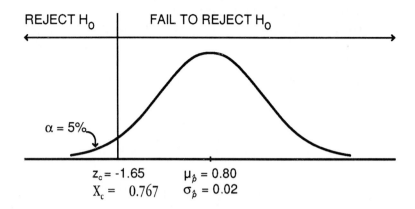

Figure 12.6

The decision rule is:

f) Reject H_o if the sample proportion (the experimental outcome), $\hat{p}$, is less than 0.767.

Step 4. *Conduct the experiment.*

Using the survey data, compute the experimental outcome, i.e. the sample proportion, $\hat{p}$.

$$\hat{p} = \frac{\text{number of women within the survey who watch the soap}}{\text{sample size}}$$

$$\hat{p} = \frac{311}{400}$$

$$\hat{p} = 0.7775$$

Step 5. *Determine the conclusion.*

Since the experimental outcome (sample proportion) $\hat{p} = 0.7775$ is **not** less than the critical value of 0.767, then we fail to reject H_o at $\alpha = 5\%$. Therefore, the ad agency *should not* change the theme of the commercials for *The Nights of Our Lives*. ∎

Example 12.7

The IRS stated that last year 20% of the Federal Income Tax Returns contained arithmetic errors. A random sample of 500 of this year's Federal Returns found 130 returns with arithmetic errors. Can the IRS conclude that the proportion of Federal Returns containing arithmetic errors has significantly changed at $\alpha = 5\%$?

Solution:

Step 1. *Formulate the hypotheses.*

H_o: For this year, the population proportion of Federal Income Tax Returns with arithmetic errors is 0.20.

In symbols, we have:

$$H_o: p = 0.20$$

H_a: For this year, the population proportion of Federal Income Tax Returns with arithmetic errors is **not** 0.20.
In symbols, we have:

$$H_a: p \neq 0.20$$

Step 2. *Design an experiment to test the null hypothesis, H_o.*

Under the assumption H_o is true and using Theorem 12.1, the expected results are:
a) The Sampling Distribution of The Proportion is approximately a normal distribution since both np and n(1-p) are greater than 5.

b) Since p = 0.20, then the mean is: $\mu_{\hat{p}} = 0.20$.

c) For n = 500, p = 0.20 and 1-p = 0.80, then the standard deviation is:

$$\sigma_{\hat{p}} = \sqrt{\frac{p(1-p)}{n}}$$

$$\sigma_{\hat{p}} = 0.0179$$

Figure 12.7 illustrates the Sampling Distribution of The Proportion which serves as the Hypothesis Testing model.

Sampling Distribution of The Proportion
(is approximately a normal distribution)

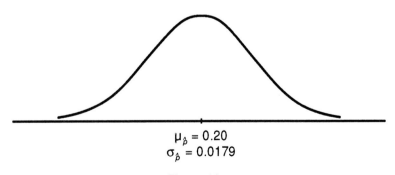

$\mu_{\hat{p}} = 0.20$
$\sigma_{\hat{p}} = 0.0179$

Figure 12.7

Step 3. *Formulate the decision rule.*

a) The alternative hypothesis is **nondirectional** since the IRS is trying to determine if the proportion of Federal Returns containing arithmetic errors for this year is **significantly different** from 0.20.
b) The type of hypothesis test is two-tailed.
c) The significance level is: $\alpha = 5\%$.

d) The critical z scores, z_{LC} and z_{RC}, for a two-tailed test with $\alpha = 5\%$ are $z_{LC} = -1.96$ and $z_{RC} = +1.96$.

Left critical value, X_{LC}, is:

$X_{LC} = \mu_{\hat{p}} + (z_{LC})(\sigma_{\hat{p}})$
$X_{LC} = 0.20 + (-1.96)(0.0179)$
$X_{LC} = 0.1649$

Right critical value, X_{RC}, is:

$X_{RC} = \mu_{\hat{p}} + (z_{RC})(\sigma_{\hat{p}})$
$X_{RC} = 0.20 + (1.96)(0.0179)$
$X_{RC} = 0.2351$

e) Figure 12.8 illustrates the decision rule for this hypothesis test.

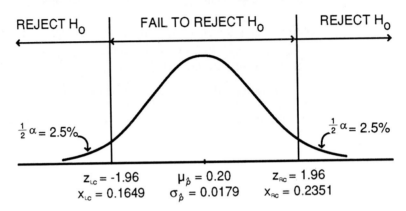

Figure 12.8

The decision rule is:

f) Reject H_o if the sample proportion, $\hat{p}$, is either *less than* 0.1649 or *greater than* 0.2351.

Step 4. *Conduct the experiment.*

Compute the experimental outcome, (the sample proportion), $\hat{p}$.

$\hat{p} = \dfrac{\text{number of returns in the sample containing arithmetic errors}}{\text{sample size}}$

$\hat{p} = \dfrac{130}{500}$

$\hat{p} = 0.26$

Step 5. *Determine the conclusion.*

Since the experimental outcome (sample proportion), $\hat{p} = 0.26$, is *greater than* the right critical outcome, $X_{RC} = 0.2351$, we reject H_o and accept H_a at $\alpha = 5\%$.

Therefore, the IRS can conclude that the proportion of this year's Federal Returns containing arithmetic errors has significantly changed at $\alpha = 5\%$. ∎

Example 12.8

A biochemist has developed a new drug which he claims is **more effective** in the treatment of a skin disorder than the existing drug. Using the existing drug only 30 percent of the people contracting this skin disorder recover completely. To test his claim, he administers this new drug to a sample of 300 people who have contracted this skin disorder. The biochemist determines that 140 patients in his sample show a complete recovery.

Based upon this sample result, can the biochemist conclude at $\alpha = 1\%$ that his new drug is more effective than the existing drug in the treatment of this skin disorder?

Solution:

Step 1. *Formulate the hypotheses.*

H_o: The population proportion of people who completely recover from this skin disorder using the new drug is 0.30.

In symbols, we have:

$$H_o: p = 0.30$$

H_a: The population proportion of people who completely recover from this skin disorder using the new drug is *greater than* 0.30.

In symbols, we have:

$$H_a: p > 0.30$$

Step 2. *Design an experiment to test the null hypothesis, H_o.*

Under the assumption H_o is true, and using Theorem 12.1, the expected results are:

a) The Sampling Distribution of The Proportion is approximately a normal distribution since both np and n(1-p) are greater than 5.

b) Since $p = 0.30$ the mean is: $\mu_{\hat{p}} = 0.30$.

c) For n = 300, p = 0.30 and 1-p = 0.70, the standard deviation is:

$$\sigma_{\hat{p}} = \sqrt{\frac{p(1-p)}{n}}$$

$$\sigma_{\hat{p}} = 0.0265$$

Figure 12.9 illustrates the Sampling Distribution of The Proportion Hypothesis Testing model.

Sampling Distribution of The Proportion
(is approximately a normal distribution)

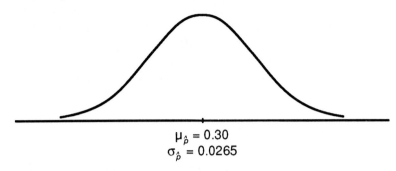

$\mu_{\hat{p}} = 0.30$
$\sigma_{\hat{p}} = 0.0265$

Figure 12.9

Step 3. *Formulate the decision rule.*

a) the alternative hypothesis is *directional* since the biochemist wants to determine if the proportion of patients completely recovering using the new drug is *greater than* 0.30.
b) the type of hypothesis test is one-tailed on the right side.
c) the significance level is: $\alpha = 1\%$.
d) the critical score, z_c, for a one-tailed test on the right side with $\alpha = 1\%$ is 2.33.

The critical value, X_c, is:

$$X_c = \mu_{\hat{p}} + (z_c)(\sigma_{\hat{p}})$$
$$X_c = 0.30 + (2.33)(0.0265)$$
$$X_c = 0.3617$$

e) Figure 12.10 illustrates the decision rule for this hypothesis test.

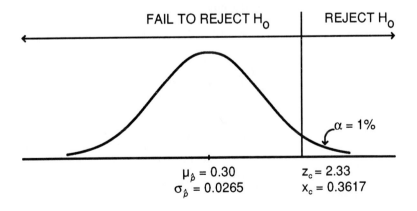

Figure 12.10

The decision rule is:

f) Reject H_0 if the sample proportion (experimental outcome), $\hat{p}$, is greater than 0.3617.

Step 4. *Conduct the experiment.*

Compute the experimental outcome, i.e. the sample proportion, $\hat{p}$.

$$\hat{p} = \frac{\text{number of people in the sample who recover using new drug}}{\text{sample size}}$$

$$\hat{p} = \frac{140}{300}$$

$$\hat{p} = 0.4667$$

Step 5. *Determine the conclusion.*

Since the experimental outcome (sample proportion), $\hat{p} = 0.4667$, is greater than the critical outcome, $X_c = 0.3617$, we reject H_o and accept H_a at $\alpha = 1\%$.

Therefore, the biochemist can conclude that his new drug is more effective than the existing drug in the treatment of this skin disorder at a significance level of 1%. ∎

CASE STUDY 12.2

Examine the information contained in the USA Snapshot entitled *Percentage of female workers who are executives or managers* shown in Figure CS12.2.

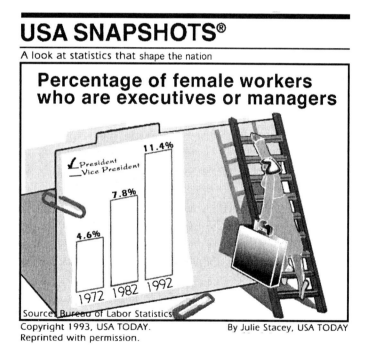

Figure CS12.2

a) What parameter is represented by the graph in Figure CS12.2?
b) What type of graph is used in Figure CS12.2 to present the percentage values?
c) Define the population described in Figure CS12.2?
d) What variable is represented by the graph?

e) Is this variable numeric or categorical? Explain.
f) According to Figure CS12.2, what proportion of female workers were executives or managers during 1972?
g) According to Figure CS12.2, what was the population proportion of female executives or managers for 1982?
h) According to Figure CS12.2, what is the population proportion of female executives or managers for 1992?

Using the information for 1992, design a hypothesis test assuming you believe that the trend shown in Figure CS12.2 will continue to the present time. For your test, state:

i) the null hypothesis, using the information pertaining to 1992 as the status quo.
j) an alternative hypothesis.
k) Is the alternative hypothesis directional or nondirectional? Is this a 1TT or a 2TT? Explain.
l) Which population parameter are you assuming you know to perform this hypothesis test? What is the value of this parameter?
m) State the name of the sampling distribution that will serve as the model for this test? Is this sampling distribution a normal or a t distribution? Explain.
n) What is the mean of this sampling distribution? What is the standard deviation of the sampling distribution?
o) Which tail of the curve would you place the critical outcome for your hypothesis test?
p) If you decide to use a significance level of 5%, then write the critical value formula required to perform the test and determine the value of the critical outcome.
q) If the experimental outcome (or sample outcome) falls to the left of the critical outcome of your test, what is the conclusion to this test? Could you agree that the proportion of the population of female executives or managers significantly changed? If you can, then explain the type of significant change you can conclude from the test. What can you conclude about the population proportion?
r) If the experimental outcome (or sample outcome) falls to the right of the critical outcome of your test, what is the conclusion to this test? Could you agree that the proportion of the population of female executives or managers has significantly changed? If you can, then explain the type of significant change you can conclude from the test. What can you conclude about the population proportion?
s) If the p-value of the hypothesis test is 4% (or 0.04), then can you reject the null hypothesis at $\alpha = 5\%$? at $\alpha = 1\%$? For each significant level, does this indicate that the percentage of the female executives or managers has significantly changed? Explain what you can conclude regarding the population proportion for each significant level.
t) If the p-value of the hypothesis is 0.90% (or 0.0090) and $\alpha = 1\%$, could you conclude that the percentage of female executive or managers has significantly changed? Explain. ∎

Summary of the Hypothesis Test Involving a Population Proportion

The following outline is a summary of the 5 step hypothesis testing procedure for a hypothesis test involving one population propoprtion.

Step 1. *Formulate the two hypotheses, H_o and H_a.*
Null Hypothesis, H_o: $p = p_o$
Alternative Hypothesis, H_a:
has one of the following three forms:
H_a: $p < p_o$ or
H_a: $p > p_o$ or
H_a: $p \neq p_o$

Step 2. *Design an experiment to test the null hypothesis, H_o.*
Under the assumption H_o is true, the expected results are:
a) the Sampling Distribution of The Proportion is approximately normal when **both** np and n(1 - p) are greater than 5.
b) the mean of the Sampling Distribution of The Proportion, denoted $\mu_{\hat{p}}$, is:
$$\mu_{\hat{p}} = p$$
c) the standard deviation of the Sampling Distribution of The Proportion, denoted $\sigma_{\hat{p}}$, is:
$$\sigma_{\hat{p}} = \sqrt{\frac{p(1-p)}{n}}$$

Step 3. *Formulate the decision rule.*
a) alternative hypothesis: directional or nondirectional.
b) type of test: 1TT or 2TT.
c) significance level: $\alpha = 1\%$ or $\alpha = 5\%$.
d) critical value formula is:
$$X_c = \mu_{\hat{p}} + (z_c)(\sigma_{\hat{p}})$$
e) draw hypothesis testing model.
f) state the decision rule.

Step 4. *Conduct the experiment.*
Compute the experimental outcome (or sample proportion), $\hat{p}$, using the formula:
$$\hat{p} = \frac{\text{number of occurrences possessing particular characteristic}}{\text{sample size}}$$

Step 5. *Determine the conclusion.*
Compare the sample proportion, $\hat{p}$, (the experimental outcome) to the critical value(s), X_c, of the decision rule and draw one of the following conclusions, either:
(a) **Reject H_o and Accept H_a at α**
or
(b) **Fail to Reject H_o at α**

Table 12.1 summarizes the hypothesis testing procedure for a population proportion.

Table 12.1

SUMMARY OF HYPOTHESIS TEST FOR TESTING THE VALUE OF A POPULATION PROPORTION			
FORM OF THE NULL HYPOTHESIS	CONDITIONS OF TEST	CRITICAL VALUE FORMULA	SAMPLING DISTRIBUTION IS:
$p = p_o$	$np > 5$ and $n(1-p) > 5$	$X_c = \mu_{\hat{p}} + (z_c)(\sigma_{\hat{p}})$ where: $\sigma_{\hat{p}} = \sqrt{\dfrac{p(1-p)}{n}}$	Approximately a Normal Distribution

❄ 12.4 Using MINITAB

In this chapter we have explored the sampling distribution of the proportion. Section 12.3 discussed the hypothesis testing procedure for a population proportion, p. When conducting hypothesis testing for the population proportion using a random sample of size n, where both np and n(1-p) are greater than five, the normal distribution is used as a model for the test.

To use MINITAB to perform a hypothesis test about a population proportion p, the standard deviation of the distribution of the population proportion must be calculated.

To calculate the standard deviation of the proportion use the formula:

Standard Deviation of the Proportion = $\sqrt{p(1-p)}$.

To illustrate this formula, let's use a population proportion of p=0.7. The standard deviation of the proportion is calculated as

$$\text{Standard Deviation of the Proportion} = \sqrt{p(1-p)}$$
$$= \sqrt{0.7(1-0.7)}$$
$$\approx 0.4583$$

To **calculate** this in MINITAB, use the **LET** command.
    LET K1 = SQRT(0.7 * 0.3)

To **display** the value of K1, the **PRINT** command is used.
    PRINT K1

Using MINITAB for which p=0.7 we have
    MTB › LET K1=SQRT(0.7*0.3)
    MTB › PRINT K1
    K1   0.458258

The value K1 is the standard deviation of the proportion. When using MINITAB the value of K1 will be labelled SIGMA.

In MINITAB to conduct an hypothesis test of a population proportion, p, we can use the ZTEST command.

To use the ZTEST command, it is assumed that the sampling distribution of the proportion is modelled by a normal distribution. Depending on the type of alternative hypothesis, (i.e. whether it is nondirectional or directional) the ZTEST command takes on one of three forms:

for a **two-tailed hypothesis test**:
ZTEST MU=(numerical value) SIGMA=(numerical value) C1

for a **one-tailed** hypothesis test, where the alternative hypothesis, H_a, is on the left is:
ZTEST MU=(numerical value) SIGMA=(numerical value) C1;
ALTERNATIVE= -1.

for a **one-tailed** hypothesis test, where the alternative hypothesis, H_a, is on the right is:
ZTEST MU=(numerical value) SIGMA=(numerical value) C1;
ALTERNATIVE= 1.

In each form the notation assumes that the population proportion p, is MU, the standard deviation of the proportion is SIGMA, and, that the random sample used to test the population proportion is in a MINITAB worksheet in column C1.

For example, to use the ZTEST command of MINITAB for testing that the population proportion p =.7 where a sample of data values is in C1 we have:

MTB > ZTEST MU=0.7 SIGMA=.4583 C1

The output for this illustration is:

TEST OF MU = 0.7000 VS MU N.E. 0.7000
THE ASSUMED SIGMA = 0.458

	N	MEAN	STDEV	SE MEAN	Z	P VALUE
C1	40	0.5500	0.5038	0.0725	-2.07	0.039

For this output what is the null hypothesis? the alternative hypothesis? Notice that the sample contained 40 data values and that the p-value for the sample is 0.039. Is the null hypothesis rejected?

Sample data used in these types of tests **must** be coded using 0 and 1. Thus, to use MINITAB to analyze a hypothesis test about a population proportion, the sample results should be entered into C1 as a sequence of 0 and 1. To illustrate this consider the following hypotheses:

H_o: Statistics is used by 80% of the population.

You believe this is **too high** and randomly sample 400 from the population. The outcome of your random sample is that of the 400 surveyed 312 use statistics. Does the random sample support your claim? (α=5%).

Since your claim is that the population proportion, p= .80 is too high, the ZTEST command is for a 1TT on the left. Thus, it takes on the form:

ZTEST MU=(numerical value) SIGMA=(numerical value) C1;
ALTERNATIVE= -1.

But in order to apply the ZTEST
   (1) SIGMA must be calculated, and
   (2) the sample results must be entered into a MINITAB worksheet.

To calculate sigma we use the LET statement
MTB > LET K1=SQRT(0.8*0.2)

The value is then printed.

MTB > PRINT K1
K1   0.400000

Thus, SIGMA is 0.4.

To enter the sample results into a MINITAB worksheet, we could use the SET command.

Since the random sample of 400 contained an outcome of 312 affirmative responses (i.e. those that use statistics), then 312 ones must be entered into the worksheet in column C1. To complete the column for the remaining 88 entries, 88 zeros are entered into C1.

MTB > SET C1

DATA > 312(1) 88(0)

DATA > END.

Since the value of SIGMA and the entries of C1 have been completed, the hypothesis test can be conducted using the MINITAB ZTEST.

MTB > ZTEST MU = 0.8 SIGMA = 0.4 C1;
SUBC> ALTERNATIVE=-1.

The output is:

TEST OF MU = 0.8000 VS MU L.T. 0.8000
THE ASSUMED SIGMA = 0.400

	N	MEAN	STDEV	SE MEAN	Z	P VALUE
C1	400	0.7800	0.4148	0.0200	-1.00	0.16

Examine the output. Do the sample results in C1 indicate that the alternative hypothesis is accepted?

GLOSSARY

Term	Section
Population Proportion, p	12.1
Sample Proportion, $\hat{p}$	12.1
Theorem 12.1: The Sampling Distribution of the Proportion	12.2
The Sampling Distribution of The Proportion	12.2
The Mean of The Sampling Distribution of The Proportion, $\mu_{\hat{p}}$	12.2
The Standard Deviation of The Sampling Distribution of The Proportion, $\sigma_{\hat{p}}$	12.2
Hypothesis Testing Procedure Involving a Population Proportion	12.3
Expected Results	12.3
Experimental Outcome (or Sample outcome)	12.3

Exercises

PART I Fill in the blanks.

1. To perform a hypothesis test involving a population proportion, the Sampling Distribution of The _____ is used as the hypothesis testing model.
2. The symbol p represents the proportion of the _____ possessing the particular characteristic.
3. The symbol $\hat{p}$ represents the proportion of the _____ and is computed using the formula:
   $\hat{p}$ = # of occurrences possessing particular characteristic / sample size
4. The Sampling Distribution of The Proportion has the following characteristics:
   a) the mean of the Sampling Distribution of The Proportion, $\mu_{\hat{p}}$, is given by the formula:
   $\mu_{\hat{p}}$ = _____
   b) the standard deviation of the Sampling Distribution of The Proportion is given by the formula:
   $\sigma_{\hat{p}}$ = _____
   c) the Sampling Distribution of The Proportion is approximately normal when both ___ and ___ are greater than 5.
5. The null hypothesis of a hypothesis test involving a population proportion is stated as:
   The population proportion, symbolized ___, is ___ to the proportion p_0. In symbols, H_0: p ___ p_0.
6. The alternative hypothesis of a hypothesis test involving a population proportion has one of the following three forms:
   a) the population proportion, symbolized ___, is greater than the proportion p_0. In symbols, H_a: p ___ p_0.
   b) the _____ proportion, p, is less than the proportion p_0. In symbols, H_a: p ___ p_0.
   c) The population _____, p, is not equal to the proportion p_0. In symbols, H_a: p ___ p_0.
7. When performing a hypothesis test about a population proportion, the expected results are determined under the assumption that _____.
8. The hypothesis testing model for a hypothesis test involving a population proportion is the Sampling Distribution of The _____.
9. The formula for the critical value, X_c, for a hypothesis test involving a population proportion is: X_c = _____.
10. To determine the conclusion of a hypothesis test involving a population proportion, the experimental outcome, (i.e. the sample proportion) is computed and compared to the _____ outcome(s).

PART II Multiple choice questions.

1. If 14 out of every 35 people in the population interviewed answered yes to the survey question, then the population proportion, p, would be:
   a) 0.14 b) 0.20 c) 0.50 d) 0.40
2. If random samples of size 40 are selected from an infinite binomial population with a proportion of women equal to 0.25, what would be the expected standard deviation of sample proportions of women?
   a) 0.0610 b) 0.0801 c) 0.0685 d) 0.0488
3. If random samples of size 80 are selected from an infinite binomial population with a proportion of minority members of 0.37, then what would be the expected mean proportion of minorities of these samples?
   a) 0.37 b) 0.80 c) 0.50 d) cannot determine
4. If the sample size is increased, then the standard deviation of the Sampling Distribution of The Proportion will:
   a) decrease   b) increase
   c) remain the same   d) can't tell
5. When performing a hypothesis test involving a population proportion the null hypotheses, H_0, is always stated as:
   a) H_0: p ≠ p_0 b) H_0: p > p_0 c) H_0: p < p_0 d) H_0: p = p_0
6. If the experimenter believes that the claimed population proportion value p_0 is too high, then his alternative hypothesis is stated as:
   a) H_a: p ≠ p_0 b) H_a: p > p_0
   c) H_a: p < p_0 d) H_a: p=p_0
7. The Sampling Distribution of The Proportion is approximately a normal distribution when:
   a) n > 30   b) always
   c) np > 5 and n(1-p) > 5   d) np > 5 or n(1-p) > 5
8. The formula for $\mu_{\hat{p}}$ is given by:
   a) $\mu_{\hat{p}}$ = np   b) $\mu_{\hat{p}}$ = p
   c) $\mu_{\hat{p}}$ = p/n   d) $\mu_{\hat{p}}$ = n(1-p)
9. If the alternative hypothesis is H_a: p > p_0, then the experimenter is trying to show:
   a) p is less than p_0 b) p equals p_0
   c) p is not equal to p_0 d) p is greater than p_0
10. In a hypothesis test the value used for p in calculating $\mu_{\hat{p}}$ and $\sigma_{\hat{p}}$ is obtained from:
    a) the experimental outcome
    b) the alternative hypothesis
    c) the null hypothesis
    d) the standard deviation

PART III Answer each statement True or False.

1. The population proportion is the number of occurrences of a particular characteristic in a population.
2. The sample proportion, $\hat{p}$, is the ratio of the number of occurrences of a particular characteristic to the sample size n.
3. The Sampling Distribution of The Proportion has properties similar to the Binomial Distribution.
4. The mean of the Sampling Distribution of The Proportion is determined from the sample proportion, $\hat{p}$.
5. The Sampling Distribution of The Proportion is approximately normal when either np or n(1-p) is greater than 5.
6. In calculating $\mu_{\hat{p}}$, we use the formula np where n is the sample size and p is the sample proportion.
7. When formulating the null hypothesis involving a population proportion p, it is assumed that $p = p_o$ where p_o represents the claimed population proportion.
8. If the alternative hypothesis is expressed as $p < p_o$, then the rejection of H_o will statistically indicate that the population proportion is significantly less than p_o.
9. The value for $\mu_{\hat{p}}$ used in the hypothesis testing procedure is based upon the assumption that H_o is true.
10. When doing a hypothesis test involving a population proportion, p, the sample proportion, $\hat{p}$, is referred to as the experimental outcome.

PART IV Problems. *Roundoff ALL answers to four decimal places.*

1. Determine the sample proportion, $\hat{p}$, for each of the following statements.
   a) Thirty out of 100 incoming freshman do not meet the minimum college requirements in reading.
   b) A nationwide survey of 1200 people found that 800 favored the death penalty for murder.
   c) A family counselor finds that 180 out of 200 patients are lonely because they have forgotten how to have fun.
   d) A government agency has determined that 1850 out of a sample of 5000 student borrowers have defaulted on their federal student loans.
   e) Four out of 37 stock investors purchase stocks based upon presumed inside information (i.e. a hot tip).
2. The following table summarizes the responses of a cross section of 9000 Americans aged 16 and over regarding the following two questions:

Question #1: "If a man doesn't want his wife to take a job, should she respect his wishes?"
Question #2: "Do you think it is or is not all right for a man and a woman to live together without getting married?"

	Question #1		Question #2	
	YES	NO	ALL RIGHT	NOT All RIGHT
TOTAL	4790	4210	4388	4612
REGION				
Northeast	710	1090	1026	774
Southeast	990	810	728	1072
Midwest	970	830	791	1009
Southwest	1210	590	87	1113
Far West	910	890	1156	644
SEX				
Women	2365	2060	1946	2479
Men	2425	2150	2442	2133
MARITAL STATUS				
Single	902	1363	1593	672
Married	1243	1042	915	1370
Widowed	1482	643	396	1729
Separated/Divorced	1163	1162	1484	841
EDUCATION				
High School Incomplete	1422	793	778	1437
High School Graduate	1343	892	915	1320
Some beyond H. S.	1122	1133	1262	993
College Graduate	903	1392	1433	862

Determine the sample proportion, $\hat{p}$, for each of the following:

a) Northeast Americalns responding yes to question #1.
b) Southwestern Americans responding yes to question #1.
c) Single Americans responding no to question #1.
d) Widowed Americans responding no to question #1.
e) Those Americans who did not complete high school responding yes to question #1.
f) College graduates responding yes to question #1.
g) Midwest Americans responding not all right to question #2.
h) Far West Americans responding all right to question #2.
i) Women responding all right to question #2.
j) Married Americans responding not all right to question #2.
k) Separated/divorced Americans responding not all right to question #2.
l) College graduates responding all right to question #2.

3. Determine the mean and standard deviation of the Sampling Distribution of The Proportion and whether or not the Sampling Distribution of The Proportion can be approximated by a normal distribution, if:

a) p = 0.40, and n = 50
b) p = 0.10, and n = 30
c) the population proportion is 0.65 and the sample size is 200.
d) the population proportion is 0.80 and the sample size is 100.
e) a random sample of 400 was selected from a population composed of 70% Democratic.
f) a random sample of 150 was selected from a population composed of 20% unemployed people.

4. If the population proportion is 0.25 and the sample size is 500, then determine the probability that the sample proportion, $\hat{p}$, is:
   a) greater than 0.29
   b) less than 0.27
   c) between 0.22 and 0.30
   d) at most 0.20
   e) at least 0.26

5. If the population proportion is 0.7 and the sample size is 300, then determine the probability that the sample proportion, $\hat{p}$, is:
   a) greater than 0.67
   b) between 0.65 and 0.73
   c) less than 0.74
   d) at least 0.64
   e) at most 0.69

6. A large New York law firm states that 80% of their cases are settled out of court. What is the probability that out of the next 200 cases:
   a) at least 75% will be settled out of court?
   b) at least 90% will be settled out of court?
   c) at least 25% will **not** be settled out of court?

7. The medical officer at a Southwestern University states that 40% of all dental students have contracted the H-2 virus. What is the probability that for a random sample of twenty students:
   a) that at least 50% will have contracted the virus?
   b) that less than 25% will have contracted the virus?
   c) that between 30% to 60% will have contracted the virus?

   For problems 8 to 25:
   a) state the Null and Alternative Hypotheses.
   b) calculate the expected results for the hypothesis test assuming the null hypothesis is true and determine the hypothesis test model.
   c) formulate the decision rule.
   d) determine the experimental outcome.
   e) determine the conclusion and answer the question(s) posed in the problem, if any.
   f) round off all calculations to four decimal places.

8. A car manufacturer claims that less than 10% of their cars have major defective parts. A skeptical consumer believes the claim is too low and randomly surveys 100 owners of the manufacturer's cars and finds 16 cars had major defective parts. Would this indicate that the manufacturer's claim was incorrect at $\alpha = 5\%$?

9. A gasoline lawn mower manufacturer claims that their power lawn mowers start up on the first try 95% of the time. A consumer's protection group feels the manufacturer claim is too high and randomly selects 200 of the manufacturer's mowers and finds that 178 start on the first try. Would this indicate that the manufacturer was indeed exaggerating at $\alpha = 1\%$?

10. The production manager claims to have devised a new production process which will significantly lower the percent of defective parts manufactured. The present process produced defective parts at a rate of 10%. If a trial run using the new process produces 70 defects in a sample of 800, can one conclude that this new process is more effective at $\alpha = 5\%$?

11. A survey conducted by an Institute for Social Research claims that 40% of the American people say life in the United States is getting worse. Would one disagree with this claim at $\alpha = 5\%$ if an independent poll of 400 Americans shows that 175 believe life in the U.S. is getting worse?

12. Past experience has shown that 40% of the students fail a University entrance exam in English. If 50 out of 100 students from a certain city failed, would one be justified in concluding that the students from this city are inferior in English using $\alpha = 5\%$?

13. In 1986, 18% of the fire alarms in Hyde Park were false alarms. This year in a random sample of 200 fire alarms 30 were false alarms. Using a 5% significance level has the rate of false alarms decreased since 1986?

14. Based on a recent survey, the New York Off-Track Betting Corporation (O.T.B.) stated that 57% of its single-male bettors earned over $18,000 per year. If 300 randomly selected single male O.T.B. bettors were interviewed and 178 earned over $18,000 per year, would one disagree with the O.T.B.'s claim at $\alpha = 1\%$?

15. According to a study released by a New York newspaper 25% of Long Island shoppers are being charged sales tax on non-taxable items in supermarkets because of the complexity of the state's tax law. Mr. Jessel, a consumer advocate, decides to test this claim against his suspicion that the claim is too low. In a random survey of supermarket shoppers, Mr. Jessel finds that 65 out of 200 shoppers were improperly charged tax on their purchases. Does this sample result support Mr. Jessel's claim at $\alpha = 1\%$?

16. The Federal Trade Commission claims that Presidential Nuts, Inc., located in Plains, Ga., puts too many peanuts in their mixed nuts. Presidential Nuts Inc. claims that their mixed nuts contain only 20% peanuts. To test this claim, the Federal Trade Commission sampled 400 nuts at random and found 102 were peanuts. Does this finding support the claim made by the Federal Trade Commission at $\alpha = 1\%$?

17. A study conducted by the National Cancer Institute found that 15% of all cancer patients treated with synthetic human interferon had their tumors shrink to less than half their original size. Medical researchers at a Western Medical School decide to test this claim at $\alpha = 1\%$. If the researchers find that interferon caused the tumors to shrink to less than half their previous size in 14 of 80 patients, then can one reject the National Cancer Institutes claim?

18. An educational psychologist claims that 80% of the recent high school graduates that enter college do so primarily because of parental pressure. Researchers from the California University system believe this claim is too high when applied to entering freshman in the California University system. The researchers randomly sample 1600 of their most recent entering freshman and find that 1248 entered college primarily because of parental pressure. Does this sample result significantly support the researcher's belief at $\alpha = 1\%$?

19. The personnel director for a Florida based insurance company believes that her new hiring procedure will result in a better employee retainment. The present hiring technique resulted in a 60% employee retainment for those employees that have been with the company for more than six months. If the company samples 200 employees that have been hired under the new procedure and 126 are still with the company after six months does this indicate that the personnel director's claim is correct using $\alpha = 5\%$?

20. A census bureau study of divorce rates reports that 68% of the men who wed before age 22 were divorced within 20 years. The Institute of Family Relations decided to test the validity of this report by randomly selecting and reviewing 250 marriages where the men wed before age 22. If 186 of these marriages ended in divorce within 20 years, can the Institute of Family Relations conclude the census bureau's claim is incorrect at $\alpha = 5\%$?

21. The market research component of a shaver company will not allow mass production of a new shaver unless they can get a test market acceptance rate of at least 85%. If the test market survey of their latest five blade shaver received 320 favorable responses out of a random sample of 400 interviews, should the company mass produce their new five blade shaver at $\alpha = 1\%$?

22. A recent survey indicated that 60% of the people who own a personal computer use it to play video games. HAL, a personal computer manufacturer, decides to test this claim against the suspicion that it is lower than 60%. If from a random sample of 100 personal computer owners, 52 indicate they use their computer to play video games. Does this sample support the manufacturer's claim at $\alpha = 1\%$?

23. An avid football fan decides to test her claim that the home team has an advantage in a *pick'em* game. To test the claim that the proportion of pick'em games the home team wins is 0.50, the fan randomly selects 40 football games that are rated *pick'em* and finds that the home team was the winning team 23 times. Can the football fan conclude that her claim is correct at $\alpha = 5\%$?

24. According to the latest National Crime Poll 75% of all women fear walking alone at night in their neighborhoods. A sociologist feels that this proportion is too high for women living in a middle class suburb on Long Island. To test her claim, the sociologist randomly samples 250 women living in this Long Island suburb and determines that 162 fear walking alone at night in their neighborhoods. Does this sample data support the sociologist's claim at $\alpha = 5\%$?

25. A national survey on "The Evolving Role of the Secretary in the Information Age" reported that 60% of the secretaries experienced eye strain when using word processors with a CRT display screen. An executive secretary decided to test this claim. She randomly surveys fifty secretaries who work with a CRT and determines that 35 experience eye strain. Based upon this sample data can the executive secretary reject the national claim at $\alpha = 1\%$?

PART V What Do You Think?

1. What Do You Think?

Examine the pie chart entitled *The projected 2005 U.S. labor force* which presents the group of workers which will comprise the U. S. labor force in 2005 according to the U. S. Bureau of Labor Statistics.

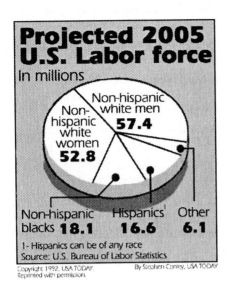

a) Define the population.
b) What variable is presented by the pie chart?
c) Is this variable numeric or categorical? Explain
d) What proportion of the labor force projected for 2005 is non-hispanic white men?
e) What proportion of the labor force projected for 2005 is non-hispanic blacks?
f) What proportion of the labor force projected for 2005 is Hispanic?
g) What proportion of the labor force projected for 2005 is listed as other?

According to the U.S. Bureau of Labor Statistics, the present proportion of U. S. workers is 43% is white men; 35% is white female: 11% is black; 8% is hispanic, and 3% is asians, native americans and others.

h) What is the percentage shift in the labor force for the white male worker?
i) What is the percentage shift in the labor force for the black worker?
j) What is the percentage shift in the labor force for the white female worker?
k) What is the percentage shift in the labor force for the hispanic worker?
l) What is the percentage shift in the labor force for the worker categorized as other?
m) Which group within the labor force will have greatest percentage increase?
n) Which group within the labor force will have smallest percentage increase?
o) Which group within the labor force will have smallest percentage decrease?

p) What parameter is represented by the pie chart for each group?

For each group of U.S. worker presented within the pie chart, design a hypothesis test for the present that includes:

q) a null hypothesis. Use the information pertaining to the present labor force as the status quo.
r) an alternative hypothesis. Use the pie chart information as your guide as to what you believe to be the direction of the proportion with regard to the present value of the true population parameter.
s) Is the alternative hypothesis directional or nondirectional? Is this a 1TT or a 2TT? Explain.
t) Which population parameter are you assuming you know to perform this hypothesis test? What is the value of this parameter?
u) State the name of the sampling distribution that will serve as the model for this test? Is this sampling distribution a normal or a t distribution? Explain.
v) What is the mean of this sampling distribution? What is the standard deviation of the sampling distribution?
w) Which tail of the curve would you place the critical outcome for your hypothesis test?
x) If you decide to use a significance level of 5% for each hypothesis test, then write the critical value formula required to perform the test and determine the value of the critical outcome.
y) If the experimental outcome (or sample outcome) falls to the right of the critical outcome for the U.S. hispanic worker hypothesis test, what is the conclusion to this test? Could you agree that the proportion of the population of U. S. hispanic workers has significantly changed at $\alpha = 5\%$? If you can, then explain the type of significant change you can conclude from the test.
z) If the p-value of the hypothesis test is 6% (or 0.06), then can you reject the null hypothesis for non-hispanic white men at $\alpha = 5\%$? Does this indicate that the percentage of the non-hispanic white male worker within the U. S. labor force has significantly changed? Explain what you can conclude regarding the population proportion? If the p-value of the hypothesis is 2.90% (or 0.0290) and $\alpha = 2\%$, could you conclude that the percentage of the non-hispanic white male worker within the U. S. labor force has significantly changed? Explain.

2. **What Do You Think?**

Within the following USA article entitled *Agency says 59% of us buckle up* a graph is presented beneath the title *More using seat belts* which is shown in the following figure.

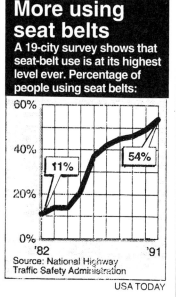

Agency says 59% of us buckle up

Safety campaigns have helped boost national seat belt use to 59%, the Transportation Department reported. Safety belt use ranges from 28% in Rhode Island to 85% in Hawaii. A survey of 19 cities showed seat belt use increased to 54% from 50% after a summer-long campaign. Belt use in cities is usually lower. "We are beginning to bring 'hard-core' non-users into the fold," said Transportation Secretary Samuel Skinner. He said each percentage point gain represents nearly 2 million people. Seat belts saved an estimated 4,800 lives in 1990.

a) Explain why the percentage figure within the title of the article (59%) is different from the percentage figure (54%) shown in the graph for the year 1991?
b) What variable is presented by the graph?
c) Is this variable numeric or categorical? Explain.
d) Is this variable discrete or continuous? Explain.

Using the stated percentage values within the article (59%, 28%, 85%, and 54%) as representing the status quo values for each population, and assuming your position is that you believe that the stated values are too high for the values 59% and 85%, while your belief is that the values of 28% and 54% are too low, then:

e) define the population for each stated percentage value.
f) state the null and alternative hypotheses.
g) state the type of alternative hypothesis.
h) state whether this hypothesis test is a 1TT or a 2TT.
i) state the name of the sampling distribution which will serve as the model for each test. Is this sampling distribution a normal or a t distribution? Explain.
j) indicate the population parameter you are assuming is true to perform your hypothesis test. State the value of each population parameter.
k) state the mean of each sampling distribution. Determine the standard deviation of each sampling distribution using a sample size of 1500 for each individual state and a total sample size of 28,500 for the 19 city survey.
l) explain which tail of the curve that the critical outcome of the test is placed for the hypothesis test. If you decide to use a significance level of 5% for each hypothesis test, then write the critical value formula required to perform the test and determine the value of the critical outcome.
m) If the experimental outcome (or sample outcome) falls to the right of the critical outcome for the hypothesis test pertaining to the Rhode Island percentage, what is the conclusion to this test? Could you agree that the proportion of the population has significantly changed at $\alpha = 5\%$? If you can, then explain the type of significant change you can conclude from the test.
n) If the p-value of the hypothesis test pertaining to the 59 percentage value is 5.3% (or 0.053), then can you reject the null hypothesis for this test at $\alpha = 5\%$? Could you conclude that the proportion of people using seat belts for this population has significantly changed? Explain.
o) If the p-value of the hypothesis test pertaining to the 54 percentage value is 1.90% (or 0.0190) and $\alpha = 2\%$, could you conclude that the percentage of people within the population that uses seat belts has significantly changed? Explain.

PART VI Exploring DATA With MINITAB

1. For the following population proportion, p, calculate SIGMA.

   a) p = .2   f) p = .55
   b) p = .25   g) p = .6
   c) p = .3   h) p = .7
   d) p = .4   i) p = .8
   e) p = .5   j) p = .9

   Calculate each SIGMA using the MINITAB command LET.
   To display SIGMA on the CRT screen use the PRINT command.

2. For each of the following samples, write a MINITAB command that inputs the sample results into a MINITAB worksheet that uses the code 1 for the outcome and 0 for the remaining sample values.

   a) n= 400 outcome= 350   e) n= 100 outcome= 58
   b) n= 200 outcome= 50   f) n= 160 outcome= 78
   c) n= 150 outcome= 75   g) n= 600 outcome= 152
   d) n= 50 outcome= 35   h) n= 600 outcome= 448

3. Use MINITAB to test the population proportion, p, given the following information:
   a) H_o: p= 0.7 H_a: p< 0.7   n= 400   outcome= 260.
   b) H_o: p= 0.6 H_a: p> 0.6   n= 200   outcome= 123.
   c) H_o: p= 0.8 H_a: p≠0.8   n= 100   outcome= 84.
   d) H_o: p= 0.3 H_a: p< 0.3   n= 50   outcome= 12.
   e) H_o: p= 0.2 H_a: p> 0.2   n= 36   outcome= 8.
   f) H_o: p= 0.1 H_a: p≠0.1   n= 64   outcome= 6.
   g) H_o: p= 0.75 H_a: p< 0.75 n= 400   outcome= 280.
   h) H_o: p= 0.67 H_a: p> 0.67 n= 200   outcome= 143.
   i) H_o: p= 0.23 H_a: p≠0.23 n= 100   outcome= 26.
   j) H_o: p= 0.95 H_a: p< 0.95 n= 400   outcome= 375.
   k) H_o: p= 0.07 H_a: p> 0.07 n= 200   outcome= 18.
   l) H_o: p= 0.61 H_a: p≠0.61 n= 400   outcome= 169.

PART VII   Projects.

1. Hypothesis Testing Project.
   A. Selection of a topic
      1. Select a reported fact or claim about a population proportion made in a local newspaper, magazine or other source.
      2. Suggested topics:
         a) Home team advantage: Randomly select any two weeks of a football, baseball, basketball or hockey season and record the following:
            1. total number of games played during the two weeks selected.
            2. the number of games the home team won.

   Test the null hypothesis: p = 1/2 against the alternative:

   $p > 1/2$ (*i.e. home team advantage*).
         b) Males equal females? Select a random sample of at least 30 families. For these families record the number of male and female children. Test the null hypothesis: p = 1/2 against the alternative hypothesis: p ≠ 1/2.
         c) In the first national election where eighteen year olds were allowed to vote, the United States Census Bureau claimed that 48% of the eligible voters in the 18-20 year age group voted. Perform a binomial hypothesis test to check if this claim is still valid.
   B. Use the following procedure to test the claim selected in Part A.
      1. State the claim, and identify the population referred to in the statement of the claim, also indicate the source.
      2. State your opinion regarding the claim (i.e. do you feel it is too high, too low, or simply don't agree with the claim) and clearly identify the population you plan to sample to test your opinion.
      3. State the null and alternative hypotheses.
      4. Develop and state a procedure for selecting a random sample from your population. Indicate the technique you are going to use to obtain your sample data.
      5. Compute the expected results, indicate the hypothesis testing model, and choose an appropriate level of significance.
      6. Calculate the critical value(s) and state the decision rule.
      7. Perform the experiment and calculate the experimental outcome, $\hat{p}$ (sample proportion).
      8. Construct a hypothesis testing model and place the experimental outcome on the model.
      9. Formulate the appropriate conclusion and interpret the conclusion with respect to your opinion (as stated in step 2).

2. Read a research article from a professional journal in a field such as medicine, education, marketing, science, social science, etc. Write a summary of the research article that includes the following:
   A. Name of article, author and journal (include volume number and date).
   B. Statement of the research problem.
   C. The null and alternative hypotheses.
   D. How the experiment was conducted (give the sample size; and the sampling technique).
   E. Determine the experimental outcome and the level of significance.
   F. State the conclusion in terms of the null and alternative hypotheses.

PART VIII   Database.

The following exercises refer to the file DATABASE listed in Appendix A. We have indicated the appropriate MINITAB commands that are necessary to answer each exercise.

1. Using the MINITAB commands: RETRIEVE, LET, SQRT, PRINT and ZTEST. Retrieve the file DATABASE.MTW and identify the column number for the variable name: SEX.

   The chairman of the math department believes that the proportion of female students is 0.5.

   Use the ZTEST command to test the claim that the proportion of female students is *not* 0.5.

   State the null and alternative hypothesis for this test.

Is the null hypothesis rejected? Identify the p-value and interpret.

State the conclusion.

2. Using the MINITAB commands: RETRIEVE, LET, SQRT, PRINT and ZTEST. Retrieve the file DATABASE.MTW and identify the column number for the variable name: MAJ.

The Vice President of Finance believes that the proportion of Liberal Arts students is at least 60%. Use the ZTEST command to test this claim if you believe that the claim is *too high*.

State the null and alternative hypothesis for this test.

Is the null hypothesis rejected? Identify the p-value and interpret.

State the conclusion.

CHAPTER 13
ESTIMATION

❄ 13.1 INTRODUCTION

"What time is it?"

The response to this question is an *estimate* of the actual time. Estimates like this are made everyday. In fact, occasions arise in business, in social science, in medicine and in science when it becomes necessary to *estimate* **a population parameter** such as the population mean, the population proportion or the population standard deviation.

For instance:

- a sociologist is interested in **the mean age** of women when delivering her first child.
- the census bureau is trying to determine **the proportion of American workers** who use computers at work.
- a quality control engineer wants to monitor **the variability of the sugar content** of soft drinks.
- an environmentalist needs to know **the mean amount of toxins** in the city's water supply.
- a student is interested in **the proportion of students** who pass a statistics course with a particular instructor.
- a commuter wishes to know **the variability of the amount of time** necessary to travel to work using mass transit.
- a company president wants to determine **the proportion of people** that would purchase a new line of shavers.

In each of these examples, the researcher is trying to determine the *true* or *actual population parameter*. In practical applications, the *true population parameter* is unattainable since populations are large and it would be either impractical or impossible to obtain the entire population data to calculate the true population parameter. Thus, it becomes necessary to select a random sample from the population and **use the sample data to *estimate* the population parameter**. The process of using sample data to make an inference about a population parameter is called ***statistical inference***. In this chapter, we will discuss one major area of statistical inference called ***estimation***.

❄ 13.2 POINT ESTIMATE OF THE POPULATION MEAN AND THE POPULATION PROPORTION

In this section, we will introduce procedures that enable us to estimate the population mean or the population proportion using information from a sample. A procedure that assigns a numerical value to a population parameter based upon sample information is called *estimation*.

> **Definition: Estimation.** Estimation is the statistical procedure where sample information is used to estimate the value of a population parameter such as the population mean, population standard deviation or population proportion.

Suppose a sociologist is interested in determining the population mean age of southern California women when delivering their first child. How would the sociologist obtain an *estimate* of the mean age of this population?

The sociologist could *randomly* select a sample of the birth records of first-time mother's from the southern California area. Using this sample of records, the sample mean age of the first-time mothers can be calculated. This *sample mean* is used to *estimate* **the population mean age** of first-time mothers. Such an estimate is called a *point estimate* of the population mean.

> **Definition: Point Estimate.** A point estimate is a sample estimate of a population parameter, such as, a population mean or population proportion. The point estimate is expressed by a single number.

CASE STUDY 13.1

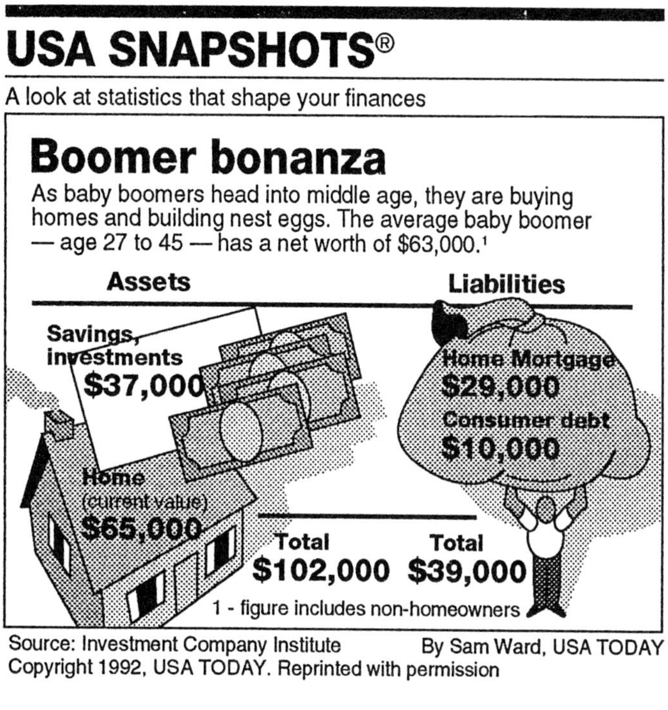

USA SNAPSHOTS®
A look at statistics that shape your finances

Boomer bonanza
As baby boomers head into middle age, they are buying homes and building nest eggs. The average baby boomer — age 27 to 45 — has a net worth of $63,000.[1]

Assets
- Savings, investments $37,000
- Home (current value) $65,000
- **Total $102,000**

Liabilities
- Home Mortgage $29,000
- Consumer debt $10,000
- **Total $39,000**

1 - figure includes non-homeowners

Source: Investment Company Institute   By Sam Ward, USA TODAY
Copyright 1992, USA TODAY. Reprinted with permission

Figure 13.1

Examine Figure 13.1. Notice that the *estimates* for the **average** assets and **average** liabilities for baby boomers are listed. These averages represent *point estimates* of the *true* population assets and liabilities for a baby boomer.

We can estimate that a baby boomer has assets of:
- $37,000 in savings and investments
- $65,000 in home value

or a total average asset of $102,000.

On the other hand, a baby boomer has liabilities of:
- $29,000 for home mortgage
- $10,000 for consumer debt

or a total average liability of $39,000.

Thus, the average baby boomer has a net worth of $63,000, which represents a point estimate of the mean net worth of **all baby boomers**. ∎

CASE STUDY 13.2

Examine the pie chart shown in Figure 13.2. The graph provides an *estimate* of the proportion of U.S. adults who consider themselves: *Republicans*, *Democrats*, or *Independents*, along with those who responded as *Don't Know*.

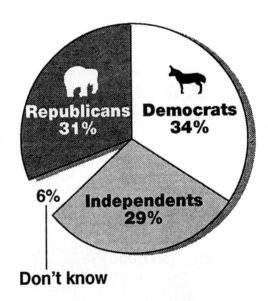

Figure 13.2

According to the information contained within Figure 13.2, these *point estimates* are based on a **sample** of 14,695 adults. Thus, the *point estimates* of the **proportion** of U.S. adults who consider themselves Democrats is 34%, Republicans is 31% and Independents is 29%. ∎

In both Case Studies, a **point estimate** was used to estimate a population parameter. In Case Study 13.1, the population mean was estimated by a sample mean. While in Case Study 13.2, the population proportion was estimated by a sample proportion. A **sample mean**, $\bar{X}$, is one method that is used to obtain a *point estimate* of the **population mean, μ**. While, the **sample proportion**, $\hat{p}$, is a method that is used to obtain a *point estimate* of the **population proportion, p. Both the sample mean and the sample proportion are called *estimators*.** Any sample statistic that is used to estimate a population parameter is called an estimator.

> **Definition: Estimator.** An estimator is a method or formula used in estimating a population parameter.

It is important to distinguish between an *estimate* and an *estimator*. If a sample mean is used to **estimate** the population mean, then the **numerical value obtained** from the sample mean formula is the **estimate** while the **estimator** is the **sample mean** *formula* which was used to compute this point estimate.

Example 13.1

The Student Government Association (SGA) at Nassau Community College is interested in estimating the mean number of hours that all full-time students work during a school week. If the SGA randomly samples 200 students and calculates the sample mean number of hours students work to be 22.4 hours, then determine the:
a) population.
b) population parameter being estimated.
c) estimator used to estimate the population parameter.
d) point estimate for the population parameter.

Solution:

a) The population is all full-time students attending Nassau Community College.
b) The population parameter to be estimated is the **population mean** number of hours that a full-time student works during a school week.
c) The estimator is the *sample mean formula* used to calculate the **mean number of hours** a full-time student works during a school week for the **sample of 200 students**.
d) The point estimate for the population mean number of hours is 22.4 hours. ∎

A question that needs to be addressed at this point is:

How do we know that the sample mean, $\overline{X}$, is a good estimator for the population mean, μ?

The criteria for a *good estimator* are:

- Unbiasedness
- Efficiency
- Consistency
- Sufficiency

We will **only** consider the criteria of *unbiasedness*. An estimator of a population parameter is said to be *unbiased* **if the mean of the point estimates obtained from the independent samples selected from the population will *approach* the true value of the population parameter as more and more samples are selected.**

Based on our discussion in Chapter 8, Theorem 8.1, the sample mean is an unbiased estimator of the population mean because the mean of all the sample means is equal to the mean of the population. It can also be shown that the sample proportion, $\hat{p}$, possesses the four criteria discussed above to be a **good estimator** of the population proportion, p.

13.3 INTERVAL ESTIMATION

In the previous section, we considered the concept of a point estimate. In practice, we will only select **one** sample to compute a point estimate. However, we must realize that the sample we select is **only one** of the many possible samples that could have been selected from the population.

For example, if you were interested in estimating the mean GPA of the population of 26,000 community college students attending Nassau Community College, you would *randomly* select a sample of students and calculate the mean GPA of the sample. Suppose you determine that the mean GPA for the selected sample is 2.63. Although this **point estimate** of 2.63 would represent our **best** guess for the true population mean GPA, you should realize that this estimate of 2.63 is *probably* **not going to be exactly equal to the true population mean!** In fact, if you selected a second random sample of students and calculated the mean GPA of this second sample, you probably would have arrived at a result that is *different* from the first sample mean GPA of 2.63.

Since samples **do vary**, one **major disadvantage** of using one sample mean as a point estimate of the population mean is that we **don't know how close or far** the point estimate is from the *true* population mean. Thus, when using a point estimate to estimate a population parameter, we are not sure of the **extent of the** *error* involved in the estimate.

An **estimate** of a population parameter would be **more useful** if we could provide a **measure of the error associated with the estimate** of a population parameter. In fact many opinion polls contain a measure of the extent of error associated with estimating a population parameter. Case Study 13.3 illustrates the idea of the error associated with a point estimate of a population proportion.

CASE STUDY 13.3

The Opinion Poll conducted by an Independent Poll in Figure 13.3 indicates that 67% is a point estimate for the proportion of the people who say that they feel that politicians in power take advantage of the general population.

*Opinion poll result from a sample size of 983. The margin of error is plus or minus three percentage points. Adapted from USA SNAPSHOTS, USA TODAY 6/17/92 by Marty Baumann. © 1992, USA TODAY. Reprinted with permission.

Figure 13.3

Notice the statement attached to the point estimate of 67% includes a warning that the sample result of 67% in estimating the true population proportion has a **margin of error** *of plus or minus 3 percentage points*. This "margin of error" is due to "sampling error" because not everyone in the population has been polled! The sampling error means that if a second random sample were selected from this population and a point estimate of the population proportion is determined, we would **expect** this second sample proportion to be different from 67%.

Since the proportion of a sample can vary from sample to sample, then the point estimates determined from different samples can yield **different estimates for the population proportion**. Thus, the margin of error of ± 3 percentage points reflects the sampling error in estimating the true population proportion. So, we interpret the opinion poll as stating that the true population proportion lies within the interval of: 67% ± 3%. The interval is estimating that the true population proportion will lie between:

(67% - 3%) to (67% + 3%)
or,
64% to 70%

This **range of percentage values from 64% to 70% is referred to as an interval estimate of the true population proportion**. Notice, an interval estimate uses a point estimate in constructing a range of values within which one can be reasonably sure that the true population parameter will lie. In particular, the interval estimate consists of two components:

1) **67%**: the value of the sample proportion which represents a point estimate of the population proportion.

2) **3 percentage points**: the margin of error which represents the range allowed for the anticipated sampling error. ∎

From Case Study 13.3, we can see that an *interval estimate* is a **technique** that **uses a point estimate** along with an **associated error** to construct an interval to estimate a population parameter.

Definition: Interval Estimate. An interval estimate is an estimate that specifies a *range of values* that the population parameter is likely to fall within.

Going back to the previous example of trying to estimate the population mean GPA of Nassau Community College students using the point estimate of 2.63, if we instead stated our estimate as: *the mean GPA may lie between the values: 2.43 and 2.83*, then the **range of values from 2.43 to 2.83 would represent an interval estimate of Nassau's mean GPA**.

In comparison to a point estimate which uses a single value to estimate the population parameter and is very unlikely to be exactly equal to the population parameter, an **interval estimate uses a range of values** to estimate the population parameter. A *probability* level can be **assigned** to the interval estimate which will *indicate* how *confident* we are that the interval will **include** the population parameter. The probability level that is associated with an interval estimate is referred to as the **confidence level**. This type of interval estimate is called **a confidence interval**. In Section 13.4, we will examine how to construct a confidence interval for estimating the mean of a population.

13.4 INTERVAL ESTIMATION: CONFIDENCE INTERVALS FOR THE POPULATION MEAN

Let's illustrate the procedure to construct an interval estimate of a population mean by considering an example. Suppose we want to determine an interval estimate for the mean GPA of Nassau Community College students.

The **first step** of the procedure in constructing an interval estimate of a population mean is to *select a random sample and calculate the sample mean,* $\overline{X}$. Suppose a random sample of 36 students was selected from the population of Nassau students and the sample mean is determined to be: $\overline{X} = 2.63$. Remember this **sample mean** is **only one of many** *possible* **sample means** that *could have been determined.* That is, this is **one** of the possible sample means within **the Sampling Distribution of the Mean.** Since the sample size, n, is greater than 30 (for this example, n = 36), then according to Theorem 8.2: The Central Limit Theorem, the Sampling Distribution of the Mean is **approximately** normal with:

$$\mu_{\overline{X}} = \mu$$

$$\sigma_{\overline{X}} = \frac{\sigma}{\sqrt{n}}$$

The Sampling Distribution of the Mean is illustrated in Figure 13.4.

Sampling Distribution of the Mean
(is approximately a Normal Distribution)

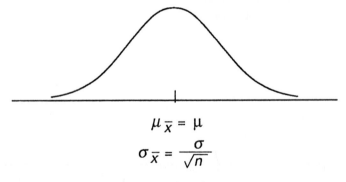

$$\mu_{\overline{X}} = \mu$$
$$\sigma_{\overline{X}} = \frac{\sigma}{\sqrt{n}}$$

Figure 13.4

Within this normal distribution of sample means, we would expect *95% of the sample means,* $\overline{X}$, *to lie within 1.96 standard deviations[1] of the Population Mean.*

Therefore, we expect 95% of the sample means to lie **between the values:**

(Population Mean) − (1.96)(Standard Deviation)
and
(Population Mean) + (1.96)(Standard Deviation)

---

[1] *Please note: The standard deviation used in this statement refers to the standard deviation of the sampling distribution of the mean,* $\sigma_{\overline{X}}$.

This can be expressed as:

$$\mu - (1.96)\sigma_{\bar{x}} \quad \text{and} \quad \mu + (1.96)\sigma_{\bar{x}}$$

Using $\sigma_{\bar{x}} = \dfrac{\sigma}{\sqrt{n}}$, we can then rewrite the values within which we would expect 95% of the sample means to fall between. These values are expressed as:

$$\mu - (1.96)\dfrac{\sigma}{\sqrt{n}} \quad \text{and} \quad \mu + (1.96)\dfrac{\sigma}{\sqrt{n}}$$

These two values are illustrated in Figure 13.5.

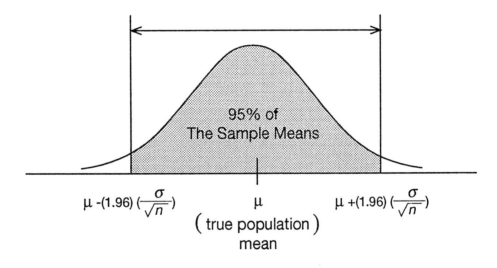

Figure 13.5

If the sample mean, $\bar{X}$, that we select from the population happens to be one of the sample means that falls within the 95% region illustrated in Figure 13.5, and we construct an interval around the sample mean by:

subtracting 1.96 standard deviations *from the sample mean,* $\bar{X}$,

expressed as: $\bar{X} - (1.96)\dfrac{\sigma}{\sqrt{n}}$

and

adding 1.96 standard deviations *to the sample mean,* $\bar{X}$,

expressed as: $\bar{X} + (1.96)\dfrac{\sigma}{\sqrt{n}}$

then the **interval, written as:**

$$\bar{X} - (1.96)\dfrac{\sigma}{\sqrt{n}} \quad \text{to} \quad \bar{X} + (1.96)\dfrac{\sigma}{\sqrt{n}}$$

will contain the *true* population mean, μ. This is illustrated in Figure 13.6.

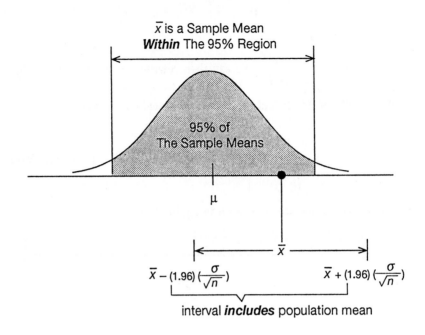

Figure 13.6

However, if the sample mean, $\overline{X}$, **that we select is a sample mean that comes from one of the regions outside** the 95% region illustrated in Figure 13.6, then the interval constructed using the formula:

$$\overline{X} - (1.96)\frac{\sigma}{\sqrt{n}} \quad \text{to} \quad \overline{X} + (1.96)\frac{\sigma}{\sqrt{n}}$$

will not contain the true population mean. This is illustrated in Figure 13.7.

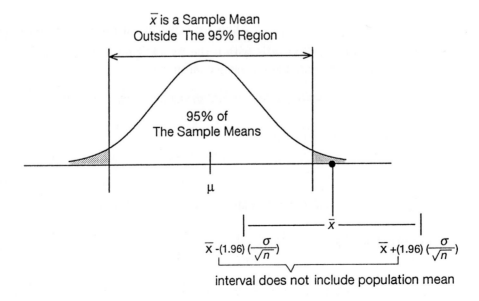

Figure 13.7

Therefore, if all possible samples of size n (for n greater than 30) are selected, and an interval is constructed using the formula:

$$\overline{X} - (1.96)\frac{\sigma}{\sqrt{n}} \text{ to } \overline{X} + (1.96)\frac{\sigma}{\sqrt{n}}$$

for each possible sample mean, then **95% of such intervals are expected to contain the true population mean, μ**. Such an interval is referred to as a **95% Confidence Interval for the Population Mean**, and is defined as follows.

Definition: 95% Confidence Interval for the Population Mean, μ, where the population standard deviation is known.

A 95% Confidence Interval for the population mean, μ, (with known population standard deviation) is constructed by the formula:

$$\overline{X} - (1.96)\frac{\sigma}{\sqrt{n}} \text{ to } \overline{X} + (1.96)\frac{\sigma}{\sqrt{n}}$$

where:
- $\overline{X}$ is the sample mean,
- σ is the population standard deviation,
- n is the sample size greater than 30

Thus, constructing a 95% Confidence Interval to **estimate the population mean GPA of Nassau students**, for which the sample mean, $\overline{X} = 2.63$ for n=36 students, is obtained using the formula:

$$\overline{X} - (1.96)\frac{\sigma}{\sqrt{n}} \text{ to } \overline{X} + (1.96)\frac{\sigma}{\sqrt{n}}$$

If we **assume** that the population standard deviation, σ, is 0.6, then a 95% confidence interval is:

$$\overline{X} - (1.96)\frac{\sigma}{\sqrt{n}} \text{ to } \overline{X} + (1.96)\frac{\sigma}{\sqrt{n}}$$

$$(2.63) - (1.96)\frac{0.6}{\sqrt{36}} \text{ to } 2.63 + (1.96)\frac{0.6}{\sqrt{36}}$$

$$(2.63) - (1.96)\frac{0.6}{6} \text{ to } 2.63 + (1.96)\frac{0.6}{6}$$

$$(2.63) - (1.96)(0.1) \text{ to } 2.63 + (1.96)(0.1)$$

$$2.43 \text{ to } 2.83$$

Therefore, a 95% Confidence Interval is: 2.43 to 2.83. That is, we estimate the mean GPA of all Nassau Community College students to fall within the interval: 2.43 to 2.83.

The interval 2.43 to 2.83 is called the *confidence interval*, and the endpoints of the interval are called the *confidence limits*. The smaller endpoint of the confidence interval is called the *lower confidence limit*, while the larger endpoint of the confidence interval is referred to as the *upper confidence limit*.

> **Definition: Confidence Limits.** The endpoints of a confidence interval are called the confidence limits of the interval.
> The smaller endpoint of the interval is referred to as the **lower confidence limit**. The larger endpoint of the interval is referred to as the **upper confidence limit**.

For this interval, the lower confidence limit is 2.43, while 2.83 represents the upper confidence limit. **The probability value expressed as a percentage which is associated with the confidence interval is called the *confidence level* or *degree of confidence*.**

> **Definition: Confidence Level.** The confidence level of a confidence interval is the probability value, expressed as a percentage, that is associated with an interval estimate. The probability value represents the chance that the procedure used to construct the confidence interval will give an interval that will include the population parameter.
> Thus, the higher the probability value, the greater the confidence that the interval estimate will include the population parameter.

The confidence level for the interval: 2.43 to 2.83 is 95%. This confidence level of 95% indicates that: **in the long run, the percentage of all intervals constructed using this procedure will *probably contain* the population mean 95% of the time**. In fact, if we were to repeat this process of constructing such an interval for all possible samples, we would be *confident that 95% of all possible intervals constructed would include the population mean*.

Because of this, **the interval 2.43 to 2.83 is referred to as a *95% Confidence Interval*. This statement doesn't indicate whether *one* particular interval, like the interval, 2.43 to 2.83, will actually contain the true population parameter**. It indicates that we are **95% confident** that the true population mean is contained within the interval 2.43 to 2.83.

Furthermore, it is interpreted to mean: that **if a 95% confidence interval for the population mean was constructed for each possible sample of size n selected from the population, we would expect that 95% of these confidence intervals to include the population mean and 5% of these confidence intervals not to include the population mean**.

In Figure 13.8, a 95% confidence interval is shown for four different possible samples. Notice from Figure 13.8 that if a sample mean falls within the 95% shaded area (like $\overline{X}_1$ and $\overline{X}_3$) then the confidence interval will contain the population mean. While if the sample mean falls outside this 95% region (like $\overline{X}_2$ and $\overline{X}_4$), then the confidence interval will not contain the population mean. Again, keep in mind that **a 95% confidence interval *doesn't* tell you whether *a particular interval will contain* the population mean**. But rather it means that: **in the long run, the percentage of all intervals constructed using this procedure will *probably contain* the population mean 95% of the time**.

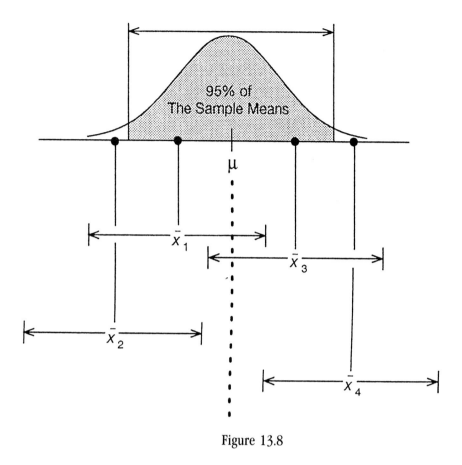

Figure 13.8

Although, any confidence level can be chosen to construct a confidence interval, the most common confidence levels are: 90%, 95% and 99%. The formulas to construct a 90% and 99% Confidence Intervals are similar to a 95% Confidence Interval formula except for the value of the z score which is used in the formula.

In a 90% Confidence Interval, the z score: 1.65 is used to construct the interval estimate, while in a 99% Confidence Interval, the appropriate z score is: 2.58 in constructing this interval estimate. Thus, formulas for a 90% and 99% Confidence Intervals are defined as follows.

Definition: 90% Confidence Interval for the Population Mean, μ, where the population standard deviation is known.

A 90% Confidence Interval for the population mean, μ, (with known population standard deviation) is constructed by the formula:

$$\overline{X} - (1.65)\frac{\sigma}{\sqrt{n}} \text{ to } \overline{X} + (1.65)\frac{\sigma}{\sqrt{n}}$$

where:
$\overline{X}$ is the sample mean
σ is the population standard deviation
n is the sample size greater than 30.

Similarly, we define a *99% Confidence Interval for the Population Mean* to be:

Definition: 99% Confidence Interval for the Population Mean, μ, where the population standard deviation is known.

A 99% Confidence Interval for the population mean, μ, (with known population standard deviation) is constructed by the formula:

$$\overline{X} - (2.58)\frac{\sigma}{\sqrt{n}} \quad \text{to} \quad \overline{X} + (2.58)\frac{\sigma}{\sqrt{n}}$$

where:
  $\overline{X}$ is the sample mean
  σ is the population standard deviation
  n is the sample size greater than 30

Notice that in the formula for a 90% confidence interval, the values, -1.65 and 1.65, represent the appropriate z scores associated with a 90% Confidence Interval, while the z scores, -2.58 and +2.58, are used to construct a 99% Confidence Interval. The z scores, **-1.65 and +1.65**, are used for a **90% confidence interval**, since the Sampling Distribution of the Mean, which is approximately a normal distribution for a sample size greater than 30 and a known population standard deviation, has approximately 90% of the sample means within 1.65 standard deviations of the true population mean. Similarly, **99% of the sample means are expected to lie within 2.58 standard deviations** of the population mean. This is illustrated in Figure 13.9.

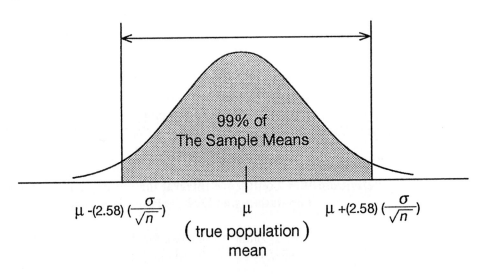

Figure 13.9

Let's calculate a 99% confidence interval for the population mean GPA for Nassau students, using $\overline{X} = 2.63$, n = 36 and $\sigma = 0.6$. From the definition of a 99% confidence interval, we have:

$$\overline{X} - (2.58)\frac{\sigma}{\sqrt{n}} \quad \text{to} \quad \overline{X} + (2.58)\frac{\sigma}{\sqrt{n}}$$

Substituting the sample mean, $\bar{X} = 2.63$, sample size, n = 36 and the standard deviation, $\sigma = 0.6$, we have:

$$(2.63) - (2.58)\frac{0.6}{\sqrt{36}} \text{ to } (2.63) + (2.58)\frac{0.6}{\sqrt{36}}$$

$$(2.63) - (2.58)\frac{0.6}{6} \text{ to } (2.63) + (2.58)\frac{0.6}{6}$$

$$(2.63) - (2.58)(0.1) \text{ to } (2.63) + (2.58)(0.1)$$

$$2.37 \text{ to } 2.89$$

Thus, a 99% confidence interval is 2.37 to 2.89.

That is, we are **99% confident that the mean GPA of all Nassau Community College students falls within the interval:**

2.37 to 2.89

In summary: we have computed both a 95% confidence interval and a 99% confidence interval for the population mean GPA of the Nassau Community College students to be:

95% Confidence Interval: 2.43 to 2.83	Width of Interval is: 2.83 − 2.43 = 0.40
99% Confidence Interval: 2.37 to 2.89	Width of Interval is: 2.89 − 2.37 = 0.52

You should notice that as the **confidence level increases** from 95% to 99%, the **width** of the confidence interval also **increases**.

In general, as the *confidence level increases*, the *width of the confidence interval also increases*.

CONSTRUCTING A CONFIDENCE INTERVAL FOR A POPULATION MEAN: WHEN THE POPULATION STANDARD DEVIATION IS UNKNOWN

In constructing a 90%, 95% or a 99% Confidence Interval for a population mean, it was assumed that the standard deviation of the population was known. In most practical applications, the population standard deviation is rarely known. Therefore, a more realistic approach in constructing an interval estimate of the true population mean is to consider the situation where the **population standard deviation is unknown**.

When the population standard deviation, σ, is unknown, the sample standard deviation, s, is used to estimate the population standard deviation, σ. Under the condition that the population standard deviation is unknown and the population from which the sample is selected is approximately a normal distribution (or bell-shaped), then the shape of the

Sampling Distribution of the Mean is **no longer** a normal distribution. Its shape approximates a *t* **Distribution** under these conditions.

Thus, the *t* **Distribution** rather than the normal distribution is used to construct confidence intervals for the population mean under the following conditions.

Conditions to Use The *t* Distribution When Constructing a Confidence Interval for the Mean of a Population

The *t* **Distribution** is used to construct a confidence interval for the population mean when the following conditions hold:

- The population standard deviation, σ, is unknown, and is estimated by the sample standard deviation, s.

- The population from which the sample is selected is approximately a normal distribution (or bell-shaped).

When the *t* **Distribution** is used to construct a confidence interval for the population mean, then the **formulas** for a 90%, 95% and a 99% confidence interval will include :

- the *sample standard deviation, s*, rather than the population standard deviation, σ, and
- *t scores* rather than z scores.

To determine the **specific** *t* score to use for a confidence interval formula, the **degrees of freedom**, *df*, associated with the *t* **score must be determined using the formula:** *df* = *n* − 1. Knowing the degrees of freedom, *df*, and the confidence level, then the appropriate *t* score for a confidence interval is found using TABLE III: Critical values for the *t* Distribution found in Appendix D.

Substituting s for σ, and replacing the z score by the appropriate *t* score, we can now define a **90%, 95% or a 99% Confidence Interval for the Population Mean,** *where the population standard deviation is unknown*.

Definition: 90% Confidence Interval for the Population Mean, μ , where Population Standard Deviation is Unknown.

A 90% Confidence Interval for the population mean, μ , when the population standard deviation is *unknown*, is constructed by the formula:

$$\overline{X} - (t_{95\%}) \frac{s}{\sqrt{n}} \quad \text{to} \quad \overline{X} + (t_{95\%}) \frac{s}{\sqrt{n}}$$

where:

$\overline{X}$ is the sample mean
n is the sample size
$t_{95\%}$ is the **positive** critical *t* score with *df* = n-1 that corresponds to a **one-tailed test at** α=5%
s is the sample standard deviation.

Definition: 95% Confidence Interval for the Population Mean, μ, where Population Standard Deviation is Unknown.

A 95% Confidence Interval for the population mean, μ, when the population standard deviation is *unknown*, is constructed by the formula:

$$\overline{X} - (t_{97.5\%}) \frac{s}{\sqrt{n}} \text{ to } \overline{X} + (t_{97.5\%}) \frac{s}{\sqrt{n}}$$

where:
- $\overline{X}$ is the sample mean
- n is the sample size
- $t_{97.5\%}$ is the **positive** critical t score with df = n-1 that corresponds to a **one-tailed test at** α=5%
- s is the sample standard deviation.

Definition: 99% Confidence Interval for the Population Mean, μ, where Population Standard Deviation is Unknown.

A 99% Confidence Interval for the population mean, μ, when the population standard deviation is *unknown*, is constructed by the formula:

$$\overline{X} - (t_{99.5\%}) \frac{s}{\sqrt{n}} \text{ to } \overline{X} + (t_{99.5\%}) \frac{s}{\sqrt{n}}$$

where:
- $\overline{X}$ is the sample mean
- n is the sample size
- $t_{99.5\%}$ is the **positive** critical t score with df = n-1 that corresponds to a **one-tailed test at** α=5%
- s is the sample standard deviation.

Notice that the t score for a 90% confidence interval is found in TABLE III under the column listed $t_{95\%}$, while the appropriate t score for a 95% confidence interval is found under the column heading $t_{97.5\%}$, and the t score for a 99% confidence interval is found under the column for $t_{99.5\%}$. The subscript for each of these t scores refers to the area under the t distribution that is to the LEFT of each positive t score. This is illustrated in Figure 13.10.

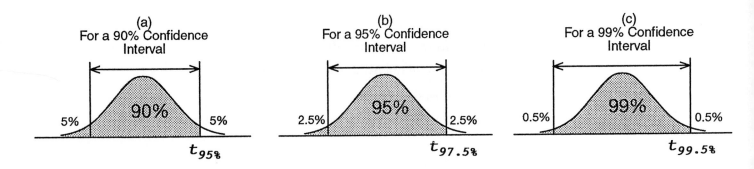

Figure 13.10

In Figure 13.10a, for a 90% confidence interval, the area to the left of the positive *t* score is 95%, therefore, this *t* score is referred to as: $t_{95\%}$. From Figure 13.10b, for a 95% confidence interval, the area to the left of the positive *t* score is 97.5%, therefore, this *t* score is referred to as: $t_{97.5\%}$. Examining Figure 13.10c, for a 99% confidence interval, the area to the left of the positive *t* score is 99.5%, therefore, this *t* score is referred to as: $t_{99.5\%}$.

Example 13.2

A random sample of 16 students was selected from the student body at a local college to determine the average amount of money that students carry with them.

If the sample mean, $\overline{X}$, was equal to $28, with a sample standard deviation of s = 6, then construct a:
a) 95% Confidence Interval for estimating the mean amount of money all students carry with them.
b) 99% Confidence Interval for estimating the mean amount of money all students carry with them.

Solution:

a) Since the population standard deviation is **unknown** and is estimated by s, then the formula to compute a 95% Confidence Interval for the population mean is:

$$\overline{X} - (t_{97.5\%}) \frac{s}{\sqrt{n}} \quad \text{to} \quad \overline{X} + (t_{97.5\%}) \frac{s}{\sqrt{n}}$$

To determine a 95% Confidence Interval, we need to determine the appropriate *t* score by finding the degrees of freedom using the formula: df = n - 1. For a sample size of 16, we have: df = 16 - 1 = 15. Using TABLE III, the appropriate *t* score is found under the column $t_{97.5\%}$ and along the row for 15 degrees of freedom. Thus, the t score, $t_{97.5\%}$, for 15 degrees of freedom is: 2.13. Using the following values:

$$\overline{X} = 28, \quad t_{97.5\%} \text{ (with } df = 15\text{)} = 2.13, \quad s = 6 \text{ and } n = 16$$

in the formula for a 95% Confidence Interval, we have:

$$(28) - (2.13)\frac{6}{\sqrt{16}} \text{ to } 28 + (2.13)\frac{6}{\sqrt{16}}$$

$$(28) - (2.13)\frac{6}{4} \text{ to } 28 + (2.13)\frac{6}{4}$$

$$(28) - (2.13)(1.50) \text{ to } 28 + (2.13)(1.50)$$

$$(28) - (3.20) \text{ to } 28 + (3.20)$$

$$24.81 \text{ to } 31.20$$

Thus, a 95% Confidence Interval for the mean amount of money all students carry with them is: $24.81 to $31.20.

b) To construct a 99% Confidence Interval for the population mean, where the population standard deviation is **unknown**, use the formula:

$$\overline{X} - (t_{99.5\%})\frac{s}{\sqrt{n}} \text{ to } \overline{X} + (t_{99.5\%})\frac{s}{\sqrt{n}}$$

To determine a 99% Confidence Interval, we need to determine the appropriate *t* score by finding the degrees of freedom using the formula: $df = n - 1$. For a sample size of 16, we have: $df = 16 - 1 = 15$. Using TABLE III, the appropriate *t* score is found under the column $t_{99\%}$ and along the row for 15 degrees of freedom. Thus the t score, $t_{99\%}$, for 15 degrees of freedom is: 2.95. Using the following values:

$$\overline{X} = 28, \quad t_{99\%} \text{ (with } df = 15) = 2.95, \text{ s} = 6 \text{ and n} = 16$$

in the formula for a 99% Confidence Interval, we have:

$$(28) - (2.95)\frac{6}{\sqrt{16}} \text{ to } 28 + (2.95)\frac{6}{\sqrt{16}}$$

$$(28) - (2.95)\frac{6}{4} \text{ to } 28 + (2.95)\frac{6}{4}$$

$$(28) - (2.95)(1.50) \text{ to } 28 + (2.95)(1.50)$$

$$(28) - (4.43) \text{ to } 28 + (4.43)$$

$$\$23.58 \text{ to } \$32.43$$

Thus, a 99% Confidence Interval for the mean amount of money all students carry with them is: $23.58 to $32.43. ∎

Example 13.3

A professor in the English department would like to estimate the average number of typing errors per page in term papers for all students enrolled in liberal arts English courses. For a random sample of 36 term papers, the professor found the mean number of typing errors per page was 4.6 with s = 1.4. Find the 90% Confidence Interval for the mean number of typing errors per page for all students enrolled in liberal arts English courses.

Solution:

Since the population standard deviation is **unknown** and is estimated by s, then the formula to compute a 90% Confidence Interval for the population mean is:

$$\overline{X} - (t_{95\%}) \frac{s}{\sqrt{n}} \text{ to } \overline{X} + (t_{95\%}) \frac{s}{\sqrt{n}}$$

To determine a 90% Confidence Interval, we need to determine the appropriate *t* score by finding the degrees of freedom using the formula: *df* = n - 1. For a sample size of 36, we have: *df* = 36 - 1 = 35. Using TABLE III, the appropriate *t* score is found under the column $t_{95\%}$.

Since *df* = 35 does **not** appear in the table and is exactly halfway between *df* = 30 and *df* = 40, we will choose the *t* score value that is farthest from zero. Thus in this case we will choose the *t* score for 30 degrees of freedom. Thus, the t score, $t_{95\%}$, for 30 degrees of freedom is: 1.70. Using the following values:

$$\overline{X} = 4.6, \quad t_{95\%} \text{ (with } df = 30\text{)} = 1.70, \text{ s} = 1.4 \text{ and n} = 36$$

in the formula for a 90% Confidence Interval, we have:

$$(4.6) - (1.70)\frac{1.4}{\sqrt{36}} \text{ to } 4.6 + (1.70)\frac{1.4}{\sqrt{36}}$$

$$(4.6) - (1.70)\frac{1.4}{6} \text{ to } 4.6 + (1.70)\frac{1.4}{6}$$

$$(4.6) - (1.70)(0.23) \text{ to } 4.6 + (1.70)(0.23)$$

$$(4.6) - (0.39) \text{ to } 4.6 + (0.39)$$

$$4.21 \text{ to } 4.99$$

Thus, a 90% Confidence Interval for the mean number of typing errors per page for all students enrolled in liberal arts English courses is: 4.21 to 4.99. ∎

❄ 13.5 INTERVAL ESTIMATION: CONFIDENCE INTERVALS FOR THE POPULATION PROPORTION

Very often, you see the results of an opinion poll stated in a newspaper or a magazine. Polls are conducted to determine the proportion or percentage of people who favor a particular issue like gun control, the proportion of people who favor a certain candidate, the percentage of people who prefer a particular soft drink, the proportion of people who don't drink and drive. In each instance, the purpose of the opinion poll is to estimate the proportion or percentage of a population with a particular characteristic or opinion.

For example, suppose an opinion poll is conducted to determine the proportion of **all American adults** who disapprove of the way Congress is handling its job. If an opinion poll of 1500 American adults indicates that 1170 adults disapprove of the way Congress is doing its job, then how can we use this sample information to estimate the proportion of all American adults who disapprove of the way Congress is doing its job? An estimate of the proportion of all American adults disapproving of the type of job Congress is doing can be calculated using the sample proportion, $\hat{p}$. This sample proportion of people is determined by the formula:

$$\hat{p} = \frac{\text{number of adults disapproving of the way Congress is doing its job}}{\text{total number of adults polled}}$$

Thus,

$$\hat{p} = \frac{1170}{1500}$$

or

$$\hat{p} = 0.78 \text{ or } 78\%$$

Since this opinion poll has determined that 78% of the american adults polled disapprove of the way Congress is handling its job, then we can use this sample proportion as a *point estimate* of the proportion of **all** american adults who disapprove of the way Congress is handling its job. Thus, the question becomes: is it reasonable to conclude from this sample proportion, $\hat{p}$, that 78% of **all american adults** disapprove of the way Congress is doing its job? That is, how **confident** can one be that this poll of 1500 american adults reflects the true feelings of the **entire population of american adults** or the **population proportion, p**?

Suppose a second opinion poll was conducted for another 1500 adults, would one expect the **same** percentage of american adults to disapprove of the way Congress is doing its job? **Probably, not!** In fact, if more and more samples were selected, the sample proportion, $\hat{p}$, would vary from sample to sample. Some of these sample proportions will be smaller than the population proportion, p, and some will be larger. Therefore, this sample proportion, $\hat{p}$, which serves as a point estimate of the population proportion, p, **does not** indicate how accurate this estimate is.

Thus, in order to provide a level of confidence in our estimate of the population proportion, we need to construct an interval estimate of the population proportion, p. To construct an interval estimate of the population proportion, we need to examine the distribution of all possible sample proportions. This distribution is referred to as the Sampling Distribution of the Proportion.

In Chapter 12, the characteristics of the Sampling Distribution of the Proportion were discussed. Within Chapter 12, Theorem 12.1 stated the **Characteristics of the Sampling Distribution of the Proportion, p,** as:

- the Sampling Distribution of The Proportion is **approximately normal** when n is **large**, that is, when **both** np and $n(1 - p)$ are greater than 5.
- the mean of the Sampling Distribution of The Proportion, denoted $\mu_{\hat{p}}$, is:
$$\mu_{\hat{p}} = p$$
- the standard deviation of the Sampling Distribution of The Proportion, denoted $\sigma_{\hat{p}}$, is:
$$\sigma_{\hat{p}} = \sqrt{\frac{p(1-p)}{n}}$$

In the same way that the sample mean was used to construct a confidence interval for the population mean, the sample proportion, $\hat{p}$, will be used to construct 90%, 95% and a 99% confidence intervals for the population proportion, p.

When constructing an interval estimate of the population proportion, the actual value of the population proportion, p, is **not known**. Thus, the standard deviation of the Sampling Distribution of the Proportion, $\sigma_{\hat{p}}$, cannot be calculated since the value of p used in the formula for $\sigma_{\hat{p}}$ is **not** known. The best estimate that we have for the population proportion, p, is the sample proportion, $\hat{p}$. Therefore, in constructing a confidence interval for the population proportion, p, we will use the sample proportion, $\hat{p}$, as an estimate of the population proportion, p, in the formula for $\sigma_{\hat{p}}$.

When we replace the population proportion, p, by the sample proportion, $\hat{p}$, in the formula for the standard deviation of the Sampling Distribution of the Proportion, $\sigma_{\hat{p}}$, then we have an **estimate of the standard deviation of the Sampling Distribution of the Proportion** and it is denoted by the symbol: $s_{\hat{p}}$.

Definition: The Estimate of Standard Deviation of the Sampling Distribution of The Proportion, $s_{\hat{p}}$.

The value of $s_{\hat{p}}$, which is an estimate of the standard deviation of the Sampling Distribution of the Proportion, $\sigma_{\hat{p}}$, is calculated by the formula:

$$s_{\hat{p}} = \sqrt{\frac{\hat{p}(1-\hat{p})}{n}}$$

where:

$\hat{p}$ is the sample proportion.

n is the sample size

Example 13.4

Determine an estimate of the standard deviation of the Sampling Distribution of the Proportion, $\sigma_{\hat{p}}$, using the formula for $s_{\hat{p}}$, if a random sample of 49 data values has a sample proportion of: $\hat{p} = 0.4$.

Solution:

An estimate of $\sigma_{\hat{p}}$ is found using the formula for $s_{\hat{p}}$, which is:

$$s_{\hat{p}} = \sqrt{\frac{\hat{p}(1-\hat{p})}{n}}$$

Substituting the value of the sample proportion: $\hat{p} = 0.4$ and the sample size: $n = 49$ into the formula for $s_{\hat{p}}$, we have:

$$s_{\hat{p}} = \sqrt{\frac{(0.4)(0.6)}{49}}$$

$$s_{\hat{p}} = \sqrt{0.0049}$$

$$s_{\hat{p}} = 0.07 \blacksquare$$

Under the assumption that the Sampling Distribution of the Population Proportion, p, is approximately a normal distribution, $\hat{p}$ serves as a *point estimate of the population proportion*, and $s_{\hat{p}}$ serves as an estimate of the standard deviation of the Sampling Distribution of the Proportion, then we can use the reasoning that was developed in constructing confidence intervals for population means to construct confidence intervals for the population proportion. Thus, we can define the *90%, 95% and 99% confidence intervals for the population proportion, p,* as follows.

Definition: 90% Confidence Interval for the Population Proportion, p, where n is large.

A 90% Confidence Interval for the Population Proportion, p, is constructed by the formula:

$$\hat{p} - (1.65)\, s_{\hat{p}} \quad \text{to} \quad \hat{p} + (1.65)\, s_{\hat{p}}$$

where:

$\hat{p}$ is the point estimate of the population proportion,

$s_{\hat{p}}$ is the estimate of the standard deviation of the sampling distribution of the proportion

---

> **Definition: 95% Confidence Interval for the Population Proportion, p, where n is large.**
> A 95% Confidence Interval for the Population Proportion, p, is constructed by the formula:
>
> $$\hat{p} - (1.96)\, s_{\hat{p}} \quad \text{to} \quad \hat{p} + (1.96)\, s_{\hat{p}}$$
>
> where:
>
> $\hat{p}$ is the point estimate of the population proportion.
>
> $s_{\hat{p}}$ is the estimate of the standard deviation of the sampling distribution of the proportion.

> **Definition: 99% Confidence Interval for the Population Proportion, p, where n is large.**
> A 99% Confidence Interval for the Population Proportion, p, is constructed by the formula:
>
> $$\hat{p} - (2.58)\, s_{\hat{p}} \quad \text{to} \quad \hat{p} + (2.58)\, s_{\hat{p}}$$
>
> where:
>
> $\hat{p}$ is the point estimate of the population proportion.
>
> $s_{\hat{p}}$ is the estimate of the standard deviation of the sampling distribution of the proportion.

Example 13.5

A sociologist reports that from a study of 1000 respondents, 65% of the respondents believe that the best place to meet a "suitable" companion is by taking an evening class. (source: Bruskin-Goldring Research Poll 2-13-92 USA TODAY)

Find a:
a) 95% confidence interval for the population proportion, p.
b) 99% confidence interval for the population proportion, p.

Solution:

a) To find a confidence interval for a population proportion, p, we must first calculate $s_{\hat{p}}$ using the formula:

$$s_{\hat{p}} = \sqrt{\frac{\hat{p}(1-\hat{p})}{n}}$$

Since the point estimate of the sample proportion, $\hat{p} = 0.65$, and n = 1000, then

$$s_{\hat{p}} = \sqrt{\frac{(0.65)(0.35)}{1000}}$$

$$s_{\hat{p}} = 0.015$$

A 95% Confidence Interval for the population proportion p is found by:

$$\hat{p} - (1.96)\, s_{\hat{p}} \quad \text{to} \quad \hat{p} + (1.96)\, s_{\hat{p}}$$

Substituting the values for $\hat{p}$ and $s_{\hat{p}}$ into the formula, we have:

$$(0.65) - (1.96)(0.015) \quad \text{to} \quad (0.65) + (1.96)(0.015)$$

$$0.620 \quad \text{to} \quad 0.679$$

Therefore, a 95% confidence interval for the population proportion, p, that believe the best place to meet a "suitable" companion is by taking an evening class is:

$$62\% \text{ to } 67.9\%$$

That is, we are 95% confident that the *true* population proportion, p, is between 62% and 68%.

b) A 99% Confidence Interval for the population proportion, p, is found by:

$$\hat{p} - (2.58)\, s_{\hat{p}} \quad \text{to} \quad \hat{p} + (2.58)\, s_{\hat{p}}$$

Substituting the values for $\hat{p}$ and $s_{\hat{p}}$ into the formula, we have:

$$(0.65) - (2.58)(0.015) \quad \text{to} \quad (0.65) + (2.58)(0.015)$$

$$0.611 \quad \text{to} \quad 0.689$$

Therefore, a 99% confidence interval for the population proportion, p, that believe the best place to meet a "suitable" companion is by taking an evening class is:

$$61.1\% \text{ to } 68.9\%$$

That is, we are 99% confident that the *true* population proportion, p, is between 61.1% and 68.9%. ∎

13.6 DETERMINING SAMPLE SIZE AND THE MARGIN OF ERROR

In the previous two sections, we developed the formulas to construct a confidence interval for a population mean and a population proportion. In those sections, we arbitrarily chose a sample size. In practice, it is important to determine an **appropriate sample size** for a desired confidence level. Selecting a sample size larger than necessary will involve a greater cost. Thus, "how does one decide on how large a sample size, n, is required?"

One factor that affects the sample size, n, is the *level of precision* that is required. If a high level of precision (that is, a better estimate or more accuracy) is required for the confidence interval, then the sample size must be increased since a larger sample will decrease the sampling error. Therefore, the more precision you need for a confidence interval, the larger the sample size you will be required to take.

Let's examine some procedures that are used to determine the sample size for a specified precision level and confidence level. First, we will consider the procedure for determining the sample size for estimating a population mean.

SAMPLE SIZE FOR ESTIMATING A POPULATION MEAN, µ

A confidence interval for the population mean, µ, where the population standard deviation, σ, is known has the form:

$$\overline{X} \pm (z\text{ score}) \left(\frac{\sigma}{\sqrt{n}} \right)$$

Thus, the confidence interval for a specific level of confidence has two components:

1) $\overline{X}$: which represents a point estimate of the population mean.

2) $(z\text{ score})\left(\frac{\sigma}{\sqrt{n}}\right)$: which represents the **margin of error**.

This **margin of error** is the maximum possible error of estimate for the population mean using the sample mean, $\overline{X}$. We will denote the margin of error by the capital letter E.

Definition: Margin of Error for Estimating the Population Mean, where the Population Standard Deviation is Known.

The margin of error, denoted by E, for the population mean is the maximum possible error of estimate for the population mean, µ, using the sample mean, $\overline{X}$. The formula to determine this margin of error is:

$$E = (z)\left(\frac{\sigma}{\sqrt{n}}\right)$$

where:
    E = margin of error
    z = z score corresponding to the level of confidence
    σ = the population standard deviation
    n = the sample size which is greater than 30

The **margin of error, E, represents one-half the width of the confidence interval** as illustrated in Figure 13.11.

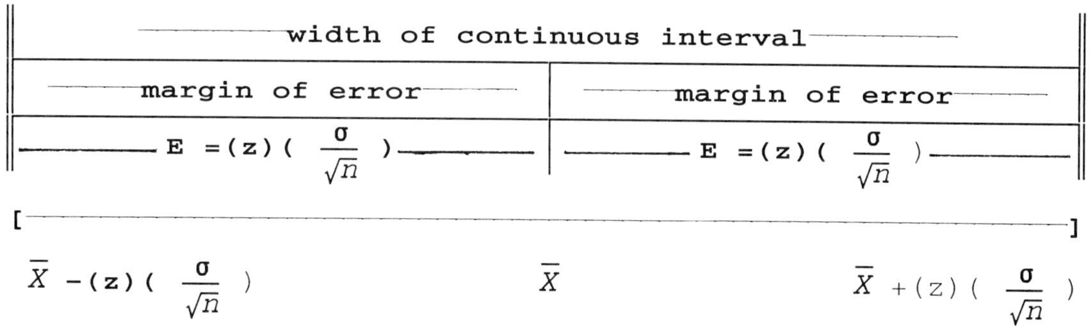

```
|——————————width of continuous interval——————————|
|————margin of error————|————margin of error————|
|————E = (z)(σ/√n)——————|————E = (z)(σ/√n)——————|
[                                                 ]
X̄ − (z)(σ/√n)           X̄              X̄ + (z)(σ/√n)
```

Figure 13.11

For a fixed sample size and population standard deviation, as the level of confidence is **increased**, the margin of error also **increases**.

This is illustrated in Table 13.1.

Table 13.1

Level of Confidence	z score	Margin of Error
90%	1.65	$(1.65)(\frac{\sigma}{\sqrt{n}})$
95%	1.96	$(1.96)(\frac{\sigma}{\sqrt{n}})$
99%	2.58	$(2.58)(\frac{\sigma}{\sqrt{n}})$

The size of the margin of error determines the precision or accuracy of the estimate of the population mean. **A precise estimate has a small margin of error.** We would like our estimate of the population mean to be precise and also to have a high level of confidence. However, as shown in Table 13.1, as the confidence level increases from 90% to 99%, the margin of error increases which causes a decrease in the precision level.

In examining the margin of error formula, you should notice that there are three quantities within the formula. These quantities are the z score, the population standard deviation, and the sample size. All three have an effect on the margin of error, but, since the population standard deviation represents the variability of **all** the data values, it is a fixed value over which we have no control. On the other hand, the z score and the sample size are quantities over which we have control.

Our objective will be to reduce the margin of error without reducing the confidence level, so that the precision of the estimate of the population mean can be improved. A simple way to reduce the margin of error, if we hold the confidence level fixed, is to increase the sample size. This is true because the sample size is in the denominator of the margin of error formula and so as the sample size increases the margin of error will decrease. This concept is illustrated in Example 13.6.

Example 13.6

During the past year, the flight times between two cities were reported to have a population standard deviation of 20 minutes. Determine the margin of error for estimating the mean flight time between the two cities for a confidence level of 95% for each of the following sample sizes:
a) 36
b) 64
c) 100
d) Compare the results of the previous three parts and state what happens to the margin of error as the sample size is increased.

Solution:

To calculate the margin of error, we use the formula:

$$E = (z)\left(\frac{\sigma}{\sqrt{n}}\right)$$

Since the population standard deviation is given to be 20 minutes, then $\sigma = 20$. For a 95% level of confidence, the z score is: 1.96. Substituting these values into the margin of error formula, we have:

$$E = (1.96)\left(\frac{20}{\sqrt{n}}\right)$$

a) For a sample size of 36, the margin of error is:

$$E = (1.96)\left(\frac{20}{\sqrt{36}}\right)$$

$$E \approx (1.96)(3.33)$$

$$E = 6.53 \text{ minutes}$$

b) For a sample size of 64, the margin of error is:

$$E = (1.96)\left(\frac{20}{\sqrt{64}}\right)$$

$$E = (1.96)(2.5)$$

$$E = 4.9 \text{ minutes}$$

c) For a sample size of 100, the margin of error is:

$$E = (1.96)\left(\frac{20}{\sqrt{100}}\right)$$

$$E = (1.96)(2)$$

$$E = 3.92 \text{ minutes}$$

d) Notice from the previous parts, as the sample size increased from 36 to 100, the margin of error, E, decreased from 6.53 minutes to 3.92 minutes. Thus, *a smaller margin of error can be achieved by increasing the sample size.* ∎

As illustrated in Example 13.6, **selecting a larger sample size for a fixed confidence level will reduce the margin of error**. *As the margin of error is decreased, the width of the confidence interval will also be reduced.* This will result in an increase in the precision or accuracy of estimating the population mean.

Since the margin of error and the size of a sample are related, we can develop a formula that determines the required sample size, n, for a predetermined margin of error (or precision requirement) and confidence level. Using the margin of error formula and solving for n, we can obtain a formula that will determine the sample size for a given margin of error.

The Formula to Determine the Sample Size for a Given Margin of Error in Estimating The Population Mean, μ

For a given confidence level, a known population standard deviation, and a given margin of error, the formula to determine the sample size, n, to satisfy these requirements in estimating the population mean is:

$$n = \left[\frac{z\sigma}{E}\right]^2$$

If the resulting value of n is not a whole number, the next largest whole number should be taken for the required sample size.

where:

n = the required sample size for the specified conditions
E = predetermined margin of error
z = the z score corresponding to the level of confidence
σ = the population standard deviation

Example 13.7

A light-bulb manufacturer wants to determine how large a random sample that the quality control department should take to be 99% confident that the sample mean will be within 25 hours of estimating the population mean.

Determine the sample size required to satisfy these requirements if we assume that the population standard deviation is 250 hours.

Solution:

To determine the sample size, n, for the given conditions, we need to use the formula:

$$n = \left[\frac{z\sigma}{E}\right]^2$$

For a confidence level of 99%, the z score is 2.58. The population standard deviation is given to be 250 hours, therefore, $\sigma = 250$. Since the quality control department wants to be within 25 hours of the population mean, then the margin of error is 25 hours. Substituting this information into the sample size formula, we have:

$$n = \left[\frac{(2.58)(250)}{25}\right]^2$$

$$n = 665.64$$

Thus, the sample size required to meet the requirements of the quality control department is 666 (rounded off to the next largest whole number). ■

If the population standard deviation, σ, is unknown, then the conventional procedure is to take a preliminary sample (of any size greater than 30) and to find the sample standard deviation, s. This value of s is used to replace the value of σ in the formula for the sample size. However, it should be noted that when s is used to estimate the value of σ, then the sample size determined using s may yield a margin of error that can be either smaller or larger than the predetermined margin of error. Why?

SAMPLE SIZE FOR ESTIMATING A POPULATION PROPORTION, p

A confidence interval for the population proportion, p, where the sample size is large, and the population proportion, p, is unknown, and is estimated by the sample proportion, $\hat{p}$, has the form:

$$\hat{p} \pm (z \text{ score})\left(\sqrt{\frac{\hat{p}(1-\hat{p})}{n}}\right)$$

Thus, the confidence interval for a specific level of confidence has two components:

1) $\hat{p}$: which represents a point estimate of the population proportion

2) $(z \text{ score})\left(\sqrt{\frac{\hat{p}(1-\hat{p})}{n}}\right)$: which represents the **margin of error**

To determine this **margin of error**, we need to know the value of the sample proportion, $\hat{p}$, and the z score corresponding to the confidence level. This margin of error is the maximum possible error of estimate for the population proportion using the sample proportion, $\hat{p}$.

We will denote the margin of error by the capital letter E. Opinion Polls usually express the point estimate, the sample proportion, of the poll as a percentage and the margin of error of the poll as percentage points. Care must be taken to distinguish between percentages and percentage points. *Percentage refers to a proportion value multiplied by 100%, while percentage points represent units on the percent scale.* The margin of error for opinion polls is always stated as a certain number of percentage points, and not as a percentage.

Definition: Margin of Error for Estimating the Population Proportion for a Large Sample.
The margin of error, denoted by E, for the population proportion is the maximum possible error of estimate for the population proportion, p, using the sample proportion, $\hat{p}$. The formula to determine the margin of error is:

$$E = (z)\left(\sqrt{\frac{\hat{p}(1-\hat{p})}{n}}\right)$$

where:
  E = margin of error
  z = z score corresponding to the level of confidence
  $\hat{p}$ = a sample proportion
  n = a large sample size

This **margin of error, E, represents one-half the width of the confidence interval** as illustrated in Figure 13.12.

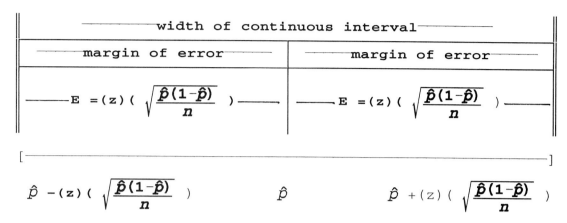

Figure 13.12

Example 13.8

The Nielsen television ratings used a random sample of 2,000 homes to estimate with 95% confidence that 45% of the proportion of all homes watched a football game on Monday Night Football. Determine the margin of error for the Nielsen's estimate of the proportion of all homes watching this football game on Monday Night Football.

Solution:

To calculate the margin of error, we use the formula:

$$E = (z)\sqrt{\frac{\hat{p}(1-\hat{p})}{n}}$$

For a 95% level of confidence, the z score is: 1.96. Since a random sample of 2,000 homes was selected, then n = 2,000. Given that the sample proportion of homes watching the football game is 45%, then $\hat{p} = 0.45$, and $1 - \hat{p} = 1 - 0.45 = 0.55$. Substituting these values into the margin of error formula, we have:

$$E = (1.96)\left(\sqrt{\frac{(0.45)(0.55)}{2,000}}\right)$$

$$E = (1.96)\left(\sqrt{\frac{(0.2475)}{2,000}}\right)$$

$$E = (1.96)(\sqrt{0.0001238})$$

$$E \approx (1.96)(0.0111)$$

$$E \approx 0.0218$$

Thus, the margin of error is approximately 0.0218. This can also be expressed as approximately **2.18 percentage points**. ∎

Case Study 13.4

The following New York Times article entitled *Method Used in Taking Survey in Phone Interview* describes the technique used in selecting the sample for a New York Times/CBS News Public Opinion Poll.

Method Used in Taking Survey in Phone Interview

The latest New York Times/CBS News Poll is based on telephone interviews conducted from last Wednesday through last Sunday with 1,536 adult men and women around the United States. Of this total, 542 said that they were Democrats, 495 said they were Republicans and 499 said they were independents.

The sample of telephone exchanges called was selected by a computer from a complete list of exchanges in the country. The exchanges were chosen in a way that would insure that each region of the country was represented in proportion to its population. For each exchange, the telephone numbers were formed by random digits, thus permitting access to both listed and unlisted residential numbers.

The results have been weighted to take account of household size and to adjust for variations in the sample relating to region, race, sex, age and education.

Higher Rate for Republicans

Republicans were sampled at a higher rate than others to insure a sufficiently large number from that party. The groups of voters, by party identification, were then weighted to reflect their proper proportion in the voting population.

In theory, one can say with 95 percent certainty that the results based on the entire sample differ by no more than 3 percentage points in either direction from what would have been obtained by interviewing all adult Americans. The error for smaller subgroups is larger, depending on the number of sample cases in the subgroup.

The theoretical errors do not take into account a margin of additional error resulting from the various practical difficulties in taking any survey of public opinion.

Figure CS13.1.

---

a) How many people were selected for this poll?
b) Name some procedures that the pollsters did when selecting the sample to insure that it was representative of the population?
c) Explain what is meant by the statement: "in theory, one can say with 95% certainty that the results based on the entire sample differ by no more than 3 percentage points in either direction from what would have been obtained by interviewing all adult americans"? What confidence interval are they referring to within this statement? What is the margin of error for this poll? What is the population of people for this poll?
d) What additional errors are the pollsters referring to in the statement: "the theoretical errors do not take into account a margin of additional errors resulting from the various practical difficulties in taking any survey of public opinion"? ■

As shown earlier in this section, when estimating the mean of a population, **the margin of error decreases as the sample size increases**. This also holds true when estimating a population proportion. *As the margin of error is decreased, then the precision or accuracy of estimating the population proportion will increase.*

Since the margin of error and the size of a sample are related, we can develop a formula that determines the required sample size, n, for a predetermined margin of error (or precision requirement) and confidence level. Using the margin of error formula and solving for n, we can obtain a formula that will determine the sample size for a given margin of error when estimating a population proportion, p. When the margin of error formula is solved for n, we get:

$$n = \frac{z^2(\hat{p})(1-\hat{p})}{E^2}$$

The Formula to Determine the Sample Size for a Given Margin of Error in Estimating The Population Proportion, p

For a given confidence level, a point estimate of the population proportion, $\hat{p}$, and a given margin of error, the formula to determine the sample size, n, to satisfy these requirements in estimating the population proportion is:

$$n = \frac{z^2(\hat{p})(1-\hat{p})}{E^2}$$

If the resulting value of n is not a whole number, the next largest whole number should be taken for the required sample size.

where:
    n = the required sample size for the specified conditions
    E = predetermined margin of error
    z = the z score corresponding to the level of confidence
    $\hat{p}$ = a point estimate of the population proportion

Thus, to determine the sample size, n, we need to find the values for z, E and $\hat{p}$. The value for z is the z score that corresponds to the confidence level used in estimating the population proportion, while the value for E is the **predetermined margin of error** for estimating the population proportion. However, the value of $\hat{p}$, the sample proportion, is *unknown* since we would need to select a sample to determine $\hat{p}$. Thus, our problem is: the formula for the sample size requires that we have a value for $\hat{p}$, but in order for us to determine this value for $\hat{p}$, we need to decide upon how large a sample to select which is the purpose of using this sample size formula.

Consequently, we don't have a value for $\hat{p}$, and thus, we cannot use the formula for the sample size, n.

We can alleviate this problem in one of two ways:
1) *take a pilot sample (of any sample size that is sufficiently large) and calculate $\hat{p}$ for this sample, or use the value of $\hat{p}$ from a previous sample. This value of $\hat{p}$ is used to find the sample size, n.*
2) *determine the maximum possible product of $\hat{p}$ and $1 - \hat{p}$, and simply replace $\hat{p}$ and $1 - \hat{p}$ with values that produce the maximum possible product of $\hat{p}$ and $1 - \hat{p}$. This procedure is referred to as the* **conservative estimate of the sample size**.

To help find this conservative estimate of the sample size, we need to determine the maximum product of $\hat{p}$ and $1 - \hat{p}$. Table 13.2 displays the product of $\hat{p}$ and $1 - \hat{p}$ for values of $\hat{p}$ from 0.1 to 0.9.

Table 13.2

Table of Values of $\hat{p}$, $(1-\hat{p})$, and Their Product

$\hat{p}$	$(1-\hat{p})$	$(\hat{p})(1-\hat{p})$
0.1	0.9	0.09
0.2	0.8	0.16
0.3	0.7	0.21
0.4	0.6	0.24
0.5	0.5	0.25
0.6	0.4	0.24
0.7	0.3	0.21
0.8	0.2	0.16
0.9	0.1	0.09

From Table 13.2, you should notice that the maximum value of the product of $\hat{p}$ and $1-\hat{p}$ occurs when $\hat{p}$ is equal to 0.5. Thus, the maximum possible product of $\hat{p}$ and $1-\hat{p}$ will be 0.25. Therefore, $(\hat{p})(1-\hat{p})$ will always be less than $(0.5)(0.5) = 0.25$. Substituting 0.25 for the product: $(\hat{p})(1-\hat{p})$ into the formula for the sample size, n, we obtain:

$$n = \frac{z^2(0.25)}{E^2}$$

This formula for the sample size, n, based upon using 0.5 for the value of $\hat{p}$ will give a **conservative estimate of the sample size**. This is referred to as a conservative estimate because it gives the **maximum sample size** regardless of the possible value of $\hat{p}$.

Conservative Estimate of Determining the Sample Size for a Given Margin of Error in Estimating The Population Proportion, p

For a given confidence level, and a given margin of error, the formula to determine the conservative sample size, n, to satisfy these requirements in estimating the population proportion is:

$$n = \frac{z^2(0.25)}{E^2}$$

If the resulting value of n is not a whole number, the next largest whole number should be taken for the required sample size.

where:
  n = the required sample size for the specified conditions
  E = predetermined margin of error
  z = the z score corresponding to the level of confidence

Example 13.9

An automobile club would like to estimate the proportion of all drivers who wear seat belts while driving. If the automobile club wants the margin of error to be within 2 percentage points of the true population proportion for a 99% confidence interval, then determine:

a) a conservative estimate for the sample size.
b) an estimate of the sample size if a previous study showed that 65% wear seat belts.

Solution:

a) To determine a conservative estimate for the sample size of drivers who wear seat belts, we use the formula:

$$n = \frac{z^2(0.25)}{E^2}$$

To calculate the sample size, n, using this formula, we need to determine the z score and a margin of error for these requirements. A confidence level of 99%, uses the z score, z = 2.58.

Since the automobile club wants to be within 2 percentage points, then the margin of error is: E = 0.02.

Substituting these values into the formula for n, we have:

$$n = \frac{(2.58)^2(0.25)}{(0.02)^2}$$

$$n = \frac{1.6641}{0.0004}$$

$$n = 4160.25$$

Since the resulting value of n is not a whole number, the next largest whole number is taken, which is: 4161.

Thus, the automobile club's conservative estimate of the sample size is 4161 drivers. Therefore, if the club wants to be within two percentage points of the population proportion of drivers who wear seat belts, and 99% confident of their estimate, they should take a sample of 4161 drivers.

b) To determine an estimate of the sample size, n, using a sample proportion, $\hat{p}$, determined from a previous study, we use the formula:

$$n = \frac{z^2(\hat{p})(1-\hat{p})}{E^2}$$

To calculate the sample size, n, using this formula, we need to determine the z score, a margin of error for these requirements, and a sample proportion, $\hat{p}$. A confidence level of 99%, uses a z score, z = 2.58. Since the automobile club wants to be within 2 percentage points, then the margin of error is: E = 0.02. From the previous study, we can use 0.65 for the value for $\hat{p}$, so $\hat{p}$ = 0.65 and 1 - $\hat{p}$ = 0.35. Substituting these values into the formula for the sample size, n, we have:

$$n = \frac{(2.58)^2 (0.65)(0.35)}{(0.02)^2}$$

$$n = \frac{1.5143}{0.0004}$$

$$n = 3785.75$$

Since the resulting value of n is not a whole number, the next largest whole number is taken, which is: 3786.

Thus, if the automobile club takes a sample of 3786 drivers, then the estimate of the population of all drivers that wear seat belts will be within 2 percentage points of the population proportion. ∎

In Example 13.9, it is important to realize that the sample size of 3786 drivers was determined using the sample proportion, $\hat{p} = 0.65$, and that the automobile club wanted to be within 2 percentage points of the population proportion. However, you should realize that if the sample proportion for the new sample of 3786 drivers happens to be greater than 0.65, the margin of error will not be within the requirement of 2 percentage points.

Therefore, if you want to be cautious and more conservative in your estimate, you may want to select a sample size that is greater than 3786 drivers. For example, perhaps you would use the sample size of 4161 that was determined in *part a* of Example 13.9.

❄ 13.7 USING MINITAB

In this chapter we have studied the process of constructing confidence intervals about population parameters and have learned about three types:

(1) Confidence intervals about a population mean, μ, where the population standard deviation, σ, is **known**.
(2) Confidence intervals about a population mean, μ, where the population standard deviation, σ, is **unknown**.
(3) Confidence intervals about a population proportion, p, where n $\hat{p}$ > 5 and

n(1- $\hat{p}$) > 5.

The MINITAB commands, ZINTERVAL and TINTERVAL are designed to construct a confidence interval for the population mean or proportion for any level of confidence.

When a population standard deviation, σ, is known then the ZINTERVAL command is used. Its form is ZINTERVAL [(1-α)% confidence] sigma=K, for C,...,C.

For example, suppose data values are collected and inputed into column C1 of a MINITAB worksheet. If the population standard deviation, σ, is 10, then the MINITAB command for a 95% confidence interval for the data in C1 is

MTB› ZINTERVAL 95 SIGMA=10 C1

MINITAB output for this command could be: THE ASSUMED SIGMA =10.0

	N	MEAN	STDEV	SE MEAN	95.0 PERCENT C.I.
C1	100	50.52	9.92	1.00	(48.56, 52.48)

If you examine the output, you should notice that a 95% confidence interval for a population mean for the sample of 100 data values is listed under the heading 95.0 PERCENT C.I.. The values 48.56 and 52.48 are the interval values for the 95% confidence interval.

Thus, the 95% confidence interval for the population mean estimated by the sample within C1 is: 48.56 to 52.48. Similar confidence intervals can be constructed for the 90% or 99% levels of confidence using the ZINTERVAL command.

The TINTERVAL command is used to estimate the population mean when the population standard deviation, σ, is unknown. It is also used to estimate the population proportion. The form for the TINTERVAL command is:

TINTERVAL [with (1-α) percent confidence] for data in C,...,C

Let's use the TINTERVAL command to construct a 90% confidence interval for a population mean for sample data stored within column C1 of a MINITAB worksheet.

The command

    MTB >   TINTERVAL 90 C1

produces the output:

	N	MEAN	STDEV	SE MEAN	90.0 PERCENT C.I.
C1	16	3.0352	0.2870	0.0717	(2.9094, 3.1610)

Examine the output. What is the 90% confidence interval. How many data values are in the sample? What is the sample mean? the sample standard deviation? the standard error of the mean? How is the SE MEAN calculated?

Let's use the TINTERVAL command to estimate a population proportion from sample data. Suppose 80% of a random sample of 350 voters believe in the President's foreign trade policies. Use the TINTERVAL command to construct the 99% confidence interval for the population proportion.

To use the TINTERVAL command we must be sure that the sample results are contained within a MINITAB worksheet. Let's use the command SET to put the point estimate 80% into column C1.

Since there are 350 voters in the sample, 80% of 350 is 280. Thus, using the SET command

MTB> SET C1

DATA>   280(1) 70(0)

DATA>   END

followed by

MTB >   TINTERVAL 99% C1

yields the output

	N	MEAN	STDEV	SE MEAN	99.0 PERCENT C.I.
C1	350	0.8000	0.4006	0.0214	(0.7445, 0.8555)

Examine the output. What is the 99% confident interval for the population proportion, p? What is the margin of error? What is the point estimate?

SUMMARY OF CONFIDENCE INTERVALS

Table 13.3 contains a summary of the confidence intervals and the appropriate conditions.

Table 13.3

\multicolumn{5}{c}{SUMMARY OF CONFIDENCE INTERVALS}				
PARAMETER BEING ESTIMATED	**CONDITIONS OF ESTIMATE**	**POINT ESTIMATE**	**CONFIDENCE INTERVAL**	**SAMPLING DISTRIBUTION IS A:**
μ	Known σ	$\bar{x}$	$\bar{x} - (z)\left(\frac{\sigma}{\sqrt{n}}\right)$ to $\bar{x} + (z)\left(\frac{\sigma}{\sqrt{n}}\right)$ where: $z = 1.65$ for 90% $z = 1.96$ for 95% $z = 2.58$ for 99%	Approximately a Normal Distribution
μ	Unknown σ	$\bar{x}$	$\bar{x} - (t)\left(\frac{s}{\sqrt{n}}\right)$ to $\bar{x} + (t)\left(\frac{s}{\sqrt{n}}\right)$ where: $t = t_{95\%}$ for 90% $t = t_{97.5\%}$ for 95% $t = t_{99.5\%}$ for 99%	t Distribution with: df = n − 1
p	$n\hat{p} > 5$ and $n(1-\hat{p}) > 5$	$\hat{p}$	$\hat{p} - (z)(s_{\hat{p}})$ to $\hat{p} + (z)(s_{\hat{p}})$ where: $z = 1.65$ for 90% $z = 1.96$ for 95% $z = 2.58$ for 99%	Approximately a Normal Distribution

Glossary

Term	Section
Estimate	13.1, 13.2
Estimation	13.1
Point Estimate of the Population Mean, μ	13.2, 13.3
Point Estimate of the Population Proportion, p	13.2, 13.4
Estimator	13.2
Unbiased	13.2
Good Estimator	13.2
Interval Estimate	13.3
Margin of Error	13.3
90% Confidence Interval for Population Mean, μ	13.3
95% Confidence Interval for Population Mean, μ	13.3
99% Confidence Interval for Population Mean, μ	13.3
Sampling Distribution of the Proportion	13.4
Standard Deviation of the Point Estimate of the Sample Proportion, $s_{\hat{p}}$	13.4
90% Confidence Interval for Population Proportion, p	13.4
95% Confidence Interval for Population Proportion, p	13.4
99% Confidence Interval for Population Proportion, p	13.4
Precision	13.6
Conservative Estimate of the Sample Size	13.6

Exercises

PART I Fill in the blanks.

1. The mean, proportion or standard deviation of a population is called a population _____.
2. To estimate a population parameter, we select a ___ from the population.
3. The statistical procedure where sample information is used to estimate the value of a population parameter such as the population mean, population standard deviation or population proportion is called _____.
4. A sample estimate using a single number to estimate a population parameter, such as, a population mean or population proportion is called a ___ estimate.
5. A sample mean, $\bar{X}$, is one method that is used to obtain a ___ estimate of the population mean, μ.
6. A method that is used to obtain a point estimate of the population proportion, p, is the ___.
7. Any sample statistic that is used to estimate a population parameter is called an ___.
8. The numerical value obtained from the sample mean (or sample proportion) formula is referred to as an ___ of the population mean (or sample proportion). The sample mean (or sample proportion) formula which is used to find a point estimate of the population mean (or population proportion) is called an _____.
9. An estimator of a population parameter is said to be ___, if the mean of the point estimates obtained from the independent samples selected from the population will approach the true value of the population parameter as more and more samples are selected.
10. The "margin of error" usually given in an opinion poll is due to ___ error because not everyone in the population was selected for the poll.
11. An ___ estimate is an estimate that specifies a range of values that the population parameter is likely to fall within.
12. The first step of the procedure in constructing an interval estimate of a population parameter is to select a random sample and calculate a ___ estimate.
13. The 95% Confidence Interval for the population mean, μ, where the population standard deviation is known and the sample size is greater than 30, is constructed by the formula: ___ to ___.
14. The 90% Confidence Interval for the population mean, μ, where the population standard deviation is known and the sample size is greater than 30, is constructed by the formula: ___ to ___.
15. The 99% Confidence Interval for the population mean, μ, where the population standard deviation is known and the sample size is greater than 30, is constructed by the formula: ___ to ___.
16. The 90% Confidence Interval for the population mean, μ, when the population standard deviation is unknown, is constructed by the formula: ___ to ___.
17. The 95% Confidence Interval for the population mean, μ, when the population standard deviation is unknown, is constructed by the formula: ___ to ___.
18. The 99% Confidence Interval for the population mean, μ, when the population standard deviation is unknown, is constructed by the formula: ___ to ___.
19. The value of $s_{\hat{p}}$, which is an estimate of the standard deviation of the population proportion, $\sigma_{\hat{p}}$, is determined by the formula: $s_{\hat{p}} =$ ___.
20. The 90% Confidence Interval for the Population Proportion, p, is constructed by the formula: ___ to ___.
21. The 95% Confidence Interval for the Population Proportion, p, is constructed by the formula: ___ to ___.
22. The 99% Confidence Interval for the Population Proportion, p, is constructed by the formula: ___ to ___.
23. The confidence level of a confidence interval is the ___ value, expressed as percentage, that is associated with an interval estimate, which represents the chance that the ___ used to construct the confidence interval will give an interval that will include the population ___. Thus, the higher the probability value, the greater the ___ that the interval estimate will include the population parameter.
24. A 95% confidence interval for a population mean doesn't tell you whether a particular ___ will contain the population mean. But rather it means that: in the long run, the percentage of ___ intervals constructed using this procedure will probably contain the population mean ___ of the time.
25. In general, as the confidence level increases, the width of the confidence interval ___.
26. The t Distribution is used to construct a confidence interval for the population mean when the following conditions hold.
    a) The population standard deviation, σ, is ___, and is estimated by the ___.
    b) The population from which the sample is selected is approximately a ___ distribution.
27. To increase the precision of a confidence interval with a fixed confidence level, you must ___ the sample size.

28. A confidence interval for a specific level of confidence consists of two components: a point estimate of the ____, and the ____ of error.
29. The margin of error for a confidence interval to estimate the population mean represents the ____ possible error of estimate for the population mean using the ____. The margin of error is denoted by the symbol ____. The margin of error represents one-half the ____ of the confidence interval. As the level of confidence is increased, where the sample size and population standard deviation remain fixed, the margin of error ____. To reduce the margin of error while holding the confidence level fixed, then the sample size must be ____.
30. The formula used to calculate the margin of error when estimating the population mean, where the population standard deviation is known, is: E =____.
31. The margin of error formula used to determine the maximum possible error of estimate for the population proportion, p, using the sample proportion, $\hat{p}$, is: E = ____.
32. The width of a confidence interval is twice the ____.
33. For a given confidence level, a known population standard deviation, and a given margin of error, the formula to determine the sample size, n, to satisfy these requirements in estimating the population mean is: n = ____.
34. For a given confidence level, and a given margin of error, the formula to determine the conservative sample size, n, to satisfy these requirements in estimating the population proportion is: n = ____.
35. For a given confidence level, a point estimate of the population proportion, $\hat{p}$, and a given margin of error, the formula to determine the sample size, n, to satisfy these requirements in estimating the population proportion is: n = ____.

PART II Multiple choice questions.

1. Which symbol represents a Point Estimate of the Population Mean?
   a) p b) $\mu_{\hat{p}}$ c) $\hat{p}$ d) $\bar{X}$ e) Eμ
2. Which symbol represents a Point Estimate of the Population Proportion?
   a) p b) $\mu_{\hat{p}}$ c) $\hat{p}$ d) $\bar{X}$ e) Eμ
3. Which formula is used in the calculation of a confidence interval when the population standard deviation is **known**?
   a) $\frac{\sigma}{\sqrt{n}}$ b) $\frac{s}{\sqrt{n}}$ c) $\sqrt{np(1-p)}$ d) $\sqrt{\frac{\hat{p}(1-\hat{p})}{n}}$ e) np

4. Which formula is used in the calculation of a confidence interval when the population standard deviation is **unknown**?
   a) $\frac{\sigma}{\sqrt{n}}$ b) $\frac{s}{\sqrt{n}}$ c) $\sqrt{np(1-p)}$ d) Eμ e) np
5. Which formula is used to calculate the Standard Deviation of the Point Estimate of the Sample Proportion?
   a) $\frac{\sigma}{\sqrt{n}}$ b) $\frac{s}{\sqrt{n}}$ c) $\sqrt{np(1-p)}$ d) $\sqrt{\frac{\hat{p}(1-\hat{p})}{n}}$ e) np
6. Which formula is used in the calculation of a confidence interval for the Population Proportion?
   a) p b) $\mu_{\hat{p}}$ c) $\hat{p}$ d) $\bar{X}$ e) Eμ
7. When the sampling distribution of the population proportion p is approximated by a normal distribution, which z score is used to construct the 95% Confidence Interval?
   a) 0.95 b) 1.65 c) 1.96 d) 2.33 e) 2.58
8. When the sampling distribution of the population proportion p is approximated by a normal distribution, which z score is used to construct the 99% Confidence Interval?
   a) 0.95 b) 1.65 c) 1.96 d) 2.33 e) 2.58
9. When constructing a 90% confidence interval where the population standard deviation is **unknown**, the t score with df = n-1 corresponds to:
   a) a two-tailed test at α = 5%
   b) a one-tailed test at α = 5%
   c) a two-tailed test at α = 1%
   d) a one-tailed test at α = 1%
   e) a z score
10. When constructing a 99% confidence interval where the population standard deviation is **unknown**, the t score with df = n-1 corresponds to:
    a) a two-tailed test at α = 5%
    b) a one-tailed test at α = 5%
    c) a two-tailed test at α = 1%
    d) a one-tailed test at α = 1%
    e) a z score

PART III Answer each statement True or False.

1. Random samples can be used to estimate population parameters.
2. A point estimate is a procedure that is used to estimate the value of a population parameter.

3. An example of an estimator is the formula for the Standard Deviation of the Point Estimate of the Sample Proportion.
4. An estimator is considered **unbiased** because as more and more samples are selected from the population, the mean of the point estimates obtained from independent random samples approach the true value of the population parameter.
5. One major disadvantage of using a sample mean point estimate to estimate the true population mean is that we are never sure how far the point estimate is from the true population mean.
6. An interval estimate uses a range of values and a probability level to indicate how confident one is that the interval will **not** contain the population parameter.
7. The 95% Confidence Interval for the population mean suggests that 5% of these confidence intervals constructed for all possible sample means will **not** contain the true population mean.
8. When constructing a 99% Confidence Interval for the population mean where the population standard deviation is **not** known, the sample standard deviation is needed to complete the construction.
9. When constructing a 99% Confidence Interval for the population proportion p for a sample size n, it is assumed that both np and n(1-p) are greater than 5.
10. A margin of error is due to the inaccuracy of a sample to be random.
11. When determining the margin of error, the sample standard deviation is the one measure you can control.
12. As the sample size increases, the margin of error increases.

PART IV Problems.

1. Determine a 90%, 95% and 99% Confidence Intervals and the margin of error for the following sample data:
   a) $\bar{X}$ = 78, s = 4 and n = 16
   b) $\bar{X}$ = 78, s = 4 and n = 36
   c) $\bar{X}$ = 78, s = 4 and n = 49
   d) $\bar{X}$ = 78, s = 4 and n = 100
   e) What is happening to the width of each confidence interval and the margin of error as the sample size increases?

2. On the basis of a random sample of 64 couples, the mean length of dating before marriage is 540 days with a sample standard deviation is 90 days.
   a) Construct a 90% confidence interval to estimate the mean number of dating days for the population.
   b) State the lower and upper confidence limits for this interval.

3. Using the information: a random sample of 36 students averaged 9.06 hours of sleep on the weekend with a sample standard deviation of 1.08, construct:
   a) a 90% confidence interval for the population mean amount of time students sleep on the weekend
   b) a 95% confidence interval for the population mean amount of time students sleep on the weekend
   c) a 99% confidence interval for the population mean amount of time students sleep on the weekend
   d) Determine the width of each confidence interval by subtracting the lower confidence limit from the upper confidence limit.
   e) Compare and discuss the relationship of the width of each confidence interval with the confidence level.

4. A sociology professor is interested in estimating the mean number of hours that a day student spends per week studying. The professor randomly selects 400 day students and calculates that the sample mean number of hours spent on studying is 18.5 hrs with s = 4. Determine the:
   a) population
   b) population parameter being estimated
   c) estimator used to estimate the population parameter
   d) point estimate
   e) margin of error for the 90%, 95%, and 99% confidence level
   f) 90% Confidence Interval for the Population Mean, μ
   g) 95% Confidence Interval for the Population Mean, μ
   h) 99% Confidence Interval for the Population Mean, μ

5. The Student Government Association President wants to estimate the mean amount of dollars spent on textbooks and supplies per semester by a full-time student. The President randomly selects 625 full-time students and calculates the mean amount of dollars they spent on textbooks and supplies per semester to be $247 with s = 35. Determine the:
   a) population
   b) population parameter being estimated
   c) estimator used to estimate the population parameter
   d) point estimate
   e) margin of error for the 90%, 95%, and 99% confidence level

f) 90% Confidence Interval for the Population Mean, μ
g) 95% Confidence Interval for the Population Mean, μ
h) 99% Confidence Interval for the Population Mean, μ

6. A special *fat reducing* diet is tested on 28 middle aged males. The results of the test indicated that the mean amount of weight loss was 14 lbs with s = 2.5. Determine a 99% Confidence Interval for the diet. What is the margin of error associated with this estimate?

7. The amount of gasoline consumption of a particular make of car is known to be approximately normally distributed. If a random sample of 25 cars had a mean of 31.7 mpg for highway driving with s = 2.9 mpg. Construct a 95% confidence interval for the mean gasoline consumption of this make of car.

8. The registration time for students at a community college is approximately normally distributed. If a random sample of 49 students has a mean registration time of 92 minutes with a standard deviation of 11.2 minutes, construct a 90% confidence interval for the mean registration time for this community college.

9. A sociologists is interested in estimating the educational level of a community. She randomly interviews 25 adults residents of the community and determines the number of years they attended school. Her sample had a mean of 12.5 years with s = 3.8 years. Use a 95% confidence interval to estimate the educational level of all the adults of this community.

10. A random sample of 16 college students living within a dorm had a mean monthly telephone bill of $52.79 with s = $8.64. Find a 90% confidence interval for the mean monthly telephone bill for all students living in the college dorm.

11. The mean cholesterol level of a random sample of 36 adults of age 45 was 215 milligrams of cholesterol with s = 19.84. Construct a 95% confidence interval for the mean cholesterol level of all adults aged 45.

12. A manufacturer of a new light bulb selects a random sample of 64 of these new bulbs to estimate the mean amount of hours that the new light bulb will last. Construct a 95% confidence interval to estimate the mean life of all these new bulbs if the sample mean is 1405 hours with s = 137 hours.

13. A random sample of n students is selected from the student body of college students to estimate the mean IQ score of the student population. Assuming that the sample mean is 108, and the population standard deviation is equal to 15, construct a 95% confidence interval for samples of size:
a) 36
b) 100
c) 400
d) Determine the width of the confidence intervals found in parts a, b & c.
e) What is the effect on the width of the confidence interval as the sample sized is increased while the confidence level remains the same?

14. A random sample of 100 American workers showed that the average number of hours an employee works per week is 36.3 with the sample standard deviation equal to 4.2 hours. Find a 95% confidence interval for the mean number of hours that all american employees work per week.

15. A survey of 40 NFL players found the average weight to be 230.7 pounds with s = 15.7 lbs, with an average height of 6 feet 2 inches with s = 3.8 inches. Find:
a) a 95% confidence interval to estimate the mean weight of all NFL players.
b) a 90% confidence interval to estimate the mean height of all NFL players.

For problems 16 to 27, when computing $s_{\hat{p}}$, round off to four decimal places.

16. Determine a 90%, 95% and 99% Confidence Intervals for the following sample data:
a) $\hat{p}$ = 0.30, and n=100
b) $\hat{p}$ = 0.30, and n=1600
c) $\hat{p}$ = 0.30, and n=100,000
d) What is happening to the width of each confidence interval and the margin of error as the sample size increases?

17. DPT Enterprises, Inc. is about to market a new kind of shaver, a shaver with six blades. In order to determine if they should proceed with their manufacturing plans, the marketing department of the company decides to survey a cross section of the potential buyers as to their feelings regarding their purchasing of the new shaver. The results of the survey were:
152 potential buyers said they would buy the shaver, sample size = 7569
Determine the:
a) population
b) population parameter being estimated
c) estimator used to estimate the population parameter
d) point estimate
e) margin of error for the 90%, 95%, and 99% confidence level

f) 90% Confidence Interval for the Population Proportion p
g) 95% Confidence Interval for the Population Proportion p
h) 99% Confidence Interval for the Population Proportion p

18. Suppose you are the polling consultant for a state senator. You find that a *straw poll* of 625 registered voters indicates that 263 registered voters name drugs within the schools as the most important issue facing the inner city schools. Determine a 95% Confidence Interval for the proportion of all voters who hold this opinion. What is the margin of error associated with this estimate?

19. *PC DAY* computing magazine recently published that Microhard word processing software now has 24% of the nation's word processing market. Determine the:
    a) population
    b) population parameter being estimated
    c) estimator used to estimate the population parameter
    d) point estimate
    e) 90% Confidence Interval and margin of error for the Population Proportion if the sample size used is:
        1) 100
        2) 400
        3) 1024
    f) 95% Confidence Interval and margin of error for the Population Proportion if the sample size used is:
        1) 100
        2) 400
        3) 1024
    g) 99% Confidence Interval and margin of error for the Population Proportion if the sample size used is:
        1) 100
        2) 400
        3) 1024.
    h) Discuss what happens to the margin of error as the sample size increases for each confidence level.
    i) Discuss what happens to the margin of error as both the sample size and confidence level increase.

20. According to an opinion poll of 500 economists, 43% stated that they felt the stock market would get worse before the end of the year. Construct a 99% confidence interval to estimate the population proportion of economists who feel that the stock market will get worse before the end of the year.

21. A recent magazine poll determined that 1278 of 1800 American adults interviewed were in favor of tougher laws to combat crime. Find a 99% confidence interval for the true population proportion of American adults who are in favor of tougher laws to combat crime.

22. According to an independent poll of 1160 people, the Internal Revenue Service (IRS) provided the wrong answers to 198 people who called to ask a question pertaining to their tax return during the recent tax season. Construct a 90% confidence interval to estimate the proportion of all times that the IRS fails to provide a correct answer to a tax question.

23. According to a nationwide survey of 1254 adults, 915 adults said they suffered from headache pain, and 702 reported that they suffered from backache pain. Find:
    a) a 90% confidence interval for the true proportion of all adults that suffer from headache pain.
    b) a 95% confidence interval for the true proportion of all adults that suffer from backache pain.

24. A Department of Transportation survey of 1480 American drivers stated that 873 drivers wear seat belts.
    Construct:
    a) a 95% confidence interval to estimate the true proportion of all American drivers that wear seat belts.
    b) a 99% confidence interval to estimate the true proportion of all American drivers that wear seat belts.

25. A telephone survey of 1050 married adults indicated that 724 adults preferred to celebrate their anniversary by eating at a romantic restaurant. Construct a 90% confidence interval to estimate the true proportion of all married adults who prefer to celebrate their anniversary by eating at a romantic restaurant.

26. A random sample of 1350 men aged 18 years and older found that 486 men were single, while a random sample of 1560 women aged 18 years and older showed that 624 were single.
    Construct:
    a) a 95% confidence interval to estimate the true proportion of all single women that are 18 years or older.
    b) a 99% confidence interval to estimate the true proportion of all single men that are 18 years or older.

27. A National Health Survey of 1640 women determined that 574 women indicated that they are trying to lose weight at any given time, while 468 out of 1560 men stated that they are constantly trying to maintain their weight.

Construct:
a) a 95% confidence interval to estimate the true proportion of all women who are trying to lose weight at any given time.
b) a 90% confidence interval estimate the true proportion of all men who are constantly trying to maintain their weight.

28. Determine a *conservative estimate* of the sample size for estimating the population proportion using a 95% Confidence Interval where the margin of error is to be:
a) 0.01
b) 0.10
c) What advantage is there in computing this?

29. A manufacturing company wants to estimate the proportion of defective squash balls that are produced by a machine to be within 0.03 of the population proportion p for a 99% Confidence level. How large of a sample size, n, must the company select to accomplish this using the conservative estimate of the sample size formula?

30. a) How large a sample is required if one would like to be 95% confident that the estimate of the population mean height of Americans will be within 0.5 inches of the population mean? (Assume that the population standard deviation is 6 inches.)
b) How large a sample is required for a 99% confidence interval of the population mean height of Americans? (Assume that the population standard deviation is 6 inches.)

31. A random sample of 49 college male students indicates that the mean pulse rate, while at rest, is 68.3 beats per minute. Assuming the population standard deviation is 9.6 beats per minute, determine:
a) a 95% confidence interval for the mean pulse rate, while at rest, for the population of college male students.
b) the width of this 95% confidence interval.
c) the margin of error for this 95% confidence interval.
d) how large a sample of college students would one need to select to be 95% confident that the margin of error is within 1 heart beat of the mean pulse rate for all college male students?

32. The quality-control department of an electrical manufacturer estimated the mean lifetime of their new miser 60 watt bulb by randomly selecting 64 bulbs and determined the sample mean lifetime to be: $\bar{x}$ = 1817 hours. If the population standard deviation of these miser bulbs is 240 hours, then:
a) determine the margin of error for a 99% confidence interval.
b) what is the width of this 99% confidence interval?
c) within how many hours can the manufacturer conclude that they are 99% confident that the sample mean lifetime of 1817 hours is from the population mean lifetime of these new miser bulbs?

33. A medical student is interested in estimating with a 90% confidence level the proportion of babies that are born between the hours of: 6AM and 6PM, called a "morning baby". If the student wants the margin of error to be within 4 percentage points, determine:
a) a conservative estimate for the sample size required to meet these requirements.
b) an estimate of the sample size required to satisfy these conditions, if a previous sample showed the proportion of morning babies to be 0.62.

34. How many times would a fair coin have to be randomly tossed to be 99% confident that between 40% and 60% of the tosses will land heads?

35. An opinion poll estimates the percentage of voters who favor candidate P to be 48% with a margin of error of 3 percentage points. Candidate P is not pleased with this poll and decides to take another poll that will have a higher precision level of 2 percentage points for the margin of error. If the candidate is willing to accept a 95% confidence level, then determine:
a) how many voters should be interviewed for this new poll.
b) the cost of the new poll, if the cost per interview is $1.50.

36. A political activist wants to estimate the proportion of all 18-year-olds who are registered to vote. Determine:
a) a conservative estimate of the sample size that would produce a margin of error of 3 percentage points for a 95% confidence interval.
b) an estimate for the size of a sample for a 95% confidence interval, if a previous sample indicated that the proportion of 18-year-olds who are registered to vote was 28%.

37. Given the statement: *a 95% confidence interval for the proportion of people owning an answering machine is: 45% ± 3 percentage points*, explain why the following interpretation of this 95% confidence interval statement is incorrect.

This confidence interval is interpreted to mean that you are 95% certain that the true proportion of all people that own an answering machine is within 3 percentage points of 45%.

PART V What Do You Think?

1. **What Do You Think?**

   The following article entitled *Survey Finds High Fear of Crime* discusses the results of a Gallup Poll on crime.

   ## Survey Finds High Fear of Crime

   Fear of crime continues to pervade American society, especially in urban areas, where 76 percent of women fear walking alone at night in their neighborhood, according to the latest Gallup Poll.

   Over all, 45 percent of Americans are afraid to go out alone at night within a mile of their homes, the researchers reported. Thirteen percent had that fear in the day. Sixty-two percent of the women surveyed had that fear, as did 26 percent of the men.

   The survey also found that crime had affected an average of 25 percent of American households in the last 12 months.

   Comparing surveys over the last decade, the poll reported that the incidence of crime and fear of crime had remained about the same.

   Another statistic that has remained constant is the number of people who do not feel safe at home at night. Sixteen percent said they felt unsafe, as did 16 percent in 1991, 15 percent in 1987 and 20 percent in 1985. The difference in these figures is statistically insignificant because the survey's margin of error is plus or minus three percentage points.

   The one area in which a marked change was found was the number of people who thought there was "more crime in this area than there was a year ago". In 1992, 47 percent felt there was more crime than there had been a year before, in contrast to 37 percent in the latest poll.

   The poll said its findings showed that "the actual crime situation in this country is more serious than official Government figures" indicate because "many incidents are not reported to the police."

   The findings were based on interviews conducted in person with 1,555 adults from Jan. 28-31.

   a) Define the population for this survey.
   b) What type of estimate does the percentage value of 76% represent? What population parameter is being estimated by this result? Explain.
   c) What was the margin of error for this survey? How would you interpret this margin of error?
   d) State an interval that you believe might contain the true population proportion of americans who are afraid to go out alone at night within one mile of their homes. How did you construct such an interval?
   e) What is meant by the statement "the difference in these figures is statistically insignificant because the survey's margin of error is plus or minus three percentage points"?
   f) State a sample point estimate for the proportion of women who are afraid to go out alone at night within a mile of their homes. State an interval estimate for the proportion of women who are afraid to go out alone at night within a mile of their homes. What is the difference between these two estimates? Explain.

2. **What Do You Think?**

   The following New York Times section entitled *Pulse/Working Mothers* presents the results of four questions conducted by the Gallup Poll.

   a) Define the population for this survey.
   b) In question #1, what type of estimate does the percentage value of 67% represent? What population parameter is being estimated by this result? Explain.
   c) Calculate the margin of error for question #1 using a z score of 1.96. How does this compare to the margin of error for the poll?
   d) Construct a 95% confidence interval for the yes response to question #1. How would you interpret this result? What is the width of this confidence interval? How does this compare to the margin of error?
   e) Determine a 95% confidence interval for the population proportion of women who feel it is very important to share child-rearing responsibilities with another adult. What is the width of this confidence interval? How does this compare to the margin of error?
   f) Give a point estimate of the women who feel that conditions for working mothers would improve if more women were elected to prominent Federal government positions. Determine a 95% confidence interval using this point estimate. What is the width of this interval?
   g) State an interval that you believe, with 95% confidence, might contain the true population proportion of women who feel that most women who work outside the home generally seem to make worse mothers than women who do not work outside the home. What is the width of this interval? How does this width compare to the margin of error?

PULSE | Working Mothers

Responses from a Gallup Poll conducted Dec. 9, 1992, of 506 women in the Northeast 20 to 55 years old with children under 18 years old living at home. Margin of sampling error is four percentage points. The Northeast region includes Connecticut, Delaware, Maryland, Maine, New Hampshire, New Jersey, New York, Pennsylvania, Rhode Island, Vermont and Virginia.

Source: Gallup Poll

❶ Do you work outside the home?

❷ Would you say that most women who work outside the home generally seem to make better or worse mothers than women who do not work outside the home or is there no difference?

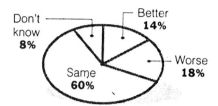

❸ Do you feel that conditions for working mothers would improve if more women were elected to prominent Federal government positions?

❹ How important is it to be able to share child-rearing responsibilities with another adult?

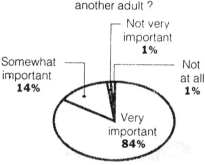

h) State a point estimate for the proportion of women who feel it is somewhat important to share child-rearing responsibilities with another adult? Determine a 95% and a 99% confidence interval using this point estimate. What is the width of each of these intervals? What can you conclude about the relationship between the width of the confidence intervals and the degree of confidence?

PART VI Exploring DATA With MINITAB

1. Use MINITAB to find the 90%, 95%, and 99% confidence intervals for the population mean for the sample data randomly selected from the population whose standard deviation is 20, if the data values are:
104 75 59 73 71 74 39 93
84 80 62 120 50 80 56 62
What is the point estimate?
For each confidence interval determine the margin of error.

2. Use MINITAB to find the 90%, 95%, and 99% confidence intervals for the population mean for the random sample of data values selected from the population.

   2.91 2.34 1.94 3.33 3.95 2.55 2.67 2.87 2.95
   2.78 3.56 3.68 2.88 3.05 2.68 3.55 2.98 1.94
   2.47 2.88

   What is the point estimate? For each confidence interval determine the margin of error.

3. Use MINITAB to find the 90%, 95%, and 99% confidence intervals for the population proportion for a random sample of 400 data values selected from a population that yields a point estimate of 0.6.
   For each confidence interval determine the margin of error.

4. Suppose that you randomly sample 20 times from a population whose data values are approximately normal with a mean of 500 and a standard deviation of 100.
   a) If you were to form a 95% confidence interval for your 20 random samples, how many of the confidence intervals would you expect to include the true population mean value of 500?
   b) What would be the margin of error for your simulation if the sample size was 400?
   c) Use the sequence of MINITAB commands:

      MTB> RANDOM 400 C1-C20;

      SUBC> NORMAL MU=500 SIGMA=100 C1-C20.

      to simulate the selection of 20 random samples of size 400 from a normal distribution whose mean is 500 and whose standard deviation is 100.
   d) For each of the 20 random samples developed in part (c), calculate the 95% confidence interval using the command

      MTB> ZINTERVAL SIGMA=100, C1-C20

   e) How many of the 20 confidence intervals contained the true population mean of 500? How does this compare to your answer to part (a)? Explain.

PART VII Projects.

1. Find an example of an **interval estimate** that is published in a newspaper or magazine. Identify the:
   a) population
   b) population parameter being estimated
   c) estimator used to estimate the population parameter
   d) point estimate
   e) sample size
   f) margin of error

2. Poll your fellow students on an issue of concern for your school or local. For your poll, identify the:
   a) population
   b) population parameter being estimated
   c) estimator used to estimate the population parameter
   d) point estimate
   e) margin of error
   f) 90% Confidence Interval for the Population Proportion
   g) 95% Confidence Interval for the Population Proportion
   h) 99% Confidence Interval for the Population Proportion

3. Randomly select a page from this text and count the number of words on the page.
   a) Estimate the number of words in this book using this information.
   b) Construct a 95% confidence interval using this information.

4. Use a random sample of students at your school to estimate the population of students at the school that are:
   a) female.
   b) male.
   Using this information, construct a 95% confidence interval for the percentage of all students that are:
   c) female.
   d) male.

PART VIII Database.

The following exercises refer to the file DATABASE listed in Appendix A. We have indicated the appropriate MINITAB commands that are necessary to answer each exercise.

1. Using the MINITAB commands: RETRIEVE, TINTERVAL, and PRINT Retrieve the file DATABASE.MTW and identify the column number: for the variable names: GPA, AVE, WRK, STY, GFT. Construct 90%, 95% and 99% Confidence Intervals for each of the variables. Interpret each of the confidence intervals in terms of the population means to which they are inferring.

2. Using the MINITAB commands: RETRIEVE, TINTERVAL, and PRINT Retrieve the file DATABASE.MTW and identify the column number: for the variable names: SEX, EYE, HR, MAJ. Construct 90%, 95% and 99% Confidence Intervals for each portion of the population proportion contained within each variable. Interpret each of the confidence intervals in terms of the population proportion to which they are inferring.

CHAPTER 14
CHI-SQUARE

❄ 14.1 INTRODUCTION

A population can be divided many ways. For example, it can be divided according to gender, age, religion, marital status, annual income, political affiliation, or education. Divisions such as these are usually referred to as ***classifications*** or ***factors*** of the population.

In this chapter, we will present a technique that determines whether two classifications of a population are ***independent*** (i.e., the classifications are not related) or ***dependent*** (i.e., the classifications are related). For example,

- are annual income and education level dependent? If the classifications of annual income and education level **are dependent**, then one might *expect* to find individuals with a low education level to have a low annual income, and, those with a high education level to have a high annual income. On the other hand, if the classifications **are independent**, then knowledge of one's education level is *not* indicative of the individual's annual income level.

- are movie preference and an individual's gender dependent? If the classifications of movie preference and gender **are dependent** then a relationship between the classifications might be that females generally prefer romance type movies and males generally prefer adventure type movies. Whereas, if the classifications **are independent** then knowing one's gender would *not* give you a clue to their movie preference.

- are male ages and incidence of heart attack dependent? If the classifications of age and incidence of heart attack **are dependent** then we might expect to find older males more susceptible to heart attack then younger males. If the classifications **are independent** then one's age would have no relation to the incidence of heart attack.

These examples involve the analysis of **categorical** or **count** data to determine if there is a statistically significant relationship between the population classifications. That is, are the classifications ***dependent***?

A statistical technique used to investigate whether the classifications of a population are related is the ***chi-square test of independence***. To apply the chi-square test, a random sample is selected from the population and the sample count data is separated into the different categories for each classification of the population. The ***frequency*** (or number) of responses falling into the distinct categories for each classification are recorded in a table. These sample count results are called ***observed frequencies***.

Table 14.1 illustrates the observed frequencies for a random sample of 200 co-eds selected from a mid-western university. The sample is classified by gender and movie preference. This table is referred to as a ***contingency table***.

> **Definition: Contingency Table.** A contingency table is an arrangement of sample count data into a two-way classification.

Table 14.1
CONTINGENCY TABLE

Gender	Movie Preference					Row Total
	Romance	Comedy	Drama	Adventure	Mystery	
Male	a 10	b 20	c 5	d 40	e 15	90
Female	f 50	g 25	h 5	i 15	j 15	110
Column Total	60	45	10	55	30	200

Table 14.1 is a contingency table with the row classification (i.e., the GENDER classification) separated into two categories and the column classification (i.e., the MOVIE PREFERENCE classification) separated into five categories. A contingency table with 2 row classifications and 5 column classifications is referred to as a **2 by 5 contingency table.**

In general, an **r by c contingency table** is a table where the row classification has been separated into **r** rows and the column classification has been separated into **c** columns.

Each individual box in the table is referred to as a **cell**. The number within each cell is the **observed cell frequency**. For example, in cell d of Table 14.1 the observed cell frequency is 40. That is, the number of males within this sample that prefer adventure movies is 40. Similarly, in cell f the observed frequency is 50 which represents the number of females within the sample that prefer a romantic movie.

To determine whether the two classifications, *gender and movie preference*, are related we will compare the observed frequencies of the random sample to the frequencies we would **expect** if there is **no relationship** between the classifications *gender and movie preference*. If the differences between the **observed** and these **expected** frequencies are *small*, then we will conclude that the two classifications are **not related** or **independent**. If the differences between the observed and expected frequencies are *large*, then we will conclude that the two classifications are **related** or **dependent**. The statistic used to determine whether the differences between the observed and expected frequencies for all the cells of the contingency table are **large** or **small** is known as **Pearson's chi-square statistic, denoted by X^2** (where X is the capital Greek letter Chi, pronounced kye which rhymes with hi).

The Definition Formula For Computing Pearson's Chi-Square Statistic, X^2

Pearson's Chi-Square Statistic = $\sum_{all\ cells} \frac{(Observed - Expected)^2}{Expected}$

This formula is symbolized as:

$$X^2 = \sum_{all\ cells} \frac{(O-E)^2}{E}$$

where:
  O = Observed Cell Frequency
  E = Expected Cell Frequency

If the classifications are *truly independent*, then this would be indicated by a situation where each expected cell frequency, symbolized by E, is *close* or *equal* to its corresponding observed cell frequency, symbolized by O. Consequently, in the case of independence, the numerical value of X^2 would be *small* or *zero*.

When the observed and expected frequencies are the same (i.e., there are *no differences* between the observed and expected frequencies), the value of X^2 is zero. Whereas, *any difference* between the observed and expected frequencies will result in a positive value for X^2. As the differences between the observed and expected frequencies get larger, the value of X^2 will increase.

It then follows that a *large* value of X^2 would tend to indicate a **significant** relationship exists between the two classifications. Thus, we need to determine how *large* a value of X^2 would indicate that the two classifications are dependent. To determine when a value of Pearson's chi-square statistic is considered to be large, we need to examine a distribution that consists of all the possible values of the chi-square statistic. This distribution is called the Sampling Distribution of Pearson's Chi-Square Statistic. The Sampling Distribution of Pearson's Chi-Square Statistic is approximated by the **Chi-Square Distribution**, denoted by χ^2 **Distribution** (where χ is the lowercase Greek letter chi, pronounced kye and rhymes with hi). Let us now examine the properties of the **Chi-Square, χ^2, Distribution**.

❋ 14.2 Properties of the Chi-Square Distribution

Figure 14.1 illustrates a typical chi-square distribution.

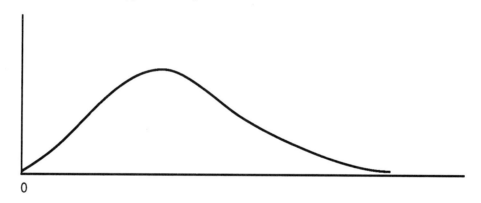

Figure 14.1

Properties of the Chi-Square Distribution

1. The value of χ^2 is always non-negative; it is zero or positively valued.
2. The graph of a chi-square distribution is *not symmetric*, it is *skewed* to the right and extends infinitely to the right of zero.
3. There is a family of chi-square distributions similar to the *t* distribution as illustrated in Figure 14.2. Each chi-square distribution is dependent upon the number of degrees of freedom, *df*, determined by the formula:

$$df = (r-1)(c-1),$$

where: r represents the number of row classifications within the contingency table, and
c represents the number of column classifications within the contingency table.

As the number of degrees of freedom, df, increases, the shape of the chi-square distribution approaches the shape of the normal distribution.

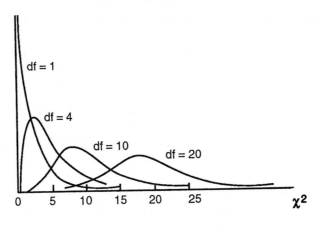

Figure 14.2

14.3 CHI-SQUARE HYPOTHESIS TEST OF INDEPENDENCE

Let's apply the general five step hypothesis testing procedure to test the independence of two population classifications. This hypothesis testing procedure is called the **Chi-Square Test of Independence**.

Step 1. *Formulate the two Hypotheses, H_o and H_a.*
These hypotheses have the following form:

Null Hypothesis Form

H_o: The two population classifications are **independent**.

This means that the population classifications are *not* related.

Alternative Hypothesis Form

H_a: The two population classifications are **dependent**.

This means that the population classifications *are* related.

By stating the null hypothesis as "the two population classifications are independent" we are making the assumption that there is *no relationship* between the classifications. Consequently, by *rejecting* the null hypothesis we can conclude that the two population classifications are **dependent**. This infers that *there is a relationship between the two population classifications*.

Step 2. *Design an experiment to test the null hypothesis, H_o.*

Under the assumption that H_o *is true*, calculate the expected cell frequencies. **To calculate the expected frequency for a particular cell, we multiply its row total (RT) by its column total (CT) and divide this product by the sample size (n).**

Expected Frequency Cell Formula

The expected frequency formula for a particular cell is:

$$\text{expected cell frequency} = \frac{(RT)(CT)}{n}$$

where:
- RT is the row total for the cell
- CT is the column total for the cell
- n is the sample size

Step 3. *Formulate the decision rule.*

All chi-square tests of independence will be one-tailed tests on the right, since a significant Pearson's X^2 statistic will always be a large positive number.

To determine the critical chi-square value necessary to formulate the decision rule, we must perform the following steps:

Procedure to formulate the decision rule

a) Identify the significance level, $\alpha = 1\%$ or $\alpha = 5\%$.

b) Calculate the degrees of freedom, *df*, using:

  df = (number of rows minus one)(number of columns minus one).

  In symbols, $df = (r-1)(c-1)$.

c) Determine the critical χ^2 value, denoted χ^2_α, using TABLE IV: Critical Values for the Chi-Square Distribution, χ^2_α, found in Appendix D: Statistical Tables, and draw a chi-square distribution model with the χ^2_α value appropriately placed on the curve. This hypothesis testing model is illustrated in Figure 14.3.

d) State the decision rule.

  The decision rule is: *reject H_o if X^2 statistic is greater than* χ^2_α.

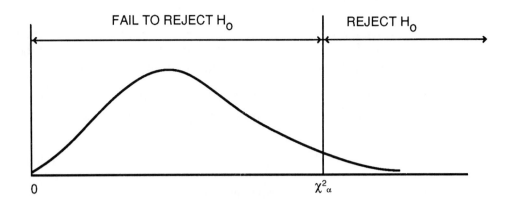

Figure 14.3

Step 4. *Calculate Pearson's X² statistic.*

There are two equivalent formulas, the definition formula and the computational formula, that are used to calculate Pearson's X² statistic. Let's examine each of these two formulas for the X² statistic.

Pearson's Chi-Square Statistic Formulas

The definition formula for Pearson's chi-square statistic is:

$$X^2 = \sum_{all\ cells} \frac{(O-E)^2}{E}$$

The computational formula for Pearson's chi-square statistic is:

$$X^2 = \sum_{all\ cells} \left(\frac{O^2}{E}\right) - n$$

where: O represents the observed frequency of a particular cell,
E represents the expected frequency for the same cell,
n represents the sample size.

Step 5. *Determine the conclusion.*

Compare the value of Pearson's chi-square statistic, X^2, to the critical value of chi-square, χ^2_α, and choose the appropriate conclusion.

If the value of the X^2 statistic is greater than or equal to the critical value of chi-square, χ^2_α, then the conclusion is: **Reject H_o and Accept H_a at α and we would conclude that there is a significant relationship between the two population classifications.** This means that **the two population classifications are dependent.**

Similarly, if the value of the X^2 statistic is less than the critical value of chi-square, χ_α^2, then the conclusion is:

Fail to Reject H_o at α and we would conclude that there is no significant relationship between the two population classifications. This means that **the two population classifications are independent.**

In summary, we have:

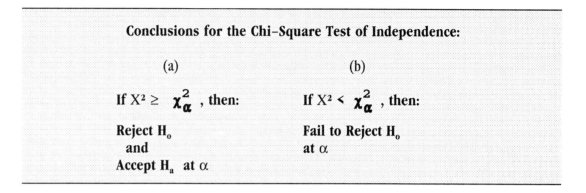

These conclusions are illustrated in Figure 14.4.

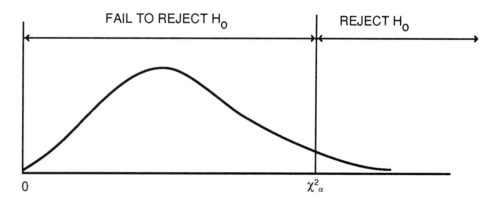

Figure 14.4

Example 14.1

A study was conducted at the 1% level of significance to determine if there is a relationship between gender and one's view on capital punishment. A random sample of 100 adults was selected and asked their opinion on the question:

"Do you believe that the death penalty should be given to those convicted of first degree murder?"

The sample results which represent the observed frequencies are summarized in Table 14.2.

Table 14.2

Table of Observed Frequencies
View on Capital Punishment

Gender	View on Capital Punishment			Row Totals
	Favor	Against	Undecided	
Male	a 20	b 25	c 15	60
Female	d 10	e 25	f 5	40
Column Totals	30	50	20	100

Based on these sample results can we conclude that there is a relationship between gender and one's view on capital punishment at $\alpha = 1\%$?

Solution:

Step 1. *Formulate the two hypotheses, H_o and H_a.*

H_o: Gender and one's view on capital punishment are **independent.**
H_a: Gender and one's view on capital punishment are **dependent.**

Step 2. *Design an experiment to test the null hypothesis, H_o.*

Table 14.2 is a contingency table of the observed frequencies. *To test H_o the expected frequencies for each cell must be computed.* **To compute these expected frequencies, we assume H_o is true.** That is, **we assume that the classifications of gender and one's view on capital punishment are independent.** This means that since 30 of the 100 people sampled favor capital punishment then we would *expect* 30% of the 60 males to favor capital punishment. Similarly, we would also *expect* 30% of the 40 females to favor capital punishment. Therefore, the expected number of males favoring capital punishment is

$$\frac{(30)(60)}{(100)} = 18$$

and the expected number of females favoring capital punishment is

$$\frac{(30)(40)}{(100)} = 12$$

Table 14.3 represents the *expected frequencies* for each of the cells a through f. Notice that the value in cell (a) in Table 14.3 is 18. This cell value represents the number of males that are *expected* to favor capital punishment. Also the value in cell (d) in Table 14.3 is 12. This cell value represents the number of females that are *expected* to favor capital punishment.

Each of the remaining cell values are obtained based upon the assumption that there is *no relationship* between gender and one's view on capital punishment. The expected frequency for a particular cell may be calculated by multiplying its row total by its column total and then dividing the product by the sample size. Therefore, the value in cell (a), the expected frequency for cell (a), could have been obtained by the formula:

$$\text{expected frequency for cell a} = \frac{\text{(row total for cell a)} \text{(column total for cell a)}}{(100)}$$

$$\text{expected frequency for cell a} = \frac{(60)(30)}{100}$$

$$\text{expected frequency for cell a} = 18$$

Therefore, each of the remaining expected cell values can be obtained using the formula:

$$\text{expected frequency for a cell} = \frac{\text{(row total for the cell)} \text{(column total for the cell)}}{\text{sample size}}$$

Table 14.3
Table of Expected Frequencies
View on Capital Punishment

Gender	View on Capital Punishment			
	Favor	Against	Undecided	Row Totals
Male	a 18	b 30	c 12	60
Female	d 12	e 20	f 8	40
Column Totals	30	50	20	100

The observed and expected frequencies are conveniently summarized in Table 14.4. In each cell, the expected frequency is shown in the upper right corner and the observed cell frequency is in the center of the cell.

Table 14.4
Table of Observed and Expected Frequencies
View on Capital Punishment

Gender	View on Capital Punishment			
	Favor	Against	Undecided	Row Totals
Male	a 20 [18]	b 25 [30]	c 15 [12]	60
Female	d 10 [12]	e 25 [20]	f 5 [8]	40
Column Totals	30	50	20	100

Step 3. *Formulate the decision rule.*

a) The significance level is $\alpha = 1\%$.
b) The degrees of freedom is found using the formula:

$$df = (r - 1)(c - 1)$$

For this contingency table the row classification of **gender** has a row value of 2; the column classification of **view on capital punishment** has a column value of 3.

$$df = (2 - 1)(3 - 1)$$

$$df = 2$$

Thus, this 2 by 3 contingency table has **two degrees of freedom.**

c) The critical value, χ^2_α, is found in TABLE IV. For $\alpha = 1\%$ and $df = 2$,

$$\chi^2_\alpha = 9.21$$

The hypothesis testing model is a chi-square distribution with $df = 2$ and $\chi^2_\alpha = 9.21$.

d) The decision rule is : reject H_o if the X^2 statistic is greater than 9.21. This is illustrated in Figure 14.5.

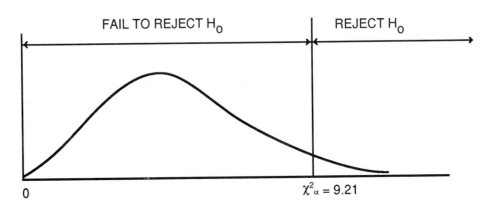

Figure 14.5

Step 4. *Calculate the X^2 statistic.* For this example we will use the definition formula to calculate Pearson's chi-square statistic. The definition formula is:

$$X^2 = \sum_{all\ cells} \frac{(O-E)^2}{E}$$

Table 14.5 has been constructed to aid in the computation of this formula. The observed and expected cell frequency values were obtained from Table 14.4.

Table 14.5

Cell	Observed Frequency (O)	Expected Frequency (E)	(O - E)	(O - E)²	$\frac{(O-E)^2}{E}$
a	20	18	2	4	0.22
b	25	30	-5	25	0.83
c	15	12	3	9	0.75
d	10	12	-2	4	0.33
e	25	20	5	25	1.25
f	5	8	-3	9	1.13

$$X^2 = \sum_{all\ cells} \frac{(O-E)^2}{E} = 4.51$$

To determine the X^2 statistic *sum the values* in the last column labelled: $\frac{(O-E)^2}{E}$.

Thus, $X^2 = 4.51$

Pearson's X^2 statistic could have been calculated using the computational formula:

$$X^2 = \sum_{all\ cells} (\frac{O^2}{E}) - n$$

We will illustrate the use of the computational formula in Example 14.2.

Step 5. *Determine the conclusion.*

Since the X^2 statistic value of 4.51 is less than χ^2_α value of 9.21 we **fail to reject the null hypothesis at** $\alpha = 1\%$. Thus, *there is no statistically significant relationship between gender and one's view on capital punishment.* (i.e., the two classifications are independent). This is illustrated in Figure 14.6.

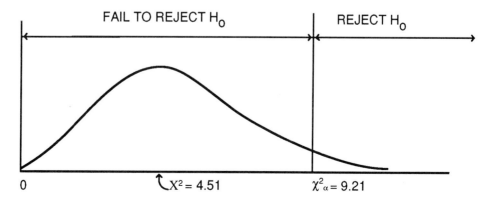

Figure 14.6 ∎

Example 14.2

A study was conducted to determine if there is a significant relationship between the frequency of times meat is served as a main meal per month for individuals living in eastern, central, and western United States. To help conduct the study, a questionnaire was administered to a random sample of 300 families The results of the questionnaire are summarized in Table 14.6.

Table 14.6

Living Area	Frequency of Meat as Main Meal Per Month			Row Totals
	Less Than 11	11 to 20	More Than 20	
Eastern US	a 46	b 36	c 20	102
Central US	d 40	e 40	f 26	106
Western US	g 16	h 38	i 38	92
Column Totals	102	114	84	300

Do these sample results indicate that there is a significant relationship between the frequency of meat served as a main meal per month and geographic living area at $\alpha = 5\%$?

Solution:

Step 1. *Formulate the two hypotheses, H_o and H_a.*

H_o: The frequency of meat served as a main meal per month and geographic living area are independent.

H_a: The frequency of meat served as a main meal per month and geographic living area are dependent.

Step 2. *Design an experiment and collect the data to test the null hypothesis, H_o.*

The results of the experiment are summarized in Table 14.6. To calculate the chi-square statistic, an expected frequency table representing the expected frequency for each cell must be developed. The expected frequency table is based upon the assumption that the null hypothesis is true. Each of the cell entries is found by the formula:

$$\text{expected frequency for a cell} = \frac{(\text{row total for the cell})(\text{column total for the cell})}{\text{sample size}}$$

The expected and observed cell frequencies are summarized in Table 14.7.

Table 14.7

Living Area	Frequency of Meat as Main Meal Per Month			Row Totals
	Less Than 11	11 to 20	More Than 20	
Eastern US	a 46 34.68	b 36 38.76	c 20 28.56	102
Central US	d 40 36.04	e 40 40.28	f 26 29.68	106
Western US	g 16 31.28	h 38 34.96	i 38 25.76	92
Column Totals	102	114	84	300

Step 3. *Formulate the decision rule.*

a) The significance level is $\alpha = 5\%$.
b) The degrees of freedom, *df*, is found using the formula:
$df = (r - 1)(c - 1)$
The degrees of freedom for a 3 by 3 contingency table is:
$df = (3 - 1)(3 - 1)$
Thus,
$df = 4$

c) Using TABLE IV, the critical χ^2 value for $\alpha = 5\%$ and $df = 4$ is:

$$\chi^2_\alpha = 9.49$$

The hypothesis testing model is a chi-square distribution with $df = 4$ and $\chi^2_\alpha = 9.49$.

d) The decision rule is: reject H_o if the X² statistic is greater than or equal to 9.49. Figure 14.7 illustrates the decision rule for this hypothesis test.

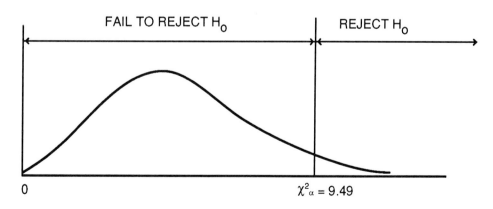

Figure 14.7

Step 4. *Calculate the X² statistic.*

For this example we will use the computational formula for the chi-square statistic.

$$X^2 = \sum_{all\ cells} \left(\frac{O^2}{E}\right) - n$$

Table 14.8 has been constructed to aid in the computation of this formula.

Table 14.8

Cell	O	O^2	E	$\dfrac{O^2}{E}$
a	46	2116	34.68	61.01
b	36	1296	38.76	33.44
c	20	400	28.56	14.01
d	40	1600	36.04	44.40
e	40	1600	40.28	39.72
f	26	676	29.68	22.78
g	16	256	31.28	8.18
h	38	1444	34.96	41.30
i	38	1444	25.76	56.06

$$n = 300 \qquad \sum_{allcells} \left(\dfrac{O^2}{E}\right) = 320.90$$

To compute the X^2 statistic using the computational formula: $X^2 = \sum_{allcells}\left(\dfrac{O^2}{E}\right) - n$

1) Sum the values in the last column labelled $\dfrac{O^2}{E}$ in Table 14.8.

Thus, $\sum_{allcells}\left(\dfrac{O^2}{E}\right) = 320.90$

2) The sample size, n, is 300.

Therefore,

$$X^2 = \sum_{allcells}\left(\dfrac{O^2}{E}\right) - n$$

$$= 320.90 - 300$$

$$X^2 = 20.90$$

Step 5. *Determine the Conclusion.*

Since the X^2 statistic value of 20.90 is greater than the χ^2_α value of 9.49, we reject H_o and accept H_a at $\alpha = 5\%$. These results are illustrated in Figure 14.8.

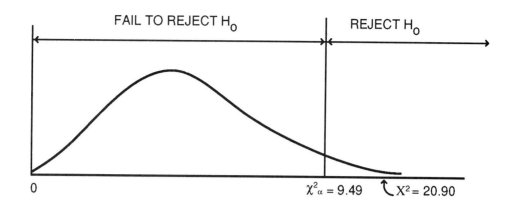

Figure 14.8

Thus, *there is a **statistically significant** relationship between the frequency of times meat is served as a main meal and one's geographic living location* at $\alpha = 5\%$. ∎

Case Study 14.1

Examine the USA Snapshot entitled *Music for the ages* in Figure CS14.1, which presents the percentage of tapes, CDs and records for the indicated age groups.

Figure CS14.1

a) Identify the population classifications presented within Figure CS14.1.
b) If you were to perform a hypothesis test to determine if there is a relationship between the population classifications, how would the null hypothesis be stated?
c) How would the alternative hypothesis be stated?
d) When the null hypothesis is assumed to be true, what does that indicate about the

relationship between the two population classifications?

e) How many age categories are presented within the USA snapshot? If for each age category, we classify the people as either purchasing or not purchasing music, then what would be the size of the contigency table for this type of chi-square test?

f) Assuming 2,000 people were randomly selected for the information contained within the USA snapshot, where 800 people were aged 10 - 19 years, 1000 were 20 - 39 and 200 were 40 or older, then determine the number of people that would fall within the categories of the contingency table.

g) What are these values within the contingency table called?

h) Determine the expected frequencies for this hypothesis test.

i) What is the number of degrees of freedom for this test?

j) Determine the critical value for the 5% and 1% level of significance.

k) State a decision rule for each level of significance.

l) Calculate Pearson's Chi-Square statistic for this test.

m) What is the conclusion for this test, if $\alpha = 5\%$? Can the null hypothesis be rejected at $\alpha = 5\%$? Can you conclude that the two population classifications are independent?

n) Is there a statistically significant relationship between the population classifications at $\alpha = 1\%$? Can the null hypothesis be rejected at $\alpha = 1\%$? Can you conclude that the two population classifications are dependent?

o) Can you conclude that the p-value of this hypothesis test is less than 5%? Explain.

p) Can you conclude that the p-value of this hypothesis test is less than 1%? Explain.

14.4 Assumptions Underlying the Chi-Square Test

When using the χ^2 distribution to perform a hypothesis test, several conditions must be satisfied. These conditions are referred to as the **assumptions underlying the chi-square test**. Let's now discuss these assumptions.

Assumptions Underlying the Chi-Square Test

- Each of the observations must be **independent of all other observations**.

- Each observation must be **classified into one and only one cell**. Care must be exercised to avoid designing classifications that allow an observation to fall into more than one cell.

- Perhaps the most important condition necessary to perform a X^2 test is that the sample size, n, be sufficiently large. To be sufficiently large means that the **smallest expected cell frequency is at least five**.

Summary of The Chi-Square Hypothesis Test of Independence

The following outline is a summary of the 5 step hypothesis testing procedure for The Chi-Square Test of Independence.

Step 1. *Formulate the two hypotheses, H_o and H_a.*
H_o: The two population classifications are independent.
H_a: The two population classifications are dependent.

Step 2. *Design an experiment to test the null hypothesis, H_o.*
Under the assumption that H_o is true, calculate the expected cell frequencies using the formula:

expected cell frequency $= \dfrac{(RT)(CT)}{n}$

Step 3. *Formulate the decision rule.*
All chi-square tests of independence will be one-tailed tests on the right, since a significant Pearson's X^2 statistic will always be a large positive number. To determine the critical chi-square value, perform the following steps:
a) Identify the significance level, $\alpha = 1\%$ or $\alpha = 5\%$.
b) Calculate the degrees of freedom: $df = (r - 1)(c - 1)$.
c) Determine the critical χ^2 value, denoted χ^2_α, using TABLE IV: Critical Values for the Chi-Square Distribution, χ^2_α, found in Appendix D: Statistical Tables, and draw a chi-square distribution model with the χ^2_α value appropriately placed on the curve.
d) State the decision rule.
The decision rule is: reject H_o if X^2 statistic is greater than χ^2_α.

Step 4. *Conduct the experiment.*
Calculate Pearson's X^2 statistic, using either of the following formulas:

$$X^2 = \sum_{\text{all cells}} \frac{(O-E)^2}{E} \quad \text{or}$$

$$X^2 = \sum_{\text{all cells}} \left(\frac{O^2}{E}\right) - n$$

Step 5. *Determine the conclusion.*
Compare the value of Pearson's chi-square statistic, X^2, to the critical value of chi-square, χ^2_α, either:

a) **Reject H_o and Accept H_a at α,** if X^2 is greater than or equal to the critical value of chi-square, χ^2_α.

or

b) **Fail to Reject H_o at α,** if X^2 is less than the critical value of chi-square, χ^2_α.

14.5 Test of Goodness-of-Fit

Another important application of the chi-square distribution is the determination of how well sample data fits a particular theoretical probability model. This type of Chi-Square Test is referred to as a **Goodness-of-Fit Test**.

The Goodness-of-Fit Test is used to determine whether a particular population can be described by a known theoretical distribution model. Throughout the text, we have always assumed that the distribution we were working with fit a particular theoretical distribution. For example, in several chapters, we assumed that certain distributions were normal. Using the Goodness-of-Fit Test, we can use the chi-square distribution to determine whether a particular distribution is approximately normal by comparing a random sample of data selected from the distribution to a theoretical normal distribution.

The chi-square statistic is used to compare the sample data to the theoretical distribution model, like a normal distribution, to see if the model provides a "good fit" for the data. This type of statistical test is called a goodness-of-fit test. The procedure to perform a Goodness-of-Fit-test involves the same five step hypothesis testing procedure that we've used in the preceding chapters with a few modifications. Example 14.3 outlines the procedure for a goodness-of-fit test.

Example 14.3

A market researcher wants to determine if the color of a clients' packaging material is an important factor when a consumer is selecting their product. To answer this question the researcher selects 800 consumers at random and gives them four different colored packages of the same product. However, the consumers are told that the four packages have different chemical properties and that the different color wrappers are to be used only for identification. Six weeks later each person in the sample is asked to complete an order form indicating which one of the four packages they preferred. To insure that the consumer selects the color they truly prefer, each consumer will be sent a free months supply of the product of their choice. The results of this survey are given in Table 14.9.

Table 14.9
Survey Results

Color of Wrapper	Pink	Blue	Brown	Grey
Number of consumers selecting wrapper	250	300	100	150

Remark: In this example the market researcher is trying to determine if the consumer prefers a particular colored package. If color is *not* a factor in the consumer's selection then we would expect an *equal* number of consumers choosing each color package. Whenever a statistician expects an equal number of observations in each category, the theoretical distribution model is called the ***uniform distribution***. The Goodness-of-Fit test can now be used to determine if the observed data in Table 14.9 fits the theoretical uniform distribution.

Solution:

We will modify the five step hypothesis testing procedure used to perform a chi-square test of independence to outline the Goodness-of-Fit test.

Step 1. *Formulate the two hypotheses H_o and H_a.*
In a Goodness-of-Fit test, the **null hypothesis** is stated as:

Null Hypothesis Form

H_o: the sample data fits a particular distribution

In this example we are trying to determine if there is an equal preference for each color. Therefore, the null hypothesis, H_o, is stated:

H_o: The sample data fits a uniform distribution.

In a *Goodness-of-Fit* test, the **alternative hypothesis** is stated as:

Alternative Hypothesis Form

H_a: the sample data does *not* fit the particular distribution

---

If color is a significant factor in consumer selection of this product then the alternative hypothesis, H_a, is:

H_a: The sample data does not fit a uniform distribution.

Step 2. *Design an experiment to test the null hypothesis, H_o*
To calculate the expected frequency for each cell, we assume the null hypothesis to be true. Thus **if we assume that the sample data fits a uniform distribution, then all the expected frequencies would be equal.** Thus, the expected frequency for each cell of the four cells is 200, since 800 consumers responded to the survey. The observed and expected frequencies are summarized in Table 14.10.

Table 14.10
Observed and Expected Frequencies

Color of Wrapper	Pink	Blue	Brown	Grey	Total
Number of consumers selecting wrapper	a 250 200	b 300 200	c 100 200	d 150 200	800

Step 3. *Formulate the decision rule.*

a) In this example we will use a significance level $\alpha = 1\%$.

b) In this example the sample data is only classified by color. This is referred to as a single classification case. **In a Goodness-of-Fit test, the degrees of freedom are dependent upon the number of categories within a single classification case.**

> The degrees of freedom, *df*, for a single classification case is:
>
> $$df = k - 1$$
>
> where k equals the number of categories within the single classification.

Since there are four categories of color, then k = 4.
Thus,
$$df = 4 - 1$$
$$df = 3$$

c) The critical χ^2 value for this test can be found in TABLE IV where $\alpha = 1\%$ and $df = 3$.

The critical χ^2 value is:

$$\chi^2_\alpha = 11.34$$

d) The decision rule is: reject H_o if the X^2 statistic is greater than 11.34. The hypothesis testing model is illustrated in Figure 14.9.

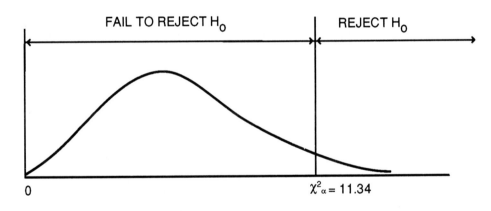

Figure 14.9

Step 4. *Calculate the X^2 statistic using the definition formula:*

$$X^2 = \sum_{all\,cells} \frac{(O-E)^2}{E}$$

Table 14.11 is used for the computation of the X^2 statistic.

Table 14.11

Cell	Observed Frequency (O)	Expected Frequency (E)	(O - E)	(O - E)²	$\frac{(O - E)^2}{E}$
a	250	200	50	2500	12.50
b	300	200	100	10000	50.00
c	100	200	-100	10000	50.00
d	150	200	-50	2500	12.50

$X^2 = 125.00$

Step 5. *Determine the conclusion.*

Since the X^2 statistic value of 125.00 is greater than χ^2_α value of 11.34, we reject H_o and accept H_a at $\alpha = 1\%$. **Thus the sample data does not fit a uniform distribution.** This conclusion is illustrated in Figure 14.10.

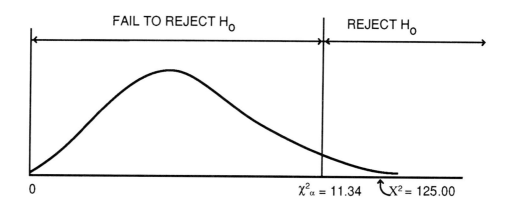

Figure 14.10

Therefore, color is a significant factor in consumer selection of this product. ■

Example 14.4

Table 14.12 indicates the proportion of viewers per network that watched the six o'clock news last year.

Table 14.12
Percent of Viewers

Network	Percent of Viewers
ABC	30
CBS	25
NBC	30
Independent	15

This year a random sample of 400 viewers was conducted to determine if the number of viewers watching the six o'clock news on each of the networks has significantly changed from last year's results. The results of the sample data is listed in Table 14.13.

Table 14.13
Sample of Viewers

Network	Number of Viewers
ABC	110
CBS	95
NBC	130
Independent	65

Based on the results of this sample data can we conclude that this year's viewing audience has significantly changed from last year's? Use $\alpha = 5\%$.

Solution:

The goodness-of-fit test will be used to test whether this year's sample data fits last year's viewing distribution.

Step 1. *Formulate the two hypotheses H_o and H_a.*

H_o: The sample data fits last year's viewing distribution.

H_a: The sample data does not fit last year's viewing distribution.

Step 2. *Design an experiment to test the null hypothesis, H_o.*

To calculate the expected frequency for each cell, we assume the null hypothesis to be true. Thus if we assume that the sample data fits last year's viewing distribution, then the expected frequencies are calculated from last year's viewing audience. For example, 30 percent of last year's viewing audience watched the six o'clock news on the ABC network. Thus, for this year's sample of 400 viewers, we would expect 30 percent of the viewers to have watched the six o'clock news on ABC network. Therefore we would expect 120 viewers to watch ABC's six o'clock news. The expected and the observed frequencies are summarized in Table 14.14.

Table 14.14
Observed and Expected Frequencies

Network	ABC	CBS	NBC	Independent	Total
Number of Viewers	a 110 120	b 95 100	c 130 120	d 65 60	400

Step 3. *Formulate the decision rule.*

a) In this example we will use a significance level $\alpha = 5\%$.

b) In this example the sample data is only classified by network. The degrees of freedom, df, for a single classification case is $df = k - 1$, where k is the number of categories within the single classification.
Since there are four networks, then k = 4.
Thus,
$$df = 4 - 1$$
$$df = 3$$

c) The critical χ^2 value for this test can be found in TABLE IV where α = 5% and df = 3.

The critical χ^2 value is: $\chi^2_\alpha = 7.82$

d) The decision rule is: reject H_0 if the X^2 statistic is greater than 7.82. The hypothesis testing model is illustrated in Figure 14.11.

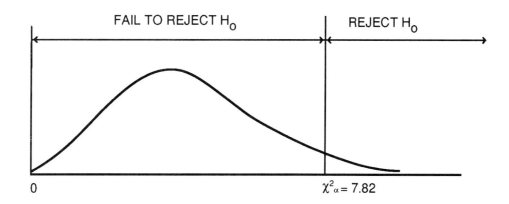

Figure 14.11

Step 4. *Calculate the X^2 statistic using the definition formula:*

$$X^2 = \sum_{all\,cells} \frac{(O-E)^2}{E}$$

Table 14.15 is used for the computation of the X^2 statistic.

Table 14.15

Cell	Observed Frequency (O)	Expected Frequency (E)	(O - E)	(O - E)²	$\frac{(O-E)^2}{E}$
a	110	120	-10	100	0.83
b	95	100	-5	25	0.25
c	130	120	10	100	0.83
d	65	60	5	25	0.25

$X^2 = 2.16$

Step 5. *Determine the conclusion.*

Since the X^2 statistic (2.16) is less than χ^2_α (7.82) we fail to reject H_0 at $\alpha = 5\%$. This conclusion is illustrated in Figure 14.12.

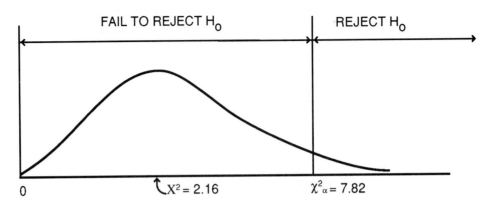

Figure 14.12

Therefore, we can conclude that this year's viewing distribution has not significantly changed from last year's. ∎

Case Study 14.2

A research article entitled *The Rekurring Kase Of The Special K* examined whether the letter K occurs with more frequency as the initial letter in top brand names than it does as a first letter in our language in general. This study appeared in the October/November 1990 issue of the Journal of Advertising Research. One aspect of this research was to examine the eight most frequently occurring initial letters in top name brands which were: **C, S, A, B, M, P, D and T**. The researcher is trying to determine if they occur more frequently as the initial letter in top brand names than they appear as a first letter in our language in general.

These eight most frequently occurring letters were selected by the researchers from a composite list of unduplicated top brand names. From this list, only unique names were selected. Using only unique names eliminated overrepresentation of certain letters. The expected frequency for each of these letters were based on Webster's Ninth New Collegiate Dictionary(1987) model. The result of this chi-square test is shown in Figure CS14.2.

The second aspect of this research was to determine what happens to the analysis when the letter K is added to this list. This aspect of the hypothesis test will be examined in the: *PART V What Do You Think?* question set of the exercise section at the end of this chapter.

Chi-Square Analysis of Most Frequently Occurring Initial Letters of Top Brand Names

Initial letter	Observed frequency	Expected* frequency	Cumulative percent Observed	Cumulative percent Expected
C	51	52.4	9.43	9.68
S	46	67.3	17.93	22.12
A	42	32.2	25.69	28.08
B	42	29.2	33.45	33.47
M	40	29.5	40.84	38.92
P	40	47.5	48.23	47.70
D	30	28.2	53.78	52.91
T	29	32.0	59.14	58.83

$X^2 = 20.677$ df = 7 $p < .005$

Figure CS14.2

*Based on Webster's Ninth New Collegiate Dictionary (1987), distribution of words beginning with each letter.

a) Define the population for this test.
b) What variable is being examined within this test?
c) Is this variable numeric or categorical? Explain.
d) What type of chi-square test is being illustrated in this research?
e) What distribution serves as the theoretical distribution?
f) What serves as the sample data for this test?
g) What would serve as the null hypothesis for this test?
h) What would serve as the alternative hypothesis?
i) What assumption was used to calculate the expected frequencies within the table? Which hypothesis was used for this assumption?
j) How many classifications are there for this test? State the classification(s) for this test. What does each category within a classification represent for this test?
k) What formula was used to calculate the number of degrees of freedom for this test? What are the degrees of freedom for this test?
l) What is the value of $\dfrac{(O - E)^2}{E}$ for each initial letter of the information contained within the table of Figure CS14.2, where O is the Observed frequency and E is the Expected frequency? What does the sum of the values of $\dfrac{(O - E)^2}{E}$ represent?
m) What is the critical chi-square value for this test at $\alpha = 5\%$ and $\alpha = 1\%$?
n) What is the conclusion of this hypothesis test for each of these significance levels? Can you reject the null hypothesis for any of these significance levels? Explain.
o) What is the conclusion to this test for a significance level of 0.75%? What information did you use to answer this question? Explain.
p) What is the smallest level of significance that you can use to obtain a significant result?
q) State the conclusion to the chi-square test shown in Figure CS14.2.
r) Was the result of the chi-square significant for the hypothesis test shown in Figure CS14.2? Explain.

14.6 USING MINITAB

In this chapter we have examined the chi-square statistic as a technique to analyze categorical data. Two hypothesis tests were discussed:

(1) Chi-Square Test of Independence, and (2) Test of Goodness-of-Fit. Let's use MINITAB to perform these tests.

In Example 14.2, a contingency table was given that summarized the results for a random sample of 300 families that were administered a questionnaire to explore the relationship between the frequency of times meat is served as a main meal per month and a family's living area. The table for this example is repeated here.

Table 14.16

Living Area	Frequency of Meat as Main Meal Per Month			Row Totals
	Less Than 11	11 to 20	More Than 20	
Eastern US	a 46	b 36	c 20	102
Central US	d 40	e 40	f 26	106
Western US	g 16	h 38	i 38	92
Column Totals	102	114	84	300

Do these sample results indicate that there is a significant relationship between frequency of meat served as a main meal per month and geographic living area? The answer to this question was explored within example 14.2 using the chi-square test of independence.

Before MINITAB can calculate the X^2 statistic, additional commands must be applied in order to complete the analysis. Thus, in order to use MINITAB to conduct a X^2 hypothesis test of independence the following sequence of steps should be followed:

(1) Enter the observed frequencies using a READ command.

Since there are three columns of observed data the READ command is

```
MTB> READ C1 C2 C3
DATA> 46 36 20
DATA> 40 40 26
DATA> 16 38 38
DATA> END
```

(2) Use the CHISQUARE command applied to the columns for which the data values are contained.

```
MTB> CHISQUARE C1 C2 C3
```

The output for these commands is:

```
Expected counts are printed below observed counts

            C1       C2       C3     Total
   1        46       36       20       102
         34.68    38.76    28.56

   2        40       40       26       106
         36.04    40.28    29.68

   3        16       38       38        92
         31.28    34.96    25.76

Total      102      114       84       300

ChiSq =  3.695 +  0.197 +  2.566 +
         0.435 +  0.002 +  0.456 +
         7.464 +  0.264 +  5.816  =  20.895
df = 4
```

If you examine the output you will notice that **both** the observed and the expected cell values are contained within the contingency table. Compare this output to the contingency table in Table 14.16. Notice that the value of X^2 is given and is 20.895 along with the degrees of freedom, df = 4. However, unlike other MINITAB testing analysis, what is missing is the p-value. Thus, additional steps must be completed to interpret the significance of the X^2 statistic. This could be done by using a table of significant X^2 values or by using MINITAB. Let's use MINITAB.

(3) Find the p-value of the X^2 statistic using the sequence:

```
MTB> CDF (X² statistic) K1;
SUBC> CHISQUARE (df).
MTB> LET K2=1-K1
MTB> PRINT K2
```

Thus for this illustration we have:

    MTB>   CDF 20.895 K1;
    SUBC> CHISQUARE 4.
    MTB>   LET K2=1-K1
    MTB>   PRINT K2

Using this sequence will produce the p-value. The p-value is indicated on the output as K2:

K2   0.000299990

Thus the p-value for the X^2 test statistic is 0.00029990.
Is the test statistic significant? What is the maximum allowable significance level?

Oftentimes, data values are included in a file from which many of the variables are categorical. It is possible to apply the chi-square hypothesis test to these categorical variables by using the TABLE command followed by the CHISQUARE subcommand. Output that is similar to the output from the previous illustration will be produced.

The second application of the chi-square statistic discussed in this chapter was concerned with a Test of the Goodness-of-Fit. In Example 14.4 a random sample of 400 viewers was explored to investigate if the number of viewers watching the six o'clock news on each of the networks significantly changed from last year's results. The null and alternative hypotheses for this goodness-of-fit test is:

H_o: The sample data fits last year's viewing distribution.

H_a: The sample data does not fit last year's viewing distribution.

The results of the sample data along with the expected values calculated from the previous year's data is listed in Table 14.17.

Table 14.17
Observed and Expected Frequencies

Network	ABC	CBS	NBC	Independent	Total
Number of Viewers	a 110 120	b 95 100	c 130 120	d 65 60	400

Unfortunately, MINITAB does not have a goodness-of-fit test. However, we can use Pearson's Chi-Square test to determine the goodness-of-fit. The illustration that we are examining is comparing last year's viewing numbers to this year's. Table 14.17 contains the listing for the number of viewers this year. These numbers are the observed frequencies within the table. The MINITAB SET command is used to input these numbers in a MINITAB worksheet.

MTB >   SET C1
DATA>   110 95 130 65
DATA>   END

Similarly, the SET command is used to input the expected frequencies of Table 14.17.

MTB >   SET C2
DATA>   120 100 120 60
DATA>   END

The LET command is used to calculate the values for $\frac{(O-E)^2}{E}$ for each cell pair of O and E.

MTB > LET C3=(C1-C2)**2/C2

Next the NAME command is used to name columns C1, C2, and C3 as 'OBS' for observed frequency, 'EXP' for expected frequency, and 'RSD' for $\frac{(O-E)^2}{E}$.

MTB > NAME C1 'OBS' C2 'EXP' C3 'RSD'

Adding the values in 'RSD' we can calculate the chi-square statistic, X^2.

```
MTB >  SUM('RSD') K1
SUM    =    2.3333
```

Thus, X^2 is 2.333. As in the previous illustration, the p-value for this test statistic must be found.

Therefore, the p-value for a chi-square value of 2.333 with $df = 3$ is found using the MINITAB commands:

```
MTB >   CDF 2.333 K1;
SUBC>   CHISQUARE 3.
MTB >   LET K2=1-K1
MTB >   PRINT K2
K2         0.506798
```

The p-value, listed as K2, is 0.506798, for $X^2 = 2.333$. Is this p-value significant? What is the conclusion regarding the goodness-of-fit? Do these sample data fit last year's viewing distribution?

GLOSSARY

TERM	SECTION
Classifications	14.1
Independent	14.1
Dependent	14.1
Qualitative (Count) data	14.1
Chi-Square Test of Independence	14.1
Frequency	14.1
Observed Cell Frequency	14.1
Expected Cell Frequency	14.1, 14.3
Contingency Table	14.1
Cell	14.1
Pearson's Chi-Square Statistic, X^2	14.1
Chi-Square Distribution	14.2
Symmetric	14.2
Skewed	14.2
Degrees of Freedom df	14.2, 14.3, 14.5
Definition Formula for Pearson's X^2	14.3
Computational Formula for Pearson's X^2	14.3
Critical Chi-Square, χ^2_α	14.3
Goodness-of-Fit Test	14.5
Uniform Distribution	14.5

Exercises

PART I Fill in the blanks.

1. A population can be divided by gender, age, religion, marital status, etc. These divisions are called ____.
2. A statistical technique used to investigate whether the classifications of a population are related is the Chi-Square Test of ____.
3. In a Chi-Square Test of Independence the sample data represents qualitative or ____ data.
4. The sample count data falling into each of the cells of the population classifications are called the ____ frequencies.
5. An arrangement of sample count data into a two-way classification is called a ____ table.
6. Each individual box in the contingency table is referred to as a ____.
7. A 4 x 3 contingency table has ____ rows and ____ columns.
8. a) The definition formula for computing Pearson's Chi-Square statistic is: $X^2 =$ ____
   b) The computational formula for computing Pearson's Chi-Square statistic is: $X^2 =$ ____.
9. The Pearson's Chi-Square statistic is computed by comparing the observed frequencies to the ____ frequencies of each cell within the contingency table.
10. The value of the Chi-Square is always ____.
11. The graph of a Chi-Square distribution is not symmetric, but it is skewed to the ____.
12. There are a family of Chi-Square distributions where each chi-square distribution is dependent upon the number of ____. The formula to calculate df is given by: $df =$ ____.
13. In a Chi-Square test of Independence, the null hypothesis is stated as: the two population classifications are ____.
    The alternative hypothesis is stated as: the two population classifications are ____.
14. In a Chi-Square test of Independence, the expected frequency for a particular cell is calculated by multiplying its row ____ by its ____ total and dividing this product by the ____.
15. In formulating the conclusion of a Chi-Square test of Independence when:
    a) the value of the X^2 statistic is less than the critical value of chi-square, χ^2_α, then the conclusion is ____.
    b) the value of the X^2 statistic is greater than or equal to the critical value of chi-square, χ^2_α, then the conclusion is ____.
16. The assumptions underlying the Chi-Square test of Independence are:
    a) Each of the observations must be ____ of all the other observations.
    b) Each observation must be classified into one and only one ____.
    c) The sample size must be sufficiently large so that the smallest expected cell frequency is at least ____.
17. The Goodness-of-Fit Chi-Square Test is used to determine if the sample data fits a particular ____ distribution model.
18. In a Goodness-of-Fit Test, the null hypothesis is stated as:
    the ____ data fits a particular theoretical distribution model.
19. In a Goodness-of-Fit Test, the degrees of freedom of a single classification case with k categories is: $df =$ ____.

PART II Multiple choice questions.

1. How many degrees of freedom does a 3 by 4 contingency table have? a) 3 b) 12 c) 6 d) 4
2. What is the critical value for a χ^2 test of independence for a 2 by 5 contingency table using $\alpha = 1\%$?
   a) 13.28 b) 11.07 c) 5.99 d) 9.49
3. What are the degrees of freedom for a Goodness-of-Fit test with a single classification of six categories?
   a) 4 b) 5 c) 6 d) 3
4. If the value of Pearson's chi-square statistic is 6.79 and $\chi^2_\alpha = 9.49$, what is the conclusion of this test?
   a) reject H_o and accept H_a
   b) fail to reject H_o
   c) reject H_o and reject H_a d) fail to reject H_a
5. In a χ^2 test for independence the null hypothesis is always
   a) the distribution is normal
   b) the population classifications are independent
   c) the population classifications are dependent
   d) the distribution is uniform

6. When doing a χ^2 test what must be true about the sample size?
   a) at least 50
   b) no more than 300
   c) large enough so that the smallest expected cell frequency is at least 5
   d) large enough so that each cell in the expected frequency is at least 30

7. What type of test is used to determine if the sample data conforms to a theoretical distribution model?
   a) sample mean hypothesis test
   b) goodness-of-fit test
   c) chi-square test of independence
   d) a normal distribution test

8. The expected frequency for a particular cell is calculated by multiplying its row total by its column total and dividing by
   a) the degrees of freedom b) the critical χ^2
   c) the sample size d) number of classifications

9. In designing a contingency table, care must be taken so that each observation falls into
   a) at least one cell b) no more than two cells
   c) only one cell d) none of these choices

10. For a sample of 200, what is the expected frequency for a cell whose row total is 70 and its column total is 30?
    a) 10.5 b) 21   c) 105 d) 2100

PART III Answer each statement True or False.

1. Chi-Square tests of independence involve the analysis of quantitative continuous data.
2. The shape of the chi-square distribution is similar to the shape of the normal distribution.
3. All chi-square tests of independence are one-tailed tests on the right.
4. If the value of the X² statistic is greater than or equal to the critical value of chi-square, χ_α^2, then the conclusion is to reject H_o and accept H_a at α.
5. If the conclusion is to reject H_o and accept H_a at α, then the chi-square test of independence has indicated that the population classifications are independent.
6. If the conclusion is that the classifications are independent then the Pearson's chi-square statistic, X², is greater than the χ_α^2.

7. A calculated value of X² is less than the critical χ_α^2.

   The conclusion is: Fail to reject the null hypothesis at α.

8. The test of Goodness-of-Fit refers to a type of chi-square test that can be used to determine how well sample data fits the normal distribution.
9. A Goodness-of-Fit test is used to determine if sample data fits a uniform distribution. The conclusion is to reject H_o and accept H_a at α. This conclusion means that the sample data does not fit a uniform distribution.
10. A Goodness-of-Fit test is used to determine if sample data fits a uniform distribution. The conclusion is fail to reject H_o at α. This means that the data fits a normal distribution.

PART IV Problems.

Problems 1 - 15 use the chi-square test of independence. Problems 16 - 23 use the test of Goodness-of-Fit. For all of the problems:
   a) Formulate the hypotheses.
   b) State the decision rule.
   c) Calculate Pearson's chi-square statistic.
   d) Determine the conclusion, and answer the question(s) posed in the problem, if any.

1. A study was undertaken to determine whether or not membership in campus fraternity groups is related to grades. A random sample of 100 students was selected and the results are summarized in the following table:

Fraternity Group	Grade Point Average (GPA)			Row Totals
	Less Than 2.00	From 2.00 to 3.00	Greater Than 3.00	
Member	a 16	b 17	c 7	40
Non-member	d 9	e 28	f 23	60
Column Totals	25	45	30	100

Based on these sample results can we conclude that there is a relationship between membership in a campus fraternity group and GPA at $\alpha = 1\%$?

2. A study was conducted to determine if gender is related to the number of compact discs (CDs) purchased per year. A random sample of 200 CD buyers were interviewed and their responses are recorded in the following table:

Gender	Number of Compact Discs Purchased Per Year			
	0 - 20	21 - 40	41 and Higher	Row Totals
Male	a 40	b 50	c 30	120
Female	d 20	e 50	f 10	80
Column Totals	60	100	40	200

Based on the sample data can we conclude that gender is related to the number of CDs purchased per year at $\alpha = 5\%$?

3. A study was conducted to determine if a relationship exists between gender and movie preference. Two hundred students were questioned regarding their movie preference. The results of the survey are listed in the following table:

Gender	Movie Preference					
	Romance	Comedy	Drama	Adventure	Mystery	Row Totals
Male	a 10	b 20	c 5	d 40	e 15	90
Female	f 50	g 25	h 5	i 15	j 15	110
Column Totals	60	45	10	55	30	200

Based on the sample data can we conclude that gender is related to movie preference at $\alpha = 5\%$?

4. A quality control engineer wants to compare the production process of parts manufactured by three different companies. The engineer randomly samples a total of 270 parts from the three companies and summarizes the results in the following table.

Part Quality	Different Companies			
	Company A	Company B	Company C	Row Totals
Defective	a 15	b 10	c 10	35
Non-Defective	d 70	e 80	f 85	235
Column Totals	85	90	95	270

Based upon the results of the sample, can the quality control engineer conclude that part quality and manufacturer are independent at $\alpha = 5\%$?

5. A consumer research agency was asked to determine if the consumer preference in packaging material was independent of marital status. Three different types of packaging material (clear, white and dark plastic) were tested. The research agency randomly selected and interviewed 500 individuals to determine their packaging preference. These results are summarized in the following table:

Marital Status	Plastic Packaging Material			
	Clear	White	Dark	Row Totals
Married	a 200	b 75	c 25	300
Not-Married	d 50	e 75	f 75	200
Column Totals	250	150	100	500

Do these sample data indicate that marital status and preference in packaging material are independent at $\alpha = 1\%$?

6. In an experiment designed to compare the effectiveness of different brands of pain relievers, a group of 225 individuals who complained of chronic pain were randomly assigned to three daily treatments: three tablets of Brand A (prescription drug), three tablets of Brand B(non-prescription drug), or three placebos.

After the experiment, each subject was questioned to determine the effectiveness of their treatment. Their responses are classified in the following table.

Result of Treatment	Drug Treatment			Row Totals
	Brand A	Brand B	Placebo	
Relief from Pain	a 17	b 15	c 13	45
SomeRelief from Pain	d 33	e 34	f 38	105
No Relief from Pain	g 25	h 26	i 24	75
Column Totals	75	75	75	225

Based upon these sample results can one conclude that drug treatment and relief from pain are independent at $\alpha = 5\%$?

7. A Nassau County study was conducted to determine the relationship between a person's educational level and their selection to serve on a grand jury. A random sample of 240 people were interviewed for grand jury service and the selection results were classified with respect to the person's educational level. These results are summarized in the following table.

Educational Level	Grand Jury		Row Totals
	Selected	Not Selected	
Elementary	a 30	b 20	50
Secondary	c 63	d 47	110
Some College	e 15	f 31	46
College Degree	g 12	h 22	34
Column Totals	120	120	240

Do the results of the study indicate at $\alpha = 1\%$ that a person's selection to serve on the grand jury is dependent on educational level?

8. A financial investment broker believes that Dow Jones Average is related to the prime interest rate. She studies the Dow Jones Average and the prime interest rate for a three year period. Her results are summarized in the following table.

Prime Rate	Dow Jones Average			Row Totals
	Increase	Decrease	No Change	
Increase	a 17	b 15	c 13	45
Decrease	d 33	e 34	f 38	105
No Change	g 25	h 26	i 24	75
Column Totals	75	75	75	225

Do the results of the study indicate that the Dow Jones Average and the Prime Interest Rate are dependent at $\alpha = 1\%$?

9. John, a mathematics counselor, is interested in the relationship between math anxiety and the successful completion of the introductory college math course. A random sample of 300 college students enrolled in the introductory college math course were given a test to measure their anxiety level on the first day of the course. The results of the anxiety test and their final grade in the math course are given in the following table.

Grade	Anxiety Level		
	Low	Medium	High
Passed	a 59	b 32	c 29
Failed or Withdrew	d 31	e 58	f 91

Based upon the sample data, can you conclude that anxiety level and successful completion of the math course are independent at $\alpha = 1\%$?

10. Kate and Allie advertising would like to determine if gender is related to the number of compact discs purchased per year. A random sample of 300 people were interviewed and their responses are summarized in the following table.

Gender	Number of Compact Discs Purchased Per Year		
	0 - 20	21 - 40	Above 40
Male	a 50	b 55	c 35
Female	d 40	e 70	f 50

Based on the sample data can the advertising firm conclude that the number of compact discs purchased per year is based upon gender? (Use $\alpha = 5\%$)

11. A sociologist is researching the question: "Is there a relationship between the level of education and job satisfaction?". For a random sample of 270 workers she determines each individual's education level and level of job satisfaction. The results are recorded in the following table:

Education Level	Job Satisfaction Level		
	LOW	MEDIUM	HIGH
College	a 20	b 70	c 30
High School	d 15	e 60	f 25
Grade School	g 15	h 25	i 10

Perform a chi-square test of independence to determine if educational level and job satisfaction are dependent at $\alpha = 5\%$.

12. A statistics instructor decides to investigate the relationship between a student's grade in the course and class attendance. The instructor randomly samples the records of 120 students and the results are listed in the following table.

Number of Absences	Job Satisfaction Level		
	F - D$^+$	C - B	B$^+$ - A
0 - 1	a 10	b 18	c 12
2 - 5	d 25	e 17	f 8
above 5	g 20	h 5	i 5

Perform a chi-square test of independence to determine if a student's grade in the statistics course and class attendance are dependent at $\alpha = 1\%$?

13. Tom the Greek, a serious basketball handicapper, wants to determine if an NBA team's winning home record is independent of their division. He uses the results of last season's NBA games as his random sample and records the following information.

DIVISION	Home Record		DIVISION	Home Record	
Atlantic	Wins	Losses	Central	Wins	Losses
Team #1	35	6	Team #1	37	4
Team #2	30	11	Team #2	37	4
Team #3	32	9	Team #3	33	8
Team #4	30	11	Team #4	31	10
Team #5	17	24	Team #5	30	11
Team #6	12	29	Team #6	20	21

DIVISION	Home Record		DIVISION	Home Record	
Midwest	Wins	Losses	Pacific	Wins	Losses
Team #1	34	6	Team #1	35	6
Team #2	31	11	Team #2	35	6
Team #3	35	9	Team #3	31	10
Team #4	24	11	Team #4	29	12
Team #5	18	24	Team #5	28	13
Team #6	12	29	Team #6	21	20
			Team #7	17	24

a) Using this information, complete the following contingency table.

Home Record	NBA Division			
	Atlantic	Central	Midwest	Pacific
Wins	a	b	c	d
Losses	e	f	g	h

b) Based upon the sample data, can Tom conclude that a team's winning home record is independent of their NBA division at $\alpha = 1\%$?

14. A study was conducted on a random sample of 500 married couples to determine if there is a significant relationship between the frequency of times per year a married couple goes to the movies and their educational background. The results of the study are shown below:

Frequency of Movie Attendance Per Year	Educational Background (Highest Level Achieved)			
	Neither Went To College	Each Has Some College	Only One Grad College	Both Grad College
0 - 6	a 55	b 20	c 35	d 17
6 - 12	e 65	f 30	g 30	h 8
Above 12	i 55	j 75	k 45	l 65

Do these sample results indicate at $\alpha = 5\%$ that there is a significant relationship between frequency of times married couples go to the movies and their educational background?

15. A record company conducted a random survey of 500 people to determine if there was a relationship between a person's age and their favorite music category. The results of their survey are summarized in the following table.

Music Category	Age				
	Under 20 years	20 to 29	30 to 39	40 to 49	Over 49
Rock	a 45	b 32	c 29	d 14	e 10
Jazz	f 5	g 7	h 18	i 19	j 21
Country	k 7	l 11	m 22	n 15	o 15
Classical	p 5	q 8	r 11	s 24	t 32
Contemporary	u 23	v 35	w 34	x 33	y 25

Based on the survey results, can the record company conclude that a person's age and their favorite music category are dependent at $\alpha = 5\%$?

16. A marketing research agency conducted a random survey to determine the opinions of prospective customers for a new product. The results of the survey were:

Opinion	Number
outstanding	97
excellent	98
good	88
fair	67
not interested	70

If the opinions are uniformly distributed among the five categories then the marketing agency will not recommend the manufacturer to mass produce the new product. Test the null hypothesis that the sample data (opinions of the customers) fits a uniform distribution at $\alpha = 5\%$. Should the marketing agency recommend mass production of the new product to the manufacturer?

17. A market researcher wants to determine if the shape of a client's packaging material is an important factor is a consumer's selection of a product. The researcher randomly selects 2000 consumers an gives each of them five differently shaped packages of the same product. One month later each consumer is asked to indicate a preference for one of the shaped packages. The results of the survey are given in the table below.

Number of Consumers Selecting Package	Shape of Package				
	#1	#2	#3	#4	#5
	430	450	385	365	370

Use a goodness-of-fit test to test the null hypothesis that sample data fits a uniform distribution at $\alpha = 1\%$. Can you conclude that the sample data indicates that there is a preference for packaging shape?

18. "Let it Ride" Larry decides to test the fairness of a die. He rolls it 120 times with the results shown in the following table:

Number Rolled	Frequency
1	24
2	18
3	27
4	17
5	16
6	18

Use a Goodness-of-Fit test to determine if the die is fair at $\alpha = 5\%$.

19. According to recently published census figures the socio-economic stratification for a small northeast town is: 20% upper, 50% middle and 30% lower. One hundred and fifty residents of this town are supposedly chosen at random. To check this the sample data is sorted and summarized to be:

Socio-Economic Stratification	Percentage of Residents
Lower	37
Middle	86
Upper	27

Should we conclude that the sample was not randomly selected from the town using $\alpha = 1\%$?

20. A mathematics professor has analyzed the results of her statistics final exam for the past few years. She has determined that the grade distribution for her final exam is:

Grade	Percent of Students Receiving
A	10%
B	15%
C	40%
D	25%
F	10%

This year the professor administers the exam to 200 students. The number of students receiving the letter grades A, B, C, D, and F are 24, 20, 89, 40, and 27 respectively. Use the goodness-of-fit test to determine if this year's final exam results fit her previous grade distribution at $\alpha = 5\%$.

21. The quality control engineer of a parts manufacturing company wants to determine if more defective parts are produced on a particular day of the six day work week. A random sample of the production records indicate the following data:

Day of the Week	Number of Defective Parts
Monday	26
Tuesday	15
Wednesday	24
Thursday	27
Friday	12
Saturday	16

The engineer states the null hypothesis as: the sample data (number of rejected parts) fits a uniform distribution. Perform a goodness-of-fit test at $\alpha = 5\%$ and determine if the engineer can conclude that the production of defective parts is not evenly distributed throughout the workweek.

22. The student government at Fernworth College has been told that the student population consists of 58% female and 42% male. A random sample of 400 students consists of 54% female and 46% male. Are the sample data consistent with the population proportions at an $\alpha = 5\%$?

23. The Pit Boss at the Golden Flamingo casino wants to determine if a certain pair of dice is "loaded". She tosses the dice 1080 times and records the sum for each toss. The data for her experiment is listed below:

Sum of the Pair of Dice	Number of Times
2	24
3	70
4	77
5	128
6	156
7	156
8	148
9	136
10	111
11	48
12	26

Can the Pit Boss conclude that the dice are indeed "loaded" using $\alpha = 5\%$?

The following problem outlines a Goodness-of-Fit test to determine if the sample data fits a normal distribution.

24. It has been claimed that the number of hours worked per week by a full-time college student is normally distributed. To test this claim a random sample of 600 full-time college students were selected. It was determined that the sample data had a mean number of hours worked per week per student of 21.7 hrs with $s = 6.7$ hrs. Six intervals were used to summarize the data and the sample results are given in the following table.

Interval Boundaries	Frequency
Below 12.32	31
12.32 - 17.01	96
17.01 - 21.70	173
21.70 - 26.39	147
26.39 - 31.08	112
Above 31.08	41

Complete the following steps which outline a Goodness-of-Fit test to test the null hypothesis that the number of hours worked per week by a full-time student is normally distributed at $\alpha = 5\%$.

a) **Step 1.** *Formulate the two hypotheses H_o and H_a.*
   H_o _____

   H_a _____

b) **Step 2.** *Design an experiment to test the null hypothesis, H_o.*
   To calculate the expected frequency for each interval we assume the null hypothesis to be true, that is, we assume a normal distribution with
   $\mu \approx$ sample mean = 21.70 and $\sigma \approx s = 6.70$.
   The expected frequencies for each of the interval boundaries are calculated by:
   1) converting each boundary into a z score.
   2) obtaining the corresponding probabilities for the z scores from TABLE II: The Normal Distribution Table.
   3) calculating each expected frequency using these percentages and the sample size.

   Complete the following table using this procedure.

Interval Boundaries	z score	Probability	Expected Frequency
Below 12.32	-1.40	8.08%	48.48
12.32 - 17.01			
17.01 - 21.70			
21.70 - 26.39			
26.39 - 31.08			
Above 31.08			

c) **Step 3.** *Formulate the decision rule.*
   1) In this example we will use a significance level $\alpha =$ ___.
   2) **The degrees of freedom, *df*, for a single classification case is $df = k - 1$, where k is the number of categories within the single classification.**

   However, since the mean and standard deviation were estimated then the degrees of freedom must be reduced by one for each estimated parameter used.

   Thus,
   $df = k - 3$
   $df =$ ___

3) The critical chi-square value for this test can be found in TABLE IV where $\alpha = 5\%$ and $df = 3$. The critical chi-square value is: $\chi^2_\alpha = $ ____

4) The decision rule is: _____

d) **Step 4**. *Calculate the Pearson's chi-square statistic.*

$X^2 = $ ____

e) **Step 5**. *Determine the conclusion* and state whether the number of hours worked per week by a full-time student fits a normal distribution.

PART V What Do You Think?

1. **What Do You Think?**

Examine the USA Snapshot entitled *Would you date a co-worker?* in the following figure.

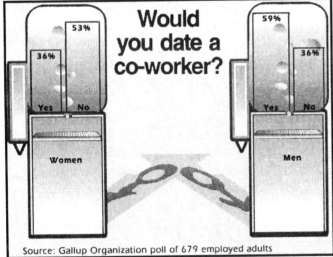

Source: Gallup Organization poll of 679 employed adults
Copyright 1993, USA TODAY. By Julie Stacey, USA TODAY
Reprinted with permission.

a) If you wanted to determine if there was a relationship between the sex of a worker and the response to the question: *Would you date a co-worker?*, then identify the population classifications presented within the USA Snapshot.

b) If you were to perform a hypothesis test to determine if there is a relationship between the population classifications, how would the null hypothesis be stated?

c) How would the alternative hypothesis be stated?

d) When the null hypothesis is assumed to be true, what does that indicate about the relationship between the two population classifications?

e) If we classify the response to the question as either: **yes**, **no** and **undecided**, then what would be the size of the contingency table for this type of chi-square test?

f) What is the sample size for this information contained within the USA snapshot? If three hundred people of the sample were women, then determine the number of people that would fall within each category of the contingency table.

g) What are these values within the contingency table called?

h) Determine the expected frequencies for this hypothesis test.

i) What is the number of degrees of freedom for this test?

j) Determine the critical value for the 5% and 1% level of significance.

k) State a decision rule for each level of significance.

l) Calculate Pearson's Chi-Square statistic for this test.

m) What is the conclusion for this test, if $\alpha = 5\%$? Can the null hypothesis be rejected at $\alpha = 5\%$? Can you conclude that the two population classifications are independent?

n) Is there a statistically significant relationship between the sex of a worker and the response to the question: *Would you date a co-worker?* at $\alpha = 1\%$? Can the null hypothesis be rejected at $\alpha = 1\%$? Can you conclude that the two population classifications are dependent?

o) Can you conclude that the p-value of this hypothesis test is less than 5%? Explain.

p) Can you conclude that the p-value of this hypothesis test is less than 1%? Explain.

2. **What Do You Think?**

A research article entitled *Research On Attitudes Of Professionals Toward Advertising: A Methodological Caveat* appeared in June/July 1990 issue of the Journal of Advertising Research. One aspect of this article investigates the sampling frames actually used to represent the professional population to determine if they represent the population under study. This research stresses the importance of selecting a sampling frame which is a valid representation of the population under study since the authors believe that researchers may be making inaccurate predictions from surveys in which they have relied on an inappropriate sampling frame. The authors of the study state the choice of a sampling frame leaves one important question unanswered: **Are all members of a profession (which depicts the population) equally represented in the various sampling frames?** Given that professional association membership lists are commonly used in studies of professionals, **the researchers decided to compare the responses of members and non-members of a professional state association.** Using a sampling frame for a study which omitted the non-members of a professional association might raise serious doubts about the results of the study.

In their study, the researchers selected a random sample of 600 dentists from a list of currently licensed Oregon dentists provided by the Oregon Board of Dentistry. This sampling frame was used to be certain that all Oregon dentists in practice during the study had the opportunity of being selected for the study. To allow a thorough comparison of the members and non-members of the professional association, questions were designed to span the range of issues that typically addressed in surveys of professional providers - their values, attitudes, behaviors, and demographic characteristics. The questionnaire used in the study contained the items listed in the following table.

Questions Used to Measure Values, Attitudes, Behaviors and Demographic Traits

Values
- V1. I am satisfied with dentistry as a career.
- V2. I encourage young people to consider dentistry as a career.
- V3. The government should guarantee the availability of dental service to all Americans.

Attitudes
- A1. Advertising my services would be beneficial to me personally.
- A2. The advertising of fees would adversely affect the public image of dentists.
- A3. Dentists should be allowed to advertise without restrictions.

Behaviors
- B1. What was the average number of patients that you had per day in 1987?
- B2. Number of weekdays taken as vacation.
- B3. How frequently do you attend religious services?

Demographic
- D1. Net Income
- D2. Gross Income
- D3. Gender
- D4. Age
- D5. Self-Employed
- D6. Practice Type

Values and attitudes were measured using a five-point Likert-type scale, while behavior questions used ordinal categories except for 'days on vacation' which was ratio scaled. The income question used a six-point ordinal scale; practice type was a three category nominal measure, gender and "employee" were (nominal) dichotomous measures, and age was actual (ratio-scaled).

The results of the comparisons are shown in the following table. You should notice that chi-square tests were used for the comparison between the members and the non-members for each of the measures listed in the following table with the exception of "age" and "days of vacation" for which a t-test was performed.

TESTS OF DIFFERENCES BETWEEN THOSE INCLUDED FROM TYPICAL SAMPLING FRAMES
Professional Organization Members versus Non-Members

Question	Statistic[1]	p ≤	df	N[*]
V1	$X^2 = 10.92$	.05	4	414
V2	$X^2 = 13.90$	.01	4	414
V3	$X^2 = 6.16$	.18	4	412
A1	$X^2 = 9.80$	.05	4	408
A2	$X^2 = 11.58$	.05	4	411
A3	$X^2 = 3.88$	.42	4	411
B1	$X^2 = 21.71$	.001	5	411
B2	$t = -3.17$	.002	192	402
B3	$X^2 = 6.09$	.10	3	406
D1	$X^2 = 24.49$	.001	5	393
D2	$X^2 = 40.01$	.001	5	384
D3	$X^2 = 1.65$	.19	1	410
D4	$t = -2.77$	.01	397	399
D5	$X^2 = 30.90$	.001	1	413
D6	$X^2 = 66.15$	.001	2	413

[1] A significant statistic is evidence that the groups are different.
[*] Variations in numbers reflect non-response to a question.

a) What is the null and alternative hypothesis for the chi-square test shown for question V1? Was the result of this test significant at $\alpha = 5\%$? at $\alpha = 1\%$? What is the degrees of freedom for this test? How did they arrive at this value?

b) What is the null and alternative hypothesis for the chi-square test shown for question V3? Can the null hypothesis be rejected at $\alpha = 5\%$? at $\alpha = 1\%$? What is the degrees of freedom for this test? What is the smallest level of significance that would result in the rejection of the null hypothesis?

c) Which chi-square test shown for Attitudes A1, A2, and A3 would not lead to the rejection of the null hypothesis at $\alpha = 10\%$? Explain.

d) What is the null and alternative hypothesis for the chi-square test shown for question B1? Can the alternative hypothesis be accepted at $\alpha = 5\%$? at $\alpha = 1\%$? What is the degrees of freedom for this test? What is the smallest level of significance that would result in the rejection of the null hypothesis?

e) Why wasn't a chi-square test used to perform a hypothesis test to detect a difference between members and non-members with regard to the number of weekdays taken as vacation? Explain.
f) What is the null and alternative hypothesis for the chi-square test shown for question D1? Can the null hypothesis be rejected at $\alpha = 5\%$? at $\alpha = 1\%$? What is the degrees of freedom for this test? What is the smallest level of significance that would result in the rejection of the null hypothesis? What is the degrees of freedom for this test? Based on this number of degrees of freedom, what can you conclude regarding the number of categories for net income? Explain.
g) Why wasn't a chi-square test used to perform a hypothesis test to detect a difference between members and non-members with regard to the age? What type of test was used to detect this difference? What is the degrees of freedom for test? How was this determined? Explain. Was this difference significant at $\alpha = 5\%$? at $\alpha = 1\%$? What is the smallest level of significance that would result in the rejection of the null hypothesis for this test?
h) For which of these chi-square tests would the null hypothesis test be rejected at a 0.5% level of significance?
i) For which questions, was there no significant difference between the members and non-members of the professional association?
j) Can you conclude that there is a significant difference between the members and non-members with regard to private practice at $\alpha = 1\%$? Explain.
k) Can you conclude that self-employment and membership in the professional association are independent at $\alpha = 1\%$? Explain.
l) Can you conclude that advertising without restrictions and membership in the professional association are dependent at $\alpha = 5\%$? Explain.
m) Based on these results, can you conclude that the dentists who are members of the professional association are significantly different from the non-members? Explain.

3. **What Do You Think?**

A research article entitled *The Rekurring Kase Of The Special K* studied whether the letter K occurs with more frequency as the initial letter in top brand names than it does as a first letter in our language in general which appeared in the October/November 1990 issue of the Journal of Advertising Research. The first aspect of this research examined the eight most frequently occurring initial letters: **C, S, A, B, M, P, D and T** in top name brands to determine if they occur more frequently as the initial letter in top brand names than they appear as a first letter in our language in general. The results of this study were examined in Case Study 14.2.

The second aspect of this research was to determine what happens to the analysis when the letter K was added to the list. This aspect of the test will be discussed within this question. The following table shows the results of a chi-square test with the letter K added to the previous list of the eight most frequently occurring initial letters. The expected frequency for each of these letters were based on Webster's Ninth New Collegiate Dictionary (1987) model.

Chi-Square Analysis of Most Frequently Occurring Initial Letters Plus the Letter K

Initial letter	Observed frequency	Expected* frequency	Cumulative percent Observed	Cumulative percent Expected
C	51	52.4	9.43	9.68
S	46	67.3	17.93	22.12
A	42	32.2	25.69	28.08
B	42	29.2	33.45	33.47
M	40	29.5	40.84	38.92
P	40	47.5	48.23	47.70
D	30	28.2	53.78	52.91
T	29	32.0	59.14	58.83
K	16	4.6	62.10	59.68

$X^2 = 48.927 \quad df = 8 \quad p < .001$

*Based on Webster's Ninth New Collegiate Dictionary (1987), distribution of words beginning with each letter.

a) Read Case Study 14.2, and discuss the objective of the hypothesis test along with the results of the chi-square analysis.
b) Define the population for this test.
c) What variable is being examined within this test?
d) Is this variable numeric or categorical? Explain.
e) What type of chi-square test is being illustrated in this research?
f) What distribution serves as the theoretical distribution?
g) What serves as the sample data for this test?
h) What would serve as the null hypothesis for this test?
i) What would serve as the alternative hypothesis?
j) What assumption was used to calculate the expected frequencies within the table? Which hypothesis was used for this assumption? Explain.
k) How many classifications are there for this test? State the classification(s) for this test. What does each category within a classification represent for this test?

l) What formula was used to calculate the number of degrees of freedom for this test? What were the degrees of freedom for this test? Why is the degrees of freedom for this test different from the degrees of freedom for the test in Case Study 14.2?

m) What is the value of $\dfrac{(O - E)^2}{E}$ for each initial letter of the information contained within the table of Figure CS14.2, where O is the Observed frequency and E is the Expected frequency? What does the sum of the values of $\dfrac{(O - E)^2}{E}$ represent?

n) What is the critical chi-square value for this test at $\alpha = 5\%$ and $\alpha = 1\%$?

o) What is the conclusion of this hypothesis test for each of these significance levels? Can you reject the null hypothesis for any of these significance levels? Explain.

p) What is the conclusion to this test for a significance level of 0.30%? What information did you use to answer this question? Explain.

q) What is the smallest level of significance that you can use to obtain a significant result?

r) State the conclusion to the chi-square test shown in Figure CS14.2.

s) Was the result of the chi-square significant for the hypothesis test shown in Figure CS14.2? Explain.

t) What happened to the value of the chi-square statistic when the letter K was added to the initial list of eight most frequently occurring letters? What do you think this result indicates about the letter K?

u) Which letter do you think occurred significantly more often as an initial letter on the brand-name list than it did for words in the dictionary? Explain your reasoning in answering this question.

PART VI Exploring DATA With MINITAB

1. Use the MINITAB commands to conduct Chi-Square Hypothesis Tests on problems numbered 1-15 in PART IV. Compare your results to those indicated at the back of the textbook.

2. Use the MINITAB commands to conduct Tests of Goodness-of-Fit on problems numbered 16-23 in PART IV. Compare your results to those indicated at the back of the textbook.

3. MINITAB can be used to determine if sample data is approximately normally distributed using the Goodness-of-Fit test.

Suppose a sample of 1000 data values are *standardized* using the z score formula and the *number* of z scores within a given interval is tallied. These tallies are indicated in the column of observed data. Furthermore, for 1000 data values the *expected number* in a *normal distribution* within each z score interval is calculated.

z interval	Observed	Expected
$z \le -2$	27	22.7
$-2 < z \le -1$	23	135.9
$-1 < z \le 0$	355	341.3
$0 < z \le 1$	325	341.3
$1 < z \le 2$	145	135.9
$2 < z$	25	22.7

Use the Goodness-of-Fit hypothesis test to determine if the sample of 1000 scores is approximately normally distributed.

PART VII Projects:

1. Chi-Square Test of Independence Hypothesis Testing Project.
    A. Selection of a topic
        1. Select an article from a local newspaper, magazine or other source that can be used to perform a chi-square test of independence.
        2. Suggested topics:
            a) Test the claim that there is a relationship between student views on course requirements for there major and the major area of concentration.
            b) Test the claim that there is a relationship between gender and student views on capital punishment.
    B. Use the following procedure to test the claim selected in part A.
        1. Identify the classifications of the population referred to in the statement of the claim and indicate where the claim was found.
        2. State the null and alternative hypotheses.
        3. Develop and state a procedure for selecting a random sample from your population. Indicate the technique you are going to use to obtain your sample data.
        4. Indicate the hypothesis testing model, and choose an appropriate level of significance.
        5. Perform the experiment and summarize the observed frequencies in a contingency table.

6. Find the degrees of freedom and determine the appropriate critical chi-square value.
7. Calculate the expected frequencies and determine Pearson's chi-square statistic.
8. Construct a chi-square distribution model and place Pearson's chi-square statistic on the model.
9. Formulate the appropriate conclusion and interpret the conclusion with respect to your opinion (as stated in step 2).

2. Read a research article from a professional journal in a field such as: medicine, education, marketing, science, social science, etc that illustrates the use of the chi-square test of independence. Write a summary of the research article that includes the following:
   a) Name of article, author, and journal (include volume number and date).
   b) Statement of the research problem.
   c) The null and alternative hypotheses.
   d) How the experiment was conducted. (include the sample size and sampling technique)
   e) Determine contingency table, the level of significance, the value of Pearson's chi-square statistic and the critical chi-square value.
   f) State the conclusion(s) in terms of the null and alternative hypotheses.

3. Get hold of a pair of dice from any board game. Do an experiment by tossing the dice against 1000 times and recorded the number of times each sum appears. Compare the experimental results to the results that one would expect if the dice were fair. Use an α = 5%.

PART VIII Database.

The following exercises refer to the file DATABASE listed in Appendix A. We have indicated the appropriate MINITAB commands that are necessary to answer each exercise.

1. Using the MINITAB commands:
   RETRIEVE, TABLE, CHISQUARE, CDF, LET, and PRINT
   Retrieve the file DATABASE.MTW and identify the columns names for the variables Hair color and Eye color.
   a) Use the TABLE command to produce a table for the variables 'HR' and 'EYE'.
   b) Add the subcommand CHISQUARE to the TABLE command to direct MINITAB to calculate and print the CHI-SQUARE value along with the expected and the observed frequencies in each cell.
   c) Determine if the CHI-SQUARE value is significant at α=5% by using the CDF command and calculating the p-value.
   d) Interpret the results in terms of the variables Hair color and Eye color.

2. Using the MINITAB commands:
   RETRIEVE, TABLE, CHISQUARE, CDF, LET, and PRINT
   Retrieve the file DATABASE.MTW and identify the columns names for the variables SEX and MAJ.
   a) Use the TABLE command to produce a table for the variables 'SEX' and 'MAJ'.
   b) Add the subcommand CHISQUARE to the TABLE command to direct MINITAB to calculate and print the CHI-SQUARE value along with the expected and the observed frequencies in each cell.
   c) Determine if the CHI-SQUARE value is significant at α=5% by using the CDF command and calculating the p-value.
   d) Interpret the results in terms of the variables SEX and MAJ.

CHAPTER 15
LINEAR CORRELATION AND REGRESSION ANALYSIS

❄ 15.1 Introduction

Most of us have examined a chart in a Doctor's office or health magazine that relates height to weight. Imagine the following scenario in a *Doctor's examination room* as illustrated in Figure 15.1.

The Patient a rather **plump** individual about **5 feet 2 inches**, is standing on a Doctor's scale, facing the Doctor. The Doctor, dressed in a white examination jacket, is facing the patient.

Figure 15.1

DOCTOR: "Now let's see... you **weigh about 160 lbs**. How **tall** are you?"
PATIENT: "**Six feet.**"
DOCTOR: "You are **not** anywhere near six feet tall!"
PATIENT: "But Doctor, then I would be ***too*** short for my weight!

The scenario illustrates the two major concepts that are presented in this chapter. The first concept is known as ***correlation analysis.*** *Correlation is a technique that measures the strength (or the degree) of the relationship between two variables.* For example, in this scenario, the height and weight of an individual represents the two variables.

Since, we are familiar with the notion that there is strong relationship between the height and weight of an individual, then using the statistical concept of correlation, we could measure the strength of this relationship. We will examine how to determine this measure in a later section.

Remember the chart you use to estimate your "ideal" weight given your respective height? Such charts have been developed using a statistical technique known as ***regression***. Essentially, regression is a statistical technique that produces a model of the relationship (correlation) between the two variables.

The height/weight chart used in the previous scenario was generated by a regression formula. This chart represents the model for the relationship between the height and weight of an individual. When there is a strong correlation between two variables, such as height and weight, then regression analysis can be applied to produce a formula (or rule) to model the relationship between the two variables. In this chapter, we will explore these important statistical concepts: *linear correlation* and *linear regression analysis*.

15.2 THE SCATTER DIAGRAM

As illustrated in Section 15.1, there are times when we are interested in determining whether a relationship exists between two variables. In Chapter 14, we examined whether or not a relationship exists between two classifications where the data was expressed as count data. Now, we will examine the relationship between two variables where the data represents a measurement rather than count data.

To help us begin our examination of the relationship between two variables, let's consider studying the relationship between height and weight for female students attending Bermuda Community College. We will randomly select a sample of 12 female students from a statistics class at this community college and measure each female's height and weight. Thus, **two measurements** are recorded for **each female student**. These two measurements represent the value of each of the two variables, height and weight, for each female student.

If we label the height as the x variable and the weight as the y variable, then for each female student we will have a pair of numbers. The sample data representing the pair of numbers for the twelve female students has been recorded and summarized in Table 15.1.

Table 15.1

female student number	x Height (inches)	y Weight (lbs)
1	62	123
2	58	102
3	64	110
4	69	137
5	61	145
6	70	132
7	59	108
8	60	112
9	63	124
10	72	155
11	71	170
12	68	140

These pairs of numbers can also be written in the following form: **(x,y)**, which is called an **ordered pair** since the value of the x variable is always written first within the parentheses. For example, the ordered pair for female student number 1 is written: (62, 123), where the first number, 62, within the parentheses represents the height of student number 1 and the second number, 123, within the parentheses represents the weight of student number 1.

Using the sample information contained in Table 15.1, we can begin to analyze the relationship between a female's height and weight by constructing a graph of these ordered pairs. This type of graph is called a **scatter diagram**.

Definition: Scatter Diagram. A scatter diagram is a graph representing the ordered pairs of data on a set of axes.

To construct a scatter diagram, we begin by drawing two lines, a horizontal and a vertical line, to represent the two axes. The horizontal line is called the x axis and represents the x values. In this case the heights of the female students represents the x values. Thus, the x axis is labeled heights. The vertical line is called the y axis and represents the y values, in this case the weights of the female students. Thus, the y axis is labeled weights. This is illustrated in Figure 15.2.

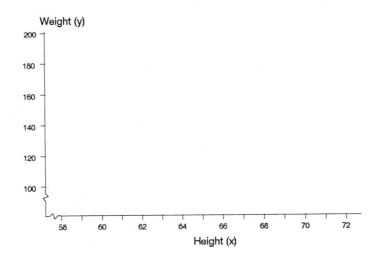

Figure 15.2

We use a dot to represent each ordered pair of measurements. The dot is placed directly above the female's height and directly to the right of the female's weight. Thus, for the first female student with a height of 62 inches and a weight of 123 pounds, a dot is placed above 62 on the axis labeled height and to the right of 123 on the axis labeled weight. This dot represents the height and weight of female student number 1. This same procedure is used to plot all the ordered pairs of the sample data of Table 15.1 and Figure 15.3 illustrates the scatter diagram for this sample data.

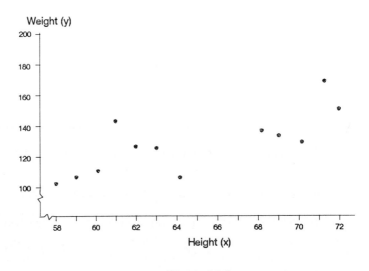

Figure 15.3

Visual examination of a scatter diagram can help to determine whether there is an apparent relationship or correlation between the two variables and, if one does exist, to determine the type of correlation that exists. For example, it appears that a relationship does exist between a female's height and her weight as we examine the scatter diagram in Figure 15.3.

Let's examine the type of relationship that we notice in Figure 15.3. The scatter diagram of Figure 15.3 seems to show that **as a female's height, represented by the x values, increases** then **her weight, represented by the y values, also increases**. This type of correlation is called a **positive linear correlation**.

> **Definition: Positive Linear Correlation.** A positive linear correlation between two variables, x and y, occurs when high measurements on the x variable tend to be associated with high measurements on the y variable and low measurements on the x variable tend to be associated with low measurements on the y variable.

You will also notice that the dots in the scatter diagram tend to follow a **straight-line pattern or model**. This type of correlation is called a **linear correlation**. This can be better visualized by drawing a straight line through the scatter diagram of Figure 15.3, and noticing that the points tend to lie very close to the line. This is illustrated in Figure 15.4.

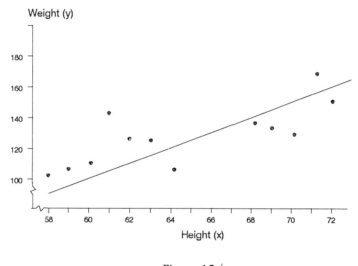

Figure 15.4

From Figure 15.4, we notice that there is a **positive linear correlation between the two variables, height and weight.**

Another type of correlation that can exist between two variables is a **negative linear correlation**.

> **Definition: Negative Linear Correlation.** A negative linear correlation between two variables, x and y, occurs when high measurements on the x variable tend to be associated with low measurements on the y variable and low measurements on the x variable tend to be associated with high measurements on the y variable.

Let's examine this idea of negative correlation using the scatter diagram in Figure 15.5. Notice in Figure 15.5, as the values of the x variable increase, the value of the y variable tend to decrease. The two variables tend to vary in opposite directions. Thus, the relationship between the two variables is a **negative linear correlation**.

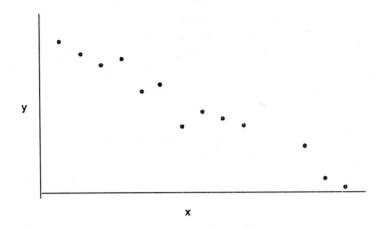

Figure 15.5

It is also possible that when examining the relationship between two variables, you may notice that there is no linear relationship between the two variables. This is referred to as **no linear correlation**.

> **Definition: No linear Correlation.** No linear correlation means there is no linear relationship between the two variables. That is, high and low measurements on the two variables are not associated in any predictable straight line pattern.

Figure 15.6 illustrates a scatter diagram that shows no linear correlation between the two variables.

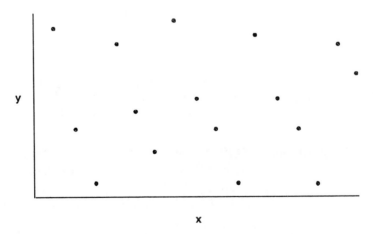

Figure 15.6

Remark: We will only examine linear (or straight line) correlations in this chapter. Although, it is possible that two variables can be related to each other in a curvilinear relationship, this type of correlation is beyond the scope of this textbook. An example of a curvilinear relationship is illustrated by the scatter diagram in Figure 15.7.

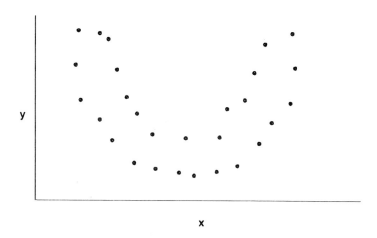

Figure 15.7

Let's discuss the procedure to construct a scatter diagram and explain the type of linear correlation that may exist.

Example 15.1

Using the sample data in Table 15.2, construct a scatter diagram and indicate the type of linear correlation, if any exist.

Table 15.2

x	y
1	4
2	5
3	5
5	6
8	9

Solution:

To construct a scatter diagram, perform the following steps:

Step 1. Draw a horizontal axis and label it as the x axis. Now draw a vertical axis and label it as the y axis.

Step 2. Plot each ordered pair (x,y) of the sample data from Table 15.2 in the appropriate position on the set of axes. Figure 15.8 represents the scatter diagram for the sample data of Table 15.2.

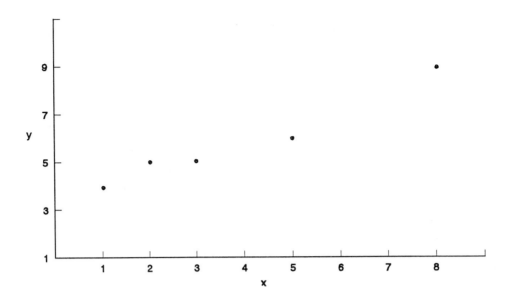

Figure 15.8

To determine the type of linear correlation that may exist, we must examine the pattern that is shown by the sample data pairs in Figure 15.8. As you examine the scatter diagram from left to right, you will notice the pattern of the points is going in an upward direction. That is, as the x values *increase*, the y values also *increase*. Thus, the scatter diagram indicates a pattern that corresponds to a *positive* linear correlation. ■

Example 15.2

Using the sample data in Table 15.3, construct a scatter diagram and indicate the type of linear correlation, if any exist.

Table 15.3

x	y
2	11
3	9
5	8
6	6
8	6
10	3

Solution:

Step 1. Draw a horizontal axis and label it as the x axis. Now draw a vertical axis and label it as the y axis.

Step 2. Plot each ordered pair (x,y) of the sample data from Table 15.3 in the appropriate position on the set of axes. Figure 15.9 represents the scatter diagram for the sample data of Table 15.3.

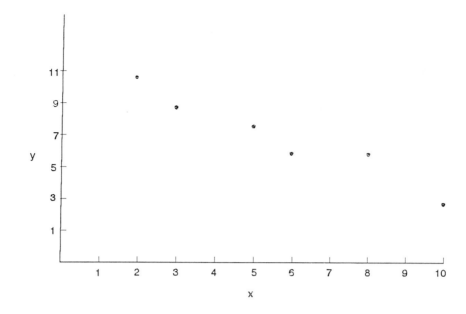

Figure 15.9

To determine the type of linear correlation that may exist, we must examine the pattern that is shown by the sample data pairs in Figure 15.9. As you examine the scatter diagram from left to right, you will notice the pattern of the points is going in a downward direction. That is, as the x variable *increases*, the y variable *decreases*. This scatter diagram indicates a **negative linear correlation** for the sample data listed in Table 15.3. ■

Example 15.3

Using the sample data in Table 15.4, construct a scatter diagram and indicate the type of linear correlation, if any exist.

Table 15.4

x	y
3	1
4	3
5	2
7	5
5	5
3	4
6	4
7	2
8	1
6	3

Solution:

Step 1. Draw a horizontal axis and label it as the x axis. Now draw a vertical axis and label it as the y axis.

Step 2. Plot each ordered pair (x,y) of the sample data from Table 15.4 in the appropriate position on the set of axes. Figure 15.10 represents the scatter diagram for the sample data of Table 15.4.

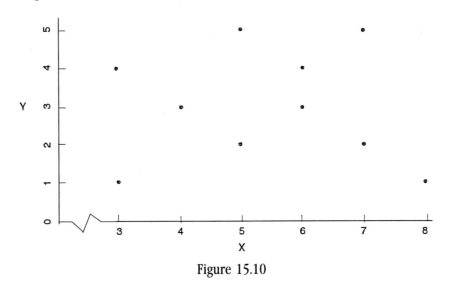

Figure 15.10

To determine the type of linear correlation that may exist, we must examine the pattern that is shown by the sample data pairs in Figure 15.10 As you examine the scatter diagram from left to right, you will notice that there is no definite pattern for the points. This scatter diagram indicates that we have ***zero or no linear correlation*** for the sample data listed in Table 15.4. ∎

❄ 15.3 THE COEFFICIENT OF LINEAR CORRELATION

When a scatter diagram seems to indicate that there is a linear correlation between two variables, our next step is to determine the **strength of the relationship that exist between the two variables.** By a **linear correlation**, we mean how closely the points of a scatter diagram approximates a straight-line pattern.

The closer the points of a scatter diagram approximate a straight-line pattern, the **stronger the correlation between the two variables.** The **strength** of the correlation between the two variables can be numerically measured by **Pearson's correlation coefficient, r.**

Definition: **Pearson's Correlation Coefficient, r.** Pearson's Correlation Coefficient, r, measures the strength of a linear relationship between two variables for a sample. This sample correlation coefficient, r, is calculated by the following formula:

$$r = \frac{n\Sigma(xy) - (\Sigma x)(\Sigma y)}{\sqrt{n(\Sigma x^2) - (\Sigma x)^2}\sqrt{n(\Sigma y^2) - (\Sigma y)^2}}$$

where: **r** is Pearson's Correlation Coefficient
x represents the data values for the *first variable*.
y represents the data values for the *second variable*.
n represents the *number of pairs of data values*

Pearson's correlation coefficient, r, will always be a **numerical value between -1 and +1**. That is,

$$-1 \leq r \leq +1$$

Let's interpret what the different values of r mean before we examine how to calculate Pearson's correlation coefficient.

Interpreting the values of r

A value of **r = +1** represents the strongest positive correlation possible and it indicates a perfect positive correlation. This means that **all the points of the scatter diagram will lie on a straight line which is sloping upward from left to right.**

A value of **r = -1** represents the strongest negative correlation possible and it indicates a perfect negative correlation. This means that **all the points of the scatter diagram will lie on a straight line which is sloping downward from left to right.**

A value of **r = 0 indicates no linear correlation between the two variables.**

The closer the value of the correlation coefficient, r, is to -1 or to +1 the stronger the linear relationship between the two variables. For example, if we say that height, represented by variable one, is **strongly correlated** to weight, represented by variable two, then **we would expect the numerical value of r to be either close to -1 or close to +1. However, a value of r close to zero indicates a weak linear relationship between the two variables.**

Let's examine how to use Pearson's correlation coefficient formula to calculate the strength of the correlation for the sample data in Example 15.4.

Example 15.4

Using the sample data in Table 15.5, calculate Pearson's correlation coefficient, r.

Table 15.5

x	y
1	4
2	5
3	5
5	6
8	9

Solution:

Use the correlation coefficient formula:

$$r = \frac{n\Sigma(xy)-(\Sigma x)(\Sigma y)}{\sqrt{n(\Sigma x^2)-(\Sigma x)^2}\sqrt{n(\Sigma y^2)-(\Sigma y)^2}}$$

to calculate r, we need to determine the values for n, $\Sigma(xy)$, Σx, Σx^2, Σy and Σy^2. To obtain the sums needed to calculate r, we will construct a table that has a column for each of the quantities: **x, y, x², y²** and **xy**. This is the first step in calculating r.

Step 1. *Construct a table that lists the quantities: x, y, x², y² and xy.*

Now, perform the calculation for each of these quantities and list the result in the appropriate column. These results are listed in Table 15.6.

Table 15.6

x	y	x²	y²	xy
1	4	1	16	4
2	5	4	25	10
3	5	9	25	15
5	6	25	36	30
8	9	64	81	72

Step 2. *Using the information in Table 15.6, calculate $\Sigma(xy)$, Σx, Σx^2, Σy and Σy^2.*

Adding the values within each column, we obtain the following results:

$$\Sigma x = 19 \quad \Sigma y = 29 \quad \Sigma x^2 = 103 \quad \Sigma y^2 = 183 \quad \Sigma(xy) = 131$$

Remembering that n is the **number of pairs of x and y values** and since there are five pairs of x and y values, then **n** equals 5.

Step 3. *Substitute the calculated values of n, $\Sigma(xy)$, Σx, Σx^2, Σy and Σy^2 into the formula and calculate r.*

$$r = \frac{5(131)-(19)(29)}{\sqrt{5(103)-(19)^2}\sqrt{5(183)-(29)^2}}$$

$$r = \frac{655-551}{\sqrt{515-361}\sqrt{915-841}}$$

$$r = \frac{104}{\sqrt{154}\sqrt{74}}$$

$$r \approx \frac{104}{106.73}$$

$$r \approx 0.97$$

Thus, Pearson's correlation coefficient for the sample data is: r = 0.97. ■

You should notice that the sign associated with an r = 0.97 is *positive*. Whenever the sign associated with a correlation coefficient is positive, this indicates that there is a positive correlation between the two variables. Similarly, whenever the sign associated with a correlation coefficient is *negative*, the correlation between the two variables is *negative*.

Example 15.5

Using the sample data in Table 15.7, calculate Pearson's correlation coefficient, r.

Table 15.7

x	y
2	11
3	9
5	8
6	6
8	6
10	3

Solution:

Step 1. *Construct a table that lists the quantities: x, y, x^2, y^2 and xy.*

Now, perform the calculation for each of these quantities and list the result in the appropriate column. These results are listed in Table 15.8.

Table 15.8

x	y	x^2	y^2	xy
2	11	4	121	22
3	9	9	81	27
5	8	25	64	40
6	6	36	36	36
8	6	64	36	48
10	3	100	9	30

Step 2. *Using the information in Table 15.8 calculate n, $\Sigma(xy)$, Σx, Σx^2, Σy and Σy^2.*

Adding the values within each column, we obtain the following results:

$$\Sigma x = 34 \quad \Sigma y = 43 \quad \Sigma x^2 = 238 \quad \Sigma y^2 = 347 \quad \Sigma(xy) = 203$$

Since there are six pairs of x and y values, then n = 6.

Step 3. *Substitute the calculated values of n, $\Sigma(xy)$, Σx, Σx^2, Σy and Σy^2 into the formula and calculate r.*

$$r = \frac{n\Sigma(xy)-(\Sigma x)(\Sigma y)}{\sqrt{n(\Sigma x^2)-(\Sigma x)^2}\sqrt{n(\Sigma y^2)-(\Sigma y)^2}}$$

$$r = \frac{6(203)-(34)(43)}{\sqrt{6(238)-(34)^2}\sqrt{6(347)-(43)^2}}$$

$$r = \frac{1218-1462}{\sqrt{1428-1156}\sqrt{2082-1849}}$$

$$r = \frac{-244}{\sqrt{272}\sqrt{233}}$$

$$r \approx \frac{-244}{251.76}$$

$$r \approx -0.97$$

Thus, Pearson's correlation coefficient is: $r = -0.97$. ∎

Procedure[1] to Calculate Pearson's Correlation Coefficient, r

Step 1. Construct a table that contains the quantities: x, y, x^2, y^2 and xy. Perform the calculation for each of these quantities and list the result in the appropriate column of the table.

Step 2. Using the information in the table, calculate:
n, $\Sigma(xy)$, Σx, Σx^2, Σy and Σy^2.

Step 3. Substitute the calculated values of n, $\Sigma(xy)$, Σx, Σx^2, Σy and Σy^2 into the formula and calculate r.

To calculate r, use Pearson's Correlation Coefficient formula:

$$r = \frac{n\Sigma(xy)-(\Sigma x)(\Sigma y)}{\sqrt{n(\Sigma x^2)-(\Sigma x)^2}\sqrt{n(\Sigma y^2)-(\Sigma y)^2}}$$

When Pearson's correlation coefficient is greater than zero, that is: r > 0, then there is a positive linear correlation between the two variables.

When Pearson's correlation coefficient is less than zero, that is: r < 0, then there is a negative linear correlation between the two variables.

❄ 15.4 More on the Relationship Between Correlation Coefficient and the Scatter Diagram

We have previously stated that the correlation coefficient, r, measures the strength of a 0relationship between two variables. Furthermore, we have indicated that $-1 \leq r \leq +1$. Carefully examine Figures 15.10a, 15.10b and 15.10c.

---

[1] At the end of this chapter, we have provided you with a general procedure to calculate the correlation coefficient, r, using a statistical calculator.

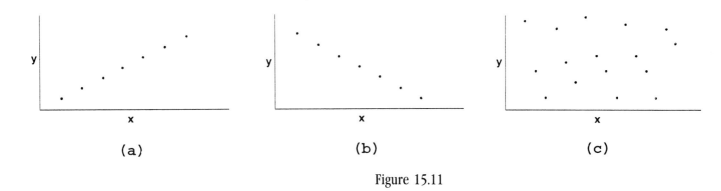

Figure 15.11

The scatter diagrams illustrated in Figures 15.11a or 15.11b, have correlation coefficients equal to +1 and -1 respectively. In each case such a value of r would mean that a **perfect linear relationship exists between the x and y variables.** If the value of r is +1, then all the points of the scatter diagram would lie on a line that is sloping upward from left to right as illustrated in Figure 15.11a and the relationship between the variables would be referred to as a **perfect positive linear relationship.**

If the value of r is -1, then all the points of the scatter diagram would lie on a line that is sloping downward from left to right as illustrated in Figure 15.11b and the relationship between the variables would be referred to as a **perfect negative linear relationship.**

On the other hand if r = 0, then the scatter diagram would indicate no linear correlation as illustrated in Figure 15.11c. Such a scatter diagram would suggest **no linear** relationship exists between the variables x and y.

In practice, Pearson correlation coefficient values of r that are +1, or -1, or 0 rarely occur. In the next section we will consider one way to interpret the correlation coefficient specifically for values other than +1, -1, and 0.

15.5 TESTING THE SIGNIFICANCE OF THE CORRELATION COEFFICIENT

There are many practical situations where it is necessary to determine whether a *significant* correlation exists between two variables. For example:

- Is there a significant correlation between High School average and success in college? or,

- Is there a significant correlation between the number of hours of television a five year old watches per week and his/her IQ score? or,

- Is there a significant correlation between a person's age and their blood pressure? or,

- Is there a significant correlation between a company's advertising expenditures and sales for one of their new products?

To help us answer these questions, we would need to determine the population correlation coefficient for the two variables. The population correlation coefficient is symbolized by the lower case Greek letter rho: ρ, pronounced as row. However, it is impractical to collect the entire population data for the two variables, consequently, we would have to use a random sample to help us estimate the population correlation coefficient, ρ. We use Pearson's correlation coefficient, r, to calculate the correlation coefficient for the sample data. **The correlation coefficient, r, for the sample is used as an estimate of the population correlation coefficient, ρ.**

After we calculate the correlation coefficient, r, for the sample, we need to determine if this sample value of r indicates that the unknown population correlation coefficient, ρ, is a significant correlation. That is, the value of the population correlation coefficient, ρ, is significantly different from a correlation of zero (i.e. no correlation).

In this section, we will **examine how to test the correlation coefficient for statistical significance.** The statistical procedure needed to perform this test involves a modification of the general five step hypothesis testing procedure that was developed in Chapter 9. Let's develop this procedure to test the significance of a population correlation coefficient.

Procedure to Test the Significance of the Correlation Coefficient

This test is based upon the assumption that both variables x and y are normally distributed.

Step 1. *Formulate the two hypotheses, H_o and H_a.*

Null Hypothesis:
In general, the null hypothesis for testing the significance of the population correlation coefficient has the following form.

Null Hypothesis Form:

H_o: The population correlation coefficient is equal to zero. That is, there is no linear correlation between the two variables.

Thus, the null hypothesis is symbolized as:

$$H_o: \rho = 0$$

Alternative Hypothesis:
Since the alternative hypothesis, H_a, can be stated as the population correlation coefficient is either: *greater than, less than,* or *not equal to zero*, then the alternative hypothesis can have one of three forms:

Form (a): Positive or Greater Than Form For H_a	Form (b): Negative or Less Than Form For H_a	Form (c): Not Equal Form For H_a
H_a: The population correlation coefficient, ρ, is claimed to be greater than zero. That is, there is a positive linear correlation between the two variables. This is symbolized as: $H_a: \rho > 0$	H_a: The population correlation coefficient, ρ, is claimed to be less than zero. That is, there is a negative linear correlation between the two variables. This is symbolized as: $H_a: \rho < 0$	H_a: The population correlation coefficient, ρ, is claimed to be not equal to zero. That is, there is a significant linear correlation between the two variables. This is symbolized as: $H_a: \rho \neq 0$

Step 2. *Design an experiment to test the null hypothesis, H_o.*

> An important consideration to the design is the assumption that the variables x and y are both normally distributed. Under the assumption the null hypothesis is true, that is: $\rho = 0$, the sampling distribution of r values is symmetric about $r = 0$ and is referred to as **the Sampling Distribution of the Correlation Coefficients.** This distribution will be used as the model for the hypothesis test. Figure 15.12 illustrates different sampling distributions of the correlation coefficients for various choices of n *pairs* of data.

HYPOTHESIS TESTING MODEL
The Sampling Distribution of the Correlation Coefficients

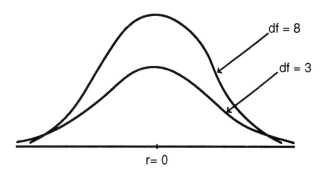

Figure 15.12

Step 3. *Formulate the decision rule.*
The decision rule is based on the following information.

a) Determine the type of alternative hypothesis (i.e., is the alternative hypothesis **directional** or **nondirectional**?)
b) Determine the type of hypothesis test (i.e., is the test **1TT** or **2TT**?)
c) Identify the significance level (i.e., is $\alpha = 1\%$ or $\alpha = 5\%$?)
d) Calculate the critical value(s) using **TABLE V: Critical Values of the Correlation Coefficient, r_α, found in Appendix D: Statistical Tables.**

> **TABLE V** contains the **Critical Values of the Correlation Coefficient, r_α.** To determine the appropriate critical value of r, denoted as r_α, from TABLE V for the decision rule, we need to determine the degrees of freedom for testing the correlation coefficient.
> **Degrees of freedom, *df*.**
> The degrees of freedom, df, for testing the correlation coefficient is calculated by subtracting two from the number of pairs of sample data, n.
> Thus, the formula for *df* is: **df = n − 2**

e) Construct the appropriate **hypothesis testing model**.

Figure 15.13 illustrates the appropriate sampling distribution of the correlation coefficients that is used as the hypothesis testing model. Notice that the actual model chosen is dependent upon the **form of the alternative hypothesis**.

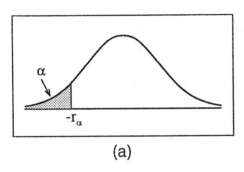
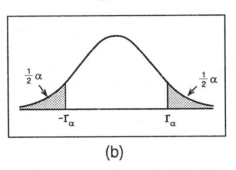
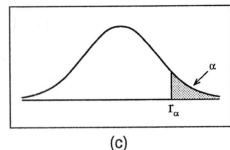

Figure 15.13

Step 4. *Conduct the experiment.*

Randomly select the necessary data from the population to **calculate Pearson's correlation coefficient, r, for the sample data**. This r value is called **the experimental outcome**. To calculate Pearson's correlation coefficient, r, use the formula:

$$r = \frac{n\Sigma(xy) - (\Sigma x)(\Sigma y)}{\sqrt{n(\Sigma x^2) - (\Sigma x)^2}\sqrt{n(\Sigma y^2) - (\Sigma y)^2}}$$

Step 5. *Determine the conclusion.*

Compare the **experimental outcome (i.e., Pearson's correlation coefficient r)** to the **critical value, r_α**, of the decision rule and **draw one** of the following conclusions:

a) **Reject H_o and Accept H_a at α**

b) **Fail to reject H_o at α**

In summary, the general procedure used to perform a hypothesis test about a ***population correlation coefficient***, ρ, is similar to the general hypothesis testing procedure developed in Chapter 9. However, there were some refinements to this general procedure. They were:
- The variables x and y are each normally distributed and the ***sampling distribution of the correlation coefficients*** which is symmetric about r = 0 when $\rho = 0$ is used as the appropriate hypothesis testing model.
- The ***experimental outcome*** is calculated by using ***Pearson's correlation coefficient, r***, for the sample data. The ***critical value, r_α***, for the decision rule is found in TABLE V using degrees of freedom, ***df***, calculated by the formula: ***df* = n − 2**, and identifying the α value and the type of hypothesis test.

Example 15.6 illustrates the use of this hypothesis testing procedure to test the significance of a correlation coefficient.

Example 15.6

A Parents' Association in a Mission Viejo, California school district interviewed and collected data from a random sample of twenty parents whose children attend the elementary school. The data indicated the number of hours per week that each child watches television and the number of hours spent doing homework. The Association believes there is a negative correlation between the number of hours per week a child watches television and the number of hours per week the child spends doing homework. **Pearson's correlation coefficient, r,** was calculated to be -0.52 for the sample. Test the hypothesis that there is a negative relationship between the number of hours a child watches television and spends doing homework. Use $\alpha = 5\%$.

Solution:

Step 1. *Formulate the hypotheses.*

Null Hypothesis:

H_o: **The population correlation coefficient for the number of hours a child watches TV and spends doing homework is zero.**

This is symbolized as: $H_o: \rho = 0$.

A population correlation coefficient of zero is interpreted to mean that there is **no linear** relationship between the number of hours per week a child watches television and the number of hours per week the child spends doing homework.

Alternative Hypothesis:

Since the association **believes that there is a negative correlation** between the number of hours per week a child watches television and the number of hours per week the child spends doing homework, then they are trying to show that the population correlation coefficient is *less than* zero.

H_a: **The population correlation coefficient for the number of hours a child watches TV and spends doing homework *is less than* zero.**

This is symbolized as: $H_a: \rho < 0$.

Step 2. *Design an experiment to test the null hypothesis, H_o.*

To test the null hypothesis, the association randomly selected 20 parents whose children attend the elementary school and determined the number of hours per week each child watched television and the number of hours per week each child dedicated to homework. To test the null hypothesis the sampling distribution of the correlation coefficients is used as the hypothesis testing model and both variables are assumed to be normally distributed.

Assuming the null hypothesis is true, the *expected value for the population correlation coefficient is:*

$$\rho = 0.$$

Step 3. *Formulate the decision rule.*

a) The alternative hypothesis is directional, since the association is trying to show that ρ **is less than zero.**
b) The type of hypothesis test is **one-tailed on the left side.**
c) The significance level is: α = 5%.
d) To determine the critical r value, r_α, we need to compute the degrees of freedom, df, using: $df = n - 2$.

For n = 20, we have: $df = 20 - 2 = 18$. Thus, the critical r value, r_α, for a 1TT on the left with $df = 18$ is found in TABLE V. From TABLE V, the critical value is: $r_\alpha = -0.38$.

e) Thus, the decision rule is: *Reject H_o*, if the experimental outcome, r, is *less than* -0.38.

Figure 15.14 illustrates the decision rule for this hypothesis test.

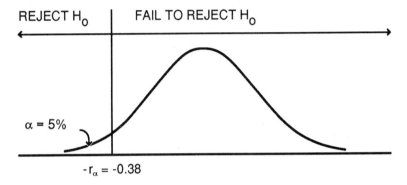

Figure 15.14

Step 4. The Parents' Association calculated the *experimental outcome* using Pearson's correlation coefficient formula for the sample of twenty children. The calculated r value for the experimental outcome was: r = -0.52.

Step 5. Since the experimental outcome, r = -0.52, is *less than* the critical value, $r_\alpha = -0.38$, then the conclusion is to *Reject H_o and Accept H_a* at α = 5%. Therefore, the Parents' Association found that *there is a significant negative linear correlation between the number of hours per week a child watches television and the number of hours per week the child spends doing homework at α = 5%.* ∎

Example 15.7

A personnel director for ECHO Publishing, Inc. believes that there is a significant correlation between the distance (in miles) an employee travels to work and the number of minutes per month the employee is late. Data from a random sampling of 10 employees obtained a sample correlation coefficient of: r = 0.46. Test the significance of the population correlation coefficient at α = 1%.

Solution:

> **Step 1.** *Formulate the hypotheses.*
>
> **Null Hypothesis:**
> H_o: The population correlation coefficient for the distance an employee travels to work and number of minutes the employee is late is zero. This is symbolized as: $H_o: \rho = 0$.
>
> **Alternative Hypothesis:**
>   Since the personnel director did not indicate a direction as to his belief about the relationship between the number of miles an employee travels to work and the number of minutes per month the employee is late, the alternative hypothesis is *nondirectional*.
>
> Therefore, the alternative hypothesis is:
>
> H_a: The population correlation coefficient for the distance an employee travels to work and number of minutes the employee is late is *not equal* to zero.
> This is symbolized as: $H_a: \rho \neq 0$.
>
> **Step 2.** *Design an experiment to test the null hypothesis, H_o.*
>
> The personnel director will randomly select 10 employees and determine the number of miles each employee travels to work and the number of minutes per month the employee is late. The distribution of correlation coefficients is used as the hypothesis testing model and both variables are assumed to be normally distributed. Under the assumption that the null hypothesis is true, the expected value for the population correlation coefficient is: $\rho = 0$.
>
> **Step 3.** *Formulate the decision rule.*
>
> a) The alternative hypothesis is nondirectional, since the director is trying to show that ρ **is not equal to zero.**
> b) The type of hypothesis test is **two-tailed**.
> c) The significance level is: $\alpha = 1\%$.
> d) To determine the critical r values, r_α, we need to compute the degrees of freedom, *df*, using: *df* = n - 2.
>
> For n = 10, we have: *df* = 10 - 2 = 8. Thus, the critical r value, r_α, for a 2TT with *df* = 8 and $\alpha = 1\%$ is found in TABLE V. From TABLE V, the critical values are:
> $$r_\alpha = \pm\, 0.76.$$
>
> e) Thus, the decision rule is to Reject H_o, if the experimental outcome, r, is either: less than the critical r value, $r_\alpha = -0.76$ or r is greater than the critical value, $r_\alpha = 0.76$.
> Figure 15.15 illustrates the appropriate hypothesis testing model.

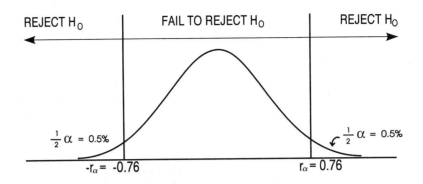

Figure 15.15

Step 4. *Conduct the experiment.*

For this study, ten employees were surveyed. Pearson's correlation coefficient, r, was calculated to be 0.65 for this sample. Thus, the experimental outcome is: r = 0.65.

Step 5. *Determine the conclusion.*

Since the experimental outcome, r = 0.65, is neither less than the critical value of $r_\alpha = -0.76$ nor greater than the critical value of $r_\alpha = 0.76$, then the personnel director *fails to reject H_o* at $\alpha = 1\%$. Thus, the correlation between the number of miles travelled to work and the number of minutes late per month is *not statistically significant at $\alpha = 1\%$*. ∎

15.6 THE COEFFICIENT OF DETERMINATION

An important statistic measure that can be calculated from the correlation coefficient, r, is called the **Coefficient of Determination**. This statistical measure is used to explain the degree of influence that one variable called the independent variable, usually denoted by x, has on the other variable called the dependent variable, usually denoted by y.

> **Definition: Coefficient of Determination, r^2.** The coefficient of determination measures the proportion of the variance of the dependent variable **y** that can be accounted for by the variance of the independent variable **x**. It is calculated by squaring the correlation coefficient, r.
> Thus,
>
> **Coefficient of Determination = r^2**

Example 15.8

In Example 15.6, it was shown that a significant linear correlation, r = -0.52, was found between the number of hours per week a child watches television (the independent variable) and the number of hours per week the child spends doing homework (the dependent variable). Using this information, then:
a) calculate the coefficient of determination, r^2.
b) interpret the meaning of r^2.

Solution:

a) To calculate the coefficient of determination, the correlation coefficient, r, must be determined. Since r was found to be -0.52, the coefficient of determination is found by squaring r. Thus, the coefficient of determination is:
$$r^2 = (-0.52)^2$$
$$r^2 = 0.27$$

b) To convert the coefficient of determination to a percent multiply r^2 by 100%. Thus, in percent form, the coefficient of determination is 27%. This means that 27% of the variance in the number of hours per week a child spends doing homework (the dependent variable) can be accounted for by the variance in the number of hours per week the child watches television (the independent variable). ■

The coefficient of determination can also be used to suggest the existence of **other** reasons for the variation of one variable due to the variation in the other. Looking back at the previous example, one may conclude that the variation in the hours a child spends doing homework that **cannot be accounted for,** (i.e., 100% - 27% = 73%) may be attributed to a number of reasons. For example, a child may not be assigned homework on a regular basis. Or, the child is inconsistent in doing homework regularly. Since the coefficient of determination is 27%, we may conclude that there is significant amount of variance in the number of hours per week the child spends doing homework that is **not yet explained. Thus, 73% reflects this unexplained amount.**

CASE STUDY 15.1

The research article entitled *Dropping Out And Absenteeism in High School* appeared in Psychological Reports in the June 1990 issue. The study investigated the relationships among dropping out, absenteeism days of school year, and size of school enrollment. The data were collected from the records of a North Central Kansas High School District. The table in Figure CS15.1 represents the Pearson correlations for the variables.

Pearson Correlations For Variables

	1	2	3	4	5	6	7
1. Days of school in the year							
2. Enrollment	.80*						
3. Total days students absent less semester test days	.81*	.92*					
4. Number of students dropping out	.53†	.74*	.48†				
5. Percentage of dropouts							
6. Total number of graduating seniors	.07	.72*	.66*	.29			
7. Total number of seniors not graduating	−.74†	−.15	−.37	−.18		.09	

*p<.01. †p<.05.

a) Identify the different variables that were investigated within the study.
b) If a scatter diagram was constructed for the variables enrollment and total number of seniors not graduating, what would you expect the general pattern to look like?
c) If a scatter diagram was constructed for the variables enrollment and total days students absent less semester test days, what would you expect the general pattern to look like?
d) As enrollment increases, what would you expect would happen to the value of the variable number of students dropping out?

e) Interpret the meaning of 0.80 within the second row and first column of the table? What variables are being compared? Is this a positive or negative relationship between the variables? Explain. Is this a significant relationship at α = 5%? at α = 1%? Explain.

f) Interpret the meaning of 0.07 within the sixth row and first column? What variables are being compared? Is this a positive or negative relationship between the variables? Explain. Is this a significant result at 5%? at α = 1%? Explain.

g) What happens to the value of the variables on enrollment, days of school of the school year, and absences as the value of the variable on the number of students dropping out increases? Are these relationships positive or negative?

h) Can the null hypothesis be rejected for the variables number of students dropping out and days of the school year at α = 5%? at α = 1%? Explain. What is the smallest level of significance that would cause the rejection of the null hypothesis? Explain.

i) What is the proportion of the variance in enrollment that can be accounted for by the variance in the number of students dropping out? What formula is used to determine this proportion, and what is this statistical measure called?

j) What is the proportion of the variance in total number of seniors not graduating that can be accounted for by the variance in the days of school in the year?

15.7 LINEAR REGRESSION ANALYSIS

Once a significant linear correlation has been established between two variables, a **linear model** can be developed that is used to **predict** a value for the dependent variable, usually denoted as the y variable, given a value for the independent variable, usually denoted as the x variable.

For example, if a significant linear correlation has been established between a father's height (the independent variable) and his son's height (the dependent variable), then it would be useful to develop a linear model that could reasonably *predict* the son's height *given* his father's height. This linear model could be used to **predict** a son's height simply by using the father's height.

You are probably aware that many colleges and universities successfully use a student's standardized test results as a main criteria for acceptance into the school. The schools are able to predict success with *a high degree of confidence* because research studies have established that there is a *high correlation* between the standardized test results and success at the college, as measured by the student's first year grade point average (GPA). To accomplish this, a model is statistically developed that can be used to **predict** a student's first year GPA from his/her standardized test scores.

These examples show how statistical modeling can be applied to the relationship between two variables. This type of modeling involves regression analysis. Regression analysis enables us to develop models so we can make these types of predictions. *Linear regression analysis provides us with a linear model or equation that can be used to predict the value of the y or dependent variable given the value of the x or independent variable.* The value of y that is predicted by this model is usually not the exact value, however it is a "close" estimate of the actual y value. To determine the linear model that will generate a "close" estimate of the actual y value, we obtain the line that **"best fits"** all the sample points of the scatter diagram. In regression analysis, the technique used to obtain this **"best fitting"** line is called the **method of least squares**. Essentially, this technique selects as the **"best fitting"** line the line which **minimizes the sum of the squared distances between the predicted y values and the actual y values.** This idea is illustrated in Figure 15.16.

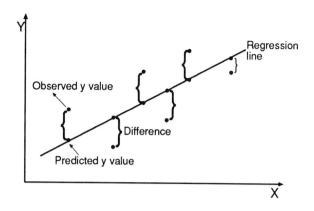

Figure 15.16

The **"best fitting"** line is called the **regression line**. It is used to predict the dependent variable (y) based upon the value of the independent variable (x). Let's examine how to obtain the regression line by examining the regression line formula.

Regression Line Formula

$$y' = a + bx$$

where: **y' is the predicted value of y, the dependent variable, given the value of x, the independent variable.**

and

a and b are the regression coefficients obtained by the formulas:

$$b = \frac{n\Sigma(xy) - \Sigma x \Sigma y}{n(\Sigma x^2) - (\Sigma x)^2}$$

$$a = \frac{\Sigma y - b\Sigma x}{n}$$

Example 15.9 illustrates how to calculate the regression line.

Example 15.9

A psychologist has established that there is a significant negative correlation between the number of sunny days per year for a given region and the number of adult suicides in that region. The data pairs listed in Table 15.9 represent the random sample used by the psychologist.

Table 15.9

region number	number of sunny days per year	number of suicides per year
1	250	28
2	330	10
3	210	34
4	342	08
5	265	30
6	227	41
7	280	19
8	278	21
9	321	14
10	338	09

Use the sample data listed in Table 15.9 to develop a regression line to predict the number of suicides per year given the number of sunny days per year.

Solution:

Step 1. *Choose the independent and dependent variables.*

To find the regression line, y' = a + bx, we must first choose the dependent variable (y). The dependent variable is the variable you want to predict. Thus, the *number of suicides per year* is the dependent variable. Therefore, the independent variable (x), is the *number of sunny days per year*. This choice of the variable names will produce a regression line model that will *predict* the number of suicides per year, y', *given* the number of sunny days, x.

Step 2. *Calculate the regression constants, a and b.*

Of the two regression constants, b **must** be computed **first**, since the calculated value of b is *used* in the formula to calculate a.

To calculate b, use the formula:

$$b = \frac{n\Sigma(xy) - \Sigma x \Sigma y}{n(\Sigma x^2) - (\Sigma x)^2}$$

Since the calculation of b needs the quantities Σx, Σy, Σx^2, Σxy and n, use Table 15.10 to help organize the calculations.

Table 15.10

x	y	xy	x²
250	28	7,000	62,500
330	10	3,300	108,900
210	34	7,140	44,100
342	08	2,736	116,964
265	30	7,950	70,225
227	41	9,307	51,529
280	19	5,320	78,400
278	21	5,838	77,284
321	14	4,494	103,041
338	09	3,042	114,244

From Table 15.10, we obtain the following sums:
$\Sigma x = 2,841 \quad \Sigma y = 214 \quad \Sigma x^2 = 827,187 \quad \Sigma xy = 56,127$

Since there are 10 pairs of sample data, n = 10.

To calculate the regression coefficient b, substitute the above values into the formula for b:

$$b = \frac{n\Sigma(xy) - \Sigma x \Sigma y}{n(\Sigma x^2) - (\Sigma x)^2}$$

$$b = \frac{10(56127) - (2841)(214)}{10(827187) - (2841)^2}$$

$$b = \frac{561,270 - 607,974}{8,271,870 - 8,071,281}$$

$$b = \frac{-46,704}{200,589}$$

$$b \approx -0.2328$$

To calculate regression coefficient a, use the formula and the value of b:

$$a = \frac{\Sigma y - b\Sigma x}{n}$$

$$a = \frac{214 - (-0.2328)(2,841)}{10}$$

$$a = \frac{214 + 661.38}{10}$$

$$a = \frac{875.38}{10}$$

$$a \approx 87.54$$

Step 3. *Substitute the values of a and b into the equation for the regression line:*

y' = a + bx

y' = 87.54 + (−0.2328)x ∎

Let's summarize the procedure to calculate the regression line: **y' = a + bx.**

Procedure to Determine the Regression Line

Step 1. Choose the independent and dependent variables. The dependent variable is the variable you are going to predict.

Step 2. Calculate the regression coefficients, a and b. Calculate b first. Use the formula:

$$b = \frac{n\Sigma(xy) - \Sigma x \Sigma y}{n(\Sigma x^2) - (\Sigma x)^2}$$

Calculate a. Use the formula:

$$a = \frac{\Sigma y - b \Sigma x}{n}$$

Step 3. Substitute the values of a and b into the equation for the regression line:

y' = a + bx

Example 15.10

In Example 15.9 we found the regression line to be y' = 87.54 + (−0.2328)x. This formula represents a *linear* model of the *best fitting line* for the sample data in Table 15.9. Use this regression line model to predict the number of suicides in a year where the number of sunny days is: a) 200 b) 300 c) 0.

Solution:

Since we want to predict the number of suicides per year, this represents the dependent variable y. Thus, the **number of sunny days per year** represents the independent variable x. The regression line equation,

y' = 87.54 + (−0.2328)x,

is a model to predict the number of suicides per year (y) given the number of sunny days per year (x).

a) If x = 200 then
$$y' = 87.54 + (-0.2328)(200)$$
$$y' = 40.98 \text{ suicides per year.}$$

Thus, the predicted number of suicides per year for a region with 200 sunny days per year is 40.98.

b) If x = 300 then

$$y' = 87.54 + (-0.2328)(300)$$
$$y' = 17.70 \text{ suicides per year}$$

Thus, the predicted number of suicides per year for a region with 300 sunny days per year is 17.70.

c) If x = 0 then

$$y' = 87.54 + (-0.2328)(0)$$
$$y' = 87.54 \text{ suicides per year}$$

Thus, the predicted number of suicides per year for a region with zero sunny days per year is 87.54.

Example 15.11

Use the regression line equation and the sample data pairs from Example 15.9 to graph, on the *same* axes, the scatter diagram of the sample data and the regression line. The regression line model equation from Example 15.9 is:

$$y' = 87.54 + (-0.2328)x$$

and the sample data is listed in Table 15.11.

Table 15.11

region number	number of sunny days per year	number of suicides per year
1	250	28
2	330	10
3	210	34
4	342	08
5	265	30
6	227	41
7	280	19
8	278	21
9	321	14
10	338	09

Solution:

The scatter diagram for these data pairs, where the independent variable x represents the number of sunny days per year and the dependent variable y represents the number of suicides per year, is shown in Figure 15.17. The regression line has also been drawn. Notice that the regression line has the property that it **"best fits"** the data.

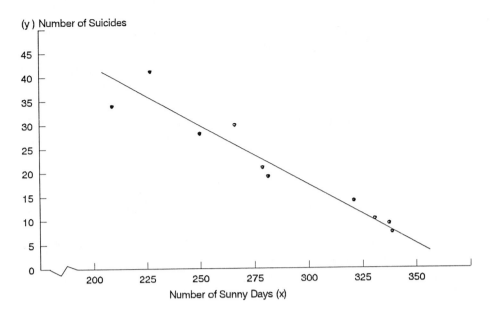

Figure 15.17

CASE STUDY 15.2

The research article entitled *Suicide Rates, Handgun Control Laws, and Sociodemographic Variables* appeared in Psychological Reports of the June 1990 issue. One aspect of the study examined the relationships between suicide rates, and sex, age, percent black, percent metropolitan population, population density, and rates of population change (increase or decrease), divorce, crime, and unemployment. The Pearson correlation coefficient obtained among suicide rates, handgun control laws (identified as Attribute I: Restrictions on the seller of handguns and Attribute II: Restrictions on the buyer of handguns), and all sociodemographic variables are presented in the table of Figure CS15.2.

Pearson Correlations (N = 51)

Variable	1	2	3	4	5	6
1. Suicide rate, 1985						
2. Percent male, 1980	.55‡					
3. Percent ages 35-64, 1986	−.29*	−.35†				
4. Percent black, 1985	−.42†	−.52‡	.23			
5. Percent metropolitan, 1985	−.31*	−.35†	.54‡	.32*		
6. Percent per square mile, 1985	−.37†	−.42†	.21	.67‡	.32*	
7. Percent population change, 1980-1985	.52‡	.63‡	−.17	−.14	.03	−.15
8. Divorce rate, 1985	.69‡	.41†	−.05	−.12	−.08	−.11
9. Crime rate, 1985	.19	.02	.19	.26	.60‡	.34*
10. Unemployment rate, 1985	.15	.08	−.20	.22	−.18	.04
11. Restrictions on seller (Attribute I), 1987	−.45‡	−.25	.45‡	−.02	.48‡	−.07
12. Restrictions on buyer (Attribute II), 1987	−.54‡	−.30*	.30*	.37†	.35†	.44‡
M	12.80	48.90	31.80	10.80	64.10	351.30
SD	3.20	1.10	1.90	12.50	22.50	1386.40

	7	8	9	10	11	12
8. Divorce rate, 1985	.58‡					
9. Crime rate, 1985	.51‡	.32*				
10. Unemployment rate, 1985	−.01	.22	.03			
11. Restrictions on seller (Attribute I), 1987	−.40†	−.35†	−.06	−.09		
12. Restrictions on buyer (Attribute II), 1987	−.31*	−.29*	.18	−.03	.21	
M	6.00	5.20	48.20	7.10	1.30	0.50
SD	5.70	1.80	13.60	1.90	1.20	0.90

*$p<.05$. †$p<.01$. ‡$p<.001$.

Figure CS15.2.

a) Examine the table in Figure CS15.2 and identify all the variables examined within this study.
b) What type of relationship exists between the suicide rate and the restrictions of the seller? Is this a significant relationship at $\alpha = 5\%$? at $\alpha = 1\%$? Can one conclude from this relationship that high suicide rates had significantly less stringent handgun control laws? Explain.

 What proportion of the variance in the suicide rate can be accounted for by the variance in the restrictions of the seller? What formula is used to determine this proportion, and what is this statistical measure called?
c) What type of relationship exists between the suicide rate and the percent of male? Is this a significant relationship at $\alpha = 5\%$? at $\alpha = 1\%$? What proportion of the variance in the suicide rate can be accounted for by the variance in the percent of males? What formula is used to determine this proportion, and what is this statistical measure called?
d) If a scatter diagram were constructed for the variables suicide rate and divorce rate, what would you expect the general pattern to look like?
e) As suicide rates increase, what would you expect to happen to the variable persons per square mile?
f) As crime rates increase, what would you expect to happen to the variable persons per square mile?
g) With what variables would suicide rates be significantly correlated? Complete the following statements:

 high suicide rates had significantly _____ percent of blacks.
 high suicide rates had significantly _____ divorce rates.
 high crime rates had significantly _____ persons per square mile.
 For each of the other variables, write a similar statement.
h) Can the null hypothesis be rejected for the variables suicide rate and percentage of persons in the 35 to 64 age group at $\alpha = 5\%$? at $\alpha = 1\%$? Explain. What is the smallest level of significance that would cause the rejection of the null hypothesis? Explain. ■

Summary of Hypothesis Testing Procedure to Test the Significance of a Correlation Coefficient

The following outline is a summary of the 5 step hypothesis testing procedure for the test of a correlation coefficient.

Step 1. *Formulate the two hypotheses, H_o and H_a.*
 Null Hypothesis: H_o: $\rho = 0$
 Alternative Hypothesis, H_a: can have one of following three forms:
 H_a: $\rho > 0$ or H_a: $\rho < 0$ or H_a: $\rho \neq 0$

Step 2. *Design an experiment to test the null hypothesis, H_o.*
 Under the assumption the null hypothesis is true, that is: $\rho = 0$, the sampling distribution of r values is symmetric about r=0 and is referred to as the:
 the Sampling Distribution of the Correlation Coefficients.

Step 3. *Formulate the decision rule.*
 a) type of alternative hypothesis: **directional** or **nondirectional**
 b) type of test: **1TT** or **2TT**
 c) significance level: $\alpha = 1\%$ or $\alpha = 5\%$
 d) determine the critical value, r_α, using **TABLE V** and the **degrees of freedom:** $df = n - 2$
 e) construct the hypothesis testing model
 f) state the decision rule.

Step 4. *Conduct the experiment.*
 Calculate Pearson's correlation coefficient, r.

Step 5. *Determine the conclusion.*
 Compare the **experimental outcome, (i.e., Pearson's correlation coefficient, r)** to the **critical value, r_α**, and state the conclusion:
 either: a) **Reject H_o and Accept H_a at α**
 or   b) **Fail to reject H_o at α**

15.8 Assumptions for Linear Regression Analysis

Beside the assumption of a **linear relationship** between the two variables **x** and **y**, the following three assumptions must be satisfied in order to apply the linear regression model.

For each value of x under consideration, there exists a population of y values and they must conform to the following conditions:

- these y populations must have a normal distribution.
- the means of all these populations must lie on a straight line called the population regression line.
- the standard deviation of all the y populations must be equal.

Furthermore, when selecting a value for the independent variable x, this x value must be within the range of the sample x data.

Some Cautions Regarding the Interpretation of Correlation Results

There are some common mistakes that people make when they interpret correlation results. To help you avoid these pitfalls, we will discuss two cautions regarding the use of correlation.

Caution #1:
Don't Overlook the Possibility of a Non-Linear Relationship

The correlation coefficient, r, only measures the **linear** relationship between two variables. It is possible for two variables to have a linear correlation near zero, and, yet they could have a *significant non-linear relationship*. Be careful not to interpret a non-significant linear correlation as meaning there doesn't exist any relationship between the variables.

Caution #2:
Correlation Doesn't Indicate a Cause-And-Effect Relationship

You must be careful not to interpret a significant linear relationship between two variables to imply there exists a ***cause-and-effect*** relationship. Simply because two variables are correlated does not guarantee a cause-and-effect relationship. Two seemingly non-correlated variables may often be highly correlated.

For example, the number of storks nesting in various European towns in the early 1900's and the number of human babies born in the same towns during this period had a very high correlation. However, we can't conclude that an increase in the number of storks will cause an increase in the number of babies.

Therefore, a significant correlation between two variables should not be interpreted to mean that a change in one variable *causes* a change in the other variable, but rather only that changes in one variable are associated with or accompanied by changes in the other variable.

USING A CALCULATOR TO EVALUATE PEARSON'S CORRELATION COEFFICIENT, r, AND THE REGRESSION LINE QUANTITIES: a AND b

Statistical calculators have built-in programs that enable users to determine the correlation coefficient, r, and the regression line quantities: a and b. The procedure to obtain the correlation coefficient, r, and the regression quantities: a and b requires the user to simply enter the x and y data values using special keystrokes, and then retrieve the results from the statistical memories through a combination of keystrokes. However, the specific procedure to determine these quantities is dependent upon the calculator manufacturer and the model you own.

The general generic procedure is:

Step 1. Set the calculator in statistical mode for linear regression.

Step 2. Enter each pair of data values (the x and y values) one at a time using the appropriate keystrokes.

The general sequence of steps to enter the x and y values are:

1) Key in x value
2) Press special key to store this x value.
3) Key in y value
4) Press a second special key to enter this pair of x and y data values.

This sequence of four steps is repeated until all the pairs of x and y data values are entered.

The special keys used in the 2nd and 4th step of the previous sequence is dependent upon the manufacturer of the calculator.

For Texas Instruments Calculators, the two special keys usually are:

the $X \rightleftarrows Y$ key and the $\Sigma+$ key.

The $X \rightleftarrows Y$ key is pressed after the x value is keyed in to store the x value, while the $\Sigma+$ key is pressed after the y value is keyed in to enter the pair of data values.

For Casio Calculators, the two special keys usually are:
the X_D, Y_D key and the **DATA** key.

The X_D, Y_D key is pressed after the x value is keyed in to store the x value, while the **DATA** key is pressed after the y value is keyed in to enter the pair of data values.

For Sharp Calculators, the two special keys usually are:
the **(X,Y)** key and the **DATA** key. The **(X,Y)** key is pressed after the x value is keyed in to store the x value, while the **DATA** key is pressed after the y value is keyed in to enter the pair of data values.

Depending on the calculator, as the pairs of data values are entered, the display will either show the number of data values entered or the last data value entered.

Step 3. Retrieve the statistical results for: r, a and b from memory using the appropriate keystrokes outlined in the manual for the calculator.

Although, the previous procedure is generic and not dependent upon any specific calculator, Appendix B contains specific Calculator Instructions for some models of the calculator manufacturers: Texas Instruments, Casio, Sharp and Radio Shack.

15.9 USING MINITAB

Table 15.1 listed the heights and respective weights for a sample of 12 female students. Let's use MINITAB to the find the correlation between the variables height and weight; the scatter diagram; and, the regression equation predicting a female weight given a height.

Using the data values in Table 15.1 let's enter these data into a MINITAB worksheet using the READ command.

```
MTB >   READ C1 C2
DATA>   62 123
DATA>   58 102
DATA>   64 110
DATA>   69 137
DATA>   61 145
DATA>   70 132
DATA>   59 108
DATA>   60 112
DATA>   63 124
DATA>   72 155
DATA>   71 170
DATA>   68 140
DATA>   END
```

To make the output easier to interpret we will name and print the variables in columns C1 and C2 to correspond to the characteristic of each variable.

```
MTB >   NAME C1 'HT(IN)' C2 'WT(LB)'
MTB >   PRINT C1 C2
```

The output for variable height and weight is:

```
ROW    HT(IN)    WT(LB)
 1       62       123
 2       58       102
 3       64       110
 4       69       137
 5       61       145
 6       70       132
 7       59       108
 8       60       112
 9       63       124
10       72       155
11       71       170
12       68       140
```

To find the correlation, use the CORRELATION command.

```
MTB >   CORRELATION C1 C2
```

The correlation output is:

```
Correlation of HT(IN) and WT(LB) = 0.789
```

A scatter diagram for these variables is displayed using the PLOT command. Notice that when the PLOT command is used, the first variable listed corresponds to the horizontal axis on the graph.

```
MTB >   PLOT C2 C1
```

```
MTB >  PLOT C2 C1

         175+                                                               *
WT(LB)     -
           -                                                                     *
           -
         150+
           -           *
           -                              *
           -                           *
           -                                   *
         125+       * *
           -
           -   *  *         *
           -
         100+ *
           -
             --------+---------+---------+---------+---------+--------HT(IN)
                   60.0      62.5      65.0      67.5      70.0
```

To find the regression equation, use the MINITAB command REGRESSION C2 1 C1. The form of this command requires that the dependent variable is listed first, followed by the number of independent variables, and, finally the listing of the independent variable(s). For simple linear regression, that is, regression with **one** independent variable, a numeral one is indicated between the dependent variable and the independent variable. Thus to determine the regression equation we have

MTB > REGRESSION C2 1 C1

The output for these variables is:

```
The regression equation is
WT(LB) = - 81.1 + 3.26 HT(IN)

Predictor         Coef    Stdev   t-ratio      p
Constant        -81.12    52.14    -1.56    0.151
HT(IN)          3.2580   0.8031     4.06    0.002

s = 13.30    R-sq = 62.2%    R-sq(adj) = 58.4%

Analysis of Variance
SOURCE      DF       SS       MS        F        p
Regression   1    2911.0   2911.0    16.46    0.002
Error       10    1768.7    176.9
Total       11    4679.7

Unusual Observations
Obs.HT(IN)  WT(LB)     Fit  Stdev.Fit  Residual   St.Resid
  5   61.0  145.00   117.62      4.88     27.38     2.21R

R denotes an obs. with a large st. resid.
```

As you can see there is a lot of output generated by the REGRESSION command. Let's considered the output as blocks of information. The first line of the output gives the regression equation. Next you'll see information on the regression equation as well as the variables used. In the last block of the output there is listed information about potential outliers. Can you find where the constants "a" and "b", that are used in the regression equation, are listed? Of course the regression equation is one place to look. Can you find those values for "a" and "b" anywhere else?

Notice that the p-value of 0.002 is given for the line labelled 'HT'. This indicates that the correlation between HT and WT has a p-value of 0.002. Would you consider HT and WT to be significantly correlated?

An interesting command of MINITAB is the MPLOT command. This command graphs multiple scatter diagrams on the same axes.

Suppose we plot both the original data values in our illustration and a set of data values that are generated by prediction using the regression equation that was found for the data.

To accomplish this, the LET command is used for the regression equation such that the predicted values are generated into a separate column, C3. The NAME and PRINT commands are also used to help understand what is happening.

```
MTB >   NAME C3 'PRDWT'
MTB >   LET C3=3.26*C1 + - 81.1
MTB >   PRINT C1-C3
```

The MINITAB output is:

```
ROW   HT(IN)   WT(LB)   PRDWT
 1      62      123     121.02
 2      58      102     107.98
 3      64      110     127.54
 4      69      137     143.84
 5      61      145     117.76
 6      70      132     147.10
 7      59      108     111.24
 8      60      112     114.50
 9      63      124     124.28
10      72      155     153.62
11      71      170     150.36
12      68      140     140.58
```

Next the MPLOT command is used yielding two graphs on the plot.

```
MTB >   MPLOT C2 C1 C3 C1
```

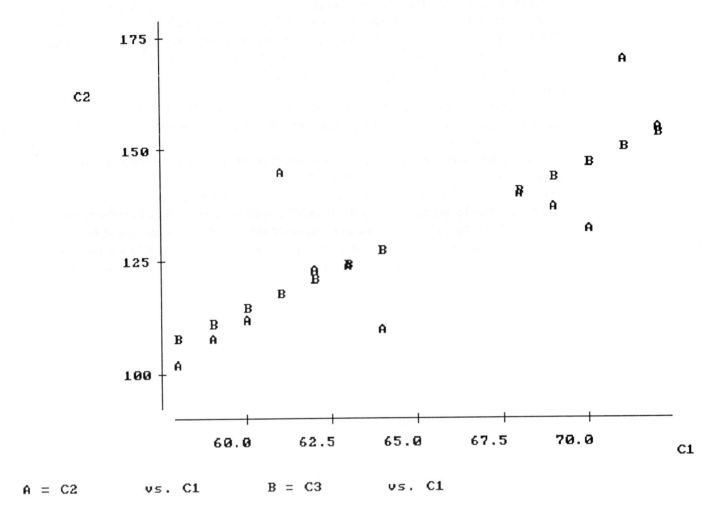

The MPLOT command's output lists the original scatter diagram's points as A, while the predicted points are listed as B. The numeral 2, indicates where both the original and predicted ordered pairs coincide. Try using a ruler to line up the B's. This is the plot of the regression line also known as the least squares line.

Glossary

TERM	SECTION
Linear correlation	15.1
Ordered pair	15.2
Scatter diagram	15.2
Positive linear correlation	15.2
Negative linear correlation	15.2
No linear correlation	15.2
Pearson's correlation coefficient, r	15.3
Population correlation coefficient, ρ	15.5
Sampling distribution of correlation coefficients	15.5
Critical value of the correlation coefficient, r_α	15.5
Coefficient of determination, r^2	15.6
Linear regression	15.7
Best fitting line	15.7
Regression line	15.7
Method of least squares	15.7
Predicted value of y, y'	15.7
Independent variable	15.7
Dependent variable	15.7
Regression coefficients, a, b	15.7

Exercises

PART I Fill in the blanks.

1. Correlation is a technique that measures the strength (or the degree) of the relationship between ___ ___.
2. In correlation analysis, two variables represent data that is a _____ not count data.
3. A pair of numbers that is written in the following form: (x,y) is called an _____ pair since the value of the x variable is always written _____ within the parentheses.
4. A graph representing the ordered pairs of sample data on a set of axes is called a ___ ___ .
5. A _____ correlation is a correlation between two variables, x and y, that occurs when high measurements on the x variable tend to be associated with high measurements on the y variable and low measurements on the x variable tend to be associated with low measurements on the y variable.
6. If the dots in the scatter diagram tend to follow a straight-line path, then this is type of correlation is called a ___.
7. A _____ correlation is a correlation between two variables, x and y, that occurs when high measurements on the x variable tend to be associated with low measurements on the y variable and low measurements on the x variable tend to be associated with high measurements on the y variable.
8. A _____ correlation means there is no linear relationship between the two variables. That is, high and low measurements on the two variables are __ associated in any predictable way.
9. The closer the points of a scatter diagram approximate a straight-line pattern, the _____ the correlation between the two variables.
10. The strength of the correlation between the two variables of a sample can be numerically measured by _____.
11. Pearson's correlation coefficient, r, will always be a numerical value between ___ and ___.
12. A value of r = +1 represents the _____ positive correlation possible and it indicates a _____ positive correlation. This means that all the points of the scatter diagram will lie on a straight line which is sloping ___ from left to right.
13. A value of r = -1 represents the _____ negative correlation possible and it indicates a _____ negative correlation. This means that all the points of the scatter diagram will lie on a straight line which is sloping ___ from left to right.
14. A value of r = 0 represents no _____ correlation between the two variables.
15. The closer the value of the correlation coefficient, r, is to _____ or to _ the stronger the linear relationship between the two variables. However, a value of r close to zero indicates a _____ linear relationship between the two variables.
16. When Pearson's correlation coefficient is greater than zero, that is: r > 0, then there is a _____ linear correlation between the two variables. When Pearson's correlation coefficient is less than zero, that is: r < 0, then there is a _____ linear correlation between the two variables.
17. The population correlation coefficient is symbolized by the lower case Greek letter ρ, pronounced as row.
18. We use Pearson's correlation coefficient, r, to calculate the correlation coefficient for the _____ data. The correlation coefficient, r, for the sample is used as an _____ of the population correlation coefficient, ρ.
19. a) The null hypothesis for testing the significance of the population correlation coefficient is stated as: the population correlation coefficient is equal to _____. That is, there is __ linear correlation between the two variables. This is symbolized as: $H_o: \rho = \_\_$.
    b) The alternative hypothesis can be stated in one of the following three forms:
    form a: The population correlation coefficient, ρ, is claimed to be greater than zero. That is, there is a _____ linear correlation between the two variables. This is symbolized as: $H_a: \rho \_\_ 0$.
    form b: The population correlation coefficient, ρ, is claimed to be less than zero. That is, there is a _____ linear correlation between the two variables. This is symbolized as: $H_a: \rho \_ 0$
    form c: The population correlation coefficient, ρ, is claimed to be not equal to zero. That is, there is a _____ linear correlation between the two variables. This is symbolized as: $H_a: \rho \_ 0$.
20. The test of the significance of the correlation coefficient is based upon the assumption that both variables x and y are _____ distributed. This test uses as its hypothesis testing model: the sampling distribution of the _____ _____.
21. The critical value of the correlation coefficient is

denoted as ____. To determine this critical r value, you need to calculate the degrees of freedom for the test using the formula: df = ____.

22. To calculate Pearson's correlation coefficient, we use the formula r = ____.

23. The coefficient of determination measures the proportion of the variance of the _____ variable that can be accounted for by the variance of the _____ variable. To calculate the coefficient of determination, you need to square the value of _____.

24. In linear regression analysis, the technique used to obtain the best fitting line is called the method of ____ ____. The best fitting line is called the ____ line.

25. The regression line formula is given by: _____.
To calculate the regression coefficients a and b, we use the following formulas: b = ____ and a = ____. The regression line is used to predict the ____ variable given the value of the ____ variable.

PART II   Multiple choice questions.

1. Which type of correlation is indicated by a scatter diagram whose points are generally lower as you move from left to right?
   a) positive b) none c) negative
2. Which type of correlation is indicated by a scatter diagram whose points appear to be randomly placed in no particular pattern?
   a) positive b) none c) negative
3. When testing the significance of the correlation coefficient, if there are ten pairs of data, $\alpha = 5\%$ and it's a 1TT, the r_α = ____.
   a) 0.54 b) 0.83 c) 0.62 d) 0.71
4. If r = 0.8 then the coefficient of determination is?
   a) 0.88 b) 0.64 c) 1.6 d) 0.40
5. A negative value of r indicates:
   a) no relationship   b) weak relationship
   c) positive relationship   d) negative relationship
6. When testing the significance of the correlation coefficient, r, the null hypothesis, H_o, is always symbolized as:
   a) $\rho > 0$ b) $\rho < 0$ c) $\rho = 0$ d) $\rho \neq 0$

7. The coefficient of determination, r^2, measures the proportion of variance of:
   a) the independent variable which can be accounted for by dependent variable.
   b) the dependent variable which can be accounted for by the independent variable.
   c) neither of these
8. When testing the significance of r the degrees of freedom is given by:
   a) n   b) n - 2   c) n - 1   d) 2n
9. If x and y are perfectly correlated then r must equal:
   a) 1   b) -1   c) 0   d) 1 or -1
10. If y increases as x decreases this indicates that x and y are correlated:
    a) positively b) not a all c) negatively

PART III   Answer each statement True or False.

1. The linear correlation between two variables, x and y, is a number between -1 and +1.
2. A linear correlation that is close to zero can be interpreted to mean that no relationship exists between the variables, x and y.
3. Negative correlations are usually not significant.
4. A linear correlation between the two variables is a line drawn between the variables.
5. Scatter diagrams are graphs that indicate the degree of variation between x and y.
6. When the coefficient of determination is zero, this means that there is a perfect linear correlation between the x and y variables.
7. A coefficient of determination of 0.5, means that there is a 50% chance that a linear relationship exists between variables x and y.
8. When the correlation coefficient is significant, then the coefficient of determination is greater than 50%.
9. Linear regression can be used to predict value for the dependent variable from a value of the independent variable.
10. A linear regression model produces a line that best fits the data pairs from a sample.

PART IV   Problems.

Many problems in this section require calculations that are long and tedious. Therefore, we recommend that you use a calculator or a computer with statistical application software to help you perform these calculations. In problems requiring the testing of the significance of the correlation coefficient, assume that both the x and y variables are normally distributed.

1. In each of the following scatter diagrams, if a linear correlation exists between the two variables, then state whether the type of linear correlation is positive, negative or none.

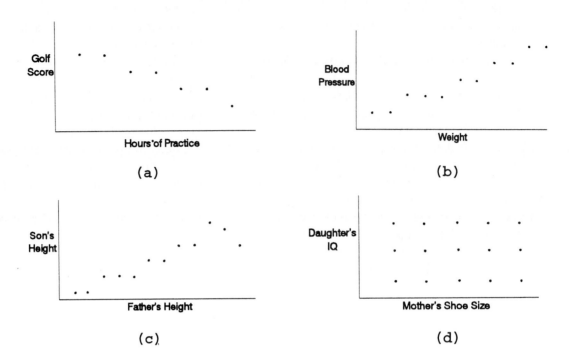

2. For each of the following scatter diagrams, choose the statement that best describes the type of relationship that exists between the two variables.
   1) strong linear correlation.
   2) moderate or weak linear correlation.
   3) no linear correlation.

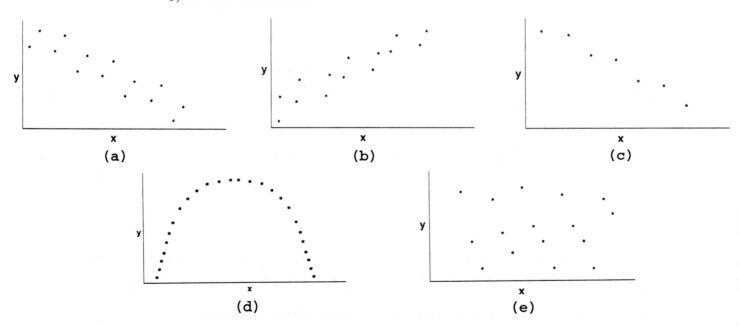

3. For the following scatter diagrams:
   a) identify those that you believe represent a linear relationship between x and y.
   b) for those that you have identified as representing a linear relationship between x and y, determine which are positive relationships and which are negative.

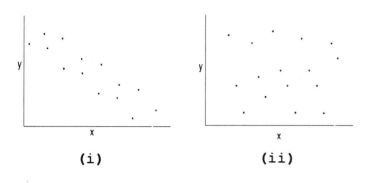
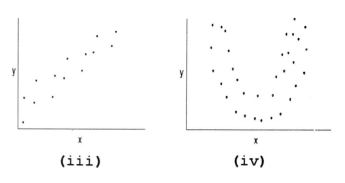

(i)   (ii)   (iii)   (iv)

4. For each of the following two variables, state whether you would expect a positive, negative or no correlation.
   a) an individual's IQ score and annual salary
   b) an individual's age and blood pressure
   c) the number of days a student misses class and final grade
   d) a mother's birthday and her daughter's birthday
   e) an individual's driving speed and gasoline mileage
   f) an individual's annual income and education level
   g) average daily temperature during the summer months and the number of air conditioners sold during these months
   h) an adult's shoe size and IQ score
   i) amount of fertilizer applied to tomato plants and the number of tomatoes per plant
   j) number of hours practicing golf and golf score
   k) a husband's age and his wife's age
   l) number of hours practicing bowling and bowling score
   m) amount of alcohol consumed and reaction time
   n) mother's birth weight and her child's birth weight
   o) a person's age and number of days of sick-leave
   p) a person's age and life insurance premium
   q) high school average and college grade-point average
   r) amount of grams of fat consumed daily and cholesterol level

5. Rank the following values of Pearson's correlation coefficient, r, from the strongest to weakest linear association.
   −0.95, 0.05, +1.00, 0.69, −0.69, 0, 0.50

6. Explain the difference between the following pair of r values.
   a) r = 0.55 and r = 0.87
   b) r = −0.60 and r = 0.60
   c) r = 0.05 and r = 0.95

7. Using the Pearson correlation coefficient formula, calculate the value of r for each of the following data pairs.

a) x	y	b) x	y	c) x	y
2	4	5	11	4	9
3	7	7	8	5	11
4	8	9	1	6	13
5	11	10	6	7	15
6	11	16	2	8	17

d) x	y	e) x	y	f) x	y
2	3	11	12	72	222
2	5	16	9	62	145
3	5	15	9	58	125
6	6	21	5	68	165
7	9	23	4	68	172
4	3	16	8	63	135
1	4	26	4	61	120
5	8	29	3	73	185
3	6	34	1	75	210
9	10	28	5	66	158

8. In the table below is a list of sample information along with Pearson correlation coefficients that have been calculated for various sample data pairs. For each collection of sample information, use the information in the table and TABLE V to determine r_α, and to test the significance.
   a) n = 15, r = 0.60, 2TT, α = 1%
   b) n = 25, r = −0.60, 1TT, α = 1%
   c) n = 30, r = 0.60, 1TT, α = 1%
   d) df = 10, r = 0.42, 2TT, α = 5%
   e) df = 20, r = −0.42, 1TT, α = 5%
   f) df = 28, r = 0.42, 1TT, α = 5%

9. In the table below is a list of sample information along with Pearson correlation coefficients that have been calculated for various sample data pairs. For each collection of sample information, use the information in the table and TABLE V to determine r_α, and to test the significance.

a) $n=18$, $r=0.35$, $H_a: \rho \neq 0$, $\alpha=1\%$
b) $n=28$, $r=+0.47$, $H_a: \rho > 0$, $\alpha=1\%$
c) $n=30$, $r=-0.46$, $H_a: \rho < 0$, $\alpha=1\%$
d) $df=19$, $r=0.58$, $H_a: \rho \neq 0$, $\alpha=5\%$
e) $df=25$, $r=-0.34$, $H_a: \rho < 0$, $\alpha=5\%$
f) $df=13$, $r=0.40$, $H_a: \rho > 0$, $\alpha=5\%$

10. Audrey, an anthropologist, has made a study of the linear relationship between the leg length and total height of prehistoric animals. Using a random sample of 17 prehistoric animal fossils, she obtained a correlation coefficient of 0.48.
    a) Can Audrey conclude that there is a positive correlation between the length and total height of prehistoric animals at $\alpha = 5\%$?
    b) What percent of variability in total height can be accounted for by the variability in the leg length of the prehistoric animals?

11. Carol, a nutritionist, has conducted a study to determine the linear relationship between oat bran consumption and cholesterol level.
    a) If Carol obtains a correlation coefficient of -0.56 from a random sample of 24 people, can she conclude that there is a negative correlation between the amount of oat bran consumed and cholesterol level at $\alpha = 1\%$?
    b) What percent of the variability in cholesterol level can be accounted for by the variability in the amount of oat bran consumed?

12. Matthew, a medical researcher, wants to determine if a positive linear relationship exists between the number of pounds a male is overweight and his blood pressure. He randomly selects a sample of 10 males and records the number of pounds that the individual is over his "ideal" weight according to a medical weight chart and his systolic blood pressure. Matthew's sample data is listed in the following table.

male subject	number of lbs. above ideal weight	systolic blood pressure
1	12	150
2	8	142
3	20	165
4	17	152
5	14	147
6	23	158
7	6	135
8	4	128
9	15	135
10	19	153

   a) Construct a scatter diagram.
   b) Calculate the Pearson correlation coefficient, r.
   c) Determine if r is significant at $\alpha = 1\%$. If r is significant at $\alpha = 1\%$, then do parts d, e, f.
   d) Find r^2 and interpret its meaning.
   e) Determine the regression equation, y'.
   f) Using the regression equation, predict a male's systolic blood pressure if the individual is 10 pounds over his ideal weight.

13. Craig, a marketing executive for a brewery company, wants to determine if there is a positive linear relationship between advertising expenditures and sales for their new lite beer product, Less Ale. He randomly samples 7 sales regions and records the sample data in the following table, where the x variable represents the advertising expenditure in thousands of dollars and the y variable represents the beer sales in million dollars.

region	x	y
1	35	38
2	47	35
3	65	42
4	92	50
5	55	40
6	25	35
7	82	48

   a) Construct a scatter diagram.
   b) Calculate the Pearson correlation coefficient, r.
   c) Determine if r is significant at $\alpha = 1\%$. If r is significant at $\alpha = 1\%$, then do parts d, e, f.
   d) Find r^2 and interpret its meaning.
   e) Determine the regression equation, y'.
   f) Using the regression equation, predict the beer sales for a region advertising budget expenditure of $50,000.

14. I.M. Smart, a statistics nerd, wants to investigate if there is a positive linear relationship between the number of hours a student studies for a statistics exam and the student's test grade. He randomly samples 12 students that are taking statistics at a community college and records the sample data in the following table.

student	hours studying for exam	test grade
1	10	80
2	8	60
3	12	78
4	20	90
5	7	65
6	4	60
7	9	70
8	15	85
9	11	75
10	6	70
11	2	45
12	1	50

a) Construct a scatter diagram.
b) Calculate the Pearson correlation coefficient, r.
c) Determine if r is significant at α = 1%. If r is significant at α = 1%, then do parts d, e, f.
d) Find r^2 and interpret its meaning.
e) Determine the regression equation, y'.
f) Using the regression equation, predict a student's statistics grade if the student spent 14 hours studying for the exam.

15. The grounds keeper at the Sunview Golf course wants to know if there is a positive linear relationship between the density of seed spread on a newly landscaped fairway and the density of new grass six months later. The following data pairs represent some sample plantings made by the grounds keeper.

seed density (lbs/400ft²) x	grass density (seedlings/ft²) y
1.0	170
2.0	200
3.0	240
4.0	300
5.0	310
6.0	290
7.0	290

a) Construct a scatter diagram.
b) Calculate the Pearson correlation coefficient, r.
c) Determine if r is significant at α = 5%. If r is significant at α = 5%, then do parts d,e,f.
d) Find r^2 and interpret its meaning.
e) Determine the regression equation, y'.
f) Using the regression equation, predict the grass density given that grass seed was applied at a rate of 3.5 (lbs/400ft²).

16. The scientists at the Microbiology Laboratory in Woods Hale, MA. want to determine if there is a linear relationship between the amount of rainfall in May and the number of mosquitoes. For each of the selected years, a random sample of data pairs have been collected and are listed in the table that follows:

Year	amount of rainfall in inches	Mosquito population index
1990	5.3	7.7
1989	2.7	5.4
1987	3.9	4.8
1985	2.4	3.5
1983	5.7	6.9
1981	3.5	4.5

a) Construct a scatter diagram.
b) Calculate the Pearson correlation coefficient, r.
c) Determine if r is significant at α = 5%. If r is significant at α= 5%, then do parts d, e, f.
d) Find r^2 and interpret its meaning.
e) Determine the regression equation, y'.
f) Using the regression equation, predict the mosquito population index if we have 4.3 inches of rain in May.

17. Grace, the personnel director at Weight Lookers International, believes there is a negative linear relationship between the number of sick days taken per year by an employee and the percent of yearly salary increase the employee receives. A random sample of 20 employees yielded the following data pairs.

x number of sick days per year	y % of yearly salary increase
4	12
16	8
11	8
14	7
6	14
20	0
9	10
12	9
13	7
1	14
0	12
17	2
18	2
7	13
9	10
15	7
15	8
13	9
10	8
5	12

a) Construct a scatter diagram.
b) Calculate the Pearson correlation coefficient, r.
c) Determine if r is significant at α = 5%. If r is significant at α = 5%, then do parts d, e, f.
d) Find r^2 and interpret its meaning.
e) Determine the regression equation, y'.
f) Using the regression equation, predict an employee's yearly percentage of salary increase if over a given year the employee was out sick eight times.

18. Professor DePorto believes there is a positive linear relationship between Midterm examination grades and Final examination grades for students taking his elementary introduction to statistics course. A random sample of 26 students who have taken both exams produced the following data pairs.

Midterm Exam Grade	Final Exam Grade
39	37
28	28
32	30
35	24
30	25
42	39
34	26
36	25
22	19
33	32
22	19
38	30
29	25
50	42
16	13
41	37
43	35
44	40
39	21
30	34
24	11
16	17
29	29
33	15
44	29
39	35

a) Construct a scatter diagram.
b) Calculate the Pearson correlation coefficient, r.
c) Test the significant of r at $\alpha = 5\%$. If r is significant at $\alpha = 5\%$, the do parts d, e, f.
d) Find r^2 and interpret its meaning.
e) Determine the regression equation, y'.
f) Linda received a 24 on her Midterm exam. Use the regression equation and predict Linda's Final exam grade.

PART V What Do You Think?

1. **What Do You Think?**
   Examine the following table which represents the correlations between the SAT (Scholastic Aptitude Test) scores and the GRE (Graduate Record Examination) scores for 22,923 subjects in a study entitled *The Differential Impact Of Curriculum On Aptitude Test Scores* which appeared in the Journal of Educational Measurement in 1990.

   The SAT is a test administered by the College Board to college-bound high school students. The results are used by college undergraduate admission departments in making decisions regarding high school applicants. While the Graduate Record Examination (GRE), a College Board administered exam, is taken by college students preparing to apply for admission to graduate schools. The purpose of the study was to determine the correlation between the SAT and GRE scores of students who took these exams at the normal times during their academic careers with the typical number of years between exams.

   a) What type of correlation exists between the SAT Verbal and the GRE Verbal? Interpret this relationship. That is, in general, what would you expect to happen to the value of the GRE Verbal scores as the scores on the SAT Verbal increases? If a scatter diagram were constructed for the variables SAT Verbal and the GRE Verbal, what would you expect the general pattern to look like?

   b) What proportion of the variance in the SAT Verbal scores can be accounted for by the variance in the GRE Verbal scores? What formula is used to determine this proportion, and what is this statistical measure called?

The population of interest for the study consisted of examinees who took the SAT and also the GRE General Test at the normal times in their academic careers, with the typical number of years intervening.

Intercorrelations Between SAT and GRE Scores for the Total Study Sample
n = 22,923

	SAT Verbal	SAT Mathematical	GRE Verbal	GRE Quantitative	GRE Analytical	Mean	Standard Deviation
SAT Verbal	1.000	.628	.858	.547	.637	518.8	104.7
SAT Mathematical	.628	1.000	.598	.862	.734	556.0	110.2
GRE Verbal	.858	.598	1.000	.560	.649	510.1	107.7
GRE Quantitative	.547	.862	.560	1.000	.730	573.4	125.6
GRE Analytical	.637	.734	.649	.730	1.000	579.7	117.6

c) What does a value of 1.000 indicate about the relationship between SAT Verbal and SAT Verbal? What type of a relationship is this called?
d) For the SAT Verbal Column, what row variable, other than the SAT Verbal, has the strongest relationship with the SAT Verbal? Explain. What do you believe might cause this to happen?
e) For the SAT Mathematical Column, what row variable, other than the SAT Mathematical, has the strongest relationship with the SAT Mathematical? Explain. What do you think might lead to this strong relationship?
f) What proportion of the variance in the SAT Mathematical scores can be accounted for by the variance in the GRE Analytical? What proportion of the variance in the SAT Mathematical scores can be accounted for by the variance in the GRE Quantitative? Explain what the difference in these two results mean?
g) The authors of the study state: "the correlations between SAT Verbal and GRE Verbal and between the SAT Mathematical and GRE Quantitative, both of which are 0.86, indicate that the linear relationship between SAT and GRE scores explains almost three fourths of the variance in GRE Verbal and GRE Quantitative scores taken four years later." Explain the meaning of this statement and how do you think the authors arrived at this three fourths value? What is an equivalent percentage value for this three fourths figure? What do the authors mean by "linear relationship"? Explain.
h) For the GRE Analytical Column, what row variable has the weakest relationship with the GRE Analytical? Explain. What do you think might lead to this weak relationship?
i) In words, describe the type of relationship and the strength of the relationship between the variables GRE Quantitative and SAT Mathematical.
j) Interpret the information that the mean and standard deviation indicate about the SAT and GRE test results. Which distribution of test scores has the greatest variability?
k) If the distribution of each of these test scores is approximately normal, then what is the median and modal test score for each test?
l) If the distribution of each of these test scores is approximately normal, then what percent of the SAT Mathematical scores are within 110.2 points of 556? within 220.4 points of 556? within 330.6 points of 556?
m) As the scores of the SAT Mathematical increase, what would you expect to happen to the scores on the variable GRE Quantitative? Explain.

2. **What Do You Think?**
A research article from the Journal of Social and Behavior Personality entitled *Effort and reward in college: A replication of some puzzling findings* presented the following intercorrelations table. The table represents the correlations between the overall grade point average (GPA) and predictor variables for students in three sections of an Introduction to Sociology class taught by the researcher.

Intercorrelations (r) among GPA and Predictor Variables for Three Sections of Introduction to Sociology[a]

	(1)	(2)	(3)	(4)	(5)	(6)	(7)	(8)
GPA (1)	-	.45***	.65***	.72***	.70***	.00	.30**	-.39**
HSGPA (2)		-	.52***	.32***	.37***	.06	.16	-.01
SYNGRADE (3)			-	.86***	.81***	.02	.29*	-.50***
FINAL (4)				-	.83***	-.11	.31*	-.51***
EXAM1 (5)					-	-.13	.26*	-.37**
STUDYDAY (6)						-	.17	-.01
STUDYEND (7)							-	.00
ABSENCES (8)								-

[a]Abbreviations: GPA = reported overall college Grade Point Average
HSGPA = reported High School Grade Point Average
SYNGRADE = semester grade in Sociology course
FINAL = score on comprehensive Final in Sociology
EXAM1 = score on first exam in Sociology
STUDYDAY = estimated average number of hours studied during weekdays
STUDYEND = estimated average number of hours studied on weekends
ABSENCES = actual number of absences in the Sociology course

Correlations involving SYNGRADE are based on N = 58; correlations involving FINAL on N = 55; All other correlations based on N = 60.

*.05 level of significance; **.01 level of significance; ***.001 level of significance.

Use this table to answer the following questions.
a) Which variable has the greatest significant correlation with the GPA variable? The smallest correlation? Interpret these results.
b) In words, describe the type and strength of the relationship between the variables *number of absences in the Sociology course* and *semester grade in the course*.
c) If a student has missed a large number of classes, then what might you predict about this student's performance on the comprehensive final exam in Sociology? Explain. What did you use to make this prediction?
d) If you performed a hypothesis test to determine if there exists a positive linear relationship between the grade on the first exam and the grade on the final exam at a significance level of 1%, then could you conclude that there existed a significant positive linear correlation? Explain.
e) If you performed a hypothesis test to determine if there exists a negative linear relationship between the grade on the first exam and the number of absences in the course at a significance level of 1%, then could you conclude that there existed a significant negative linear correlation? Explain.
f) If you performed a hypothesis test to determine if there exists a negative linear relationship between the grade point average and the number of absences in the course at a significance level of 5%, then can you reject the null hypothesis for this test? Explain.
g) Which variable is not a significant predictor of the other variables? How might you interpret this result?
h) If the p-value for the hypothesis test to determine if there exists a positive linear relationship between the grade on the first exam and grade point average is 0.020, then could you conclude that there existed a significant positive linear correlation? Explain.
i) If the p-value for the hypothesis test to determine if there exists a positive linear relationship between the semester grade for the Sociology course and grade point average is 0.002, then could you reject the null hypothesis for this test? Explain.
j) What proportion of the variance in the GPA can be accounted for by the variance in the scores on the comprehensive Sociology? Which formula is used to determine this proportion, and what is this statistical measure called?
k) If a scatter diagram were constructed for the variables GPA and Absences, what would you expect the general pattern to look like? Explain.
l) If a scatter diagram were constructed for the variables GPA and EXAM1, what would you expect the general pattern to look like? Explain.
m) As GPA increases, what would you expect to happen to the variable FINAL? Explain.

3. **What Do You Think?**
The following are excerpts from different studies.
(1) *In a study of 748 pregnant women, maternal birth weight was significantly related to baby's birth weight. The lower the maternal birth weight, the lower the baby's birth weight.*
(2) *The College Board which administers the SAT's stated that there is a correlation between family income and test performance with low-income students not doing as well on the SAT's as students from high-income families.*
(3) *A research finding indicated that tall children under age 8 tend to do better on intelligence tests than short children.*
(4) *The results of a study support the notion that a child's drive is related to the age of the father. The higher the scores of a student, the younger the father. Older fathers were defined as those over age 30 at the time of their offspring's birth.*
(5) *A researcher cites that time spent on homework is positively related to achievement.*
   a) For each excerpt, state the variables within the study. Identify which variable is the independent and the dependent variable.
   b) Indicate the type of relationship that exists between the two variables.
   c) If a scatter diagram was constructed for each of the variables, what would you expect the general pattern to look like? Explain.

PART VI **Exploring DATA With MINITAB**

1. For problems numbered 12-18 in PART IV of the Exercise section, use MINITAB to do all calculations.
2. For each of the exercises in problem 1 use the NAME and PRINT commands to construct a table of the original variables x and y. Use the LET command to calculate and add a column, C3, of predicted y values alongside the x and y variables. Use the MPLOT to construct a scatter diagram for the original variables and a graph of the regression equation on the same axes. How many data pairs are not on the regression line? Is this unusual? Explain.

PART VII Projects.

1. Randomly collect the heights to the nearest inch (x) and the weights to the nearest pound (y) of 20 students at your school.
   a) Construct a scatter diagram.
   b) Find r for your random sample.
   c) What type of linear correlation, if any, do you observe?
   d) Test H_o: $\rho = 0$, against H_a: $\rho > 0$ at $\alpha = 5\%$.
   e) If the test is significant at $\alpha = 5\%$, determine the regression line model and plot it on your scatter diagram.
   f) Estimate the weight of a student if his/her height is 74 inches.
   g) Determine the percent of variance in weight that is accounted for by the variance in height.
2. Randomly select 25 students at your school and ask them the following questions:
   I. How many hours per week do you work?
   II. What is your most recent cumulative average?
   With the data you obtain do the following: (let x represent the number of hours the student works per week)
   a) Construct a scatter diagram.
   b) Find r for your random sample.
   c) What type of linear correlation, if any, do you observe?
   d) Test H_o: $\rho = 0$, against H_a: $\rho \neq 0$ at $\alpha = 5\%$.
   e) If the test is significant at $\alpha=5\%$, determine the regression line model and plot it on your scatter diagram.
   f) Estimate the cumulative average of a student who works 35 hours per week.
   g) Determine the percent of variance in cumulative hours that is accounted for by the variance in hours worked.
3. Select two classifications that you believe are related and randomly sample the selected population. Using your sample data follow the procedures set forth in projects 1 and 2 above.

PART VIII Database.

The following exercises refer to the file DATABASE listed in Appendix A. We have indicated the appropriate MINITAB commands that are necessary to answer each exercise.

1. Using the MINITAB commands:
   RETRIEVE, PLOT, PRINT, CORRELATION and REGRESS
   a) Retrieve the file DATABASE.MTW and identify the column numbers for the variable names: AVE and GPA
   b) Print the columns AVE and GPA.
   c) Plot a scattergram for the variables AVE and GPA.
   d) Calculate the correlation between AVE and GPA. Determine the strength of the correlation between the two variables. Determine if the correlation coefficient is significant at $\alpha = 5\%$.
   e) Determine the regression equation and r^2. Which variable is considered the independent variable? Which is considered the dependent variable?
   f) What other things appear on the printout?
2. Using the MINITAB commands:
   RETRIEVE, PLOT, PRINT, CORRELATION and REGRESS
   a) Retrieve the file DATABASE.MTW and identify the column numbers for the variable names: SHT and PHT
   b) Print the columns SHT and PHT.
   c) Plot a scattergram for the variables SHT and PHT.
   d) Calculate the correlation between SHT and PHT. Is there a significant correlation between SHT and PHT at $\alpha = 1\%$? What can you conclude about the relationship between SHT and PHT?
   e) Determine the regression equation and r^2. How might you use the results on the regression equation?
   f) What other things appear on the printout?

ANSWERS

CHAPTER 1

Part I

1. Raw
2. Statistics
3. Descriptive; Inferential
4. Descriptive
5. Population
6. Sample
7. Population
8. Inferential; Representativeness
9. Statistic
10. Parameter
11. Pictograph
12. non-representative or biased
13. Statistic
14. Descriptive
15. Inferential
16. a. Sample
    b. Population
    c. Statistic
17. a. Parameter
    b. Sample
18. a. Working women in San Francisco, CA
    b. 250 working women
    c. Average annual Salary
19. a. People living in the suburb of Miami
    b. 400 people surveyed
    c. Sample
20. a. 10,000 families living in the community.
    b. 100 families randomly selected.
    c. Average annual income to measure the socio-economic level.
    d. The average annual income of the 100 families

Part IV

1. a) inferential; Population is individuals that watch TV; the sample size is 1994; the percentage of men who change channels is greater than the percentage of women.
   b) inferential; Population is the American people; the sample size is 1047; Two out of three Americans favor the death penalty for murder.
   c) not inferential
   d) inferential; Population is women that are mothers; the sample size is 84,193 women; The percentage of mothers that breast feed is increasing.

Part II

1. b
2. a
3. b
4. a
5. b
6. a
7. c
8. b
9. a

Part III

1. False
2. True
3. True
4. True
5. False
6. True
7. True
8. True
9. False
10. True

2a.

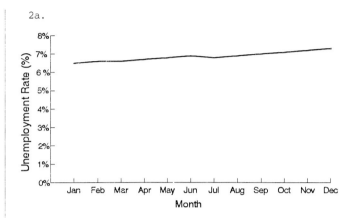

2b.

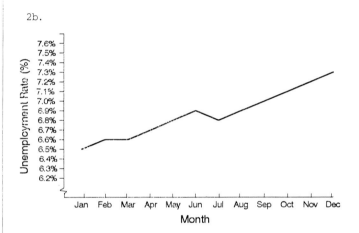

3a.

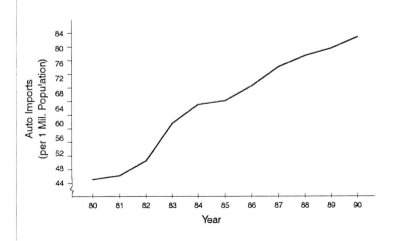

3b.

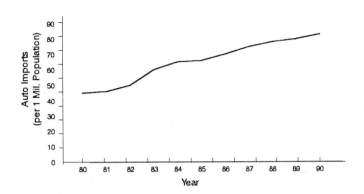

4. a) Statement does not indicate what pure means.
   b) We are not told anything about the survey. For example, how many doctors were surveyed, how was the survey taken, what questions were the doctors asked, etc.
   c) Similar to answer b.
   d) The word "juicier" is not defined. What does it mean for tobacco to be juicier?
   e) There are no details about the recent study. Such as who was surveyed, how was it surveyed, when did the survey take place, was it a random survey, etc.
   f) The pictograph of the money bags gives a visual impression that the oil profits have more than doubled.
   g) What does the concept of "the plain truth" mean? The graph is not labeled and gives a distorted impression by presenting a bar graph with a portion of the bars missing.

4. h) The graph is not labeled so we do not know by how much the circulation increased. The Journal's graph is also inferring that the circulation will continue to increase as indicated by the arrow head on the graph.
   i) The graph is not labeled and it implies that the population everywhere in the US is bulging at the seams.
   j) The pictographs give a visual impression that distorts the growth. There is no connection between the statistics presented and land investments.
   k) The slogan does not indicate the time period the statistics were compiled, and the civilian statistics include the two age groups (infants & senior citizens) with the highest mortality rates.
   l) The line graph indicates a greater visual increase through the years than the actual increase of $280.
5. Discuss this with your instructor
6. Some of the problems are:
   1. A bias sample is used by the manufacturer in their claims.
   2. False or misleading conclusions drawn from the sample.
   3. Small samples are used in the claims.
7. a) The axes aren't numbered, so you can't determine amount of moisture or amount of time.
   b) It is difficult to interpret what "ten times better" and "1000 times better" really means.
8. a) 50 cents
   b) Approx. 3.3%
   c) Yes, if you don't realize the actual amount of your contribution to the environmental groups.

CHAPTER 2

Part I

1. variable
2. value
3. categorical; numerical
4. categories
5. numerical; data
6. continuous; discrete
7. continuous
8. discrete
9. exploratory data analysis
10. distribution
11. stem-and-leaf
12. 34; 6
13. outlier
14. stem
15. zero; four; five; nine
16. classes; frequency
17. categories
18. significant
19. mark; upper; lower
20. mark
21. width
22. midpoint
23. relative; total number
24. one
25. percent; 100; 100%
26. sum; less than
27. sum; greater than
28. discrete; categorical
29. time
30. vertical; horizontal
31. frequency or relative frequency or relative percentage
32. omission
33. variable; gaps; breaks
34. marks; continuous
35. ogive
36. pie
37. pictograph
38. center; decreasing
39. skewed
40. right; low; high
41. left; high; low
42. same
43. extreme end; center
44. one; opposite
45. two

Part II

1. b
2. a
3. b
4. c
5. b
6. d
7. b
8. d
9. b
10. a
11. b
12. a
13. a
14. d
15. d

Part III

1. false
2. true
3. true
4. false
5. true
6. true
7. false
8. true
9. true
10. true
11. false
12. true
13. true
14. true
15. false
16. true

Part IV

1. (a) numerical; continuous
   (b) numerical; discrete
   (c) categorical
   (d) numerical; continuous
   (e) numerical; discrete
   (f) categorical
   (g) numerical; discrete
   (h) numerical; continuous
   (i) numerical; continuous
   (j) categorical
   (k) numerical; discrete
   (l) numerical; continuous
   (m) numerical; continuous
   (n) categorical

2. a)

variable	type
sex	categorical
status	categorical
age	numerical: continuous
height	numerical: continuous
weight	numerical: continuous
frat. membership	categorical
high school avg	numerical: continuous
grade point avg	numerical: continuous

   b) student #9857:
   sex: female; status: freshman; age: 18; height: 62; weight: 105; frat: no; havg: 93; gpa: 3.45

3. a)
   ```
   1  9 9
   2  4 6 8 9 9
   3  0 1 2 2 3 4 4 5 7 7 9
   4  1 1 4 6 6 9 9
   5  0 1 2 2 4 6 9
   6  1 3 3 6
   7  3 4 6
   8  4
   ```
   b) skewed to the right   c) 30
   d) 25-39 years

4. a)
   ```
   0  3 4 5 5 5 7 8 8
   1  0 2 3 5 5 8 9
   2  0 0 1 3 5 5
   3  0 4 5 8
   4  0 0 5
   5  0
   6  4
   ```
   b) skewed to the right   c) 5-15 miles

5. a)
   ```
   1   0
   2
   3   0 2 5 7 8 9
   4   2 5 5 5 7 8 8 9 9
   5   0 0 3 5 8
   6   1 2 4
   7   0 0 1 2 2 2 2 2 3 4 5 5 5 5 7 7 7 9 9 9
   8   0
   9   5
   10
   11
   12  0
   ```
   b) bimodal
   c) yes; there are two outliers: 10 and 120 minutes.
   d) 66%   e) 2/50 or 4%

6. a) 13.8, 13.9, 14.0, 14.3, 14.5, 14.7, 15.1, 15.1, 15.4, 15.4, 15.8, 16.1, 16.2, 16.9, 17.0, 17.2, 18.0, 18.1, 18.3, 19.0, 19.1, 19.4, 19.4, 19.5, 20.0, 20.5, 20.6, 20.6, 20.8, 21.1, 21.1, 21.5, 21.5, 22.0, 22.1, 22.2, 22.2, 22.3, 23.1, 23.4, 23.7, 23.8, 24.1, 24.3, 28.4

   b) 28.4; an outlier

6. c)

classes	frequency
13.8-17.4	16
17.5-21.1	15
21.2-24.8	13
24.9-28.5	1

7. a)
```
4  0 5 7 7 8
5  0 0 3 4 5 5 7 9
6  0 0 0 4 5
7  0 1 2 3 5 5 5 7 9 9
8  0 2 2 3 5 6 8 9 9
9  2 3 3 5
```

b) bimodal

c)

classes	frequency	class boundaries
40-49	5	39.5-49.5
50-59	8	49.5-59.5
60-69	5	59.5-69.5
70-79	10	69.5-79.5
80-89	9	79.5-89.5
90-99	4	89.5-99.5

8. b)

classes	freq.	relative frequency	class mark	class boundaries
18-24	24	21.82%	21	17.5-24.5
25-31	25	22.73%	28	24.5-31.5
32-38	16	14.55%	35	31.5-38.5
39-45	13	11.82%	42	38.5-45.5
46-52	9	8.18%	49	45.5-52.5
53-59	11	10.00%	56	52.5-59.5
60-66	7	6.36%	63	59.5-66.5
67-73	5	4.55%	70	66.5-73.5

8. c1)

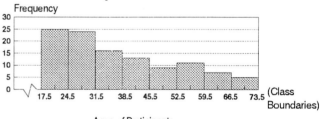

c2)

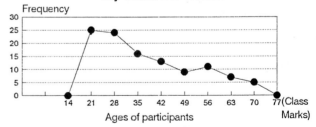

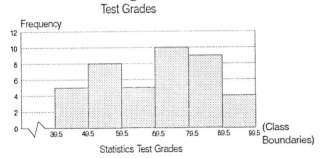

7. d) 4   e) 31.71%   f) 43.90%   g) 24.39%.

8. a)
```
1  8 8 8 9 9 9 9 9
2  0 0 1 1 1 2 2 2 2 2 3 3 3 3 4 4 5 6 6 6 7 7 7 7 7 7 8 8 8 9 9 9 9 9 9 9
3  0 0 1 1 1 2 2 3 3 4 4 4 5 5 5 6 7 7 7 7 8
4  0 1 1 1 2 3 3 3 3 4 4 4 5 8 8 9 9
5  0 0 1 2 2 4 4 5 5 6 6 8 8 9 9
6  0 0 0 0 1 4 5 7 9 9
7  1 2
```

8. d) skewed to the right

9. a)
```
          9 8 7 5 5  10  9
    9 9 8 7 5 4 3 3 2  11  3 7
    8 7 6 5 5 3 1 0 0  12  4 6 6 7 8
                  0 1  13  5 5 6 6 7 7 8 8 8 9
                       14  2 3 5 7 8 8 9
```

b) yes

c) experimental: symmetric bell-shaped;
control: skewed to the left

10. a)

```
                                        smoking              non-smoking
                                    4 1  3
                                         4
                   12 12 9 9 7 5 4 3 3 2 2  5  9 11 11 13 15
                   9 9 8  8 8 6 6 6 5 5 2 2 1 1  6  5 6 6 6 8 8 9 10 12 14 14 14
                   8 8 7  6  6 5 5 5 4 4 3 2 2 0  7  0 1 1 2 3 3 5  6  6  8  9  9 10 11 12 14 15
                                   3 3 2 2 0 0  8  1 2 4 4 5 6 8 9
                                         3 2 1  9  0 3 3 4 8
                                                10  1 5
                                                11  2
```

b) skewed;   10. c) yes

11. a) auto teller service table:

classes	cumulative frequency
less than 123	0
less than 162	10
less than 201	26
less than 240	36
less than 279	43
less than 318	47
less than 357	48
less than 396	49
less than 435	50

regular service table:

classes	cumulative frequency
less than 123	0
less than 162	3
less than 201	13
less than 240	23
less than 279	32
less than 318	42
less than 357	45
less than 396	48
less than 435	50

11. b) 72%   c) 46%

12. b)

classes	men's pressures	women's pressures
less than 104	0	0
less than 115	18	28
less than 126	43	52
less than 137	64	70
less than 148	71	74
less than 159	75	75

12. c) 85.33%   d) 93.33%

12. e)

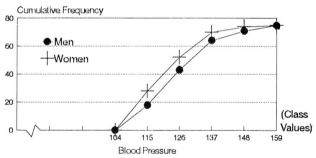

Ogive
Men's and Women's Blood Pressure

12. a)

```
                        men's leaves    stem   women's leaves
                      6 6 6 6 5 4 4 0   10  4 4 5 5 6 7 8 8 8
              8 8 8 8 6 6 5 4 2 2 2 2 0 0 0 0 0 0   11  0 0 0 0 0 0 0 0 2 2 2 2 2 4 4 4 4 6 6 6 6 6 8 8 8 8 8 9
  8 8 6 6 6 6 4 2 0 0 0 0 0 0 0 0 0 0 0 0 0   12  0 0 0 0 0 2 2 2 4 4 5 6 6 6 6 8 8 8 8 9
                        6 6 4 4 2 0 0 0 0 0 0 0 0 0   13  0 0 0 0 4 4 5 8 8 9
                              6 6 5 5 3 2 0   14  0
                                    6 6 0 0   15  6
```

13. a)

classes	class mark	class boundaries
44-50	47	43.5-50.5
51-57	54	50.5-57.5
58-64	61	57.5-64.5
65-71	68	64.5-71.5
72-78	75	71.5-78.5
79-85	82	78.5-85.5
86-92	89	85.5-92.5
93-99	96	92.5-99.5

13. b)

classes	class mark	class boundaries
44-51	47.5	43.5-51.5
52-59	55.5	51.5-59.5
60-67	63.5	59.5-67.5
68-75	71.5	67.5-75.5
76-83	79.5	75.5-83.5
84-91	87.5	83.5-91.5
92-99	95.5	91.5-99.5

13. c) 8 classes   d) no

14. a)

classes	class mark	class boundaries
1-8	4.5	0.5-8.5
9-16	12.5	8.5-16.5
17-24	20.5	16.5-24.5
25-32	28.5	24.5-32.5
33-40	36.5	32.5-40.5

14. b)

classes	class mark	class boundaries
1-7	4	0.5-7.5
8-14	11	7.5-14.5
15-21	18	14.5-21.5
22-28	25	21.5-28.5
29-35	32	28.5-35.5
36-42	39	35.5-42.5

15. a) yes b) no c) yes d) no e) no

16. a) 15; 12.5% b) 59; 49.17% c) not possible
    d) not possible e) not possible
    f) 21; 17.5%

17. a)

classes	frequency
2-3	7
4-5	22
6-7	7
8-9	3
10-11	1

17. b)

classes	frequency
2-4	17
5-7	19
8-10	4

18. a)

Classes (IQ Scores)	freq.	class mark	class boundary	rel. freq.	relative percentage
92-97	12	94.5	91.5-97.5	0.10	10%
98-103	11	100.5	97.5-103.5	0.09	9%
104-109	17	106.5	103.5-109.5	0.14	14%
110-115	19	112.5	109.5-115.5	0.16	16%
116-121	22	118.5	115.5-121.5	0.18	18%
122-127	19	124.5	121.5-127.5	0.16	16%
128-133	8	130.5	127.5-133.5	0.07	7%
134-139	4	136.5	133.5-139.5	0.03	3%
140-145	5	142.5	139.5-145.5	0.04	4%
146-151	3	148.5	145.5-151.5	0.03	3%

18. b) sum of rel. freq. = 1.00; sum of rel. freq. percentages = 100%
   c) (1) 49.17% (2) 6.67% (3) 32.50%
      (4) 33.34% (5) 10%

19. a) class width = 3
    b)

classes (gallons of gasoline)	freq.	class mark	class boundaries	relative percentages
5-7	17	6	4.5-7.5	5.67%
8-10	33	9	7.5-10.5	11.00%
11-13	42	12	10.5-13.5	14.00%
14-16	88	15	13.5-16.5	29.33%
17-19	61	18	16.5-19.5	20.33%
20-22	38	21	19.5-22.5	12.67%
23-25	21	24	22.5-25.5	7.00%

19. c)

classes	cumulative frequency
less than 5	0
less than 8	17
less than 11	50
less than 14	92
less than 17	180
less than 20	241
less than 23	279
less than 26	300

Ogive
Purchase of Gasoline

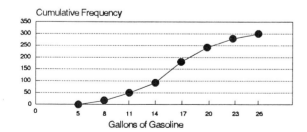

19. d) 50 customers; 16.67%   e) 6%
    f) 191 customers; 63.67%

20. a) & b)

classes (preference)	frequency	relative frequency	relative percentage
SIPEPS	20	0.40	40%
COAKS	24	0.48	48%
NEITHER	6	0.12	12%

20. c)

Pie Chart
Preference for Soft Drink

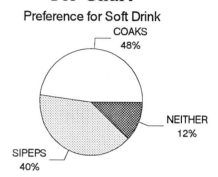

21. a)

Finished Shopping By:	Percent of Shoppers
November 30	6%
December 15	52%
December 23	17%
December 25	25%

21. b)

Date Completed Christmas Shopping

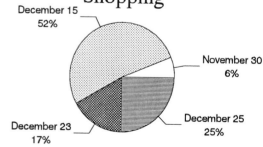

21. c) December 15   d) 75%   e) 25%

22. a) class width = 10
    b)

class mark	class boundaries	relative percentages for males	relative percentages for females
29.5	24.5-34.5	0.11%	0.08%
39.5	34.5-44.5	0.94%	0.42%
49.5	44.5-54.5	4.05%	2.08%
59.5	54.5-64.5	11.63%	8.27%
69.5	64.5-74.5	26.31%	24.22%
79.5	74.5-84.5	56.96%	64.94%

c) males: sum of rel. percents = 100%;
   females: sum of rel. percents = 100.01%

22. d)

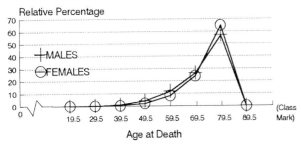

23. a)

Park Size in Acres	Cumulative Frequency	Relative Percentages
Less than 1	0	0
Less than 10	17	6
Less than 50	51	19
Less than 100	103	37
Less than 200	150	55
Less than 500	182	66
Less than 750	207	75
Less than 1000	275	100

23. b)

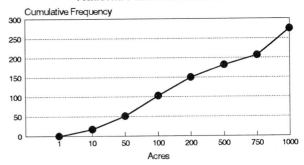

23. c) see table
    d) 1) 6%; 2) 55%; 3) 75%
    e) 50-750 acres

24. a)

classes (grades)	Frequency	class boundaries	class mark
44-55	4	43.5-55.5	49.5
56-67	6	55.5-67.5	61.5
68-79	6	67.5-79.5	73.5
80-91	11	79.5-91.5	85.5
92-103	3	91.5-103.5	97.5

24. b)

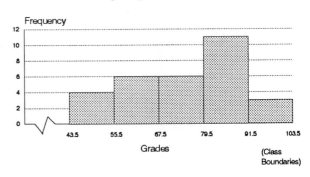

24. c)

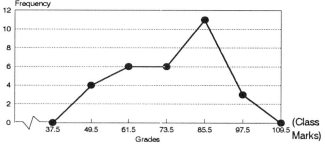

25. a)

classes (age)	freq.	class mark	class boundaries	relative frequency
18-22	27	20	17.5-22.5	0.13
23-27	54	25	22.5-27.5	0.26
28-32	55	30	27.5-32.5	0.26
33-37	31	35	32.5-37.5	0.15
38-42	21	40	37.5-42.5	0.10
43-47	10	45	42.5-47.5	0.05
48-52	7	50	47.5-52.5	0.03
53-57	3	55	52.5-57.5	0.01
58-62	2	60	57.5-62.5	0.01

25. b) shape: skewed to the right

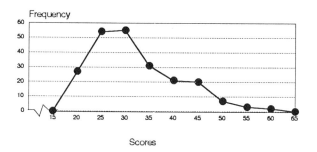

25. c)

classes	cumulative frequency
less than 18	0
less than 23	27
less than 28	81
less than 33	136
less than 38	167
less than 43	188
less than 48	198
less than 53	205
less than 58	208
less than 63	210

25. d)

classes	cumulative frequency
18 or more	210
23 or more	183
28 or more	129
33 or more	74
38 or more	43
43 or more	22
48 or more	12
53 or more	5
58 or more	2
63 or more	0

25. e)

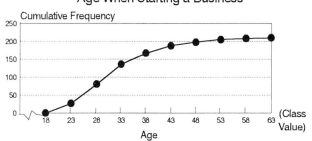

25. f) 65% g) 10% h) 52%

26. a)

classes (Indices)	Frequency	class boundaries	class mark
30-37	6	29.5-37.5	33.5
38-45	16	37.5-45.5	41.5
46-53	18	45.5-53.5	49.5
54-61	31	53.5-61.5	57.5
62-69	14	61.5-69.5	65.5
70-77	10	69.5-77.5	73.5
78-85	5	77.5-85.5	81.5

26. b)

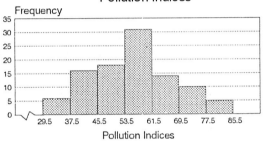

26. c)

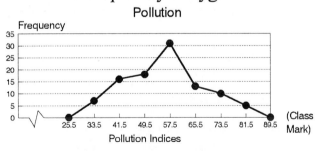

27. a)

Number of Minutes to Register for Class	Class Boundaries	Number of Students
0-15	-0.5-15.5	20
16-30	15.5-30.5	45
31-45	30.5-45.5	90
46-60	45.5-60.5	110
61-75	60.5-75.5	30
76-90	75.5-90.5	5

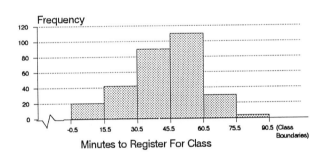

27. b)

Classes	Cumulative Frequency
Less Than 0	0
Less Than 16	20
Less than 31	65
Less than 46	155
Less than 61	265
Less Than 76	295
Less Than 91	300

Ogive
Minutes to Register for Class

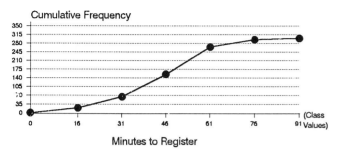

27. c) 55 min

28. a)

Classes (IQ Scores)	class boundaries	BOYS Rel. Freq.	GIRLS Rel. Freq.	THIRD GRADERS Rel. Freq.
60-79	59.5-79.5	11.43%	2.5%	6.67%
80-99	79.5-99.5	38.57%	43.75%	41.33%
100-119	99.5-119.5	32.86%	18.75%	25.33%
120-139	119.5-139.5	14.29%	31.25%	23.33%
140-159	139.5-159.5	2.86%	3.75%	3.33%

28. b1)

Histogram
Third Grade Boys

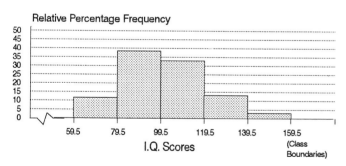

28. b2)

Histogram
Third Grade Girls

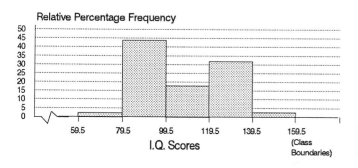

28. b3)

Histogram
Third Grade Students

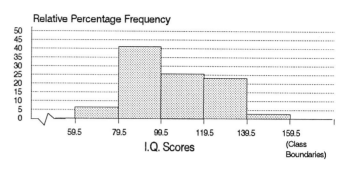

28. c) (1) 53.75% (2) 50.01% (3) 51.99% (4) 93.33% (5) 6.67%

29. a)

Classes (SAT Scores)	Class Mark	Class Boundaries	VERBAL		MATHEMATICS	
			% of Males	% of Females	% of Males	% of Females
200-299	249.5	199.5-299.5	8.57%	5.13%	5.71%	12.82%
300-399	349.5	299.5-399.5	25.71%	15.38%	17.14%	28.21%
400-499	449.5	399.5-499.5	20.00%	15.38%	17.14%	17.95%
500-599	549.5	499.5-599.5	25.71%	30.77%	31.43%	23.08%
600-699	649.5	599.5-699.5	5.71%	7.69%	8.57%	5.13%
700-799	749.5	699.5-799.5	14.29%	25.64%	20.00%	12.82%

29. b1)

Histogram
Male Verbal SAT Scores

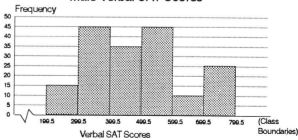

29. b2)

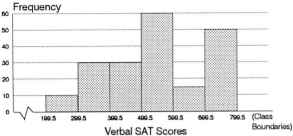

29. c1)

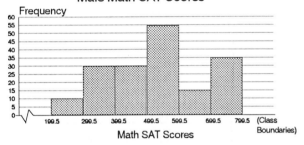

29. c2)

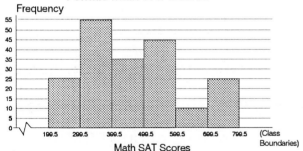

29. d1)

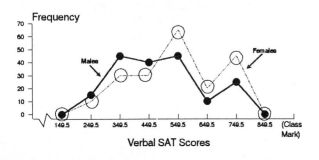

29. d2)

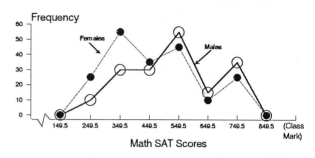

e) (1) 79.48% (2) 65.71% (3) 17.95% (4) 28.57%

30.

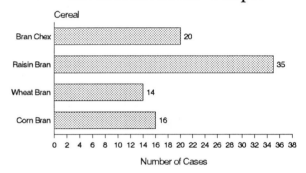

31.

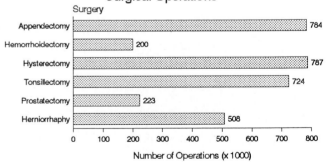

32.

CHAPTER 2   ANSWERS   •   751

33.

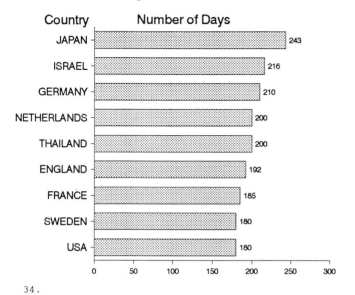

34.

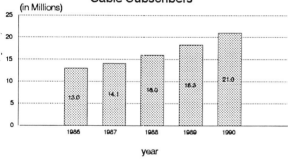

35. a)

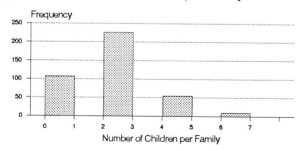

35. b)

classes	frequency
less than 0	0
less than 2	102
less than 4	330
less than 6	387
less than 8	400

35. c) 74.5%   d) 82.5%   e) approx. 11 homes

36. a)

Classes	Cumulative Frequency
less than 0.5	0
less than 1.0	26
less than 1.5	74
less than 2.0	134
less than 2.5	200
less than 3.0	273
less than 3.5	358
less than 4.0	458
less than 4.5	500

36. b)

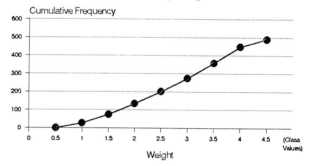

36. c) 55 %

37

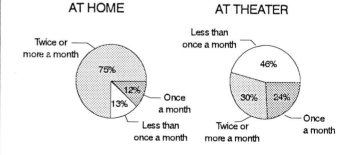

38.

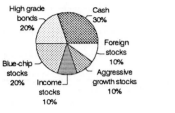
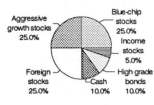

39. a)

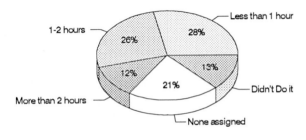

39. b) 38%;   c) 67%

40.

Classes (Level of Football)	Number of Injuries
High School	152
Professional	144
College	56
Youth	48

41. a) 18 b) 20% c) 20% d) 36% e) 36% f) yes g) 4%

CHAPTER 3

PART I

1. mean, median, mode
2. center
3. adding, number
4. μ
5. σ
6. N
7. $\mu = \dfrac{\Sigma x}{N}$
8. (a) 13
   (b) -1
   (c) ½
   (d) 21
9. 10
10. (a) 45
    (b) 3
    (c) 5
    (d) 20
11. 0
12. -5
13. mean
14. (c) $(X_1 + X_2 + \ldots + X_N)^2$
    (d) $(X_1-2) + (X_2-2) + \ldots + (X_N-2)$
    (e) $(X_1-2)^2 + (X_2-2)^2 + \ldots + (X_N-2)^2$
    (f) $[(X_1-2) + (X_2-2) + \ldots + (X_N-2)]^2$
15. middle
16. mean
17. equals
18. 75
19. mode
20. bimodal
21. nonmodal
22. $\bar{x}$
23. $\bar{x} = \dfrac{\Sigma x}{n}$
24. population
25. sample
26. population
27. statistic
28. extreme; outliers
29. is not
30. tail
31. resistant
32. mode
33. outlier
34. right; extreme
35. center

PART II

1. a
2. b
3. c
4. b
5. b
6. a
7. c
8. d
9. a
10. a
11. a
12. b
13. b

PART III

1. true
2. true
3. false
4. false
5. false
6. false
7. false
8. true
9. false
10. false
11. false
12. false
13. false
14. true
15. true

PART IV

1. (a) (1) 4
       (2) 4
       (3) no mode
       (4) 0
       (5) 190
       (6) 1600
       (7) 30
       (8) 0
       (9) 174
   (b) (1) 10
       (2) 9.5
       (3) 9
       (4) 0
       (5) 1034
       (6) 10,000
       (7) 34
       (8) 0
       (9) 934
2. (a) 10
   (b) 34
   (c) 340
   (d) 100
   (e) 1156
   (f) 30
   (g) 294
   (h) 90
   (i) 1

PART IV (continued)

3. (a) 12
   (b) 21
   (c) 252
   (d) 144
   (e) 441
   (f) 34
   (g) 91
   (h) 53
   (i) 0.68
4. (a) μ = $170; median = $165; mode = $165
   (b) they all indicate a fair price
5. (a) μ = 26; median = 26.5; mode = 28
   (b) mode, it best supports the ad campaign
6. (a) mean = $597,100; median = $507,500
       mode = $425,000
   (b) they are not representative because the mean is too high and the mode is too low.
7. (a) mean = 65.1; median = 80.5; mode = 82 (b) mean
8. (a) mean = 4.30; median = 4; mode = 4 (b) mean
9. (a) mean= $1,565,833.30; median= $960,000
       mode= $640,000
   (b) the mode is a poor measure of central tendency because it's at the extreme low end of the distribution.
10. (a) μ = 89.5
    (b) the student is basing his remark just on the mean. he is not taking into consideration the fact that approximately 46% of the grades are above 90.
11. (a) 140; 50   (c) 50; 50
    (b) median   (d) equally good
    (e) median   (f) mean
12. (a) M(FrChk);T(Spag);W(MtLf);Th(Bf);F(Pz)
    (b) Spaghetti and Meat Balls
13. mode, because the data is categorical
14. (a) mean = 34; mode = 28; median = 30
    (b) mean
15. (a) 6.41 (b) 6.25 (c) 5½ (d) mode
16. (a) 28.67 mpg (b) $\bar{x}$
17. (a) mean = 5.24 mgs; median = 4.95 mgs
    (b) 4.95 mgs
18. $31.60
19. (a) 59,896.88 (b) mode
20. (a) 5 lbs
    (b) no, since 14 of the bags weigh less than five pounds.
21. (a) 24
    (b) if the sample is representative of the class then one can see that very few of the students are 24 or older.
22. (a) I: μ = 6.20
        II: μ = 10.70
        III: μ = 13.30
    (b) 10.07
    (c) mean length using all the survey data:10.07
23. No! They should divide the sum of their hits by the sum of their AB's over the 10 years.
24. No! This logic is not correct, since he doesn't take into consideration that each semester's avg is for a different number of credits.
25. 107
26. (a) (1) Brown
        (2) Brown
        (3) Brown hair; Brown Eyes
        (4) Brown hair; Blue Eyes
    (b) Mode, because the data is categorical
27. (a) $13,875 (c) $19,500
    (b) $7,500   (d) $3,406.25
28. (a) 888.5
    (b) 0.25 lbs
    (c) 7.5 lbs
29. (a) 2440
    (b) 60 (c) yes
30. see instructor
31. mode = X   median = Y   mean = Z
32. mode = 20 median = 19 mean = 18
33. 1) skewed right; mean
    2) bell shaped; mean=median
    3) skewed left; median
    4) skewed right; mean
    5) bell shaped; mean=median
    6) skewed left; median
    7) skewed right; mean

CHAPTER 4

Part I

1. range, standard deviation, variance
2. dispersion or spread
3. smallest; largest
4. 2
5. 5
6. 18
7. mean, μ
8. $\dfrac{\Sigma(x-\mu)^2}{N}$
9. 500
10. 200
11. variance
12. σ (sigma)
13. $\sqrt{\dfrac{\Sigma(x-\mu)^2}{N}}$; $\sqrt{\dfrac{\Sigma x^2 - \dfrac{(\Sigma x)^2}{N}}{N}}$
14. 12
15. 14, 14, 14, 0, 0, 0
16. $\Sigma(x-\mu)^2$
17. (1) below   (2) above   (3) m
18. 13
19. (a) 3 (b) 3 (c) 1/2 (d) 15 (e) 12
20. 1/2
21. y; x
22. 55, 60, 65, 45, 40, 35
23. σ
24. $\sqrt{\dfrac{\Sigma(x-\bar{x})^2}{n-1}}$; $\sqrt{\dfrac{\Sigma x^2 - \dfrac{(\Sigma x)^2}{n}}{n-1}}$
25. Chebyshev's; Empirical
26. Chebyshev's Theorem
27. Symmetric Bell-shaped
28. standard deviations
29. a) 68% b) 95% c) all; 99%; 100%
30. two
31. relative; absolute
32. Coefficient of Variation = $\dfrac{\text{standard deviation}}{\text{mean}}$ (100%)
33. standard deviation
34. different; means

Part II

1. a   11. a
2. b   12. b
3. c   13. a
4. b   14. b
5. b   15. a
6. c
7. a
8. a
9. c
10. b

Part III

1. true   2. true
3. false
4. true
5. false
6. false
7. false
8. true
9. true
10. true
11. false
12. true
13. false
14. false
15. false
16. true
17. false
18. true
19. false
20. false
21. false
22. true

Part IV

1. (a) 13
   (b) 21.20
   (c) 4.60
   (d) 5.15
2. (a) 18   (b) 41.60
   (c) 6.45 (d) 7.21
3. Resort 1: s=20.26
   Resort 2: s=5.54
   Resort 2 has more consistent weather
4. Ms. Sue
5. (a) R=42; σ=14
   (b) R=42; σ=14
   (c) R=168; σ=56
6. store one
7. Sarabella: μ= $159,000
                 σ= $139,685
   20th Century: μ= $184,500
                 σ= $124,878
8. (a) μ=14; σ=4   (b) 18
   (c) 10   (d) 10;12;14;16;18
   (e) none   (f) 5
9. Joann: μ= 6;   σ= 4.52
   Maryann: μ= 17; σ= 2.26
   Amy: μ= 10.38; σ= 6.02
   Kathleen: μ= 10; σ= 4.52
   Debbie: μ= 20; σ= 4.41
   Ellen: μ= 11; σ= 4.84
   Grace: μ= 13; σ= 5.95
   Maryann receives the award
10. (a) New wages under: Plan A:
    $31.30;$30.05;$31.80;$30.55;$30.80
    Plan B: $31.35; $29.98; $31.90; $30.53; $30.80
    Plan C: $31.31; $30.19; $31.76; $30.63; $30.86
    (b) Plan A: R = $1.75; σ = $0.60
        Plan B: R = $1.92; σ = $0.66
        Plan C: R = $1.57; σ = $0.54
11. (a) $15.94   (b) $0.69
12. σ_x = 6.68; σ_y = 4.12
    team y has the more consistent defense
13. (a) $\bar{x}$ = 0.083 (8.3%) (c) 11 (e) 100%
    (b) s = 0.013 (1.3%) (d) 73.3%
14. (a) & (b)
    stock $\bar{x}$ s
    x   68   1.83
    y   68   2.71
    z   68   5:31
    (c) stock z is more speculative
        stock x is more conservative
15. a) 67.6 for both machines b) s_1= 0.17 oz; s_2= 0.27 oz
    machine one is more consistent in its operation.
16. drug $\bar{x}$ s
    Afferin   48 min   8.04 min
    Banacin   48 min   6.64 min
    Banacin was more consistent
17. (a) $\bar{x}$ = 16.82 ppm; s = 0.90 ppm
    (b) none exceeded the standard of 18.62 ppm
18. Drug companies   Cosmetic companies
        R & D   Marketing   R & D   Marketing
    (a) μ = 3.11   μ = 2.40   μ = 1.39   μ = 3.68
    (b) σ = 0.88   σ = 0.61   σ= 0.53   σ = 0.75
    (c) The drug companies spend more on R & D relative to
        the cosmetic companies. While, the cosmetic companies
        spend more on marketing than the drug companies.
19. (a) $\bar{x}$ = 5.67 ; s = 4.29 (b) $\bar{x}$ = 5.2 ; s = 0.56
    (c) no; yes
20. (a) (1) 75% (2) 89% (3) 94% (b) 11%
21. (a) μ = $67.84 ; σ = $39.88 (b) 93.33% (c) 93.33%
        (in billion dollars)
22. (a) 95% (b) 68% (c) 81.5% (d) approx 97.5%
23. (a) 95% (b) 84% (c) approx 0%
    (d) manufacturer's claim is too high because the
        chance of a tire lasting 71,250 miles is
        virtually impossible.
24. (a) the percentage can vary from 0% to 25% (b) 2.5%
25. a) μ=3.35 σ=1.70 b) 100% c) 100%
    d) shape is approximately uniform.
       stem-and-leaf of die 1   N = 40
       Leaf Unit = 0.10
        7   1   0000000
       15   2   00000000
       (7) 3   0000000
       18   4   000000
       12   5   000000
        6   6   000000
    e) the results of both parts (b) and (c) are
       consistent with Chebyshev's Theorem.
26. a) μ=6.93 σ=2.35 b)95% c)100%
    d) the results of both parts (b) and (c) are
       consistent with Chebyshev's Theorem.
27. a) μ=7.60 σ=1.71 b)90% c)100%
    d) the results of both parts (b) and (c) are
       consistent with Chebyshev's Theorem.
28. a) approximately bell-shaped. b) yes. 75% c) 96%
    d) yes. Since the Empirical Rule states that approx-
       imately 95% of the data values will lie within
       two standard deviations of the mean and we have
       calculated 96%, the results are very close to
       what they should be.
29. Although officer B has more absolute variability
    (standard deviation=15) than officer A (std. dev.=5),
    the relative variation for officer B (coefficient of
    variation=10%) is less than the relative variation
    for officer A (coefficient of variation=12.5%).
30. Training program A, coefficient of variability = 24%.
31. a) Fund C   b) Fund A   c) Fund C
32. a) Stock D b) Because the means are far apart.
33. a) Mickey Mantle   b) Because the data values are
       expressed in different units.

CHAPTER 5

Part I

1. z score or percentile rank
2. z score
3. $\frac{x-\mu}{\sigma}$
4. less
5. +1
6. -1.5
7. mean
8. less
9. greater
10. (a) 0
    (b) -2.5
    (c) 108
    (d) 106
11. m + zs
12. 3; greater
13. 2; less
14. 4
15. 25
16. 0
17. 90; 10
18. less; equal; distribution
19. whole
20. p
21. percentile
22. quartile
23. quartile
24. 110
25. decile
26. data value; 0; 100
27. exploratory; center; spread; shape
28. first; third
29. whisker
30. first; third; box; 50%

Part IV

1. (a) 1
   (b) -2
   (c) -1.33
   (d) 1.30
   (e) 1.50
4. (a) A: greater than 95
      B: from 87 to 95
      C: between 79 and 87
      D: from 71 to 79
      F: less than 71
   (b) A: greater than 97
      B: from 79 to 97
      C: between 61 and 79
      D: from 43 to 61
      F: less than 43
5. auto mechanic (z = +2.83)
6. Lou
7. Lorenzo's grade = 94
   Michael's grade = 87
   Lorenzo had the higher grade
8. Sprinters 100m run(z=-2)
9. (a) $\bar{x}$ = 6.04, s = 0.13
   (b) z = -1.5, x = 5.85
       z = 1.5, x = 6.24
   (c) Yes, the parts do meet the conditions

2. (a) 35
   (b) 38
   (c) 147.5
   (d) 267
   (e) 129.2
3. (a) m = 76.5
   (b) s = 9.98
   (c) temp   z
       80   0.35
       90   1.35
       69   -0.75
       60   -1.65
       85   0.85
       75   -0.15

Part II

1. c
2. a
3. c
4. b
5. c
6. a
7. b
8. c
9. d
10. c

Part III

1. false
2. true
3. false
4. false
5. true
6. false
7. false
8. false
9. false
10. false
11. true
12. false
13. true
14. true
15. false
16. false
17. true
18. true

10. Golfer(z=-1.57)
11. a) Condo A
    b) Condo M
12. one year adjustable
13. (a) 10
    (b) 25
    (c) 50
    (d) 70
    (e) 75
    (f) 81
14. (a) 115
    (b) 119
    (c) 105
    (d) 87
    (e) 105
    (f) 91
15. (a) 372
    (b) 433
    (c) 500
    (d) 525
    (e) 567
    (f) 85%
    (g) 40%
    (h) 50%
    (i) 134 point difference
16. (a) 20 seconds
    (b) 28 seconds
    (c) 85%
    (d) 16 seconds
    (e) 12 seconds
    (f) 36 seconds
17. student   PR
    L   25
    A   30
    W   50
    Y   75
    E   81
    R   90
18. Barbara ; Mike ; Alice
19. (a) $13,345
20. (a) I; III
    (b) III; I
    (c) II; III
21. (a) 25%
    (b) 25%
    (c) 77%
    (d) 40%
    (e) 50%
    (f) 40%
22. (a) 490
    (b) 510
    (c) 570
    (d) 500
    (e) 500
    (f) 750
23. Luanna: her produce had largest z score
24. max.buying price: $12.73
    min. selling price: $44.07
25. (a) 164 to 188
    (b) 152.48 to 199.52
    (c) 152 to 200
    (d) 145.04 to 206.96
    (e) 140 to 212
26. Jim
27. (a) 100   (b) 116
    (c) 16   (d) 94%
    (e) 124   (f) 34%
    (g) 4%

28. a) Roger Maris:1st quartile = 11, median = 19.5; 3rd quartile = 30.5
    Babe Ruth: 1st quartile = 34.5, median = 46; 3rd quartile = 51.5
    b)
    d) larger variability: Roger Maris; smaller variability: Babe Ruth IQR for Roger Maris: 19.5; IQR for Babe Ruth: 17
    e) potential outlier: 61 for Roger Maris z score for 61 is 2.49
    f) Roger Maris: skewed to the right Babe Ruth: skewed to the left

29. a) A:51   B:67
    b) A:67   B:87
    c) A:57   B:79
    d) A:43,93; B:45,95
    e) A:16   B:20
    f) A:skewed to the right
       B:skewed to the left
    g) 75   h) 50   i) 50
    k) A: mininimum knowledge;
       A: top students; B: easier test

30. a) 6.5 b) 6.65 c) 6.58
    d) 6.24; 6.82 e) 0.15

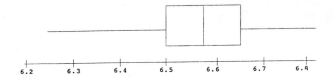

g) approximately symmetric bell-shaped i) yes; 6.24 is an outlier since it is more than 1.5 (IQR) less than 1st quartile.

31. a) 106, 142, 175, 186, 272
    b)

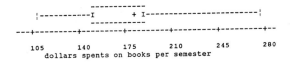

c) skewed left
d) IQR=44 (1) 272 (2) none
e) 272; student may have need to buy technical texts that are often very expensive relative to other standard texts.

32. a) 4, 5.5, 8.5, 10, 120
    b) MINITAB output for Boxplot:

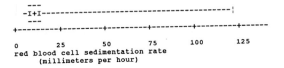

c) approximately symmetric, with the possibility of outliers.
d) IQR= 4.5; 105, 120; 105, 120.
e) Data values 105 and 120 are considered to be outliers since they are more than 3 (IQR). These values may have occured because of an inaccuracy in measurement from the test of red blood cell sedimentation or perhaps they are an indication of an abnormality in the patient.

33. (a)

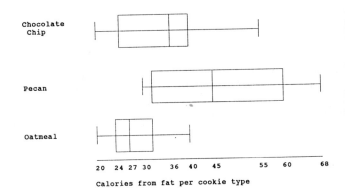

Chocolate Chip: approximately skewed to the left Pecan: approximately symmetric bell-shaped
Oatmeal: skewed to the right

(b) The number of calories from fat for the Oatmeal cookies are consistently less and less spread out than the other cookie types. Also, the number of calories from fat for the pecan cookies are more spread out than the other cookie types.

CHAPTER 6

Part I

1. experiment
2. sample space
3. 216
4. event
5. more; HHHT; HTHH; HHTH; HHHH
6. definite
7. factorial; 4;3;2;1;24
8. permutations; $_nP_n$; n!
9. 720
10. 24
11. $_nP_s$; n(n-1)(n-2)...(n-s+1)
12. 20
13. $\dfrac{N!}{(n_1!)(n_2!)\ldots(n_k!)}$
14. 420
15. order
16. $\binom{n}{s}$; $\dfrac{_nP_s}{s!}$ or $\dfrac{n!}{(n-s)!\,s!}$
17. 35
18. 56
19. $\dfrac{\text{number of outcomes satisfying event E}}{\text{total number of outcomes in the sample space}}$
20. (a) 1/12; (b) 3/12 = 1/4; (c) 0
21. 0; 1
22. 1
23. A; minus; same
24. queen; red card; queen and a red card; queen; red card; queen and a red card
25. 0.60 + 0.40 − 0.30 = 0.70
26. mutually exclusive
27. P(A); P(B)
28. (a) 0; (b) P(C) + P(D)
29. mutually exclusive; same; ace; king; 4/52; 4/52; 8/52
30. mutually exclusive; 4; 5
31. not mutually exclusive; 4; even number; 4
32. 1; E; E'; 1
33. P(getting 4 heads)
34. independent
35. independent; and; times; A and B; A; B
36. a white marble; times; a white marble; independent; 4/7; 4/7; 16/49
37. dependent; conditional
38. dependent; B; occurred; A and B; B|A
39. dependent; 6; B|A; 1/4; 1/3; 1/12
40. a) P(B|A)
    b) probability of B, given that event A has occurred
    c) P(A|B)
    d) probability of A, given that event B has occurred
41. and; B; A|B; B
42. conditional; first shot is a blank; 20/30; 2/5
43. no two people have the same birthday; 0.814
44. (a) fixed (b) independent (c) success; failure (d) success
45. independent; successes; failures; ways; success; failure
46. 20;3;6;20
47. 4; HHHT; HHTH; HTHH; THHH
48. 6; 7; sssssfs; sssssfss; sssfsss; ssfssss; sfsssss; fssssss
49. (a) 10; 1/27; 4/9; 40/243
    (b) 1; 1/243; 1; 1/243

Part II

1. c   6. d
2. b   7. a
3. c   8. c
4. a   9. b
5. d

Part III

1. False
2. True
3. False
4. True
5. True
6. True
7. False
8. False
9. True
10. False
11. False

50. 5/32
51. 7; 5; 1/4; 3/4; 21; 2; 5; 21
52. 1/5

Part IV

1. a) R, R, R, R, Y, Y, Y, B, B, B
   b) BB ** AA
      B* *B A*
      BA *A AB
   c) HHH
      HHT
      HTH
      HTT
      THH
      THT
      TTH
      TTT
   d) same as answer 1c
   e) BYE
      BEY
      EBY
      EYB
      YBE
      YEB
   f) GOODOGOD
      GDOOOGDO
      OODGODOG
      OOGDODGO
      DOOGDOGO
      DGOOGODO

2. a) RY   b) RR RB   c) JQ QJ KJ AJ   d) JQ QJ KJ AJ
      RB   BB BR   JK QK KA AQ   JK QK KA AQ
      YR   YY YB   JA QA KQ AK   JA QA KQ AK
      YB   RY BY   JJ QQ KK AA
      BB   YR
      BR
      BY

3. a) 1/52 b) 1/9999 c) 1/12 d) 4/5 e) 7/8
4. a) 1/13 b) 10000/99999 c) 1/52 d) 1/365
   e) 1/5   f) 1/8
5. a) 9/169 b) $\dfrac{132}{2652} = \dfrac{33}{663}$ c) 25/169 d) 95/663

6. a)

outcome	2	3	4	5	6	7	8	9	10	11	12
probability	$\dfrac{1}{36}$	$\dfrac{2}{36}$	$\dfrac{3}{36}$	$\dfrac{4}{36}$	$\dfrac{5}{36}$	$\dfrac{6}{36}$	$\dfrac{5}{36}$	$\dfrac{4}{36}$	$\dfrac{3}{36}$	$\dfrac{2}{36}$	$\dfrac{1}{36}$

   b) $\dfrac{8}{36}$ c) $\dfrac{28}{36}$ d) $\dfrac{26}{36}$ e) $\dfrac{18}{36}$

7. a) $\dfrac{15}{36}$ b) $\dfrac{15}{36}$

8. a) $\dfrac{6}{24}$ b) $\dfrac{2}{24}$ c) $\dfrac{1}{24}$

9. a) BBBB GGGG
      BBBG GGGB
      BBGB GGBG
      BBGG GGBB
      BGBB GBGG
      BGBG GBGB
      BGGB GBBG
      BGGG GBBB

9. b) $\frac{4}{16}$ c) $\frac{15}{16}$ d) $\frac{11}{16}$ e) $\frac{5}{16}$ f) $\frac{1}{16}$

10. a) $\frac{3}{8}$ b) $\frac{3}{8}$ c) $\frac{4}{8}$

11. a) $\frac{18}{30}$ b) $\frac{12}{30}$ c) $\frac{5}{30}$ d) $\frac{16}{30}$ e) $\frac{10}{30}$ f) $\frac{14}{30}$

12. a) 11 21 31 41
       12 22 32 42
       13 23 33 43
       14 24 34 44
    b) $\frac{4}{16}$ c) $\frac{4}{16}$ d) $\frac{8}{16}$

13. a) $\frac{1}{1000}$ b) $\frac{6}{1000}$ c) $\frac{100}{1000}$ d) $\frac{10}{1000}$ e) $\frac{6}{1000}$

14. P PND
    N PDQ
    D PNQ
    Q NDQ
    PN PNDQ
    PD
    PQ
    ND
    NQ
    DQ

    a) $\frac{11}{15}$ b) $\frac{11}{15}$ c) $\frac{7}{15}$ d) $\frac{8}{15}$ e) $\frac{14}{15}$

15. a) $\frac{21}{26}$ b) $\frac{5}{26}$ c) $\frac{14}{26}$ d) $\frac{2}{26}$ e) $\frac{17}{26}$ f) $\frac{5}{26}$
    g) $\frac{21}{26}$ h) $\frac{6}{26}$ i) $\frac{4}{26}$ j) $\frac{1}{26}$

16. a) $\frac{1}{24}$ b) $\frac{6}{24}$ c) $\frac{10}{24}$ d) $\frac{18}{24}$ e) $\frac{20}{24}$ f) 0

17. a) $\frac{16}{30}$ b) $\frac{12}{30}$ c) $\frac{2}{30}$ d) $\frac{28}{30}$

18. a) $\frac{30}{56}$ b) $\frac{24}{56}$ c) $\frac{2}{56}$

19. a) $\frac{3}{6}$ b) $\frac{1}{6}$ c) 0 d) $\frac{2}{6}$

20. $\frac{4}{10}$

21. 60
22. a) 240 b) 11,520
23. a) 1 b) 49 c) 343 d) 2401
24. a) 2 b) 24 c) 12 d) 24 e) 210 f) 1,663,200
    g) 95,040
25. a) DOG b) 6
       DGO
       OGD
       ODG
       GOD
       GDO
26. a) PIG b) 6
       PGI
       GIP
       GPI
       IPG
       IGP

27. 40,320
28. 120
29. 32
30. 840
31. 1320
32. a) 362,880 b) 4,151,347,200
33. a) 10,000 b) 5040 c) 9630
34. $2^8 = 256$
35. a) 40,320 = 8! b) 24 = 4! c) 384 = (4! 2)
    d) 1152 = (4!4!·2)
36. a) MISS b) 12
       ISSM
       SISM
       SSIM
       SIMS
       SMIS
       SSMI
       ISMS
       IMSS
       MSSI
       SMSI
       SMIS
37. 60
38. 30
39. a) 453,600 b) 3780 c) 1680 d) 3360
40. a) 676 b) 650 c) 6760 d) 20,280
41. a) 17,576,000 b) 308,915,780 c) 1,000,000
    d) 165,765,600 e) 151,200

42. $4200 = \frac{10!}{4!\,3!\,3!}$

43. $378{,}378{,}000 = \frac{15!}{2!\,4!\,3!\,2!\,3!}$

44. 1/100

45. a) $\frac{1}{15{,}600}$ b) $\frac{1}{7{,}893{,}600}$

46. a) 120 b) 60 c) $\frac{1}{120}$

47. a) 720 b) 120 c) $\frac{1}{720}$

48. $\frac{6}{21}$

49. a) 24 b) 1/24 c) 6 d) 1/6 e) 12 f) 1/12

50. a) $\frac{330}{19{,}487{,}171} = \frac{\binom{11}{7}}{11^7}$ b) $\frac{1}{1{,}771{,}561} = \frac{1}{11^6}$

51. a) 1. $5 = {}_5P_1$ 2. $20 = {}_5P_2$ 3. $60 = {}_5P_3$
       4. $120 = {}_5P_4$ 5. $120 = {}_5P_5$

    b) 1. $5 = \binom{5}{1}$ 2. $10 = \binom{5}{2}$ 3. $10 = \binom{5}{3}$
       4. $5 = \binom{5}{4}$ 5. $1 = \binom{5}{5}$

    c) In part (a) permutations consider the order of
       the objects whereas in part (b) the order is
       completely irrelevant when dealing with
       combinations. Hence different permutations
       can represent the same combination. Therefore
       the answers in part (a) are larger.

52. a) 20 b) 10 c) 210 d) 5,200,300

52. e) 133,784,560
53. a) 4   b) 21   c) 495   d) 1326
    e) 635,013,560,000

54. $99,884,400 = \binom{50}{7}$

55. $45 = \binom{10}{2}$

56. $21 = \binom{7}{2}$

57. $75,287,520 = \binom{100}{5}$

58. $3,838,380 = \binom{40}{6}$

59. $1365 = \binom{15}{4}$

60. $56 = \binom{8}{3}$

61. a) $5005 = \binom{15}{9}$ b) $1890 = \binom{9}{5} \cdot \binom{6}{4}$

62. $280 = \binom{4}{3} \cdot \binom{8}{4}$

63. $150 = \binom{5}{3} \cdot \binom{6}{4}$

64. a) $\dfrac{1,069,263}{2,598,960} = \dfrac{\binom{13}{1} \cdot \binom{39}{4}}{\binom{52}{5}}$

    b) $\dfrac{2,023,203}{2,598,960} = 1 - \dfrac{\binom{39}{5}}{\binom{52}{5}}$

65. $\dfrac{182}{210} = 1 - \dfrac{\binom{2}{0} \cdot \binom{8}{6}}{\binom{10}{6}}$

66. a) $\dfrac{3744}{2,598,960} = \dfrac{\binom{13}{1} \cdot \binom{4}{2} \cdot \binom{12}{1} \cdot \binom{4}{3}}{\binom{52}{5}}$

    b) $\dfrac{624}{2,598,960} = \dfrac{\binom{13}{1} \cdot \binom{4}{4} \cdot \binom{12}{1} \cdot \binom{4}{1}}{\binom{52}{5}}$

    c) $\dfrac{68,640}{2,598,960} = \dfrac{\binom{13}{1} \cdot \binom{4}{2} \cdot \binom{12}{3} \cdot \binom{4}{1}}{\binom{52}{5}}$

    d) $\dfrac{247,104}{2,598,960} = \dfrac{\binom{13}{1} \cdot \binom{4}{2} \cdot \binom{12}{1} \cdot \binom{4}{2} \cdot \binom{11}{1} \cdot \binom{4}{1}}{\binom{52}{5}}$

    e) $\dfrac{5,148}{2,598,960} = \dfrac{\binom{13}{5} \cdot \binom{4}{1}}{\binom{52}{5}}$

67. $\dfrac{64}{1326} = \dfrac{\binom{4}{1} \cdot \binom{16}{1}}{\binom{52}{2}}$

68. a) $635,013,560,000 = \binom{52}{13}$

    b) $1 = \binom{13}{13} \cdot \binom{39}{0}$

    c) $\dfrac{2,613,754}{635,013,560,000} = \dfrac{\binom{13}{10} \cdot \binom{39}{3}}{\binom{52}{13}}$

69. a) $4 = \binom{4}{3}$ b) $60 = \binom{4}{1} \cdot \binom{6}{2}$

    c) $36 = \binom{4}{2} \cdot \binom{6}{1}$ d) $120 = \binom{10}{3}$

70. a) $\dfrac{120}{1001} = \dfrac{\binom{5}{1} \cdot \binom{3}{1} \cdot \binom{4}{1} \cdot \binom{2}{1}}{\binom{14}{4}}$

70. b) $\frac{90}{1001} = \frac{\binom{5}{3}\cdot\binom{9}{1}}{\binom{14}{4}}$

  c) $\frac{270}{1001} = \frac{\binom{4}{2}\cdot\binom{10}{2}}{\binom{14}{4}}$

71. a) $\frac{1}{31}$ b) $\frac{812}{900}$ c) $\frac{88}{900}$

72. a) $\frac{1}{5}$ b) $\frac{4}{25}$ c) $\frac{16}{125}$ d) $\frac{369}{625}$

73. 2/3

74. a) $\frac{1}{3}$ b) $\frac{2}{3}$ c) $\frac{1}{2}$

75. a) $\frac{58}{100}$ b) $\frac{29}{53}$ c) $\frac{18}{42}$

76. a) $\frac{1}{30}$ b) $\frac{1}{7}$ c) $\frac{1}{7}$ d) $\frac{1}{30}$

77. a) $\frac{5}{90}$ b) $\frac{20}{25}$

78. 0.26

79. a) 0.10 b) 0.68 c) 0.22

80. a) 0.75
    b) The manager should bring in the relief pitcher.

81. a) $\frac{8}{36}$ b) $\frac{4}{36}$ c) $\frac{3}{9}$ d) $\frac{4}{10}$ e) $\frac{5}{11}$ f) 0.49293

82. a) The missionary should place only one blue berry in one of the coconuts and the remaining nineteen berries in the other coconut.
    b) 0.74

83. a) 3   b1) 7   b2) 15   b3) 31
    c) For N discs, the least number of moves is $2^N - 1$.
    d) $2^{64} - 1$ e) 5.85×10^{11} years

84. a) $\frac{1}{3}$ b) $\frac{2}{3}$

85. a) $\frac{1}{243}$ b) $\frac{51}{243}$ c) $\frac{11}{243}$

86. a) $\frac{16}{1024}$ b) $\frac{918}{1024}$ c) $\frac{243}{1024}$

87. a) $\frac{10}{32}$ b) $\frac{5}{32}$ c) $\frac{6}{32}$

88. a) $\frac{256}{10,000}$ b) $\frac{1296}{10,000}$ c) $\frac{1792}{10,000}$

89. 0.017

90. a) $\frac{625}{1296}$ b) $\frac{1}{1296}$ c) $\frac{1125}{1296}$

91. a) $\frac{106}{1024}$ b) $\frac{243}{1024}$ c) $\frac{405}{1024}$

92. $\frac{1}{1024}$

93. $\frac{30618}{262,144}$

94. a) 0.03 b) 0.53 c) 0.17

95. 0.10

96. a) 0.01 b) 0.08 c) 0.34

97. a) 0.70 b) 0

98. a) 0.01 b) yes, it was helpful.

99. 0.23

100. a) $\frac{16}{625}$ b) $\frac{17}{625}$ c) $\frac{1}{625}$ d) $\frac{512}{625}$
     e) $\frac{608}{625}$ f) $\frac{81}{625}$ g) $\frac{96}{625}$

101. a) $\frac{1024}{3125}$ b) $\frac{2944}{3125}$ c) $\frac{2304}{3125}$

102. 0.15

103. a) $\frac{23040}{100,000}$ b) $\frac{1024}{100,000}$ c) $\frac{31744}{100,000}$ d) $\frac{7776}{100,000}$

104. a) 0.03 b) 0.97 c) 0.97

105. a) 0.10 b) 0.16 c) 0.16

CHAPTER 7

Part I

1. mean(median or mode)
2. center
3. 50.00
4. 100
5. (a) 68.26
   (b) 95.44
   (c) 99.74
6. 68.26
7. 34.13; 95(or 55)
8. +0.53
9. 0;2
10. 15.87; 84.13
11. P_{16}
12. median(mean or mode); zero
13. 85.2
14. normal
15. 5
16. (a) np
    (b) $\sqrt{np(1-p)}$
17. (a) 50
    (b) 5
18. (a) No
    (b) No
    (c) Yes

Part IV

1. (a) 99.38%
   (b) 68.16%
   (c) 90.32%
2. (a) 4.01%
   (b) 12.3%
   (c) 14.98%
3. (a) z = +1.65
   (b) z = -1.18
   (c) z = +0.84 & z_x = -0.84
   (d) z = -0.52
   (e) z = -0.67
4. (a) P_{98}
   (b) P_{16}
   (c) P_{69}
5. (a) 85.77%
   (b) 65.89%
   (c) 25.14%
6. (a) 28.43%
   (b) 12.71%
   (c) 41.92%
7. (a) P_{96}
   (b) P_{77}
   (c) P_{11}
8. (a) X = 341.6
   (b) X = 284.4
   (c) X = 320.8
   (d) X = 248.8
   (e) X = 326.8
   (f) X = 310
9. (a) 9.68%
   (b) approx. 27 games
   (c) 34.46%
10. (a) 13.57%
    (b) 59.47%
    (c) 48,400 families
    (d) $36,700
    (e) $29,150
    (f) $34,450
    (g) $17,000

Part II

1. b
2. c
3. a
4. d
5. c
6. a
7. d
8. b
9. c
10. a

Part III

1. True
2. True
3. False
4. True
5. False
6. True
7. True
8. False
9. False
10. True

(h) $49,000
(i) $18,450
11. (a) 10.56%
    (b) 37.21%
    (c) approx. 906 males
    (d) 158
12. (a) 106.5 minutes
    (b) 102.8 minutes
    (c) 113.3 minutes
13. (a) 99
    (b) 170
    (c) 168
    (d) 166
    (e) 162
    (f) 168
    (g) 163
    (h) 165; 83.84%
14. 10.56%
15. (a) 15.87%
    (b) 6.68%
    (c) 15.85%
16. (a) 95.25%
    (b) 99.62%
    (c) 2.29%
17. (a) 0.62%
    (b) approx. 1543 students
    (c) 72.62%
18. (a) 15.87%
    (b) 2.28%
    (c) 30.72%
19. $87,800
20. 86.12 inches
21. 10.26 months
22. (a) 22.36%
    (b) 23.89%
    (c) 85.20%
    (d) 10.74 seconds
23. approx. 730 tax returns

24. (a) 8.08%
    (b) 0.47%
    (c) 7.52%
25. 754 hours
26. approx 45 months
27. m = 67.4 ounces
28. m = 82.95
29. (a) 97.72%
    (b) 50%
    (c) 30.85%
30. 2.28%
31. Tina; Michael; Allison Mary Ann
32. (a) 1. 1.79%
       2. 1.08%
       3. 2.87%
       4. 98.92%
    (b) 1. 13.5%
       2. 12.1%
       3. 87.9%
       4. 100%
    (c) 1. 6.55%
       2. 6.81%
       3. 15.62%
       4. 0.49%
       5. 77.08%
33. (a) 4.95%
    (b) 30.62%
    (c) 3.98%
    (d) 20.75%
34. 8.78%
35. 0.10%
36. (a) 16.35%
    (b) 99.41%
    (c) 1.60%
37. (a) 0.55%
    (b) 4.36%
38. (a) 1.02%
    (b) 0.57%
    (c) 61.79%
    (d) 69.50%
39. (a) 4.65%
    (b) 99.60%
    (c) 0.81%
40. (a) 11.70%
    (b) 0.52%
    (c) 46.88%
41. (a) 0.62%
    (b) 32.43%
    (c) 76.31%
42. (a) 47%
    (b) 24.2%
    (c) 2.68%
    (d) betting strategy (a)
43. (a) 3.59%
    (b) 53.42%
    (c) 0.01%
44. (a) 1.19%
    (b) 2.33%
45. (a) 1.92%
    (b) 0.07%
46. (a) 12.92%
    (b) 3.01%
    (c) 8.38%
47. (a) 0.35%
    (b) 0.07%
48. (a) 0.87%
    (b) 12.92%
49. (a) 99.99%
    (b) 0.25%

50. (a) 0.89%
    (b) 6.06%
51. (a) 0.73%
    (b) 11.70%
52. (a) 1.50%
    (b) 9.68%
53. (a) 0.03%
    (b) 12.10%
54. 95.35%
55. (a) 10.03%
    (b) 99.32%
56. (a) 320
    (b) 99.48%
    (c) 11.70%
    (d) $6,183

CHAPTER 8

Part I

1. Sampling Distribution
2. sample
3. representative
4. equal
5. all
6. a) mean; μ
   b) size; $\dfrac{\sigma}{\sqrt{n}}$
7. a) x
   b) μ
   c) $\mu_{\overline{x}}$
   d) $\sigma_{\overline{x}}$
   e) σ
8. size; decreases
9. normal; large; 30; shape
10. a) normal
    b) 130; 130
    c) 1; 1
       Central Limit
11. normal

Part II

1. d
2. b
3. a
4. d
5. a
6. d
7. b
8. a
9. b
10. a

Part III

1. T
2. F
3. F
4. F
5. T
6. F
7. T
8. F
9. T
10. F

Part IV

1. a) not random
   b) not random
   c) random
2. a) $\mu_{\overline{x}} = 100$; $\sigma_{\overline{x}} = 3.0$
   b) $\mu_{\overline{x}} = 100$; $\sigma_{\overline{x}} = 1.0$
   c) $\mu_{\overline{x}} = 100$; $\sigma_{\overline{x}} = 0.9$
   d) $\mu_{\overline{x}} = 100$; $\sigma_{\overline{x}} = 0.45$
3. a) 1) normal
      2) 160
      3) 1
   b) 1) normal
      2) 16
      3) 0.33
   c) 1) unknown
      2) 4000
      3) 40
   d) 1) normal
      2) 1200
      3) 100
   e) 1) normal
      2) 32
      3) 0.5
   f) 1) normal
      2) 500
      3) 5
   g) 1) normal
      2) 20,000
      3) 300
   h) 1) unknown
      2) 8
      3) 0.4
   i) 1) normal
      2) 1000
      3) 50
   j) 1) unknown
      2) 24,000
      3) 2000
4. a) 3.59%
   b) 0.35%
   c) 0.13%
5. a) 9.18%
   b) 97.34%
   c) 0.05%
6. a) $\mu_{\overline{x}} = 72"$, $\sigma_{\overline{x}} = 0.25"$
   b) 68.26%
   c) 0.47%
   d) 71.79"
7. a) 97.72%
   b) 84.13%
   c) 81.85%
8. a) 97.72%
   b) 50%
   c) 0%
9. a) 74 years
   b) 1 year
   c) yes
   d) 97.72%; 0%
   e) 57.93%; 27.43%
   f) because the standard deviations are not equal.
10. a) 150 minutes
    b) 1 minute
    c) yes
    d) 0%
    e) 10.56%
    f) because the standard deviations are not equal.
11. a) 84.13%
    b) 2.28%
    c) 97.72%
12. 95.25%
13. 93.32%
14. 90.50%
15. 99.01%
16. a) 00.13%
    b) yes
17. 1791.2 hours to 1908.8 hours
18. a) 106.70 to 113.30
    b) 106.08 to 113.92
    c) 104.84 to 115.16
19. a) 1.52%
    b) 15.87%
    c) 18.15%
20. a) 30.85%
    b) 13.36%

CHAPTER 9

Part I

1. true; reject H_o
2. null; H_a
3. I
4. II
5. expected; significant
6. alternative
7. I
8. α
9. decreases; increases
10. β
11. one
12. two
13. equally; 0.5
14. mean; z_c; standard deviation
15. -1.65; +1.65
16. -1.96; +1.96
17. -2.33; +2.33
18. -2.58; +2.58
19. reject; accept
20. fail to reject
21. smallest
22. 0.89
23. less than or equal to; null
24. very; significant; marginally or not; not
25. yes; 3%
26. twice

Part II

1. c
2. a
3. b
4. b
5. a
6. c
7. a
8. c
9. b
10. a

Part III

1. True
2. True
3. True
4. True
5. False
6. False
7. True
8. True
9. True
10. False
11. False
12. True
13. True
14. True

Part IV Problems:

1. H_o: The mean weight loss is 12 lbs for the first ten days.
   H_a: The mean weight loss is less than 12 lbs for the first ten days.
   directional test.
2. H_o: The mean pollutant level is 13 ppm.
   H_a: The mean pollutant level is greater than 13 ppm.
   directional test.
3. H_o: The percent of asthmatic children whose mothers smoked during pregnancy is 10%.
   H_a: The percent of asthmatic children whose mothers smoked during pregnancy is not 10%.
   nondirectional test.
4. H_o: 11% of all licensed drivers wear seat belts.
   H_a: More than 11% of all licensed drivers wear seat belts.
   directional test.
5. H_o: 20% of the wives who hold full-time jobs earn more than their husbands.
   H_a: Less than 20% of the wives who hold full-time jobs earn more than their husbands.
   directional test.
6. H_o: The mean salary of public school teachers in the union is $25,000.
   H_a: The mean salary of public school teachers in the union is less than $25,000.
   directional test.
7. H_o: The mean consumption of electricity per day of the energy saving refrigerator is 4.9 k.w.h.
   H_a: The mean consumption of electricity per day of the energy saving refrigerator is more than 4.9 k.w.h.
   directional test.
8. H_o: The percent of women between the ages of 16 and 40 years who have a form of herpes virus is 30%.
   H_a: The percent of women between the ages of 16 and 40 years who have a form of herpes virus is not 30%.
   nondirectional test.
9. H_o: The mean time of an automatic opening device for the manufacturer's parachutes is 6 seconds.
   H_a: The mean time of an automatic opening device for the manufacturer's parachutes is not 6 seconds.
   nondirectional test.
10. H_o: The average daily rate of attendance is 94%.
    H_a: The average daily rate of attendance is greater than 94%.
    directional test.
11. a) Reject H_o if the experimental outcome is either less than 80.4 or greater than 119.6
    b) Yes
    c) Yes
12. a) Reject H_o if the experimental outcome is greater than 58.25.
    b) Yes
    c) No
13. a) Reject H_o if the experimental outcome is less than 97.69.
    b) No c) No d) Yes e) Yes
14. a) Reject H_o if the experimental outcome is less than 10.35.
    b) Yes c) Yes d) No e) No
15. a) Reject H_o if the experimental outcome is either less than 59.36 or greater than 100.64
    b) Fail to reject H_o
    c) Reject H_o, accept H_a
    d) Reject H_o, accept H_a
16. a)

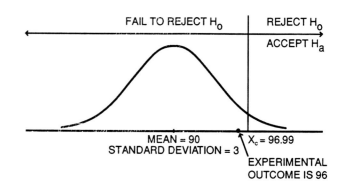

   b) Fail to reject H_o
   c) No

17. a)

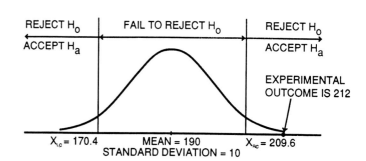

   b) Reject H_o, accept H_a
   c) Yes

18. a)

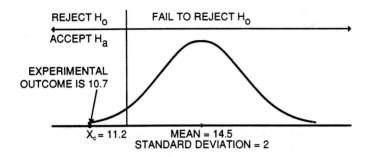

b) Reject H_o, accept H_a
c) Yes

19. a)

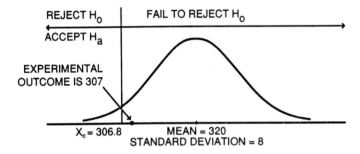

b) Fail to reject H_o
c) No

20. a)

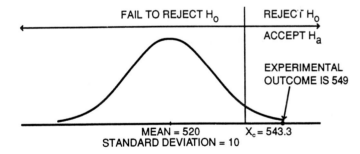

b) Reject H_o, accept H_a
c) Yes

21. a) z of 80.4 = -1.96 ; z of 119.6 = 1.96
    b) 5%
22. a) z of 58.25 = 1.65
    b) 5%
23. a) z of 97.69 = -2.33
    b) 1%
24. a) z of 10.35 = -1.65
    b) 5%
25. a) z of 59.36 = -2.58 ; z of 100.64 = 2.58
    b) 1%
26. a) z of 96.99 = 2.33
    b) 1%
27. a) z of 170.4 = -1.96 ; z of 209.6 = 1.96
    b) 5%
28. a) z of 11.2 = -1.65
    b) 5%
29. a) z of 306.8 = -1.65
    b) 5%
30. a) z of 543.3 = 2.33
    b) 1%
31. a) H_o: The mean age of recipients of unemployment benefits is 37 years.
       H_a: The mean age of recipients of unemployment benefits is not 37 years.
    b)

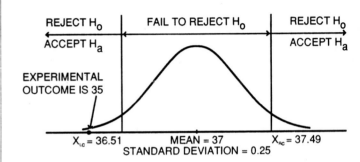

    c) p-value=0.0000 or 0% d) Reject H_o, accept H_a.
32. a) H_o: Mean expenditure of an American family of four for medical care is $2000.
       H_a: Mean expenditure of an American family of four for medical care is less than $2000.
    b)

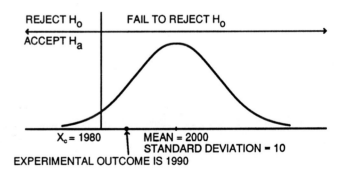

    c) p-value=0.1586 or 15.86% or 0%
    d) Fail to reject H_o.
33. a) H_o: 70% of the subscribers to *Playperson* magazine are aged 21 through 35 years.
       H_a: Less than 70% of the subscribers to *Playperson* magazine are aged 21 through 35 years.

b)

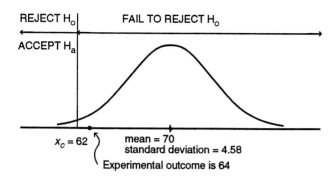

c) 0.0951 or 9.51%
d) Fail to reject H_o

34. a) H_o: 50% of U.S. families earn more than $20,000 per year.
    H_a: More than 50% of U.S. families earn more than $20,000 per year.
    b)

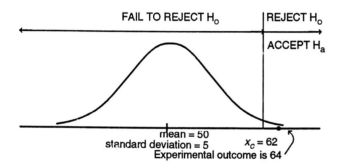

c) p-value = 0.0026 or 0.26%
d) Reject H_o, accept H_a.

35.

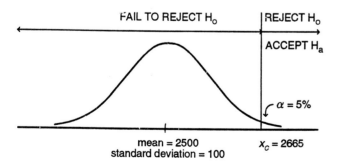

Decision rule: Reject H_o if the experimental outcome is greater than 2665.
a) (1) Reject H_o, accept H_a
   (2) Fail to reject H_o
   (3) Fail to reject H_o
   (4) Fail to reject H_o
b) 2715

36.

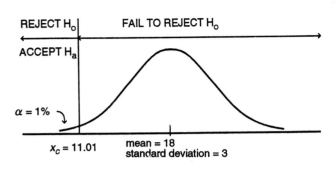

Decision rule: Reject H_o if the experimental outcome is less than 11.01
a) (1) Fail to reject H_o
   (2) Fail to reject H_o
   (3) Reject H_o, accept H_a
   (4) Fail to reject H_o
b) 11

37.

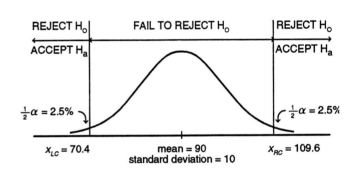

Decision rule: Reject H_o if the experimental outcome is less than 70.4 or greater than 109.6
a) (1) Fail to reject H_o
   (2) Fail to reject H_o
   (3) Reject H_o, accept H_a
   (4) Fail to reject H_o
b) 110

38.

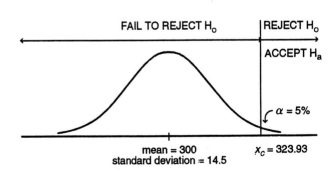

Decision rule: Reject H_o if the experimental outcome is greater than 323.93
a) (1) Fail to reject H_o
   (2) Reject H_o, accept H_a
   (3) Fail to reject H_o
   (4) Fail to reject H_o
b) 330

39.

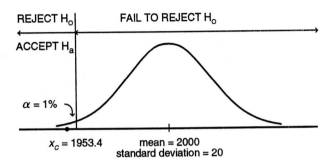

Decision rule: Reject H_o if the experimental outcome is less than 1953.4
a) (1) Reject H_o, accept H_a
   (2) Reject H_o, accept H_a
   (3) Fail to reject H_o
   (4) Fail to reject H_o
b) 1874 ; 1884

40.

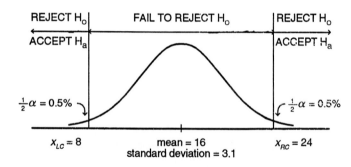

Decision rule: Reject H_o if the experimental outcome is less than 8 or greater than 24.
a) (1) Reject H_o, accept H_a
   (2) Fail to reject H_o
   (3) Fail to reject H_o
   (4) Reject H_o, accept H_a
b) 25 ; 2

41. a) H_o: 20% of the babies born are delivered by Cesarean Section.
H_a: More than 20% of the babies born are delivered by Cesarean section.
b) X_c = 26.6 babies

Decision rule: Reject H_o if the experimental outcome is more than 26.6 babies.

c)

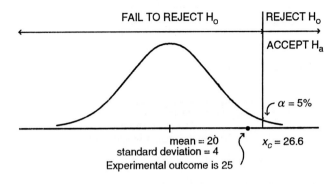

d) p–value = 0.1056 or 10.56%
e) *Conclusion*: Fail to reject H_o.
f) The experimental outcome is not statistically significant.

42. a) H_o: The unemployment rate is 8.7%.
H_a: The unemployment rate is lower than 8.7%.

b) X_c = 10.82 residents.
Decision rule: Reject H_o if the experimental outcome is less than 10.82 residents.

c)

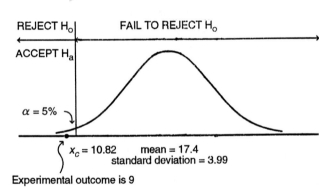

d) p–values = 0.0174 or 1.74%
e) *Conclusion*: Reject H_o, accept H_a.
f) The experimental outcome is statistically significant.

43. a) H_o: The mean achievement test score for fifth graders is 88.
H_a: The mean achievement test score for fifth graders is higher than 88.
b) X_c = a test score of 90.80
Decision Rule: Reject H_o if the experimental outcome is more than the test score of 90.80.

c)

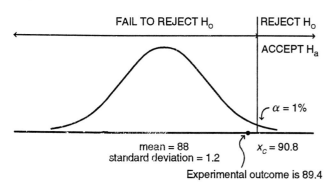

d) p-value 0.1210 or 12.10%
e) *Conclusion:* Fail to reject H_o.
f) The experimental outcome is not statistically significant.

44. a) H_o: The mean amount of soda dispensed by the machine is 2.0 liters.
    H_a: The mean amount of soda dispensed by the machine is not 2.0 liters.
    b) $X_{LC} = 1.95$ liters and $X_{RC} = 2.05$ liters.
    Decision Rule: Reject Ho if the experimental outcome is either less than 1.95 liters or greater than 2.05 liters.
    c)

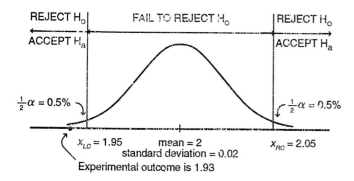

d) p-value 0.0004 or 0.04%
e) *Conclusion:* Reject H_o, accept H_a.
f) The experimental outcome is statistically significant.

45. a) H_o: The antibiotic is 90% effective against type A bacteria.
    H_a: The antibiotic is more than 90% effective against type A bacteria

b) $X_c = 373.98$ individual
   Decision Rule: Reject H_o if the experimental outcome is greater than 373.98 individuals.
c)

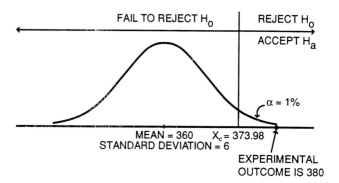

d) p-value 0.0005 or 0.05%
e) *Conclusion:* Reject H_o, accept H_a
f) The experimental outcome is statistically significant.

46. a) H_o: The mean reaction time is 0.20 seconds.
    H_a: The mean reaction time is less than 0.20 seconds.
    b) $X_c = 0.18$ seconds
    Decision rule: Reject H_o if the experimental outcome is less than 0.18 seconds.
    c)

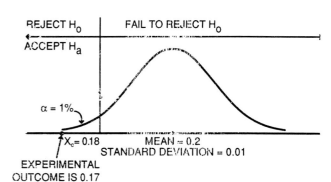

d) p-value 0.0013 or 0.13%
e) *Conclusion:* Reject H_o, accept H_a
f) The experimental outcome is statistically significant.

47. a) 0.0287 or 2.87%   b) No   c) 0.0287 or 2.87%
48. a) 0.0080 or 0.80%   b) Yes   c) 0.0080 or 0.80%

CHAPTER 10

Part I

1. directional
2. nondirectional
3. directional
4. true
5. estimate
6. $s_{\bar{x}}$; $s/\sqrt{n}$
7. a) normal
   b) μ_0
   c) $\sigma/\sqrt{n}$
8. the mean
9. population; normal
10. a) bell shaped
    b) mean
    c) degrees of freedom
11. one; n-1
12. equal; =
13. a) μ_0 ; >
    b) mean; μ_0 ; <
    c) population; ≠
14. H_0
15. a) z_c ; $\sigma_{\bar{x}}$
    b) t_c ; $s_{\bar{x}}$
16. 24

Part II

1. d
2. b
3. c
4. d
5. a
6. d
7. c
8. a
9. a
10. a

Part III

1. true
2. false
3. false
4. false
5. false
6. true
7. true
8. false
9. true
10. false

Part IV

1. a) -1.71
   b) -2.72
   c) 1.72
   d) 2.57
   e) -2.02; 2.02
   f) -2.66; 2.66
2. a) -1.80
   b) 1.70
   c) 2.65
   d) -2.57
   e) -2.01; 2.01
   f) -2.63; 2.63
3. a) -2.13; -1.83; -1.68; -1.66; -1.65
   b) -1.65
   c) as n increases, t_c approaches critical z score, z_c.
4. a) H_0: $\mu=120$; H_a: $\mu<120$
   b) $\mu_{\bar{x}}=120$; $\sigma_{\bar{x}}=3$; the sampling distribution of the mean is a normal distribution.
   c) reject H_0 if the sample mean is less than 115.05.
   d) $\bar{x}=114$
   e) reject H_0, accept H_a at $\sigma=5\%$; yes
5. a) H_0: $\mu=90$; H_a: $\mu>90$
   b) $\mu_{\bar{x}}=90$; $\sigma_{\bar{x}}=0.75$; the sampling distribution of the mean is a normal distribution.
   c) reject H_0 if the sample mean is greater than 91.75.
   d) $\bar{x}=93$
   e) reject H_0, accept H_a at $\alpha=1\%$; yes
6. a) H_0: $\mu=37$; H_a: $\mu\neq37$
   b) $\mu_{\bar{x}}=37$; $\sigma_{\bar{x}}=0.25$; the sampling distribution of the mean is a normal distribution.
   c) reject H_0 if the sample mean is either less than 36.36 or greater than 37.65.
   d) $\bar{x}=35$
   e) reject H_0, accept H_a at $\alpha=1\%$; yes
7. a) H_0: $\mu=6$; H_a: $\mu\neq6$
   b) $\mu_{\bar{x}}=6$; $\sigma_{\bar{x}}=0.16$; the sampling distribution of the mean is a normal distribution.
   c) reject H_0 if the sample mean is either less than 5.69 or greater than 6.31.
   d) $\bar{x}=5.73$
   e) fail to reject H_0 at $\sigma=5\%$; no
8. a) H_0: $\mu=112$; H_a: $\mu>112$
   b) $\mu_{\bar{x}}=112$; $\sigma_{\bar{x}}=2$; the sampling distribution of the mean is a normal distribution
   c) reject H_0 if the sample mean is greater than 116.66.
   d) $\bar{x}=116$
   e) fail to reject H_0 at $\alpha=1\%$; no
9. a) H_0: $\mu=90$; H_a: $\mu<90$
   b) $\mu_{\bar{x}}=90$; $\sigma_{\bar{x}}=2$; the sampling distribution of the mean is a normal distribution.
   c) reject H_0 if the sample mean is less than 85.34.
   d) $\bar{x}=84$
   e) reject H_0, accept H_a at $\alpha=1\%$ yes
10. a) H_0: $\mu=125,000$; H_a: $\mu>125,000$
    b) $\mu_{\bar{x}}=125,000$; $\sigma_{\bar{x}}=666.67$; the sampling distribution of the mean is a normal distribution.
    c) reject H_0 if the sample mean is greater than 126,100.01
    d) $\bar{x}=126,000$
    e) fail to reject H_0 at $\alpha=5\%$; no
11. a) H_0: $\mu=10$; H_a: $\mu>10$
    b) $\mu_{\bar{x}}=10$; $\sigma_{\bar{x}}=0.47$; the sampling distribution of the mean is a normal distribution.
    c) reject H_0 if the sample mean is greater than 10.78.
    d) $\bar{x}=12$
    e) reject H_0, accept H_a at $\alpha=5\%$; yes
12. a) H_0: $\mu=5$; H_a: $\mu<5$
    b) $\mu_{\bar{x}}=5$; $\sigma_{\bar{x}}=0.075$ hours, the sampling distribution of the mean is a normal distribution.
    c) reject H_0 if the sample mean is less than 4.83 hours.
    d) $\bar{x}=4.75$
    e) reject H_0, accept H_a at $\alpha=1\%$; yes
13. a) H_0: $\mu=45$; H_a: $\mu<45$
    b) $\mu_{\bar{x}}=45$; $\sigma_{\bar{x}}=1$; the sampling distribution of the mean is a normal distribution.
    c) reject H_0 if the sample mean is less than 43.35.
    d) $\bar{x}=42$
    e) reject H_0, accept H_a at $\alpha=5\%$; yes
14. a) H_0: $\mu=1236.45$; H_a: $\mu<1236.45$
    b) $\mu_{\bar{x}}=1236.45$; $\sigma_{\bar{x}}=15.88$; the sampling distribution of the mean is a normal distribution.
    c) reject H_0 if the sample mean is less than 1210.25.
    d) $\bar{x}=1217.69$
    e) fail to reject H_0 at $\alpha=5\%$; no
15. a) H_0: $\mu=36$; H_a: $\mu<36$
    b) $\mu_{\bar{x}}=36$; $\sigma_{\bar{x}}=0.5$; the sampling distribution of the mean is a normal distribution.
    c) reject H_0 if the sample mean is less than 35.18.
    d) $\bar{x}=33$
    e) reject H_0, accept H_a at $\alpha=5\%$; yes
16. a) H_0: $\mu=25$; H_a: $\mu\neq25$
    b) $\mu_{\bar{x}}=25$; $s_{\bar{x}}=0.1$; the sampling distribution of the mean is a t distribution.
    c) reject H_0 if the experimental outcome is either less than 24.8 or greater than 25.2.
    d) $\bar{x}=25.4$
    e) reject H_0, accept H_a at $\alpha=5\%$; yes

17. a) H_o: m=14.3; H_a: m>14.3
    b) $\mu_{\bar{x}}$=14.3; $s_{\bar{x}}$=0.3; the sampling distribution of the mean is a *t* distribution.
    c) reject H_o if the experimental outcome is greater than 14.81.
    d) $\bar{x}$=15.4
    e) reject H_o, accept H_a at a=5%; yes
18. a) H_o: m=12; H_a: m>12
    b) $\mu_{\bar{x}}$=12; $s_{\bar{x}}$=0.25; the sampling distribution of the mean is a *t* distribution.
    c) reject H_o if the experimental outcome is greater than 12.65.
    d) $\bar{x}$=13
    e) reject H_o, accept H_a at a=1%; yes
19. a) H_o: m=35; H_a: m¹35
    b) $\mu_{\bar{x}}$=35; $s_{\bar{x}}$=0.8; the sampling distribution of the mean is a *t* distribution.
    c) reject H_o if the experimental outcome is either less than 32.76 or greater than 37.24.
    d) $\bar{x}$=36
    e) fail to reject H_o at a=1%; no
20. a) H_o: m=4.5; H_a: m¹4.5
    b) $\mu_{\bar{x}}$=4.5; $s_{\bar{x}}$=0.05; the sampling distribution of the mean is a *t* distribution.
    c) reject H_o if the experimental outcome is either less than 4.37 or greater than 4.63.
    d) $\bar{x}$=4.3
    e) reject H_o, accept H_a at a=1%; yes
21. a) H_o: m=80,000; H_a: m<80,000
    b) $\mu_{\bar{x}}$=80,000; $s_{\bar{x}}$=50; the sampling distribution of the mean is a *t* distribution.
    c) reject H_o if the experimental outcome is less than 79,870 lbs.
    d) $\bar{x}$=79,880
    e) fail to reject H_o at a=1%; yes
22. a) H_o: m=5000; H_a: m<5000
    b) $\mu_{\bar{x}}$=5000; $s_{\bar{x}}$=50; the sampling distribution of the mean is a *t* distribution.
    c) reject H_o if the experimental outcome is less than 4907.
    d) $\bar{x}$=4930
    e) fail to reject H_o at a=5%; yes
23. a) H_o: m=27; H_a: m>27
    b) $\mu_{\bar{x}}$=27; $s_{\bar{x}}$=0.7; the sampling distribution of the mean is a *t* distribution.
    c) reject H_o if the experimental outcome is greater than 28.23 months.
    d) $\bar{x}$=28.05
    e) fail to reject H_o at a=5%; no
24. a) H_o: m=100; H_a: m<100
    b) $\mu_{\bar{x}}$=100; $s_{\bar{x}}$=1; the sampling distribution of the mean is a *t* distribution.
    c) reject H_o if the experimental outcome is less than 97.60.
    d) $\bar{x}$=97.45
    e) reject H_o, accept H_a at a=1%; yes
25. a) H_o: m=9; H_a: m<9
    b) $\mu_{\bar{x}}$=9; $s_{\bar{x}}$=0.58; the sampling distribution of the mean is a *t* distribution.
    c) reject H_o if the experimental outcome is less than 7.96.
    d) $\bar{x}$=7
    e) reject H_o, accept H_a at a=5%; yes
26. a) H_o: m=4.4; H_a: m<4.4
    b) $\mu_{\bar{x}}$=4.4; $s_{\bar{x}}$=0.1; the sampling distribution of the mean is a *t* distribution.
    c) reject H_o if the experimental outcome is less than 4.16.
    d) $\bar{x}$=4.2
    e) fail to reject H_o at a=1%; no
27. a) H_o: m=21; H_a: m<21
    b) $\mu_{\bar{x}}$=21; $s_{\bar{x}}$=0.69; the sampling distribution of the mean is a *t* distribution.
    c) reject H_o if the experimental outcome is less than 19.97.
    d) $\bar{x}$=20
    e) fail to reject H_o at a=5%; no
28. a) H_o: m=93; H_a: m>93
    b) $\mu_{\bar{x}}$=93; $s_{\bar{x}}$=0.55; the sampling distribution of the mean is a *t* distribution.
    c) reject H_o if the experimental outcome is greater than 93.96.
    d) $\bar{x}$=94.31
    e) reject H_o, accept H_a at a=5%; yes
29. a) H_o: m=850; H_a: m¹850
    b) $\mu_{\bar{x}}$=850; $s_{\bar{x}}$=7.5; the sampling distribution of the mean is a *t* distribution.
    c) reject H_o if the experimental outcome is either less than $830.27 or greater than $869.73.
    d) $\bar{x}$=872
    e) reject H_o, accept H_a at a=1%; yes
30. a) H_o: m=20; H_a: m<20
    b) $\mu_{\bar{x}}$=20; $s_{\bar{x}}$=0.5; the sampling distribution of the mean is a *t* distribution.
    c) reject H_o if the experimental outcome is less than 19.15.
    d) $\bar{x}$=19.14
    e) reject H_o, accept H_a at a=5%; yes
31. a) H_o: m=90; H_a: m¹90
    b) $\mu_{\bar{x}}$=90; $s_{\bar{x}}$=0.94; the sampling distribution of the mean is a *t* distribution.
    c) reject H_o if the experimental outcome is either less than 87.99 or greater than 92.01.
    d) $\bar{x}$=91.53
    e) fail to reject H_o at a=5%; no
32. a) H_o: m=84; H_a: m>84
    b) $\mu_{\bar{x}}$=84; $s_{\bar{x}}$=0.53; the sampling distribution of the mean is a *t* distribution.
    c) reject H_o if the experimental outcome is greater than 85.28.
    d) $\bar{x}$=89.23
    e) reject H_o, accept H_a at a=1%; yes
33. a) fail to reject H_o
    b) fail to reject H_o
    c) no
    d) m=120
    e) m=122
    f) the same population would have two different means.

CHAPTER 11

Part I

1. Two means
2. Sample means, mean, mean
3. a) $\mu_{\bar{x}_1-\bar{x}_2}$, μ_1, μ_2, population, population
   b) $s_{\bar{x}_1-\bar{x}_2}$, size, population 1, size, population 2,
   $$\sqrt{\frac{(n_1-1)s_1^2 + (n_2-1)s_2^2}{n_1 + n_2 - 2} \cdot \left(\frac{1}{n_1} + \frac{1}{n_2}\right)}$$
   c) t, $n_1 + n_2 - 2$
4. No, mean, mean, $\mu_1 - \mu_2$
5. a) mean of population 2,
   $\mu_{\bar{x}_1-\bar{x}_2} > 0$
   b) mean, population 2,
   $\mu_{\bar{x}_1-\bar{x}_2} < 0$
   c) population, mean, 2,
   $\mu_{\bar{x}_1-\bar{x}_2} \neq 0$
6. H_o
7. t_c, $s_{\bar{x}_1-\bar{x}_2}$, 0, t_c, $s_{\bar{x}_1-\bar{x}_2}$
8. $\bar{x}_1 - \bar{x}_2$
9. 34
10. B, A
11. 1.16
12. 1.09

Part II

1. a
2. b
3. d
4. b
5. a
6. b
7. a
8. c
9. c
10. c

Part III

1. True
2. False
3. False
4. True
5. True
6. True
7. False
8. True
9. False
10. False

Part IV

1. Pop 1: urban sixth grade students
   Pop 2: rural sixth grade students
   a) *Hypotheses*:
      H_o: There is no difference between the mean score of the urban sixth grade students and the mean score of the rural sixth grade students on the standardized reading comprehension test: $\mu_1 - \mu_2 = 0$.
      H_a: The mean score of the urban sixth grade students is greater than the mean score of the rural sixth grade students on the standardized reading comprehension test: $\mu_1 - \mu_2 > 0$.
   b) *Expected Results*:
      $\mu_{\bar{x}_1-\bar{x}_2} = 0$
      $\sigma_{\bar{x}_1-\bar{x}_2} \approx 2.08$
      Hypothesis test model: The Sampling Distribution of The Difference Between Two Means is a Normal Distribution.
   c) *Decision rule*:
      Reject H_o if $(\bar{x}_1 - \bar{x}_2)$ is greater than 3.43.
   d) *Experimental Outcome*:
      $(\bar{x}_1 - \bar{x}_2)$ is 4.5.
   e) *Conclusion*:
      Reject H_o, Accept H_a at $\alpha = 5\%$.
      Question:
      The educational researcher can conclude that the urban sixth grade students performed significantly better on the standardized test at $\alpha = 5\%$.

2. Pop 1: female college students
   Pop 2: male college students
   a) *Hypotheses*:
      H_o: There is no difference between the mean score of the female college students and the mean score of the male college students on the IQ test: $\mu_1 - \mu_2 = 0$.
      H_a: There is a difference between the mean score of the female college students and the mean score of the male college students on the IQ test: $\mu_1 - \mu_2 \neq 0$.
   b) *Expected Results*:
      $\mu_{\bar{x}_1-\bar{x}_2} = 0$
      $\sigma_{\bar{x}_1-\bar{x}_2} \approx 1.79$
      Hypothesis test model: The Sampling Distribution of The Difference Between Two Means is a Normal Distribution.
   c) *Decision rule*:
      Reject H_o if $(\bar{x}_1 - \bar{x}_2)$ is less than -4.62 or greater than 4.62.
   d) *Experimental Outcome*:
      $(\bar{x}_1 - \bar{x}_2)$ is -3.9.
   e) *Conclusion*:
      Fail to Reject H_o at $\alpha = 1\%$.
      Question:
      The psychologist can't conclude that there is a significant difference between the two groups on the IQ test at $\alpha = 1\%$.

3. Pop 1: manufacturer's new light bulbs.
   Pop 2: competitor's light bulbs.
   a) *Hypotheses*:
      H_o: There is no difference between the mean life of the new light bulbs and the mean life of the competitor's light bulbs: $\mu_1 - \mu_2 = 0$.
      H_a: The mean life of the new light bulbs is greater than the mean life of the competitor's light bulbs: $\mu_1 - \mu_2 > 0$.
   b) *Expected Results*:
      $\mu_{\bar{x}_1-\bar{x}_2} = 0$
      $s_{\bar{x}_1-\bar{x}_2} \approx 19.03$
      Hypothesis test model: The Sampling Distribution of The Difference Between Two Means is a *t* Distribution.
   c) *Decision rule*:
      Reject H_o if $(\bar{x}_1 - \bar{x}_2)$ is greater than 31.59 hours.
   d) *Experimental Outcome*:
      $(\bar{x}_1 - \bar{x}_2)$ is 50 hours.
   e) *Conclusion*:
      Reject H_o, Accept H_a at $\alpha = 5\%$.
      Question:
      Sample results support the manufacturer's claim.

4. Pop 1: effective life of the new insecticide.
   Pop 2: effective life of the present insecticide.
   a) *Hypotheses*:
      H_o: There is no difference between the effective life of the new insecticide and the present insecticide: $\mu_1 - \mu_2 = 0$.
      H_a: The effective life of the new insecticide is greater than the effective life of the present insecticide: $\mu_1 - \mu_2 > 0$.
   b) *Expected Results*:
      $\mu_{\bar{x}_1-\bar{x}_2} = 0$
      $s_{\bar{x}_1-\bar{x}_2} \approx 1.48$
      Hypothesis test model: The Sampling Distribution of The Difference Between Two Means is a *t* Distribution.

c) *Decision Rule*:
   Reject H_o if $(\bar{x}_1-\bar{x}_2)$ is greater than 3.49 minutes.
d) *Experimental Outcome*:
   $(\bar{x}_1-\bar{x}_2)$ is 3 minutes.
e) *Conclusion*:
   Fail to reject H_o at $\alpha=1\%$.
   Answer:
   The sample results don't support manufacturer's claim.

5. Pop 1: final grades of the female students.
   Pop 2: final grades of the male students.
   a) *Hypotheses*:
      H_o: There is no difference between the mean final grade of the female and male students: $\mu_1-\mu_2=0$.
      H_a: The mean final grade of the female students is less than the mean final grade of the male students: $\mu_1-\mu_2<0$.
   b) *Expected Results*:
      $\mu_{\bar{x}_1-\bar{x}_2}=0$
      $s_{\bar{x}_1-\bar{x}_2} \approx 1.17$
      Hypothesis test model: The Sampling Distribution of The Difference Between Two Means is a *t* Distribution.
   c) *Decision Rule*:
      Reject H_o if $(\bar{x}_1-\bar{x}_2)$ is less than -1.93.
   d) *Experimental Outcome*:
      $(\bar{x}_1-\bar{x}_2)$ is -2.
   e) *Conclusion*:
      Reject H_o, Accept H_a at $\alpha=5\%$.
      Answer:
      The results support the student government's claim.

6. Pop 1: Treatment group.
   Pop 2: Control group.
   a) *Hypotheses*:
      H_o: There is no difference in the mean stamina level of the treatment and control groups: $\mu_1-\mu_2=0$.
      H_a: The mean stamina level of the treatment group is greater than the mean stamina level of the control group: $\mu_1-\mu_2>0$.
   b) *Expected Results*:
      $\mu_{\bar{x}_1-\bar{x}_2}=0$
      $s_{\bar{x}_1-\bar{x}_2} \approx 2.78$
      Hypothesis test model: The Sampling Distribution of The Difference Between Two Means is a *t* Distribution.
   c) *Decision rule*:
      Reject H_o if $(\bar{x}_1-\bar{x}_2)$ is greater than 4.73.
   d) *Experimental Outcome*:
      $(\bar{x}_1-\bar{x}_2)$ is 8.
   e) *Conclusion*:
      Reject H_o, Accept H_a at $\alpha=5\%$.
      Answer:
      The results support the nutritionist's belief.

7. Pop 1: Treatment group.
   Pop 2: Control group.
   a) *Hypotheses*:
      H_o: There is no difference between the mean number of cavities of the treatment group and the control group: $\mu_1-\mu_2=0$.
      H_a: The mean number of cavities of the treatment group is less than the mean number of cavities of the control group: $\mu_1-\mu_2<0$.
   b) *Expected Results*:
      $\mu_{\bar{x}_1-\bar{x}_2}=0$
      $s_{\bar{x}_1-\bar{x}_2} \approx 0.09$

Hypothesis test model: The Sampling Distribution of The Difference Between Two Means is a *t* Distribution.
   c) *Decision Rule*:
      Reject H_o if $(\bar{x}_1-\bar{x}_2)$ is less than -0.21.
   d) *Experimental Outcome*:
      $(\bar{x}_1-\bar{x}_2)$ is -0.2.
   e) *Conclusion*:
      Fail to Reject H_o at $\alpha=1\%$.
      Answer:
      The results don't indicate that DPT significantly reduces cavities.

8. Pop 1: typing speeds of the community college graduates.
   Pop 2: typing speeds of the secretarial student graduates.
   a) *Hypotheses*:
      H_o: There is no difference in the mean typing speed of the community college graduates and the secretarial student graduates: $\mu_1-\mu_2=0$.
      H_a: There is a significant difference between the mean typing speed of the community college graduates and the secretarial student graduates: $\mu_1-\mu_2 \neq 0$.
   b) *Expected Results*:
      $\mu_{\bar{x}_1-\bar{x}_2}=0$
      $s_{\bar{x}_1-\bar{x}_2} \approx 3.10$
      Hypothesis test model: The Sampling Distribution of The Difference Between Two Means is a *t* Distribution.
   c) *Decision Rule*:
      Reject H_o if $(\bar{x}_1-\bar{x}_2)$ is either less than -6.17 or greater than 6.17.
   d) *Experimental Outcome*:
      $(\bar{x}_1-\bar{x}_2)$ is 4.
   e) *Conclusion*:
      Fail to reject H_o at $\alpha=5\%$.
      Answer:
      The personnel director cannot conclude that there is a significant difference in the mean typing speed of the community college graduate and secretarial school graduates.

9. Pop 1: number of accidents per year for male drivers under age 25 years.
   Pop 2: number of accidents per year for male drivers 25 years of age or older.
   a) *Hypotheses*:
      H_o: There is no difference between the mean number of accidents per year of male drivers under age 25 and male drivers 25 years of age or older: $\mu_1-\mu_2=0$.
      H_a: The mean number of accidents per year for male drivers under age 25 years is greater than the mean number of accidents per year for male drivers 25 years of age or older: $\mu_1-\mu_2>0$.
   b) *Expected Results*:
      $\mu_{\bar{x}_1-\bar{x}_2}=0$
      $s_{\bar{x}_1-\bar{x}_2} \approx 0.04$
      Hypothesis test model: The Sampling Distribution of The Difference Between Two Means is a *t* Distribution.
   c) *Decision Rule*:
      Reject H_o if $(\bar{x}_1-\bar{x}_2)$ is greater than 0.07.
   d) *Experimental Outcome*:
      $(\bar{x}_1-\bar{x}_2)$ is 0.26.
   e) *Conclusion*:
      Reject H_o, Accept H_a at $\alpha=5\%$.
      Answer:
      The insurance company's claim is correct.

10. Pop 1: tomato yield using special fertilizer.
    Pop 2: tomato yield using the usual fertilizer.
    a) *Hypotheses*:
       H_o: There is no difference in the average bushel yield per acre using the special fertilizer and the usual fertilizer: $\mu_1-\mu_2=0$.
       H_a: The average bushel yield per acre using the special fertilizer is greater than the average bushel yield per acre using the usual fertilizer: $\mu_1-\mu_2>0$.
    b) *Expected Results*:
       $\mu_{\bar{x}_1-\bar{x}_2}=0$
       $s_{\bar{x}_1-\bar{x}_2}\approx 5.38$
       Hypothesis test model: The Sampling Distribution of The Difference Between Two Means is a *t* Distribution.
    c) *Decision rule*:
       Reject H_o if $(\bar{x}_1-\bar{x}_2)$ is greater than 12.59.
    d) *Experimental Outcome*:
       $(\bar{x}_1-\bar{x}_2)$ is 25.
    e) *Conclusion*:
       Reject H_o, accept H_a at $\alpha=1\%$.
       Answer:
       The results support the farmer's claim.

11. Pop 1: married graduate students
    Pop 2: unmarried graduate students
    a) *Hypotheses*:
       H_o: There is no difference between the mean cumulative average of married and unmarried graduate students: $\mu_1-\mu_2=0$
       H_a: The mean cumulative average of married graduate students is lower than the mean cumulative average of unmarred graduate students: $\mu_1-\mu_2<0$
    b) *Expected Results*:
       $\mu_{\bar{x}_1-\bar{x}_2}=0$
       $s_{\bar{x}_1-\bar{x}_2}\approx 0.16$
       Hypothesis test model: The Sampling Distribution of The Difference Between Two Means is a *t* Distribution.
    c) *Decision rule*:
       If the experimental outcome is less than -0.4, then reject H_o.
    d) *Experimental Outcome*:
       $\bar{x}_1-\bar{x}_2=-0.21$
    e) *Conclusion*:
       Fail to reject H_o at $\alpha=1\%$
       Answer:
       We cannot agree with the sociologists claim.

12. Pop 1: American League Pitchers
    Pop 2: National League Pitchers
    a) *Hypotheses*:
       H_o: There is no difference between the mean E.R.A. of American League pitchers and National League pitchers: $\mu_1-\mu_2=0$
       H_a: The mean E.R.A of American League pitchers is lower than the mean E.R.A of National League pitchers: $\mu_1-\mu_2<0$
    b) *Expected Results*:
       $\mu_{\bar{x}_1-\bar{x}_2}=0$
       $s_{\bar{x}_1-\bar{x}_2}\approx 0.07$
       Hypothesis test model: The Sampling Distribution of The Difference Between Two Means is a *t* distribution.
    c) *Decision rule*:
       If the experimental outcome is less than -0.17, reject H_o.

12. d) *Experimental Outcome*:
       $\bar{x}_1-\bar{x}_2=-0.10$
    e) *Conclusion*:
       Fail to reject H_o at $\alpha=1\%$
       Answer:
       The sportswriter's claim is not justified.

13. Pop 1: male journalists
    Pop 2: female journalists
    a) *Hypotheses*:
       H_o: There is no difference between the mean salary of male and female journalists: $\mu_1-\mu_2=0$
       H_a: The mean salary of male journalists is greater than the mean salary of female journalists: $\mu_1-\mu_2>0$
    b) *Expected Results*:
       $\mu_{\bar{x}_1-\bar{x}_2}=0$
       $s_{\bar{x}_1-\bar{x}_2}\approx 347.31$
       Hypothesis test model: The Sampling Distribution of The Difference Between Two Means is a *t* Distribution.
    c) *Decision rule*:
       If the experimental outcome is greater than 590.43, reject H_o.
    d) *Experimental Outcome*:
       $\bar{x}_1-\bar{x}_2=1,100$
    e) *Conclusion*:
       Reject H_o, Accept H_a at $\alpha=5\%$
       Answer:
       The Affirmative Action Officer can conclude that there is a significant difference in their salaries.

14. Pop 1: statistics class with class projects
    Pop 2: statistics class without class projects
    a) *Hypotheses*:
       H_o: There is no difference between the mean performance of the statistics class with class projects and the statistics class without class projects: $\mu_1-\mu_2=0$
       H_a: The mean performance of the statistics class with class projects is greater than the mean performance of the statistics class without class projects: $\mu_1-\mu_2>0$
    b) *Expected Results*:
       $\mu_{\bar{x}_1-\bar{x}_2}=0$
       $s_{\bar{x}_1-\bar{x}_2}\approx 2.63$
       Hypothesis test model: The Sampling Distribution of The Difference Between Two Means is a *t* Distribution.
    c) *Decision rule*:
       If the experimental outcome is greater than 4.42, reject H_o.
    d) *Experimental Outcome*:
       $\bar{x}_1-\bar{x}_2=3$
    e) *Conclusion*:
       Fail to reject H_o at $\alpha=5\%$
       Answer:
       The statistics instructor can't conclude that there is a significant difference in the mean performance between the two statistics classes.

15. Pop 1: group using both eyes on perception test
    Pop 2: group using one eye covered on perception test
    a) *Hypotheses*:
       H_o: There is no difference between the mean length on the perception test of group using both eyes and group using one eye covered: $m_1-m_2=0$
       H_a: The mean length on the perception test of the group using both eyes is significantly different than the mean length of the group using one eye covered: $\mu_1-\mu_2\neq 0$

b) *Expected Results*:
   $\mu_{\bar{x}_1-\bar{x}_2}=0$
   $s_{\bar{x}_1-\bar{x}_2}\approx 0.87$
   Hypothesis test model: The Sampling Distribution of The Difference Between Two Means is a *t* Distribution.
c) *Decision rule*:
   If the experimental outcome is less than -2.47 or greater than 2.47, reject H_o.
d) *Experimental Outcome*:
   $\bar{x}_1-\bar{x}_2=-1.75$
e) *Conclusion*:
   Fail to reject H_o at $\alpha=1\%$
   Answer:
   Psychologist can't conclude that there is a significant difference on the perception between the two groups.

Pop 1: control group
Pop 2: treatment group
a) *Hypotheses*:
   H_o: There is no difference between the mean amount of Vitamin A stored in mice between the control and treatment groups: $\mu_1-\mu_2=0$
   H_a: The mean amount of Vitamin A stored in the mice of the control group is greater than the mean amount of Vitamin A stored in the mice of the treatment group: $\mu_1-\mu_2>0$
b) *Expected Results*:
   $\mu_{\bar{x}_1-\bar{x}_2}=0$
   $s_{\bar{x}_1-\bar{x}_2}\approx 139.63$
   Hypothesis test model: The Sampling Distribution of The Difference Between Two Means is a *t* Distribution.
c) *Decision rule*:
   If the experimental outcome is greater than 344.89, reject H_o.
d) *Experimental Outcome*:
   $\bar{x}_1-\bar{x}_2=893.33$
e) *Conclusion*:
   Reject H_o, Accept H_a at $\alpha=1\%$
   Answer:
   The researcher can conclude that the mean amount of Vitamin A stored in mice on the diet deficient diet is significantly less.

Pop 1: treatment group
Pop 2: control group
a) *Hypotheses*:
   H_o: There is no difference between the mean number of flowers per plant of the treatment and control groups: $\mu_1-\mu_2=0$
   H_a: The mean amount of flowers per plant of the treatment group is greater than the mean amount of flowers per plant of the control group: $\mu_1-\mu_2>0$
b) *Expected Results*:
   $\mu_{\bar{x}_1-\bar{x}_2}=0$
   $s_{\bar{x}_1-\bar{x}_2}\approx 0.81$
   Hypothesis test model: The Sampling Distribution of The Difference Between Two Means is a *t* Distribution.
c) *Decision rule*:
   If the experimental is greater than 1.39, reject H_o.
d) *Experimental Outcome*:
   $\bar{x}_1-\bar{x}_2=4.83$
e) *Conclusion*:
   Reject H_o, Accept H_a at $\alpha=5\%$
   Answer:
   The botanist can conclude that the plants treated with "Miraculous Growth" had a greater number of flowers per plant.

18. I.
   Pop 1: Ivy League College
   Pop 2: Military Academy
   a) *Hypotheses*:
      H_o: There is no difference between the mean verbal S.A.T. score of the IVY League College and the Military Academy: $\mu_1-\mu_2=0$
      H_a: The mean verbal S.A.T. score of the IVY League College is greater than the mean verbal S.A.T. score of the Military Academy: $\mu_1-\mu_2>0$
   b) *Expected Results*:
      $\mu_{\bar{x}_1-\bar{x}_2}=0$
      $s_{\bar{x}_1-\bar{x}_2}\approx 11.05$
      Hypothesis test model: The Sampling Distribution of The Difference Between Two Means is a *t* Distribution.
   c) *Decision rule*:
      If the experimental outcome is greater than 26.07, reject H_o.
   d) *Experimental Outcome*:
      $\bar{x}_1-\bar{x}_2=40$
   e) *Conclusion*:
      Reject H_o, Accept H_a at $\alpha=1\%$
      Answer:
      The educational researcher can conclude that the IVY League College has a higher mean verbal S.A.T.

18. II.
   Pop 1: IVY League College
   Pop 2: Military Academy
   a) *Hypotheses*:
      H_o: There is no difference between the mean mathematics S.A.T. score of the IVY League College and the Military Academy: $\mu_1-\mu_2=0$
      H_a: The mean mathematics S.A.T. score of the IVY League College is lower than the mean mathematics S.A.T. score of the Military Academy: $\mu_1-\mu_2<0$
   b) *Expected Results*:
      $\mu_{\bar{x}_1-\bar{x}_2}=0$
      $s_{\bar{x}_1-\bar{x}_2}\approx 12.17$
      Hypothesis test model: The Sampling Distribution of The Difference Between Two Means is a *t* Distribution.
   c) *Decision rule*:
      If the experimental outcome is less than -28.71, reject H_o.
   d) *Experimental Outcome*:
      $\bar{x}_1-\bar{x}_2=-35$
   e) *Conclusion*:
      Reject H_o, Accept H_a at $\alpha=1\%$
      Answer:
      The educational researcher can conclude the mean math S.A.T. score is greater for the Military Academy.

18. III.
   Pop 1: IVY League College
   Pop 2: Military Academy
   a) *Hypotheses*:
      H_o: There is no difference between the mean combined S.A.T. score of the IVY League College and the Military Academy: $\mu_1-\mu_2=0$
      H_a: There is a significant difference between the mean combined S.A.T score of the IVY League College and the Military Academy: $\mu_1-\mu_2 \neq 0$
   b) *Expected Results*:
      $\mu_{\bar{x}_1-\bar{x}_2}=0$
      $s_{\bar{x}_1-\bar{x}_2}\approx 16.42$

Hypothesis test model: The Sampling Distribution of The Difference Between Two Means is a *t* Distribution.

c) *Decision rule*:
If the experimental outcome is less than -43.18 or greater than 43.18, reject H_o.

d) *Experimental Outcome*:
$\bar{x}_1 - \bar{x}_2 = 5$

e) *Conclusion*:
Fail to reject H_o at $\alpha = 1\%$
Answer:
The educational researcher can't conclude that the combined mean S.A.T. score is significantly different for the IVY League College and the Military Academy.

CHAPTER 12

Part I

1. Proportion
2. population
3. sample, number of occurrences in the sample possessing the particular characteristic
4. a) p
   b) $\sqrt{\dfrac{p(1-p)}{n}}$
   c) np, n(1-p)
5. p, equal, =
6. a) p, >
   b) population, <
   c) proportion, ≠
7. H_o = true
8. Proportion
9. $\mu_{\hat{p}} \pm (z_c)(\sigma_{\hat{p}})$.
10. critical

Part II

1. d
2. c
3. a
4. a
5. d
6. c
7. c
8. b
9. d
10. c

Part III

1. False
2. True
3. True
4. False
5. False
6. False
7. True
8. True
9. True
10. True

Part IV

1. a) 0.30
   b) 0.67
   c) 0.90
   d) 0.37
   e) 0.11
2. a) 0.39
   b) 0.67
   c) 0.60
   d) 0.30
   e) 0.64
   f) 0.39
   g) 0.56
   h) 0.64
   i) 0.44
   j) 0.60
   k) 0.27
   l) 0.62
3. a) $\mu_{\hat{p}}$ = 0.40
      $\sigma_{\hat{p}}$ = 0.07
      Yes
   b) $\mu_{\hat{p}}$ = 0.10
      $\sigma_{\hat{p}}$ = 0.0548
      No
   c) $\mu_{\hat{p}}$ = 0.65
      $\sigma_{\hat{p}}$ = 0.0337
      Yes
   d) $\mu_{\hat{p}}$ = 0.8
      $\sigma_{\hat{p}}$ = 0.04
      Yes
   e) $\mu_{\hat{p}}$ = 0.7
      $\sigma_{\hat{p}}$ = 0.0229
      Yes
   f) $\mu_{\hat{p}}$ = 0.2
      $\sigma_{\hat{p}}$ = 0.0327
      Yes
4. a) 1.97%
   b) 84.85%
   c) 93.45%
   d) 0.49%
   e) 30.15%
5. a) 87.08%
   b) 84.14%
   c) 93.45%
   d) 98.81%
   e) 35.20%
6. a) 96.16%
   b) 0.02%
   c) 3.84%
7. a) 18.14%
   b) 8.53%
   c) 78.50%
8. a) H_o: p = 0.1 ; H_a: p > 0.1
   b) $\mu_{\hat{p}}$ = 0.1; $\sigma_{\hat{p}}$ = 0.03; Sampling Distribution of The Proportion
   c) Reject H_o if the experimental outcome is greater than 0.1495
   d) $\hat{p}$ = 0.16
   e) Reject H_o, accept H_a at α = 5%; Yes
9. a) H_o: p = 0.95; H_a: p < 0.95
   b) $\mu_{\hat{p}}$ = 0.95; $\sigma_{\hat{p}}$ = 0.0154; Sampling Distribution of The Proportion
   c) Reject H_o if the experimental outcome is less than 0.9141
   d) $\hat{p}$ = 0.89
   e) Reject H_o; accept H_a at α = 1%; Yes
10. a) H_o: p = 0.1; H_a: p < 0.1
    b) $\mu_{\hat{p}}$ = 0.1; $\sigma_{\hat{p}}$ = 0.0106; Sampling Distribution of The Proportion
    c) Reject H_o if the experimental outcome is less than 0.0825
    d) $\hat{p}$ = 0.0875
    e) Fail to reject H_o at α = 5%; No
11. a) H_o: p = 0.40; H_a: p ≠ 0.40
    b) $\mu_{\hat{p}}$ = 0.40; $\sigma_{\hat{p}}$ = 0.0245; Sampling Distribution of The Proportion
    c) Reject H_o, if the experimental outcome is either less than 0.3520 or greater than 0.4480
    d) $\hat{p}$ = 0.4375
    e) Fail to reject H_o at α = 5%; No
12. a) H_o: p = 0.4; H_a: p > 0.4
    b) $\mu_{\hat{p}}$ = 0.4; $\sigma_{\hat{p}}$ = 0.0490; Sampling Distribution of The Proportion
    c) Reject H_o if the experimental outcome is greater than 0.4809

d) $\hat{p} = 0.5$
e) Reject H_o, accept H_a at $\alpha = 5\%$; Yes

13. a) $H_o: p = 0.18$; $H_a: p < 0.18$
    b) $\mu_{\hat{p}} = 0.18$; $\sigma_{\hat{p}} = 0.0272$; Sampling Distribution of The Proportion
    c) Reject H_o, if the experimental outcome is less than 0.1351
    d) $\hat{p} = 0.15$
    e) Fail to reject H_o at $\alpha = 5\%$; No

14. a) $H_o: p = 0.57$; $H_a: p \neq 0.57$
    b) $\mu_{\hat{p}} = 0.57$; $\sigma_{\hat{p}} = 0.0286$, Sampling Distribution of The Proportion
    c) Reject H_o, If the experimental outcome is either less than 0.4962 or greater than 0.6438
    d) $\hat{p} = 0.5933$
    e) Fail to reject H_o at $\alpha = 1\%$; No

15. a) $H_o: p = 0.25$; $H_a: p > 0.25$
    b) $\mu_{\hat{p}} = 0.25$; $\sigma_{\hat{p}} = 0.0306$; Sampling Distribution of The Proportion
    c) Reject H_o, if the experimental outcome is greater than 0.3213
    d) $\hat{p} = 0.325$
    e) Reject H_o, accept H_a at $\alpha = 1\%$; Yes

16. a) $H_o: p = 0.2$; $H_a: p > 0.2$
    b) $\mu_{\hat{p}} = 0.2$; $\sigma_{\hat{p}} = 0.02$; Sampling Distribution of The Proportion
    c) Reject H_o, if the experimental outcome is greater than 0.2466
    d) $\hat{p} = 0.255$
    e) Reject H_o, accept H_a at $\alpha = 1\%$; Yes

17. a) $H_o: p = 0.15$; $H_a: p \neq 0.15$
    b) $\mu_{\hat{p}} = 0.15$; $\sigma_{\hat{p}} = 0.0399$, Sampling Distribution of The Proportion
    c) Reject H_o, if the experimental outcome is either less than 0.0471 or greater than 0.2529
    d) $\hat{p} = 0.175$
    e) Fail to reject H_o at $\alpha = 1\%$; No

18. a) $H_o: p = 0.80$; $H_a: p < 0.80$
    b) $\mu_{\hat{p}} = 0.8$; $\sigma_{\hat{p}} = 0.01$, Sampling Distribution of The Proportion
    c) Reject H_o, if the experimental outcome is less than 0.7767
    d) $\hat{p} = 0.78$
    e) Fail to reject H_o at $\alpha = 1\%$; No

19. a) $H_o: p = 0.6$; $H_a: p > 0.6$
    b) $\mu_{\hat{p}} = 0.6$; $\sigma_{\hat{p}} = 0.0346$, Sampling Distribution of The Proportion
    c) Reject H_o, if the experimental outcome is greater than 0.6571
    d) $\hat{p} = 0.63$
    e) Fail to reject H_o at $\alpha = 5\%$; No

20. a) $H_o: p = 0.68$; $H_a: p \neq 0.68$
    b) $\mu_{\hat{p}} = 0.68$ $\sigma_{\hat{p}} = 0.0295$, Sampling Distribution of The Proportion
    c) Reject H_o, if the experimental outcome is either less than 0.6222 or greater than 0.7378
    d) $\hat{p} = 0.744$
    e) Reject H_o, accept H_a at $\alpha = 5\%$; Yes

21. a) $H_o: p = 0.85$; $H_a: p < 0.85$
    b) $\mu_{\hat{p}} = 0.85$; $\sigma_{\hat{p}} = 0.0179$, Sampling Distribution of The Proportion
    c) Reject H_o, if the experimental outcome is less than 0.8083
    d) $\hat{p} = 0.800$
    e) Reject H_o, accept H_a at $\alpha = 1\%$; No

22. a) $H_o: p = 0.6$; $H_a: p < 0.6$
    b) $\mu_{\hat{p}} = 0.6$; $\sigma_{\hat{p}} = 0.0490$, Sampling Distribution of The Proportion
    c) Reject H_o, if the experimental outcome is less than 0.4858
    d) $\hat{p} = 0.52$
    e) Fail to reject H_o at $\alpha = 1\%$; No

23. a) $H_o: p = 0.5$; $H_a: p > 0.5$
    b) $\mu_{\hat{p}} = 0.5$ $\sigma_{\hat{p}} = 0.0791$; Sampling Distribution of The Proportion
    c) Reject H_o, if the experimental outcome is greater than 0.6305
    d) $\hat{p} = 0.575$
    e) Fail to Reject H_o at $\alpha = 5\%$; No

24. a) $H_o: p = 0.75$; $H_a: p < 0.75$
    b) $\mu_{\hat{p}} = 0.75$; $\sigma_{\hat{p}} = 0.0274$, Sampling Distribution of The Proportion
    c) Reject H_o, if the experimental outcome is less than 0.7048
    d) $\hat{p} = 0.648$
    e) Reject H_o, accept H_a at $\alpha = 5\%$; Yes

25. a) $H_o: p = 0.6$; $H_a: p \neq 0.6$
    b) $\mu_{\hat{p}} = 0.6$; $\sigma_{\hat{p}} = 0.0693$; Sampling Distribution of The Proportion
    c) Reject H_o, if the experimental outcome is either less than 0.4212 or greater than 0.7788
    d) $\hat{p} = 0.7$
    e) Fail to reject H_o at $\alpha = 1\%$; No

CHAPTER 13

Part I

1. parameter
2. random sample
3. estimation
4. point
5. point
6. sample proportion, $\hat{p}$
7. estimator
8. estimate; estimator
9. unbiased
10. sampling
11. interval
12. point
13. $\bar{x} - (1.96)\frac{\sigma}{\sqrt{n}}$; $\bar{x} + (1.96)\frac{\sigma}{\sqrt{n}}$
14. $\bar{x} - (1.65)\frac{\sigma}{\sqrt{n}}$; $\bar{x} + (1.65)\frac{\sigma}{\sqrt{n}}$
15. $\bar{x} - (2.58)\frac{\sigma}{\sqrt{n}}$; $\bar{x} + (2.58)\frac{\sigma}{\sqrt{n}}$
16. $\bar{x} - (t_{95\%})\frac{s}{\sqrt{n}}$; $\bar{x} + (t_{95\%})\frac{s}{\sqrt{n}}$
17. $\bar{x} - (t_{97.5\%})\frac{s}{\sqrt{n}}$; $\bar{x} + (t_{97.5\%})\frac{s}{\sqrt{n}}$
18. $\bar{x} - (t_{99.5\%})\frac{s}{\sqrt{n}}$; $\bar{x} + (t_{99.5\%})\frac{s}{\sqrt{n}}$
19. $\sqrt{\frac{\hat{p}(1-\hat{p})}{n}}$
20. $\hat{p} - (1.65){}^s\hat{p}$; $\hat{p} + (1.65){}^s\hat{p}$
21. $\hat{p} - (1.96){}^s\hat{p}$; $\hat{p} + (1.96){}^s\hat{p}$
22. $\hat{p} - (2.58){}^s\hat{p}$; $\hat{p} + (2.58){}^s\hat{p}$
23. probability; procedure; parameter; confidence
24. interval; all; 95%
25. increases
26. unknown; sample standard deviation, s; normal
27. increase
28. population parameter; margin
29. maximum; sample mean, $\bar{x}$; E; width; increases; increased
30. $(z)\left(\frac{\sigma}{\sqrt{n}}\right)$
31. $(z)\left(\sqrt{\frac{\hat{p}(1-\hat{p})}{n}}\right)$
32. margin of error
33. $\left[\frac{z\sigma}{E}\right]^2$
34. $\frac{z^2(0.25)}{E^2}$
35. $\frac{z^2(\hat{p})(1-\hat{p})}{E^2}$

Part II

1. d
2. c
3. a
4. b
5. d
6. c
7. c
8. e
9. a
10. c

Part III

1. T
2. T
3. T
4. T
5. T
6. F
7. T
8. T
9. T
10. F
11. F
12. F

Part IV

1) a) 90%:76.25 to 79.75; margin of error= 1.75
      95%:75.87 to 80.13; margin of error= 2.13
      99%:75.05 to 80.95; margin of error= 2.95
   b) 90%:76.87 to 79.13; margin of error= 1.13
      95%:76.64 to 79.36; margin of error= 1.36
      99%:76.17 to 79.83; margin of error= 1.83
   c) 90%:77.04 to 78.96; margin of error= 0.96
      95%:76.85 to 79.15; margin of error= 1.15
      99%:76.47 to 79.53; margin of error= 1.53
   d) 90%:77.34 to 78.66; margin of error= 0.66
      95%:77.21 to 78.79; margin of error= 0.79
      99%:76.95 to 79.05; margin of error= 1.05
   e) The width of the confidence intervals and the margin of error at each confidence level decreases.
2) a) 521.21 to 558.79 days
   b) LCL = 521.21 days; UCL = 558.79 days
3) a) 8.76 to 9.36 hrs
   b) 8.70 to 9.42 hrs
   c) 8.62 to 9.50 hrs
   d) 0.60; 0.72; 0.88
   e) as the confidence level increases, the width of the confidence interval increases.
4) a) day students
   b) population mean
   c) sample mean
   d) 18.5 hrs/wk
   e) 90%: margin of error= 0.33 hrs/wk
      95%: margin of error= 0.39 hrs/wk
      99%: margin of error= 0.52 hrs/wk
   f) 90%: 18.17 to 18.83 hrs/wk
   g) 95%: 18.11 to 18.89 hrs/wk
   h) 99%: 17.98 to 19.02 hrs/wk
5) a) full-time students
   b) population mean
   c) sample mean
   d) $247
   e) 90%: margin of error= $2.31
      95%: margin of error= $2.74
      99%: margin of error= $3.61
   f) 90%: 244.69 to 249.31 dollars
   g) 95%: 244.26 to 249.74 dollars
   h) 99%: 243.39 to 250.61 dollars
6) 99%: 12.7 to 15.3 lbs
   margin of error= 1.30 lbs
7) 30.51 to 32.89 mpg
8) 89.31 to 94.69 minutes
9) 10.93 to 14.07 years
10) $49.01 to $56.57
11) 208.25 to 221.75 milligrams
12) 1370.75 to 1439.25 hours
14) 35.47 to 37.13 hours
15) a) 225.69 to 235.71 lbs
    b) 6' 0.99" to 6' 3.01"
16) a) 90%:0.2244 to 0.3756; margin of error= 0.0756
       95%:0.2102 to 0.3898; margin of error= 0.0898
       99%:0.1818 to 0.4182; margin of error= 0.1182
    b) 90%:0.2810 to 0.3190; margin of error= 0.0190
       95%:0.2775 to 0.3225; margin of error= 0.0225
       99%:0.2703 to 0.3297; margin of error= 0.0297
    c) 90%:0.2977 to 0.3023; margin of error= 0.0023
       95%:0.2973 to 0.3027; margin of error= 0.0027
       99%:0.2964 to 0.3036; margin of error= 0.0036
    d) The width of each confidence interval and the margin of error at each confidence level decreases.

17) a) potential buyers of shavers
    b) population proportion
    c) sample proportion
    d) 0.0201
    e) 90%: margin of error= 0.0026
       95%: margin of error= 0.0031
       99%: margin of error= 0.0041
    f) 90%: 0.0175 to 0.0227
       95%: 0.0170 to 0.0232
       99%: 0.0160 to 0.0242
18) 95%: 0.3822 to 0.4594
    margin of error= 0.0386
19) a) nation's WP market
    b) population proportion
    c) sample proportion
    d) 0.24
    e) 1) 90%: margin of error= 0.0705
          0.1695 to 0.3105
       2) 90%: margin of error= 0.0353
          0.2047 to 0.2753
       3) 90%: margin of error= 0.0219
          0.2181 to 0.2619
    f) 1) 95%: margin of error= 0.0837
          0.1563 to 0.3237
       2) 95%: margin of error= 0.0419
          0.1981 to 0.2819
       3) 95%: margin of error= 0.0261
          0.2139 to 0.2661
    g) 1) 99%: margin of error= 0.1102
          0.1298 to 0.3502
       2) 99%: margin of error= 0.0552
          0.1848 to 0.2952
       3) 99%: margin of error= 0.0343
          0.2057 to 0.2743

20) 0.3730 to 0.4970
21) 0.6824 to 0.7376
22) 0.1525 to 0.1889
23) a) 0.7091 to 0.7503
    b) 0.5324 to 0.5872
24) a) 0.5648 to 0.6150
    b) 0.5569 to 0.6229
25) 0.6659 to 0.7131
26) a) 0.3757 to 0.4243
    b) 0.3262 to 0.3938
27) a) 0.3269 to 0.3731
    b) 0.2809 to 0.3191
28) a) n=9604
    b) n=97
    c) To determine a conservative estimate of the minimum sample size needed to meet the requirements of the confidence level and margin of error.
29) n=1849
30) a) 554
    b) 959
31) a) 65.61 to 70.99
    b) 5.38
    c) 2.69
    d) 355
32) a) 77.4
    b) 154.8
    c) 77.4
33) a) 426
    b) 401
34) 167
35) a) 2398
    b) $3597
36) a) 1068
    b) 861

CHAPTER 14

Part I

1. classification
2. independence
3. count
4. observed
5. contingency
6. cell
7. 4;3
8. a) $X^2 = \sum \frac{(O-E)^2}{E}$
   b) $X^2 = \sum \frac{O^2}{E} - n$
9. expected
10. non-negative
11. right
12. degrees of freedom; $(r-1)(c-1)$
13. independent; dependent
14. total; column; sample size
15. a) fail to reject H_o
    b) reject H_o and accept H_a
16. a) independent; b) cell; c) five
17. theoretical
18. sample
19. $k-1$

Part II

1. c
2. a
3. b
4. b
5. b
6. c
7. b
8. c
9. c
10. a

Part III

1. false
2. false
3. true
4. true
5. false
6. false
7. true
8. true
9. false
10. false

Part IV

1. (a) *Formulate the two hypotheses H_o and H_a.*
   H_o: Grade point average and membership to a fraternity group are independent.
   H_a: Grade point average and membership to a fraternity group are dependent.
   (b) For $\alpha = 1\%$ and $df = 2$, $\chi_\alpha^2 = 9.21$.
   Decision Rule: reject H_o if X^2 is greater than 9.21.
   (c) $X^2 = 9.57$
   (d) *Conclusion*: reject H_o and accept H_a at $\alpha = 1\%$. Yes, we can conclude that there is a relationship between membership in a campus fraternity and GPA.

2. (a) *Formulate the two hypotheses H_o and H_a.*
   H_o: Gender and the number of compact discs purchased per year are independent.
   H_a: Gender and the number of compact discs purchased per year are dependent.
   (b) For $\alpha = 5\%$ and $df = 2$, $\chi_\alpha^2 = 5.99$.
   Decision Rule: reject H_o if X^2 is greater than 5.99.
   (c) $X^2 = 9.03$
   (d) *Conclusion*: reject H_o and accept H_a at $\alpha = 5\%$. Yes, we can conclude that gender and the number of CD's purchased per year are related.

3. (a) *Formulate the two hypotheses H_o and H_a.*
   H_o: Gender and movie preference are independent.
   H_a: Gender and movie preference are dependent.
   (b) For $\alpha = 5\%$ and $df = 4$, $\chi_\alpha^2 = 9.49$.
   Decision Rule: reject H_o if X^2 is greater than 9.49.
   (c) $X^2 = 36.96$
   (d) *Conclusion*: reject H_o and accept H_a at $\alpha = 5\%$. Yes, we can conclude that there is a relationship between gender and movie preference.

4. (a) *Formulate the two hypotheses H_o and H_a.*
   H_o: Part quality and manufacturer are independent.
   H_a: Part quality and manufacturer are dependent.
   (b) For $\alpha = 5\%$ and $df = 2$, $\chi_\alpha^2 = 5.99$.
   Decision Rule: reject H_o if X^2 is greater than 5.99.
   (c) $X^2 = 2.43$
   (d) *Conclusion*: fail to reject H_o at $\alpha = 5\%$. Yes, the quality control engineer can conclude that part quality and manufacturer are independent at $\alpha = 5\%$.

5. (a) *Formulate the two hypotheses H_o and H_a.*
   H_o: Preference in packaging material and marital status are independent.
   H_a: Preference in packaging material and marital status are dependent.
   (b) For $\alpha = 1\%$ and $df = 2$, $\chi_\alpha^2 = 9.21$.
   Decision Rule: reject H_o if X^2 is greater than 9.21.
   (c) $X^2 = 98.96$
   (d) *Conclusion*: reject H_o and accept H_a at $\alpha = 1\%$. No, marital status and preference in packaging material are not independent at $\alpha = 1\%$.

6. (a) *Formulate the two hypotheses H_o and H_a.*
   H_o: Drug treatment and result of the treatment are independent.
   H_a: Drug treatment and result of the treatment are dependent.
   (b) For $\alpha = 5\%$ and $df = 4$, $\chi_\alpha^2 = 9.49$.
   Decision Rule: reject H_o if X^2 is greater than 9.49.
   (c) $X^2 = 1.02$
   (d) *Conclusion*: fail to reject H_o at $\alpha = 5\%$. Yes, Drug treatment and relief from pain are independent at $\alpha = 5\%$.

7. (a) *Formulate the two hypotheses H_o and H_a.*
   H_o: Educational level and the selection to serve on the grand jury are independent.

H_a: Educational level and the selection to serve on the grand jury are dependent.
(b) For $\alpha = 1\%$ and $df = 3$, $\chi_\alpha^2 = 11.34$.
Decision Rule: reject H_o if X^2 is greater than 11.34.
(c) $X^2 = 12.83$
(d) Conclusion: reject H_o and accept H_a at $\alpha = 1\%$.
Yes, the educational level and the selection to serve on a grand jury are dependent at an $\alpha = 1\%$.

8. (a) Formulate the two hypotheses H_o and H_a.
H_o: The Dow Jones average and the prime interest rate are independent.
H_a: The Dow Jones average and the prime interest rate are dependent.
(b) For $\alpha = 1\%$ and $df = 4$, $\chi_\alpha^2 = 13.28$.
Decision Rule: reject H_o if X^2 is greater than 13.28.
(c) $X^2 = 1.01$
(d) Conclusion: fail to reject H_o at $\alpha = 1\%$.
No, the Dow Jones average and the Prime Interest Rate are not dependent at $\alpha = 1\%$.

9. (a) Formulate the two hypotheses H_o and H_a.
H_o: Math anxiety and the successful completion of the introductory math course are independent.
H_a: Math anxiety and the successful completion of the introductory math course are dependent.
(b) For $\alpha = 1\%$ and $df = 2$, $\chi_\alpha^2 = 9.21$.
Decision Rule: reject H_o if X^2 is greater than 9.21.
(c) $X^2 = 37.77$
(d) Conclusion: reject H_o and accept H_a at $\alpha = 1\%$. No, Math anxiety and the successful completion of the introductory math course are not independent at $\alpha = 1\%$.

10. (a) Formulate the two hypotheses H_o and H_a.
H_o: The number of compact discs purchased per year and gender are independent.
H_a: The number of compact discs purchased per year are dependent.
(b) For $\alpha = 5\%$ and $df = 2$, $\chi_\alpha^2 = 5.99$.
Decision Rule: reject H_o if X^2 is greater than 5.99.
(c) $X^2 = 4.24$
(d) Conclusion: fail to reject H_o at $\alpha = 5\%$.
No, the number of compact discs purchased per year is not based on gender at $\alpha = 5\%$.

11. (a) Formulate the two hypotheses H_o and H_a.
H_o: Level of education and job satisfaction are independent.
H_a: Level of education and job satisfaction are dependent.
(b) For $\alpha = 5\%$ and $df = 4$, $\chi_\alpha^2 = 5.99$.
Decision Rule: reject H_o if X^2 is greater than 9.49.
(c) $X^2 = 5.49$
(d) Conclusion: fail to reject H_o at $\alpha = 5\%$.
Level of education and job satisfaction are independent at $\alpha = 5\%$.

12. (a) Formulate the two hypotheses H_o and H_a.
H_o: Course grade and the number of class absences are independent.
H_a: Course grade and the number of class absences are dependent.
(b) For $\alpha = 1\%$ and $df = 4$, $\chi_\alpha^2 = 13.28$.
Decision Rule: reject H_o if X^2 is greater than 13.28.
(c) $X^2 = 13.38$
(d) Conclusion: reject H_o and accept H_a at $\alpha = 1\%$. Course grade and the number of class absences are dependent.

13. (1)

Home Record	NBA Division			
	Atlantic	Central	Midwest	Pacific
Wins	a 156	b 188	c 154	d 196
Losses	e 90	f 58	g 92	h 91

(2a) Formulate the two hypotheses H_o and H_a.
H_o: A team's home winning record and NBA division are independent.
H_a: A team's home winning record and NBA division are dependent.
(b) For $\alpha = 1\%$ and $df = 3$, $\chi_\alpha^2 = 11.34$.
Decision Rule: reject H_o if X^2 is greater than 11.34.
(c) $X^2 = 13.60$
(d) Conclusion: reject H_o and accept H_a at $\alpha = 1\%$. A team's home winning record and NBA division are dependent.

14. (a) Formulate the two hypotheses H_o and H_a.
H_o: Educational background and the frequency of movie attendance per year are independent.
H_a: Educational background and the frequency of movie attendance are dependent.
(b) For $\alpha = 5\%$ and $df = 6$, $\chi_\alpha^2 = 12.59$.
Decision Rule: reject H_o if X^2 is greater than 12.59.
(c) $X^2 = 54.32$
(d) Conclusion: reject H_o and accept H_a at $\alpha = 5\%$. Yes, educational background and the frequency of movie attendance per year are dependent at $\alpha = 5\%$.

15. (a) Formulate the two hypotheses H_o and H_a.
H_o: Age and music category are independent.
H_a: Age and music category are dependent.
(b) For $\alpha = 5\%$ and $df = 16$, $\chi_\alpha^2 = 26.30$.
Decision Rule: reject H_o if X^2 is greater than 26.30.
(c) $X^2 = 91.66$
(d) Conclusion: reject H_o and accept H_a at $\alpha = 5\%$. Yes, age and music category are dependent at $\alpha = 5\%$.

16. (a) Formulate the two hypotheses H_o and H_a.
H_o: sample data fits a uniform distribution.
H_a: sample data doesn't fit a uniform distribution.
(b) For $\alpha = 5\%$ and $df = 4$, $\chi_\alpha^2 = 9.49$.
Decision Rule: reject H_o if X^2 is greater than 9.49.
(c) $X^2 = 10.30$
(d) Conclusion: reject H_o and accept H_a at $\alpha = 5\%$. The marketing agency should recommend that the manufacturer mass produce the new product.

17. (a) Formulate the two hypotheses H_o and H_a.
H_o: sample data fits a uniform distribution.
H_a: sample data doesn't fit a uniform distribution.
(b) For $\alpha = 1\%$ and $df = 4$, $\chi_\alpha^2 = 13.28$.
Decision Rule: reject H_o if X^2 is greater than 13.28.
(c) $X^2 = 14.37$
(d) Conclusion: reject H_o and accept H_a at $\alpha = 1\%$. We can conclude that there is a preference for packaging shape.

18. (a) Formulate the two hypotheses H_o and H_a.
H_o: sample data fits a uniform distribution.
H_a: sample data doesn't fit a uniform distribution.

(b) For α=5% and df = 5, χ^2_α = 11.07.
    Decision Rule: reject H_o if X^2 is greater than 11.07.
(c) X^2 = 4.90
(d) Conclusion: fail to reject H_o at α = 5%. The die is fair.

19. (a) Formulate the two hypotheses H_o and H_a.
    H_o: sample data fits census figures distribution.
    H_a: sample data doesn't fit census figures distribution.
    (b) For α=1% and df = 2, χ^2_α = 9.21.
    Decision Rule: reject H_o if X^2 is greater than 9.21.
    (c) X^2 = 10.44
    (d) Conclusion: reject H_o and accept H_a at α = 1%. We should conclude that the sample wasn't randomly selected from the town.

20. (a) Formulate the two hypotheses H_o and H_a.
    H_o: sample data fits previous year's professor's grade distribution.
    H_a: sample data doesn't fit previous year's professor's grade distribution.
    (b) For α=5% and df = 4, χ^2_α = 9.49.
    Decision Rule: reject H_o if X^2 is greater than 9.49.
    (c) X^2 = 9.59
    (d) Conclusion: reject H_o and accept H_a at α = 5%. This year's final exam results don't fit previous year's grade distribution.

21. (a) Formulate the two hypotheses H_o and H_a.
    H_o: sample data fits a uniform distribution.
    H_a: sample data doesn't fit a uniform distribution.
    (b) For α=5% and df = 5, χ^2_α = 11.07.
    Decision Rule: reject H_o if X^2 is greater than 11.07.
    (c) X^2 = 10.30
    (d) Conclusion: fail to reject H_o at α = 5%. The quality control engineer can conclude that the production of defective parts is evenly distributed throughout the workweek.

22. (a) Formulate the two hypotheses H_o and H_a.
    H_o: sample data fits the population proportions.
    H_a: sample data doesn't fit the population proportions.
    (b) For α=5% and df = 1, χ^2_α = 3.84.
    Decision Rule: reject H_o if X^2 is greater than 3.84.
    (c) X^2 = 2.62
    (d) Conclusion: fail to reject H_o at α = 5%. The sample data is consistent with the population proportions.

23. (a) Formulate the two hypotheses H_o and H_a.
    H_o: Dice are fair.
    H_a: Dice are loaded.
    (b) For α=5% and df = 10, χ^2_α = 18.31.
    Decision Rule: reject H_o if X^2 is greater than 18.31.
    (c) X^2 = 18.71
    (d) Conclusion: reject H_o and accept H_a at α = 5%. The pit boss can conclude that the dice are loaded.

24. (a) Formulate the two hypotheses H_o and H_a.
    H_o: The number of hours worked per week fits a normal distribution.
    H_a: The number of hours worked per week doesn't fit a normal distribution.
    (b)

Interval Boundaries	z score	Probability	Expected Frequency
below 12.32	below -1.40	8.08%	48.48
12.32 - 17.01	-1.40 to -0.70	16.12%	96.72
17.01 - 21.70	-0.70 to 0.0	25.80%	154.80
21.70 - 26.39	0.0 to 0.70	25.80%	154.80
26.39 - 31.08	0.70 to 1.40	16.12%	96.72
above 31.08	above 1.40	8.08%	48.48

(c) For α=5% and df = 3, χ^2_α = 7.82
Decision Rule: reject H_o if X^2 is greater than 7.82
(d) X^2 = 12.36
(e) reject H_o and accept H_a at α = 5%. The number of hours worked per week doesn't fit a normal distribution.

CHAPTER 15

Part I

1. two variables
2. measurement
3. ordered; first
4. scatter diagram
5. positive
6. linear correlation
7. negative
8. zero or no; not
9. stronger
10. Pearson's correlation coefficient, r.
11. -1; +1
12. strongest; perfect; upward
13. strongest; negative; downward
14. linear
15. -1;+1; weak
16. positive; negative
17. rho: ρ
18. sample; estimate
19. a) zero; no; 0
    b) positive; >; negative; <; significant; ≠
20. normally; correlation coefficients
21. r_α; n-2
22. $r = \dfrac{n\Sigma(xy) - (\Sigma x)(\Sigma y)}{\sqrt{n(\Sigma x^2) - (\Sigma x)^2}\sqrt{n(\Sigma y^2) - (\Sigma y)^2}}$
23. dependent; independent; r^2
24. least squares; regression
25. $y' = a + bx$; $\dfrac{n\Sigma(xy) - \Sigma x \Sigma y}{n(\Sigma x^2) - (\Sigma x)^2}$; $\dfrac{\Sigma y - b\Sigma x}{n}$
    dependent; independent

Part II

1. c   6. c
2. b   7. b
3. a   8. b
4. b   9. d
5. d   10. c

Part III

1. true   6. false
2. true   7. false
3. false   8. false
4. false   9. true
5. false 10. true

Part IV

1. a) negative
   b) positive
   c) positive
   d) no correlation
2. a) 2
   b) 1
   c) 1
   d) 3
   e) 3
3. a) i; iii
   b) i is negative, iii positive
4. a) positive
   b) positive or no correlation
   c) negative
   d) no correlation
   e) negative
   f) positive
   g) positive
   h) no correlation
   i) positive
   j) negative
   k) positive
   l) positive
   m) positive
   n) positive
   o) positive
   p) positive
   q) positive
   f) positive
5. highest 1, -0.95, 0.69 and -0.69, 0.5, 0.05
6. a) r= 0.87 is a stronger positive correlation
   b) although the r values represent a different type of relationship they have the same strength
   c) r= 0.95 represents a very strong relationship, whereas, r= 0.05 represents a very weak relationship
7. a) r=0.97 b) r=-0.74 c) r=1.00
   d) r=0.84 e) r=-0.95 f) r=0.93
8.
	r_α	significant (Yes,No)
a)	±0.64	no
b)	-0.46	yes
c)	0.42	yes
d)	±0.58	no
e)	-0.36	yes
f)	0.31	yes
9.
	r_α	significant (Yes,No)
a)	0.59	no
b)	0.44	yes
c)	-0.42	yes
d)	0.43	yes
e)	-0.32	yes
f)	0.44	no
10. a) yes
    b) 23.04%
11. a) yes
    b) 31.36%
12. a)

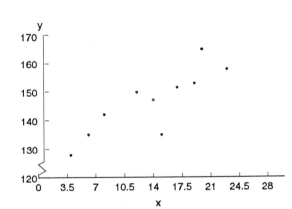

    b) r=0.85
    c) yes
    d) 72.25% of the variance in blood pressure that can be accounted for by the variance in pounds above ideal weight.
    e) y'= 1.54x + 125.25
    f) y' ≈ 141 (lbs)
13. a)

    b) r=0.94
    c) yes

d) 88.36% of the variance in advertising expenditures that can be accounted for by the variance in beer sales.
e) y' = 0.23x + 27.86
f) y' ≈ $39,360,000.00

14. a)

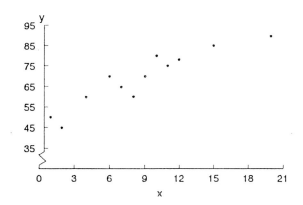

b) r=0.93
c) yes
d) 86.49% of the variance in test grades that can be accounted for by the variance in hours studying for exam.
e) y' = 2.35x + 48.47
f) y' ≈ 81.32

15. a)

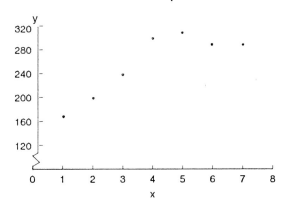

b) r=0.86
c) yes
d) 73.96% of the variance in grass density that can be accounted for by the variance in seed density.
e) y' = 21.8x + 169.94
f) y' ≈ 246 seedlings/ft²

16. a)

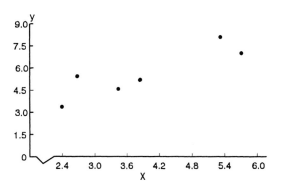

b) r=0.87
c) yes
d) 74.83% of the variance in mosquito population that can be accounted for by the variance in the amount of rainfall in may.
e) y' = 1.01x + 1.52
f) y' ≈ 5.86

17. a)

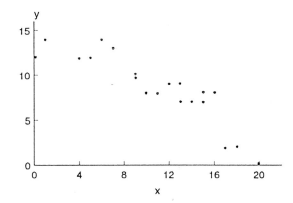

b) r=-0.89
c) yes
d) 79.21% of the variance in percent of yearly salary increase that can be accounted for by the variance in the number of sick days per year.
e) y' = -0.62x + 15.27
f) y' = 10.31% increase in yearly salary.

18. a)

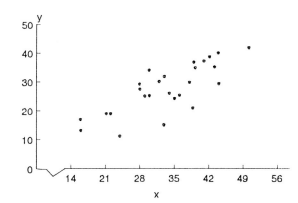

b) r=0.78
c) yes
d) 60.84% of the variance in final exam grade that can be accounted for by the variance in midterm exam grade.
e) y' = 0.77x + 1.87
f) y' ≈ 20 final exam grade.

APPENDICES

APPENDIX A: DATABASE

- Database layout, variable names, variable descriptions and variable values
- Database file: DATABASE.MTW

APPENDIX B: CALCULATOR INSTRUCTIONS

- For mean, standard deviation, Σx, and Σx^2
- For correlation coefficient, & regression quantities: a and b

APPENDIX C: SUMMARY

- FORMULAS
- HYPOTHESIS TESTS
- CONFIDENCE INTERVALS

APPENDIX D: STATISTICAL TABLES

- TABLE I: SUCCESS OCCURRENCE TABLE
- TABLE II: NORMAL CURVE AREA TABLE
- TABLE III: CRITICAL VALUES FOR THE t DISTRIBUTIONS
- TABLE IV: CRITICAL VALUES FOR THE CHI-SQUARE DISTRIBUTIONS
- TABLE V: CRITICAL VALUES OF THE CORRELATION COEFFICIENT

APPENDIX A: DATABASE

The following database, DATABASE.MTW, represents a sample of 75 student records. The MINITAB record layout for each student is as follows:

MINITAB Column	Variable Name	Variable Description
C1	ID	identification number
C2	AGE	age in years
C3	SEX	gender
C4	MAJ	academic area of concentration
C5	AVE	high school average
C6	HR	hair color
C7	EYE	eye color
C8	WRK	hours worked per week
C9	STY	hours studied per week
C10	MAT	matriculation status
C11	GPA	grade point average
C12	NAP	hour of the day a nap is taken
C13	SHT	student height in inches
C14	PHT	same sex parent's height in inches
C15	AM	ideal age to marry
C16	MOB	month of birth
C17	LK	preference of softdrink: Coke, Pepsi or neither
C18	HND	handedness
C19	GFT	amount in dollars spent on gift
C20	PH	number of hours spent on phone per week
C21	NB	response to question: Choose a number: 1, 2, 3, 4.
C22	AST	response to question: Do you believe in astrology?

Data Variable Values

Variable Name	Variable Value
ID	4 digit numeric
AGE	2 digit numeric
SEX	0=male, 1=female
MAJ	1=bus, 2=cmp, 3=fpa, 4=hth, 5=lib, 6=sci
AVE	2 digit numeric
HR	1=blonde, 2=black, 3=brown, 4=red
EYE	1=blue, 2=brown, 3=green
WRK	2 digit numeric
STY	2 digit numeric
MAT	0=yes, 1=no
GPA	3 digit 2 decimal numeric
NAP	standard military hours
SHT	2 digit numeric
PHT	2 digit numeric
AM	2 digit numeric
MOB	2 digit numeric, 01-12, month
LK	1=n , 2=p 3=c
HND	0=left, 1=right
GFT	3 digit numeric
PH	2 digit numeric
NB	1 digit numeric
AST	0=yes, 1=no

C1	C2	C3	C4	C5	C6	C7	C8	C9	C10	C11	C12	C13	C14	C15	C16	C17	C18	C19	C20	C21	C22
1618	19	0	5	81	3	1	16	11	0	1.87	1300	74	71	25	06	2	1	100	02	4	0
7290	20	1	5	87	1	1	12	19	0	3.05	2000	60	65	25	04	3	1	120	04	2	1
5506	21	0	1	85	3	3	15	10	1	3.62	1500	74	72	33	08	1	1	050	01	4	1
8496	19	1	2	88	3	3	14	09	0	3.67	1700	61	63	25	06	3	0	050	10	3	1
2832	18	1	1	72	3	3	08	23	0	2.25	0700	69	64	23	07	1	1	200	10	3	0
3407	22	0	5	83	3	1	14	20	0	2.70	1400	74	72	30	12	2	1	025	01	4	1
8573	19	1	5	78	3	2	15	17	1	2.30	1500	60	62	25	04	2	1	150	05	4	0
6265	20	0	6	76	3	2	22	16	1	2.51	0900	67	67	26	03	1	0	100	02	4	0

4821	19	0	1	74	1	1	18	22	1	2.10	2200	74	68	30	03	3	1	100	01	4	1
9029	19	0	1	79	1	1	10	12	0	1.92	1600	69	71	25	10	3	1	200	01	1	0
3001	19	1	5	83	3	1	10	20	1	3.15	1500	64	66	24	02	2	1	050	10	2	1
8010	23	1	1	78	3	2	20	11	0	2.00	1700	65	66	30	07	3	1	025	02	3	0
4699	19	1	5	74	1	3	26	09	0	1.95	1400	65	64	25	12	3	1	050	07	2	1
8298	24	0	1	88	3	3	12	19	1	2.88	1600	70	75	24	03	2	1	100	10	4	1
1896	18	1	5	81	3	1	18	09	1	2.65	1800	68	66	26	01	2	1	050	10	2	0
4789	19	1	5	84	3	1	17	08	0	1.98	0900	65	65	25	12	3	1	050	05	3	0
3592	27	1	1	83	3	1	15	19	1	3.21	2200	65	63	25	11	3	1	050	02	4	0
7679	21	0	1	87	1	1	16	07	0	1.95	1600	76	74	30	02	3	1	035	03	2	0
0209	18	1	6	80	3	1	22	10	1	2.96	1400	66	66	24	10	3	1	070	10	4	0
4489	21	0	5	84	1	3	14	08	0	2.95	1000	71	69	24	07	3	1	050	02	4	1
0880	18	0	1	90	3	2	08	18	0	2.00	1500	71	75	26	03	3	1	075	02	4	1
7168	37	1	5	87	3	2	11	16	1	2.82	2100	61	60	30	04	3	1	050	02	1	0
5709	28	1	1	85	2	2	07	10	1	1.95	1700	65	67	32	01	3	1	100	25	4	0
7520	20	1	5	77	2	2	25	12	1	3.01	1400	62	65	23	01	3	1	040	07	3	0
2757	20	1	4	72	1	1	27	15	0	2.70	1300	63	62	25	08	2	1	100	03	2	1
3016	19	1	5	83	3	3	20	11	0	2.67	1100	60	65	28	04	3	0	175	07	4	0
7298	21	0	2	77	3	3	12	09	0	2.81	1500	68	68	26	11	3	1	100	02	4	1
5677	18	0	6	85	3	1	16	27	1	3.76	1600	66	65	25	10	2	1	040	04	4	0
0406	21	1	2	70	3	2	14	10	1	2.67	1200	62	64	23	03	1	1	050	07	3	1
7673	19	1	4	77	3	2	16	07	0	2.32	1500	67	66	26	04	3	1	125	04	3	0
3480	20	1	1	83	4	3	24	08	0	2.09	1600	64	63	23	04	3	1	100	05	2	1
6994	18	1	5	86	3	3	19	09	0	2.42	1700	65	61	27	01	2	1	100	02	3	0
4009	18	1	5	84	3	2	10	10	1	1.90	1300	62	64	26	01	1	1	075	02	4	0
7879	20	0	6	77	2	2	08	07	1	1.98	1900	69	68	26	02	3	1	090	02	1	1
2021	19	1	5	82	1	3	20	12	0	3.15	1000	65	64	26	01	2	1	100	10	1	0
0044	19	1	4	90	1	2	00	14	0	3.40	1600	67	70	24	02	3	1	050	09	3	0
4508	19	1	1	84	1	3	15	10	1	2.87	2100	65	65	24	07	3	1	075	06	2	1
1085	23	1	5	91	3	2	10	28	1	3.51	1700	67	64	26	07	2	1	040	10	2	0
8289	20	1	1	78	2	2	22	08	1	1.95	1700	63	62	26	09	3	1	050	15	1	0
5664	18	0	5	75	3	1	26	06	0	2.43	1600	71	67	25	10	2	1	050	01	3	1
1284	19	1	5	81	2	2	12	08	0	2.47	1500	66	63	23	07	1	1	100	20	2	0
9406	19	0	6	79	2	2	14	10	0	2.66	1300	68	69	25	06	3	1	100	03	1	1
8022	19	0	1	88	1	3	00	12	1	3.28	1600	71	69	28	09	3	1	050	01	2	1
0954	25	1	5	87	3	2	09	15	1	3.41	1600	63	63	27	04	1	1	050	01	4	1
5012	20	0	5	75	2	2	10	21	1	2.20	2100	77	71	35	03	2	0	150	02	3	0
0013	19	1	5	80	3	1	30	11	1	2.32	1700	63	61	25	08	2	1	175	25	3	0
8809	22	1	4	88	1	2	08	20	1	2.97	1600	65	69	25	09	3	1	100	05	2	1
3697	20	1	1	93	3	2	16	29	1	3.75	1500	62	66	26	08	3	1	075	12	2	1
7807	18	1	5	77	3	2	18	20	0	2.78	1300	64	62	25	10	3	1	050	15	4	1
2058	30	1	4	82	3	2	20	15	1	2.84	1600	61	62	28	06	2	1	030	03	3	0
9801	18	1	4	79	3	2	08	10	1	3.10	1400	65	64	28	04	1	1	040	15	3	1
5464	18	1	5	75	1	1	24	06	0	2.49	1500	65	67	30	07	3	0	075	10	1	1
8002	24	1	4	88	3	2	10	10	1	3.24	1300	69	67	28	09	1	1	100	05	3	0
1919	18	0	1	81	3	1	09	21	1	2.95	1100	69	66	28	04	3	1	050	08	3	1
6728	19	0	1	86	3	2	17	23	1	2.81	1900	72	69	23	01	3	1	150	20	2	1
9603	26	1	4	79	2	2	16	10	0	2.36	1600	70	66	24	10	3	1	100	02	4	1
8871	22	0	2	88	3	2	08	17	1	3.10	2000	67	67	27	03	3	1	200	01	3	1
6587	19	1	5	86	3	2	15	28	1	3.78	2200	61	59	25	03	3	1	100	03	3	1
0250	19	0	5	90	3	2	12	15	0	3.08	1900	71	70	28	10	3	1	100	13	2	0
1010	19	0	1	81	3	2	20	11	1	3.07	1100	69	70	25	12	2	1	150	03	4	1
2721	22	1	5	92	1	2	17	12	0	3.15	1600	60	63	25	07	1	1	040	03	2	0
5256	18	1	5	95	1	2	12	15	1	3.65	1500	63	64	26	01	3	1	050	06	3	1
9847	18	0	1	89	3	2	08	14	1	3.70	1000	66	66	24	09	2	1	200	20	2	1
1617	20	1	5	81	2	1	16	21	1	2.73	1500	65	64	24	03	3	1	075	01	2	0
8077	20	0	6	88	3	1	10	17	1	3.75	1300	73	75	27	01	3	1	050	09	4	1
0001	19	1	5	90	3	2	10	20	1	3.17	1600	68	65	26	06	3	1	040	02	4	0
7755	31	0	1	87	3	2	37	25	0	3.56	1300	67	67	40	09	3	1	075	07	4	0
4607	18	1	3	84	3	2	16	30	0	3.75	1700	65	63	25	04	1	1	075	05	4	0
6930	20	1	1	76	3	2	28	23	1	2.05	1600	65	64	27	09	2	1	030	10	1	0
9788	19	0	5	89	3	2	17	28	1	3.85	1500	69	68	24	02	3	1	050	01	3	0
4697	22	0	5	74	3	2	16	09	0	2.74	1600	70	65	25	03	3	1	100	10	1	0
6796	18	1	1	86	3	2	17	09	1	2.64	1300	68	65	26	11	2	1	075	01	3	1
3529	23	1	3	73	3	2	15	12	0	2.93	1700	64	63	26	09	1	1	070	10	3	1
9244	19	0	5	79	1	1	12	24	0	2.47	1500	72	68	26	03	2	1	100	01	2	1
2581	24	0	3	82	3	2	25	08	1	2.18	1800	71	70	28	09	2	0	025	01	2	1

APPENDIX B: CALCULATOR INSTRUCTIONS

For mean, standard deviation, Σx, and Σx^2

SHARP EL-512
1) To place in correct mode: Press 2nd F & STAT
2) Enter DATA using M+ KEY
3) To retrieve mean:
   $\overline{X}$: Press 2nd F & 4 KEY
4) To retrieve sample standard deviation:
   S_X : Press 2nd F & 5 KEY
5) To retrieve population standard deviation:
   σ_X : Press 2nd F & 6 KEY
6) To get number of scores:
   n: Press (KEY
7) To obtain sum of scores:
   ΣX : Press) KEY
8) To retrieve sum of the squared scores:
   ΣX^2 : Press 2nd F &) KEY
   To clear: TURN OFF

SHARP EL-506D
1) To place in correct mode: Press MODE 4
2) Then hit 0 when 0 ~ 6 appears
3) Enter DATA using M+ KEY
4) Use a OR b Keys to scroll through quantities, then press = key to retrieve result.

SHARP EL-513C
1) Press 2nd F & STAT to place in correct mode
2) Enter DATA using M+ KEY
3) To retrieve mean:
   $\overline{X}$: Press X→M KEY
4) To retrieve sample standard deviation:
   S : Press RM KEY
5) To retrieve population standard deviation:
   σ : Press 2nd and RM KEY
6) To get number of scores:
   n: Press) KEY
7) To obtain sum of the scores:
   ΣX : Press 2nd and) KEY
8) To retrieve sum of the squared scores:
   ΣX^2 : Press 2nd and X→M KEY
   To clear: TURN OFF

SHARP EL-5150
1) To place in correct mode: Press 2nd F & STAT
2) Enter DATA using M+ KEY
3) To retrieve mean:
   $\overline{X}$: Press 2nd F & 4 KEY
4) To retrieve sample standard deviation:
   S_X : Press 2nd F & 5 KEY
5) To retrieve population standard deviation:
   σ_X : Press 2nd F & 6 KEY
6) To get number of scores:
   n: Press 0 KEY
7) To obtain sum of the scores:
   ΣX : Press) KEY
8) To retrieve sum of the squared scores:
   ΣX^2 : Press (-) KEY
   To clear: TURN OFF

CASIO FX-570AV; FX-991D FX-330V; FX-115V
1) To clear: Press Shift and AC KEY
2) Enter DATA using M+ KEY
3) To retrieve mean:
   $\overline{X}$: Press SHIFT and 1 KEY
4) To retrieve population standard deviation:
   σ : Press SHIFT and 2 KEY
5) To retrieve sample standard deviation:
   S : Press SHIFT and 3 KEY
6) To get sum of the squared scores:
   ΣX^2 : Press K OUT & 1 KEY
7) To obtain sum of the scores:
   ΣX : Press K OUT and 2 KEY
8) To get number of scores:
   n: Press K OUT & 3 KEY
   To clear: PRESS Shift AND AC KEY

RADIO SHACK EC-4031; EC-4035
1) a) For EC-4031 model: Press MODE and 3: SD MODE is indicated.
   b) For EC-4035 model: Press Mode and + Key: SD MODE is indicated.

The following procedure is the same for both calculators:
2) Enter DATA using M+ KEY
3) To retrieve mean:
   $\overline{X}$: Press SHIFT and 1 KEY
4) To retrieve population standard deviation:
   $X\sigma_n$: Press SHIFT & 2 KEY
5) To retrieve sample standard deviation:
   $X\sigma_{n-1}$: Press SHIFT & 3 KEY
6) To get sum of the squared scores:
   ΣX^2 : Press K OUT & 1 KEY
7) To obtain sum of the scores:
   ΣX : Press K OUT and 2 KEY
8) To get number of scores:
   n: Press K OUT and 3 KEY
   To clear: Shift & KAC KEY

TEXAS INSTRUMENTS: TI-60
1) Enter DATA using Σ+ KEY
2) To retrieve mean:
   mean: Press 2nd & $X \rightleftarrows Y$ KEY
3) To retrieve sample standard deviation:
   σ_{n-1} : Press 2nd and (KEY
4) To retrieve population standard deviation:
   σ_n : Press 2nd and) KEY
   To clear: Press 2nd & CSR KEY

TI-35X & TI-36X
1) To place in correct mode: Press 3rd and STAT 1
2) Enter DATA using Σ+ KEY
3) To retrieve mean:
   $\overline{X}$: Press 2nd and x^2 KEY
4) To retrieve sample standard deviation:
   σ_{xn-1} : Press 2nd and $\sqrt{X}$ KEY
5) To retrieve population standard deviation:
   σ_{xn} : Press 2nd and ÷ KEY
   To clear: Press 2nd & CSR KEY

TI-30 SLR+
1) Enter DATA using Σ+ KEY
3) To retrieve mean:
   $\overline{X}$: Press 2nd and STO KEY
4) To retrieve sample standard deviation:
   σ_{n-1} : Press 2nd & RCL KEY
5) To retrieve population standard deviation:
   σ_n : Press 2nd and SUM KEY
6) To get sum of the scores:
   ΣX : Press 2nd and) KEY
7) To obtain sum of the squared scores:
   ΣX^2 : Press 2nd and (KEY
   To clear: Press 2nd & CSR KEY

APPENDIX B: CALCULATOR INSTRUCTIONS:
For correlation coefficient, & regression quantities: a and b

SHARP EL-512
To place in LR Mode
Press 2nd F
Press Mode
Press 3 then 1
To enter the x & y data values
1) Key in x value
2) Press (x,y) [RM Key]
3) Key in y value
4) Press Data [M+ Key]
   Repeat these 4 steps until all the pairs of data values are entered.
To retrieve r:
Press S. Calc Key [x→M Key]
Press r Key [9 Key]
To retreive a:
Press S. Calc Key [x→M Key]
Press 2nd F and a Key [7 Key]
To retreive b:
Press S. Calc Key [x→M Key]
Press 2nd F and b Key [8 Key]

SHARP EL-506D
To place in LR Mode
Press Mode
Press 4 then 1
To enter the x & y data values
1) Key in x value
2) Press (x,y) [RM Key]
3) Key in y value
4) Press Data [M+ Key]
   Repeat these 4 steps until all the pairs of data values are entered.
To retrieve the results, use the Scroll Keys which are the:
△ [a Key] or ▽ [b Key]
To retrieve r:
Press either Scroll Key until you arrive at: r =
Press = Key to obtain r value
To retreive a:
Press either Scroll Key until you arrive at: a =
Press = Key to obtain a value
To retreive b:
Press either Scroll Key until you arrive at: b =
Press = Key to obtain b value

SHARP EL-512H
To place in LR Mode
Press 2nd F
Press DEC [X Key]
To enter the x & y data values
1) Key in x value
2) Press (x,y) Key
3) Key in y value
4) Press Data Key
   Repeat these 4 steps until all the pairs of data values are entered.
To retrieve r:
Press Alpha Key
Press r Key [(Key]
To retreive a:
Press Alpha Key
Press a Key [X Key]
To retreive b:
Press Alpha Key
Press b Key [+ Key]

CASIO:
FX 570AV; FX 991 D;
FX 115V; FX 115D; FX 330V
To place in LR Mode
Press Mode
Press 2
To enter the x & y data values
1) Key in x value
2) Press X_D, Y_D Key {[(--- Key}
3) Key in y value
4) Press Data Key [M+ Key]
   Repeat these 4 steps until all the pairs of data values are entered.
To retrieve r:
Press Shift and r Key [a Key]
To retreive a:
Press Shift and a Key [7 Key]
To retreive b:
Press Shift and b Key [8 Key]

TEXAS INSTRUMENTS: TI-60
For this model, it is not necessary to place in LR mode.
To enter the x & y data values
1) Key in x value
2) Press $X \rightleftarrows Y$ KEY
3) Key in y value
4) Press Σ+ KEY
   Repeat these 4 steps until all the pairs of data values are entered.
To retrieve r:
Press 2nd & Corr Key [÷ Key]
To retreive a:
Press 2nd & Intep Key [7 Key]
To retreive b:
Press 2nd & Slope Key [8 Key]

TI-35X; TI-36X
To place in LR Mode
Press 3rd and STAT 2
To enter the x & y data values
1) Key in x value
2) Press $X \rightleftarrows Y$ KEY
3) Key in y value
4) Press Σ+ KEY
   Repeat these 4 steps until all the pairs of data values are entered.
To retrieve r:
Press 3rd & Corr Key [4 Key]
To retreive a:
Press 2nd & Itc Key [4 Key]
To retreive b:
Press 2nd & Slp Key [5 Key]

RADIO SHACK:
EC4031; EC4035
To place in LR Mode
Press Mode and - Key
To enter the x & y data values
1) Key in x value
2) Press X_D, Y_D Key {[(Key}
3) Key in y value
4) Press DATA Key [M+ Key]
   Repeat these 4 steps until all the pairs of data values are entered.
To retrieve r:
Press Shift & r [9 Key]
To retreive a:
Press Shift & A [7 Key]
To retreive b:
Press Shift & B [8 Key]

RADIO SHACK: EC4026
To place in LR Mode
Press Mode and 2
To enter the x & y data values
1) Key in x value
2) Press , Key [comma Key]
3) Key in y value
4) Press M+ Key
   Repeat these 4 steps until all the pairs of data values are entered.
To retrieve r:
Press 2nd F & r [(Key]
Then EXE Key
To retreive a:
Press 2nd F & a [STO Key]
Then EXE Key
To retreive b:
Press 2nd & b [RCL Key]
Then EXE Key

APPENDIX C: SUMMARY FORMULAS

CHAPTER 3

Population Mean:

$$\mu = \frac{\Sigma x}{N}$$

Median is the middle value (after the data values have been arranged in numerical order)

position of median $= \frac{N+1}{2}$ th data value

Mode is most frequent data value

Sample Mean:

$$\bar{x} = \frac{\Sigma x}{n}$$

CHAPTER 4

Range = largest value − smallest value

Population Variance $= \frac{\Sigma(x-\mu)^2}{N}$

OR

$$\sigma^2 = \frac{\Sigma(x-\mu)^2}{N}$$

Definition Formula for the Population Standard Deviation:

$$\sigma = \sqrt{variance}$$

OR

$$\sigma = \sqrt{\frac{\Sigma(x-\mu)^2}{N}}$$

Computational Formula for the Population Standard Deviation:

$$\sigma = \sqrt{\frac{\Sigma x^2 - \frac{(\Sigma x)^2}{N}}{N}}$$

Definition Formula for the Sample Standard Deviation:

$$s = \sqrt{\frac{\Sigma(x-\bar{x})^2}{n-1}}$$

Computational Formula for the Sample Standard Deviation:

$$s = \sqrt{\frac{\Sigma x^2 - \frac{(\Sigma x)^2}{n}}{n-1}}$$

CHAPTER 5

z Score Formula:

$$z = \frac{x - \mu}{\sigma}$$

Raw Score Formula:

$$x = \mu + z\sigma$$

Percentile Rank Formula:

$$PR \text{ of } x = \frac{[B + (\tfrac{1}{2})E]}{N}(100)$$

CHAPTER 6

Counting Rule 1: Permutation Rule.
The number of permutations (arrangements) of n different objects taken altogether, denoted $_nP_n$, is:

$$_nP_n = n!$$

COUNTING RULE 2: Permutation Rule for n objects taken s at a time.
The number of permutations of *n* different objects taken *s* at a time, denoted $_nP_s$, is:

$$_nP_s = n(n-1)(n-2) \cdots (n-s+1)$$

COUNTING RULE 3: Permutation Rule of N objects with k alike objects
Given N objects where n_1 are alike, n_2 are alike, ..., n_k are alike, then the number of permutations of these N objects is:

$$\frac{N!}{(n_1!)(n_2!)\cdots(n_k!)}$$

COUNTING RULE 4: Number of Combinations of n objects taken s at a time
The number of combinations of n objects taken s at a time, is:

$$\binom{n}{s} = \frac{_nP_s}{s!}$$

The formula for $\binom{n}{s}$ in factorial notation is written as:

$$\binom{n}{s} = \frac{n!}{s!(n-s)!}$$

CHAPTER 6

$$P(\text{Event } E) = \frac{\text{number of outcomes satisfying event } E}{\text{total number of outcomes in sample space}}$$

The Addition Rule:
$$P(A \text{ or } B) = P(A) + P(B) - P(A \text{ and } B)$$

The Complement Rule:
$$P(E) + P(E') = 1$$

Multiplication Rule for Independent Events:
$$P(A \text{ and } B) = P(A) \cdot P(B)$$

Multiplication Rule for dependent Events:
$$P(A \text{ and } B) = P(A) \cdot P(B|A)$$

Conditional Probability Formula
$$P(A|B) = \frac{P(A \text{ and } B)}{P(B)}$$

Binomial Probability Formula
$$P(s \text{ successes in } n \text{ trials}) = \binom{n}{s} p^s q^{(n-s)}$$

CHAPTER 7

z Score Formula:
$$z = \frac{x - \mu}{\sigma}$$

Raw Score Formula:
$$X = \mu + (z)\sigma$$

Mean of the Binomial Distribution
$$\mu_s = np$$

Standard Deviation of the Binomial Distribution
$$\sigma_s = \sqrt{np(1-p)}$$

CHAPTER 8

Mean of the Sampling Distribution of the Mean:
$$\mu_{\bar{x}} = \mu$$

Standard Deviation of the Sampling Distribution of the Mean:
$$\sigma_{\bar{x}} = \frac{\sigma}{\sqrt{n}}$$

CHAPTER 9

z SCORE Formula:
$$z_x = \frac{x - \mu}{\sigma}$$

CRITICAL VALUE Formula:

$$X_c = \text{mean} + (z_c)(\text{standard deviation})$$

CHAPTER 10

mean of the sampling distribution of the mean is:
$$\mu_{\bar{x}} = \mu$$

standard deviation of the sampling distribution of the mean is:
$$\sigma_{\bar{x}} = \frac{\sigma}{\sqrt{n}}$$

estimate of the standard deviation of the sampling distribution of the mean
$$s_{\bar{x}} = \frac{s}{\sqrt{n}}$$

critical value formula:
a) for normal distribution:
$$X_c = \mu_{\bar{x}} + (z_c)(\sigma_{\bar{x}})$$

b) for t distribution:
$$X_c = \mu_{\bar{x}} + (t_c)(s_{\bar{x}})$$

To find t_c, need to determine degrees of freedom using formula:
$$df = n-1$$

CHAPTER 11

The mean of the Sampling Distribution of the Difference Between Two Means, is:
$$\mu_{\bar{x}_1 - \bar{x}_2} = \mu_1 - \mu_2$$

The standard deviation of the Sampling Distribution of the Difference Between Two Means is:
$$\sigma_{\bar{x}_1 - \bar{x}_2} = \sqrt{\frac{\sigma_1^2}{n_1} + \frac{\sigma_2^2}{n_2}}$$

The estimate of the standard deviation of the Sampling Distribution of The Difference Between Two Means is:
$$s_{\bar{x}_1 - \bar{x}_2} = \sqrt{\frac{(n_1-1)s_1^2 + (n_2-1)s_2^2}{n_1+n_2-2} \cdot \left(\frac{1}{n_1} + \frac{1}{n_2}\right)}$$

The Sampling Distribution of The Difference Between Two Means is approximated by a t distribution with degrees of freedom which is given by:
$$df = n_1 + n_2 - 2.$$

CHAPTER 11

The critical value formula is:

$$X_c = t_c(s_{\bar{X}_1 - \bar{X}_2})$$

The difference between the two sample means is: $d = \bar{X}_1 - \bar{X}_2$

CHAPTER 12

The population proportion is:

$$p = \frac{x}{N}$$

The sample proportion is:

$$\hat{p} = \frac{x}{n}$$

the mean of the Sampling Distribution of The Proportion is:

$$\mu_{\hat{p}} = p$$

the standard deviation of the Sampling Distribution of The Proportion is:

$$\sigma_{\hat{p}} = \sqrt{\frac{p(1-p)}{n}}$$

the critical value, X_c, formula is:

$$X_c = \mu_{\hat{p}} + (z_c)(\sigma_{\hat{p}})$$

CHAPTER 13

A 90% Confidence Interval for the population mean, μ, (with known population standard deviation) is:

$$\bar{X} - (1.65)\frac{\sigma}{\sqrt{n}} \text{ to } \bar{X} + (1.65)\frac{\sigma}{\sqrt{n}}$$

A 95% Confidence Interval for the population mean, μ, (with known population standard deviation) is:

$$\bar{X} - (1.96)\frac{\sigma}{\sqrt{n}} \text{ to } \bar{X} + (1.96)\frac{\sigma}{\sqrt{n}}$$

A 99% Confidence Interval for the population mean, μ, (with known population standard deviation) is:

$$\bar{X} - (2.58)\frac{\sigma}{\sqrt{n}} \text{ to } \bar{X} + (2.58)\frac{\sigma}{\sqrt{n}}$$

A 90% Confidence Interval for the population mean, μ, when the population standard deviation is *unknown*, is:

$$\bar{X} - (t_{95\%})\frac{s}{\sqrt{n}} \text{ to } \bar{X} + (t_{95\%})\frac{s}{\sqrt{n}}$$

A 95% Confidence Interval for the population mean, μ, when the population standard deviation is *unknown*, is:

$$\bar{X} - (t_{97.5\%})\frac{s}{\sqrt{n}} \text{ to } \bar{X} + (t_{97.5\%})\frac{s}{\sqrt{n}}$$

A 99% Confidence Interval for the population mean, μ, when the population standard deviation is *unknown*, is:

$$\bar{X} - (t_{99.5\%})\frac{s}{\sqrt{n}} \text{ to } \bar{X} + (t_{99.5\%})\frac{s}{\sqrt{n}}$$

A 90% Confidence Interval for the Population Proportion, p, is:

$$\hat{p} - (1.65)s_{\hat{p}} \text{ to } \hat{p} + (1.65)s_{\hat{p}}$$

A 95% Confidence Interval for the Population Proportion, p, is:

$$\hat{p} - (1.96)s_{\hat{p}} \text{ to } \hat{p} + (1.96)s_{\hat{p}}$$

A 99% Confidence Interval for the Population Proportion, p, is:

$$\hat{p} - (2.58)s_{\hat{p}} \text{ to } \hat{p} + (2.58)s_{\hat{p}}$$

The margin of error for estimating a population mean, where the population standard is known, is:

$$E = (z)\left(\frac{\sigma}{\sqrt{n}}\right)$$

CHAPTER 13

The sample size formula when estimating the population mean is:

$$n = \left[\frac{z\sigma}{E}\right]^2$$

The margin of error for estimating a population proportion is:

$$E = (z)\left(\sqrt{\frac{\hat{p}(1-\hat{p})}{n}}\right)$$

The sample size formula when estimating the population proportion using $\hat{p}$, a point estimate of the population proportion, is:

$$n = \frac{z^2(\hat{p})(1-\hat{p})}{E^2}$$

The conservative sample size formula in estimating the population proportion is:

$$n = \frac{z^2(0.25)}{E^2}$$

CHAPTER 14

Pearson's Chi-Square Statistic

$$X^2 = \sum_{all\ cells} \frac{(O-E)^2}{E}$$

or

$$X^2 = \sum_{all\ cells} \left(\frac{O^2}{E}\right) - n$$

Chi-Square Distribution: degrees of freedom:

$$df = (r-1)(c-1)$$

$$\text{expected cell frequency} = \frac{(RT)(CT)}{n}$$

CHAPTER 15

Pearson's Correlation Coefficient, r.

$$r = \frac{n\sum(xy) - (\sum x)(\sum y)}{\sqrt{n(\sum x^2) - (\sum x)^2}\sqrt{n(\sum y^2) - (\sum y)^2}}$$

degrees of freedom, df, for testing the correlation coefficient is:

$$df = n - 2$$

Coefficient of Determination = r^2

Regression Line Formula: $y' = a + bx$

where:
y' is the predicted value of y, the dependent variable, given the value of x, the independent variable,

and

a and b are the regression coefficients obtained by the formulas:

$$b = \frac{n\sum(xy) - \sum x \sum y}{n(\sum x^2) - (\sum x)^2}$$

$$a = \frac{\sum y - b\sum x}{n}$$

SUMMARY OF HYPOTHESIS TESTS

SUMMARY OF HYPOTHESIS TESTS FOR TESTING THE VALUE OF A POPULATION PARAMETER			
FORM OF THE NULL HYPOTHESIS	CONDITIONS OF TEST	CRITICAL VALUE FORMULA	SAMPLING DISTRIBUTION IS A:
$\mu = \mu_0$	KNOWN σ and $n > 30$	$X_c = \mu_{\bar{x}} + (z_c)(\sigma_{\bar{x}})$ where: $\sigma_{\bar{x}} = \dfrac{\sigma}{\sqrt{n}}$	Normal Distribution
$\mu = \mu_0$	UNKNOWN σ	$X_c = \mu_{\bar{x}} + (t_c)(s_{\bar{x}})$ where: $s_{\bar{x}} = \dfrac{s}{\sqrt{n}}$	t Distribution where: df = n - 1
$p = p_0$	$np > 5$ and $n(1-p) > 5$	$X_c = \mu_{\hat{p}} + (z_c)(\sigma_{\hat{p}})$ where: $\sigma_{\hat{p}} = \sqrt{\dfrac{p(1-p)}{n}}$	Normal Distribution

SUMMARY OF HYPOTHESIS TESTS INVOLVING TWO POPULATION MEANS			
FORM OF THE NULL HYPOTHESIS	CONDITIONS OF TEST	CRITICAL VALUE FORMULA	SAMPLING DISTRIBUTION IS A:
$\mu_1 - \mu_2 = 0$	KNOWN: σ_1 and σ_2 and n_1 and n_2 are both greater than 30	$X_c = (z_c)(\sigma_{\bar{x}_1 - \bar{x}_2})$ where: $\sigma_{\bar{x}_1 - \bar{x}_2} = \sqrt{\dfrac{\sigma_1^2}{n_1} + \dfrac{\sigma_2^2}{n_2}}$	Normal Distribution
$\mu_1 - \mu_2 = 0$	UNKNOWN: σ_1 and σ_2 and $\sigma_1 \approx \sigma_2$	$X_c = (t_c)(s_{\bar{x}_1 - \bar{x}_2})$ where: $s_{\bar{x}_1 - \bar{x}_2} = \sqrt{\dfrac{(n_1-1)s_1^2 + (n_2-1)s_2^2}{n_1 + n_2 - 2} \cdot \left(\dfrac{1}{n_1} + \dfrac{1}{n_2}\right)}$	t Distribution where: df = $n_1 + n_2 - 2$

SUMMARY OF CONFIDENCE INTERVALS

CONFIDENCE INTERVALS FORMULAS AND CONDITIONS

PARAMETER BEING ESTIMATED	CONDITIONS OF ESTIMATE	POINT ESTIMATE	CONFIDENCE INTERVAL	SAMPLING DISTRIBUTION IS A:
μ	Known σ	$\bar{x}$	$\bar{x} - (z)\left(\dfrac{\sigma}{\sqrt{n}}\right)$ to $\bar{x} + (z)\left(\dfrac{\sigma}{\sqrt{n}}\right)$ where: $z = 1.65$ for 90% $z = 1.96$ for 95% $z = 2.58$ for 99%	Normal Distribution
μ	Unknown σ	$\bar{x}$	$\bar{x} - (t)\left(\dfrac{s}{\sqrt{n}}\right)$ to $\bar{x} + (t)\left(\dfrac{s}{\sqrt{n}}\right)$ where: $t = t_{95\%}$ for 90% $t = t_{97.5\%}$ for 95% $t = t_{99.5\%}$ for 99%	t Distribution with: df = n - 1
p	$n\hat{p} > 5$ and $n(1-\hat{p}) > 5$	$\hat{p}$	$\hat{p} - (z)(s_{\hat{p}})$ to $\hat{p} + (z)(s_{\hat{p}})$ where: $z = 1.65$ for 90% $z = 1.96$ for 95% $z = 2.58$ for 99%	Normal Distribution

APPENDIX D: STATISTICAL TABLES

TABLE I: The Success Occurrence Table, $\binom{n}{s}$

The entries in the table give the number of ways to get s successes in n trials, which is represented by $\binom{n}{s}$ [1]

n \ s	0	1	2	3	4	5	6	7	8	9	10	11	12	13	14	15	16	17	18	19	20	n
0	1																					0
1	1	1																				1
2	1	2	1																			2
3	1	3	3	1																		3
4	1	4	6	4	1																	4
5	1	5	10	10	5	1																5
6	1	6	15	20	15	6	1															6
7	1	7	21	35	35	21	7	1														7
8	1	8	28	56	70	56	28	8	1													8
9	1	9	36	84	126	126	84	36	9	1												9
10	1	10	45	120	210	252	210	120	45	10	1											10
11	1	11	55	165	330	462	462	330	165	55	11	1										11
12	1	12	66	220	495	792	924	792	495	220	66	12	1									12
13	1	13	78	286	715	1287	1716	1716	1287	715	286	78	13	1								13
14	1	14	91	364	1001	2002	3003	3432	3003	2002	1001	364	91	14	1							14
15	1	15	105	455	1365	3003	5005	6435	6435	5005	3003	1365	455	105	15	1						15
16	1	16	120	560	1820	4368	8008	11440	12870	11440	8008	4368	1820	560	120	16	1					16
17	1	17	136	680	2380	6188	12376	19448	24310	24310	19448	12376	6188	2380	680	136	17	1				17
18	1	18	153	816	3060	8568	18564	31824	43758	48620	43758	31824	18564	8568	3060	816	153	18	1			18
19	1	19	171	969	3876	11628	27132	50388	75582	92378	92378	75582	50388	27132	11628	3876	969	171	19	1		19
20	1	20	190	1140	4845	15504	38760	77520	125970	167960	184756	167960	125970	77520	38760	15504	4845	1140	190	20	1	20

---

[1] The values for $\binom{n}{s}$ can also be obtained by using the formula: $\binom{n}{s} = \dfrac{n!}{s!\,(n-s)!}$.

This formula was discussed in Section 6.3 as Counting Rule 4.

TABLE II: The Normal Curve Area Table

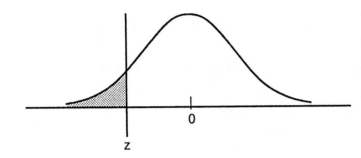

THE ENTRIES IN TABLE II represent the percent of area under the Standard Normal Distribution that is to the left of the z score.

z	0	1	2	3	4	5	6	7	8	9
-3.	00.13	00.10	00.07	00.05	00.03	00.02	00.02	00.01	00.01	00.00[2]
-2.9	00.19	00.18	00.17	00.17	00.16	00.16	00.15	00.15	00.14	00.14
-2.8	00.26	00.25	00.24	00.23	00.23	00.22	00.21	00.21	00.20	00.19
-2.7	00.35	00.34	00.33	00.32	00.31	00.30	00.29	00.28	00.27	00.26
-2.6	00.47	00.45	00.44	00.43	00.41	00.40	00.39	00.38	00.37	00.36
-2.5	00.62	00.60	00.59	00.57	00.55	00.54	00.52	00.51	00.49	00.48
-2.4	00.82	00.80	00.78	00.75	00.73	00.71	00.69	00.68	00.66	00.64
-2.3	01.07	01.04	01.02	00.99	00.96	00.94	00.91	00.89	00.87	00.84
-2.2	01.39	01.36	01.32	01.29	01.25	01.22	01.19	01.16	01.13	01.10
-2.1	01.79	01.74	01.70	01.66	01.62	01.58	01.54	01.50	01.46	01.43
-2.0	02.28	02.22	02.17	02.12	02.07	02.02	01.97	01.92	01.88	01.83
-1.9	02.87	02.81	02.74	02.68	02.62	02.56	02.50	02.44	02.39	02.33
-1.8	03.59	03.52	03.44	03.36	03.29	03.22	03.14	03.07	03.01	02.94
-1.7	04.46	04.36	04.27	04.18	04.09	04.01	03.92	03.84	03.75	03.67
-1.6	05.48	05.37	05.26	05.16	05.05	04.95	04.85	04.75	04.65	04.55
-1.5	06.68	06.55	06.43	06.30	06.18	06.06	05.94	05.82	05.71	05.59
-1.4	08.08	07.93	07.78	07.64	07.49	07.35	07.22	07.08	06.94	06.81
-1.3	09.68	09.51	09.34	09.18	09.01	08.85	08.69	08.53	08.38	08.23
-1.2	11.51	11.31	11.12	10.93	10.75	10.56	10.38	10.20	10.03	09.85
-1.1	13.57	13.35	13.14	12.92	12.71	12.51	12.30	12.10	11.90	11.70
-1.0	15.87	15.62	15.39	15.15	14.92	14.69	14.46	14.23	14.01	13.79
-0.9	18.41	18.14	17.88	17.62	17.36	17.11	16.85	16.60	16.35	16.11
-0.8	21.19	20.90	20.61	20.33	20.05	19.77	19.49	19.22	18.94	18.67
-0.7	24.20	23.89	23.58	23.27	22.96	22.66	22.36	22.06	21.77	21.48
-0.6	27.43	27.09	26.76	26.43	26.11	25.78	25.46	25.14	24.83	24.51
-0.5	30.85	30.50	30.15	29.81	29.46	29.12	28.77	28.43	28.10	27.76
-0.4	34.46	34.09	33.72	33.36	33.00	32.64	32.28	31.92	31.56	31.21
-0.3	38.21	37.83	37.45	37.07	36.69	36.32	35.94	35.57	35.20	34.83
-0.2	42.07	41.68	41.29	40.90	40.52	40.13	39.74	39.36	38.97	38.59
-0.1	46.02	45.62	45.22	44.83	44.43	44.04	43.64	43.25	42.86	42.47
-0.0	50.00	49.60	49.20	48.80	48.40	48.01	47.61	47.21	46.81	46.41

---

[2] Please note: to two decimal places, the percent of area to the left of z = -3.9 is *approximately* zero.

TABLE II: The Normal Curve Area Table

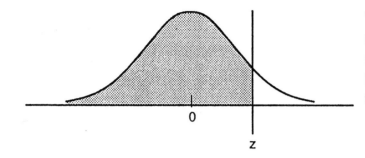

THE ENTRIES IN TABLE II represent the percent of area under the Standard Normal Distribution that is to the left of the z score.

z	0	1	2	3	4	5	6	7	8	9
0.0	50.00	50.40	50.80	51.20	51.60	51.99	52.39	52.79	53.19	53.59
0.1	53.98	54.38	54.78	55.17	55.57	55.96	56.36	56.75	57.14	57.53
0.2	57.93	58.32	58.71	59.10	59.48	59.87	60.26	60.64	61.03	61.41
0.3	61.79	62.17	62.55	62.93	63.31	63.68	64.06	64.43	64.80	65.17
0.4	65.54	65.91	66.28	66.64	67.00	67.36	67.72	68.08	68.44	68.79
0.5	69.15	69.50	69.85	70.19	70.54	70.88	71.23	71.57	71.90	72.24
0.6	72.57	72.91	73.24	73.57	73.89	74.22	74.54	74.86	75.17	75.49
0.7	75.80	76.11	76.42	76.73	77.04	77.34	77.64	77.94	78.23	78.52
0.8	78.81	79.10	79.39	79.67	79.95	80.23	80.51	80.78	81.06	81.33
0.9	81.59	81.86	82.12	82.38	82.64	82.89	83.15	83.40	83.65	83.89
1.0	84.13	84.38	84.61	84.85	85.08	85.31	85.54	85.77	85.99	86.21
1.1	86.43	86.65	86.86	87.08	87.29	87.49	87.70	87.90	88.10	88.30
1.2	88.49	88.69	88.88	89.07	89.25	89.44	89.62	89.80	89.97	90.15
1.3	90.32	90.49	90.66	90.82	90.99	91.15	91.31	91.47	91.62	91.77
1.4	91.92	92.07	92.22	92.36	92.51	92.65	92.78	92.92	93.06	93.19
1.5	93.32	93.45	93.57	93.70	93.82	93.94	94.06	94.18	94.29	94.41
1.6	94.52	94.63	94.74	94.84	94.95	95.05	95.15	95.25	95.35	95.45
1.7	95.54	95.64	95.73	95.82	95.91	95.99	96.08	96.16	96.25	96.33
1.8	96.41	96.49	96.56	96.64	96.71	96.78	96.86	96.93	96.99	97.06
1.9	97.13	97.19	97.26	97.32	97.38	97.44	97.50	97.56	97.61	97.67
2.0	97.72	97.78	97.83	97.88	97.93	97.98	98.03	98.08	98.12	98.17
2.1	98.21	98.26	98.30	98.34	98.38	98.42	98.46	98.50	98.54	98.57
2.2	98.61	98.64	98.68	98.71	98.75	98.78	98.81	98.84	98.87	98.90
2.3	98.93	98.96	98.98	99.01	99.04	99.06	99.09	99.11	99.13	99.16
2.4	99.18	99.20	99.22	99.25	99.27	99.29	99.31	99.32	99.34	99.36
2.5	99.38	99.40	99.41	99.43	99.45	99.46	99.48	99.49	99.51	99.52
2.6	99.53	99.55	99.56	99.57	99.59	99.60	99.61	99.62	99.63	99.64
2.7	99.65	99.66	99.67	99.68	99.69	99.70	99.71	99.72	99.73	99.74
2.8	99.74	99.75	99.76	99.77	99.77	99.78	99.79	99.79	99.80	99.81
2.9	99.81	99.82	99.82	99.83	99.84	99.84	99.85	99.85	99.86	99.86
3.	99.87	99.90	99.93	99.95	99.97	99.98	99.98	99.99	99.99	100.00[3]

---

[3] Please note: to two decimal places, the percent of area to the left of z = 3.9 is *approximately* 100.

TABLE III: Critical values for the t distributions
FOR HYPOTHESIS TESTING

The entries in the table give the critical t values for the specified number of degrees of freedom (df) and the level of significance (α) for a one tailed or a two tailed test.

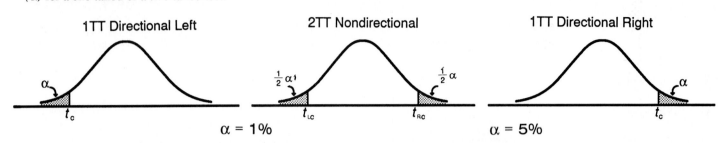

	Level of significance α = 1%				Level of significance α = 5%			
	ONE TAIL		TWO TAIL		ONE TAIL		TWO TAIL	
degrees of freedom	critical t left (t_C)	critical t right (t_C)	critical t left (t_{LC})	critical t right (t_{RC})	critical t left (t_C)	critical t right (t_C)	critical t left (t_{LC})	critical t right (t_{RC})
	FOR CONFIDENCE INTERVALS							
df	$t_{01\%}$	$t_{99\%}$	$t_{0.5\%}$	$t_{99.5\%}$	$t_{5\%}$	$t_{95\%}$	$t_{2.5\%}$	$t_{97.5\%}$
1	−31.82	31.82	−63.66	63.66	−6.31	6.31	−12.71	12.71
2	−6.96	6.96	−9.92	9.92	−2.92	2.92	−4.30	4.30
3	−4.54	4.54	−5.84	5.84	−2.35	2.35	−3.18	3.18
4	−3.75	3.75	−4.60	4.60	−2.13	2.13	−2.78	2.78
5	−3.36	3.36	−4.03	4.03	−2.02	2.02	−2.57	2.57
6	−3.14	3.14	−3.71	3.71	−1.94	1.94	−2.45	2.45
7	−3.00	3.00	−3.50	3.50	−1.90	1.90	−2.36	2.36
8	−2.90	2.90	−3.36	3.36	−1.86	1.86	−2.31	2.31
9	−2.82	2.82	−3.25	3.25	−1.83	1.83	−2.26	2.26
10	−2.76	2.76	−3.17	3.17	−1.81	1.81	−2.23	2.23
11	−2.72	2.72	−3.11	3.11	−1.80	1.80	−2.20	2.20
12	−2.68	2.68	−3.06	3.06	−1.78	1.78	−2.18	2.18
13	−2.65	2.65	−3.01	3.01	−1.77	1.77	−2.16	2.16
14	−2.62	2.62	−2.98	2.98	−1.76	1.76	−2.14	2.14
15	−2.60	2.60	−2.95	2.95	−1.75	1.75	−2.13	2.13
16	−2.58	2.58	−2.92	2.92	−1.75	1.75	−2.12	2.12
17	−2.57	2.57	−2.90	2.90	−1.74	1.74	−2.11	2.11
18	−2.55	2.55	−2.88	2.88	−1.73	1.73	−2.10	2.10
19	−2.54	2.54	−2.86	2.86	−1.73	1.73	−2.09	2.09
20	−2.53	2.53	−2.84	2.84	−1.72	1.72	−2.09	2.09
21	−2.52	2.52	−2.83	2.83	−1.72	1.72	−2.08	2.08
22	−2.51	2.51	−2.82	2.82	−1.72	1.72	−2.07	2.07
23	−2.50	2.50	−2.81	2.81	−1.71	1.71	−2.07	2.07
24	−2.49	2.49	−2.80	2.80	−1.71	1.71	−2.06	2.06
25	−2.48	2.48	−2.79	2.79	−1.71	1.71	−2.06	2.06
26	−2.48	2.48	−2.78	2.78	−1.71	1.71	−2.06	2.06
27	−2.47	2.47	−2.77	2.77	−1.70	1.70	−2.05	2.05
28	−2.47	2.47	−2.76	2.76	−1.70	1.70	−2.05	2.05
29	−2.46	2.46	−2.76	2.76	−1.70	1.70	−2.04	2.04
30	−2.46	2.46	−2.75	2.75	−1.70	1.70	−2.04	2.04
40	−2.42	2.42	−2.70	2.70	−1.68	1.68	−2.02	2.02
50	−2.40	2.40	−2.68	2.68	−1.68	1.68	−2.01	2.01
60	−2.39	2.39	−2.66	2.66	−1.67	1.67	−2.00	2.00
80	−2.37	2.37	−2.64	2.64	−1.66	1.66	−1.99	1.99
100	−2.36	2.36	−2.63	2.63	−1.66	1.66	−1.98	1.98
200	−2.34	2.34	−2.60	2.60	−1.65	1.65	−1.97	1.97
500	−2.33	2.33	−2.59	2.59	−1.65	1.65	−1.96	1.96
(normal distribution)	−2.33	2.33	−2.58	2.58	−1.65	1.65	−1.96	1.96

TABLE IV: Critical Values for the Chi-Square Distributions, χ_α^2

The entries in the table give the critical chi-square values (χ_α^2) for the specified number of degrees of freedom (df) and the level of significance (α).

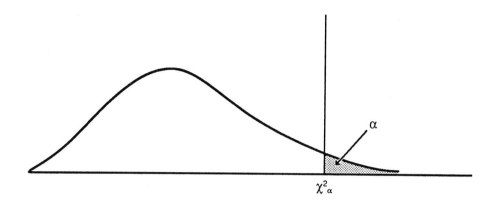

degrees of freedom	significance level	
df	$\alpha = 5\%$	$\alpha = 1\%$
1	3.84	6.64
2	5.99	9.21
3	7.82	11.34
4	9.49	13.28
5	11.07	15.09
6	12.59	16.81
7	14.07	18.48
8	15.51	20.09
9	16.92	21.67
10	18.31	23.21
11	19.68	24.72
12	21.03	26.22
13	22.36	27.69
14	23.68	29.14
15	25.00	30.58
16	26.30	32.00
17	27.59	33.41
18	28.87	34.80
19	30.14	36.19
20	31.41	37.57

APPENDIX D: STATISTICAL TABLES

TABLE V: Critical Values of the Correlation Coefficient, r_α

1TT Directional Left use a negative critical r value, $-r_\alpha$	2TT Nondirectional use a positive critical r value, r_α and a negative critical r value, $-r_\alpha$	1TT Directional Right use a positive critical r value, r_α
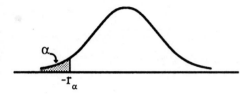	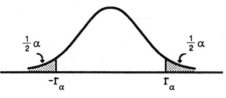	

	$\alpha = 5\%$		$\alpha = 1\%$	
df	one tail	two tail	one tail	two tail
1	0.99	1.00	1.00	1.00
2	0.90	0.95	0.98	0.99
3	0.81	0.88	0.93	0.96
4	0.73	0.81	0.88	0.92
5	0.67	0.75	0.83	0.87
6	0.62	0.71	0.79	0.83
7	0.58	0.67	0.75	0.80
8	0.54	0.63	0.72	0.76
9	0.52	0.60	0.69	0.73
10	0.50	0.58	0.66	0.71
11	0.48	0.53	0.63	0.68
12	0.46	0.53	0.61	0.66
13	0.44	0.51	0.59	0.64
14	0.42	0.50	0.57	0.61
15	0.41	0.48	0.56	0.61
16	0.40	0.47	0.54	0.59
17	0.39	0.46	0.53	0.58
18	0.38	0.44	0.52	0.56
19	0.37	0.43	0.50	0.55
20	0.36	0.42	0.49	0.54
21	0.35	0.41	0.48	0.53
22	0.34	0.40	0.47	0.52
23	0.34	0.40	0.46	0.51
24	0.33	0.39	0.45	0.50
25	0.32	0.38	0.45	0.49
26	0.32	0.37	0.44	0.48
27	0.31	0.37	0.43	0.47
28	0.31	0.36	0.42	0.46

To determine the critical value of the correlation coefficient, r_α, when the *df* is greater than 28, use the formula:

$$r_\alpha = \frac{t_c}{\sqrt{t_c^2 + (n-2)}}$$

where:
  t_c is the corresponding critical value of t for $(n-2)$ degrees of freedom

INDEX

Alpha level, α (see Significance level
        or Type I Error) 441-459
    definition one 441
    definition two 444
Addition rule 260-264
    definition 260
    mutually exclusive events 262-264
Alternative hypothesis 424-426
    definition 424
    directional 425
    introduction 424
    nondirectional 426
Average 116
Bar graph 65-68
    definition 66
    horizontal 66
    vertical 66
    time-series 67
Best fitting line 712-716
Beta, β (see Type II Error) 447-448
    definition 447
Bimodal
    definition 126
    distribution 88
Binomial distribution 310-313, 365-382
    mean 366
    normal approximation conditions 368
    normal approximation procedure 371
    standard deviation 366
    using MINITAB 310-313, 379-382
Binomial experiment 292-295
    definition 292
Binomial probability 292-313
    calculator procedure 299
    formula 294
    normal approximation to 365-382
    using MINITAB 310-313
Box-and-Whisker plot (Box plot) 212-222
    definition 212
    interpretation 213
    procedure to construct 216
    using MINITAB 220-222
Calculator procedure
    for binomial probability 299
    for correlation coefficient 721-722
    for mean 130
    for standard deviation 161, 169
    for regression coefficients 721-722
Categorical data 43-44
    analysis of 647-649
Categorical variable 31-33
    definition 32
Central limit theorem 401-406
    definition 401
Central tendency 116
Chebyshev's theorem 171-174
Chi-Square 647-674
    assumptions 662
    classifications 647
    degrees of freedom 649, 651
    goodness-of-fit 664-667
    test of independence 650-653
    using MINITAB 671-674

Chi-Square distribution 649
    properties 649
Chi-Square goodness-of-fit 664-667
Chi-Square statistic 648-649, 652
    critical value 651
    definition 648
    formulas 648, 652
    Pearson's 648
Chi-Square test of independence
    introduction 650-653
    summary 663
    using MINITAB 671-674
Class boundary
    definition 50
    normal approximation to binomial 370-371
Class limit 47-56
    definition 47
    lower 47-56
    upper 47-56
Class mark
    definition 49
Classifications 647-650
    chi-square 647-650
    data 31-33
    dependent 647-650
    independent 647-650
    population 647-650
Coefficient of determination 710-711
    definition 710
Coefficient of variation 178-180
    definition 178
    when to use 178
Combination 240-250
    counting rule 4 248
    definition 247
    using Table I 250
Complement of event 264-267
    definition 264
    rule 265
Conditional probability 278-291
    definition 286
    formula 286
    multiplication rule 279, 284
Confidence interval 606-624
    90%: for Population Mean 611, 614
    90%: for Population Proportion 621
    95% confidence interval 609
    95%: for Population Mean 609, 615
    95%: for Population Proportion 622
    99%: for Population Mean 612, 615
    99%: for Population Proportion 622
    Confidence level 610
    Confidence limits 610
    population standard deviation
        unknown 613-616
    summary 636
    t distribution 614
    using MINITAB 634-635
    width 613, 625, 629
Confidence intervals 606-624
    degrees of freedom 614
    summary of conditions 636
Confidence level. 610, 613
    definition 610

Confidence limits
    definition 610
Conservative estimate of sample size 631-632
Contingency table 647-648
    cell 648
    definition 648
    observed frequencies 647-649
Continuity Correction 370-371
Correlation
    analysis 699
    calculate 698-700
    cautions 720-721
    coefficient of determination 710-711
    coefficient of linear correlation 698-699
    correlation coefficient 698-700
    critical values 705-706
    degrees of freedom 705
    experimental outcome 706
    hypothesis testing model 705
    interpreting 693-695, 703-706
    interpreting r 698-699, 720-721
    introduction 689-690
    linear 690, 693-695, 702-703, 712-713, 720-721
    negative linear 693-694, 697
    no linear 694, 698, 721
    ordered pair 691
    Pearson's correlation coefficient 698-699
    population correlation coefficient 703-706
    positive linear 692-693, 699, 703
    procedure to calculate 702
    relationship with scatter diagram 702-703
    sampling distribution of correlation
        coefficients 705
    scatter diagram 690-698, 723-726
    significance of 721
    strength 699, 710-711
    test of significance 704
    using a calculator 721-722
    using MINITAB 722-726
    weak linear 699
    zero 694-695, 699
Critical value(s)
    chi-square statistic 651-653
    correlation coefficient 705-706
    differences between population means 537
    formula for 449
    introduction 440-449
    left critical, X_{LC} 451-453
    population mean 493, 505
    population proportion 569-570
    right critical, X_{RC} 451-453
    t distribution 501-504
    two sample t test 537
Critical t score 501-504
    left, t_{LC} 502
    right, t_{RC} 503
Critical z score 449-453
    introduction 449
    left, z_{LC} 452
    right, z_{RC} 452
Data 2
    categorical 31-33
    classifications of 31-33
    continuous 32
    discrete 32
    exploring 2, 34-43
    outlier 36-37, 218, 221
    raw 2

Deciles 209-212
Decision rule 427-460
    procedure 427, 457
    table for 5% 458
    table for 1% 459
Degrees of freedom
    chi-square 649, 651
    confidence intervals 614
    correlation coefficient 705
    difference between two means 537
    one sample hypothesis test 500-501
    two sample t test 537
    t distribution 500-501
Dependent
    classifications 647-650
Deviation score 199
Difference between two means
    characteristics of sampling distribution 532-533
    pooled standard deviation 534
    standard deviations known 533
    standard deviations unknown 534
    t distribution 534
Directional alternative hypothesis
    definition 425
Distribution
    bell-shaped 87, 132, 331-332
    bimodal 88
    center 37
    characteristics of 37
    chi-square 649-650
    definition 34
    differences between two means 532-534
    frequency 43
    mean 399
    normal 331-382
    rectangular 87
    reverse J-shaped 88
    sampling 395
    sampling distribution of correlation coefficients shape 37, 86-89, 401-402, 500, 533-534
    skewed 37, 87, 132-133
    spread 37
    symmetric 37, 87, 132
    t 498-501
    u-shaped 88
    uniform 87, 664-665
Empirical rule 174-175
Equally likely 251-256
    definition 251
Estimate
    good estimator 603
    point 601-602, 604-605
Estimate of the population mean
    point estimate 601-602, 604
    interval estimate 606-619
Estimate of the population proportion
    interval estimate 604-605, 619-622
    point estimate 600-602
Estimate of the population standard
    deviation 166-171
Estimation
    definition 601
    estimator 602-603
    introduction 600-605
    using MINITAB 634-635
Estimator 602-603
    definition 603
    good 603
Event 239-240

definition 239
Experiment 236-240
    binomial 292-295
    definition 236
Experimental outcome 236-240, 427, 429
    correlation coefficient 706
    difference between two means 538
    mean 493
    proportion 578
Exploratory data analysis 2, 34-43, 212-222
    box-and-whisker plot 212-222
    stem-and-leaf display 34-43
5-Number summary 213-220
    definition 213
    procedure to construct 215
Factors (see classifications)
Frequencies
    expected 648-649, 651-652
    observed 647-649, 651-652
Frequency 34, 43
Frequency distribution table 44-65
    class boundary 50
    class mark 49
    cumulative frequency 59-65
    definition 43
    guidelines to construct 46
    less than cumulative 59-61
    more than cumulative 61-65
    numerical data 32
    procedure to construct 52
    relative frequency 56-59
Frequency polygon 76-78
    definition 76
    procedure to construct 77
General multiplication rule 284-286
    definition 284
Goodness-of-Fit test
    introduction 664-667
Graph(s) 13-15
    bar 14, 65-68
    box-and-whisker 212-222
    frequency polygon 76-78
    histogram 69-75
    interpreting 89-95
    ogive 80
    pictograph 14, 85
    pie chart 2, 81-84
    scatter diagram 690-698, 723-726
    shapes 86-89
    stem-and-leaf 34-43
    time-series 67
Histogram 69-75
    procedure to construct 71
    using MINITAB 96
Hypothesis test
    application to Judicial System 427
    alternative hypothesis 424
    chi-square test of independence 650-653
    control groups 545-546
    correlation coefficient 704-706
    critical value(s) 440-449
    critical z-score 449-453
    decision rule 427-460
    double-blind study 545-546
    errors 427-428, 444-448
    general procedure 427, 460
    introduction 424-431
    level of significance, α 441-459
    model 427, 450, 453
    nondirectional hypothesis 426
    null hypothesis 424
    one-tailed test, 1TT 450
    p-value 466-473
    population correlation coefficient 704-706
    population mean 490-491, 516-517
    population standard deviation is known 492, 533
    population proportion 575-578
    population standard deviation
        unknown 498-501, 535-538
    single-blind study 545-546
    summary chi-square test of independence 663
    summary correlation test of significance 720
    summary one population mean 516-517
    summary two population means 551-552
    summary one population proportion 587-588
    t distribution 504-517, 534-538
    treatment group 545-549
    two population means 535-538
    two sample t test 535-538
    two-tailed test, 2TT 453
    type I error 428
    type II error 428
    using MINITAB 473-477, 552-556
Independent
    classifications 647
Independent Events 268-278, 288-295
    definition 268
    multiplication rule 268
Interquartile range, IQR 218-220
    definition 218
Interval Estimate 604-624
    central limit theorem 606
    confidence interval 606-624
    definition 605
    error 604
    margin of error 605, 624-634
    population mean 604-619
    population proportion 619-624
    sampling distribution of the mean 606-609, 614
    sampling distribution of the proportion 620
Level of significance, α 441-459
    definition one 441
    definition two 444
Linear correlation 690, 693-695, 702-703, 712-713, 720-721
Linear model 712
Linear regression analysis 712-722
    assumption for 720
    best fitting 713
    introduction 712
    linear model 712
    method of least squares 712
    prediction from 713
    procedure for regression line 716
    regression line formula 713
Margin of error 604-605, 624-634
Mean   binomial distribution 366
    calculator 130
    population 117-118,
    properties 122-123,
    sample 129
    sampling distribution 399
    relationship between median and mode 131-134
    trimmed mean 137-138
    using MINITAB 136
Median 124-125, 213
    definition 124
    position formula 125

relationship between median and mode 131-134
Method of least squares 712
Mode 126-128
    bimodal 126
    definition 126
    nonmodal 127
    relationship between median and mode 131-134
Multiplication rule
    independent events 268
    dependent events 279
    general 284
Mutually exclusive events 262-264
    definition 262
Negative linear correlation 693-694, 697
    definition 693
    scatter diagram 694
Nondirectional alternative hypothesis
    definition 426
Nonmodal
    definition 127
Normal curve
    see normal distribution
Normal curve area
    using Table II 335-340
Normal distribution
    applications of 341-347
    approximating binomial probabilities 365-382
    continuity correction 370-371
    introduction 331-332
    percentile rank 347-360
    percentiles 347-360
    probability 360-365
    properties of 332-335
    using MINITAB 379-382
Normal distribution approximation of binomial probabilities 365-382
    continuity correction 370-371
Null hypothesis 424-425
    definition 424
Numerical variable
    definition 32
Ogive 80-81
    definition 80
    procedure to construct 81
One-tailed test, 1TT
    definition 450
Ordered pair 691
Outlier 36-37, 136, 218, 221
    definition 37
    Tukey's rule 218
p-value 466-472
    definition 468
    interpretation 470
    procedure to calculate 471-472
    rule to reject H_o 470
    using MINITAB 473-477
Parameter 8
    mean 395
    median 395
    standard deviation 395
Pearson's correlation coefficient, r
    definition 698
    formula 698
    interpreting values of 699
    procedure to calculate 702
    strength of 699, 710-711
    test of significance 704-706, 720-721
Percentile rank 206-212
    computation of 207
    definition 207
    meaning of 207, 209
    normal distribution 347-360
Percentile(s) 208-212,
    deciles 209-212
    definition 208
    meaning of 209
    normal distribution 347-348
    quartiles 209-212
Permutation 240-247
    counting rule 1 242
    counting rule 2 243
    Counting rule 3 247
    definition 241
Pictograph 85-86
    definition 85
Pie chart 81-84
    definition 81
    procedure to construct 84
Point estimate 600-606
    definition 601
    sample mean 601-604, 606-619
    sample proportion 602, 604-605, 619-623, 628-634
Pooled
    standard deviation 534, 553
Population 6-8,
    definition 6
    mean 117-123
    median 124-125
    standard deviation 154-166
Population correlation coefficient 703-706
Population mean 117-118
    definition 117
    estimate of 601-602
    hypothesis test 490-493, 504-505
    procedure for hypothesis test 490-493, 516-517
    formula 117
    using calculator 130
    using MINITAB 517-519
Population parameter
    estimate of 600-605, 619-620
Population proportion 569-570, 619-620
    definition 569
    estimate of 619-620
    hypothesis test 575-578
    using MINITAB 588-590
Population standard deviation 154-166
    applications of 171-180
    computational formula 159
    definition 157
    properties of 163-165
    using calculator 161, 169
Population variance 155-159
    definition 156
Positive linear correlation 692-693, 699, 703
    definition 692
    scatter diagram 693
Probability
    addition rule 260-264
    binomial 292-313
    birthday problem 288
    classical probability definition 252
    combinations 240-250
    complement of event 264-267
    conditional 278-291
    conditional probability formula 286
    definition 251, 252,
    equally likely 251

event 252-277
experiment 236-240
fundamental counting principle 238-239
fundamental rules 257-277
general multiplication rule 284
independent events 268-278, 288-295
multiplication rule dependent events 279
multiplication rule independent events 268
mutually exclusive events 262-264
normal distribution 360-382
permutation(s) 270-277
sample space 252
using MINITAB 309-313
Probability of an event E 252-277
definition 252
normal distribution 361
Quartiles 209-222
procedure 213-214
using MINITAB 221
Random sampling 395-397
definition 395
Range 152-154
definition 152
Raw score 204-206
formula 206
Regression
"best fit" 712-713
assumptions 720
introduction 689-690, 712-713
linear 712, 721
model 712
procedure regression line 716
regression line formula 713
scatter diagram 712-713, 724-726
using a calculator 721-722
using MINITAB 724-726
Regression analysis 712-726
regression line 713
Regression line
best fitting 712-713
Regression line formula 713, 716
calculate a 713
calculate b 713
procedure to determine 716
Relative frequency
definition 56
Relative percentage frequency
definition 57
Sample 6-8
definition 6
margin of error 624-634
non-representative 8, 15, 16
representative 7
Sample mean
definition 129
formula 129
using calculator 130
Sample proportion 570-571, 602-605, 619-620
definition 570
Sample size
conservative estimate 631-632
determining 631-632
effect on sampling distribution 407-408
margin of error 624-628, 631-632
population proportion 630-632
Sample space 236-240
definition 236
Sample standard deviation 166-171
applications 171-180

computational formula 167
confidence interval 613-616
definition formula 166
using calculator 169
using MINITAB 180-183
Sampling distribution
definition 395
Sampling distribution of the correlation
coefficients 705
distribution model 705
hypothesis testing model 705
Sampling distribution of the difference between two
means 531-534
characteristics of 533-534
hypothesis test 535-549
standard deviations known 533
standard deviations unknown 534
using MINITAB 552-556
Sampling distribution of the mean
approximately normal 401
central limit theorem 401-406
characteristics of 401-403
definition 398
effect of sample size 407-408
estimate of the standard deviation 505
hypothesis test 490-493, 504-505
mean of 399
standard deviation of 399
t distribution 498-505
using MINITAB 408-412
Sampling distribution of the proportion 572, 576-577, 620
characteristics 572
estimate of standard deviation 620
introduction 572
mean 572
normal distribution 572
properties of 572
standard deviation 572
Sampling error
introduction 604-605
Scatter diagram 690-698, 712-713, 724
definition 691
introduction 690-691
ordered pair 691
regression 712-713
relationship with r 691-695, 699, 703
Shapes of distributions 86-89, 131-134, 216-220
Significance
level of 441-459
Significant correlation coefficient 704-706
Simulation 408-412
Skewed
to the left 87, 132, 216-217
to the right 87, 133, 216-217
Standard deviation
binomial distribution 366
estimate 166-171, 498-499
pooled 534, 554
population 154-166
properties 163-165
sample 166-171,
sampling distribution of the differences between two
means 533-534
sampling distribution of the mean 399-412, 505
sampling distribution of the proportion 572
Standard deviation estimate
sampling distribution of the mean 499, 504-505
Standard deviation formula
sampling from a finite population

standard deviation of sampling distribution of the mean 399, 492
Statistic 1,
Statistical significance
    introduction 430-431
Statistics
    definition 6
    descriptive 6
    inappropriate comparisons 16-17
    inferential 7
    misuses of 12-17
    uses of 10-12
Stem-and-Leaf display 34-43
    back-to-back 40-42
    definition 34
    procedure to construct 38
    stretched 42-43
    using MINITAB 95-96
t score 499-504
    critical 501-504
    procedure to locate critical t 504
t distribution 498-504
    critical values 501-504
    degrees of freedom 500-501
    difference between two means 534
    introduction 498-501
    properties of 501
Test of significance
    correlation coefficient 703-706

Two-tailed test, 2TT
    definition 453
Type I error
    definition 428
Type II error
    definition 428
Unbiasedness
    criteria 603
Using calculator
    correlation coefficient 721-722
    binomial formula 299
    mean 130
    regression 721-722
    standard deviation 161, 169
Variability 151-184
Variable(s)
    categorical 31-33,
    definition 31
    numerical 32
Variance
    population 154-157
z score 199-206, 442
    critical 449-453
    definition 201
    formula 201
    interpreting sign of 202-204
    to raw scores 204-206
Zero correlation 694, 699, 703, 721